Lecture Notes in Computer Science 13351

Derek Groen · Clélia de Mulatier ·
Maciej Paszynski · Valeria V. Krzhizhanovskaya ·
Jack J. Dongarra · Peter M. A. Sloot (Eds.)

Computational Science – ICCS 2022

22nd International Conference
London, UK, June 21–23, 2022
Proceedings, Part II

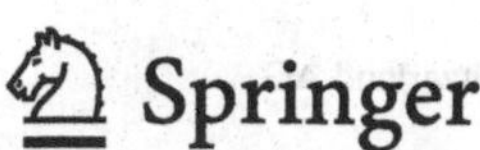

Editors
Derek Groen
Brunel University London
London, UK

Clélia de Mulatier
University of Amsterdam
Amsterdam, The Netherlands

Maciej Paszynski
AGH University of Science and Technology
Krakow, Poland

Valeria V. Krzhizhanovskaya
University of Amsterdam
Amsterdam, The Netherlands

Jack J. Dongarra
University of Tennessee at Knoxville
Knoxville, TN, USA

Peter M. A. Sloot
University of Amsterdam
Amsterdam, The Netherlands

ISSN 0302-9743 ISSN 1611-3349 (electronic)
Lecture Notes in Computer Science
ISBN 978-3-031-08753-0 ISBN 978-3-031-08754-7 (eBook)
https://doi.org/10.1007/978-3-031-08754-7

This Springer imprint is published by the registered company Springer Nature Switzerland AG
The registered company address is: Gewerbestrasse 11, 6330 Cham, Switzerland

Preface

Welcome to the 22nd annual International Conference on Computational Science (ICCS 2022 - https://www.iccs-meeting.org/iccs2022/), held during 21–23 June, 2022, at Brunel University London, UK. After more than two years of a pandemic that has changed so much of our world and daily lives, this edition marks our return to a – partially – in-person event. Those who were not yet able to join us in London had the option to participate online, as all conference sessions were streamed.

Although the challenges of such a hybrid format are manifold, we have tried our best to keep the ICCS community as dynamic, creative, and productive as always. We are proud to present the proceedings you are reading as a result of that.

Standing on the River Thames in southeast England, at the head of a 50-mile (80 km) estuary down to the North Sea, London is the capital and largest city of England and the UK. With a rich history spanning back to Roman times, modern London is one of the world's global cities, having a prominent role in areas ranging from arts and entertainment to commerce, finance, and education. London is the biggest urban economy in Europe and one of the major financial centres in the world. It also features Europe's largest concentration of higher education institutions.

ICCS 2022 was jointly organized by Brunel University London, the University of Amsterdam, NTU Singapore, and the University of Tennessee.

Brunel University London is a public research university located in the Uxbridge area of London. It was founded in 1966 and named after the Victorian engineer Isambard Kingdom Brunel, who managed to design and build a 214m long suspension bridge in Bristol back in 1831. Brunel is well-known for its excellent Engineering and Computer Science Departments, and its campus houses a dedicated conference centre (the Hamilton Centre) which was used to host ICCS. It is also one of the few universities to host a full-length athletics track, which has been used both for practice purposes by athletes such as Usain Bolt for the 2012 Olympics and for graduation ceremonies.

The International Conference on Computational Science is an annual conference that brings together researchers and scientists from mathematics and computer science as basic computing disciplines, as well as researchers from various application areas who are pioneering computational methods in sciences such as physics, chemistry, life sciences, engineering, arts, and humanitarian fields, to discuss problems and solutions in the area, identify new issues, and shape future directions for research.

Since its inception in 2001, ICCS has attracted increasing numbers of attendees and higher-quality papers, and this year – in spite of the ongoing pandemic—was not an exception, with over 300 registered participants. The proceedings series has become a primary intellectual resource for computational science researchers, defining and advancing the state of the art in this field.

The theme for 2022, "The Computational Planet," highlights the role of computational science in tackling the current challenges of the all-important quest for sustainable development. This conference aimed to be a unique event focusing on recent developments in scalable scientific algorithms, advanced software tools, computational grids, advanced numerical methods, and novel application areas. These innovative novel models, algorithms, and tools drive new science through efficient application in physical systems, computational and systems biology, environmental systems, finance, and other areas.

ICCS is well-known for its excellent lineup of keynote speakers. The keynotes for 2022 were as follows:

- Robert Axtell, George Mason University, USA
- Peter Coveney, University College London, UK
- Thomas Engels, Technische Universität Berlin, Germany
- Neil Ferguson, Imperial College London, UK
- Giulia Galli, University of Chicago, USA
- Rebecca Wade, Heidelberg Institute for Theoretical Studies, Germany

This year we had 474 submissions (169 submissions to the main track and 305 to the thematic tracks). In the main track, 55 full papers were accepted (32%), and in the thematic tracks, 120 full papers (39%). A higher acceptance rate in the thematic tracks is explained by the nature of these, where track organizers personally invite many experts in a particular field to participate in their sessions.

ICCS relies strongly on our thematic track organizers' vital contributions to attract high-quality papers in many subject areas. We would like to thank all committee members from the main and thematic tracks for their contribution to ensure a high standard for the accepted papers. We would also like to thank Springer, Elsevier, and Intellegibilis for their support. Finally, we appreciate all the local organizing committee members for their hard work to prepare for this conference.

We are proud to note that ICCS is an A-rank conference in the CORE classification.

We wish you good health in these troubled times and look forward to meeting you at the next conference, whether virtually or in-person.

June 2022

Derek Groen
Clélia de Mulatier
Maciej Paszynski
Valeria V. Krzhizhanovskaya
Jack J. Dongarra
Peter M. A. Sloot

Organization

General Chair

Valeria Krzhizhanovskaya — University of Amsterdam, The Netherlands

Main Track Chair

Clélia de Mulatier — University of Amsterdam, The Netherlands

Thematic Tracks Chair

Maciej Paszynski — AGH University of Science and Technology, Poland

Scientific Chairs

Peter M. A. Sloot — University of Amsterdam, The Netherlands | Complexity Institute NTU, Singapore

Jack Dongarra — University of Tennessee, USA

Local Organizing Committee

Chair

Derek Groen — Brunel University London, UK

Members

Simon Taylor	Brunel University London, UK
Anastasia Anagnostou	Brunel University London, UK
Diana Suleimenova	Brunel University London, UK
Xiaohui Liu	Brunel University London, UK
Zidong Wang	Brunel University London, UK
Steven Sam	Brunel University London, UK
Alireza Jahani	Brunel University London, UK
Yani Xue	Brunel University London, UK
Nadine Aburumman	Brunel University London, UK
Katie Mintram	Brunel University London, UK
Arindam Saha	Brunel University London, UK
Nura Abubakar	Brunel University London, UK

Thematic Tracks and Organizers

Advances in High-Performance Computational Earth Sciences: Applications and Frameworks – IHPCES

Takashi Shimokawabe	University of Tokyo, Japan
Kohei Fujita	University of Tokyo, Japan
Dominik Bartuschat	Friedrich-Alexander-Universität Erlangen-Nürnberg, Germany

Artificial Intelligence and High-Performance Computing for Advanced Simulations – AIHPC4AS

Maciej Paszynski	AGH University of Science and Technology, Poland

Biomedical and Bioinformatics Challenges for Computer Science – BBC

Mario Cannataro	Università Magna Graecia di Catanzaro, Italy
Giuseppe Agapito	Università Magna Graecia di Catanzaro, Italy
Mauro Castelli	Universidade Nova de Lisboa, Portugal
Riccardo Dondi	University of Bergamo, Italy
Rodrigo Weber dos Santos	Universidade Federal de Juiz de Fora, Brazil
Italo Zoppis	Università degli Studi di Milano-Bicocca, Italy

Computational Collective Intelligence – CCI

Marcin Maleszka	Wroclaw University of Science and Technology, Poland
Ngoc Thanh Nguyen	Wroclaw University of Science and Technology, Poland
Dosam Hwang	Yeungnam University, South Korea

Computational Health – CompHealth

Sergey Kovalchuk	ITMO University, Russia
Stefan Thurner	Medical University of Vienna, Austria
Georgiy Bobashev	RTI International, USA
Jude Hemanth	Karunya University, India
Anastasia Angelopoulou	University of Westminster, UK

Computational Optimization, Modelling, and Simulation – COMS

Xin-She Yang	Middlesex University London, UK
Leifur Leifsson	Purdue University, USA
Slawomir Koziel	Reykjavik University, Iceland

Computer Graphics, Image Processing, and Artificial Intelligence – CGIPAI

Andres Iglesias	Universidad de Cantabria, Spain

Machine Learning and Data Assimilation for Dynamical Systems – MLDADS

Rossella Arcucci	Imperial College London, UK

Multiscale Modelling and Simulation – MMS

Derek Groen	Brunel University London, UK
Diana Suleimenova	Brunel University London, UK
Bartosz Bosak	Poznan Supercomputing and Networking Center, Poland
Gabor Závodszky	University of Amsterdam, The Netherlands
Stefano Casarin	Houston Methodist Research Institute, USA
Ulf D. Schiller	Clemson University, USA
Wouter Edeling	Centrum Wiskunde & Informatica, The Netherlands

Quantum Computing – QCW

Katarzyna Rycerz	AGH University of Science and Technology, Poland
Marian Bubak	Sano Centre for Computational Medicine and AGH University of Science and Technology, Poland | University of Amsterdam, The Netherlands

Simulations of Flow and Transport: Modeling, Algorithms, and Computation – SOFTMAC

Shuyu Sun	King Abdullah University of Science and Technology, Saudi Arabia
Jingfa Li	Beijing Institute of Petrochemical Technology, China
James Liu	Colorado State University, USA

Smart Systems: Bringing Together Computer Vision, Sensor Networks, and Machine Learning – SmartSys

Pedro Cardoso	University of Algarve, Portugal
João Rodrigues	University of Algarve, Portugal
Jânio Monteiro	University of Algarve, Portugal
Roberto Lam	University of Algarve, Portugal

Software Engineering for Computational Science – SE4Science

Jeffrey Carver	University of Alabama, USA
Caroline Jay	University of Manchester, UK
Yochannah Yehudi	University of Manchester, UK
Neil Chue Hong	University of Edinburgh, UK

Solving Problems with Uncertainty – SPU

Vassil Alexandrov	Hartree Centre - STFC, UK
Aneta Karaivanova	Institute for Parallel Processing, Bulgarian Academy of Sciences, Bulgaria

Teaching Computational Science – WTCS

Angela Shiflet	Wofford College, USA
Nia Alexandrov	Hartree Centre - STFC, UK

Uncertainty Quantification for Computational Models – UNEQUIvOCAL

Wouter Edeling	Centrum Wiskunde & Informatica, The Netherlands
Anna Nikishova	SISSA, Italy

Reviewers

Tesfamariam Mulugeta Abuhay
Jaime Afonso Martins
Giuseppe Agapito
Shahbaz Ahmad
Elisabete Alberdi
Luis Alexandre
Nia Alexandrov
Vassil Alexandrov
Julen Alvarez-Aramberri
Domingos Alves
Sergey Alyaev
Anastasia Anagnostou
Anastasia Angelopoulou
Samuel Aning
Hideo Aochi
Rossella Arcucci
Costin Badica
Bartosz Balis
Daniel Balouek-Thomert
Krzysztof Banaś
Dariusz Barbucha
João Barroso
Valeria Bartsch
Dominik Bartuschat
Pouria Behnodfaur
Jörn Behrens
Adrian Bekasiewicz
Gebrail Bekdas
Mehmet Ali Belen
Stefano Beretta
Benjamin Berkels
Daniel Berrar
Georgiy Bobashev
Marcel Boersma
Tomasz Boiński
Carlos Bordons
Bartosz Bosak
Giuseppe Brandi
Lars Braubach
Marian Bubak

Jérémy Buisson
Aleksander Byrski
Cristiano Cabrita
Xing Cai
Barbara Calabrese
Nurullah Calik
Almudena Campuzano
Mario Cannataro
Pedro Cardoso
Alberto Carrassi
Alfonso Carriazo
Jeffrey Carver
Stefano Casarin
Manuel Castañón-Puga
Mauro Castelli
Nicholas Chancellor
Ehtzaz Chaudhry
Thierry Chaussalet
Sibo Cheng
Siew Ann Cheong
Andrei Chernykh
Lock-Yue Chew
Su-Fong Chien
Marta Chinnici
Amine Chohra
Neil Chue Hong
Svetlana Chuprina
Paola Cinnella
Noélia Correia
Adriano Cortes
Ana Cortes
Enrique Costa-Montenegro
David Coster
Carlos Cotta
Helene Coullon
Daan Crommelin
Attila Csikasz-Nagy
Javier Cuenca
António Cunha
Pawel Czarnul
Lisandro D. Dalcin
Bhaskar Dasgupta
Clélia de Mulatier
Charlotte Debus
Javier Delserlorente
Pasquale De-Luca
Quanling Deng
Vasily Desnitsky
Mittal Dhruv
Eric Dignum
Riccardo Dondi
Rafal Drezewski
Hans du Buf
Vitor Duarte
Richard Dwight
Wouter Edeling
Nasir Eisty
Kareem El-Safty
Nahid Emad
Gökhan Ertaylan
Roberto R. Expósito
Fangxin Fang
Antonino Fiannaca
Christos Filelis-Papadopoulos
Pawel Foszner
Piotr Frąckiewicz
Martin Frank
Alberto Freitas
Ruy Freitas Reis
Karl Frinkle
Kohei Fujita
Takeshi Fukaya
Wlodzimierz Funika
Takashi Furumura
Ernst Fusch
Leszek Gajecki
Ardelio Galletti
Marco Gallieri
Teresa Galvão
Akemi Galvez-Tomida
Maria Ganzha
Luis Garcia-Castillo
Bartłomiej Gardas
Delia Garijo
Frédéric Gava
Piotr Gawron
Bernhard Geiger
Alex Gerbessiotis
Philippe Giabbanelli
Konstantinos Giannoutakis

Adam Glos
Ivo Goncalves
Alexandrino Gonçalves
Jorge González-Domínguez
Yuriy Gorbachev
Pawel Gorecki
Markus Götz
Michael Gowanlock
George Gravvanis
Derek Groen
Lutz Gross
Lluis Guasch
Pedro Guerreiro
Tobias Guggemos
Xiaohu Guo
Manish Gupta
Piotr Gurgul
Zulfiqar Habib
Mohamed Hamada
Yue Hao
Habibollah Haron
Ali Hashemian
Carina Haupt
Claire Heaney
Alexander Heinecke
Jude Hemanth
Marcin Hernes
Bogumila Hnatkowska
Maximilian Höb
Jori Hoencamp
Rolf Hoffmann
Wladyslaw Homenda
Tzung-Pei Hong
Muhammad Hussain
Dosam Hwang
Mauro Iacono
David Iclanzan
Andres Iglesias
Mirjana Ivanovic
Takeshi Iwashita
Alireza Jahani
Peter Janků
Jiri Jaros
Agnieszka Jastrzebska
Caroline Jay
Piotr Jedrzejowicz
Gordan Jezic
Zhong Jin
David Johnson
Guido Juckeland
Piotr Kalita
Drona Kandhai
Epaminondas Kapetanios
Aneta Karaivanova
Artur Karczmarczyk
Takahiro Katagiri
Timo Kehrer
Christoph Kessler
Loo Chu Kiong
Harald Koestler
Ivana Kolingerova
Georgy Kopanitsa
Pavankumar Koratikere
Triston Kosloske
Sotiris Kotsiantis
Remous-Aris Koutsiamanis
Sergey Kovalchuk
Slawomir Koziel
Dariusz Krol
Marek Krótkiewicz
Valeria Krzhizhanovskaya
Marek Kubalcík
Sebastian Kuckuk
Eileen Kuehn
Michael Kuhn
Tomasz Kulpa
Julian Martin Kunkel
Krzysztof Kurowski
Marcin Kuta
Panagiotis Kyziropoulos
Roberto Lam
Anna-Lena Lamprecht
Kun-Chan Lan
Rubin Landau
Leon Lang
Johannes Langguth
Leifur Leifsson
Kenneth Leiter
Florin Leon
Vasiliy Leonenko

Jean-Hugues Lestang
Jake Lever
Andrew Lewis
Jingfa Li
Way Soong Lim
Denis Mayr Lima Martins
James Liu
Zhao Liu
Hong Liu
Che Liu
Yen-Chen Liu
Hui Liu
Marcelo Lobosco
Doina Logafatu
Marcin Los
Stephane Louise
Frederic Loulergue
Paul Lu
Stefan Luding
Laura Lyman
Lukasz Madej
Luca Magri
Peyman Mahouti
Marcin Maleszka
Bernadetta Maleszka
Alexander Malyshev
Livia Marcellino
Tomas Margalef
Tiziana Margaria
Svetozar Margenov
Osni Marques
Carmen Marquez
Paula Martins
Pawel Matuszyk
Valerie Maxville
Wagner Meira Jr.
Roderick Melnik
Pedro Mendes Guerreiro
Ivan Merelli
Lyudmila Mihaylova
Marianna Milano
Jaroslaw Miszczak
Janio Monteiro
Fernando Monteiro
Andrew Moore
Eugénia Moreira Bernardino
Anabela Moreira Bernardino
Peter Mueller
Ignacio Muga
Khan Muhammad
Daichi Mukunoki
Vivek Muniraj
Judit Munoz-Matute
Hiromichi Nagao
Jethro Nagawakar
Kengo Nakajima
Grzegorz J. Nalepa
Yves Nanfack
Pratik Nayak
Philipp Neumann
David Chek-Ling Ngo
Ngoc Thanh Nguyen
Nancy Nichols
Sinan Melih Nigdeli
Anna Nikishova
Hitoshi Nishizawa
Algirdas Noreika
Manuel Núñez
Frederike Oetker
Schenk Olaf
Javier Omella
Boon-Yaik Ooi
Eneko Osaba
Aziz Ouaarab
Raymond Padmos
Nikela Papadopoulou
Marcin Paprzycki
David Pardo
Diego Paredesconcha
Anna Paszynska
Maciej Paszynski
Ebo Peerbooms
Sara Perez-Carabaza
Dana Petcu
Serge Petiton
Frank Phillipson
Eugenio Piasini
Juan C. Pichel
Anna Pietrenko-Dabrowska
Laércio L. Pilla

Armando Pinho
Yuri Pirola
Mihail Popov
Cristina Portales
Roland Potthast
Małgorzata Przybyła-Kasperek
Ela Pustulka-Hunt
Vladimir Puzyrev
Rick Quax
Cesar Quilodran-Casas
Enrique S. Quintana-Orti
Issam Rais
Andrianirina Rakotoharisoa
Raul Ramirez
Celia Ramos
Vishwas Rao
Kurunathan Ratnavelu
Lukasz Rauch
Robin Richardson
Miguel Ridao
Heike Riel
Sophie Robert
Joao Rodrigues
Daniel Rodriguez
Albert Romkes
Debraj Roy
Katarzyna Rycerz
Emmanuelle Saillard
Ozlem Salehi
Tarith Samson
Alberto Sanchez
Ayşin Sancı
Gabriele Santin
Vinicius Santos-Silva
Allah Bux Sargano
Robert Schaefer
Ulf D. Schiller
Bertil Schmidt
Martin Schreiber
Gabriela Schütz
Franciszek Seredynski
Marzia Settino
Mostafa Shahriari
Zhendan Shang
Angela Shiflet
Takashi Shimokawabe
Alexander Shukhman
Marcin Sieniek
Nazareen Sikkandar-Basha
Robert Sinkovits
Mateusz Sitko
Haozhen Situ
Leszek Siwik
Renata Słota
Oskar Slowik
Grażyna Ślusarczyk
Sucha Smanchat
Maciej Smołka
Thiago Sobral
Isabel Sofia Brito
Piotr Sowiński
Robert Speck
Christian Spieker
Michał Staniszewski
Robert Staszewski
Steve Stevenson
Tomasz Stopa
Achim Streit
Barbara Strug
Patricia Suarez
Dante Suarez
Diana Suleimenova
Shuyu Sun
Martin Swain
Jerzy Świątek
Piotr Szczepaniak
Edward Szczerbicki
Tadeusz Szuba
Ryszard Tadeusiewicz
Daisuke Takahashi
Osamu Tatebe
Carlos Tavares Calafate
Kasim Tersic
Jannis Teunissen
Mau Luen Tham
Stefan Thurner
Nestor Tiglao
T. O. Ting
Alfredo Tirado-Ramos
Pawel Topa

Bogdan Trawinski
Jan Treur
Leonardo Trujillo
Paolo Trunfio
Hassan Ugail
Eirik Valseth
Casper van Elteren
Ben van Werkhoven
Vítor Vasconcelos
Alexandra Vatyan
Colin C. Venters
Milana Vuckovic
Shuangbu Wang
Jianwu Wang
Peng Wang
Katarzyna Wasielewska
Jaroslaw Watrobski
Rodrigo Weber dos Santos
Mei Wen
Lars Wienbrandt
Iza Wierzbowska
Maciej Woźniak
Dunhui Xiao
Huilin Xing
Yani Xue
Abuzer Yakaryilmaz
Xin-She Yang
Dongwei Ye
Yochannah Yehudi
Lihua You
Drago Žagar
Constantin-Bala Zamfirescu
Gabor Závodszky
Jian-Jun Zhang
Yao Zhang
Wenbin Zhang
Haoxi Zhang
Jinghui Zhong
Sotirios Ziavras
Zoltan Zimboras
Italo Zoppis
Chiara Zucco
Pavel Zun
Simon Portegies Zwart
Karol Życzkowski

Contents – Part II

ICCS 2022 Main Track Short Papers

Computational Collective Intelligence

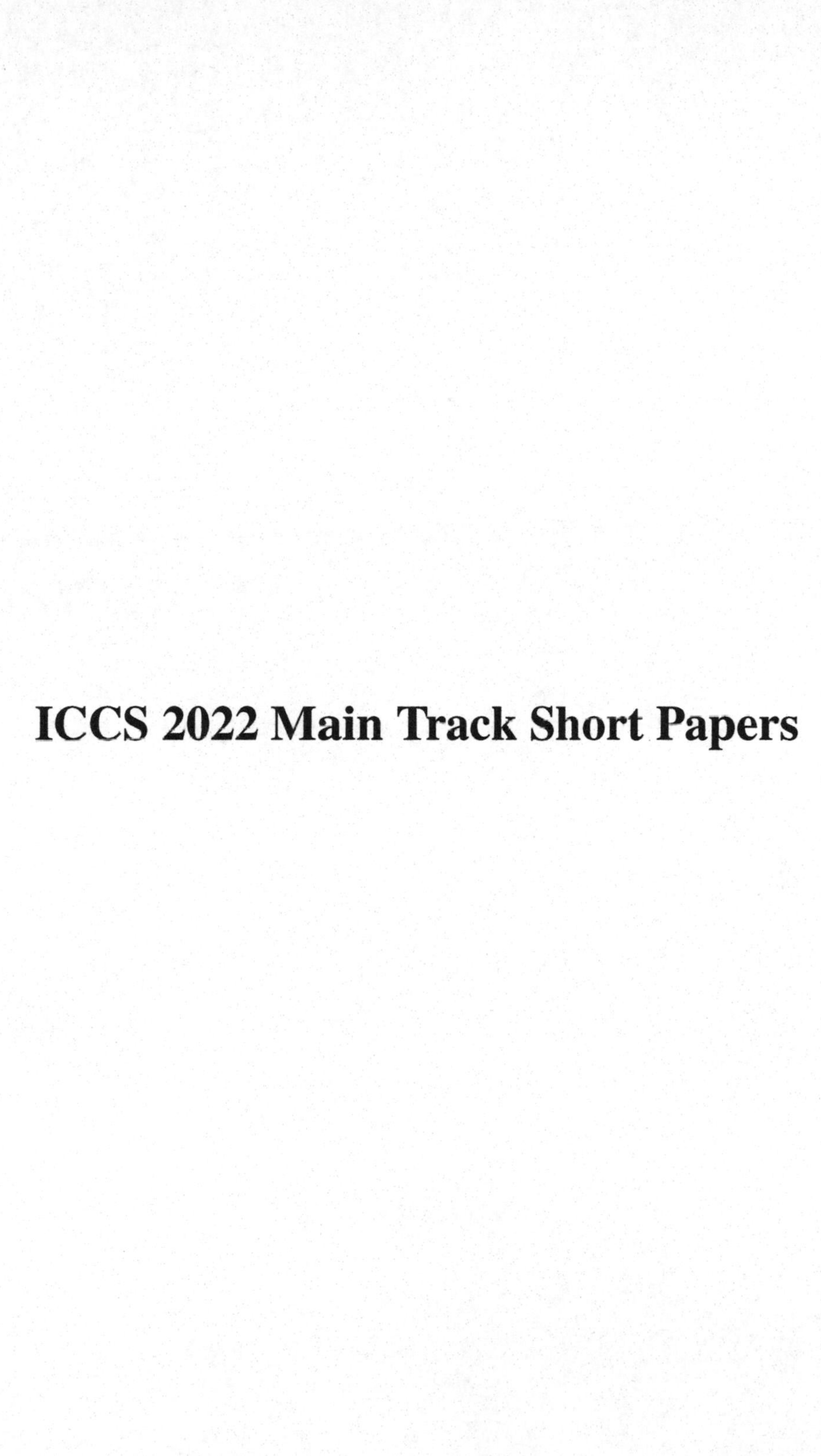

ICCS 2022 Main Track Short Papers

Neuroevolutionary Feature Representations for Causal Inference

Michael C. Burkhart[1](✉) and Gabriel Ruiz[2]

[1] University of Cambridge, Cambridge, UK
mcb93@cam.ac.uk
[2] UCLA, Los Angeles, CA, USA
ruizg@ucla.edu

Abstract. Within the field of causal inference, we consider the problem of estimating heterogeneous treatment effects from data. We propose and validate a novel approach for learning feature representations to aid the estimation of the conditional average treatment effect or CATE. Our method focuses on an intermediate layer in a neural network trained to predict the outcome from the features. In contrast to previous approaches that encourage the distribution of representations to be treatment-invariant, we leverage a genetic algorithm to optimize over representations useful for predicting the outcome to select those less useful for predicting the treatment. This allows us to retain information within the features useful for predicting outcome even if that information may be related to treatment assignment. We validate our method on synthetic examples and illustrate its use on a real life dataset.

Keywords: Causal inference · Heterogeneous treatment effects · Feature representations · Neuroevolutionary algorithms · Counterfactual inference

1 Introduction

In this note, we aim to engineer feature representations to aid in the estimation of heterogeneous treatment effects. We consider the following graphical model

$$
\begin{array}{ccc} & X & \\ \swarrow & & \searrow \\ W & \longrightarrow & Y \end{array} \tag{1}
$$

where $X \in \mathbb{R}^d$ denotes a vector of features, $W \in \{0, 1\}$ represents a boolean treatment, and $Y \in \mathbb{R}$ denotes the outcome. Suppose (X_i, W_i, Y_i) for $i = 1, \ldots, n$ are i.i.d. samples from a distribution P respecting the graph (1). Within the potential outcomes framework [10], we let $Y_i(0)$ denote the potential outcome if W_i were set to 0 and $Y_i(1)$ denote the potential outcome if W_i were set to

M.B. and G.R. were supported by Adobe Inc. (San José, Calif., U.S.A.).

D. Groen et al. (Eds.): ICCS 2022, LNCS 13351, pp. 3–10, 2022.
https://doi.org/10.1007/978-3-031-08754-7_1

1. We wish to estimate the conditional average treatment effect (CATE) defined by $\tau(x) = \mathbb{E}[Y(1) - Y(0)|X = x]$. We impose standard assumptions that the treatment assignment is unconfounded, meaning that $\{Y_i(0), Y_i(1)\} \perp W_i \mid X_i$, and random in the sense that $\epsilon < P(W_i = 1|X_i = x_i) < 1 - \epsilon$ for all i, some $\epsilon > 0$, and all x_i in the support of X_i. These assumptions jointly constitute *strong ignorability* [13] and prove sufficient for the CATE to be identifiable. Under them, there exist methods to estimate the CATE from observed data that then allow us to predict the expected individualized impact of an intervention for novel examples using only their features. Viewing these approaches as black box estimators, we seek a mapping $\Phi : \mathbb{R}^d \to \mathbb{R}^m$ such that the estimate of the CATE learned on the transformed training data $(\Phi(X_i), W_i, Y_i)$ is more accurate than an estimate learned on the original samples (X_i, W_i, Y_i). In particular, we desire a function Φ yielding a corresponding representation $\Phi(X)$ such that (1) $\Phi(X)$ is as useful as X for estimating Y, and (2) among such representations, $\Phi(X)$ is least useful for estimating W. In this way, we hope to produce a new set of features $\Phi(X)$ that retain information relevant for predicting the outcome but are less related to treatment assignment. *We propose learning Φ as a hidden layer in a neural network estimating a functional relationship of Y given X. We apply a genetic algorithm to a population of such mappings to evolve and select the one for which the associated representation $\Phi(X)$ is least useful for approximating W.*

Feature representations have previously been used for causal modeling. Johansson, et al. [4,14] viewed counterfactual inference as a covariate shift problem and learned representations designed to produce similar empirical distributions among the treatment and control populations. Li & Fu [8] and Yao, et al. [16] developed representations designed to preserve local similarity. However, we generally agree with Zhang et al.'s [17] recent argument that domain invariance often removes too much information from the features for causal inference.[1] *In contrast to most previous approaches, we develop a feature representation that attempts to preserve information useful for predicting the treatment effect if it is also useful for predicting the outcome.*

Outline. The next section describes related work. In Sect. 3, we outline our methodology. We validate our method on artificial data in Sect. 4 and on a publicly available experimental dataset in Sect. 5, before concluding in Sect. 6.

2 Related Work

In this section, we discuss meta-learning approaches for inferring the CATE and briefly introduce genetic algorithms.

Meta-learners. Meta-learning approaches leverage an arbitrary regression framework (e.g., random forests, neural networks, linear regression models) to estimate the CATE from data. The **S-learner** (single-learner) uses a standard

[1] Zhao et al. [18] make this argument in a more general setting.

supervised learner to estimate $\mu(x, w) = \mathbb{E}[Y|X = x, W = w]$ and then predicts $\hat{\tau}_S(x) = \hat{\mu}(x, 1) - \hat{\mu}(x, 0)$. The **T-learner** (two-learner) estimates $\mu_1(x) = \mathbb{E}[Y(1)|X = x]$ from treatment data and $\mu_0(x) = \mathbb{E}[Y(0)|X = x]$ from control data and then predicts $\hat{\tau}_T(x) = \hat{\mu}_1(x) - \hat{\mu}_0(x)$. The **X-learner** [6] estimates μ_1 and μ_0 as in the T-learner, and then predicts the contrapositive outcome for each training point. It then estimates $\tau_1(x) = \mathbb{E}[Y_i - \hat{\mu}_0(X_i) \mid X = x]$ and $\tau_0(x) = \mathbb{E}[\hat{\mu}_1(X_i) - Y_i \mid X = x]$ before predicting $\hat{\tau}_X(x) = g(x)\hat{\tau}_0(x) + (1 - g(x))\hat{\tau}_1(x)$ where $g : \mathbb{R}^d \to [0, 1]$ is a weight function.[2] The **R-learner** [11] leverages Robinson's decomposition that led to Robin's reformulation of the CATE function as the solution to $\tau(\cdot) = \arg\min_\tau \{\mathbb{E}_{(X,W,Y)\sim P}[|(Y - m(X)) - (W - e(X))\tau(X)|^2]\}$ in terms of the treatment propensity e and conditional mean outcome $m(x) = \mathbb{E}[Y|X = x]$.

Neuroevolutionary Algorithms. Holland introduced genetic algorithms [2] as a nature-inspired approach to optimization. These algorithms produce successive generations of candidate solutions. New generations are formed by selecting the fittest members from the previous generation and performing cross-over and/or mutation operations to produce new offspring candidates. Evolutionary algorithms encompass extensions to and generalizations of this approach including memetic algorithms that perform local refinements, genetic programming that acts on programs represented as trees, and evolutionary programming and strategies that operate on more general representations. When such methods are applied specifically to the design and training of neural networks, they are commonly called neuroevolutionary algorithms. See Stanley et al. [15] for a review.

3 Methodology

We now describe how to create our feature mapping $\Phi : \mathbb{R}^d \to \mathbb{R}^m$. Individual candidate solutions derive from a hidden layer in a network trained to predict Y from X. We then evolve cohorts of parameter sets for such maps to minimize the functional usefulness of candidate representations for predicting W.

Candidate Solutions. We consider neural networks $f_\Theta : \mathbb{R}^d \to \mathbb{R}$ of the form $f_\Theta(x) = M_2 \cdot a(M_1 \cdot x + b_1) + b_2$ for a nonlinear activation function a where the parameter set Θ denotes $M_1 \in \mathbb{R}^{m\times d}$, $M_2 \in \mathbb{R}^{1\times m}$, $b_1 \in \mathbb{R}^m$, and $b_2 \in \mathbb{R}^1$. Though f_Θ is decidedly not a deep neural network, we note that, as a neural network with a single hidden layer, it remains a universal function approximator in the sense of Hornik et al. [3]. Optimizing the network f_Θ in order to best predict Y from X seeks the solution $\Theta_* = \arg\min_\Theta \mathbb{E}\,|Y - f_\Theta(X)|^2$. For fixed Θ, we let $\Phi_\Theta : \mathbb{R}^d \to \mathbb{R}^m$ given by $\Phi_\Theta(x) = a(M_1 \cdot x + b_1)$ denote the output of the hidden layer.

---

[2] It is also possible to estimate τ from $\{(X_i, Y_i - \hat{\mu}_0(X_i))\}_{W_i=1} \cup \{(X_i, \hat{\mu}_1(X_i) - Y_i)\}_{W_i=0}$ or, using $\hat{\mu}(x, w)$ from the S-learner approach, with $\{(X_i, Y_i - \hat{\mu}(X_i, 0))\}_{W_i=1} \cup \{(X_i, \hat{\mu}(X_i, 1) - Y_i)\}_{W_i=0}$. We find that these alternate approaches work well in practice and obviate the need to estimate or fix g.

Fitness Function. For Θ near the optimum Θ_*, $\Phi_\Theta(X)$ should be approximately as useful as X for estimating Y. However, the mapped features $\Phi_\Theta(X)$ may also carry information useful for predicting W. To this end, we define $g_{\Psi,\Theta} : \mathbb{R}^d \to [0,1]$ by $g_{\Psi,\Theta}(x) = \sigma(M_4 \cdot a(M_3 \cdot \Phi_\Theta(x) + b_3) + b_4)$ for a nonlinear activation a, sigmoidal activation σ, and parameter set Ψ consisting of $M_3 \in \mathbb{R}^{k\times m}$, $M_4 \in \mathbb{R}^{1\times k}$, $b_3 \in \mathbb{R}^k$, and $b_4 \in \mathbb{R}$. We define the fitness of a parameter set Θ to be $\mu(\Theta) = \min_\Psi \mathbb{E}\,|W - g_{\Psi,\Theta}(X)|^2$. In this way, we express a preference for representations $\Phi_\Theta(X)$ that are less useful for predicting W.

Evolutionary Algorithm. Given training and validation datasets, we form an initial cohort of c candidates independently as follows. For $1 \le j \le c$, we randomly instantiate Θ_j using Glorot normal initialization for the weights and apply the Adam optimizer on training data to seek Θ_*. We use Tikhonov regularization for the weights and apply dropout after the $a(x) = \tanh(x)$ activation function[3] to prevent overfitting. For each constituent Θ_j in the cohort, we then initialize and train a network g_{Ψ,Θ_j} to seek $\Psi_j = \arg\min_\Psi \mathbb{E}\left|W - g_{\Psi,\Theta_j}(X)\right|^2$ on the training set and then evaluate $\mathbb{E}\left|W - g_{\Psi_j,\Theta_j}(X)\right|^2$ empirically on the validation set to estimate $\mu(\Theta_j)$. For each of the $\binom{\ell}{2}$ pairs formed from the ℓ fittest members of the current cohort, we apply Montana and Davis's node-based crossover [9] method to the parameters M_1 and b_1 that we use to form Φ. The next generation then consists of the fittest candidate from the previous generation, the candidates formed from cross-over, and new candidates generated from scratch.

Remarks. Due to our choice of representation Φ, after training the network $f_\Theta(x)$, we expect the relationship between the learned features $\Phi(X)$ and the outcome Y to be approximately linear. In particular, $Y \approx M_2 \cdot \Phi(X)$. For this reason, the causal meta-learners trained using a linear regression base learner may benefit more extensively from using the transformed features instead of the original features, especially in cases where the relationship between the original features and outcomes is not well-approximated as linear.

In order to use the represented features $\Phi(X_i)$ in place of the original features X_i, we require that strong ignorability holds for the transformed dataset $(\Phi(X_i), W_i, Y_i)$, $i = 1, \dots, n$. One sufficient, though generally not necessary, assumption that would imply strong ignorability is for Φ to be invertible on the support of X [14, assumption 1]. Unconfoundedness would also be guaranteed if $\Phi(X)$ satisfied the backdoor condition with respect to (W, Y) [12, section 3.3.1].

4 Ablation Study on Generated Data

Due to the fundamental challenge of causal inference (namely, that the counterfactual outcome cannot be observed, even in controlled experiments), it is common practice to compare approaches to CATE estimation on artificially generated

[3] We tested rectified and exponential linear unit activation functions for a in Φ_Θ but noticed only minor differences in subsequent performance of the causal forest.

datasets for which the CATE can be calculated. In this section, we perform experiments using Setups A and C from Nie & Wager's paper [11].[4]

Comparison Methodology. For both setups, we ran 100 independent trials. Within each trial, we randomly partitioned a simulated dataset it into training, validation, and testing subsets at a 70%-15%-15% rate. We trained causal inference methods on the training and validation sets, and predicted on the test dataset. We then developed a feature map using the training and validation data as described in the previous section, applied this map to all features, and repeated the training and testing process using the new features. To determine the impact of the fitness selection process, we also learned a feature transformation that did not make use of the fitness function at all. It simply generated a single candidate mapping and used it to transform all the features. This ablative method is referred to as "no-fitness" in Table 1. We compared the causal forest [1] with default options, and the S-, T-, and X-learners with two base learners: LightGBM [5] and cross-validated ridge regression.

Results. We report results in Table 1. For both setups, we consider a paired t-test for equal means against a two-sided alternative. For setup A, we find that the improvement in MSE from using the transformed features in place of the original features corresponds to a statistically significant difference for the following learners: the causal forest ($p < 0.001$), the S-learner with ridge regression ($p < 0.001$), the T-learner with both LightGBM ($p < 0.001$) and ridge regression ($p < 0.001$), and the X-learner with both LightGBM ($p < 0.001$) and ridge regression ($p < 0.001$). For setup C, we again find significant differences for the causal forest ($p = 0.023$), S-learner with LightGBM ($p = 0.003$), T-learner with ridge regression ($p < 0.001$) and X-learner with ridge regression ($p < 0.001$). In summary, we find that our feature transformation method improves the performance of multiple standard estimators for the CATE under two data generation models.

Table 1. Average Mean Squared Error (MSE) over 100 independent trials.

learner	features	Set. A	Set. C
Causal forest	initial	0.175	0.035
	no-fitness	0.114	0.029
	transformed	0.120	0.029
S-L. LGBM	initial	0.149	0.226
	no-fitness	0.140	0.211
	transformed	0.135	0.204
S-L. Ridge	initial	0.093	0.015
	no-fitness	0.079	0.015
	transformed	0.081	0.015
T-L. LGBM	initial	0.666	0.567
	no-fitness	0.536	0.551
	transformed	0.512	0.544
T-L. Ridge	initial	0.745	0.178
	no-fitness	0.333	0.125
	transformed	0.325	0.128
X-L. LGBM	initial	0.411	0.313
	no-fitness	0.335	0.313
	transformed	0.317	0.298
X-L. Ridge	initial	0.630	0.166
	no-fitness	0.289	0.109
	transformed	0.288	0.114

[4] Nie & Wager's paper included four setups, namely A–D; however setup B modeled a controlled randomized trial and setup D had unrelated treatment and control arms.

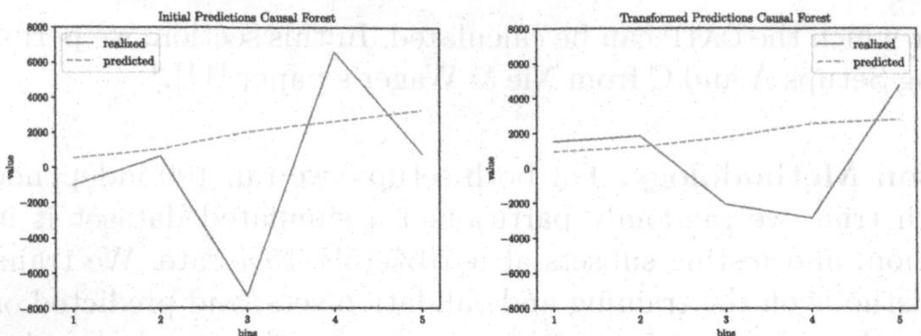

Fig. 1. We plot estimated realized and predicted average treatment effects versus the quintiles of predicted treatment effect (5 bins) for a causal forest using (left) the initial features and (right) the features transformed using our method.

5 Application to Econometric Data

In this section, we apply our feature engineering method to the LaLonde dataset [7] chronicling the results of an experimental study on temporary employment opportunities. The dataset contains information from 445 participants who were randomly assigned to either an experimental group that received a temporary job and career counseling or to a control group that received no assistance. We consider the outcome of earnings in 1978 (in $, after treatment).

We cannot determine true average treatment effects based on individual-level characteristics (i.e. the true CATE values) for real life experimental data as we can with the synthetic examples of the previous section. Instead, we evaluate performance by comparing the average realized and predicted treatment effects within bins formed by sorting study participants according to predicted treatment effect as demonstrated in Fig. 1. Applying the causal forest predictor to the original features results in a root mean square difference between the average predicted and realized treatment effects of 4729.51. Using the transformed features improves this discrepancy to 3114.82. From a practical perspective, one may learn the CATE in order to select a subset of people for whom a given intervention has an expected net benefit (and then deliver that intervention only to persons predicted to benefit from it). When we focus on the 20% of people predicted to benefit most from this treatment, we find that the estimated realized benefit for those chosen using the transformed features ($4732.89) is much greater than the benefit for those chosen using the original feature set ($816.92). This can be seen visually in Fig. 1 by comparing bin #5 (the rightmost bin) in both plots.

6 Conclusions

Causal inference, especially on real life datasets, poses significant challenges but offers a crucial avenue for predicting the impact of potential interventions. Learned feature representations help us to better infer the CATE, improving our ability to individually tailor predictions and target subsets of the general population. In this paper, we propose and validate a novel representation-based method that uses a neuroevolutionary approach to remove information from features irrelevant for predicting the outcome. We demonstrate that this method can yield improved estimates for the CATE on standard synthetic examples and illustrate its use on a real life dataset. We believe that representational learning is particularly well-suited for removing extraneous information in causal models and anticipate future research in this area.

Acknowledgements. We would like to thank Binjie Lai, Yi-Hong Kuo, Xiang Wu, and the anonymous reviewers for their insights and suggestions.

References

1. Athey, S., et al.: Generalized random forests. Ann. Statist. **47**(2), 1148–1178 (2019)
2. Holland, J.H.: Adaptation in Natural and Artificial Systems. University of Michigan Press, Ann Arbor (1975)
3. Hornik, K., et al.: Multilayer feedforward networks are universal approximators. Neural Netw. **2**(5), 359–366 (1989)
4. Johansson, F., et al.: Learning representations for counterfactual inference. In: International Conference on Machine Learning, pp. 3020–3029 (2016)
5. Ke, G., et al.: LightGBM. In: Advances in Neural Information Processing Systems, pp. 3146–3154 (2017)
6. Künzel, S.R., et al.: Metalearners for estimating heterogeneous treatment effects using machine learning. Proc. Natl. Acad. Sci. **116**(10), 4156–4165 (2019)
7. LaLonde, R.J.: Evaluating the econometric evaluations of training programs with experimental data. Am. Econ. Rev. **76**(4), 604–620 (1986)
8. Li, S.D., Fu, Y.: Matching on balanced nonlinear representations for treatment effects estimation. In: Advances in Neural Information Processing Systems, pp. 930–940 (2017)
9. Montana, D.J., Davis, L.: Training feedforward neural networks using genetic algorithms. In: International Joint Conference on Artificial Intelligence, pp. 762–767 (1989)
10. Neyman, J.: Sur les applications de la théorie des probabilités aux experiences agricoles: Essai des principes. Rocz. Nauk Rol. **10**, 1–51 (1923)
11. Nie, X., Wager, S.: Quasi-oracle estimation of heterogeneous treatment effects. Biometrika **108**(2), 299–319 (2021)
12. Pearl, J.: Causality, 2nd edn. Cambridge University Press, Cambridge (2009)
13. Rosenbaum, P.R., Rubin, D.B.: The central role of the propensity score in observational studies for causal effects. Biometrika **70**(1), 41–55 (1983)
14. Shalit, U., et al.: Estimating individual treatment effect: generalization bounds and algorithms. In: International Conference on Machine Learning, vol. 70, pp. 3076–3085 (2017)
15. Stanley, D.O., et al.: Designing neural networks through neuroevolution. Nat. Mach. Intell. **1**, 24–35 (2019)
16. Yao, L., et al.: Representation learning for treatment effect estimation from observational data. In: Advances in Neural Information Processing Systems, vol. 31, pp. 2633–2643 (2018)

17. Zhang, Y., et al.: Learning overlapping representations for the estimation of individualized treatment effects. In: International Conference on Artificial Intelligence and Statistics (2020)
18. Zhao, H., et al.: On learning invariant representations for domain adaptation. In: International Conference on Machine Learning (2019)

Compiling Linear Algebra Expressions into Efficient Code

Julien Klaus(✉), Mark Blacher, Joachim Giesen, Paul Gerhardt Rump, and Konstantin Wiedom

Friedrich Schiller University Jena, Jena, Germany
{julien.klaus,mark.blacher,joachim.giesen,paul.gerhardt.rump,konstantin.wiedom}@uni-jena.de

Abstract. In textbooks, linear algebra expressions often use indices to specify the elements of variables. This index form expressions cannot be directly translated into efficient code, since optimized linear algebra libraries and frameworks require expressions in index-free form. To address this problem, we developed *Lina*, a tool that automatically converts linear algebra expressions with indices into index-free linear algebra expressions that we map efficiently to NumPy and Eigen code.

Keywords: Linear algebra · Vectorization · Domain specific languages · Mathematics of computing

1 Introduction

In textbooks, linear algebra expressions often use indices to access the entries of vectors and matrices, or to sum over certain dimensions. These expressions can be translated directly into loops over indices in the program code, which, however, is often not efficient [1,6]. It is more efficient to map expressions with indices to highly tuned parallel linear algebra libraries like NumPy [9] or Eigen [8]. These libraries expect their input in index-free form. Therefore, in order to use efficient linear algebra libraries, expressions in index form need to be transformed into index-free form. We present an implementation of this approach, that we call *Lina*. *Lina* comprises three parts:

1. A formal input language close to textbook form (Sect. 2),
2. the transformation from index form into index-free form (Sect. 3), and
3. mappings to linear algebra libraries and frameworks (Sect. 4).

The transformation from index form to index-free form is the most challenging part and requires a good understanding of fundamental linear algebra. Consider for example the classical ridge regression problem [13]. Given a feature matrix $X \in \mathbb{R}^{n\times m}$, a label vector $y \in \mathbb{R}^n$, and hyperparameters $\beta \in \mathbb{R}^m$, $\mu, \lambda \in \mathbb{R}$ the ridge regression problem in textbook form reads as

$$\min_{\beta} \quad \sum_{i=1}^{n}\Big(y_i - \mu - \sum_{j=1}^{m} X_{ij}\beta_j\Big)^2 + \lambda \sum_{j=1}^{m} \beta_j.$$

D. Groen et al. (Eds.): ICCS 2022, LNCS 13351, pp. 11–17, 2022.
https://doi.org/10.1007/978-3-031-08754-7_2

This expression includes three summation operations, each of which become loops in the implementation. Transforming the expression into index-free form results in

$$\min_{\beta} \quad (y - \mu \cdot \mathbb{1}^n - X^\top \beta)^\top (y - \mu \cdot \mathbb{1}^n - X^\top \beta) + \lambda \cdot \beta^\top \mathbb{1}^m,$$

where $\mathbb{1}^n = (1, 1, \ldots, 1)$ is the all-ones vector. Since the index-free expression does only contain compound linear algebra operations, we can map it directly to linear algebra libraries. However, developers usually do not take the time to formulate their problems in index-free form, although using a highly optimized linear algebra library would lead to a better performance. Here, our focus is on automatically transforming expressions into index-free form. We further optimize the resulting index-free expressions before we map them to Eigen and NumPy routines. An easy-to-use implementation of our approach can be found online at https://lina.ti2.uni-jena.de.

Related Work. Various approaches already exist for mapping expressions in index-free form to linear algebra libraries [7,14,15]. Often such methods make use of additional information about the expressions' variables and parameters, for example, symmetry of matrices [2,16]. Also, multiple attempts are known to generate efficient code for expressions in index form [3,4,12]. These approaches are not transforming expressions into index-free form, but directly optimize the loops, for instance by reordering.

2 A Language for Linear Algebra Expressions

In this section, we describe a formal language for extended linear algebra expressions in index form. The notation used in this language is close to MATLAB [10]. It is rich enough to cover most classical machine learning problems, even problems not contained in standard libraries like scikit-learn [5].

⟨*expr*⟩	::=	⟨*term*⟩ {('+' | '-') ⟨*term*⟩}
⟨*term*⟩	::=	['-'] ⟨*factor*⟩ {('*' | '/') ['-'] ⟨*factor*⟩}
⟨*factor*⟩	::=	⟨*atom*⟩ ['^' ⟨*factor*⟩]
⟨*atom*⟩	::=	number | ⟨*function*⟩ '(' ⟨*expr*⟩ ')' | ⟨*variable*⟩
⟨*function*⟩	::=	'sin' | 'cos' | 'exp' | 'log' | 'sign' | 'sqrt' | 'abs' | 'sum' '[' ⟨*index*⟩ ']'
⟨*variable*⟩	::=	alpha+ ['[' ⟨*index*⟩ {',' ⟨*index*⟩} ']']
⟨*index*⟩	::=	alpha

Fig. 1. EBNF grammar for linear algebra expressions in index form. In this grammar, *number* is a placeholder for an arbitrary floating point number and *alpha* for Latin characters.

The language supports binary operations as well as unary point-wise operations like log or exp, and of course, variables and numbers. A special operation is the summation operation *sum* that has a non-optional index. This index is used to address elements of vectors or matrices. The full grammar for the language is shown in Fig. 1. In this language, the classical ridge regression example reads as

$$\mathrm{sum}[i]((y[i] - \mu - \mathrm{sum}[j](X[i,j] * \beta[j]))^2) + \lambda * \mathrm{sum}[j](\beta[j]).$$

A point worth emphasizing is that the indices always select scalar entries of a vector or matrix. This makes every operation an operation between scalars, which is different in index-free notation, where operations are on compound structures.

Expressions that follow the above grammar are parsed into an expression tree. An expression tree $G = (V, E)$ is a binary tree, where each node $v \in V$ has a specific label. This label can be either an operation, a variable name, a number, or an index. Furthermore, we assign each node a scope, containing indices. For leaf nodes, describing vectors and matrices, the scope contains the associated indices, and for all other nodes, except for the special *sum* nodes, the scope is the union of the scopes of the child nodes. Since the *sum* operation removes an index, the scope of a *sum* node removes an index from the union of their children's scopes. An expression tree for the ridge regression example is shown in Fig. 2.

3 Transformation from Index Form into Index-Free Form

The main part of *Lina* is the automatic transformation of expressions from index form into index-free form. In index form expressions, all operations are operations on scalars, whereas in index-free form expressions operations are on compound structures like vectors or matrices. For example, the multiplication $X_{ij} \cdot \beta_j$ multiplies the value of X at index (i, j) with the value of β at index j. We can collect these values into an $(m \times n)$-matrix $(X_{ij} \cdot \beta_j)_{i \in [n], j \in [m]}$. This matrix can be transformed into a point-wise product of two matrices, where X is the first matrix and the outer product $\mathbb{1}^n \beta^\top$ is the second matrix. Indeed, we compute

$$\begin{aligned}
\begin{pmatrix} X_{11} \cdot \beta_1 & \dots & X_{1m} \cdot \beta_m \\ \vdots & \ddots & \vdots \\ X_{n1} \cdot \beta_1 & \dots & X_{nm} \cdot \beta_m \end{pmatrix} &= \begin{pmatrix} X_{11} & \dots & X_{1m} \\ \vdots & \ddots & \vdots \\ X_{n1} & \dots & X_{nm} \end{pmatrix} \odot \begin{pmatrix} \beta_1 & \dots & \beta_m \\ \vdots & \ddots & \vdots \\ \beta_1 & \dots & \beta_m \end{pmatrix} \\
&= \begin{pmatrix} X_{11} & \dots & X_{1m} \\ \vdots & \ddots & \vdots \\ X_{n1} & \dots & X_{nm} \end{pmatrix} \odot \begin{pmatrix} 1 \\ \vdots \\ 1 \end{pmatrix} \begin{pmatrix} \beta_1 \\ \vdots \\ \beta_m \end{pmatrix}^\top .
\end{aligned}$$

The idea of increasing the dimension of a subexpression by an outer product to enable a point-wise operation works directly for vectors, but not for matrices. In

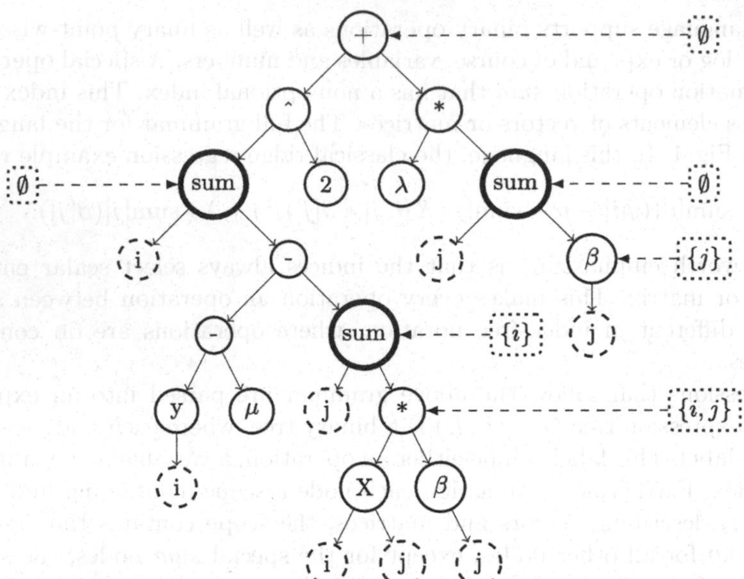

Fig. 2. Expression tree for the ridge regression problem with different node types. Bold nodes indicate sum operations, dashed nodes are indices, and all other nodes are either common operators or variables. For some nodes, we show the scope of the node in dotted rectangles.

these cases, we use the scope assigned to each node. To work with the scopes, we switch to an Einstein like notation. In this notation, each multiplication is represented by a tuple (m, l, r). The tuple (m, l, r) contains the dimension m of the result of the operation, the dimension l of the left, and the dimension r of the right operand. The following equations show how to represent the different multiplication types in Einstein-like notation as well as in linear algebra notation. Let $x \in \mathbb{R}^n$ and $y \in \mathbb{R}^n$, then

$$x^\top \cdot y = x \cdot_{(\emptyset,n,n)} y, \qquad \text{(inner product)}$$

$$x \odot y = x \cdot_{(n,n,n)} y, \qquad \text{(point-wise multiplication)}$$

$$x \cdot y^\top = x \cdot_{(nn,n,n)} y. \qquad \text{(outer product)}$$

The scope of a node and the scopes of its left and right child, respectively, directly correspond to the entries of the tuple. This notation enables us to describe the multiplication type we need during the transformation. The transformation starts at the root and recursively adjust the left and right child of nodes to satisfy their index requirements. If we encounter a node, except a *sum* node, with a left or right child that does not have the same scope, we adjust the scope by adding a multiplication node to the respective subtrees. The new multiplication nodes multiply the old subtrees with an all-ones vectors to supply the missing index. In our example, we have multiplied an all-ones vector with β to supply the dimension represented by index i. The *sum* node is special since it removes an index from the scope. This can be accomplished by an inner-product with an all-ones vector. We therefore relabel *sum* nodes as multiplication nodes, and change their left child nodes, which represent an index, into all-ones vectors with the corresponding indices.

In summary, we use outer products to transform binary operations over unequal dimensions into point wise operations over equal dimensions and inner products to reduce a dimension by *sum* operations. The transformed expression tree, for the ridge regression example, is shown in Fig. 3. There, we highlight added and adjusted nodes in gray. Note, that *sum* nodes have turned into multiplication nodes. The semantics of product nodes can be decided from their scopes.

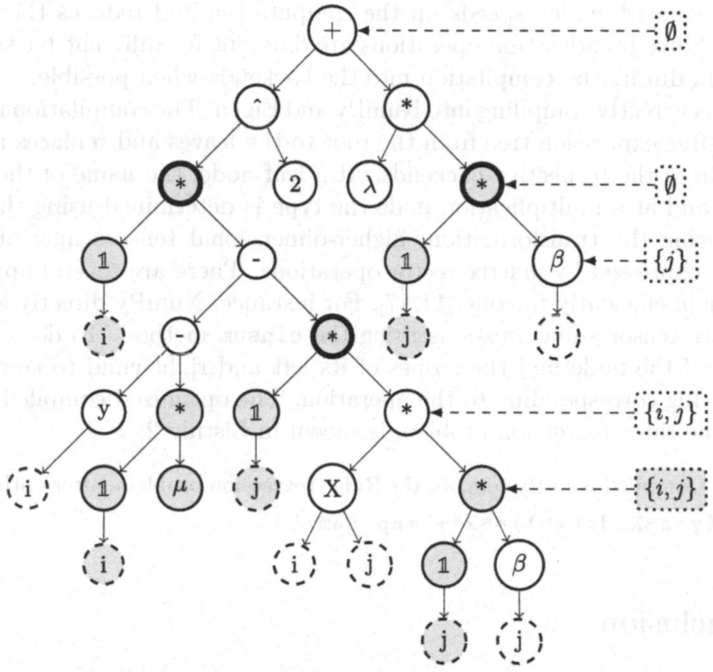

Fig. 3. Expression tree for the ridge regression problem after the transformation into index-free form. Nodes that have been transformed are highlighted in gray. For clarity, we still show the index nodes, although they are no longer needed.

4 Compilation into Multiple Backends

During the transformation from index to index-free form, the nodes of the expression tree are replaced by compound subexpressions, which can make the tree unnecessarily large. Listing 1 shows non-optimized index-free NumPy code.

Listing 1. Compiled NumPy code for the ridge regression problem given as $\sum_{i=1}^{n}(y_i - \mu - \sum_{j=1}^{m} X_{ij}\beta_j)^2 + \lambda \sum_{j=1}^{m} \beta_j$ without any optimization.

```
np.add(np.sum(np.multiply(np.ones(y_shape[0]),np.power(np.
    subtract(np.subtract(y,np.multiply(m,np.ones(y_shape[0]))
    ),np.dot(np.multiply(X,b),np.ones(X_shape[1]))),2))),np.
    multiply(l,np.sum(np.multiply(np.ones(X_shape[1]),b))))
```

Therefore, before we compile the expression tree into the different backends, we perform various optimizations that reduce the size of the tree. For instance, we perform a default constant folding, and identify common subexpressions, extract them, and link nodes in the tree to the corresponding common subexpression. Furthermore, we exploit broadcasting operations of the individual backends like NumPy or Eigen. For example, the subexpression $y - \mathbb{1}^n \cdot \mu$ in index-free form can be written as $y - \mu$ in most backends. In this case, no extra all-ones vector has to be created, which speeds up the computation and reduces the memory footprint. Since broadcasting operations are different for different backends, we apply them during the compilation into the backends when possible.

Lina is currently compiling into NumPy and Eigen. The compilation traverses the index-free expression tree from the root to the leaves and replaces nodes by methods from the respective backends. At a leaf node, the name of the node is returned, and at a multiplication node the type is determined using the nodes' scope. During the transformation, higher-dimensional tensors may arise that cannot be expressed by matrix-vector operations. There are several approaches to map them efficiently to code [11,17]. For instance, NumPy directly allows us to calculate tensor subexpressions using the `einsum` method. To do so, we use the scope of the node and the scopes of its left and right child to compute an `einsum` string corresponding to the operation. The optimized compiled NumPy code for the ridge regression problem is shown in Listing 2.

Listing 2. Compiled NumPy code for the Ridge regression problem, after optimization.

```
np.sum((y-m-X.dot(b))**2)+l*np.sum(b)
```

5 Conclusion

We have presented *Lina*, a tool for automatically compiling extended linear algebra expressions with indices into efficient linear algebra routines. Our main contribution is the transformation of expressions with indices into index-free form. We map this index-free form to NumPy and Eigen, which exploit the parallelization and vectorization capabilities of the underlying hardware. *Lina*, is available at https://lina.ti2.uni-jena.de.

Acknowledgements. This work was supported by the Carl Zeiss Foundation within the project *Interactive Inference* and from the Ministry for Economics, Sciences and Digital Society of Thuringia (TMWWDG), under the framework of the Landesprogramm ProDigital (DigLeben-5575/10-9).

References

1. Ascher, D., Dubois, P.F., Hinsen, K., Hugunin, J., Oliphant, T., et al.: Numerical Python (2001)
2. Barthels, H., Psarras, C., Bientinesi, P.: Linnea: automatic generation of efficient linear algebra programs. ACM Trans. Math. Softw. **47**(3), 1–26 (2021)
3. Baumgartner, G., et al.: Synthesis of high-performance parallel programs for a class of ab initio quantum chemistry models. Proc. IEEE **93**(2), 276–292 (2005)
4. Bilmes, J.A., Asanovic, K., Chin, C., Demmel, J.: Author retrospective for optimizing matrix multiply using PHiPAC: a portable high-performance ANSI C coding methodology. In: ACM International Conference on Supercomputing 25th Anniversary Volume. ACM (2014)
5. Buitinck, L., et al.: API design for machine learning software: experiences from the Scikit-learn project. In: ECML PKDD Workshop (2013)
6. Cai, X., Langtangen, H.P., Moe, H.: On the performance of the python programming language for serial and parallel scientific computations. Sci. Program. **13**(1) (2005)
7. Franchetti, F., et al.: SPIRAL: extreme performance portability. Proc. IEEE **106**(11), 1935–1968 (2018)
8. Guennebaud, G., Jacob, B., et al.: Eigen v3. http://eigen.tuxfamily.org (2010)
9. Harris, C.R., et al.: Array programming with NumPy. Nature **585**(7825), 357–362 (2020)
10. The Mathworks Inc, Natick, Massachusetts: MATLAB version R2021a (2021)
11. Matthews, D.A.: High-performance tensor contraction without transposition. SIAM J. Sci. Comput. **40**(1), C1–C24 (2018)
12. Nuzman, D., et al.: Vapor SIMD: auto-vectorize once, run everywhere. In: Proceedings of the CGO 2011. IEEE Computer Society (2011)
13. Owen, A.B.: A robust hybrid of lasso and ridge regression. Contemp. Math. **443**(7), 59–72 (2007)
14. Psarras, C., Barthels, H., Bientinesi, P.: The linear algebra mapping problem. arXiv preprint arXiv:1911.09421 (2019)
15. Sethi, R., Ullman, J.D.: The generation of optimal code for arithmetic expressions. J. ACM **17**(4), 715–728 (1970)
16. Spampinato, D.G., Fabregat-Traver, D., Bientinesi, P., Püschel, M.: Program generation for small-scale linear algebra applications. In: Proceedings of the 2018 International Symposium on Code Generation and Optimization. ACM (2018)
17. Vasilache, N., et al.: Tensor comprehensions: framework-agnostic high-performance machine learning abstractions. CoRR abs/1802.04730 (2018)

Interval Modification of the Fast PIES in Solving 2D Potential BVPs with Uncertainly Defined Polygonal Boundary Shape

Andrzej Kużelewski(✉), Eugeniusz Zieniuk, and Marta Czupryna

Institute of Computer Science, University of Bialystok,
Ciolkowskiego 1M, 15-245 Bialystok, Poland
{a.kuzelewski,e.zieniuk,m.czupryna}@uwb.edu.pl

Abstract. The paper presents a new modification of the fast parametric integral equations system (FPIES) by application of interval numbers and interval arithmetic in solving potential 2D boundary value problems with complex shapes. Obtained interval modified fast PIES is used to model the uncertainty of measurement data, which are necessary to define boundary shape. The uncertainty was defined using interval numbers and modelled using modified directed interval arithmetic previously developed by the authors. The reliability and efficiency of the interval modified fast PIES solutions obtained using such arithmetic were verified on 2D complex potential problems with polygonal domains. The solutions were compared with the interval solutions obtained by the interval PIES. All performed tests indicated high efficiency of the interval modified fast PIES method.

Keywords: Fast PIES · Interval numbers · Interval arithmetic · Directed intervals

1 Introduction

One of the robust numerical tools for solving boundary value problems (BVPs) is the parametric integral equations system (PIES) [1]. The method was successfully used to solve many different problems (e.g. [2,3]). The disadvantages of the PIES connected with the generation of dense non-symmetric coefficient matrices and with the method of solving the final system of algebraic equations (Gaussian elimination) were fixed by the application of the fast multipole method (FMM) [4]. Obtained fast PIES (FPIES) [5,6] significantly reduced the computation time, as well as the problem of huge random access memory (RAM) utilization.

The authors of this paper also developed the interval PIES (IPIES) [7], which is used to solve uncertainly defined problems. It is known, that in modelling and solving BVPs the shape of the boundary, boundary conditions and some parameters of the considered domain (i.e. material properties) should be defined. In

D. Groen et al. (Eds.): ICCS 2022, LNCS 13351, pp. 18–25, 2022.
https://doi.org/10.1007/978-3-031-08754-7_3

practice, to obtain these data we should measure some physical quantities. However, even the most precise measurement is not exact - inaccuracy of measurement instruments, gauge reading error or approximations of the models used in the analysis of measurements affect the accuracy of determining the physical quantity.

It should be noted, that the direct consideration of uncertainty in classical mathematical models is not possible - they required exact values of the data. However, in the literature, we can find a lot of modifications of known methods considered uncertainty (e.g. [8–10]). One of them is connected with the application of interval numbers and interval arithmetic to the method of modelling and solving uncertainly defined BVPs. Therefore, it was used in the interval finite element method (IFEM) [11] and the interval boundary element method (IBEM) [12], as well as the IPIES.

In general, either in IFEM or IBEM the uncertainty of the boundary shape is not considered (only material parameters or boundary conditions). Only in a few papers, some parameters of the shape (such as radius or beam length) were uncertainly defined. In the IPIES all uncertainties can be considered simultaneously [13]. Although the IPIES has advantages inherited from the PIES, such as the way of defining the boundary connected with a small number of interval control points, there are also some disadvantages. Unfortunately, the application of interval arithmetic and interval numbers made computations slower and utilized more RAM than in the PIES. Therefore, solving complex (large-scale) uncertainly defined problems required a combination of the IPIES and the FPIES.

The main goal of this paper is to present the interval modified fast PIES (IFPIES) applied for numerical solving of 2D potential complex BVPs with uncertainly defined boundary shapes. The application of interval arithmetic and interval numbers into the FPIES was required to obtain the new method for modelling and solving uncertainly defined problems. The efficiency and accuracy of the IFPIES are tested on the potential problems with polygonal domains.

2 Modelling Uncertainty of the Boundary Shape

Direct application of either classical [14] or directed [15] interval arithmetic for modelling boundary problems with uncertainly defined boundary shapes is very troublesome as presented in [13]. The main problem is the consideration of unrealistic problems as a result of the lack of continuity between boundary segments. Modelling the same boundary shape in different quadrants of the Cartesian coordinate system gives different results. Therefore, the authors proposed to modify the directed interval arithmetic by mapping arithmetic operators to the positive semi-axis as clearly described in [13].

In this paper, for modeling uncertainly defined boundary shape, linear segments in form of interval Bézier curves of the first degree are used:

$$\boldsymbol{S}_k(s) = (1-s)\boldsymbol{P}_{b(k)} + s\boldsymbol{P}_{e(k)}, \quad 0 \leq s \leq 1 \tag{1}$$

where $\boldsymbol{S}_k(s) = \{\boldsymbol{S}_k^{(1)}(s), \boldsymbol{S}_k^{(2)}(s)\}$, $k = \{1, 2, ..., n\}$ - the number of segments created boundary, s - variable in parametric reference system, $^{(1)}$ and $^{(2)}$ - the direction of coordinates in 2D Cartesian reference system, $\boldsymbol{P}_{b(k)}$,$\boldsymbol{P}_{e(k)}$ - interval endpoints which define interval Bézier curves as presented in Fig. 1.

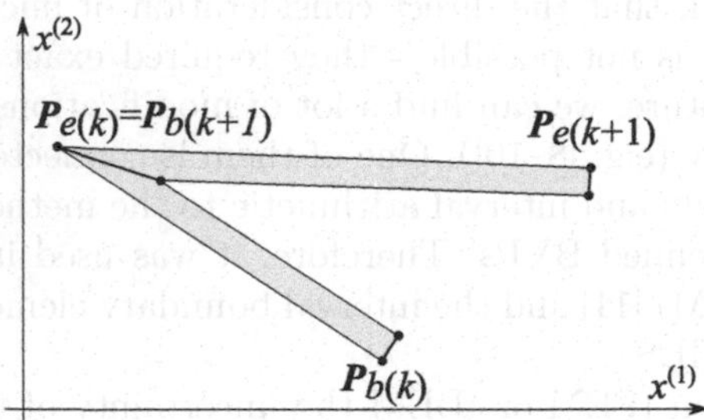

Fig. 1. The interval Bézier curve of the first degree used to define a segment of the boundary in the IPIES

3 The Interval Modified Fast PIES (IFPIES)

The FPIES for 2D potential problems [5] was obtained as the result of modification of the PIES. It includes the modification of the PIES kernels to allow for the Taylor series approximation used by the FMM. Also, the tree used by the FMM was modified to properly include the way of defining the boundary in the PIES [16]. IFPIES is obtained similarly to the FPIES. However, the application of the modified directed interval arithmetic and interval numbers is not trivial. Some variables should be defined as complex intervals, i.e. either real or imaginary part of a complex number is treated as an interval.

The basic form of the IFPIES formula is similar to the FPIES [5], however most variables are defined using interval numbers similarly to the IPIES [7]:

$$\begin{aligned} \frac{1}{2}u_l(\widehat{s}) = \sum_{j=1}^{n} \mathbb{R}\left\{ \int_{s_{j-1}}^{s_j} \widehat{\boldsymbol{U}}_{lj}^{*(c)}(\widehat{s}, s) p_j(s) \boldsymbol{J}_j^{(c)}(s) ds \right\} - \\ - \sum_{j=1}^{n} \mathbb{R}\left\{ \int_{s_{j-1}}^{s_j} \widehat{\boldsymbol{P}}_{lj}^{*(c)}(\widehat{s}, s) u_j(s) \boldsymbol{J}_j^{(c)}(s) ds \right\}, \end{aligned} \tag{2}$$

$$l = 1, 2, ..., n, \; s_{l-1} \leq \widehat{s} \leq s_l, \; s_{j-1} \leq s \leq s_j,$$

where: $\widehat{s}$ and s are defined exactly in the parametric coordinate system, s_{j-1} (s_{l-1}) correspond to the beginning and s_j (s_l) to the end of interval segment $\boldsymbol{S}_j$ ($\boldsymbol{S}_l$), n is the number of parametric segments that creates boundary of domain in 2D, $\widehat{\boldsymbol{U}}_{lj}^{*(c)}(\widehat{s}, s)$ and $\widehat{\boldsymbol{P}}_{lj}^{*(c)}(\widehat{s}, s)$ are interval kernels, $\boldsymbol{J}_j^{(c)}(s)$ is the interval Jacobian, $u_j(s)$ and $p_j(s)$ are parametric boundary functions on individual segments $\boldsymbol{S}_j$ of the interval boundary, $\mathbb{R}$ is the real part of complex function.

In the IFPIES integrals are computed using the same formulas as in the FPIES (clearly derived in [5,6]). The main difference is in the way of defining interval variables. At last, the IFPIES integrals are described as follows:

$$\int_{s_{j-1}}^{s_j} \widehat{\boldsymbol{U}}_{lj}^{*(c)}(\widehat{s},s)\,p_j(s)\,\boldsymbol{J}_j^{(c)}(s)ds = \frac{1}{2\pi}\sum_{l=0}^{N_T}(-1)^l\cdot$$
$$\cdot\left\{\sum_{k=0}^{N_T}\sum_{m=l}^{N_T}\frac{(k+m-1)!\cdot \boldsymbol{M}_k(\boldsymbol{\tau}_c)}{(\boldsymbol{\tau}_{el}-\boldsymbol{\tau}_c)^{k+m}}\cdot\frac{(\boldsymbol{\tau}'_{el}-\boldsymbol{\tau}_{el})^{m-l}}{(m-l)!}\right\}\frac{(\widehat{\boldsymbol{\tau}}-\boldsymbol{\tau}'_{el})^l}{l!},$$
$$\int_{s_{j-1}}^{s_j} \widehat{\boldsymbol{P}}_{lj}^{*(c)}(\widehat{s},s)\,u_j(s)\,\boldsymbol{J}_j^{(c)}(s)ds = \frac{1}{2\pi}\sum_{l=0}^{N_T}(-1)^l\cdot \tag{3}$$
$$\cdot\left\{\sum_{k=1}^{N_T}\sum_{m=l}^{N_T}\frac{(k+m-1)!\cdot \boldsymbol{N}_k(\boldsymbol{\tau}_c)}{(\boldsymbol{\tau}_{el}-\boldsymbol{\tau}_c)^{k+m}}\cdot\frac{(\boldsymbol{\tau}'_{el}-\boldsymbol{\tau}_{el})^{m-l}}{(m-l)!}\right\}\frac{(\widehat{\boldsymbol{\tau}}-\boldsymbol{\tau}'_{el})^l}{l!}.$$

where: N_T is the number of terms in the Taylor expansion, $\widehat{\boldsymbol{\tau}} = \boldsymbol{S}_l^{(1)}(\widehat{s}) + i\boldsymbol{S}_l^{(2)}(\widehat{s})$, $\boldsymbol{\tau} = \boldsymbol{S}_j^{(1)}(s) + i\boldsymbol{S}_j^{(2)}(s)$, complex interval points $\boldsymbol{\tau}_c$, $\boldsymbol{\tau}_{el}$, $\boldsymbol{\tau}'_c$, $\boldsymbol{\tau}'_{el}$ are midpoints of leaves obtained during tracing the tree structure (see [5,16]). Expressions $\boldsymbol{M}_k(\boldsymbol{\tau}_c)$ and $\boldsymbol{N}_k(\boldsymbol{\tau}_c)$ are called moments (and they are computed twice only) and have the form [5,16]:

$$\boldsymbol{M}_k(\boldsymbol{\tau}_c) = \int_{s_{j-1}}^{s_j}\frac{(\boldsymbol{\tau}-\boldsymbol{\tau}_c)^k}{k!}p_j(s)\,\boldsymbol{J}_j^{(c)}(s)ds,$$
$$\boldsymbol{N}_k(\boldsymbol{\tau}_c) = \int_{s_{j-1}}^{s_j}\frac{(\boldsymbol{\tau}-\boldsymbol{\tau}_c)^{k-1}}{(k-1)!}\boldsymbol{n}_j^{(c)}u_j(s)\,\boldsymbol{J}_j^{(c)}(s)\,ds. \tag{4}$$

where $\boldsymbol{n}_j^{(c)} = \boldsymbol{n}_j^{(1)} + i\boldsymbol{n}_j^{(2)}$ the complex interval normal vector to the curve created segment j.

The IPIES is solved using the pseudospectral method, therefore it is written at collocation points whose number corresponds to the number of unknowns (described in [1]). Hence, obtained interval system of algebraic equations can be compact written as $\boldsymbol{H}\mathbf{u} = \boldsymbol{G}\mathbf{p}$, where $\mathbf{u}$ and $\mathbf{p}$ are column vectors containing coefficients of approximating boundary functions $u_j(s)$ and $p_j(s)$ respectively. This system is transformed into the system of interval algebraic equations $\boldsymbol{A}\cdot\boldsymbol{x} = \boldsymbol{b}$ depending on the given type of boundary conditions. The vector $\boldsymbol{x}$ represents unknown coefficients and the column vector $\boldsymbol{b}$ contains given boundary conditions. The matrix $\boldsymbol{A}$ is dense, therefore in the IPIES direct solver in form of interval Gaussian elimination was used to solve the system.

Unlike the IPIES, the IFPIES produces the system of algebraic equations implicitly, i.e. only the result of multiplication of the matrix $\boldsymbol{A}$ by the vector of unknowns $\boldsymbol{x}$ is obtained. Therefore, an iterative GMRES solver modified by the application of directed interval arithmetic directly integrated with the FMM was applied in the IFPIES. However, the direct solver in the IPIES requires $O(N^3)$ operations to solve the interval system of algebraic equations (N is the number of equations). We also applied the GMRES solver to the IPIES to obtain a more reliable comparison.

4 Numerical Results

The first example is the gear-shaped plate presented in the Fig. 2a. The problem is described by Laplace's equation. The boundary contains 1 024 segments. Boundary conditions are also presented in Fig. 2a (where u - Dirichlet and p - Neumann boundary conditions). Tests are performed on a PC based on Intel Core i5-4590S with 16 GB RAM. Application of the IPIES and the IFPIES are compiled by g++ 7.5.0 (-O2 optimization) on 64-bit Ubuntu Linux OS (kernel 5.4.0).

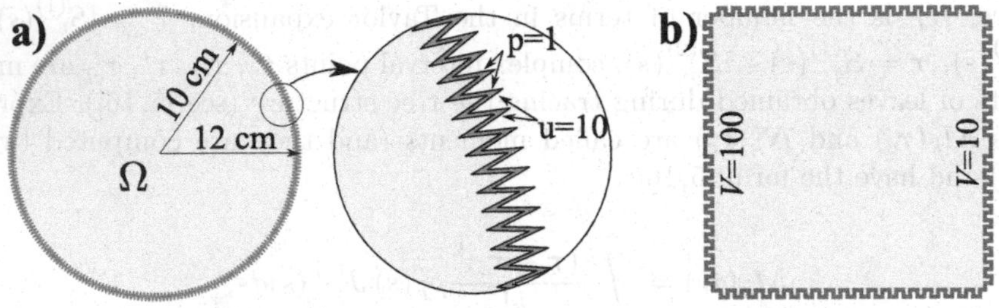

Fig. 2. Considered a) the gear-shaped, b) the square shaped boundary problem

The research focused on the CPU time, RAM utilization and the accuracy of the IFPIES compared to the IPIES. The mean square error (MSE) between infimum and supremum of the IFPIES and the IPIES solutions are computed to prove the accuracy of the proposed method.

Table 1. Comparison between the IFPIES and the IPIES

Number of		CPU time [s]		RAM utilization [MB]		MSE	
col. pts	eqs	IFPIES	IPIES	IFPIES	IPIES	inf	sup
2	2 048	12.09	37.07	28.28	197	0.0	0.0
4	4 096	49.78	160.65	83.81	775	$1.05 \cdot 10^{-13}$	$8.11 \cdot 10^{-14}$
6	6 144	119.05	395.83	168	1 741	$2.62 \cdot 10^{-10}$	$2.73 \cdot 10^{-10}$
8	8 192	239.29	768.29	316	3 122	$2.07 \cdot 10^{-11}$	$7.34 \cdot 10^{-11}$

Approximation of the modified PIES kernels uses 25 terms in the Taylor series, and the GMRES tolerance is equal to 10^{-8}. The number of collocation points is the same on all segments and equal to 2, 4, 6 or 8. Therefore, we should solve the system of 2 048, 4 096, 6 144 and 8 192 equations respectively.

As can be seen from Table 1, the IFPIES is about 3 times faster and uses up to 10 times less RAM than the IPIES. However, the mean square error (MSE) between both methods is on a very low level and does not exceed 10^{-10}. Hence, the IFPIES is as accurate as the IPIES.

The second example is the current flow through a square plate presented in Fig. 2b. The problem is also described by Laplace's equation. The boundary is composed of 16 004 segments. Potentials V (Dirichlet boundary conditions) are applied to two electrodes presented in Fig. 2b. Neumann boundary conditions in the rest of the boundary are equal to 0.

As in the previous example, 25 terms in the Taylor series, the GMRES tolerance equal to 10^{-8} and the number of collocation points from 2 to 8 are applied (the system of 32 008 to 128 032 equations is solved).

This example cannot be solved by the IPIES on a standard PC due to very high RAM utilization. Therefore, computations were carried out at the Computer Center of the University of Bialystok on Intel Xeon E5-2650v2 with 512 GB RAM. Application of the IPIES and the IFPIES are compiled by g++ 7.4.0 (-O2 optimization) on 64-bit OpenHPC and Centos Linux OS (kernel 3.10.0).

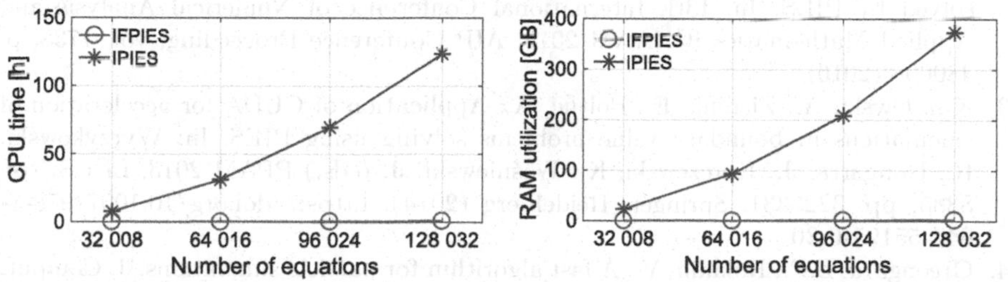

Fig. 3. The CPU time and the RAM utilization of the problem of the current flow through a square plate

As can be seen from a Fig. 3 relationship between the CPU time (the RAM utilization) and the number of equations in the IFPIES is close to linear contrary to the IPIES. The IFPIES uses about 53 min and 2.2 GB RAM contrary to 123.5 h and 373.2 GB RAM in the IPIES for the example with 128 032 equations. Hence, the IFPIES allows for solving large-scale uncertainly defined problems in a reasonable time and small RAM utilization on a standard PC.

5 Conclusions

The paper presents the IFPIES in solving 2D potential uncertainly defined boundary value problems. The FPIES was previously applied in modelling and solving 2D single- and multi-connected certainly defined potential problems. Applied fast multipole technique with a modified binary tree allows for significant reduction of CPU time, as well as RAM utilization. Also, the IFPIES allows for highly efficient solving of complex engineering problems on a standard PC in a reasonable time. However, the real power of the IFPIES is connected with the large size of solved problems and low RAM utilization. The IPIES allows solving the problems with a system up to about 25 000 equations on a standard PC with 16 GB of RAM, whilst 128 032 equations in the IFPIES use about 2.2 GB of RAM only.

Obtained results strongly suggest that the direction of research should be continued. The authors want to extend the algorithm of the IFPIES to problems with curvilinear boundary shapes, as well as modelled by other equations.

References

1. Zieniuk, E.: Hermite curves in the modification of integral equations for potential boundary-value problems. Eng. Comput. **20**(1–2), 112–128 (2003)
2. Kużelewski, A., Zieniuk, E.: OpenMP for 3D potential boundary value problems solved by PIES. In: 13th International Conference of Numerical Analysis and Applied Mathematics ICNAAM 2015, AIP Conference Proceeding, vol. 1738, p. 480098 (2016)
3. Kuzelewski, A., Zieniuk, E., Boltuc, A.: Application of CUDA for acceleration of calculations in boundary value problems solving using PIES. In: Wyrzykowski, R., Dongarra, J., Karczewski, K., Waśniewski, J. (eds.) PPAM 2013. LNCS, vol. 8385, pp. 322–331. Springer, Heidelberg (2014). https://doi.org/10.1007/978-3-642-55195-6_30
4. Greengard, L.F., Rokhlin, V.: A fast algorithm for particle simulations. J. Comput. Phys. **73**(2), 325–348 (1987)
5. Kużelewski, A., Zieniuk, E.: The fast parametric integral equations system in an acceleration of solving polygonal potential boundary value problems. Adv. Eng. Softw. **141**, 102770 (2020)
6. Kużelewski, A., Zieniuk, E.: Solving of multi-connected curvilinear boundary value problems by the fast PIES. Comput. Methods Appl. Mech. Eng. **391**, 114618 (2022)
7. Zieniuk, E., Kapturczak, M., Kużelewski, A.: Modification of interval arithmetic for modelling and solving uncertainly defined problems by interval parametric integral equations system. In: Shi, Y., et al. (eds.) ICCS 2018. LNCS, vol. 10862, pp. 231–240. Springer, Cham (2018). https://doi.org/10.1007/978-3-319-93713-7_19
8. Gouyandeh, Z., Allahviranloo, T., Abbasbandy, S., Armand, A.: A fuzzy solution of heat equation under generalized Hukuhara differentiability by fuzzy Fourier transform. Fuzzy Sets Syst. **309**, 81–97 (2017)
9. Wang, C., Matthies, H.G.: Dual-stage uncertainty modeling and evaluation for transient temperature effect on structural vibration property. Comput. Mech. **63**(2), 323–333 (2018). https://doi.org/10.1007/s00466-018-1596-3

10. Fu, C., Zhan, Q., Liu, W.: Evidential reasoning based ensemble classifier for uncertain imbalanced data. Inf. Sci. **578**, 378–400 (2021)
11. Ni, B.Y., Jiang, C.: Interval field model and interval finite element analysis. Comput. Methods Appl. Mech. Eng. **360**, 112713 (2020)
12. Zalewski, B., Mullen, R., Muhanna, R.: Interval boundary element method in the presence of uncertain boundary conditions, integration errors, and truncation errors. Eng. Anal. Boundary Elem. **33**, 508–513 (2009)
13. Zieniuk, E., Czupryna, M.: The strategy of modeling and solving the problems described by Laplace's equation with uncertainly defined boundary shape and boundary conditions. Inf. Sci. **582**, 439–461 (2022)
14. Moore, R.E.: Interval Analysis. Prentice-Hall, New York (1966)
15. Markov, S.M.: On directed interval arithmetic and its applications. J. Univ. Comput. Sci. **1**(7), 514–526 (1995)
16. Kużelewski, A., Zieniuk, E., Bołtuć, A., Szerszeń, K.: Modified binary tree in the fast PIES for 2D problems with complex shapes. In: Krzhizhanovskaya, V.V., et al. (eds.) ICCS 2020. LNCS, vol. 12138, pp. 1–14. Springer, Cham (2020). https://doi.org/10.1007/978-3-030-50417-5_1

Validation and Optimisation of Player Motion Models in Football

Moritz Renkin[1], Jonas Bischofberger[2](✉), Erich Schikuta[1], and Arnold Baca[2]

[1] Research Group Workflow Systems and Technology, Faculty of Computer Science, University of Vienna, Vienna, Austria
a11807211@unet.univie.ac.at

[2] Department of Biomechanics, Kinesiology and Computer Science in Sport, Institute of Sports Science, University of Vienna, Vienna, Austria
jonas.bischofberger@univie.ac.at

Abstract. Modelling the trajectorial motion of humans along the ground is a foundational task in the quantitative analysis of sports like association football. Most existing models of football player motion have not been validated yet with respect to actual data. One of the reasons for this lack is that such a validation is not straightforward, because the validation typically needs to be performed with respect to noisy extreme values rather than expected values.

This paper proposes a validation routine for trajectorial motion models that measures and optimises the ability of a motion model to accurately predict all possibly reachable positions by favoring the smallest predicted area that encompasses all observed reached positions up to a manually defined threshold. We demonstrate validation and optimisation on four different motion models, assuming (a) motion with constant speed, (b) motion with constant acceleration, (c) motion with constant acceleration with a speed limit, and (d) motion along two segments with constant speed. Our results show that assuming motion with constant speed or constant acceleration without a limit on the achievable speed is particularly inappropriate for an accurate distinction between reachable and unreachable locations. Motion along two segments of constant speed provides by far the highest accuracy among the tested models and serves as an efficient and accurate approximation of real-world player motion.

Keywords: Football · Positional data · Motion models · Performance analysis · Model validation · Complex systems

1 Introduction

Recently, professional association football has seen a surge in the availability of positional data of the players and the ball, typically collected by GPS, radar or camera systems. The growing availability of such data has opened up an exciting new avenue for performance analysis. High-quality measures of performance that

D. Groen et al. (Eds.): ICCS 2022, LNCS 13351, pp. 26–32, 2022.
https://doi.org/10.1007/978-3-031-08754-7_4

include positional information are invaluable for effective training, opposition scouting, and player recruitment.

The modelling of human motion is a foundational component of many performance metrics based on positional data. For example, algorithms that compute space control [7] or simulate passes [6] implicitly or explicitly make assumptions about human kinematics. These kinematic assumptions have never been verified so far, which calls the validity of these assumptions and the resulting models into question.

Rather than predicting actual human motion, many applications merely require the prediction of possibly reachable positions. This requirement essentially shifts the purpose of a motion model from predicting expected positions towards estimating the most remote reachable positions. Estimating such extreme values from real-world data can be difficult, because extreme values are typically rare and particularly likely to include a component of measurement error with an often unknown distribution that cannot easily be accounted for.

The contributions of this paper are twofold: First, we formally propose a validation routine for the quality of player motion models. Second, we use this routine to evaluate and optimise the parameters of four models of motion.

The rest of this paper is structured as follows: Sect. 2 provides some background on motion models in football and their validation. Section 3 formally presents our validation routine. Section 4 describes our exemplary model validation and optimisation based on a real data set and discusses its results. Section 5 summarises the contributions of this paper and points out possible directions of further research.

2 Motion Models in Football: State of the Art

Assumptions about the trajectorial motion of players are inherent to many performance indicators within the analysis of sports games. One example is the commonly used concept of space control which assigns control or influence over different areas on the pitch to players. It is used, for example, as a context variable to rate football actions [5] and for time series analyses [3]. Controlled space is often defined as the area that a player is able to reach before any other player, given a specific model of motion for each player. Commonly used for this purpose are motion models assuming constant and equal speed, which results in a Voronoi partition of the pitch, or accelerated movement with limited speed [7]. Spearman et al. [6] assume accelerated player motion with a limit on acceleration and velocity in the context of modeling ground passes.

Motion models have also been estimated directly from positional data [1,2]. However, such empirical models can be computationally expensive, prone to outliers and their current versions lend themselves less naturally to extreme value estimation than theoretically derived models. Attempts to validate trajectorial player motion models are rare. Notably, Caetano et al. [2] performed a validation of their space control model, and thus indirectly also the underlying motion model, by checking how many future positions of players fall within their associated controlled area for a number of time horizons.

3 Player Motion Model and Validation Procedure

We propose a validation procedure rating a player motion model on how well it fits some real positional data. In order to abstract our validation procedure from the underlying positional data, we introduce the concept of a *trail*. A trail represents a slice of a player's trajectory over some duration Δt. Formally, a trail is defined as the quadruple: $(\vec{x}_0, \vec{v}_0, \vec{x}_t, \Delta t)$

- $\vec{x}_0$: (2D) position of a given player at some arbitrary time t_0
- $\vec{v}_0$: (2D) velocity of the player at time t_0
- $\vec{x}_t$: (2D) position of the player at time $t = t_0 + \Delta t$
- Δt: time horizon (predefined)

Since every reached position is trivially contained in a large enough area, the validation function should take not only correctness but also precision of the model into account. The correctness of a motion model measures its ability to make true predictions, i.e. to predict reachable areas that contain the true target position $\vec{x}_t$. Precision refers to how well narrowed-down the predicted areas of a model are. There is a trade-off relationship between correctness and precision.

3.1 Measuring Correctness

Considering only a single trail, a motion model m makes a prediction for the reachable area using $\vec{x}_0$, $\vec{v}_0$ and Δt. If $\vec{x}_t$ is contained in the predicted reachable area, the model has made a correct prediction. Following this logic, a motion model achieves the highest possible correctness if and only if for every trail, the model predicts a reachable area in which $\vec{x}_t$ is contained. The ratio between the number of correct predictions $n_{correct}$ and the number of total predictions n_{total} of a model m for a sample of trails T will be called *hit_ratio*.

$$hit\_ratio(m, T) = \frac{n_{correct}}{n_{total}} \tag{1}$$

We can use the *hit_ratio* of a model as an indicator for its correctness. A high *hit_ratio* corresponds to a high correctness and vice-versa.

3.2 Measuring Precision

In the context of this paper, the precision of a motion model represents how much it narrows down the reachable area of a player. Smaller reachable areas imply a higher precision of the model and are generally preferable, given an equal *hit_ratio*.

To determine the precision of a model across multiple evaluated trails, we use the inverse of the mean surface area of all correctly predicted reachable areas. Incorrect predictions, where the target position $\vec{x_t}$ is not contained in the predicted reachable area are excluded from this average, since the precision of a model would otherwise increase inappropriately for very narrow, incorrect predictions. The precision of model m across a sample of trails T is given by:

$$precision(m,T) = \frac{1}{\frac{1}{n_{correct}} \cdot \sum areas_{correct}} = \frac{n_{correct}}{\sum areas_{correct}} \quad (2)$$

where $\sum areas_{correct}$ is the sum of all correctly predicted reachable areas.

3.3 Defining an Overall Validation Score

Since we aim for a single numerical value as a score for player motion models, correctness and precision have to be balanced in some way. Due to the fact that some measurement-related extreme outliers can usually be expected in positional data from football games, a model with a *hit_ratio* of 100% might not necessarily be desirable. Therefore, we introduce a minimum level of correctness hit_ratio_{min}, which represents a minimal required ratio between correct and total predictions of a model. We propose that if a motion model m satisfies the condition $hit\_ratio(m,T) \geq hit\_ratio_{min}$ for a trail sample T, the exact $hit\_ratio(m,T)$ should be indifferent for the overall validation score of m. This way, extreme outliers in the positional data caused by measurement-related errors have no influence on the validation score, as long as hit_ratio_{min} is chosen adequately.

Consequently, for a motion model m that exceeds hit_ratio_{min}, the validation score is only determined by the precision of the model (2). We define the *score* of a motion model m with the sample of trails T as:

$$score(m,T) = \begin{cases} 0 & \text{if } hit\_ratio(m,T) < hit\_ratio_{min} \\ precision(m,T) & \text{else} \end{cases} \quad (3)$$

The *score* measures how well a motion models fits a sample of positional data. hit_ratio_{min} can be considered a free parameter of this validation procedure. It should be chosen to accommodate for the error distribution of the positional data.

4 Experiment and Evaluation of Results

4.1 Data Set

For the evaluation, we use the public sample data set provided by Metrica Sports which consists of three anonymised games of football [4]. The positional data has been collected using a video-based system and is provided at a frequency 25 Hz. For this experiment, we use a constant time horizon of $\Delta t = 1s$ After visual inspection of the data, the minimal required hit ratio is set to $hit\_ratio_{min} =$ 99.975%. We evaluate the models on a random sample of $5 \cdot 10^5$ trails across all three games and all participating players.

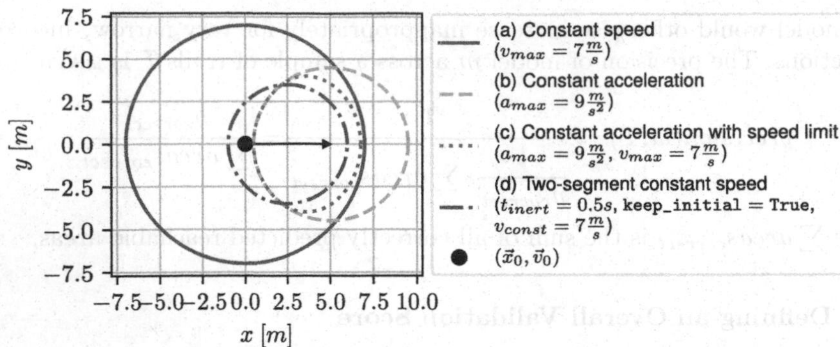

Fig. 1. Exemplary boundaries of the reachable area defined by different motion models when the player starts at $\vec{x}_0 = \vec{0}$ with velocity $\vec{v}_0 = \begin{bmatrix} 5\frac{m}{s} \\ 0 \end{bmatrix}$. The time horizon is $\Delta t = 1s$.

4.2 Preparation of Motion Models

We optimize and evaluate the following models of motion where each one defines a reachable area, depending on specific parameter values. These areas are exemplarily visualized in Fig. 1.

(a) Constant speed: Motion with constant speed v_{max} in any direction
(b) Constant acceleration: Motion with constant acceleration a_{max} in any direction
(c) Constant acceleration with speed limit: Motion with constant acceleration a_{max} in any direction until a maximal speed v_{max} is reached
(d) Two-segment constant speed: Motion along two segments of constant speed. During the first segment, the player is simulated to run in the direction of $\vec{v}_0$ with constant speed v_{seg1} which is set to either $|\vec{v}_0|$ or to some fixed value v_{const}, depending on the boolean parameter `keep_initial`. During the second segment, the speed of the player is set to either v_{const} or $\min(v_{seg1} + a_{max}t_{inert}, v_{max})$, depending on whether the parameters a_{max} and v_{max} are set.

Using the evaluation routine outlined in Sect. 3, we find the optimal parameter configuration for each model via Bayesian optimization. Discrete parameters like `keep_initial` are handled by performing one round of Bayesian optimisation for each combination of discrete parameter values and using the best score across those results.

4.3 Evaluation of Results

The performance of the optimised models (a)–(d) and their parameter values are shown in Fig. 2.

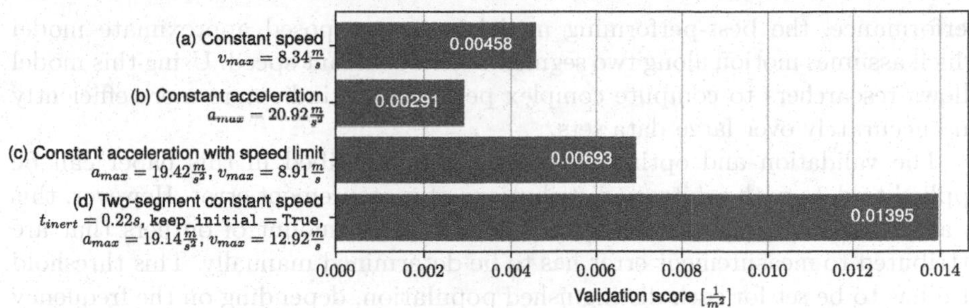

Fig. 2. Comparison of the performance of the four models (a)–(d) with their optimised parameter values.

The constant-speed model (a) unsurprisingly shows a weaker performance ($score^{-1} = 218\,\text{m}^2$) than the more sophisticated models (c) and (d), since it does not factor in the initial kinematic state of the player.

The naive constant acceleration model (b) ($score^{-1} = 344\,\text{m}^2$) performs even worse than model (a), likely because it makes the unrealistic assumption that the possible magnitude of acceleration is independent of the magnitude and direction of a player's current velocity. This implies in particular that for high speeds, the amount of reachable space in the direction that a player is moving towards will be heavily overestimated since the model assumes that the player's speed can increase unboundedly.

The model assuming constant acceleration with a speed limit (c) ($score^{-1} = 144m^2$) outperforms models (a) and (b). However, the optimised value of the maximally possible acceleration of a player a_{max} is physically unrealistic. A value of $a_{max} = 19.42\frac{m}{s^2}$ assumes that a player can accelerate from zero to the top speed $v_{max} = 8.91\frac{m}{s}(= 32.08\frac{km}{h})$ within about half a second, which is implausibly fast. Therefore, the model still overestimates the reachable area.

The two-segment constant speed model (d) ($score^{-1} = 71.7\,\text{m}^2$) is able to account for all reachable positions by predicting only about half the area of model (c). It successfully narrows down the area that a player can reach within one second to a circle with an average radius of 4.8 meters which is highly accurate. Model (d) not only achieves the best score in our evaluation, but is also mathematically simpler than model (c). For that reason, it is also computationally more efficient across the various tasks that motion models are used for, like the computation of reachable areas or the shortest time to arrive at a specific location.

5 Conclusion

We presented a novel approach to the validation and optimisation of models of trajectorial player motion in football and similar sports. We also presented an empirical comparison of the accuracy of various such models. While more sophisticated kinematic assumptions tend to be reflected in better predictive performance, the best-performing model is our proposed approximate model which assumes motion along two segments with constant speed. Using this model allows researchers to compute complex performance indicators more efficiently and accurately over large data sets.

The validation and optimisation approach described in this paper can be applied to data with arbitrary distributions of measurement error. However, this is also a disadvantage, since the threshold for the amount of outliers that are attributed to measurement error has to be determined manually. This threshold also has to be set for each distinguished population, depending on the frequency of extrema and the distribution of measurement error in the population. For example, if motion models are individualized, it would be misleading to use the same threshold for goalkeepers and outfield players, because goalkeepers produce far less positional extrema and thus outliers. As a solution, one could contextualize validation and optimization with various thresholds or derive an optimal threshold from a known error distribution.

In the future, we plan to search for motion models that further exceed the presented ones in accuracy and computational efficiency. A key towards this goal is to estimate motion models from positional data. Many problems addressed in this paper are mirrored in empirical model fitting, for example the need to exclude outliers and the lack of generalisability across populations [1]. In the context of validation, empirical models can serve as a highly informative benchmark to reveal how well theoretical models are able to approximate actual human motion.

References

1. Brefeld, U., Lasek, J., Mair, S.: Probabilistic movement models and zones of control. Mach. Learn. **108**(1), 127–147 (2018). https://doi.org/10.1007/s10994-018-5725-1
2. Caetano, F.G., et al.: Football player dominant region determined by a novel model based on instantaneous kinematics variables. Sci. Rep. **11**(1), 1–10 (2021). https://doi.org/10.1038/s41598-021-97537-4
3. Fonseca, S., Milho, J., Travassos, B., Araújo, D.: Spatial dynamics of team sports exposed by Voronoi diagrams. Hum. Mov. Sci. **31**(6), 1652–1659 (2012). https://doi.org/10.1016/j.humov.2012.04.006
4. Metrica sports sample data. GitHub repository (2021). https://github.com/metrica-sports/sample-data/commit/e706dd506b360d69d9d123d5b8026e7294b13996
5. Rein, R., Raabe, D., Memmert, D.: "Which pass is better?" novel approaches to assess passing effectiveness in elite soccer. Hum. Mov. Sci. **55**, 172–181 (2017). https://doi.org/10.1016/j.humov.2017.07.010
6. Spearman, W., Basye, A., Dick, G., Hotovy, R., Pop, P.: Physics-based modeling of pass probabilities in soccer. In: 11th MIT Sloan Sports Analytics Conference (2017)
7. Taki, T., Hasegawa, J.: Visualization of dominant region in team games and its application to teamwork analysis. In: Proceedings Computer Graphics International 2000, pp. 227–235. IEEE (2000)

Auto-scaling of Scientific Workflows in Kubernetes

Bartosz Baliś(✉), Andrzej Broński, and Mateusz Szarek

AGH University of Science and Technology, Institute of Computer Science, Krakow, Poland

balis@agh.edu.pl

Abstract. Kubernetes has gained extreme popularity as a cloud-native platform for distributed applications. However, scientific computations which typically consist of a large number of jobs – such as scientific workflows – are not typical workloads for which Kubernetes was designed. In this paper, we investigate the problem of *autoscaling*, i.e. adjusting the computing infrastructure to the current resource demands. We propose a solution for auto-scaling that takes advantage of the known workflow structure to improve scaling decisions by predicting resource demands for the near future. Such a *predictive autoscaling policy* is experimentally evaluated and compared to a regular *reactive policy* where only the current demand is taken into account. The experimental evaluation is done using the HyperFlow workflow management systems running five simultaneous instances of the Montage workflow on a Kubernetes cluster deployed in the Google Cloud Platform. The results indicate that the predictive policy allows achieving better elasticity and execution time, while reducing monetary cost.

Keywords: Scientific workflows · Auto-scaling · Kubernetes

1 Introduction

Kubernetes is a container orchestration system which has gained extreme popularity as a universal platform for management of complex distributed applications. However, scientific computations, in particular scientific workflows which are large graphs of tasks [7] – are not typical workloads for which Kubernetes was designed. We propose and evaluate a solution for auto-scaling of Kubernetes clusters tailored to scientific workflows. The main contributions of this paper are as follows: (1) two auto-scaling policies specific for scientific workflows and Kubernetes – reactive and predictive – are proposed and implemented. (2) The policies are experimentally evaluated and compared to a standard Cluster Autoscaler, using the HyperFlow Workflow Management System [1] running a workload of multiple large scientific workflows on a Kubernees cluster consisting of 12 nodes with 96 cores.

The paper is organized as follows. Section 2 presents related work. Section 3 presents the proposed solution for predictive autoscaling. Section 4 contains experimental evaluation of the solution. Section 5 concludes the paper.

D. Groen et al. (Eds.): ICCS 2022, LNCS 13351, pp. 33–40, 2022.
https://doi.org/10.1007/978-3-031-08754-7_5

2 Related Work

The basic method of scaling is *reactive* [2], wherein a scaling manager adjusts resource allocation to their actual use, based on such metrics as the number of requests per minute, or the number of active users.

Several autoscaling approaches for scientific workflows have been proposed in the context of cloud infrastructures, e.g. using AWS Spot instances [9]. Cushing et al. [3] introduce a scaling policy that relies on prediction of task execution times and estimates future demand based on the currently queued tasks. Versluis and others [10] compare several scaling policies using trace-based simulation.

Unlike the autoscaling policies for general applications, specific policies can be applied to graphs of tasks. In [6], two such policies – *Plan* and *Token* are proposed. The first one makes predictions and partial analysis of execution – so it requires knowledge of the graph structure and estimates for individual tasks. The second policy only uses information about the structure of the graph to estimate its *Level of Parallelism.* Although the quality of the estimation in the *Token* policy strongly depends on the structure of the graph, the paper [4] shows that it brings significant results for popular computational tasks. The work [5] presents an autoscaler of the *Performance-Feedback* type based on the *Token* policy, which supports many simultaneously running *workflows*, and also shows the integration architecture with *Apache Airflow.*

In summary, no existing work investigated auto-scaling of scientific workflows specifically in the context of Kubernetes using in-situ experimental evaluation.

3 Auto-scaling Scientific Workflows in Kubernetes

We adopt a solution wherein the autoscaling process can be viewed as a *MAPE* loop [8] consisting of four steps: (1) *Monitoring* – collecting information about the state of the cluster and the workflow execution state; (2) *Analysis* – predicting the future execution state and resource demands; (3) *Planning* – finding the best scaling action that accommodates the predicted workload; (4) *Execution* – performing the *scaling action.*

3.1 Monitoring

The basis of the autoscaler operation is the awareness of the current state of the cluster and the computations. We use the *SDK API* to retrieve information about *Pods* and *Nodes*, where the former determine the current demand for resources, and the latter their current supply. The demand for resources is calculated based on the *resource requests* of the Kubernetes Pods that run workflow tasks, so as to match the algorithm used by the Kubernetes scheduler. The supply of resources is the total amount of CPU and memory available on the worker nodes, reduced by the resources reserved by the Kubernetes components (e.g. Kubelet).

The workflow execution status is tracked through the events emitted by the HyperFlow workflow management system which we use to experimentally evaluate the autoscaling policies.

3.2 Analysis

The purpose of this step is to determine the demand for resources in the near future. To this end, a given time period, e.g. 5 min, is divided into smaller time frames, e.g. seconds, and each frame is assigned a specific amount of resources. To estimate how much resources are needed in the upcoming time frames, we **simulate the further execution of currently running workflow graphs** using the method described in [6], so as to predict **which tasks will be running in parallel in a given time frame**. An illustration of the analysis process is presented in Fig. 1.

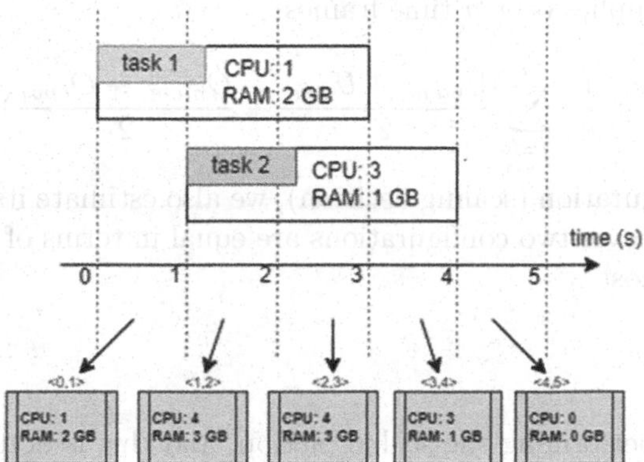

Fig. 1. Analysis of resource requests in time frames.

To calculate the demand for resources, we use the CPU and memory requests of the containers, in a similar way they are used by the Kubernetes scheduler. To take into account both CPU and memory requirements at the same time, we need to combine these two measures that have completely different units. To this end, both CPU and memory utilization are expressed as a percentage of the respective available resources and then added up.

3.3 Planning

The purpose of the planning step is to determine which action will be the most beneficial: scaling up or down by a certain number of machines, or perhaps no scaling. The outcomes of all possible decisions are checked within a given time limit, at specified time intervals (e.g. 5-frame sampling). The number of all combinations – decisions about **how to scale** and **when** – is the product of the maximum number of machines and the number of samples within the time limit. This number is sufficiently small so that all combinations can be checked.

Each combination is assigned a value of S (*score*), which represents how far the scenario deviates from an optimal match of resource demands and supplies.

The deviation from the optimal resource allocation – either in the form of underprovisioning or overprovisioning – of resource *res* is given by formula 1. There, *supply* is the value of the supply of resource *res* in the cluster taking into account the scaling action, and *demand* is the demand for resource *res*.

$$D_{res} = \left|\frac{demand_{res} - supply_{res}}{demand_{res}}\right| \tag{1}$$

Additionally, in frames where one of the resources is over- and the other is underprovisioned, the value of D is set to 0, so as to optimize runtime while accepting an additional cost.

The score S is then calculated, using formula 2, as a mean value of all resource over- and undersupplies over n time frames.

$$S = \frac{1}{n} \cdot \sum_{i=0}^{n} \left(\frac{U_{mem_i} + U_{cpu_i}}{2} + \frac{O_{mem_i} + O_{cpu_i}}{2}\right) \tag{2}$$

For each configuration (scaling decision), we also estimate its resulting **monetary cost**. Whenever two configurations are equal in terms of score, we choose one with a lower cost.

3.4 Execution

The final step is performing the scaling action, and this is done via the API of a given cloud provider. Due to the fact that scaling may take several minutes, the execution is asynchronous, i.e. we do not wait for information about the completion of scaling. To eliminate too frequent scaling attempts, after a scaling action we impose a *scaling cooldown period*, e.g. 2 min, during which no scaling actions are allowed.

3.5 Autoscaling Policies – Reactive vs. Predictive

In order to evaluate the impact of leveraging the knowledge of the workflow structure on the quality of scaling, we distinguish two *autscaling policies*. In the ***reactive policy***, we assume that the future demand for resources is identical to the current one, so the planning phase simply adapts to the current situation. With the ***predictive policy***, on the other hand, knowledge of the workflow structure is used to estimate future demand, as described in Sect. 3.2.

4 Evaluation

To evaluate the proposed predictive autoscaling algorithm, we experimentally run the same workload using three different configurations: (1) reactive policy, (2) predictive policy, (3) standard Cluster Autoscaler (which uses its own implementation of a reactive policy).

4.1 Experiment Setup

To run the experiments, we used the Google Cloud Platform (GCP), with a Kubernetes cluster consisting of one master node to run the HyperFlow components, and a pool of worker nodes, using the *n1-highcpu-8* machine type, scalable from 1 to 12 nodes (up. to 96 cores).

Let us note that when the scaling action is performed, *Kubernetes* may try to stop a *Pod* and start it on another machine. In line with our assumptions, we care first about time and secondly about the budget. Thus, we define *PodDisruptionBudget* for all tasks, which consequently blocks the removal of the machine until all previously started tasks are finished.

The test workload was the Montage (degree 2.0). The experimental workload consisted of 5 instances of this workflow running simultanoeusly. The details are shown in Table 1. The CPU requests were set to 0.5 for mDiffFit and mBackground tasks and to 1.0 for all the others. The memory request was set to 256 MiB for all tasks.

Table 1. Experimental workload – 5 instances of the Montage workflow.

Task type (agglomeration)	Count	Task type (agglomeration)	Count
mProject (3x/3 s)	1535	mImgtbl	5
mDiffFit (12x/6 s)	4310	mAdd	5
mConcatFit	5	mShrink	5
mBgModel	5	mJPEG	5
mBackground (12x/4 s)	1535		

HyperFlow supports agglomeration (clustering) of tasks to reduce the overhead of starting excessively many Pods. In the case of Montage, tasks for the three parallel stages – especially *mDiffFit* and *mBackground* – are rather short, so that Pod creation time (typically about 2 s) can introduce a significant overhead. Configuration of task agglomeration is also shown in Table 1. For example, *12x/6 s* means that HyperFlow will submit the tasks of a given type in batches of 12, but the maximum wait time to form a batch is 6 s.

4.2 Results

Figure 2 shows the visualization of the execution traces of the experimental workload, along with cluster scaling, for the React and Predict policies, respectively. Because of space limitations, a similar visualization for the CA-based execution is not shown. However, Table 2 summarizes the key metrics for all three cases (CA, React and Predict): total execution time and the cost of execution.

It can be seen that the two executions of the experimental workload, respectively driven by the React and Predict policies, are quite different. The Predict

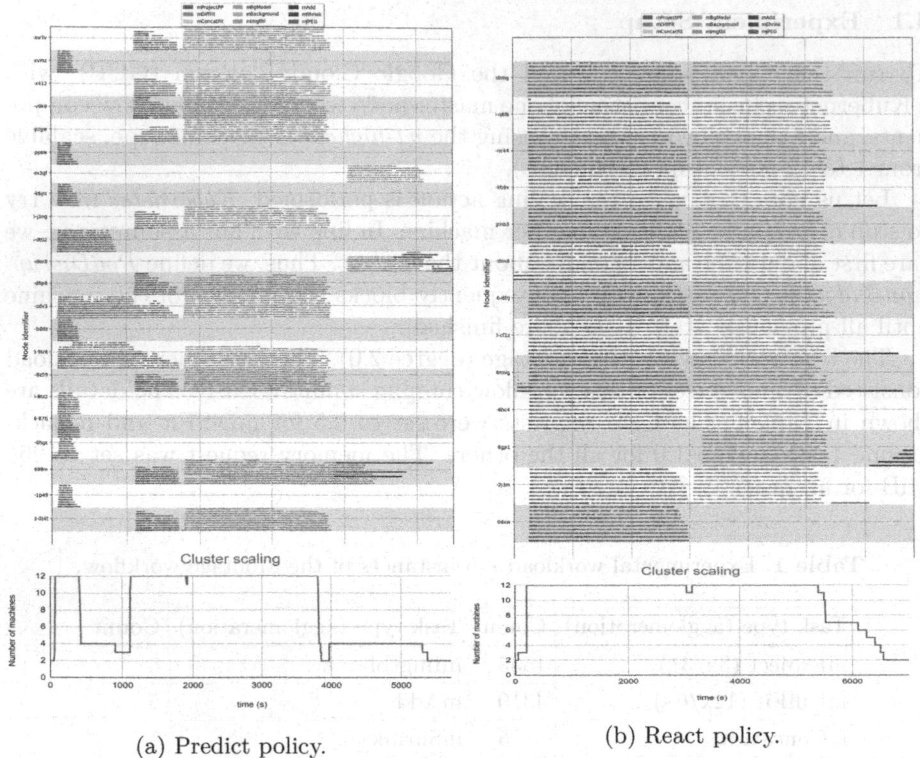

(a) Predict policy.

(b) React policy.

Fig. 2. Execution of the experimental workload.

Table 2. Experiment results summary for different auto-scaling policies.

Montage-Degree 2.0	React (CA)	Predict	React
Execution time [s]	6000	5709	6837
Cost [$]	7.05	4.53	6.95

policy results in more scaling decisions overall. The visualization of the execution trace for the Predict policy looks less 'compact' but this is only due to the fact that in total 22 different nodes are involved in the execution with the Predict policy, compared to 13 nodes for the React policy.

The results indicate that the Predict policy performed significantly better than both CA and React ones. Not only the achieved execution time was the shortest (better by 5% compared to the second best CA policy), but it resulted in the lowest cost of execution ($4.53 compared to $\sim$ $7 for the two other policies). It is reasonable to conclude that with the execution time of about 100 min, the overhead of starting and shutting down more nodes was not significant.

Another interesting observation is that our implementation of the reactive policy performed significantly worse than the policy of the standard Cluster Autoscaler. While this is not the key result of the experiment, the reasons for this will be the subject of future investigation.

5 Conclusions and Future Work

Autoscaling enables elastic resource allocation which meets the current demands, avoiding undesirable underprovisioning (hurting performance) or overprovisioning (increasing cost) of resources. We presented an autoscaling solution for running scientific workflows in a Kubernetes cluster. The proposed autoscaling policy was designed to take advantage of the knowledge of the workflow structure in order to predict the resource demands in the near future and thus, hypothetically, achieve better scaling decisions than a reactive policy. This hypothesis was confirmed experimentally.

Future work involves further experiments with different types of workflows and investigation of the impact of various factors on the execution and autoscaling, such as fine-tuning of resource demands of workflow tasks.

Acknowledgements. The research presented in this paper was partially supported by the funds of Polish Ministry of Education and Science assigned to AGH University of Science and Technology.

References

1. Balis, B.: HyperFlow: a model of computation, programming approach and enactment engine for complex distributed workflows. Futur. Gener. Comput. Syst. **55**, 147–162 (2016)
2. Chieu, T.C., Mohindra, A., Karve, A.A., Segal, A.: Dynamic scaling of web applications in a virtualized cloud computing environment. In: 2009 IEEE International Conference on e-Business Engineering. IEEE (2009)
3. Cushing, R., Koulouzis, S., Belloum, A.S., Bubak, M.: Prediction-based auto-scaling of scientific workflows. In: Proceedings of the 9th International Workshop on Middleware for Grids, Clouds and e-Science, pp. 1–6 (2011)
4. Ilyushkin, A., Ghit, B., Epema, D.: Scheduling workloads of workflows with unknown task runtimes. In: 2015 15th IEEE/ACM International Symposium on Cluster, Cloud and Grid Computing, pp. 606–616 (2015)
5. Ilyushkin, A., Bauer, A., Papadopoulos, A.V., Deelman, E., Iosup, A.: Performance-feedback autoscaling with budget constraints for cloud-based workloads of workflows. CoRR abs/1905.10270 (2019). http://arxiv.org/abs/1905.10270
6. Ilyushkin, A., et al.: An experimental performance evaluation of autoscaling policies for complex workflows. In: ICPE 2017, pp. 75–86. ACM (2017)
7. Juve, G., Chervenak, A., Deelman, E., Bharathi, S., Mehta, G., Vahi, K.: Characterizing and profiling scientific workflows. Futur. Gener. Comput. Syst. **29**(3), 682–692 (2013)
8. Lorido-Botran, T., Miguel-Alonso, J., Lozano, J.A.: A review of auto-scaling techniques for elastic applications in cloud environments. J. Grid Comput. **12**(4), 559–592 (2014)

9. Monge, D.A., Garí, Y., Mateos, C., Garino, C.G.: Autoscaling scientific workflows on the cloud by combining on-demand and spot instances. Comput. Syst. Sci. Eng. **32**(4), 291–306 (2017)
10. Versluis, L., Neacsu, M., Iosup, A.: A trace-based performance study of autoscaling workloads of workflows in datacenters. In: 2018 18th IEEE/ACM Int. Symposium on Cluster, Cloud and Grid Computing (CCGRID), pp. 223–232. IEEE (2018)

Transfer Learning Based Natural Scene Classification for Scene Understanding by Intelligent Machines

Ranjini Surendran[1], J. Anitha[1], A. Angelopoulou[2], E. Kapetanios[3], T. Chausalet[2], and D. Jude Hemanth[1](✉)

[1] Department of ECE, Karunya Institute of Technology and Sciences, Coimbatore, India
judehemanth@karunya.edu
[2] School of Computer Science and Engineering, University of Westminster, London, UK
[3] School of Physics, Engineering and Computer Science, University of Hertfordshire, Hertfordshire, UK

Abstract. Scene classification carry out an imperative accountability in the current emerging field of automation. Traditional classification methods endure with tedious processing techniques. With the advent of CNN and deep learning models have greatly accelerated the job of scene classification. In our paper we have considered an area of application where the deep learning can be used to assist in the civil and military applications and aid in navigation. Current image classifications concentrate on the various available labeled datasets of various images. This work concentrates on classification of few scenes that contain pictures of people and places that are affected in the areas of flood. This aims at assisting the rescue officials at the need of natural calamities, disasters, military attacks etc. Proposed work explains a classifying system which can categorize the small scene dataset using transfer learning approach. We collected the pictures of scenes from sites and created a small dataset with different flood affected activities. We have utilized transfer learning model, RESNET in our proposed work which showed an accuracy of 88.88% for ResNet50 and 91.04% for ResNet101 and endow with a faster and economical revelation for the application involved.

Keywords: Scene classification · Image classification · Deep learning · Transfer learning · ResNet

1 Introduction

Image classification [1] finds application in the field of automation, face detection [2], self-driving cars, robotics, aviation, civil and military applications. Many image classifications approaches classify the image based on single object or multi- object [3]. Scene classification has one of the most imperative roles in the field of scene understanding. Understanding a scene involves the steps akin to scene classification, object detection, object detection and localization and event recognition [4]. Scene consists of not merely objects, but preserves information regarding the relation and activity involved. Scenes

D. Groen et al. (Eds.): ICCS 2022, LNCS 13351, pp. 41–48, 2022.
https://doi.org/10.1007/978-3-031-08754-7_6

carry a lot information regarding the content, relation and actions within the objects present. Identifying and analyzing the scene with the meanings preserved is significant task for computer vision applications.

Deep learning [5, 6] the subset of Artificial Intelligence has taken the classification to peaks. The availability of huge dataset and high computationally powerful machines have led to the use of deep learning in classification. Many works are carried out in the field of image classification [7–11] with powerful deep learning models using publically available large indoor and outdoor datasets. Classifying images with deep learning finds many applications in our day to day life. With the inspiration from the deep architectures we tried to develop a classifying system for natural scenes. As we have seen natural disasters and calamities are affecting almost all countries in the world. We cannot not predict its effect and eliminate it, but can provide a helping hand for those affected areas. With this intension we developed a model that can classify few scenes representing the natural calamities. Here the disaster calamity of flood affected areas are showcased considering the people and vehicles if they are endangered. We have also considered scenes showing the rescue operations mainly with the aid of helicopters and boats to aid the rescuers to reach the calamity affected areas.

The dataset available were limited resource for our application. We created our own dataset consisting of few scenes. We had collected scenes of flood affected areas. We developed this model with the intention that this could serve humanity. Dataset is arranged considering scenes of people and vehicles that are drowned in flood and scenes of rescuing operation needed by boat or helicopter. Thus, our model could classify the scenes into four classes: Many people in flood, Vehicles in flood, Rescue by boat and Rescue by helicopter. These scenes could pave the way for the rescuing operations which would be hard enough for the people reaching out there. Deep models are pre-trained on large dataset with high accuracy. They are trained on millions of images. Here we have used the concept of transfer learning for modelling our system. In transfer learning we can modify the structure of already learned models to our application. Since these models have already learned from the large image dataset collections, we can use it in our work where it can learn our small scene dataset.

2 Literature Review

Classifying an image is a simple job for humans, but when it comes with the machines it is not. Human can easily visualize a scene with its objects, relations these objects are carrying with each other and the different activities involved. The machines need to be trained with the features involved within the scenes. Traditionally these features were extracted manually using machine learning algorithms like SIFT, HoG and classification algorithms like SVM [12] and fuzzy classifier [13] carried out the classification.

Convolutional Neural Network (CNN) enhanced the capability of machines to learn in the vicinity of classification. Inspired by the functionality of biological neurons in human brain, the neural layers in CNN mimics the human brain. Neural layers in the CNN extracts the features of the data, image that we are providing as input. It outperforms the traditional machine learning techniques in feature extraction. The features learned by the network layers of CNN can be further used for the classification by using any classification algorithms. Each layer in the network learns different features from the input image. Deeper the layers more the features will be learned. Deep learning is the advancement of CNN in its number of layers and has improved the efficiency of the CNN. The deep network model, ALEXNET trained on ImageNet dataset improved the efficiency and accuracy of the traditional image classification.

Transfer Learning in deep learning is an important and popular concept in the field of image classification and natural language processing. Here we choose a pretrained model and train it for a new problem. In deep models it requires many epochs or iterations to train the complete model and get higher accuracy. With pretrained models it takes only few iterations to train a model and get a desired accuracy. Thus, transfer learning saves a lot of computational power. The pretrained models are already trained with large dataset and we can use the weights of these models to train for a new problem even with lower dataset by modifying the last few layers of them. In one of paper of Manali et al. they developed a model by fine tuning the VGG16 and VGG19 model with CalTech256 and GHIM10K database. Many works are being carried out on image classification using different pretrained deep models but with state -of-art databases. In our work we have developed a pretrained model with high accuracy to classify our own data set.

3 Proposed Work

In this section we describe the methodology implemented in classifying the different scenes and the overall steps involved in our work is shown in the block diagram. Majority of the deep learning models are trained on the publically available dataset. Here we have created our own dataset which we have collected from various websites. The dataset is a customized small dataset and we are using the concept of transfer learning approach to classify our scene dataset. In our work we have used the deep ResNet transfer learning model for feature extraction and classification. Here we evaluated performance of two variants of deep ResNet model, ResNet-50 and ResNet-101 for our dataset.

4 Transfer Learning-ResNet

In deep neural network we always face a problem of vanishing gradient while updating the weights because we need to use back propagation while updating the weights and we use chain rule of calculus. The repeated multiplication will make the weights extremely small while they reach through the earlier layers. While updating the weights each layer is dependent on the previous layers. This problem of vanishing gradient is having a solution in residual network. ResNet standpoints for Residual Network. They have introduced a new concept known as skip connection in the network. Traditional network layers are learned with the output coming from each preceding layer while in residual network each layer is learned with the input applied also and we want to make input equal to output. Different variants of residual network we are discussing here are ResNet-50, ResNet-101.

ResNet-50: The value 50 means there are 50 layers in our network. The first layer is input layer having 7×7 filter and 64 such filters with stride 2. Next is a 3×3 max pooling layer with stride 2. The first block of 1 × 1 filters with 64 such filters, 64 of 3 × 3 filers and 256, 1 × 1 filters. Second block with 4 layers each 128 numbers of 1 × 1 filters, 128 of 3 × 3 filters and 512 of 1 × 1 filters (Fig. 1).

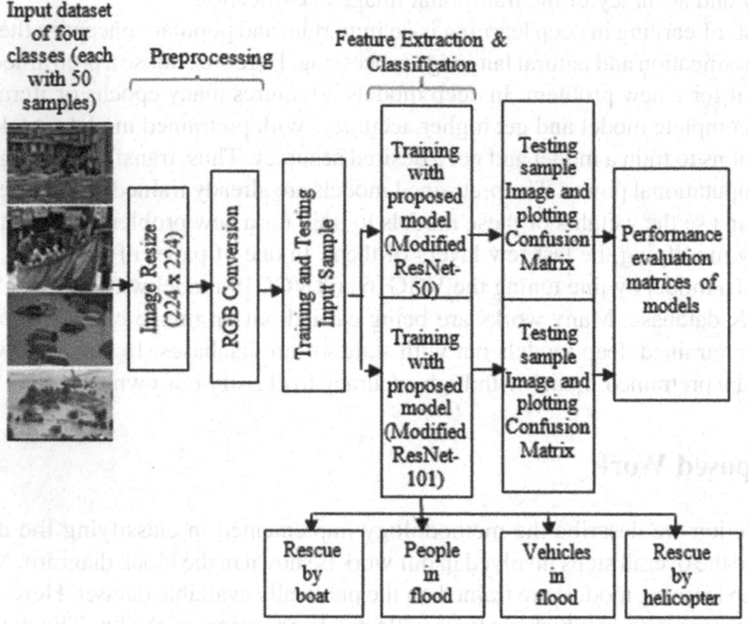

Fig. 1. Proposed model for scene classification using ResNet-50 and Resnet-101

Third block with 6 layers of 256 of 1 × 1 filters, 256 of 3 × 3 filters and 1024 1 × 1 filters. Forth block of 3 layers of 512 of 1 × 1 filters, 512 of 3 × 3 filters and 2048 of 1 × 1 filters. Next layer forms average pooling and fully connected layer. Altogether a total of 50 layers.

ResNet-101: This model architecture is having 101 layers by adding more 3 layer blocks in the network. The structure of ResNet-101 is similar to that of ResNet-50 having 24 repeated layers consisting of 256 of 1 × 1 filters, 256 of 3 × 3 filters and 1024 of 1 × 1 filters. It is here the difference lies in the architecture of Resnet-50 and ResNet-101. Remaining layers same as ResNet-101 thus forming a total of 101 layers.

5 Proposed Methodology

In our proposed methodology we have separated all the images into four separate classes. The images are then resized to 224 × 224 to be handled by the input layer of the ResNet models. These resized images are then preprocessed for RGB conversion. Preprocessed images are splitted into training and testing samples. These samples are then used to train the ResNet models. In our model the we have used ResNet-50 and Resnet-101 to extract the features from the samples. The deeper layer of the ResNet learns the sample dataset by learning many features and feature extraction is carried out by these deep models. Once the models have learned features we have to classify the images according to the learned features. For this we have made changes to the final few layers of ResNet-50 and ResNet-101. Since our classification is of four classes we have modified the final 3 layers of the ResNet. For our dataset we have replaced the last 3 layers with our own layers for carrying out the classification. The feature learned modified structured Resnet-50 and ResNet-101 are then evaluated using a test sample image. The confusion matrix is plotted for both the architectures and the performance parameters are evaluated.

5.1 Dataset

We have developed a dataset consisting of four classes of flood affected areas with 50 scenes in each class. The different types of scenes are rescuing of people in boat, people stuck in flood, Vehicles stuck in flood and rescuing by helicopter. We have collected these scenes from various sites and created our own dataset with an intension of assisting the people and the officials in finding the disaster affected areas and aid in rescue operations where there may be areas which are remote and non-reachable by people. Identifying these scenes by robots can enhance the efficiency of our application. We have trained this dataset by using the pre-trained deep ResNet architecture.

5.2 Pre-trained Models

In our work we have used pre-trained ResNet-50 and ResNet-101 for feature extraction. The input dataset is preprocessed to 224 × 224 to be handled by the input layers of ResNet. The lower layers of ResNet is freezed and kept as the same architecture. These layers which are already pre-trained with the large image dataset can now easily learn our small dataset in less time with high accuracy. In both ResNet-50 and ResNet-101 we have customized the last three layers with our layers: Fully connected layer of 4, Soft Max Layer and Classification Layer for 4 classes (Fig. 2).

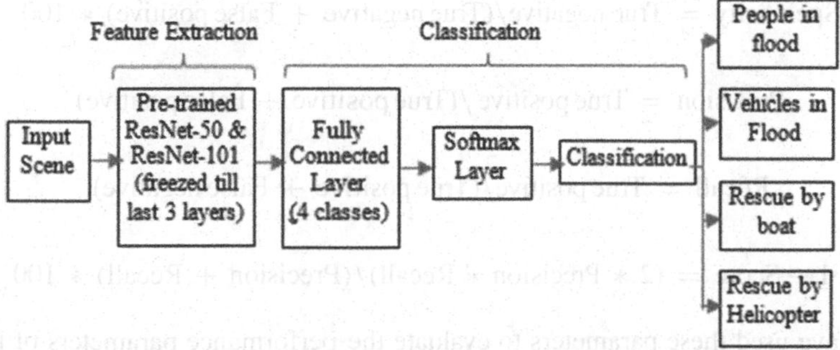

Fig. 2. Proposed architecture of ResNet-50 and ResNet-101 for scene classification

6 Experimental Results

We have done the experiment with a small dataset of 200 scenes with 50 scenes separated into four classes. Each input scene was resized to 224 × 224 and carried out RGB conversion to get suited with the input layer of ResNet architecture. 70% of preprocessed dataset divided into training set samples and remaining 30% to test set samples and trained by the pre-trained ResNet 50 and ResNet 101 models whose last layers were modified for our application. We have trained the network with batch size of 15 and for 20 epochs and each model is tested with sample input. The execution time for both models are evaluated. We have measured the efficiency of both the models by plotting the confusion matrix. The different performance matrices like accuracy, Precision, F1-Score also evaluated for both architectures.

6.1 Training Progress - ResNet50 and ResNet 101

The input samples were trained by the ResNet for a batch size of 15 and 20 epochs. Training progress monitored and the accuracy and lose model graph is plotted. After successful training the model is tested with test sample with ResNet50 having an elapsed time of 52.488234 s and ResNet 101 of 106.29 s.

6.2 Performance Metrics

We have evaluated the two proposed models with their performance matrices. We obtained the Tp (True Positive) value indicating the correct prediction of true class, Tn (True Negative) value indicates the correct prediction of false class, Fp (False Positive) and Fn (False Negative) both showing the wrong prediction of classes.

$$\text{Accuracy} = \text{True positive} + \text{True negative}/(\text{Tp} + \text{Tn} + \text{Fp} + \text{Fn}) * 100 \tag{1}$$

$$\text{Sensitivity} = \text{True positive}/(\text{True positive} + \text{False negative}) * 100 \tag{2}$$

$$\text{Specificity} = \text{True negative}/(\text{True negative} + \text{False positive}) * 100 \tag{3}$$

$$\text{Precision} = \text{True positive}\,/(\text{True positive} + \text{False positive}) \tag{4}$$

$$\text{Recall} = \text{True positive}/(\text{True positive} + \text{False negative}) \tag{5}$$

$$\text{F1} - \text{Score} = (2 * \text{Precision} * \text{Recall})/(\text{Precision} + \text{Recall}) * 100 \tag{6}$$

We have used these parameters to evaluate the performance parameters of the two models and we have compared the two models (Table 1).

Table 1. Performance evaluation of proposed ResNet50 and ResNet101

Performance matrices	Accuracy	Sensitivity	Specificity	Precision	Recall	F1-Score
Proposed Model (ResNet50)	89.28	85.71	92.85	0.92	0.85	88.88
Proposed Model (ResNet101)	91.429	87.143	95.71	0.95	0.87	91.04

7 Conclusion

We have proposed our model with an objective to classify natural scenes where we have combined different scenes. We collected 200 scenes and grouped them into four different classes indicating rescue by boat, people in flood, vehicles in flood and rescue by helicopter. We have proposed such a model to assist the robots and drones to identify the remote location affected by flood which we people may not notice. Both ResNet50 and ResNet101. Models are evaluated for their performance matrices- Sensitivity, Specificity, F1-Score etc. ResNet50 showed F1-score of 88.88 and ResNet101 showed 91.04. The execution time and accuracy of both models were also compared. Resnet50 gave an accuracy of 89.3% with an execution time of 52.4 s and Resnet101 showed the accuracy of 91.4% with a slightly larger execution time of 106.29 s. Both the proposed models gave an appreciable good response even with the small dataset considering execution time and accuracy. As a future scope we develop our model with real time scene images.

References

1. Kamavisdar, P., Saluja, S., Agrawal, S.: A survey on image classification approaches and techniques. Int. J. Adv. Res. Comput. Commun. Eng. **2**(1), 1005–1009 (2013)
2. Sharma, M., Anuradha, J., Manne, H.K., Kashyap, G.S.C.: Facial detection using deep learning. IOP Conf. Ser. Mater. Sci. Eng. **263**, 042092 (2017)
3. Bandhu, A., Sekhar Roy, S.: Classifying multi-category images using deep learning: a convolutional neural network model. In: 2nd IEEE International Conference on Recent Trends in Electronics Information & Communication Technology, pp. 1–6 (2017)
4. Regina Lourdhu Suganthi, S., Hanumanthappa, M., Kavitha, S.: Event image classification using deep learning. In: 2018 International Conference on Soft-computing and Network Security, pp. 11–17 (2018)
5. LeCun, Y., Yoshua, B., Hinton, G..: Deep learning. Nature **521**(7553), 436–444 (2015)
6. Schmidhuber, J.: Deep learning in neural networks: an overview. Neural Netw. **61**, 85–117 (2015)
7. Weng, Q., et al.: Land-use classification via extreme learning classifier based on deep convolutional features. IEEE Geosci. Remote Sens. Lett. (2017)
8. Chauhan, K., Ram, S.: Image classification with deep learning and comparison between different convolutional neural network structures using Tensorflow and Keras. Int. J. Adv. Eng. Res. 533–538 (2018)
9. Zhang, F., Du, B., Zhang, L.: Scene classification via a gradient boosting random convolutional network framework. IEEE Trans. Geosci. Remote Sens. **54**(3), 1793–1802 (2016)

10. Panigrahi, S., Nanda, A., Swarnkar, T.: Deep learning approach for image classification. In: 2018 2nd International Conference on Data Science and Business Analytics (2018)
11. Chen, Y., Jiang, H., Li, C., Jia, X., Ghamisi, P.: Deep feature extraction and classification of hyperspectral images based on convolutional neural networks. IEEE Trans. Geosci. Remote Sens. **54**(10), 6232–6251 (2016)
12. Pasolli, E., Melgani, F., Tuia, D., Pacifici, F., Emery, W.J.: SVM active learning approach for image classification using spatial information. IEEE Trans. Geosci. Remote Sens. **52**(4), 2217–2223 (2014)
13. Korytkowski, M., Rutkowski, L., Scherer, R.: Fast image classification by boosting fuzzy classifiers. Inf. Sci. **327**, 175–182 (2016)

Devulgarization of Polish Texts Using Pre-trained Language Models

Cezary Klamra[1], Grzegorz Wojdyga[1], Sebastian Żurowski[2], Paulina Rosalska[2,3], Matylda Kozłowska[4], and Maciej Ogrodniczuk[1(✉)]

[1] Institute of Computer Science, Polish Academy of Sciences, Warsaw, Poland
maciej.Ogrodniczuk@gmail.com
[2] Nicolaus Copernicus University in Toruń, Toruń, Poland
[3] Applica.ai, New York, USA
[4] Oracle, Warsaw, Poland

Abstract. We propose a text style transfer method for replacing vulgar expressions in Polish utterances with their non-vulgar equivalents while preserving the meaning of the text. We fine-tune three pre-trained language models on a newly created parallel corpus of vulgar/non-vulgar sentence pairs, then we evaluate style transfer accuracy, content preservation and language quality. To the best of our knowledge, the proposed solution is the first of its kind for Polish.

Keywords: Text style transfer · Removing obscenities · Transformer

1 Introduction

Most works on fighting with offensiveness and obscenity in language concentrated on its automatic identification. In this paper we present a mechanism for replacing vulgar expressions in Polish utterances with their non-vulgar equivalents while preserving the overall sense of the text. As a component task, we created a corpus of vulgar expressions and their equivalents.

Due to the low availability of parallel data, most previous works on this topic use non-parallel corpora. Such solutions most often separate the content of the text from its style. A style-independent representation of the content is created and used to reconstruct the text in the target style. Dos Santos et al. [14] describe methods for moderating offensive or hateful language using unsupervised text style transfer. The authors extend the standard encoder and decoder by introducing a classifier and special loss functions that allow the use of a corpus of posts from social media. Tran et al. [15] constructed an unsupervised style transfer pipeline, that uses a vocabulary of restricted words, POS tagging to locate vulgarities, RoBERTa and T5 model to create possible replacements and GPT-2 and BLEU score to select the sentence of the highest quality. Dementieva et al. [4] used a bag-of-words logistic regression model to identify toxic words and replace them using a BERT model, tuned on a toxic and non-toxic corpus.

D. Groen et al. (Eds.): ICCS 2022, LNCS 13351, pp. 49–55, 2022.
https://doi.org/10.1007/978-3-031-08754-7_7

With parallel corpora, Cheriyan et al. [2] used synonym generation, the masked language model (BERT), and the sequence-to-sequence model (BART) to rephrase offensive comments on social media. Dementieva et al. [4] used a parallel corpus and the pre-trained ruGPT-3 model to test three approaches: zero-shot, few-shot, and fine-tuning.

2 Theoretical Background for the Corpus Work

Our work concentrates on vulgarisms only, i.e. "lexical units by means of which the speaker reveals his or her emotions towards something or someone, breaking a linguistic taboo" [5]. Grochowski divides vulgar expressions into two categories: systemic and referential-customary. The former are lexical units tabooed solely because of its expressive features and often contain certain sequences of characters (e.g. *-kurw-*, *-jeb-*, *-pierd-*). The dictionary distinguishes three degrees of "strength" of systemic vulgarisms: (i) of the lowest level, generally considered vulgar among the "cultured interlocutors"; (ii) of the medium level, i.e. units commonly considered vulgar; (iii) of the highest level, i.e. regarded as very vulgar. The referential-customary vulgar expressions are tabooed because of their semantic features and the scope of their object reference; they are considered vulgar only in specific contexts. It is common for expressions of this type to cause classification problems since it can be difficult to determine what is their degree of vulgarity.

The construction of the corpus assumes the replacement of vulgarisms with euphemisms which are defined as substitute names used when direct names cannot be used because of negative connotations. The euphemistic term should instead evoke positive (or neutral) connotations. A remark made by Grochowski [5] situates euphemisms in a context: a given unit of language can only be considered euphemistic when juxtaposed with another unit of language (one that it can replace). It implies that vulgar expressions can also be interpreted as euphemisms when replacing even more vulgar ones. Based on this assumption we decide to annotate vulgarisms in context.

3 Training Data

In order to obtain representative training data we selected movie dialogues from websites aggregating subtitle translations. Unofficial Polish subtitle translations tend to preserve as much as 80% of vulgarisms present in the original script. The training corpus is based on two sources: (i) the Polish part of the OpenSubtitles corpus,[1] which mostly consists of texts prepared by non-professional translators [6]; (ii) professionally edited dialogue tracks and published scripts of two movies [7,8].

The samples were selected by annotators experienced in linguistic work who also collected and supplemented the material as necessary. The annotations collected include:

[1] see http://opus.nlpl.eu/OpenSubtitles.php, http://www.opensubtitles.org/.

- the identifier of the text from which the annotated passage came,
- the annotated vulgar expression and its context,
- the lemma of the annotated expression,
- vulgar equivalents (synonyms) of the vulgar expression,
- common equivalents (synonyms) of the expression,
- euphemistic equivalents (synonyms) of the expression.

In order to establish synonymy relations and to determine the character of particular expressions, the annotators were advised to use the Dictionary of Polish Swearwords and Vulgarisms, the Dictionary of Polish Euphemisms, or general dictionaries of Polish.

Since the corpus contains more than one substitution for some vulgar expressions, it was pre-processed to obtain pairs with different euphemisms for a given context. The grammatical forms of vulgar expressions were manually inflected so that they could be appropriately used in their contexts. The process resulted in 6691 pairs of vulgar and corresponding non-vulgar texts.

4 Devulgarization Tool

For this work, we solve the problem of substitution of vulgar expressions as a text style transfer problem defined as a reformulation of a text without affecting its content and other properties, such as sentiment, bias, degree of formality, etc. Below we compare three approaches using GPT-2, GPT-3, and T5 models. The training details have been shared online (see Sect. 6).

4.1 GPT-2-Based Solution

The first presented solution is based on GPT-2 [11], a transformer-type language representation used for language generation but also in sequence-to-sequence tasks by concatenating input and output sequences. The pre-trained model was fine-tuned on concatenated pairs of vulgar and non-vulgar texts from the training corpus (separated with the special token `<|sep|>`). We used papuGaPT-2 model pre-trained on Polish language texts, based on GPT-2 small.[2]

4.2 GPT-3-Based Solution

GPT-3 is the latest model in the GPT-n series, largely based on its predecessor but much more extensive. Although GPT-3 is only available for English, Brown et al. [1] argue that the model has some ability to process languages other than English. English words accounted for about 93% of all words in the training set, while Polish for only 0.15% but representing over 300 million occurrences nonetheless.[3]

GPT-3 has not been made publicly available but it is possible to use the model, including model tuning, through the OpenAI API.[4] The presented results

[2] https://huggingface.co/flax-community/papuGaPT2, accessed on April 12, 2022.
[3] https://github.com/openai/gpt-3/tree/master/dataset_statistics, accessed on October 5, 2021.
[4] https://beta.openai.com/, accessed on October 5, 2021.

were obtained using the second most extensive variant of the model available, the Curie model, tuned on the test corpus. The GPT-3 training adopted a similar approach as GPT-2: the training data were presented as pairs of vulgar prompts and non-vulgar completions.

4.3 T5-Based Solution

The last presented solution is based on T5 transformer-based model [12]. As a sequence-to-sequence model, T5 is well suited for tasks related to text-based language generation, such as the problem of changing the style of the text. We fine-tuned plT5, a T5-based model pre-trained on large Polish corpora.[5] The model is available in three variants (small, base, and large) differing in the number of parameters. The paper presents the results obtained for the two most extensive models, both of which were fine-tuned on the training corpus.

5 Evaluation

5.1 Evaluation Method

The models were evaluated on the test corpus consisting of 2437 vulgar sentences from the Dictionary of Polish Swearwords and Vulgarisms [5]. We used all of the examples provided in the dictionary, some of which are not strictly vulgar.

The evaluation follows the method used for text style modification [4,9,15], i.e. the quantitative quality assessment of the obtained results in three categories: (i) effectiveness of text style change (ii) preservation of the content of the original sentence, and (iii) quality of the generated language.

Text style transfer accuracy (STA) was tested using Przetak — a library for checking whether a text contains abusive or vulgar speech in Polish written in Go [3]. Przetak is able to identify offensive language with high accuracy and handles frequent misspellings and out-of-vocabulary words composed of morphemes with vulgar meaning. Evaluated sentences are assigned a score of 0 if they are classified as vulgar, and 1 otherwise.

Content preservation was checked using three metrics: (i) cosine similarity (CS) between embeddings for input and output sentences, calculated using the SBERT multilingual model [13]; (ii) word overlap (WO) between the lemmata of the original (X) and the processed (Y) sentence defined as:

$$\frac{\#(X \cap Y)}{\#(X \cup Y)} \tag{1}$$

where lemmata have been produced using spaCy; (iii) BLEU, a commonly used and well correlated with human ratings metric for assessing machine translation quality [10].

Quality of the generated language was measured with perplexity (PPL) determined using the pre-trained GPT-2 small model.

[5] https://huggingface.co/allegro/plt5-large, accessed on December 12, 2021.

As suggested by Pang and Gimpel [9], in order to compare the overall quality of the results we used the geometric mean (GM) of the corresponding metrics in the above categories, according to the formula 2.

$$GM = \Big([100 \cdot max(0, CS)] \cdot [100 \cdot max(0, STA)] \cdot max(0, \frac{1}{PPL})\Big)^{\frac{1}{3}} \quad (2)$$

5.2 Baselines

Following other studies (see [4,15]), the results achieved by presented methods are compared against two baselines:

- **Delete**—letters of words classified as vulgar by Przetak (described in Subsect. 5.1) are replaced with asterisks. The first letter of a vulgar word is not changed. This method allows preservation of the content well unless the meaning of the vulgar word is crucial for the understanding of the sentence.
- **Duplicate**—an unchanged copy of the original sentence. This baseline represents the lower bound of the models' performance.

As the models are trained to substitute some expressions which are not recognised by Przetak, metrics measuring content preservation might have disproportionately high values for the Delete baseline.

5.3 Quantitative Evaluation Results

The results of the automatic evaluation are presented in Table 1. In comparison with GPT models, T5 models generate higher-quality language, achieve better results in terms of preserving the content of the original sentence, and the corresponding values of the combined metric are higher.

Table 1. Automatic evaluation results

	Style transfer	Content preservation			Language quality	**GM**
	STA	CS	WO	BLEU	PPL	
Duplicate	0.38	1	1	1	146.86	1.78
Delete	1	0.93	0.84	0.92	246.80	4.14
GPT-2	0.90	0.86	0.71	0.86	258.44	3.71
GPT-3	0.88	0.92	0.79	0.92	359.12	3.58
T5 base	0.90	0.97	0.85	0.95	187.03	4.10
T5 large	**0.93**	**0.97**	**0.86**	**0.95**	**170.02**	**4.31**

5.4 Qualitative Evaluation Results

The presented models can, in most cases, replace vulgar words in the sentence with equivalent non-vulgar words while not changing the rest of the sentence. Replacements usually have an appropriate grammatical form and convey well the sense of the original sentence. Furthermore, the models can often cope with the replacement of the same vulgarism used in different, although homogeneous, grammatical forms or meaning. In situations where the replacement has a different grammatical form from the original (e.g. wrt. gender or number), the models can sometimes generate correct grammatical forms of the subordinate phrases. Some of the processed sentences contain swear words that have not been replaced by euphemisms.

The general effect of model inference is a decrease in the quality of language – output sentences contain numerous typos and modifications of proper names, cases, or punctuation. GPT-2 and GPT-3 models tend to modify non-vulgar parts of sentences much more frequently than T5 models, using synonyms or antonyms. In some cases, there appear words or phrases semantically unrelated to the original sentence. In individual cases, after generating such a word, the model terminates the text generation or adds a sentence ending, which is not coherent or contains numerous repetitions of certain word sequences. Such problems occur much less frequently in sentences processed by T5 models. At the same time, T5 models often perform better with sentences that are more complex, contain numerous proper names, complicated punctuation, or lower-quality language.

6 Conclusions and Future Work

This work is the first study of text devulgarization in Polish. We conducted experiments with three pre-trained models: GPT-2, GPT-3, and T5. The models were trained on a corpus of vulgar texts in Polish we created especially for this task. We developed an evaluation setup for the problem of substituting offensive language in Polish texts which aims to assess three aspects of the task — style transfer accuracy, content preservation, and language quality. Finally, we evaluated the presented methods and made all resources available.[6]

The results show that the tested approaches can be successfully used for removing offensive language, although there is room for improvement. All of the described approaches could benefit from more careful hyperparameter tuning as well as a larger training corpus. As the availability of large parallel corpora is limited, non-parallel methods for Polish could prove effective. Furthermore, we have only considered substituting profane words with euphemistic equivalents, while in some cases, simply removing the word seems to be the most appropriate strategy.

[6] http://clip.ipipan.waw.pl/DEPOTx (from *Devulgarization of Polish Texts*).

Acknowledgements. The work was financed by the European Regional Development Fund as a part of 2014–2020 Smart Growth Operational Programme, CLARIN—Common Language Resources and Technology Infrastructure, project no. POIR.04.02.00-00C002/19 and supported by the Poznan Supercomputing and Networking Center grant number 442. We kindly thank Professor Maciej Grochowski for giving us permission to use the full source material of the Dictionary of Polish Swearwords and Vulgarisms.

References

1. Brown, T.B., et al.: Language models are few-shot learners. In: Larochelle, H., Ranzato, M., Hadsell, R., Balcan, M.F., Lin, H. (eds.) Advance Neural Information Processing Systems, vol. 33, pp. 1877–1901. Curran Associates, Inc. (2020)
2. Cheriyan, J., et al.: Towards offensive language detection and reduction in four software engineering communities. In: Evaluation and Assessment in Software Engineering, pp. 254–259. Association for Computing Machinery, New York (2021)
3. Ciura, M.: Przetak: Fewer Weeds on the Web. In: Ogrodniczuk, M., Kobyliński, Ł, (eds.) Proceeding PolEval 2019 Workshop, pp. 127–133. Institute of Computer Science, Polish Academy of Sciences (2019)
4. Dementieva, D., et al.: Methods for detoxification of texts for the Russian language. Multimodal Technol. Interact. **5**, 54 (2021)
5. Grochowski, M.: Słownik polskich przekleństw i wulgaryzmów. PWN Scientific Publishers (2008)
6. Lison, P., Tiedemann, J.: OpenSubtitles2016: extracting large parallel corpora from movie and TV subtitles. In: Proceeding of Tenth International Conference on Language Resources and Evaluation, LREC 2016, pp. 923–929. ELRA, Portorož (2016)
7. Miławska, M.: Harmonia czy dysonans? O wulgaryzmach w, "Dniuświra" Marka Koterskiego. Słowo. Studia językoznawcze **4**, 188–199 (2013)
8. Osadnik, W.: Przeklinanie na ekranie, czyli o tłumaczeniu wulgaryzmów w napisach filmowych. In: P. Fast, N.S. (ed.) Tabu w przekładzie, pp. 61–71. Wydawnictwo Naukowe, "Śląsk", Katowice (2007)
9. Pang, R.Y., Gimpel, K.: Unsupervised evaluation metrics and learning criteria for non-parallel textual transfer. In: Proceedings of the 3rd Workshop on Neural Generation and Translation, pp. 138–147. ACL, Hong Kong (2019)
10. Papineni, K., et al.: Bleu: a method for automatic evaluation of machine translation. In: Proceedings of the 40th Annual Meeting of the Association for Computational Linguistics, pp. 311–318 (2002)
11. Radford, A., et al.: Language Models are Unsupervised Multitask Learners. OpenAI blog **1**(8) (2019)
12. Raffel, C., et al.: Exploring the limits of transfer learning with a unified text-to-text transformer. J. Mach. Learn. Res. **21**(140), 1–67 (2020)
13. Reimers, N., Gurevych, I.: Making Monolingual Sentence Embeddings Multilingual using Knowledge Distillation. In: Proceedings of the 2020 Conference on Empirical Methods in Natural Language Processing, EMNLP, pp. 4512–4525. ACL, Online (2020)
14. dos Santos, C.N., et al.: Fighting offensive language on social media with unsupervised text style transfer. In: Proceedings of the 56th Annual Meeting of the Association for Computational Linguistics, pp. 189–194. ACL, Melbourne (2018)
15. Tran, M., et al.: towards a friendly online community: an unsupervised style transfer framework for profanity redaction. In: Proceedings of the 28th International Conference on Computational Linguistics, pp. 2107–2114. Int. Comm. Comp. Linguist., Barcelona, Spain (Online) (2020)

Data-Driven Discovery of Time Fractional Differential Equations

Abhishek Kumar Singh[1], Mani Mehra[1(✉)], and Anatoly A. Alikhanov[2]

[1] Indian Institute of Technology Delhi, New Delhi 110016, India
mmehra@maths.iitd.ac.in

[2] North-Caucasus Center for Mathematical Research, North-Caucasus Federal University, Stavropol 355017, Russia

Abstract. In the era of data abundance and machine learning technologies, we often encounter difficulties in learning data-driven discovery of hidden physics, that is, learning differential equations/fractional differential equations via data. In [1], Schaeffer proposed a machine learning algorithm to learn the differential equation via data discovery. We extend Schaeffer's work in the case of time fractional differential equations and propose an algorithm to identify the fractional order α and discover the form of $\mathcal{F}$. Furthermore, if we have prior information regarding the set in which parameters belong to have some advantages in terms of time complexity of the algorithm over Schaeffer's work. Finally, we conduct various numerical experiments to verify the method's robustness at different noise levels.

Keywords: Fractional differential equations · Sparse optimization · Machine learning · Differential evolution

1 Introduction

The differential equations describe many phenomena in science and engineering, such as the Schrödinger equation, diffusion equation, a system of differential equations (SIR epidemic model), etc. (see [2] and references therein). The original discovery of these equations require a tremendous knowledge of physics, understanding of theories, and supportive evidence of experimental data. This is the one way to discover hidden physics. In recent years, researchers have provided the computational approach to data-driven discovery of hidden physics, that is, learning the differential equations that govern a particular phenomenon only using the data. Machine learning offers a wide range of numerical tools that efficiently learn differential equations via data discovery. Researchers in many fields have applied these tools to discover physical law from experimental data. Raissi *et al.* [3] used probabilistic machine learning to discover the linear differential equations. Again, Raissi *et al.* [4] introduced a physics-informed neural network

Supported by University Grants Commission India, SERB India and North-Caucasus Center for Mathematical Research, Russia.

D. Groen et al. (Eds.): ICCS 2022, LNCS 13351, pp. 56–63, 2022.
https://doi.org/10.1007/978-3-031-08754-7_8

to discover the non-linear partial differential equations. Due to some issues with the regular data training in the physics-informed neural network, Krishanpriyan *et al.* [5] introduced a physics-informed neural network with some advanced training techniques to discover non-linear partial differential equations. In the works mentioned above, authors know the form of the differential equation, whether its diffusion or diffusion-reaction, etc. The first time, Schaeffer [1] introduced machine learning techniques to identify the terms in underlying partial differential equations to the best of our knowledge. However, in [6], Srivastava *et al.* proposed a machine learning approach to identify the terms in underlying partial differential equations. In recent decades, the theory of fractional derivatives has been an emerging topic in physical, life, and social sciences due to its non-local properties. The non-local properties of fractional operators give a superior way to deal with the complex phenomena in physical, life and social sciences (see [7] and references therein). Gulian *et al.* [8] extend the Raissi *et al.* [3] work to find the space fractional differential equations. Pang *et al.* [9] generalized the Raissi *et al.* [4] work to learn the space-time fractional differential equations. Recently, Singh *et al.* [10,11] proposed scientific machine learning algorithm to learn the system of fractional differential equations via data.

Motivated by the above discussions, we extend the Schaeffer [1] work for time fractional differential equations. The primary approach used by Schaeffer is to convert the problem into an optimization problem and then use the indirect method to solve the optimization problem. Schaeffer used the least absolute shrinkage and selection operator (LASSO) method, which uses the L_1-norm in the regularization term, with the help of a linear system of equations [12]. In this work, we also use the LASSO method with the help of a non-linear system of Eq. and use differential evolution to solve the LASSO method.

2 Methodology

In this section, we generalized the methodology, recently given by Schaeffer [1], in the case of time fractional differential equations.

2.1 Sparse Reconstruction of Fractional Differential Equation

We will derive the method for a fractional differential equation which has the following form:

$$ {}_{C}D^{\alpha}_{0,t}\mathcal{U}(x,t) = \mathcal{F}(\mathcal{U},\mathcal{U}_x,\mathcal{U}_{xx},\cdots,\mathcal{U}_{x\cdots x}) + \mathcal{G}(x,t), \tag{1} $$

where $0 < \alpha < 1$, ${}_{C}D^{\alpha}_{0,t}$ denote the left-side Caputo fractional derivative of order α with respect to t (for definition of Caputo see the [11]). Assume that the form of the $\mathcal{F}$ and fractional order α in the above Eq. are unknown to the user, and instead user only have the data. For understanding and derivation of the methodology, here we consider the following form:

$$ {}_{C}D^{\alpha}_{0,t}\mathcal{U}(x,t) - \mathcal{G}(x,t) = \mathcal{F}(\mathcal{U},\mathcal{U}_x,\mathcal{U}_{xx}). \tag{2} $$

We can approximate the function $\mathcal{F}$ via a nth-order Taylor series expansion. This approximation can be made by the user based on some physically relevant models related to the data. Here for simplicity we take $n=2$, the second order Taylor series expansion of $\mathcal{F}$ about the origin can be written as:

$$\mathcal{F}(\mathcal{U},\mathcal{U}_x,\mathcal{U}_{xx}) \approx \mathcal{F}(0,0,0) + \mathcal{U}\frac{\partial \mathcal{F}}{\partial \mathcal{U}}(0,0,0) + \mathcal{U}_x\frac{\partial \mathcal{F}}{\partial \mathcal{U}_x}(0,0,0) + \mathcal{U}_{xx}\frac{\partial \mathcal{F}}{\partial \mathcal{U}_{xx}}(0,0,0)$$
$$+\frac{1}{2}\Big[\mathcal{U}^2\frac{\partial^2 \mathcal{F}}{\partial \mathcal{U}^2}(0,0,0) + \mathcal{U}_x^2\frac{\partial^2 \mathcal{F}}{\partial \mathcal{U}_x^2}(0,0,0) + \mathcal{U}_{xx}^2\frac{\partial^2 \mathcal{F}}{\partial \mathcal{U}_{xx}^2}(0,0,0)$$
$$+ 2\mathcal{U}\mathcal{U}_x\frac{\partial^2 \mathcal{F}}{\partial \mathcal{U}\partial \mathcal{U}_x}(0,0,0) + 2\mathcal{U}\mathcal{U}_{xx}\frac{\partial^2 \mathcal{F}}{\partial \mathcal{U}\partial \mathcal{U}_{xx}}(0,0,0) + 2\mathcal{U}_x\mathcal{U}_{xx}\frac{\partial^2 \mathcal{F}}{\partial \mathcal{U}_x\partial \mathcal{U}_{xx}}(0,0,0)\Big]. \tag{3}$$

Using the Eq. (3) in the Eq. (2), we can write

$${}_{C}D_{0,t}^{\alpha}\mathcal{U}(x,t) - \mathcal{G}(x,t) = \alpha_1 + \mathcal{U}\alpha_2 + \mathcal{U}_x\alpha_3 + \cdots + \mathcal{U}_x\mathcal{U}_{xx}\alpha_{10}, \tag{4}$$

where $\alpha_1 = \mathcal{F}(0,0,0), \alpha_2 = \frac{\partial \mathcal{F}}{\partial \mathcal{U}}(0,0,0), \alpha_3 = \frac{\partial \mathcal{F}}{\partial \mathcal{U}_x}(0,0,0), \cdots, \alpha_{10} = \frac{\partial^2 \mathcal{F}}{\partial \mathcal{U}_x \partial \mathcal{U}_{xx}}(0,0,0)$ are the coefficients of the unknown in the Eq. (3). We can write the above Eq. as:

$${}_{C}D_{0,t}^{\alpha}\mathcal{U}(x,t) - \mathcal{G}(x,t) = [1\ \mathcal{U}\ \mathcal{U}^2\ \mathcal{U}_x\ \mathcal{U}_x^2\ \mathcal{U}\mathcal{U}_x\ \mathcal{U}_{xx}\ \mathcal{U}_{xx}^2\ \mathcal{U}\mathcal{U}_{xx}\ \mathcal{U}_x\mathcal{U}_{xx}].\boldsymbol{\alpha},$$

where $\boldsymbol{\alpha} = [\alpha_1, \alpha_2, \cdots, \alpha_{10}]$. Assume that the data of the unknowns are given to us at the points $(x_j, t_k), j = 1, 2, \cdots, n, k = 1, 2, \cdots, m$. Now, we define feature vectors $f_1, f_2, f_3, \cdots, f_{10}$ as, $f_1 = [1, \ldots 1, \ldots, 1]^T$, $f_2 = [\mathcal{U}(x_1,t)\ldots\mathcal{U}(x_j,t)\ldots\mathcal{U}(x_n,t)]^T, \ldots, f_{10} = [\mathcal{U}_x(x_1,t)\mathcal{U}_{xx}(x_1,t), \ldots, \mathcal{U}_x(x_j,t)\mathcal{U}_{xx}(x_j,t), \ldots, \mathcal{U}_x(x_n,t)\mathcal{U}_{xx}(x_n,t)]^T$. Now the system of equations become,

$$V(t;\alpha) = P(t)\boldsymbol{\alpha}, \tag{5}$$

where V be the vector, which is the combination of time fractional derivative term and source term and P be the matrix of all features vector. Let $\boldsymbol{\beta} = (\alpha, \boldsymbol{\alpha})$ are the unknown parameter in the Eq. (5). In [1], author ended with the system of Eq. (5) with $\alpha = 1$, which was linear in unknown parameter. Here, we have a non-linear equation in unknown parameter $\boldsymbol{\beta}$, we can't solve the above equation with inverse or pseudo-inverse due to non-linearity. Also data may contains noise, and the problem can ill-posed. To tackle the ill-posedness, we use a regularizer with the entire data set. We use LASSO method, which is use L_1-regularization to promote the sparsity in the vector $\boldsymbol{\beta}$ is define as:

$$\min_{\boldsymbol{\beta}=(\alpha,\boldsymbol{\alpha})} \mathcal{J}(\boldsymbol{\beta}) = \frac{1}{2}\sum_{i=1}^{m} \|V(t_i,\alpha) - P(t_i)\boldsymbol{\alpha}\|_2^2 + \lambda \|\boldsymbol{\beta}\|_1, \tag{6}$$

where $\lambda > 0$ is a regularization parameter. The issue with the Schaeffer's approach how to use the method in the case of a non-linear system of equations due to fractional order. Therefore, we use differential evolution to solve the Eq. (6).

2.2 Differential Evolution

In 1996, the differential evolution method was introduced by Storn and Price [11,13]. It is a stochastic, production-based optimization method for solving non-linear optimization problems. We used differential evolution as an optimization method for our problem. Let us consider the optimization problem (6). Here we aim to find the vector $\boldsymbol{\beta}^*$, which minimizes the $\mathcal{J}$. Suppose population size is N_p. Therefore, the population matrix can be written as

$$\beta_n^g = [\beta_{n,1}^g, \beta_{n,2}^g, \cdots, \beta_{n,D}^g]. \tag{7}$$

Here D is the number of parameters, g is the generation and $n = 1, 2, \cdots, N_p$ and population matrix is generated with the help of prior information available of the parameters. If we do not have any prior information regarding parameter then we generate the population matrix with help of uniform random variable. For more details, see [11]. Now, we describes our algorithm for learning the parameter vector $\boldsymbol{\beta}$ presented in Eq. (6).

Algorithm 1. Learning the parameter involve in the Eq. (6)

Step 1 : With the help of given data, construct the feature matrix.
Step 2 : For the first generation, initialize the population matrix, and initialize all other parameters.
Step 3 : Execute the mutation and crossover steps for the given population.
Step 4 : With the help of LASSO operator evaluate error values for all the population members.
Step 5 : Execute the selection step with the help of error values.
Step 6 : Until the population members lie under a specific threshold value, repeat steps 2-5.
Step 7 : When population converges, return the final value.

---

3 Simulations and Numerical Experiments

In this Sect., we illustrate the methodology which we have discussed in Sect. 2 with the help of the following three test Examples. The first two Examples show algorithm robustness at different level of noise. The third Example shows the advantage of the algorithm in the case of prior information regarding the parameters. The simulations were run on an Intel Core i5 – 1135G7, 1.80 GHzx4 machine with 16 GB RAM.

Example 31. *Consider the time fractional diffusion equation*

$$_CD_{0,t}^{\alpha}\mathcal{U}(x,t) = \mathcal{U}_{xx}(x,t) + \mathcal{G}(x,t), \tag{8}$$

$$\mathcal{G}(x,t) = \frac{24}{\Gamma(5-\alpha)} t^{4-\alpha} \sin \pi x,\ \mathcal{U}(x,0) = 0,\ \mathcal{U}(0,t) = 0,\ \textit{and}\ \mathcal{U}(1,t) = 0.$$

Table 1. The identified terms and α for Example 31 at different level of noise.

	α	Identified term	Coefficients
No Noise	0.4942	$\mathcal{U}_{xx}$	0.9996
2% Noise	0.4879	$\mathcal{U}_{xx}$	1.0000
5% Noise	0.4851	$\mathcal{U}_{xx}$	1.0000

It can be checked that for $\alpha = 0.5$, $\mathcal{U}(x,t) = t^4 \sin \pi x$ be the exact solution of (8). Data is generated with the help of the exact solution of the equation with the 200×200 grid points and the feature matrix are obtained using third order Taylor's series expansion. Table 1 demonstrate the result of identifying the terms and fractional order in the underlying time fractional diffusion equation at different noise level. From the Table 1, one can observe that our proposed algorithm for identifying the time fractional diffusion equation is robust to noise. Figure 1 shows that the learned dynamic of $\mathcal{G}(x,t)$ is close to the original $\mathcal{G}(x,t)$ at a different noise level due to the learned fractional order.

Example 32. *Consider the time fractional Burgers equation*

$$
{}_C D^{\alpha}_{0,t}\mathcal{U}(x,t) = \mathcal{U}(x,t)\mathcal{U}_x(x,t) + 0.1\mathcal{U}_{xx}(x,t) + \mathcal{G}(x,t), \tag{9}
$$

$$
\mathcal{G}(x,t) = \frac{10}{\Gamma(2-\alpha)} t^{1-\alpha} \sin \pi x - 100t^2\pi \sin \pi x + 10t^2\pi^2 \sin \pi x,
$$

$$
\mathcal{U}(x,0) = 0 = \mathcal{U}(0,t) = \mathcal{U}(1,t).
$$

Table 2 demonstrate the result of identifying the terms and fractional order in the underlying time fractional Burgers equation at different noise level. Since the diffusion term's coefficient (viscosity) is small compared to the other term coefficients, from Table 1, one can observe that the viscosity term identifies as correctly as other terms involved in the time-fractional Burgers equation. Hence, our algorithm is efficient in identifying the terms having the coefficient with large value as well as terms having comparatively small value.

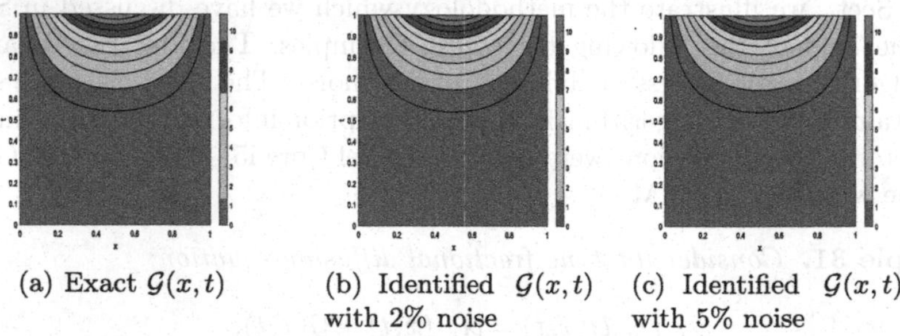

(a) Exact $\mathcal{G}(x,t)$ (b) Identified $\mathcal{G}(x,t)$ with 2% noise (c) Identified $\mathcal{G}(x,t)$ with 5% noise

Fig. 1. Exact and Identified $\mathcal{G}(x,t)$ with different level of noise.

Table 2. The identified terms and α for Example 32 at different level of noise.

	α	Identified term	Coefficient
No Noise	0.4997	$\mathcal{U}\mathcal{U}_x,\mathcal{U}_{xx}$	0.9999,0.1001
2% Noise	0.4998	$\mathcal{U}\mathcal{U}_x,\mathcal{U}_{xx}$	0.9997,0.1001
5% Noise	0.4995	$\mathcal{U}\mathcal{U}_x,\mathcal{U}_{xx}$	0.9998,0.1003

Example 33. *Consider the time fractional advection diffusion equation*

$$ {}_CD^{\alpha}_{0,t}\mathcal{U}(x,t) = \mathcal{U}_x(x,t) + \mathcal{U}_{xx}(x,t), \tag{10}$$

$$\mathcal{U}(x,0) = e^{2x},\ \mathcal{U}(0,t) = e^{6t},\ \textit{and}\ \mathcal{U}(1,t) = e^{2+6t}.$$

For simplicity we take $\alpha = 1$ and it can be checked that $\mathcal{U}(x,t) = e^{2x+6t}$ be the exact solution of (10). In this example, we will show the efficiency of our algorithm in the two cases: First, if we have prior information regarding coefficients of the terms in underlying advection-diffusion Eq., whether it is a real number or integers. Second, if we do not have any information regarding coefficients. Assuming the coefficients of the term in the underlying advection-diffusion equation are integers (one can observe from the Eq. (10)), while in the second case, we do not have any information regarding coefficients then the algorithm changes accordingly. Table 3 demonstrates the result of identifying the terms in the underlying time fractional advection diffusion equation, CPU time in second, and the number of iterations required to the convergence of the algorithm in the two cases. From the Table 3, one can observe that when we have some information regarding the coefficients of the term, our algorithm converges very fast compared to the second case. In the Schaeffer [1] approach we do not have this kind of advantage.

Table 3. The identified terms, CPU time and number of iteration required to convergence for Example 33 with two cases.

Have prior information				Do not have information			
Iteration	CPU time	Term	Coefficient	Iteration	CPU time	Term	Coefficient
14	13.52	$\mathcal{U}_x,\mathcal{U}_{xx}$	1,1	578	530.55	$\mathcal{U}_x,\mathcal{U}_{xx}$	1 ,1

4 Conclusions

To summarize, this work introduces a computational framework based on LASSO method and differential evolution to discover the time fractional differential equations. This work is an extension of the recent study done by Schaeffer [1] to the case of time fractional differential equations. In [1], the author used the indirect optimization method to solve the optimization problem. The issue with [1] is how to use the method in the case of a non-linear system of equations due to fractional order. To address this issue in this paper, we have used the direct optimization method (differential evolution). An extension of this study could be generalizing this work for time-space fractional differential equations, which remains a major conceptual challenge due to fractional Taylor series approximation and space fractional order.

Acknowledgements. The First author acknowledges the support provided by University Grants Commission (UGC), India, under the grant number 20/12/2015(ii)EU-V. The second author acknowledges the support provided by the SERB India, under the grant number SERB/F/3060/2021 – 2022. The third author acknowledges the financial support from the North-Caucasus Center for Mathematical Research under agreement number 075 – 02 – 2021 – 1749 with the Ministry of Science and Higher Education of the Russian Federation.

References

1. Schaeffer, H.: Learning partial differential equations via data discovery and sparse optimization. In: Proceedings of the Royal Society A: Mathematical, Physical and Engineering Sciences, vol. 473(2197), p. 20160446 (2017)
2. Mehra, M., Mallik, R.K.: Solutions of differential-difference equations arising from mathematical models of granulocytopoiesis. Different. Equat. Dynam. Syst. **22**(1), 33–49 (2014)
3. Raissi, M., Perdikaris, P., Karniadakis, G.E.: Machine learning of linear differential equations using gaussian processes. J. Comput. Phys. **348**, 683–693 (2017)
4. Raissi, M., Perdikaris, P., Karniadakis, G.E.: Physics-informed neural networks: a deep learning framework for solving forward and inverse problems involving non-linear partial differential equations. J. Comput. Phys. **378**, 686–707 (2019)
5. Krishnapriyan, A., Gholami, A., Zhe, S., Kirby, R., Mahoney, M.W.: Characterizing possible failure modes in physics-informed neural networks. In: Advances in Neural Information Processing Systems, vol. 34 (2021)
6. Srivastava, K., Ahlawat, M., Singh, J., Kumar, V.: Learning partial differential equations from noisy data using neural networks. J. Phys. Conf. Ser. **1655**, 012075 (2020)
7. Alikhanov, A., Beshtokov, M., Mehra, M.: The crank-nicolson type compact difference schemes for a loaded time-fractional hallaire equation. Fractional Calc. Appli. Anal. **24**(4), 1231–1256 (2021)
8. Gulian, M., Raissi, M., Perdikaris, P., Karniadakis, G.: Machine learning of space-fractional differential equations. SIAM J. Sci. Comput. **41**(4), A2485–A2509 (2019)
9. Pang, G., Lu, L., Karniadakis, G.E.: fPINNs: Fractional physics-informed neural networks. SIAM J. Sci. Comput. **41**(4), A2603–A2626 (2019)
10. Singh, A.K., Mehra, M., Gulyani, S.: A modified variable-order fractional sir model to predict the spread of covid-19 in india. Math. Meth. Appli. Sci. (2021). https://doi.org/10.1002/mma.7655
11. Singh, A.K., Mehra, M., Gulyani, S.: Learning parameters of a system of variable order fractional differential equations. In: Numerical Methods for Partial Differential Equations (2021). https://doi.org/10.1002/num.22796

12. Brunton, S.L., Proctor, J.L., Kutz, J.N.: Discovering governing equations from data by sparse identification of nonlinear dynamical systems. Proc. Natl. Acad. Sci. **113**(15), 3932–3937 (2016)
13. Storn, R., Price, K.: Differential evolution-a simple and efficient heuristic for global optimization over continuous spaces. J. Global Optim. **11**(4), 341–359 (1997)

Wide Ensembles of Neural Networks in Music Genre Classification

Daniel Kostrzewa, Wojciech Mazur, and Robert Brzeski(✉)

Department of Applied Informatics, Silesian University of Technology, Gliwice, Poland
{daniel.kostrzewa,robert.brzeski}@polsl.pl

Abstract. The classification of music genres is essential due to millions of songs in online databases. It would be nearly impossible or very costly to do this job manually. That is why there is a need to create robust and efficient methods that automatically help to do this task. In this paper, music genre recognition is implemented by exploiting the potential of wide ensembles of neural network classifiers. Creating infrequently used types of ensembles is a main contribution of authors in the development of automatic recognition of the musical genre. The paper shows how it can be done in a relatively quick and straightforward manner. The presented method can be implemented in many other use cases.

Keywords: Classification · Machine learning · Music genre recognition · Wide ensemble · Free music archive dataset · Neural network

1 Introduction

There are millions of songs available for users in online databases. Very often, we would like to listen only songs that belong to a specified music genre. It is nearly impossible to manually assign these millions songs into a music genre. One of the options is to do this automatically. For this task, machine learning methods can be used. It is possible to improve the obtained classification quality, either by pre-processing the dataset or by appropriate selection of classifiers parameters, or by creating an appropriate classifier structure (e.g., the number of layers). However, for a given classifier, at some point, the practical ability for further improving the results is limited.

In the current research, we would like to examine the comparatively simple method – the collection of ensembles. Of course, it is possible to create an ensemble with quite complex classifiers, including a convolutional neural network with many layers [6]. However, the computational cost of using such an ensemble can

This work was supported by research funds for young researchers of the Department of Applied Informatics, Silesian University of Technology, Gliwice, Poland (grant no. 02/0100/BKM22/0021 – DK) and Statutory Research funds of the Department of Applied Informatics, Silesian University of Technology, Gliwice, Poland (grant no. 02/0100/BK22/0017 – RB).

D. Groen et al. (Eds.): ICCS 2022, LNCS 13351, pp. 64–71, 2022.
https://doi.org/10.1007/978-3-031-08754-7_9

be high. Deep neural networks take a while to learn. That is why, in current research, we intend to examine the creating of wide ensembles with relatively simple classifiers. Wide ensembles are understood as built with many (dozens) base-level classifiers, in this case, multiple copies of the same classifier. This way, we can obtain another method of improving the final result of the music genres classification.

The main contribution of the paper is the way of creating wide ensembles. It can be done instantly for a neural network by creating multiple copies of the previously prepared classifier. This paper will check the change in the classification quality for this type of structure. The second contribution is checking whether the additional, late input of raw data, connected directly to the concatenation layer, which connects individual classifiers, improves classification quality. Additionally, the influence of depth of the classifiers and application of Principal Component Analysis on the final result is examined.

2 Related Work

The problem of music genre recognition (MGR) [1], as one of the sub-disciplines of music information retrieval, has become an increasingly explored issue in recent years. The article [12] can be considered the beginning of the popularity of the MGR topic [5]. The classification of musical songs can be executed using many machine learning methods. Not only the classical classifiers [2] can be used, but also newer approaches like the deep learning domain [4], with convolutional neural networks (CNN) [8] or convolutional recurrent neural networks (CRNN) [11]. Unfortunately, deep neural networks are more challenging to create, need more time for the learning process, and often give worse classification results (comparative studies are presented in Table 4) or at least there is no guarantee for obtaining a better. There are also studies in which the ensembles of various classifiers are used. Ensembles consist of base-level with a set of classifiers, as well as meta classifier [10] that tries to predict the final result based on outcomes of base-level classifiers.

3 Conditions of Experiments

The dataset that the experiment was conducted is the small subset of Free Music Archive dataset (FMA) [3]. For each excerpt, there are over 500 features in the FMA dataset. This dataset was split into three sets: training, validation, and testing in ratio 80:10:10.

The ensemble in the conducted research is built as multiple copies of the same classifier. However, each of the classifiers is learned independently. Because they have various sets of initial weights, the result of the learning process is also different for each of the classifiers. Consequently, they will generate slightly different classification outcomes. The output of individual classifiers of a given ensemble is fed to the input of the 'Concatenate' layer of the additional classifier (based on a dense neural network), which generates the final classification result. Additionally, this layer can have neurons for extra input of numerical data.

4 Experiments

4.1 Basic Dense Classifiers

At the beginning, three basic classifiers were created. Their quality (Table 1) can be treated as a benchmark for further tests.

The first one is a **simple dense classifier** (Fig. 1a) consisting of three layers – input layer and two dense layers. The results can be found in Table 1 in the 'En0' row. Accuracy of 53% is far better than a blind guess and already looks promising. Each created ensemble has a unique name (number) from En1 up to En14. The current classifier (En0) is the only one which is not an ensemble.

Table 1. Different size of ensemble, without and with raw data input.

Name	Classifiers	Raw data input	Accuracy	Precision	Recall	F1-score
En0	1 (simplest classifier)	No	0.530	0.524	0.530	0.526
En1	2	No	0.539	0.540	0.542	0.540
En2	2	Yes	0.560	0.553	0.557	0.555
En3	3	Yes	0.541	0.538	0.537	0.535
En4	5	Yes	0.586	0.584	0.590	0.585
En5	50	Yes	**0.621**	**0.618**	**0.621**	**0.617**

The second structure – En1, is the **simplest ensemble that consists of only two classifiers, without numerical input** connected by predicating layer with classifier outputs merging mechanism. Those two classifiers are just a copy of the simplest classifier.

Next structure En2, **two classifiers ensemble with numerical input** (Fig. 1b) is almost the same as the previous one. However, the difference is that a raw numerical input is attached to the merging layer.

Comparing the obtained results (Table 1), it turns out that the best approach is the usage of the two classifiers ensemble with numerical input. An additional late raw numerical input (concatenated with outcomes of base-level classifiers) to the two classifiers network ensemble, significantly increased the classification quality of a whole structure. As the results came out to be quite promising, the additional layer of numerical input will be included for the rest of the research.

4.2 Principal Component Analysis Influence

In this part of the research, the Principal Component Analysis (PCA) was introduced. The feature vector can contain not only useful data but also noise and that can lead to worsening results. One of the popular methods of preventing such a problem is feature extraction. Over five hundred features of raw data are currently fed to classifiers. Reducing that number might not only speed up processing time but also increase the accuracy of the proposed network. Time

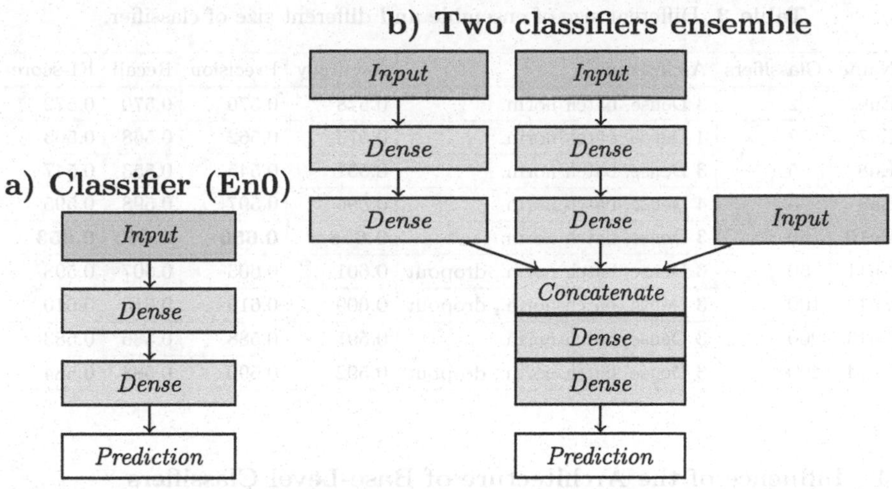

Fig. 1. a) Simplest classifier b) Ensemble with numerical input of raw data

for training for all input data without using PCA took around 6.5 s, while with PCA reduction to 300 components, training took 4 s, and with 20 components, training took only 2.8 s. The best result was achieved (Table 2) for 300 components, and that quantity will be used for the rest of the research. The reduced by PCA set of features are transferred to the classifiers' inputs and for the late raw numerical input.

Table 2. Accuracy achieved for Principal Component Analysis.

	Number of components						
Without PCA	500	400	300	200	100	50	20
0.553	0.555	0.549	**0.560**	0.553	0.548	0.521	0.486

4.3 Influence of the Number of Classifiers

In this part of the research, the influence of the number of classifiers was tested. The three (En3), five (En4), and fifty (En5) classifiers in one ensemble were taken into consideration. The achieved results are presented in Table 1.

It turns out that one additional classifier in the ensemble (En3) did not bring higher results. For another two classifiers added to the model (En4), the results are slightly improved. However, the best result is achieved for a wide ensemble of 50 base-level classifiers (En5), with an accuracy of about 62%. Nevertheless, better results also came with around 20 times longer training time than in the case of two base-level classifiers network. Additionally, the overfitting of the network can be seen. As a result, in the next test, batch normalization and dropout will be introduced.

Table 3. Different size of ensemble and different size of classifier.

Name	Classifiers	Architecture	Accuracy	Precision	Recall	F1-score
En6	2	3 Dense, batch norm.	0.578	0.576	0.570	0.572
En7	2	4 Dense, batch norm.	0.573	0.562	0.568	0.563
En8	5	3 Dense, batch norm.	0.551	0.545	0.553	0.547
En9	5	4 Dense, batch norm.	0.596	0.597	0.598	0.595
En10	50	3 Dense, batch norm.	**0.658**	**0.656**	**0.653**	**0.653**
En11	50	3 Dense, batch norm., dropout	0.601	0.603	0.607	0.598
En12	100	3 Dense, batch norm., dropout	0.609	0.610	0.613	0.610
En13	200	3 Dense, batch norm.	0.591	0.588	0.586	0.583
En14	200	3 Dense, batch norm., dropout	0.592	0.590	0.588	0.584

4.4 Influence of the Architecture of Base-Level Classifiers

This time not only the number but also the size and structure of classifiers were examined. The results of using such ensembles are presented in Table 3.

The first structure (En6) goes back to the two base-level classifiers with an additional dense and batch normalization. Comparing the obtained outcomes to the ensemble without additional layer and batch normalization (En2) shows a slight improvement in performance quality. In the next ensemble (En7), another dense layer and batch normalization were added. Apart from longer training time, there was no significant change in the classification accuracy by introducing the next dense layer. The structure of the En8 ensemble is similar to the En6, but this time with five base-level classifiers. Interestingly, the quantitative outcomes are slightly worse compared to both ensemble En6 and the earlier En4. However, the En9 ensemble, with another dense layer and batch normalization, easily beat all presented ensembles but En5 (with 50 base-level classifiers). The En10 wide ensemble is similar to the En9 one, however, this time consists of fifty classifiers. The obtained results are the best in the presented studies. A test was also carried out using additional dropout, but the results obtained in this way (En11, En12, and En14), turned out to be worse than En10. The same conclusion is for a wider ensemble (En13) with 200 classifiers but without dropout.

5 Comparison of the Outcomes

Basic classifiers and raw data input. As can be seen in Table 1 even the basic ensemble (En1) improves the results slightly in comparison to the simple classifier (En0). A much more significant improvement is obtained with a late raw numerical data input (En2).

Principal Component Analysis. Introduction of Principal Component Analysis (Table 2) was not strictly related to the model development but to the data preprocessing that the model operated on. The FMA dataset offers an overall of

518 features (for each music track), and data can be preprocessed by the dimensionality reduction method. Here, only PCA was exploited and the best result was achieved for 300 features.

Width of ensemble. Additional base-level classifiers have influenced a significant increase in quantitative results (Table 1). The actual cost of such an increase was only the time of computation. An increase of ensemble width by adding classifiers improved results significantly (up to fifty base-level classifiers). Further expanding the ensemble did not bring any advance, conversely, the outcomes have already started to worsen.

Architecture of base-level classifier. The next way of improving the accuracy of the model was the change in the structure of the base-level classifier (Table 3). Incrementing the depth of the base-level classifier results in higher accuracy values. At the same time, with more layers, the overfitting of the model became more noticeable. To reduce training accuracy spiking, batch normalization and dropout were introduced. Nonetheless, dropout did decrease overfitting but did not help with model accuracy.

Comparison of the quantitative outcomes with state-of-the-art. To compare the best result achieved in this research (wide ensemble En10 with fifty base-level classifiers for which accuracy was 0.658) with other state-of-the-art works [7,9,11,13–15] the Table 4 was created.

Table 4. Comparison of different models classifying FMA small dataset with the proposed wide ensemble En10 (all values are in %).

No.	Model	Accuracy	No.	Model	Accuracy
1	K-Nearest Neighbors [15]	36.4	12	MoER [14]	55.9
2	Logistic Regression [15]	42.3	13	FCN [13]	63.9
3	Multilayer Perceptron [15]	44.9	14	TimbreCNN [13]	61.7
4	Support Vector Machine [15]	46.4	15	End-to-end [13]	61.4
5	Original spectrogram [14]	49.4	16	CRNN [13]	63.4
6	Harmonic spectrogram [14]	43.4	17	CRNN-TF [13]	64.7
7	Percussive spectrogram [14]	50.9	18	CRNN [11]	53.5
8	Modulation spectrogram [14]	55.6	19	CNN-RNN [11]	56.4
9	MFCC [14]	47.1	20	CNN TL [7]	51.5
10	MoEB [14]	54.1	21	CNN TL [9]	56.8
11	MoEC [14]	55.6	22	C-RNN [15]	65.2
23	**Wide ensemble En10**	**65.8**	**23**	**Wide ensemble En10**	**65.8**

6 Conclusions

The results of the work are auspicious. The ensemble provided satisfying results, with the best model reaching almost 66% accuracy. It is worth mentioning that this value is in the range of state-of-the-art techniques. However, obtaining such a result is only an additional effect. The main aim of the research was to show how to relatively easily improve the originally obtained classification result. This goal has been achieved. This way, by implementing wide ensembles, we obtain another method of improving the final result of the classification without much design or programming effort. A certain limitation may be the increased computation time and the increased demand for computer resources. However, there are no initial restrictions as to the field, dataset, or nature of the research where we can try to use this method.

Summarising, if the main goal is classification quality, presented in the article methods and structures are definitely worth considering.

References

1. Aucouturier, J.J., Pachet, F.: Representing musical genre: a state of the art. J. New Music Res. **32**(1), 83–93 (2003)
2. Basili, R., Serafini, A., Stellato, A.: Z classification of musical genre: a machine learning approach. In: ISMIR (2004)
3. Defferrard, M., Benzi, K., Vandergheynst, P., Bresson, X.: FMA: a dataset for music analysis. arXiv preprint arXiv:1612.01840 (2016)
4. Kereliuk, C., Sturm, B.L., Larsen, J.: Z deep learning and music adversaries. IEEE Trans. Multim. **17**(11), 2059–2071 (2015)
5. Knees, P., Schedl, M.: A survey of music similarity and recommendation from music context data. ACM Trans. Multim. Comput. Commun. Appl. **10**(1), 1–21 (2013)
6. Kostrzewa, D., Kaminski, P., Brzeski, R.: Music genre classification: looking for the perfect network. In: Paszynski, M., Kranzlmüller, D., Krzhizhanovskaya, V.V., Dongarra, J.J., Sloot, P.M.A. (eds.) ICCS 2021. LNCS, vol. 12742, pp. 55–67. Springer, Cham (2021). https://doi.org/10.1007/978-3-030-77961-0_6
7. Lee, D., Lee, J., Park, J., Lee, K.: Z enhancing music features by knowledge transfer from user-item log data. In: ICASSP 2019–2019 IEEE International Conference on Acoustics, Speech and Signal Processing (ICASSP), pp. 386–390. IEEE (2019)
8. Lim, M., et al.: Z convolutional neural network based audio event classification. KSII Trans. Internet Inf. Syst. **12**(6) (2018)
9. Park, J., Lee, J., Park, J., Ha, J.W., Nam, J.: Z representation learning of music using artist labels. arXiv preprint arXiv:1710.06648 (2017)
10. Silla, Jr., C.N., Kaestner, C.A.A., Koerich, A.L.: Automatic music genre classification using ensemble of classifiers. In: 2007 IEEE International Conference on Systems, Man and Cybernetics, pp. 1687–1692 (2007)
11. Snigdha, C., Kavitha, A.S., Shwetha, A.N., Shreya, H., Vidyullatha, K.S.: Z music genre classification using machine learning algorithms: a comparison. Int. Res. J. Eng. Technol. **6**(5), 851–858 (2019)
12. Tzanetakis, G., Cook, P.: Musical genre classification of audio signals. IEEE Trans. Speech Audio Process. **10**(5), 293–302 (2002)
13. Wang, Z., Muknahallipatna, S., Fan, M., Okray, A., Lan, C.: Z music classification using an improved CRNN with multi-directional spatial dependencies in both time and frequency dimensions. In: 2019 International Joint Conference on Neural Networks (IJCNN), pp. 1–8. IEEE (2019)

14. Yi, Y., Chen, K.Y., Gu, H.Y.: Z mixture of CNN experts from multiple acoustic feature domain for music genre classification. In: 2019 Asia-Pacific Signal and Information Processing Association Annual Summit and Conference (APSIPA ASC), pp. 1250–1255. IEEE (2019)
15. Zhang, C., Zhang, Y., Chen, C.: Z songnet: Real-time music classification. Stanford University Press (2019)

MultiEmo: Language-Agnostic Sentiment Analysis

Piotr Miłkowski(✉), Marcin Gruza, Przemysław Kazienko, Joanna Szołomicka, Stanisław Woźniak, and Jan Kocoń

Department of Artificial Intelligence, Wrocław University of Science and Technology, Wrocław, Poland
piotr.milkowski@pwr.edu.pl

Abstract. We developed and validated a language-agnostic method for sentiment analysis. Cross-language experiments carried out on the new MultiEmo dataset with texts in 11 languages proved that LaBSE embeddings with an additional attention layer implemented in the BiLSTM architecture outperformed other methods in most cases.

Keywords: Cross-language NLP · Sentiment analysis · Language-agnostic representation · LASER · LaBSE · BiLSTM · Opinion mining · MultiEmo

1 Introduction

Two of the most important and applicable topics in natural language processing (NLP), in particular in opinion mining, include sentiment analysis [1–3] and emotion recognition [4,5]. Recently, more and more online comments are expressed in different natural languages. Consequently, there is a growing interest in new methods for sentiment analysis that are language-independent. For that purpose, appropriate language-agnostic models (embeddings) may be utilized.

In this paper, we developed and validated three language-agnostic methods for sentiment analysis: one based on the LASER model [6] and two on LaBSE [7], see Sect. 3. The latter was used in its basic version ($LaBSE_b$) and with additional attention layer ($LaBSE_a$). All of them were implemented within the bidirectional LSTM architecture (biLSTM). The experiments were performed on our new benchmark MultiEmo dataset, which is an extension of MultiEmo-Test 1.0 [8]. In the latter, only test texts were translated into other languages, whereas the MultiEmo data proposed here is fully multilingual.

This work was partially supported by the National Science Centre, Poland, project no. 2020/37/B/ST6/03806; by the statutory funds of the Department of Artificial Intelligence, Wroclaw University of Science and Technology; by the European Regional Development Fund as a part of the 2014-2020 Smart Growth Operational Programme, CLARIN - Common Language Resources and Technology Infrastructure, project no. POIR.04.02.00-00C002/19.

D. Groen et al. (Eds.): ICCS 2022, LNCS 13351, pp. 72–79, 2022.
https://doi.org/10.1007/978-3-031-08754-7_10

As the experiments revealed that LaBSE with the additional attention layer ($LaBSE_a$) performs best (Sect. 5), it was exploited in the MultiEmo web service for language-agnostic sentiment analysis: https://ws.clarin-pl.eu/multiemo. All results presented in this paper are downloadable: the MultiEmo dataset at https://clarin-pl.eu/dspace/handle/11321/798 and source codes at https://github.com/CLARIN-PL/multiemo.

2 Related Work

Recently, in the domain of sentiment analysis, most research relies on effective solutions based on deep neural networks. The currently considered state-of-the-art apply Recurrent Neural Networks and Transformers, such as BiLSTM/CNN [9,10], BERT [10,11], or RoBERTa [12,13]. The idea of knowledge transfer between domains, document types, and user biases in the context of social media was discussed in [14]. However, *Language-agnostic Sentiment Analysis* is a less considered issue. This problem goes beyond the classical approaches that rely only on a single, resource-rich language, commonly English, and focuses on other languages.

In our previous work [8], we analyzed the task of cross-language sentiment analysis. In particular, we applied vector representations not depending on a particular language directly [6], hence, transferring knowledge from one language to another appears to be quite efficient. We also proposed a benchmark dataset containing test files translated into 8 different languages. However, we have not exploited state-of-the-art methods. Therefore, the dataset published in this study was a base for our further experiments.

Another interesting approach is zero-shot learning investigated by the Slovenian CLARIN-SI team in [15]. They used it for news sentiment classification. Given the annotated dataset of positive, neutral, and negative news in Slovenia, their goal was to develop a news classification system that not only assigns sentiment categories to Slovenian news, but also to news in another language, without any additional training data. Their system was based on the multilingual BERT model [11]. At the same time, they tested different methods of processing long documents and proposed a new BERT model using emotional enrichment technology as an intermediate training step. They also evaluated the zero-sample cross-language ability of their system on the new news sentiment test set in Croatian. Due to their work, their cross-language approach is also superior to most classifiers to a large extent, and all settings without emotional richness in pretraining.

Most of the most advanced sentiment classification methods are based on supervised learning algorithms that require large amounts of manually labeled data. However, annotated resources are often differently unbalanced and of little quantity among different languages. Cross-lingual sentiment classification solves this problem by using knowledge from high-resource languages to low-resource ones. In [16], an attention-based bilingual representation learning model was proposed. It can learn the distributed semantics of documents in the source language and the target language. In each language, the authors use long short-term memory (LSTM) networks to model documents, which have proven to be very effective for word sequences. At the same time, the hierarchical attention mechanism of bilingual LSTM network (BiLSTM) was proposed. The sentence-level attention model learns which sentences in the document are more important to determine the overall sentiment, while the word-level attention model learns which words in each sentence are decisive. The proposed model achieved good results on the benchmark dataset while being trained on English and evaluated on Chinese.

We tested LASER and LaBSE embeddings on multiple NLP tasks, including primarily studies on sentiment. In this paper, however, we go further and investigate them in the multiple language environment.

3 Language-agnostic Models Combined with LSTM

LASER. Facebook Research prepared the LASER method for representing a text's sentences as a list of 1024 items long vectors supporting 93 languages [6]. It uses a common space for embeddings of sentences in any language. A single model generates embeddings for all languages. It is necessary to specify language of the input text, because sentence tokenization (dividing a document into a list of sentences) is language specific.

LaBSE. Google AI team modified multilingual BERT (M-BERT) [17] to produce language-agnostic sentence embeddings (LaBSE) for 109 languages [7]. Their work produced a state-of-the-art mask language model, allowing task-specific fine-tuning what allows to achieve the best results by one's models across all supported languages. They trained the model with the usage of a translation ranking task. It is based on bidirectional dual encoders that results in a robust combination of masked language model (MLM) and translation language model (TLM) [18].

The LaBSE model prepared during the study and subsequently released increased the average dual-text retrieval accuracy for 112 languages to 83.7%. Note that if it is compared with the 65.5% accuracy achieved by LASER on the same benchmark Tatoeba corpus, it opens up new research directions in many tasks. For example, the potential application of LaBSE includes mining parallel text from the web, however, we want to test it on sentiment analysis.

The dual-encoder architecture [19] with Additive Margin Softmax [20] essentially uses parallel encoders to encode two sequences and then to obtain a compatibility score between both encodings using a dot product. The model uses MLM and TLM pretraining to train on 17 billion single sentences and 6 billion bilingual sentence pairs. The output model is quite effective even on low-resource languages where no data is available during the training. Compared to LASER, the combination of the two makes LaBSE perform better in most cases, see e.g. [21].

Performance metrics of LaBSE have been outstanding. According to our observation, LaBSE takes less than 500ms to generate embeddings for a sentence that is close to 20 words long. It is based on Transformer model with BERT-like architecture and pretrained on over 500 k vocabulary.

In this work, we propose two LaBSE-based methods: (1) the basic usage of the LaBSE output containing plain embeddings ($LaBSE_b$) and (2) embeddings amended with the attention mask ($LaBSE_a$). The former is the classical approach often used to obtain text embeddings from transformer type models. Our attention-based variant first retrieves the token representations and then an attention mask by resizing it to match the size of the embedding output. Next, both matrices are multiplied with each other and summed up. In a later step, the summation and clamp of the attention mask are performed. The final product of this process is the division of the sum of embeddings by the prepared sum of the attention mask. This gives us an averaged embedding of tokens from the last output layer enriched with an indication of which tokens contain the most key information.

BiLSTM Architecture. A commonly used neural network with sentence embeddings is bidirectional long short-term memory (BiLSTM). A layer of such type learns bidirectional long-term dependencies between time steps of time series or sequence data. These dependencies can be useful when you want the network to learn from the complete time series at each time step. Our model operates on text's sentences encoded with LASER or LaBSE model. Overall, our deep architecture developed for the task of sentiment analysis consists of the following layers:

- The Gaussian noise layer with a standard deviation of 0.01 accepts input shapes of up to N sentences, and the vector matrix of each sentence is 1024, so the overall input shape is (N, 1024);
- a bidirectional layer with LSTM instances consisting of 1,024 hidden units, using hyperbolic tangent activation method;
- a dropout layer with a rate equal to 0.2;
- a dense layer with softmax activation (normalised exponential function) with 4 outputs representing probability of class occurrence for 4 output classes.

4 Experimental Setup

4.1 Pipeline

Model training and evaluation were done in the following stages: (1) perform training on 80% of data and validation on 10%; (2) train the model until the loss function value stops decreasing for 25 epochs; keep the lowest achieved value of loss; (3) evaluate the trained model using the test part of data – the remaining 10%. All experiments were repeated 30 times so that strong statistical tests could be performed. This removed the amount of uncertainty caused by the randomness of the neural network model learning process. If the difference between the results in our statistical tests was $p < 5\%$, they were treated as insignificantly different.

4.2 MultiEmo Dataset

We created a new kind of dataset for sentiment analysis tasks – PolEmo 2.0 [10]. Each sentence as well as the entire document are labelled with one out of the four following sentiment classes: (1) ***P***: positive; (2) ***0***: neutral; (3) ***N***: negative; (4) ***AMB***: ambivalent, i.e., there are both positive and negative aspects in the text that are balanced in terms of relevance. In all further experiments, we exploited only the labels assigned to the entire text – the document level processing. The whole MultiEmo corpus in Polish contains over 40K sentences. Since each text and sentence was manually annotated with sentiment in the 2+1 scheme, we received in total over 150 K annotations. A high value of Positive Specific Agreement (PSA) [22] equal to 0.91 for texts and 0.88 for sentences was achieved.

Additionally, the whole corpus was machine translated into different languages using DeepL (https://www.deepl.com/translator), what resulted in a new MultiEmo dataset. It provides an opportunity to train and test the model in any out of 11 languages: Polish (origin), English, Chinese, Italian, Japanese, Russian, German, Spanish, French, Dutch and Portuguese. The comprehensive profile of the MultiEmo dataset is presented in Table 1. Only the mixed-domain corpus was exploited in the experiments described in Sect. 4.3 and 5, see the last row in Table 1.

Table 1. The number of texts in the train/dev/test set of the MultiEmo corpus. The average length is calculated for the entire set (SUM).

Type	Domain	Train	Dev	Test	SUM	Average length [chars]
Mixed-domain texts (all domains)	Class P – positive	1,824	236	227	2,287	648
	Class 0 – neutral	971	128	118	1,217	854
	Class N – negative	2,469	304	339	3,112	817
	Class AMB - ambivalent	1,309	155	136	1,600	707
	All classes	**6,573**	**823**	**820**	**8,216**	**754**

4.3 Scenarios

To validate the quality of the models, we used three research scenarios, differing in the language of the texts used to train and test the models:

- **Any – > Same** – the model is both trained and tested on texts in one chosen language (e.g. Polish-Polish, English-English).

- **PL– >Any** – the model is trained only on Polish texts and tested on docs translated to any other language (e.g. Polish-English, Polish-Chinese).
- **Any– >PL** – the model is trained on texts in any language and tested only on Polish texts (e.g. English-Polish, Chinese-Polish, Dutch-Polish).

All scenarios use the same train-validation-test split, Table 1, which ensures that the model will not be trained and tested on the same translated texts.

5 Experimental Results

The results for the same language training and testing on the MultiEmo dataset (all domains mixed), which is the first scenario described in Sect. 4.3, prove that $LaBSE_a$ is better in almost all cases. There are 5 situations when $LaBSE_b$ was insignificantly better than $LaBSE_a$. It happened in English (positive and negative labels), French (positive and neutral), and Italian (neutral).

In the second scenario, the training was carried out on Polish data and testing on other languages. $LaBSE_a$ is almost always statistically better than the other models. There are only eight cases out of all 88, in which $LaBSE_b$ was insignificantly better than $LaBSE_a$. There is also one situation (Portuguese: $F1_{samples}$) where $LaBSE_b$ is insignificantly worse than $LaBSE_a$. The results aggregated over all languages separately for each of the three considered models are shown in Fig. 1a for the LASER language model, in Fig. 1b for basic LaBSE ($LaBSE_b$), and in Fig. 1c for LaBSE with the custom mean pooling ($LaBSE_a$).

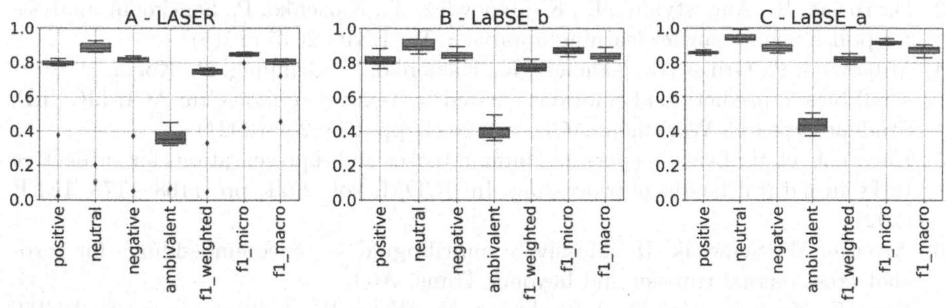

Fig. 1. Distribution of F1 scores for models learned on Polish texts and evaluated on all languages from the MultiEmo dataset (PL– >Any scenario) aggregated over all test languages. (**A**) – for the LASER embeddings; (**B**) – for the basic $LaBSE_b$ embeddings; (**C**) – for the LaBSE with attention, i.e. $LaBSE_a$ embeddings

In the third scenario, the classifier was trained on different languages but testing was performed on Polish texts only. Similarly to the previous scenarios, $LaBSE_a$ outperforms $LaBSE_b$ and LASER language models. In all scenarios, the results for the *ambivalent* class are worse by about 40%–50% than for *negative* or *positive* class meaning some documents are more controversial than others.

Rather, we should consider applying personalized reasoning to them [4,5,23, 24]. Also, the *neutral* class is poorly classified, especially for LASER and non-Latin languages (Chinese, Japanese, Russian). $LaBSE_a$ in the second scenario overcomes this problem revealing the superiority of language-agnostic solutions over language-specific ones. Languages using Latin alphabet perform almost the same.

6 Conclusion and Future Work

Language-agnostic embedding models can successfully provide valuable information for the classification of sentiment polarization. The experiments were carried out on the new multilingual MultiEmo dataset. They proved that language-agnostic representations are efficient. The best results were obtained for the LaBSE embeddings with an additional attention layer ($LaBSE_a$) and this solution was implemented in our online service. Performance for ambivalent documents may be unsatisfactory and demands other, e.g., personalized solutions. This will be investigated in future work.

References

1. Hemmatian, F., Sohrabi, M.K.: A survey on classification techniques for opinion mining and sentiment analysis. Artif. Intell. Rev.
2. Augustyniak, Ł., Szymański, P., Kajdanowicz, T., Kazienko, P.: Fast and accurate-improving lexicon-based sentiment classification with an ensemble methods
3. Bartusiak, R., Augustyniak, L., Kajdanowicz, T., Kazienko, P.: Sentiment analysis for polish using transfer learning approach. In: ENIC 2015 (2015)
4. Miłkowski, P., Gruza, M., Kanclerz, K., Kazienko, P., Grimling, D., Kocon, J.: Personal bias in prediction of emotions elicited by textual opinions. In: ACL-IJCNLP, Student Research Workshop. ACL, vol. 2021, pp. 248–259 (2021)
5. Kocoń, J., et al.: Learning personal human biases and representations for subjective tasks in natural language processing. In: ICDM, vol. 2021, pp. 1168–1173. IEEE (2021)
6. Artetxe, M., Schwenk, H.: Massively multilingual sentence embeddings for zero-shot cross-lingual transfer and beyond. Trans. ACL
7. Feng, F., Yang, Y., Cer, D., Arivazhagan, N., Wang, W.: Language-agnostic BERT sentence embedding. arXiv preprint arXiv:2007.01852 (2020)
8. Kanclerz, K., Miłkowski, P., Kocoń, J.: Cross-lingual deep neural transfer learning in sentiment analysis. Procedia Comput. Sci. **176**, 128–137 (2020)
9. Chen, T., Xu, R., He, Y., Wang, X.: Improving sentiment analysis via sentence type classification using BiLSTM-CRF and CNN. Expert Syst. Appl.
10. Kocoń, J., Miłkowski, P., Zaśko-Zielińska, M.: Multi-level sentiment analysis of PolEmo 2.0: extended corpus of multi-domain consumer reviews. In: CoNLL 2019 (2019)
11. Devlin, J., Chang, M.-W., Lee, K., Toutanova, K.: BERT: pre-training of deep bidirectional transformers for language understanding
12. Liu, Y., et al.: RoBERTa: a robustly optimized BERT pretraining approach

13. Rybak, P., Mroczkowski, R., Tracz, J., Gawlik, I.: KLEJ: comprehensive benchmark for polish language understanding. arXiv preprint arXiv:2005.00630 (2020)
14. Calais Guerra, P.H., Veloso, A., Meira Jr, W., Almeida, V.: From bias to opinion: a transfer-learning approach to real-time sentiment analysis. In: ACM SIGKDD 2011 (2011)
15. Pelicon, A., Pranjić, M., Miljković, D., Škrlj, B., Pollak, S.: Zero-shot learning for cross-lingual news sentiment classification. Appl. Sci. **10**(17), 5993 (2020)
16. Zhou, X., Wan, X., Xiao, J.: Attention-based LSTM network for cross-lingual sentiment classification. In: EMNLP 2016, pp. 247–256 (2016)
17. Pires, T., Schlinger, E., Garrette, D.: How multilingual is multilingual BERT? In: Proceedings of the 57th Annual Meeting of the ACL, 2019, pp. 4996–5001 (2019)
18. Shen, L., Xu, J., Weischedel, R.: A new string-to-dependency machine translation algorithm with a target dependency language model. In: ACL-08: HLT
19. Guo, M., et al.: Effective parallel corpus mining using bilingual sentence embeddings
20. Yang, Y., et al.: Improving multilingual sentence embedding using bi-directional dual encoder with additive margin softmax. arXiv preprint arXiv:1902.08564 (2019)
21. Gawron, K., Pogoda, M., Ropiak, N., Swędrowski, M., Kocoń, J.: Deep neural language-agnostic multi-task text classifier. In: ICDM 2021, pp. 136–142. IEEE (2021)
22. Hripcsak, G., Rothschild, A.S.: Agreement, the f-measure, and reliability in information retrieval. JAMIA **12**(3), 296–298 (2005)
23. Kocoń, J., Figas, A., Gruza, M., Puchalska, D., Kajdanowicz, T., Kazienko, P.: Offensive, aggressive, and hate speech analysis: from data-centric to human-centered approach. Inf. Process. Manag. **58**(5), 102643 (2021)
24. Kanclerz, K., et al.: Controversy and conformity: from generalized to personalized aggressiveness detection. In: ACL-IJCNLP. ACL, vol. 2021, pp. 5915–5926 (2021)

Particle Swarm Optimization Configures the Route Minimization Algorithm

Tomasz Jastrzab[1], Michal Myller[1], Lukasz Tulczyjew[1], Miroslaw Blocho[2], Wojciech Ryczko[2], Michal Kawulok[1], and Jakub Nalepa[1(✉)]

[1] Department of Algorithmics and Software, Silesian University of Technology, Gliwice, Poland
jnalepa@ieee.org
[2] Blees, Gliwice, Poland

Abstract. Solving rich vehicle routing problems is an important topic due to their numerous practical applications. Although there exist a plethora of (meta)heuristics to tackle this task, they are often heavily parameterized, and improperly tuned hyper-parameters adversely affect their performance. We exploit particle swarm optimization to select the pivotal hyper-parameters of a route minimization algorithm applied to the pickup and delivery problem with time windows. The experiments, performed on benchmark and real-life data, show that our approach automatically determines high-quality hyper-parameters of the underlying algorithm that improve its abilities and accelerate the convergence.

Keywords: Vehicle routing problem · Route minimization · Hyper-parameter selection · Particle swarm optimization · PDPTW

1 Introduction

Solving rich vehicle routing problems (VRPs) is an important research topic, as their practical applications span across different domains [12]. There are a multitude of various VRP formulations that reflect real-life constraints, including limited vehicle capacities, time windows, or the maximum waiting times [15]. Commonly, the main objective is to minimize the number of vehicles (K) serving all requests, whereas the secondary objective is to optimize the distance (T) traveled in a routing schedule. Since VRPs are often NP-hard, the most widely-used algorithms to approach such discrete optimization problems include various (meta)heuristics—they do not ensure obtaining optimal solutions, but work much faster than exact techniques and can be utilized in large-scale scheduling.

This work was supported by the European Union funds awarded to Blees Sp. z o. o. under grants POIR.04.01.01-00-0079/18-01 and UDA-RPSL.01.02.00-24-00FG/19-00. JN was supported by the Silesian University of Technology grant for maintaining and developing research potential.

D. Groen et al. (Eds.): ICCS 2022, LNCS 13351, pp. 80–87, 2022.
https://doi.org/10.1007/978-3-031-08754-7_11

Although heuristics for VRPs were shown efficient in virtually all VRPs, they are usually heavily parameterized, and the incorrect hyper-parameters lead to sub-optimal solutions. We tackle this issue and exploit particle swarm optimization (PSO) to automate the process of selecting the pivotal hyper-parameters of the route minimization algorithm (Sect. 2) applied to the pickup and delivery problem with time windows (PDPTW). It is formally defined in Sect. 1.1, whereas Sect. 1.2 presents the current state of the art in solving this VRP variant. Our experiments, performed over benchmark and real-life data, showed that PSO consistently evolves high-quality hyper-parameters that boost the abilities of the optimization technique and accelerate its convergence (Sect. 3).

1.1 Problem Formulation

The PDPTW can be defined on a directed graph $G = (V, E)$, with a set V of $B + 1$ vertices and a set of edges E. The vertices v_i, $i \in \{1, ..., B\}$, represent the locations of the requests, and v_0 indicates the depot. A set of edges $E = \{(v_i, v_{i+1}) | v_i, v_{i+1} \in V, v_i \neq v_{i+1}\}$ represents the links between particular customers. The costs $c_{i,j}$, $i, j \in \{0, 1, ..., C\}$, $i \neq j$, are equal to the distances between the travel points. Each request z_i, $i \in \{1, ..., n\}$, where $n = B/2$, is a coupled pair of pickup (P) and delivery (D) customers indicated by p_z and d_z, respectively, where $P \cup D = V \setminus \{v_0\}$ and $P \cap D = \emptyset$. For each request z_i, the amount of delivered ($q^d(z_i)$) and picked up ($q^p(z_i)$) demand is defined, where $q^d(z_i) = -q^p(z_i)$. Each customer v_i defines its demand (pickup or delivery), service time s_i ($s_0 = 0$ for the depot), and time window $[e_i, f_i]$ within which either pickup or delivery service should start (it can be completed after f_i).

In PDPTW, the fleet is homogeneous (with K denoting its size), the capacity of each vehicle is constant (Q). Each route r in the solution σ (which is a set of routes), starts and finishes at the depot. In the PDPTW, minimizing the fleet size is the primary objective, and decreasing the distance is the secondary one.

1.2 Related Work

The algorithms for minimizing the number of routes in the PDPTW include exact and approximate methods. The former techniques can find an optimal solution [11], but their main downside is that they are only applicable to small and moderate-sized problem instances [6]. On the other hand, the approximate methods allow us to find near-optimal solutions to larger problems in a short time. The heuristics fall into one of the three categories: construction, two-phase, and improvement algorithms [4]. Construction algorithms build the solution from scratch by seeding new routes when necessary. When none of the vehicles can handle a request[1], a new one is taken from the fleet. In [8], the authors seed a new route only after both *insertion* and *exchange* operations fail to introduce the request to the existing routes. Two-phase algorithms follow two general schemes: cluster-first route-second, and route-first cluster-second. The main

[1] The reasons for this inability may be capacity or time window constraint violation.

idea is to reduce the search space by combining multiple customers into clusters. Zunic et al. [16] split the set of customers into clusters based on the distance to the depot. Combining vehicles from different clusters is not allowed, but adding unclustered customers to a particular cluster is possible. Moreover, the proposed algorithm allows for hiring vehicles (if necessary) at a higher cost. The improvement algorithms start with a low-quality solution and minimize the fleet size and distance. The improvements result from relocations or exchanges of route fragments [7].

The metaheuristics involve population-based and local search approaches [3]. The first group focuses on bio-inspired techniques such as ant or bee colony optimizations, particle swarm optimization, firefly or bat algorithms [1,14], and the genetic and memetic ones [2]. The local search algorithms include tabu searches, simulated annealing, or the greedy search procedures, as well as a variety of neighborhood search methods [1,5]. Their main drawback is that they are heavily parameterized, and the parameter tuning process is time-consuming. Therefore, some techniques employ run-time adaptation through e.g., analyzing the characteristics of the problem instance and selecting the best parameters for the given problem variant. A different approach is proposed in [17], where the tuning exploits the historical data collected through GPS devices. Overall, we follow this research pathway and introduce PSO for optimizing the pivotal parameters of a guided ejection search-powered route minimization algorithm for the PDPTW [13]. This algorithm has been shown to be outperforming other approaches for minimizing the number of routes in the PDPTW, therefore we focus on it here. However, PSO is independent from the underlying route minimization technique, thus can be effectively deployed to optimize other algorithms too. Additionally, this metaheuristics is utilized because it offers high-quality performance in other tasks that involve hyper-parameters' tuning [9].

2 Configuring the Guided Ejection Search Using PSO

At each step of the guided ejection search (GES, Algorithm 1; we also report the time complexity of each step[2], where n is the number of transportation requests), inspired by [13], a random route r is drawn, its requests are inserted into the *ejection pool* (EP) (line 3), and then up to ϵ attempts to re-insert them back into the solution σ are undertaken. The *tabu pool* (TP) of maximum size α and with the maximum number of occurrences of the same entry β is initiated (line 4). The penalty counters (PCs) p indicate the re-insertion difficulty of the request. A request h_{in} is popped from the EP (line 8). If there are several feasible positions to insert h_{in} into σ, a random one is chosen (line 10). Otherwise, the inserted request violates the constraints, and the solution with h_{in} is squeezed through local moves to restore its feasibility (line 11). If squeezing fails, the request's p is increased (line 14), and h_{out} (with minimal $p[h_{out}]$) is ejected to insert h_{in} (lines 15–19). Finally, σ is mutated with a probability π using out/in-relocate

[2] Although for *Squeeze* and *Mutate*, their time complexity is fairly high, it is their worst-case complexity, and these procedures terminate much faster in practice.

and out/in-exchange moves (line 20)—the maximum and feasible number of mutation moves cannot exceed κ and λ, respectively. The optimization finishes once the maximum time has elapsed (in this work, it is two minutes).

Algorithm 1. The route minimization guided ejection search.

1: $\sigma_{best} \leftarrow \sigma_{init}$; $\sigma \leftarrow \sigma_{init}$ ▷ $\theta(n)$
2: **while not** *stop condition* **do**
3: Initialize EP with requests from a random r ▷ $O(n)$
4: Initialize tabu pool TP(α, β) ▷ $O(1)$
5: Initialize PCs as $p[h_j] := 1, j = 1, 2, ..., n$ ▷ $\theta(n)$
6: Initialize iteration counter $k := 1$ ▷ $O(1)$
7: **while** EP $\neq \emptyset$ **and** $k \leq \epsilon$ **do** ▷ max. ϵ iterations
8: $k \leftarrow k + 1$; Select and remove h_{in} from EP ▷ $O(1)$
9: **if** $S_{in}^{fe}(h_{in}, \sigma) \neq \emptyset$ **then** ▷ $O(n^2)$
10: $\sigma \leftarrow$ random $\sigma' \in S_{in}^{fe}(h_{in}, \sigma)$ ▷ $O(1)$
11: **else** $\sigma \leftarrow$ Squeeze(h_{in}, σ) ▷ $O(n^4)$
12: **end if**
13: **if** $h_{in} \notin \sigma$ **then** ▷ $O(1)$
14: $p[h_{in}] := p[h_{in}] + 1$ ▷ $O(1)$
15: Generate $S_{ej}^{fe}(h_{in}, \sigma)$ with min $p[h_{out}]$ ▷ $O(n^3)$
16: **if** $S_{ej}^{fe}(h_{in}, \sigma) \neq \emptyset$ **then** ▷ $O(1)$
17: $\sigma \leftarrow$ random $\sigma' \in S_{ej}^{fe}(h_{in}, \sigma)$ ▷ $O(1)$
18: Add the ejected request h_{out} to EP ▷ $O(1)$
19: **end if**
20: $\sigma \leftarrow$ Mutate(σ, π, λ, κ) ▷ $O(\lambda n^4)$
21: **end if**
22: **end while**
23: **if** EP $\neq \emptyset$ **then** $\sigma \leftarrow \sigma_{best}$ ▷ $\theta(n)$
24: **else** $\sigma_{best} \leftarrow \sigma$ ▷ $\theta(n)$
25: **end if**
26: **end while**
27: **return** best solution σ_{best} ▷ $O(1)$

In PSO [9], each particle's position encodes m hyper-parameters (Table 1, also in blue in Algorithm 1), and we maximize the fitness[3]: $\mathcal{F} = (0.5 \cdot (K - K_B)/K_B + 0.5 \cdot \tau_B/\tau)^{-1}$, where K and τ are the number of routes and the maximum execution time of GES with the corresponding hyper-parameters, and K_B and τ_B are the best-known number of routes for the corresponding instance in the validation set $\boldsymbol{V}$, and the best time required to converge to the solution with K routes captured during the evolution. This fitness function allows us to capture both functional and non-functional aspects of the algorithm (here, in relation to the best-known routing schedules). We evolve s random particles sampled from a uniform distribution bounded by the lower and upper parameter limits with zero initial velocity updated in each iteration for the *i-th* particle: $\boldsymbol{v_i} \leftarrow \omega \boldsymbol{v_i} + \phi_p \boldsymbol{r_p}(\boldsymbol{\lambda}_i^* - \boldsymbol{\lambda}_i) + \phi_g \boldsymbol{r_g}(\boldsymbol{\lambda}^S - \boldsymbol{\lambda}_i)$, where $\boldsymbol{r_p}, \boldsymbol{r_g}$ are from a uniform distribution $\mathcal{U}(0, 1)$, ω is the inertia, ϕ_p, ϕ_g are the acceleration coefficients, and $\boldsymbol{\lambda}_i^*$ and $\boldsymbol{\lambda}^S$ are the best *i-th* particle's and best swarm's positions visited. The *i-th* particle's position becomes $\boldsymbol{\lambda}_i = \boldsymbol{\lambda}_i + \boldsymbol{v}_i$. Once the evolution is finished, we pick the best particle that corresponds to the highest-quality parameterization.

[3] In PSO, the fitness function can be updated to reflect other aspects of the solutions.

3 Experiments

We use six Li and Lim's tests, one for the instances with clustered (c), randomized (r), and mixed (rc) requests, with short time windows and small vehicle capacities (c1, r1, and rc1), and with wider windows and larger vehicles (c2, r2, rc2). We focus on 200-request tests[4], and take the first problem instance from each group to form **V**. The max. time of calculating a fitness of a single particle amounts to 12 min (as the max. time for GES is 2 min). We pick the best hyper-parameters from each configuration for each run, and apply them to GES to solve all Li and Lim's tests (60 instances), and our 40 real-life tests of various characteristics[5]. We used the Python PSO with default parameters [9], and GES was coded in C#, and ran on Intel Xeon CPU E5-2640 v3, 2.60 Ghz, 8 GB RAM.

Table 1. The hyper-parameters of the guided ejection search.

Symbol	Range	Step	Expert's	PSO	Meaning
α	[1, 10]	1	8	3	Maximum tabu pool size
β	[2, 4]	1	2	3	Max. number of occurrences of the same entry in TP
ϵ	[1, 1000]	1	50	616	Maximum number of request ejections
π	[0.0, 1.0]	Cont	0.25	0.41	Mutation probability
λ	[1, 200]	1	100	2	Feasible number of mutations
κ	[1, 10000]	1	1000	6062	Maximum number of mutations

We keep the number of fitness evaluations ($s \cdot G_{\max}$), where $G_{\max}$ is the number of generations, constant. We investigate the impact of the swarm size s and maximum number of generations on PSO, and consider the $(s, G_{\max})$ pairs: (12, 2), (8, 3), (6, 4), (4, 6), (3, 8), and (2, 12)—each pair was run seven times. We confront PSO with Expert's (the hyper-parameters determined by an expert) and random search with 24 uniformly distributed sample points to ensure fair comparison. We always seed the *same initial solution* σ_{init} for all GES variants.

Table 2. The (a) best fitness (the best mean and median are in bold) and (b) distance traveled by the best particle in PSO.

PSO sett.→		(12, 2)	(8, 3)	(6, 4)	(4, 6)	(3, 8)	(2, 12)
(a)	Mean	0.487	0.490	0.476	**0.498**	0.481	0.451
	Std dev.	0.019	0.015	0.012	0.012	0.022	0.029
	Median	0.487	0.492	0.478	**0.502**	0.475	0.458
(b)	Mean	0.070	0.142	0.149	0.300	0.266	0.167
	Std dev.	0.024	0.060	0.041	0.059	0.168	0.166
	Median	0.078	0.159	0.174	0.317	0.243	0.082

[4] https://www.sintef.no/projectweb/top/pdptw/li-lim-benchmark/200-customers/.

[5] This set and the baseline solutions are available at https://gitlab.com/tjastrzab/iccs2022/.

Table 2 gathers the PSO results aggregated for all $(s, G_{\max})$ pairs. We can observe that balancing the swarm size and the maximum number of generations in (4, 6) leads to the largest fitness values obtained over V, hence the evolutionary process guides the particles toward high-quality parts of the solution space. The exploration capabilities of PSO are reflected in the distance traveled by the best particle in the swarm: very small swarms of two particles tend to travel larger distances, whereas larger ones, (12, 2), exploit the search space more locally, as they likely captured well-fitted random individuals in the initial population.

In Fig. 1, we confront PSO with the manual tuning process (Expert's) and random sampling. Here, we present the parameterizations that lay on the Pareto fronts, hence are not dominated by other solutions—the closer the solutions are to the point (0, 0), the better. PSO consistently delivers hyper-parameters that lead to faster convergence compared with those selected by a human expert, while maintaining low K's—the differences in K for Expert's and (4, 6) are not statistically significant (Friedman's test with Dunn's, $p < 0.05$). In Table 1, we presented the best values extracted by PSO (it was delivered by the (4, 6) configuration with the fitness of 0.52)—they are significantly different from Expert's.

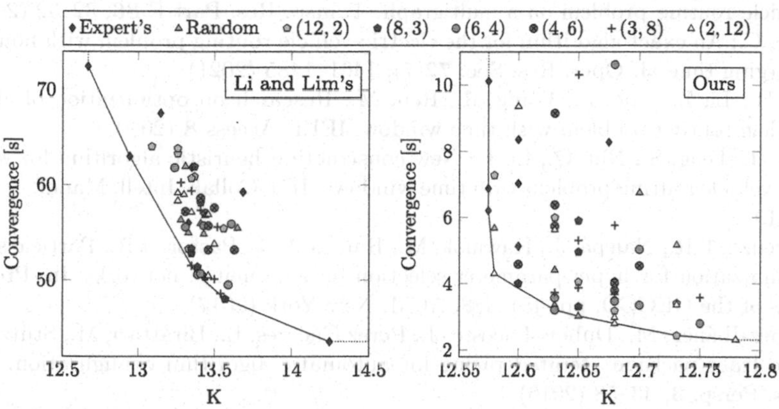

Fig. 1. The Pareto fronts (K vs. convergence in seconds, averaged across all instances in the corresponding dataset) obtained for the Li and Lim's and our test sets.

4 Conclusions

In this paper, we tackled the problem of automated parameterization of the algorithms solving rich vehicle routing problems, and employed PSO to evolve the pivotal hyper-parameters of a heuristics for minimizing routes in the PDPTW. The experiments showed that it elaborates high-quality hyper-parameters working on par with the GES parameterization delivered by a human expert, while allowing for faster convergence. Our current research efforts are focused on integrating PSO with the famous iterated racing procedures available in irace [10].

We believe that developing the solvers with the automated process of retrieving their desired parameterizations is an important step toward data-driven algorithms that will be able to solve emerging formulations of rich VRPs.

References

1. Blocho, M.: Heuristics, metaheuristics, and hyperheuristics for rich vehicle routing problems. In: Nalepa, J. (ed.) Smart Delivery Systems. Solving Complex Vehicle Routing Problems, pp. 101–156. Intelligent Data Centric Systems, Elsevier (2020)
2. Blocho, M., Nalepa, J.: LCS-based selective route exchange crossover for the pickup and delivery problem with time windows. In: Hu, B., López-Ibáñez, M. (eds.) EvoCOP 2017. LNCS, vol. 10197, pp. 124–140. Springer, Cham (2017). https://doi.org/10.1007/978-3-319-55453-2_9
3. Feng, L., et al.: Solving generalized vehicle routing problem with occasional drivers via evolutionary multitasking. In: IEEE Transactions on Cybernetics, pp. 1–14 (2019)
4. Konstantakopoulos, G., Gayialis, S., Kechagias, E.: Vehicle routing problem and related algorithms for logistics distribution: a literature review and classification. Oper. Res. (2020)
5. Lai, D., Demirag, O., Leung, J.: A tabu search heuristic for the heterogeneous vehicle routing problem on a multigraph. Transp. Res. Part E **86**, 32–52 (2016)
6. Lee, C.: An exact algorithm for the electric-vehicle routing problem with nonlinear charging time. J. Oper. Res. Soc. **72**(7), 1461–1485 (2021)
7. Li, H., Li, Z., Cao, L., Wang, R., Ren, M.: Research on optimization of electric vehicle routing problem with time window. IEEE Access **8** (2020)
8. Liu, J., Feng, S., Niu, Q., Li, L.: New construction heuristic algorithm for solving the vehicle routing problem with time windows. IET Collab. Intell. Manuf. **1**, 90–96 (2019)
9. Lorenzo, P.R., Nalepa, J., Kawulok, M., Ramos, L.S., Pastor, J.R.: Particle swarm optimization for hyper-parameter selection in deep neural networks. In: Proceedings of the GECCO, pp. 481–488. ACM, New York (2017)
10. López-Ibáñez, M., Dubois-Lacoste, J., Pérez Cáceres, L., Birattari, M., Stützle, T.: The Irace package: iterated racing for automatic algorithm configuration. Oper. Res. Persp. **3**, 43–58 (2016)
11. Mohamed, E., Ndiaye, M.: Optimal routing and scheduling in e-commerce logistics using crowdsourcing strategies. In: Proceedings of the IEEE ICITM, pp. 248–253 (2018)
12. Mor, A., Speranza, M.G.: Vehicle routing problems over time: a survey. 4OR **18**(2), 129–149 (2020)
13. Nalepa, J., Blocho, M.: Adaptive guided ejection search for pickup and delivery with time windows. J. Intell. Fuzzy Syst. **32**, 1547–1559 (2017)
14. Osaba, E., Yang, X., Fister, I., Jr., Del Ser, J., Lopez-Garcia, P., Vazquez-Pardavila, A.: A discrete and improved bat algorithm for solving a medical goods distribution problem with pharmacological waste collection. Swarm Evol. Comput. **44**, 273–286 (2019)
15. Zhang, H., Ge, H., Yang, J., Tong, Y.: Review of vehicle routing problems: models, classification and solving algorithms. Archiv. Comput. Methods Eng. **29**(1), 195–221 (2022)

16. Zunic, E., Donko, D., Supic, H., Delalic, S.: Cluster-based approach for successful solving real-world vehicle routing problems. In: Proceedings of the ACSIS, vol. 21, pp. 619–626. Springer, Cham (2020)
17. Žunić, E., Delalić, S., Donko, D.: Adaptive multi-phase approach for solving the realistic vehicle routing problems in logistics with innovative comparison method for evaluation based on real GPS data. Transport. Lett. **14**(2), 143–156 (2022)

Practical Aspects of Zero-Shot Learning

Elie Saad[1], Marcin Paprzycki[2], and Maria Ganzha[1(✉)]

[1] Warsaw University of Technology, Warsaw, Poland
{elie.saad.stud,maria.ganzha}@pw.edu.pl

[2] Systems Research Institute Polish Academy of Sciences, Warsaw, Poland
marcin.paprzycki@ibspan.waw.pl

Abstract. Zero-shot learning is applied, for instance, when properly labeled training data is not available. A number of zero-shot algorithms have been proposed. However, since none of them seems to be an "overall winner", development of a meta-classifier(s) combining "best aspects" of individual classifiers can be attempted. In this context, state-of-the-art zero-shot learning methods are compared for standard benchmark datasets. Next, multiple meta-classifiers are applied to the same datasets.

Keywords: Zero-shot learning · Meta-classifier · Benchmarking

1 Introduction and Literature Review

Many real-world applications require classifying "entities" not encountered earlier, e.g., object recognition (where every object is a category), cross-lingual dictionary induction (where every word is a category), etc. Here, one of the reasons is lack of resources to annotate available (and possibly systematically growing) datasets. To solve this problem, zero-shot learning has been proposed.

While multiple zero-shot learning approaches have been introduced ([9,19]), as of today, none of them emerged as "the best". In situations like this, meta-classifiers, which "receive suggestions" from individual classifiers and "judge" their value to select a "winner", can be explored. The assumption is that such meta-classifiers will perform better than the individual ones.

Let us start from the formal *problem formulation*. Given a dataset of *image embeddings* $\mathcal{X} = \{(x_i, y_i) \in \mathcal{X} \times \mathcal{Y} | i = 1, ..., N_{tr} + N_{te}\}$, each image is a real D-dimensional embedding vector comprised of features $x_i \in \mathbb{R}^D$, and each class label is represented by an integer $y_i \in \mathcal{Y} \equiv \{1, ..., N_0, N_0 + 1, ..., N_0 + N_1\}$ giving $N_0 + N_1$ distinct classes. Here, for generality, it is assumed that $\mathcal{X} \stackrel{\text{def}}{=} \mathbb{R}^D$. The dataset $\mathcal{X}$ is divided into two subsets: (1) training set and (2) test set. The training set is given by $X^{tr} = \{(x_i^{tr}, y_i^{tr}) \in \mathcal{X} \times \mathcal{Y}_0 | i = 1, ..., N_{tr}\}$, where $y_i^{tr} \in \mathcal{Y}_0 \equiv \{1, ..., N_0\}$, resulting in N_0 distinct training classes. The test set is given by $X^{te} = \{(x_i^{te}, y_i^{te}) \in \mathcal{X} \times \mathcal{Y}_1 | i = N_{tr} + 1, ..., N_{te}\}$, where $y_i^{te} \in \mathcal{Y}_1 \equiv \{N_0 + 1, ..., N_0 + N_1\}$ providing N_1 distinct test classes.

The goal of zero-shot learning is to train a model (on dataset X^{tr}) that performs "well" for the test dataset X^{te}. Obviously, zero-shot learning requires

D. Groen et al. (Eds.): ICCS 2022, LNCS 13351, pp. 88–95, 2022.
https://doi.org/10.1007/978-3-031-08754-7_12

auxiliary information associating labels from the training and the test sets, when $\mathcal{Y}_0 \cap \mathcal{Y}_1 = \emptyset$. The solution is to represent each class label y ($1 \leq y \leq N_0 + N_1$) by its prototype $\pi(y) = p \in \mathcal{P} \stackrel{\text{def}}{=} \mathbb{R}^M$ (semantic embedding). Here, $\pi(\cdot) : \mathcal{Y}_0 \cup \mathcal{Y}_1 \rightarrow \mathcal{P}$ is the prototyping function, and $\mathcal{P}$ is the semantic embedding space. The prototype vectors are such that any two class labels y_0 and y_1 are similar if and only if their prototype representations $\pi(y_0) = p_0$ and $\pi(y_1) = p_1$ are close in the semantic embedding space $\mathcal{P}$. For example, their inner product is large in $\mathcal{P}$, i.e., $\langle \pi(y_0), \pi(y_1) \rangle_{\mathcal{P}}$ is large. Prototyping all class labels into a joint semantic space, i.e., $\{\pi(y) | y \in \mathcal{Y}_0 \cup \mathcal{Y}_1\}$, results in labels becoming related. This resolves the problem of disjoint class sets, and the model can learn from the labels in the training set, and predict labels from the test set.

Multiple algorithms have been proposed to solve the zero-shot learning problem. Here, *DeViSE* [6], *ALE* [2], and *SJE* [3] use a bilinear compatibility function. They follow the *Stochastic Gradient Descent* (SGD), implicitly regularized by early stopping. The *ESZSL* [12] uses square loss to learn the bilinear compatibility function, and explicitly defines regularization with respect to the Frobenius norm. Kodirov et al. in [8] proposes a semantic encoder-decoder model (*SAE*), where the training instances are projected into the semantic embedding space $\mathcal{P}$, with the projection matrix W, and then projected back into the feature space $\mathcal{X}$, with the conjugate transpose of the projection matrix W^*. Another group of approaches adds a non-linearity component to the linear compatibility function [18]. Third set of approaches uses probabilistic mappings [9]. Fourth group of algorithms expresses the input image features and the semantic embeddings as a mixture of seen classes [21]. In the fifth approach, both seen and unseen classes are included in the training data [20].

In this context, a comparison of five state-of-the-art zero-shot learning approaches, applied to five popular benchmarking datasets, is presented. Next, explorations into meta-classifier for zero-shot learning are reported. Extended version of this work, with additional details and results, can be found in [14].

2 Selection of Methods and Experimental Setup

Based on the analysis of the literature, five robust zero-shot learning approaches were selected: (1) *DeViSE*, (2) *ALE*, (3) *SJE*, (4) *ESZSL*, and (5) *SAE*. Moreover, the following, popular in the literature, datasets have been picked: (a) *CUB* [17], (b) *AWA1* [9], (c) *AWA2* [19], (d) *aPY* [5], and (e) *SUN* [11]. Finally, five standard meta-classifiers have been tried: (A) Meta-Decision Tree *MDT* [16], (B) deep neural network *DNN* [7], (C) Game Theory-based approach *GT* [1], (D) Auction-based model *Auc* [1], (E) Consensus-based approach *Con* [4], and (F) a simple majority voting *MV* [13]. Here, classifiers (C), (D), (E) and (F) have been implemented following cited literature. However, the implementation of (A) differs from the one described in [16] by not applying the *weight* condition on the classifiers. However, effects of this simplification can be explored int the future. Finally, the *DNN* has two hidden layers and an output layer. All of them use the rectified linear activation function. The optimization function is the

SGD, with the mean squared error loss function. All codes and complete list of hyperparameter values for the individual classifiers and the meta-classifiers can be found in the Github repository[1]. While hyperparameter values, were obtained through multiple experiments, no claim about their optimality is made. The following standard measures have been used to measure the performance of the explored approaches: (T1) Top-1, (T5) Top-5, (LogLoss) Logarithmic Loss, and (F1) F1 score. Their definitions can be found in [10,15,19].

Separately, comparison with results reported in [19] has to be addressed. To the best of our knowledge, codes used there are not publicly available. Thus, the best effort was made to re-implement methods from [19]. As this stage, the known differences are: (1) feature vectors and semantic embedding vectors, provided in the datasets were used, instead of the calculated ones; (2) dataset splits for the written code follow the splits proposed in [19], instead of the "standard splits". Nevertheless, we believe that the results, presented herein, fairly represent the work reported in [19].

3 Experiments with Individual Classifiers

The first set of experimental results was obtained using the five classifiers applied to the five benchmark datasets. Results displayed in Table 1 show those available from [2,3,6,8,12] (in the **O** column). The **R** column represents results based on [19]. The **I** columns represents the in-house implementations of the five models. Overall, all classifiers performed "badly" when applied to the aPY dataset. Next, comparing the results between columns R and I, out of 25 results, methods based on [19] are somewhat more accurate in 15 cases. Hence, since performances are close, and one could claim that our implementation of methods from [19] is "questionable", from here on, only results based on "in house" implementations of zero-shot learning algorithms are reported.

Table 1. Individual classifier performance for the Top-1 accuracy

CLF	CUB			AWA1			AWA2		aPY			SUN		
	O	R	I	O	R	I	R	I	O	R	I	O	R	I
DeViSE	–	52	46.82	–	54.2	53.97	59.7	57.43	–	37	32.55	–	56.5	55.42
ALE	26.3	54.9	**56.34**	47.9	59.9	56.34	62.5	51.89	–	39.7	33.4	–	58.1	**62.01**
SJE	50.1	53.9	49.17	66.7	65.6	**58.63**	61.9	**59.88**	–	31.7	31.32	–	52.7	52.64
ESZSL	–	51.9	53.91	49.3	58.2	56.19	58.6	54.5	15.1	38.3	**38.48**	65.8	54.5	55.63
SAE	–	33.3	39.13	84.7	53.0	51.5	54.1	51.77	–	8.3	15.92	–	40.3	52.71

While the Top-1 performance measure is focused on the "key class", other performance measures have been tried. In [14] performance measured using *Top 5*, *LogLoss*, and *F1 score* have been reported. Overall, it can be stated

[1] https://github.com/Inars/Developing_MC_for_ZSL.

that (A) different performance measures "promote" different zero-shot learning approaches; (B) *aPY* is the "most difficult dataset" regardless of the measure; (C) no individual classifier is clear winner. Therefore, a simplistic method has been proposed, to gain a better understanding of the "overall strength" of each classifier. However, what follows "should be treated with a grain of salt". Here, individual classifiers score points ranging from 5 to 1 (from best to worst) based on the accuracy obtained for each dataset. This process is applied to all four accuracy measures. Combined scores have been reported in Table 2. Interestingly, *SAE* is the weakest method for both the individual datasets and the overall performance. The combined performance of *ALE*, *SJE*, and *EZSL* is very similar.

Table 2. Individual classifier combined performance; "winners" marked in bold font.

CLF	CUB	AWA1	AWA2	aPY	SUN	Total
DeViSE	8	8	17	16	11	60
ALE	**19**	11	7	**15**	**18**	70
SJE	12	**19**	**18**	**15**	8	72
ESZSL	17	14	14	11	17	**73**
SAE	4	8	8	7	6	33

3.1 Analysis of the Datasets

Since it became clear that the performance of the classifiers is directly related to the datasets, their "difficulty" has been explored. Hence, an instance (in a dataset) is classified as *lvl 0* if *no* classifier identified it correctly, whereas *lvl 5* if it was recognized by *all* classifiers. The results in Table 3, show how many instances (in %) belong to each category, for each dataset. Here, the *aPY* dataset has the largest percent of instances that have not been recognized at all (36.37%), or by one or two classifiers (jointly, 41.51%). At the same time, only 0.85% of its instances have been recognized by all classifiers. The *SUN* dataset is the easiest: 27.85% of instances were correctly recognized by all classifiers and about 55% of its instances are "relatively easy".

Table 3. Analysis of Instance Difficulty (represented in %)

CLF	lvl 0	lvl 1	lvl 2	lvl 3	lvl 4	lvl 5
CUB	23.86	15.98	13.11	12.4	15.23	19.41
AWA1	20.47	15.67	11.29	12.19	**19.33**	21.04
AWA2	21.88	12.95	13.46	**14.62**	12.26	22.84
aPY	**36.37**	**25.76**	**15.75**	11.75	9.53	0.85
SUN	19.31	12.78	10.69	12.5	16.88	**27.85**

Approaching the issue from different perspective, the "influence" of individual attributes has been "scored". For each correctly recognized instance, its attributes have been given +1 "points". For incorrectly recognized instances, their attributes were given −1 "points". This measure captured which attributes are the *easiest/hardest* to recognize. Obtained results can be found in Table 4. The most interesting observation is that attributes: "has eye color black" and "metal" are associated with so many instances that they are classified (in)correctly regardless if they actually "influenced" the "decision" of the classifier.

Table 4. Analysis of the datasets

CLF	Easiest attribute	Individual classifier	Hardest attribute
CUB	Has eye color black	All	Has eye color black
AWA1	Old world	DeViSE; SAE	Group
	Fast	ALE	Ocean
	Old world	SJE; ESZSL	Ocean
	Quadrupedal	Total	Swims
AWA2	Old world	DeViSE; ALE; SJE; SAE; total	Group
	Old world	ESZSL	Ocean
aPY	Metal	DeViSE; ALE; SJE; ESZSL	Metal
	Head	SAE	Metal
	Furry	Total	Metal
SUN	No horizon	All	Man-made

4 Meta-Classifiers

Let us now move to meta-classifiers. Here, note that the number of *hard* instances, found in each dataset (see, Sect. 2), establishes the hard ceiling for: *DNN*, *MDT*, and *MV*. Specifically, if not a single classifier gave a correct answer, in these approaches, their combination will also "fail". In Table 5, the performance of the six meta-classifiers is compared using the Top-1 measure, where the **Best** row denotes the best result obtained by the "winning" individual classifier, for a given dataset (see, Table 1). Results using the F1 score, can be found in [14].

Table 5. Meta-classifier performance; Top-1 accuracy

CLF	CUB	AWA1	AWA2	aPY	SUN
MV	**53.43**	**58.71**	56.56	32.72	61.94
MDT	47.89	56.43	51.89	33.40	**62.08**
DNN	48.63	57.56	54.72	**34.89**	60.63
GT	46.58	56.75	**59**	32.63	59.51
Con	46.82	53.97	57.43	32.55	55.42
Auc	47.89	56.34	51.89	33.40	62.01
Best	56.34	58.63	59.88	38.48	62.01

Comparing the results, one can see that: (a) the best individual classifier performed better than the best meta-classifier on *CUB*, *AWA2*, and *aPY* (2.91%, 0.88%, and 3.59% better); (b) the best meta-classifier performed better than the best individual classifier on *AWA1* and *SUN* datasets (0.08% and 0.08% better).

Finally, the "scoring method" was applied jointly to the meta-classifiers and the individual classifiers, for the Top-1 and the F1 score accuracy measures. Obviously, since 11 classifiers were compared, the top score was 11 points. Table 6 displays the results. It can be noticed that (1) meta-classifiers performed better than the individual classifiers (averaging 77.83 vs. 74.6 points). (2) Combining results from the individual classifiers using a simple *majority voting* algorithm brought best results. At the same time, (3) use of basic versions of more advanced meta-classifiers is not leading to immediate performance gains.

Table 6. Meta-classifier and individual classifier combined performance

CLF	CUB	AWA1	AWA2	aPY	SUN	Total
DeViSE	13	12	20	12	13	70
ALE	**21**	15	11	17	21	85
SJE	17	**20**	**22**	16	10	85
ESZSL	21	16	16	16	15	84
SAE	10	10	11	7	11	49
MV	20	21	17	**18**	20	**96**
MDT	15	16	11	**18**	**22**	82
DNN	17	18	15	17	18	85
GT	10	14	15	11	13	63
Con	13	10	13	12	13	61
Auc	15	15	11	**18**	21	80

5 Concluding Remarks

In this contribution, performance of five zero-shot learning models has been studied, when applied to popular benchmarking datasets. Moreover, the "nature of difficulty" of these datasets has been explored. Finally, six standard meta-classifiers have been experimented with. The main findings were: (1) there is no single best classifier, and results depend on the dataset and the performance measure; (2) the *aPY* dataset is the most difficult for zero-shot learning; (3) standard meta-classifiers may bring some benefits; (4) the simplest methods obtained best results (i.e., the individual classifier *ESZSL* and the meta-classifier *MV*). The obtained prediction accuracy (less than 70%) suggests that a lot of research is needed for both the individual classifiers and, possibly, the meta-classifiers. Moreover, datasets similar to the *aPY*, which are particularly difficult achieve good performance, should be used. Finally, some attention should be devoted to the role that individual attributes play in class (instance) recognition.

Acknowledgement. Research funded in part by the Centre for Priority Research Area Artificial Intelligence and Robotics of Warsaw University of Technology within the Excellence Initiative: Research University (IDUB) programme.

References

1. Abreu, M.d.C., Canuto, A.M.: Analyzing the benefits of using a fuzzy-neuro model in the accuracy of the neurage system: an agent-based system for classification tasks. In: Proceedings of IEEE International Joint Conference on NN, pp. 2959–2966. IEEE (2006)
2. Akata, Z., Perronnin, F., Harchaoui, Z., Schmid, C.: Label-embedding for image classification. IEEE Trans. Pattern Anal. Mach. Intell. **38**(7), 1425–1438 (2015)
3. Akata, Z., Reed, S., Walter, D., Lee, H., Schiele, B.: Evaluation of output embeddings for fine-grained image classification. In: Proceedings of the IEEE Conference on Computer Vision and Pattern Recognition, pp. 2927–2936 (2015)
4. Alzubi, O.A., Alzubi, J.A.A., Tedmori, S., Rashaideh, H., Almomani, O.: Consensus-based combining method for classifier ensembles. Int. Arab J. Inf. Technol. **15**(1), 76–86 (2018)
5. Farhadi, A., Endres, I., Hoiem, D., Forsyth, D.: Describing objects by their attributes. In: 2009 IEEE Conference on Computer Vision and Pattern Recognition, pp. 1778–1785. IEEE (2009)
6. Frome, A., et al.: Devise: a deep visual-semantic embedding model. Adv. Neural Inf. Proc. Sys. **26** (2013)
7. Goodfellow, I., Bengio, Y., Courville, A.: Deep Learning. MIT Press (2016)
8. Kodirov, E., Xiang, T., Gong, S.: Semantic autoencoder for zero-shot learning. In: Proceedings of the IEEE Conference on Computer Vision and Pattern Recognition, pp. 3174–3183 (2017)
9. Lampert, C., Nickisch, H., Harmeling, S.: Attribute-based classification for zero-shot learning of object categories. IEEE Trans. Pattern Anal. Mach. Intell. **36**(3) (2013)
10. Mannor, S., Peleg, D., Rubinstein, R.: The cross entropy method for classification. In: Proceedings of the 22nd International Conference on ML, pp. 561–568 (2005)

11. Patterson, G., Hays, J.: Sun attribute database: discovering, annotating, and recognizing scene attributes. In: Proceedings of IEEE Conference on Computer Vision and Pattern Recognition, pp. 2751–2758. IEEE (2012)
12. Romera-Paredes, B., Torr, P.: An embarrassingly simple approach to zero-shot learning. In: International Conference on ML, pp. 2152–2161. PMLR (2015)
13. Ruta, D., Gabrys, B.: Classifier selection for majority voting. Inf. Fusion **6**(1), 63–81 (2005)
14. Saad, E., Paprzycki, M., Ganzha, M.: Practical aspects of zero-shot learning. 10.48550/ARXIV.2203.15158, https://arxiv.org/abs/2203.15158
15. Sokolova, M., Japkowicz, N., Szpakowicz, S.: Beyond accuracy, F-Score and ROC: a family of discriminant measures for performance evaluation. In: Sattar, A., Kang, B. (eds.) AI 2006. LNCS (LNAI), vol. 4304, pp. 1015–1021. Springer, Heidelberg (2006). https://doi.org/10.1007/11941439_114
16. Todorovski, L., Džeroski, S.: Combining classifiers with meta decision trees. Mach. Learn. **50**(3), 223–249 (2003)
17. Welinder, P., et al.: Caltech-UCSD birds 200 (2010)
18. Xian, Y., Akata, Z., Sharma, G., Nguyen, Q., Hein, M., Schiele, B.: Latent embeddings for zero-shot classification. In: Proceedings of the IEEE Conference on Computer Vision and Pattern Recognition, pp. 69–77 (2016)
19. Xian, Y., Lampert, C.H., Schiele, B., Akata, Z.: Zero-shot learning-a comprehensive evaluation of the good, the bad and the ugly. IEEE Trans. Pattern Anal. Mach. Intell. **41**(9), 2251–2265 (2018)
20. Ye, M., Guo, Y.: Zero-shot classification with discriminative semantic representation learning. In: Proceedings of the IEEE Conference on Computer Vision and Pattern Recognition, pp. 7140–7148 (2017)
21. Zhang, Z., Saligrama, V.: Zero-shot learning via semantic similarity embedding. In: Proceedings of the IEEE International Conference on Computer Vision, pp. 4166–4174 (2015)

A Hypothetical Agent-Based Model Inspired by the Abstraction of Solitary Behavior in Tigers and Its Employment as a Chain Code for Compression

Khaldoon Dhou[1(✉)] and Christopher Cruzen[2]

[1] Texas A&M University Central Texas, Killeen, TX 76549, USA
kdhou@tamuct.edu

[2] University of Missouri St. Louis, St. Louis, MO 63121, USA
christopher.cruzen@mail.umsl.edu

Abstract. In this paper, we design an agent-based modeling simulation that represents the solitary behavior in tigers and utilizes it in encoding image information. Our model mainly depends on converting the digital data to a virtual environment with paths classified based on the allocation of the data in the original image. Then, we introduce virtual tigers to the environment to begin the encoding process. Tiger agents are separated from each other, and the algorithm monitors their movements and keeps them away from each other. This separation in the virtual environment allows tigers to encode information that exists in different image parts. Additionally, tigers follow a relative movement style that encodes each tiger's movement direction based on the previous one. This encoding approach allows particular movements that occur in different directions to be encoded in a similar way. After that, we apply Huffman coding to the chain of movements, the purpose of which is to reduce the size and have a new representation. The experimental findings reveal that we could obtain better results than leading standards in bi-level image compression, including JBIG family methods. Our outcomes strengthen the findings of previous studies that incorporated biological behaviors within agent-based modeling simulations and provide a new abstraction to be utilized in information processing research.

Keywords: Chain code · Agent-based model · Compression · NetLogo · Solitary behavior · Tigers · Huffman coding

1 Introduction

Swarm Intelligence (SI) is an AI research field that attempts to solve problems via applying algorithms that are inspired by nature. Mostly stimulated by biological rules, swarm intelligence computing algorithms solve numerous dilemmas such as compression and optimization problems. A recent trend of research in

D. Groen et al. (Eds.): ICCS 2022, LNCS 13351, pp. 96–102, 2022.
https://doi.org/10.1007/978-3-031-08754-7_13

swarm intelligence incorporates biological behaviors within agent-based modeling simulations. These models consider converting a piece of digital data into a two-dimensional grid that represents a virtual environment consisting of agents of different types. Each agent behaves in a particular way that simulates a biological behavior while its movements are encoded in a certain way according to the design of the simulation. For example, Mouring et al. [18] designed an ant colony algorithm that allows ants to release a pheromone to help other ants track the pieces of information. The movements of ants were further processed for additional encoding and size reduction. Similarly, Dhou and Cruzen [3] developed an agent-based model that allows agents to reproduce other agents that can further work on the problem. Other models developed consider various behaviors such beaver territories [8], predators and prey [4,6], echolocation in dolphins [7], HIV transmission [9,10], and agent modes [5].

One major issue that is likely to be encountered in some agent-based models developed for image encoding is the ability of some agents to block the way of other agents (e.g., [3,10]). This blockage can cause some delay in encoding the information. Another problem is the inclination of some agents to gather in certain areas, which can also cause encoding issues [18]. In the current article, we address these issues by utilizing the solitary behavior of tigers in encoding image information and employing the generated chains in compression. In tigers, the wild adults are considered solitary creatures that reside within large land spaces [21,22]. In our design, we define solitary behavior as the maintenance of separation between the tigers so that each works on a separate part of an image. The motivation of utilizing solitary behavior is to allow tigers to work separately while managing the blockage and gathering problems that exist in other models. Additionally, the employment of solitary behavior allows us to take advantage of relative encoding mechanisms that proved to be successful while employed in various agent-based models and other algorithms used in image encoding (e.g., [6,8,10,17]). The relative movement design is advantageous in terms of allowing multiple consecutive movements to be encoded in a similar fashion. The current approach adds to the body of knowledge and investigates a new behavior on how to capture bi-level image information and create new representations of data. Existing research shows that bi-level images are widely used by researchers from various domains such as psychology and computing (e.g., [2,11,12]).

It is essential to mention that many agent-based models for compression are stimulated by existing NetLogo models designed to solve problems in different domains such as economy, marketing, and biology. A NetLogo is an agent-based modeling programming language that is supplied with a relatively large number of existing models in various fields. The current approach is interdisciplinary, and it combines aspects from different domains, including agent-based modeling, bio-inspired computing, and image processing. However, the extensive search reveals that the closest line of research to utilizing agent-based models to encode image information is called 'chain coding.' In this approach, the edge image information is represented by passing over the contours of an image and encoding each encountered direction. Chain coding started with the Freeman Chain Code

(FCC) [13] that employs 4 or 8 codes to represent each direction depending on the design of the algorithm. After that, researchers worked on the topic from different angles, and it has been the foundation of many projects that involve information processing. For example, an enhancement was made by Bons and Kegel [1] who developed the Differential Chain Coding (DCC) approach that applies Huffman coding to the differences between subsequent codes. A further enhancement was made by Hwang et al. [15] who reduced the range of the codes in DCC by utilizing a mathematical operator. Later, chain coding research saw more advancements, including new encoding mechanisms that depend on the relative directions. Although these applications started in the image processing domain [17,23], agent-based modeling simulations designed for image encoding started to employ them from various perspectives depending on the biological behaviors embedded within a model, agent type, designed movements, and encoding directions.

2 Proposed Algorithm

The present algorithm consists of the following main steps:

- **Step 1:** Convert an image into a contour representation and then turn it into a two-dimensional virtual environment consisting of locations. Each location within the virtual environment corresponds to a pixel in the original image. The current design classifies the cells within the virtual environment into preferred and non-preferred cells. While the preferred cells are marked with green circles, each non-preferred cell is marked with a red circle that contains the 'X' sign. In addition, the areas with green circles represent the paths that are recorded by the algorithm. On the other hand, red areas can be used by the tigers within the design. However, the movements over them are not recorded by the algorithm.
- **Step 2:** The algorithm adds virtual tigers to the environment, and their job is to work in separate areas to encode image information. It is always a tiger's preference to walk over a cell marked with a green circle, and these movements are tracked by the algorithm. On the other hand, a tiger might not have the choice to identify a green cell in the neighborhood, and therefore, he chooses to walk over a red cell while hoping to transit to a green cell.
- **Step 3:** Each tiger considers a relative movement mechanism that substitutes each direction with a code that is related to the previous one. In other words, two movements can occur in different directions while still encoded similarly. For example, in Fig. 1, the movements that are marked with the same numbers are encoded in the same way regardless of the actual directions. This encoding mechanism is different from the Freeman Chain Code [13] in the way that it does not encode based on the directions themselves. Additionally, these movements occur within the context of an agent-based modeling simulation that allows incorporating behaviors and experimenting with a variable number of agents depending on a researcher's choice and the parameters of the data.

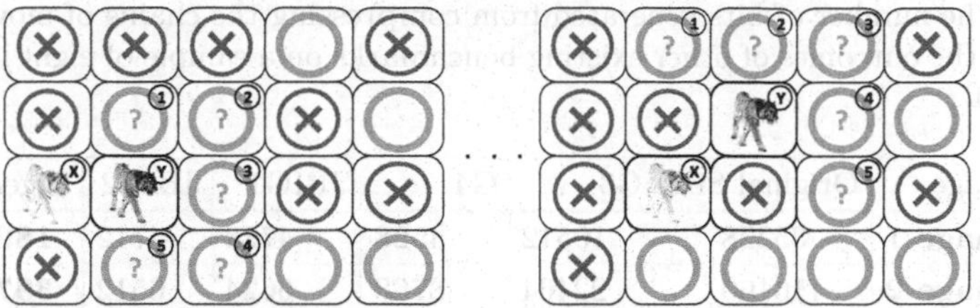

Fig. 1. Codes considered by two tigers. The codes with identical numbers are similar, although they are in different directions

- **Step 4:** The chains of movements are compressed using Huffman coding for further reduction. The results are compared with standardized benchmarks used by image processing researchers, including G3, G4, JBIG1, and JBIG2.
- **Step 5:** To reconstruct an image, all that the algorithm needs are the chains of movements and the coordinates of the first location. The algorithm is lossless, and the details of the images are maintained.

3 Experimental Results and Analysis

To evaluate the performance of the algorithm, we applied it to a sample of eight images taken from [24], and we compared the results with existing standardized algorithms: G3, G4, JBIG1, and JBIG2. Extensive search shows that these algorithms are widely used by image compression researchers, and they are used as benchmarks for comparisons. Table 1 shows the results of compressing the chains using the tiger's method, as opposed to the results obtained from other algorithms. It is evident from the table that we could outperform all the methods using all the images we employed for comparison.

The most exciting finding was that we could outperform well-known standardized benchmarks using the abstraction of solitary behavior in tigers. Another interesting finding is that the present study reveals an association between agent-based modeling, relative encoding, and bio-inspired computational behaviors. In other words, this study produced results that corroborate the findings of a great deal of the previous studies that embed bio-inspired behaviors within agent-based modeling to build encoding applications. In particular, this study explores the solitary behavior in tigers by maintaining a separation between agents and allowing them to work on encoding the information simultaneously while generating the necessary chains for representation.

These findings suggest that agent-based modeling is a promising direction to be incorporated in image processing studies. These initial results are suggestive of a link between the abstraction of solitary behavior in tigers and encoding image information. However, further testing is needed to confirm the differences between the existing results and the outcomes from other standardized benchmarks. These experiments are beyond the scope of this paper, and we leave them

Table 1. The number of bits generated from compressing the chains of movements as opposed to the outcomes of other existing benchmarks on a sample of eight images [14, 16,19,20,24]

Image	Original Size	G3	G4	JBIG1	JBIG2	Tigers
Image 1	35328	10512	4528	4728	4712	**1893**
Image 2	116160	22464	8768	6624	6512	**3974**
Image 3	96338	31936	22080	17088	16408	**15398**
Image 4	244400	60608	34800	26944	25800	**20481**
Image 5	264489	29344	12992	7704	7640	**7409**
Image 6	40000	11712	5552	5424	5208	**2425**
Image 7	91008	15888	5792	5024	4888	**2318**
Image 8	696320	56288	19776	7536	6992	**5053**
Total	**1584043**	**238752**	**114288**	**81072**	**78160**	**58951**

as future work. In fact, some of the issues emerging from this finding relate specifically to generating new codes that act as representations of different types of digital data. These can probably be utilized for various applications such as security, indexing, and document retrieval. In spite of that, more research in these areas is required to determine the practicality of agent-based modeling to be introduced as a direction in these domains.

4 Conclusion and Future Work

The aim of the present research study was to examine the practicality of solitary behavior in tigers in encoding bi-level image information and to generate new representations that can be used for compression. This was done by developing an agent-based model that simulates an abstraction of solitary behavior in tigers and designing their movements that are stimulated by existing research studies in agent-based modeling and image processing. Additionally, our model considers code replacements that examine each movement and replace it with a code to help build a new image representation to be used in compression. While these findings can be considered attractive from an academic perspective, they can also be explored by researchers from other domains such as security, and image processing.

The findings clearly indicate the effectiveness of the solitary behavior abstraction in encoding image information. Additionally, the current research study has shown that designing agent-based models that support relative encoding is significant in reducing chain lengths because they offer new chains that consist of similar codes that exist in consecutive order. These codes can also be explored for different purposes that are related to data representation. While relative encoding has been previously utilized in image processing studies (e.g. [5,6,17]), the extensive search shows that we are the first researchers to employ it in abstracting solitary behavior to be used in image processing.

The findings will be of interest to image processing researchers as they provide a deeper insight into how data can be processed and the practicality of agent-based modeling in encoding information. That is to say, the current study contributes to our understanding of solitary behaviors and how they can be designed in a way to be employed in data compression. Additionally, this work contributes to existing knowledge of agent-based models that utilize bio-inspired computing by providing a new dimension that shows the practicality of new behavior in image processing. These findings provide a new understanding of how to utilize behaviors for the sake of information processing and how to use the generated information for purposes other than the encoding itself.

References

1. Bons, J., Kegel, A.: On the digital processing and transmission of handwriting and sketching. In: Proceedings of EUROCON, vol. 77, pp. 880–890 (1977)
2. Dhou, K.: Toward a better understanding of viewers' perceptions of tag clouds: relative size judgment. Ph.D. thesis, University of North Carolina at Charlotte (2013)
3. Dhou, K., Cruzen, C.: An innovative chain coding technique for compression based on the concept of biological reproduction: an agent-based modeling approach. IEEE Internet Things J. **6**(6), 9308–9315 (2019)
4. Dhou, K.: A novel agent-based modeling approach for image coding and lossless compression based on the Wolf-Sheep predation model. In: Shi, Y., et al. (eds.) ICCS 2018. LNCS, vol. 10861, pp. 117–128. Springer, Cham (2018). https://doi.org/10.1007/978-3-319-93701-4_9
5. Dhou, K.: An innovative design of a hybrid chain coding algorithm for bi-level image compression using an agent-based modeling approach. Appl. Soft Comput. **79**, 94–110 (2019). https://doi.org/10.1016/j.asoc.2019.03.024
6. Dhou, K.: A new chain coding mechanism for compression stimulated by a virtual environment of a predator-prey ecosystem. Future Gen. Comput. Syst. **102**, 650–669 (2020)
7. Dhou, K., Cruzen, C.: A new chain code for bi-level image compression using an agent-based model of echo location in dolphins. In: The 6th IEEE International Conference on Dependability in Sensor, Cloud, and Big Data Systems and Applications (2020), In Press
8. Dhou, K., Cruzen, C.: A highly efficient chain code for compression using an agent-based modeling simulation of territories in biological beavers. Future Gen. Comput. Syst. **118**, 1–13 (2021). https://doi.org/10.1016/j.future.2020.12.016
9. Dhou, K., Cruzen, C.: An innovative employment of the NetLogo AIDS model in developing a new chain code for compression. In: Paszynski, M., Kranzlmüller, D., Krzhizhanovskaya, V.V., Dongarra, J.J., Sloot, P.M.A. (eds.) ICCS 2021. LNCS, vol. 12742, pp. 17–25. Springer, Cham (2021). https://doi.org/10.1007/978-3-030-77961-0_2
10. Dhou, K., Cruzen, C.: A creative chain coding technique for bi-level image compression inspired by the netlogo hiv agent-based modeling simulation. J. Comput. Sci. **61**, 101613 (2022). https://doi.org/10.1016/j.jocs.2022.101613

11. Dhou, K., Hadzikadic, M., Faust, M.: Typeface size and weight and word location influence on relative size judgments in tag clouds. J. Visual Lang. Comput. **44**, 97–105 (2018)
12. Dhou, K.K., Kosara, R., Hadzikadic, M., Faust, M.: Size judgment and comparison in tag clouds. In: IEEE Visualization Poster Proceedings (2013)
13. Freeman, H.: On the encoding of arbitrary geometric configurations. IRE Trans. Electron. Comput. **2**, 260–268 (1961)
14. Howard, P.G., Kossentini, F., Martins, B., Forchhammer, S., Rucklidge, W.J.: The emerging JBIG2 standard. IEEE Trans. Circuits Syst. Video Technol. **8**(7), 838–848 (1998)
15. Hwang, Y.T., Wang, Y.C., Wang, S.S.: An efficient shape coding scheme and its codec design. In: Signal Processing Systems, 2001 IEEE Workshop on, pp. 225–232. IEEE (2001)
16. ISO CCITT Recommend. T.4: Standardization of group 3 facsimile apparatus for document transmission (1980)
17. Liu, Y.K., Žalik, B.: An efficient chain code with Huffman coding. Pattern Recogn. **38**(4), 553–557 (2005)
18. Mouring, M., Dhou, K., Hadzikadic, M.: A novel algorithm for bi-level image coding and lossless compression based on virtual ant colonies. In: 3rd International Conference on Complexity, Future Information Systems and Risk, pp. 72–78. Setúbal, Portugal (2018)
19. Recommendation T6. Facsimile Coding Schemes and Coding Control Functions for Group 4 Facsimile Apparatus. International Telecommunication Union, Geneva (1988)
20. Standards/International Electrotechnical Commission, I.O., et al.: Progressive bilevel image compression. International Standard 11544 (1993)
21. Sunquist, M.: What is a tiger? ecology and behavior. In: Tilson, R., Nyhus, P.J. (eds.) Tigers of the World (Second Edition). Noyes Series in Animal Behavior, Ecology, Conservation, and Management, pp. 19–33, William Andrew Publishing, Boston, 2nd edn (2010). https://doi.org/10.1016/B978-0-8155-1570-8.00002-5, http://www.sciencedirect.com/science/article/pii/B9780815515708000025
22. Szokalski, M.S., Litchfield, C.A., Foster, W.K.: Enrichment for captive tigers (panthera tigris): current knowledge and future directions. Appl. Anim. Behav. Sci. **139**(1), 1 – 9 (2012). https://doi.org/10.1016/j.applanim.2012.02.021, http://www.sciencedirect.com/science/article/pii/S0168159112000718
23. Zahir, S., Dhou, K.: A new chain coding based method for binary image compression and reconstruction. In: Picture Coding Symposium, pp. 1321–1324 (2007)
24. Zhou, L.: A new highly efficient algorithm for lossless binary image compression. ProQuest (2007)

Analyzing the Usefulness of Public Web Camera Video Sequences for Calibrating and Validating Pedestrian Dynamics Models

Dariusz Pałka(✉), Robert Lubaś, and Jarosław Wąs(✉)

Faculty of Electrical Engineering, Automatics, IT and Biomedical Engineering, AGH University of Science and Technology, Mickiewicza 30, 30-059 Krakow, Poland
{dpalka,jaroslaw.was}@agh.edu.pl

Abstract. Calibrating and validating pedestrian dynamics models is usually conducted using data obtained in experiments with groups of people. An interesting alternative to using data obtained in such a way is using data from public web cameras. The article presents a case study of using public web cameras in the analysis of pedestrian dynamics and social behavior. We have applied YOLOv3 object detector trained on the COCO dataset in order to identify objects in the video frame and to determine their position.

Keywords: Validating pedestrian dynamics models · YOLO · Public web cameras

1 Introduction

In the process of creating pedestrian dynamics models (or their adaptation to new situations), the determination of parameters for these models and their validation are very important elements. For this to be possible, it is necessary to perform measurements in real pedestrian behavior situations [6]. Typically, in order to obtain the results of such measurements, researchers plan and conduct experiments with groups of pedestrians. An alternative method, presented in this paper, is to use data from publicly available web cameras (webcams).

This paper presents examples of extracting characteristics for a pedestrian dynamics model obtained thanks to a system we implemented that processes data from publicly available web cameras.

In times of the pandemic, methods of assessing social distance based on camera images have gained a lot of importance. In most cases when the dynamics of pedestrians is monitored using the image from cameras, we do not have additional tools (e.g. sensors indicating additional positioning) that would enable a more precise assessment of the configuration of people within the range of a camera. One can point out different approaches to crowd analysis using video cameras. Tran et al. [12] proposed a graph-based framework of grouped pedestrians based on Social Distances Model (SDM) [14] - a Cellular Automata (CA)

D. Groen et al. (Eds.): ICCS 2022, LNCS 13351, pp. 103–109, 2022.
https://doi.org/10.1007/978-3-031-08754-7_14

based model of crowd based on proxemics rules, where social distances around a pedestrian are applied. CA based methods of crowd dynamics can use data-driven paradigm and adaptable lattice [1]. One can also point out continuous Social Force [4] base crowd dynamics simulations [13]. Another trend is based on using bio inspired methods in crowd analysis [2], while the task of crowd counting and density maps estimating from videos [10] is often carried out using convolutional neural networks [5,9].

2 Application of Web Cameras

Advantages of using publicly available web cameras in comparison with conducting pedestrian experiments include: no costs needed for organizing the experiment and using the equipment, access to video data located in different locations, which gives the possibility to acquire data e.g. from different cultures and in different social situations (e.g. pandemic condition).

This approach also has significant drawbacks in the area of both planning and conducting experiments. The most important of these stem from the inability to stage a specific situation. This results in, among other things, an inability to measure parameters for situations that rarely occur in reality (such as evacuation associated with the appearance of a threat) and an inability to repeat measurements for a given situation. In addition, it is not possible to set the position and area of registration of the camera, and the measurement is limited to video data (it is not possible to obtain data from other sensors and to use the sensor fusion technique).

2.1 Object Identification

YOLOv3 [7,8] object detector trained on the *COCO* dataset was used to identify objects in the video frame and determine their position. An example of the results of recognition of different types of objects by YOLO for a webcam showing a view of Grodzka Street in Krakow are presented in Fig. 1. As can be seen, most pedestrians were correctly recognized even when they were far from the camera.

2.2 Mapping from 2D Camera Space to 3D World Space

Some characteristics related to pedestrian dynamics can be determined without knowing metric relations in 3D world space, directly from the 2D image from the camera; they include, for example:

- a number of pedestrians in the space observed by the camera
- space utilization (determines how often a location is occupied by a pedestrian)
- the location of Points of Interest (POI)

Fig. 1. The upper picture shows the view from the webcam on Grodzka Street in Krakow. The bottom picture shows the same image with the objects recognized by YOLO, where: p - person (orange frame), b - bicycle (green frame), d - dog (yellow frame), h - handbag (magenta frame), u - umbrella (brown frame), c - car (brown frame). (Color figure online)

Fig. 2. Grodzka Street relative space utilization. The upper left part of the image: the palette used to indicate the degree of space utilization: red color - the highest space utilization, blue color - the lowest. (Color figure online)

Figure 2 shows an example of calculated space utilization for the visible portion of Grodzka Street. Points of Interest can be determined on the basis of space utilization distribution. They correspond to the areas with the highest values of space utilization (i.e. places where people stay the most often and for the longest time).

There are, however, many characteristics that require the determination of metric spatial relationships, such as:

- density of people in a given area (number of people/area)
- absolute speed of movement of individuals
- distribution of absolute and relative pedestrian speeds
- minimum distances between pedestrians

In order to determine spatial relationships between objects (e.g. people), it is necessary to define the mapping from 2D camera space to 3D world space. The mapping is possible thanks to the camera calibration process, i.e. the projection matrix from 3D points (in world space) to 2D points (in camera space). Camera calibration methods can be classified into two main categories [11]:

- methods based on known calibration objects
- methods that do not depend on prior knowledge of camera scenes, so called 'camera self-calibration'

In the context of using public webcams to determine parameters of pedestrian dynamics models, a particularly useful method is camera self-calibration based on the video of walking persons [3,11].

In the case of the camera showing the view of the AGH Main Street, a procedure was carried out to determine the mapping from 2D camera space to 3D world space. The procedure was performed based on metric measurements for characteristic elements present in the image, such as street width, the height and distance between the street signs, the size of the pedestrian crossings, the sizes of the benches, etc.

The method for determining minimum distances between pedestrians is shown in Fig. 3 and 4. Figure 3 shows two pedestrians passing each other on the sidewalk. The successive pedestrian's positions are presented with an interval of 4 s (every 100 frames; the frame rate for this camera is 25 frames per second).

Figure 4 shows the moment of the greatest proximity between the pedestrians in Fig. 3. The yellow and blue frames present the area in which the pedestrians are located recognized by YOLO. The red markers are used to indicate the pedestrian center points on the ground plane. The distance in the image between the red markers (pedestrian center points) is 14 pixels, which corresponds to about 150 cm after taking into account the mapping from 3D world space to 2D camera space, while the distance between the pedestrian occupied areas (represented by the yellow and blue frames) in the image is 14 pixels, which corresponds to about 67 cm in 3D space.

Another characteristic which is important in the context of calibration and validation of pedestrian dynamics models is related to the motion paths for

Fig. 3. Two pedestrians passing each other on the sidewalk of the AGH Main Street. The image shows the frames from a video sequence recorded every 4 s.

Fig. 4. The closest approximation between two passing pedestrians. The yellow and blue frames were generated by YOLO and show the area in which the pedestrian was identified. The red markers show the calculated center point of the pedestrian on the sidewalk plane. (Color figure online)

individual pedestrians. These paths allow us, among other things, to verify if the applied model of interaction between pedestrians (based on e.g. social distances model) is correctly determined. An example of determined motion paths based on images from webcams is shown in Fig. 5.

Fig. 5. Example of the motion paths detected for two pedestrians (highlighted in red and green, respectively). The time between successive positions of the pedestrians presented in the image is 2 s. (Color figure online)

3 Conclusions

As part of the project, we created and tested an application that allows tracking the trajectory of people, determining the distances between people and indicating space utilization and some configuration patterns in the crowd. Thus, we have an important element in the data-driven modeling scheme based on the images from web-cameras. Calibrating and validating pedestrian dynamics models using data from public web cameras has many advantages. These undoubtedly include the high availability of data from different geographical locations (from different cultures) and social situations (e.g. under pandemic conditions). Additionally, thanks to widely available object detectors such as YOLO (even pre-trained on datasets such as COCO), the cost of implementing such a solution is relatively low. However, this approach also has disadvantages with respect to planning and executing experiments involving groups of individuals, the most important of which are related to the inability to plan the observed situations. Therefore, it seems that the best approach for model calibration and validation purposes is to combine these two techniques, which will allow obtaining synergistic effects.

References

1. Bazior, G., Pałka, D., Wąs, J.: Using cellular automata to model high density pedestrian dynamics. In: Krzhizhanovskaya, V.V., Závodszky, G., Lees, M.H., Dongarra, J.J., Sloot, P.M.A., Brissos, S., Teixeira, J. (eds.) ICCS 2020. LNCS, vol. 12137, pp. 486–498. Springer, Cham (2020). https://doi.org/10.1007/978-3-030-50371-0_36
2. Chrysostomou, D., Sirakoulis, G.C., Gasteratos, A.: A bio-inspired multi-camera system for dynamic crowd analysis. Pattern Recogn. Lett. **44**, 141–151 (2014). https://doi.org/10.1016/j.patrec.2013.11.020, pattern Recognition and Crowd Analysis
3. Fengjun, L.v., Zhao, T., Nevatia, R.: Camera calibration from video of a walking human. IEEE Trans. Pattern Anal. Mach. Intell. **28**(9), 1513–1518 (2006). https://doi.org/10.1109/TPAMI.2006.178
4. Helbing, D., Molnar, P.: A social force model for pedestrian dynamics. Phys. Rev. E **51**, 4284–4286 (1995)
5. Miao, Y., Han, J., Gao, Y., Zhang, B.: St-cnn: Spatial-temporal convolutional neural network for crowd counting in videos. Pattern Recogn. Lett. **125**, 113–118 (2019). https://doi.org/10.1016/j.patrec.2019.04.012
6. Porzycki, J., Schmidt-Polończyk, N., Wąs, J.: Pedestrian behavior during evacuation from road tunnel in smoke condition-empirical results. PLoS One **13**(8), e0201732 (2018). https://doi.org/10.1371/journal.pone.0201732
7. Redmon, J., Divvala, S., Girshick, R., Farhadi, A.: You only look once: unified, real-time object detection (2016)
8. Redmon, J., Farhadi, A.: Yolov3: an incremental improvement (2018). CoRR abs/1804.02767, http://arxiv.org/abs/1804.02767
9. Sun, Y., Jin, J., Wu, X., Ma, T., Yang, J.: Counting crowds with varying densities via adaptive scenario discovery framework. Neurocomputing **397**, 127–138 (2020). https://doi.org/10.1016/j.neucom.2020.02.045
10. Tadeusiewicz, R.: Electronic observation and computer monitoring of human behavior in public space. Napedy i Sterowanie **12** (2013)
11. Tang, Z., Lin, Y., Lee, K., Hwang, J., Chuang, J., Fang, Z.: Camera self-calibration from tracking of moving persons. In: 2016 23rd International Conference on Pattern Recognition, ICPR, pp. 265–270 (2016). https://doi.org/10.1109/ICPR.2016.7899644
12. Tran, K., Gala, A., Kakadiaris, I., Shah, S.: Activity analysis in crowded environments using social cues for group discovery and human interaction modeling. Pattern Recogn. Lett. **44**, 49–57 (2014). https://doi.org/10.1016/j.patrec.2013.09.015, http://www.sciencedirect.com/science/article/pii/S0167865513003516, Pattern Recognition and Crowd Analysis
13. Tytko, K., Mamica, M., Pękala, A., Wąs, J.: Simulating pedestrians' motion in different scenarios with modified social force model. In: Wyrzykowski, R., Deelman, E., Dongarra, J., Karczewski, K. (eds.) PPAM 2019. LNCS, vol. 12044, pp. 467–477. Springer, Cham (2020). https://doi.org/10.1007/978-3-030-43222-5_41
14. Wąs, J., Gudowski, B., Matuszyk, P.J.: New cellular automata model of pedestrian representation. In: El Yacoubi, S., Chopard, B., Bandini, S. (eds.) ACRI 2006. LNCS, vol. 4173, pp. 724–727. Springer, Heidelberg (2006). https://doi.org/10.1007/11861201_88

A Highly Customizable Information Visualization Framework

Luís Spínola(✉), Daniel Castro Silva, and Luís Paulo Reis

Faculty of Engineering, University of Porto (FEUP) Artificial Intelligence and Computer Science Laboratory (LIACC),
Rua Dr. Roberto Frias s/n 4200–465, Porto, Portugal
{up201405907,dcs,lpreis}@fe.up.pt

Abstract. The human brain can quickly become overwhelmed by the amounts of data computers can process. Consequently, data abstraction is necessary for a user to grasp information and identify valuable patterns. Data is usually abstracted in a pictorial or graphical format. Nowadays, users demand more personalization from the systems they use. This work proposes a user-centered framework that aims to ease creating visualizations for the developers of a platform while offering the end-user a highly customizable experience. The conceptualized solution was prototyped and tested to ensure the information about the data is transmitted to the user in a quick and effective manner. The results of a user study showed that users are pleased with the usability of the prototype and prove that they desire control over the configuration of their visualizations. This work not only confirmed the usefulness of previously explored personalization options for visual representations, but also explored promising new personalization options.

Keywords: Information visualization · Visual analytics · Visualization personalization

1 Introduction

The visualization process centers on converting data into graphics, and there are numerous methods to do so [7]. Information Visualization (InfoVis) systems conventionally follow a model offering a one-size-fits-all solution, meaning representations do not change, ignoring user preferences, capacities and overall circumstances [11]. When developing a platform to expose visualizations, there needs to be a concern with the overall design, usability and interactivity, besides the efficacy of the methods used [4,8]. However, the responsibility of selecting the best approach for each set of information usually falls under the developer.

This work emerged to explore an InfoVis process to implement graphic platforms devoted to analyzing and monitoring data. It contributes with a scalable framework offering: Visualizations of performance indicators, common graphics and means to display geo-referenced information; A method to configure and customize visualizations to cater to user preferences; Means for a developer to easily implement a visualization in a system, providing a simple workflow.

D. Groen et al. (Eds.): ICCS 2022, LNCS 13351, pp. 110–116, 2022.
https://doi.org/10.1007/978-3-031-08754-7_15

2 Background

While the amount of data keeps growing, human attention and time do not expand with that trend, instead remaining constant [2]. With larger sets of data, users need to test more attributes and experiment more before coming up with suitable abstractions [3]. Furthermore, more people are performing data analysis, and business users, while experts in their area, don't typically possess the skills to choose a suitable visual metaphor from available options [7].

Research has proven that users behave as individuals with completely different desires – their needs, abilities, and preferences have a significant impact on their performance and satisfaction when using InfoVis techniques [6]. Thus, techniques should consider the reader [12]. On that account, it is crucial to scrutinize the possibilities of new kinds of user-centered InfoVis systems. One example that supports this hypothesis is [10], where several studies were conducted: Experts were asked to choose a type of encoding for different data. Results show that there is in fact a substantial variability in visualization preferences for the same data set; The second focused on user's prior experience, visual literacy and cognitive capabilities, by trying to identify what types of graphical representations are more adequate for each type of user; A third, tried to understand how the purpose with the data could take place to change the visualization requirements. Two states were taken in consideration - analyzing and monitoring.

In [5] the concept of graphical overlays was introduced, and a system was developed to allow the dynamic addition of the proposed overlays: **Reference structures** - to assist the reader in comparing and extracting values, for instance by using grid lines; **Highlight** - to emphasize some aspect of a visualization; **Redundant encoding** - used to aid in the extraction of values or to better portray tendencies; (for example, showing data labels directly on the graphic); **Summary statistics** - helpful to portray useful statistical information about a data set. The paper present the mean, median, and maximum as examples.

A concept similar to graphical overlays was introduced (visualization aids) [11], which, much like overlays, can be included in a representation without having previous insights of user's intent. To evaluate the visual prompts, a study was conducted using two techniques, namely bar and line charts, alongside checkboxes allowing to toggle on and off five different aids. A task consisted of showing graphics to a user, alongside questions that required interpreting some data. With 40 participants and 50 tasks to be completed, the mean accuracy was 87% and showed that a user turned on an aid with an average of around 49 times. The most selected option was 'show data', followed by 'horizontal line grid'; 'vertical grid line', 'fill area' and 'dot grid' were barely used.

Similar work proposes options that are used as overlays in InfoVis techniques [5,11]. That fact makes it viable to be used in previously done visualizations, but limits the amount of overlays that can be applied and studied. This work proposes a distinctive approach that follows the trends of research on InfoVis aids by suggesting a framework for producing visualizations that offers personalizations that are proven to add value to the readability of a representation, also serving as a platform to investigate further aids.

3 Solution Conceptualization

A family of graphics is a set of visualizations serving a purpose. When seeding data to start the process, the programmer needs to specify which family the data belongs to. The families introduced so far in the prototype are: **One Numerical** - Simple categorical data (bar and pie chart); **Two Numerical** - To compare two numerical values (side-by-side comparison bar, scatter, and bar chart); **N Numerical** - Data distribution when more than two categories are present (bar, line, and area charts). **Time Series** - Data over time (line, area, and bar charts); **Performance** - Display KPIs (gauge, bullet plot); **Geo Quantities** - Values at coordinates (column and bubble map); **Geo Densities** - Density graphs (heat map, hexagonal map); **Geo Path** - Paths. A temporal sequence of coordinates traveled, with icons displaying stops. This division provides flexibility to the developer. When seeding data, the developer knows what graphics are available to the user. Families in this prototype follow the most common norms for each type of data. Still, families can be added by using existing graphics, granting them a different objective and combining them with new techniques.

The framework is divided into three simple layers. Figure 1 depicts the flow between layers. The first layer, **Data Input**, receives the data in JSON format. The metadata of the JSON object is created and appended to the data fetched. The developer queries the data and appends the metadata, which is used to specify the graphic family, as well as inform the next layer of which data key corresponds to what (for example which key corresponds to the date in a Time Series or which key corresponds to the coordinates in a georeferenced representation. The developer can also specify a user ID and a graph ID that in conjunction are unique and are needed to make user configurations persistent.

The second layer, **Graph Chooser**, uses the information from the first layer and selects the adequate graphic family for the data. User personalizations and graph selection made in the following layer are persistent. As such, this layer queries the database to find out if the visual representation is being revisited and, if so, loads previous graph selection and personalizations.

The final layer, **View/Personalization**, is the stage that handles the visualization and the process of personalization configuration by the user. This layer also communicates with the database to store individual graphic information and, if desired, can also store user metrics such as preferred graph per family and the amount of times each personalization option was used.

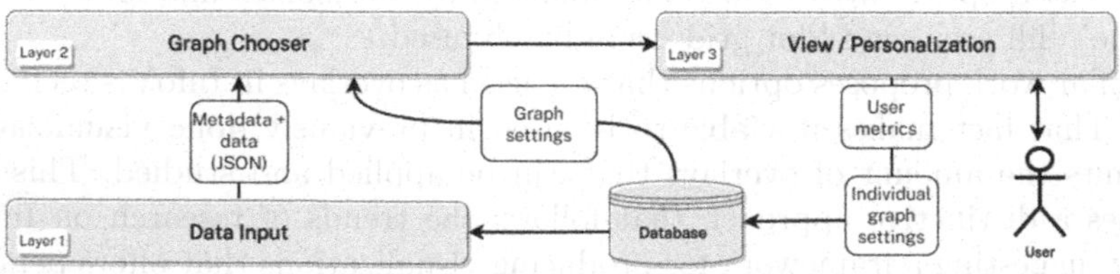

Fig. 1. Framework architecture diagram.

4 Implementation Details

Visualization on the web cannot solely offer a static representation as interaction eventually triggers some change, and although treated as different components, representation and interaction are intrinsically connected and positively contribute to the final experience [14]. Thus, all graphics provided offer interaction.

When evaluating a tool, some prominent factors are task completion time and task completion correctness [9]. JavaScript, widely used to produce interactive web-applications, was the chosen language. For the overall framework, **React**, an open-source web framework, was chosen, being used in conjunction with **Material-UI**, a set of components easing the process of the design aspects of a system. For InfoVis techniques, **Recharts** was used alongside **Deck.gl**[1].

The visualization provides the user with an icon giving access to a sidebar (see Fig. 2), which can be docked to the right or left side. The first option is a selection with the available techniques. Other options are organized in sections:

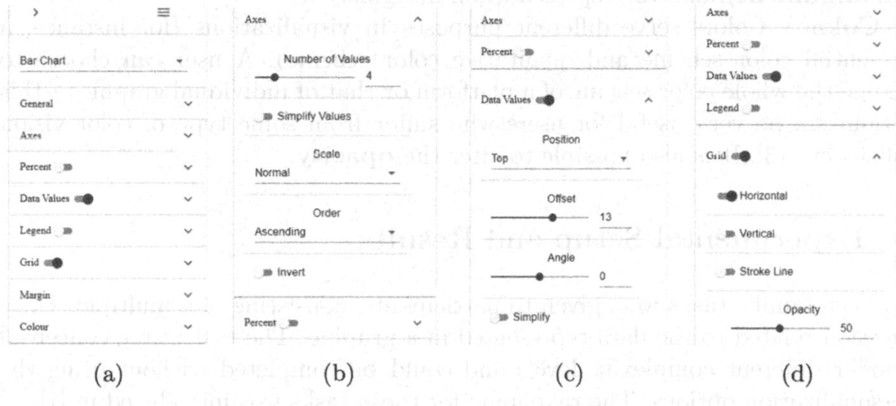

(a) (b) (c) (d)

Fig. 2. Personalization sidebar. a) Default state. b) Axes. c) Data values. d) Grid.

General – General definitions, such as **height**. An option allows data to be **grouped** in one graphic or distributed among several (as dimensions on the same graphic encourage analysis and comparison, while separation encourages monitoring [10]). For line/area charts, we can change the type of **interpolation**. For line charts, the aid of using dots and the accessibility option of tuning the **line width** exists. For bar/area plots, there is the option to display the values **stacked**. For geovisualisations, the user can change the **background map** between several options (normal, terrain, relief, satellite, or none). There's also an option to show **borders** and **text layers** (e.g. country districts' borders and names); this option can be configured according to the scope of the visualization.

[1] Offical websites: reactjs.org, material-ui.com, recharts.org, deck.gl.

Axes (for graphs contained in a Cartesian plane) – Options include the **number of values** in the numerical axis, and the option to **simplify** those values (for example showing 5500 as 5.5k). It is also possible to change the **scale** between normal and logarithmic and, when possible, to choose the **order** in which data is displayed. It is also possible to **invert axis** (for example, turning a vertical bar char into a horizontal one).

Data Values – When activated, exact values are shown directly onto the representation. Options include changing the **position** where values are displayed relative to the graphic elements and adjusting the **displacement** and **angle**. Another option is to **simplify** the displayed values.

Percentage – Changes visual representation to a percentage. Options include to display the percentage in **relation to the max value** on a representation, to display the percentage in **relation to the total value**, or to display the percentage in **relation to a custom value**.

Grid, Legend, and Margin – There is an option to toggle **grid lines**; **horizontal**, **vertical** or both, to change the **grid opacity** and to have a **stroke line**. The graphic legend (hide or show the legend and to tweak its **position** and **alignment**); and the representation margins.

Color – Colors serve different purposes in visualizations (for instance, a sequential color scheme and qualitative color scheme). A user can choose to change the whole color scheme of a platform or that of individual graphics – this option can be very useful for users who suffer from some type of color vision deficiency [13]. It is also possible to alter the **opacity**.

5 Experimental Setup and Results

For each family, tasks were given to participants, consisting of a multiple-choice question related to the data represented in a graphic. The tasks were conceived to offer different complexity levels and could be completed without using the personalization options. The taxonomy for these tasks was introduced in [1].

The test started with a demographic survey , followed by questions using a five-value scale, collecting data on participant's experience with data analysis and InfoVis techniques. After completing the tasks for each visualization, users were asked the preferred visual representation, and an open question to express opinions and problems found. There were a total of 18 tasks. After completion, a post-survey was presented to inquiry about the usability of the options, including questions about being allowed to display the same data in different graphics and asking if the user interacted with the graphic or used the personalization options to complete the tasks. The metrics collected were time, type of graphic selected, and the number of times each option was used.

The study comprised 37 participants, 21 male and 16 female, with ages ranging from 18 to 64 years old. 24 participants had a college graduation, of which 11 had a master or doctorate, and the remaining 13 had basic education. Participants rated themselves on handling data through graphic visualizations, with an average of 3.65 out of 5. They showed a strong opinion towards having the option to personalize their platforms and systems, with an average of 4.38 out of 5, and 4.24 out of 5 in relation to the personalization of graphic visualizations.

Time and Accuracy – There was an average of 15.3 successfully completed tasks (correct answers) per participant, which translates to an accuracy of 84.8%. The time taken to complete the study (not including initial and final questionnaires) was on average 21 min and 24 s.

Graph Selection – 6 participants did not change the default visualization. Of the 31 participants that did, 7 only changed it between 1 and 5 times, while the remaining 24 changed over 5 times, and 9 visited all graphics available on the study. Results show that while participants felt the default graphic was the most useful one for each task and graphic family, always being the most voted (average of 64%), other selections were also popular.

Personalization Options – 9 out of the 37 participants did not make any personalization. These 9 participants had a completion time average of 13m,57s, which is considerably lower than the global average (21m,24s). However, these participants also showed lower accuracy compared to the global average, a value of 80.3% (14.5 out of 18). Some participants opted to only use the options that directly aided on the task completion; others explored the "aesthetic" options as well. This is probably the cause of a high discrepancy between the number of personalization options used by participants. Nonetheless, the options were changed a total of 2703 times, which translates to an average of 73 changes per user. The most popular option was to show data values (29% of changes) directly on the visualization. By order of popularity, other options were: grouped (13%), percentage (11%), order (9%), number of values on an axis (9%), invert axis (6%), stacked (4%) and simplify data values (3%). Grid options were not toggled as much as expected, probably due to it being enabled by default. Users found the interface pleasing with a score of 4.16 (out of 5) and that most people would have an easy time learning to use it, with a score of 4.29, they also found the personalization options explicit and overall useful, with a score of 4.19.

6 Conclusions and Future Work

Results show that there is usefulness in giving users the option to personalize visualizations. As shown, personalization is desired by the common user. We confirmed the usefulness of some options and proposed new ones with potential. We also show that a higher successful task rate can be achieved with some options, and although users that did not use options completed tasks faster with a slightly lower accuracy rate, it did not account for the fact that users were not yet familiar with the system. This paper hopes to encourage and motivate building of systems having personalization in mind from the start. It continues research on forms of personalizing visualizations, and proposes a framework that can be included in web platforms for translating data into visualizations.

While the results are very promising, this work can be further improved by several factors. Users tend to use the same options from graphic to graphic, and on big platforms with many dashboards, personalizing each graphic might become a tedious process. Such a fact opens up the possibilities for personalization-based adaptation. The framework was conceptualized to facilitate the process of data exploration for the user, but also to ease the process of introducing visualizations by developers. So, another interesting path for future work is to conduct tests to prove the framework effectiveness from the programmer's perspective. There is also the perspective of creating mechanisms to provide the user with the power of making their own queries to the database and, in conjunction with the proposed framework, create their own dashboards and similar solutions.

Acknowledgements. This work was financially supported by Base Funding - UIDB/00027/2020 of the Artificial Intelligence and Computer Science Laboratory - LIACC - funded by national funds through the FCT/MCTES (PIDDAC).

References

1. Amar, R., Eagan, J., Stasko, J.: Low-level components of analytic activity in information visualization. In: IEEE Symposium on Information Visualization, INFOVIS 2005, pp. 111–117. IEEE (2005)
2. Chittaro, L.: Information visualization and its application to medicine. Artifi. Intell. Med. **22**, 81–88 (2001)
3. Keim, D.: Information visualization and visual data mining. IEEE Trans. Visuali. Comput. Graph. **8**, 1–8 (2002)
4. Koh, L.C., Slingsby, A., Dykes, J., Kam, T.S.: Developing and applying a user-centered model for the design and implementation of information visualization tools, pp. 90–95. IEEE, July 2011
5. Kong, N., Agrawala, M.: Graphical overlays: using layered elements to aid chart reading. IEEE Trans. Visuali. Comput. Graph. **18**, 2631–2638 (2012)
6. Lallé, S., Conati, C., Carenini, G.: Prediction of individual learning curves across information visualizations. User Model. User-Adap. Inter. **26**, 307–345 (2016)
7. Liu, S., Cui, W., Wu, Y., Liu, M.: A survey on information visualization: recent advances and challenges. The Vis. Comput. **30**, 1373–1393 (2014)
8. Moere, A.V., Purchase, H.: On the role of design in information visualization. Inf. Visuali. **10**, 356–371 (2011)
9. Nazemi, K., Burkhardt, D., Hoppe, D., Nazemi, M., Kohlhammer, J.: Web-based evaluation of information visualization. Procedia Manufact. **3**, 5527–5534 (2015)
10. Poetzsch, T., Germanakos, P., Huestegge, L.: Toward a taxonomy for adaptive data visualization in analytics applications. Front. Artifi. Intell. **3**, 9 (2020)
11. Steichen, B., Fu, B.: Towards adaptive information visualization - a study of information visualization aids and the role of user cognitive style. Front. Artifi. Intelli. **2**, 22 (2019)
12. Toker, D., Conati, C., Steichen, B., Carenini, G.: Individual user characteristics and information visualizatio, pp. 295–304. ACM, April 2013
13. Wexler, S., Shaffer, J., Cotgreave, A.: The big book of dashboards: visualizing your data using real-world business scenarios. John Wiley & Sons (2017)
14. Yi, J.S., Ah Kang, Y., Stasko, J., Jacko, J.A.: Toward a deeper understanding of the role of interaction in information visualization. IEEE Trans. Visuali. Comput. Graph. **13**, 1224–1231 (2007)

Incremental Dynamic Analysis and Fragility Assessment of Buildings with Different Structural Arrangements Experiencing Earthquake-Induced Structural Pounding

Mahmoud Miari(✉) and Robert Jankowski

Department of Construction Management and Earthquake Engineering, Faculty of Civil and Environmental Engineering, Gdańsk University of Technology, Gdańsk, Poland
mahmoud.miari@pg.edu.pl, mahmoud-miari@hotmail.com

Abstract. Structural pounding is considered as one of the most critical phenomena occurring during earthquakes. This paper presents the incremental dynamic analysis and fragility assessment of buildings experiencing earthquake-induced pounding. Three 3-D buildings with different number of storeys and under different structural arrangements have been considered. Three pounding scenarios have been taken into account, i.e. pounding between 5-storey and 7-storey buildings, pounding between 5-storey and 9-storey buildings and pounding between 7-storey and 9-storey buildings. The incremental dynamic analysis and fragility assessment has been performed for these three buildings in the three pounding scenarios as well as for the no pounding case. The results of both incremental dynamic analysis and fragility assessment illustrate that pounding can be beneficial and destructive, depending on the structural response and ground motion shift versus time. No clear relation has been observed because pounding is a highly complicated phenomenon.

Keywords: Structural pounding · Incremental dynamic analysis · Fragility assessment · Earthquakes · Buildings · Performance levels

1 Introduction

Structural pounding is defined as repeatedly observed collisions occurring between adjacent structures during earthquakes which is considered as a significant phenomenon [1–3]. It has been experienced in several earthquakes, such as the Mexico earthquake where in 40% of buildings pounding was found, and in 15% of buildings with severe damage or collapse, pounding was visible [4] where in 20–30% of them pounding was the major reason of damage [5]. Pounding was also experienced in 200 out of 500 surveyed buildings in the Loma Prieta earthquake [6]. It was also experienced in Christchurch (New Zealand, 2011) [7] and Gorkha (Nepal, 2015) [8] earthquakes.

Research on earthquake-induced pounding has been conducted for more than three decades (see [9, 10]). Pounding was found to increase the floor peak accelerations, shear

D. Groen et al. (Eds.): ICCS 2022, LNCS 13351, pp. 117–124, 2022.
https://doi.org/10.1007/978-3-031-08754-7_16

forces, and impact forces while the displacement may increase or decrease [11]. The degree of the amplification depends on the dynamic properties (mass, ductility, damping ratio, period, etc.) of colliding buildings. The properties of the ground motion also have a significant effect on the colliding structures [12]. The response of colliding buildings is substantially affected in the direction of pounding and unaffected in the other direction [13]. Crozet et al. [14, 15] also found that the frequency ratio has the largest influence on the maximum impact force and ductility demands while the frequency and mass ratios have the largest influence on the impact impulse (mass ratio is predominant for low frequency range).

The previously mentioned literature review illustrates that pounding is a substantial phenomenon and leads to severe damages during earthquakes. However, little attention has been paid to the damage state and the performance level of the colliding buildings during earthquakes. Therefore, the aim of this paper is to perform incremental dynamic analysis and fragility assessment of buildings with different structural arrangements experiencing pounding.

2 Incremental Dynamic Analysis and Fragility Assessment

The incremental dynamic analysis (IDA) and the fragility assessment method are among the modern methods to evaluate the seismic response of colliding buildings. IDA is a parametric analysis method used to estimate the structural performance of vibrating buildings under certain earthquake record (see [16] for details). It has been widely used in nonlinear dynamic analyses as well as in studying pounding phenomenon (see [17–19] for example). In this paper, the method proposed by Ibrahim and El-Shami (2011) were used to develop the fragility curves [20]. Five different performance levels have been considered by researchers, i.e. the operational performance (OP), immediate occupancy (IO), damage control (DC), life safety (LS), and collapse prevention (CP). The maximum allowable interstorey drifts for each performance level were taken into account based on Xue et al. (2008) [21] recommendations which are 0.5%, 1.0%, 1.5%, 2.0% and 2.5% for the for the OP, IO, DC, LS and CP performance levels, respectively [21].

3 Numerical Models of Buildings

Three buildings with 5, 7 and 9 storeys have been analysed. All of them have a storey height of 3 m and a width of 16 m in the x-direction and 12 m in the y-direction (the bays are 4 × 4 m and 3 × 4 m in x- and y- directions, respectively). The analysis has been performed using ETABS software. The Finite Element (FE) models of these three buildings are shown in Fig. 1.

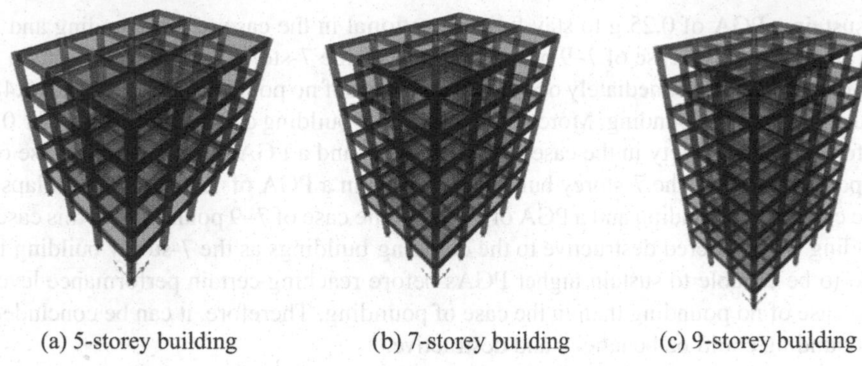

(a) 5-storey building (b) 7-storey building (c) 9-storey building

Fig. 1. FE model of the studied buildings

Three pounding scenarios have been considered, i.e. pounding between 5-storey and 7-storey buildings (5–7 pounding), pounding between 5-storey and 9-storey buildings (5–9 pounding) and pounding between 7-storey and 9-storey buildings (7–9 pounding). A gap of 4 cm has been provided between these buildings for all the cases. The soil type A (hard rock) defined in the ASCE 7–10 code [22] has been chosen in all cases of both pounding and no pounding cases. The soil type has been considered using the response spectrum concept (see [11, 23], for details). The IDA has been conducted for three earthquake records which are: San Fernando, Loma Prieta, and Imperial Valley (station: Agrarias). Then, the fragility curves of the colliding buildings have been developed based on the IDA curves.

4 IDA

In this section, the average IDA of the IDA curves of the three ground motions are presented in different pounding scenarios and compared with the no pounding case. The average IDA curves for the 5-storey, 7-storey, and 9-storey buildings are presented in Figs. 2a, 2b and 2c, respectively, under different pounding scenarios. As it can be seen from Fig. 2a, the 5-storey building can sustain a PGA of 0.27 g to stay fully operational in the case of no pounding, a PGA of 0.28 g in the case of 5–7 pounding and a PGA of 0.3 g in the case of 5–9 pounding. Indeed, the 5-storey building can sustain a PGA of 0.53 g to be immediately occupied in the case of no pounding, a PGA of 0.55 g in the case of 5–7 pounding and a PGA of 0.61 g in the case of 5–9 pounding. Moreover, the 5-storey building can sustain a PGA of 1.06 g before losing its safety in the case of no pounding, a PGA of 1.08 g in the case of 5–7 pounding and a PGA of 1.21 g in the case of 5–9 pounding. Also, the 5-storey building can sustain a PGA of 1.33 g before collapse in the case of no pounding, a PGA of 1.35g in the case of 5–7 pounding and a PGA of 1.49 g in the case of 5–9 pounding. In this case, pounding is considered beneficial to the colliding buildings as the 5-storey building is found to be capable to sustain higher PGAs before reaching certain performance level in the case of pounding than in the case of no pounding. This is referred to the fact that the pounding blocks the movement of vibrating buildings. Furthermore, as it can be seen from Fig. 2b, the 7-storey building can sustain a PGA of 0.25 g to stay fully operational in the case of no pounding and a PGA of 0.23 g in the case of 7–9 pounding. Indeed, the 7-storey building can sustain a PGA of 0.51 g to be immediately occupied in the case of no pounding and a PGA of 0.48 g in the case of 7–9 pounding. Moreover, the 7-storey building can sustain a PGA of 1.01 g before losing its safety in the case of no pounding and a PGA of 0.93 g in the case of 7–9 pounding. Also, the 7-storey building can sustain a PGA of 1.26 g before collapse in the case of no pounding and a PGA of 1.18 g in the case of 7–9 pounding. In this case, pounding is considered destructive to the colliding buildings as the 7-storey building is found to be capable to sustain higher PGAs before reaching certain performance level in the case of no pounding than in the case of pounding. Therefore, it can be concluded that pounding could be beneficial and destructive.

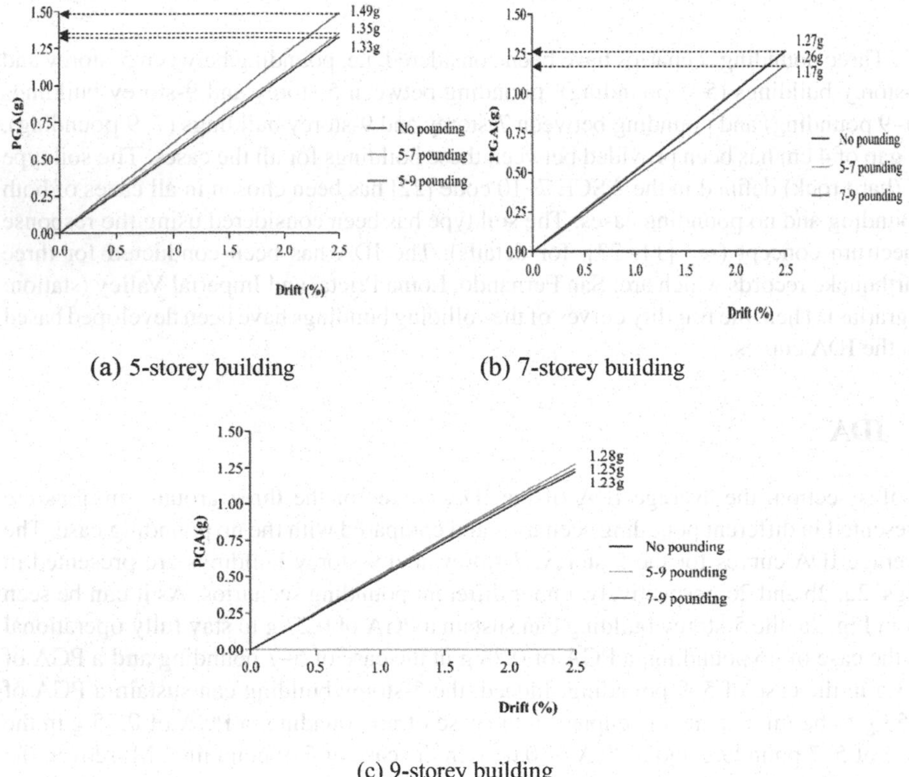

(a) 5-storey building (b) 7-storey building

(c) 9-storey building

Fig. 2. Average IDA curves of the considered buildings in different pounding scenarios

5 Fragility Assessment

In this section, the fragility curves are presented. The fragility curves have been developed based on the IDA curves presented in Sect. 4. Figure 3 presents the fragility curves for the

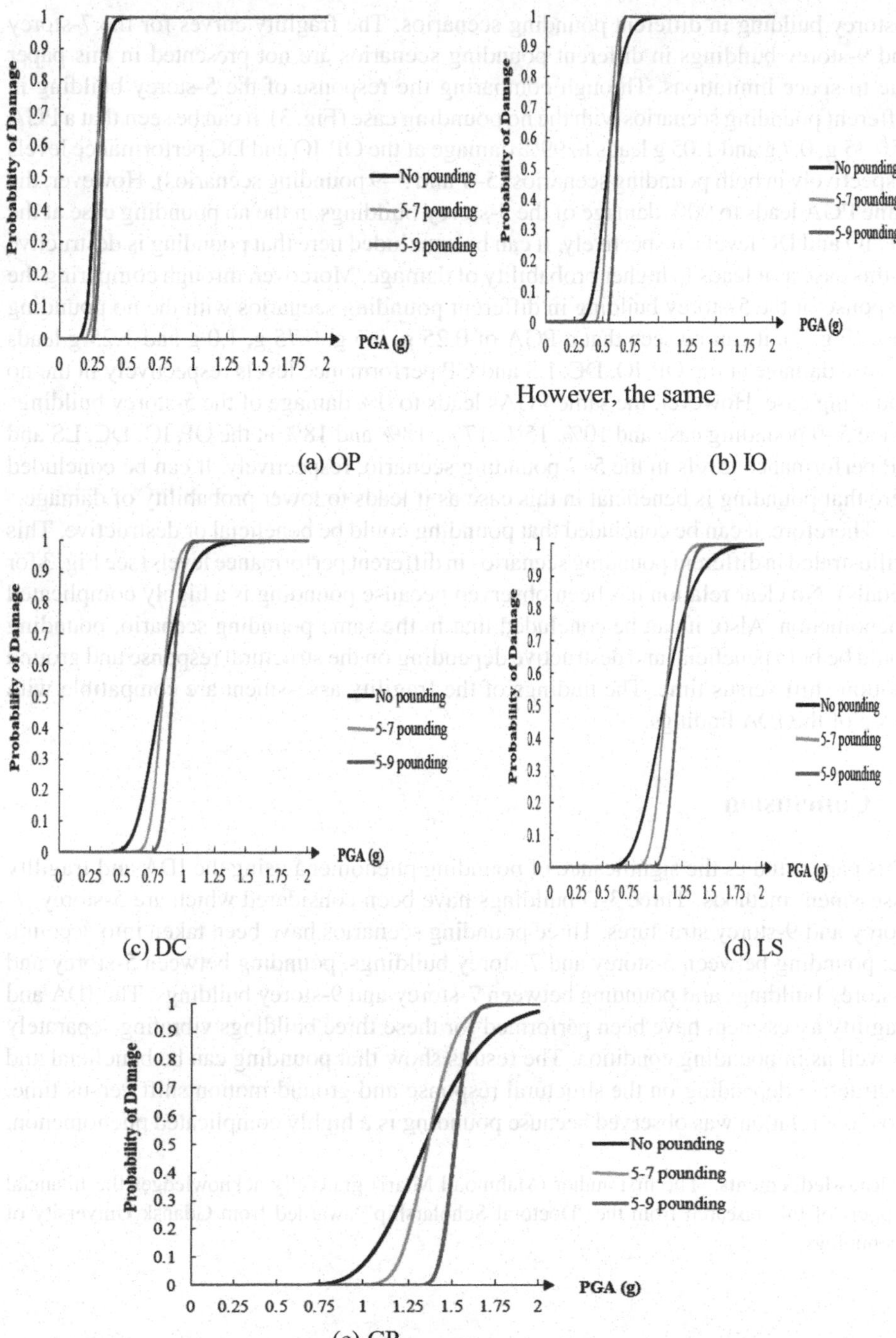

Fig. 3. Fragility curves of the 5-storey building in different pounding scenarios

5-storey building in different pounding scenarios. The fragility curves for the 7-storey and 9-storey buildings in different pounding scenarios are not presented in this paper due to space limitations. Through comparing the response of the 5-storey building in different pounding scenarios with the no pounding case (Fig. 3), it can be seen that a PGA of 0.35 g, 0.7 g and 1.05 g leads to 99% damage at the OP, IO and DC performance levels respectively in both pounding scenarios (5–7 and 5–9 pounding scenarios). However, the same PGA leads to 90% damage of the 5-storey buildings in the no pounding case at the OP, IO and DC levels, respectively. It can be concluded here that pounding is destructive in this case as it leads to higher probability of damage. Moreover, through comparing the response of the 5-storey building in different pounding scenarios with the no pounding case (Fig. 3), it can be seen that a PGA of 0.25 g, 0.5 g, 0.75 g, 1.0 g and 1.25 g leads to 36% damage at the OP, IO, DC, LS and CP performance levels respectively in the no pounding case. However, the same PGAs leads to 0% damage of the 5-storey buildings in the 5–9 pounding case and 10%, 15%, 17%, 18% and 18% at the OP, IO, DC, LS and CP performance levels in the 5–7 pounding scenario, respectively. It can be concluded here that pounding is beneficial in this case as it leads to lower probability of damage.

Therefore, it can be concluded that pounding could be beneficial or destructive. This is illustrated in different pounding scenarios in different performance levels (see Fig. 3 for details). No clear relation has been observed because pounding is a highly complicated phenomenon. Also, it can be concluded that in the same pounding scenario, pounding could be both beneficial and destructive depending on the structural response and ground motion shift versus time. The findings of the fragility assessment are compatible with those of the IDA findings.

6 Conclusion

This paper studies the significance of pounding phenomena using the IDA and fragility assessment methods. Three 3-D buildings have been considered which are 5-storey, 7-storey and 9-storey structures. Three pounding scenarios have been taken into account, i.e. pounding between 5-storey and 7-storey buildings, pounding between 5-storey and 9-storey buildings and pounding between 7-storey and 9-storey buildings. The IDA and fragility assessment have been performed for these three buildings vibrating separately as well as in pounding condition. The results show that pounding can be beneficial and destructive depending on the structural response and ground motion shift versus time. No clear relation was observed because pounding is a highly complicated phenomenon.

Acknowledgements. The first author (Mahmoud Miari) gratefully acknowledges the financial support of this research from the "Doctoral Scholarship" awarded from Gdańsk University of Technology.

References

1. Miari, M., Jankowski, R.: Pounding between high-rise buildings founded on different soil types. In: 17th World Conference on Earthquake Engineering, Sendai, Japan (2021)
2. Kazemi, F., Miari, M., Jankowski, R.: Investigating the effects of structural pounding on the seismic performance of adjacent RC and steel MRFs. Bull. Earthq. Eng. **19**, 317–343 (2021)
3. Miari, M., Jankowski, R.: Seismic gap between buildings founded on different soil types experiencing pounding during earthquakes. Earthq. Spectra (2022). Published online 07.04.2022. https://doi.org/10.1177/87552930221082968
4. Rosenblueth, E., Meli, R.: The 1985 Mexico earthquake. Concr. Int. **8**(5), 23–34 (1986)
5. Anagnostopoulos, S.: Building pounding re-examined: how serious a problem is it. In: Eleventh World Conference on Earthquake Engineering, p. 2108, Pergamon. Elsevier Science Oxford (1996)
6. Kasai, K., Maison, B.F.: Building pounding damage during the 1989 Loma Prieta earthquake. Eng. Struct. **19**(3), 195–207 (1997)
7. Cole, G.L., Dhakal, R.P., Turner, F.M.: Building pounding damage observed in the 2011 Christchurch earthquake. Earthq. Eng. Struct. Dynam. **41**(5), 893–913 (2012)
8. Sharma, K., Deng, L., Noguez, C.C.: Field investigation on the performance of building structures during the April 25, 2015, Gorkha earthquake in Nepal. Eng. Struct. **121**, 61–74 (2016)
9. Miari, M., Choong, K.K., Jankowski, R.: Seismic pounding between adjacent buildings: Identification of parameters, soil interaction issues and mitigation measures. Soil Dyn. Earthq. Eng. **121**, 135–150 (2019)
10. Miari, M., Choong, K.K., Jankowski, R.: Seismic pounding between bridge segments: a state-of-the-art review. Arch. Comput. Meth. Eng. **28**(2), 495–504 (2020)
11. Miari, M., Jankowski, R.: Analysis of pounding between adjacent buildings founded on different soil types. Soil Dyn. Earthq. Eng. **154**, 107156 (2022)
12. Abdel-Mooty, M., Al-Atrpy, H., Ghouneim, M.: Modeling and analysis of factors affecting seismic pounding of adjacent multi-story buildings. WIT Trans. Built Environ. **104**, 127–138 (2009)
13. Jameel, M., Islam, A., Hussain, R.R., Hasan, S.D., Khaleel, M.: Non-linear FEM analysis of seismic induced pounding between neighbouring multi-storey structures. Latin Am. J. Solids Struct. **10**(5), 921–939 (2013)
14. Crozet, V., Politopoulos, I., Yang, M., Martinez, J.M., Erlicher, S.: Sensitivity analysis of pounding between adjacent structures. Earthq. Eng. Struct. Dyn. **47**(1), 219–235 (2018)
15. Crozet, V., Politopoulos, I., Yang, M., Martinez, J., Erlicher, S.: Influential structural parameters of pounding between buildings during earthquakes. Procedia Eng. **199**, 1092–1097 (2017)
16. Vamvatsikos, D., Cornell, C.A.: Incremental dynamic analysis. Earthq. Eng. Struct. Dyn. **31**(3), 491–514 (2002)
17. Kazemi, F., Mohebi, B., Yakhchalian, M.: Evaluation of the P-delta effect on collapse capacity of adjacent structures subjected to far-field ground motions. Civil Eng. J. **4**(5), 1066–1073 (2018)
18. Mohebi, B., Kazemi, F., Yakhchalian, M.: Investigating the P-Delta effects on the seismic collapse capacity of adjacent structures. In: The 16th European Conference on Earthquake Engineering (16ECEE), pp. 18–21 (2018)
19. Kazemi, F., Mohebi, B., Yakhchalian, M.: Enhancing the seismic performance of adjacent pounding structures using viscous dampers. In: The 16th European Conference on Earthquake Engineering (16ECEE), pp. 18–21 (2018)

20. Ibrahim, Y.E., El-Shami, M.M.: Seismic fragility curves for mid-rise reinforced concrete frames in Kingdom of Saudi Arabia. IES J. Part A Civil Struct. Eng. **4**(4), 213–223 (2011)
21. Xue, Q., Wu, C.-W., Chen, C.-C., Chen, K.-C.: The draft code for performance-based seismic design of buildings in Taiwan. Eng. Struct. **30**(6), 1535–1547 (2008)
22. Minimum Design Loads for Buildings and Other Structures (ASCE/SEI 7-10), 078447785X (2013)
23. Miari, M., Jankowski, R.: Incremental dynamic analysis and fragility assessment of buildings founded on different soil types experiencing structural pounding during earthquakes. Eng. Struct. **252**, 113118 (2022)

PIES with Trimmed Surfaces for Solving Elastoplastic Boundary Problems

Agnieszka Bołtuć(✉) and Eugeniusz Zieniuk

Institute of Computer Science, University of Bialystok, Bialystok, Poland
{aboltuc,ezieniuk}@ii.uwb.edu.pl

Abstract. The paper presents the strategy for solving elastoplastic problems using a parametric integral equation system (PIES) and a trimming technique. It allows even complex shapes of a plastic zone to be modeled with a single surface and a set of trimming curves. New schemes for integration and approximation of solutions are developed to include changed requirements. However, both of them have kept their advantages. Some examples are solved, and the obtained results are compared with analytical solutions and those received from other numerical methods.

Keywords: PIES · Elastoplastic problems · Trimmed surfaces

1 Introduction

Next to FEM [1] and BEM [2], PIES [3,4] is used for solving boundary problems e.g. elastoplastic. The main advantage of PIES, in this case, is the elimination of discretization of the predicted yield region (by cells in BEM) or the whole domain (by finite elements in FEM). Instead, the plastic zone is modeled by a surface (e.g. a Bezier surface [5]), which requires a small number of control points to be defined. Modification of the surface is also very easy because is limited to changing the positions of some control points. Moreover, the integrals are calculated globally over the whole domain, and the approximation of plastic strains is performed in the same way. This distinguishes PIES from the approaches in which Bezier surfaces are used for FEM modeling (e.g. [6]), as FEM still requires discretization into elements for numerical integration or solution approximation.

The surfaces, however, have some limitations. If the expected yield region is complex, it is difficult to deal with it using one surface. A solution to this situation is the trimming technique, which allows modeling arbitrary regions using trimming curves. The initial domain is defined by the bilinear surface and the desired shape is created by the set of trimming curves. The proposed approach has one additional advantage, very often there is no need to determine the expected shape of the plastic area, as the entire created domain is treated as potentially plastic. This prevents cases where an incorrectly predicted shape necessitates resolving the problem. The trimming technique is often used in isogeometric FEM [7,8] since the whole domain is always defined, but as is mentioned above discretization into elements is still present.

D. Groen et al. (Eds.): ICCS 2022, LNCS 13351, pp. 125–131, 2022.
https://doi.org/10.1007/978-3-031-08754-7_17

The main aim of the paper is to develop an approach for solving elastoplastic problems with any shape of plastic regions using the trimming technique. The proposed strategy requires modifying the plastic strain approximation method along with determining the necessary interpolation points and adjusting the integration method currently used in PIES. For approximation, the inverse distance weighting (IDW) method [9] is used, performed only on points designated by the projection scheme [7], while for integration a modified transformation technique [2] for calculating singular integrals is applied. Three test examples are included, with the results confirming the effectiveness of the approach.

2 PIES for Elastoplastic Problems

The PIES formula for solving 2D elastoplastic problems using initial stress formulation can be presented in the following form [3,4]

$$0.5\dot{\boldsymbol{u}}_l(\bar{s}) = \sum_{j=1}^{n} \int_{s_{j-1}}^{s_j} \left\{ \mathbf{U}_{lj}^*(\bar{s}, s)\dot{\boldsymbol{p}}_j(s) - \mathbf{P}_{lj}^*(\bar{s}, s)\dot{\boldsymbol{u}}_j(s) \right\} J_j(s)ds + \int_{\Omega} \boldsymbol{\sigma}_l^*(\bar{s}, \boldsymbol{y})\dot{\boldsymbol{\varepsilon}}^p(\boldsymbol{y})d\Omega(\boldsymbol{y}). \quad (1)$$

The fundamental solutions for displacement $\mathbf{U}_{lj}^*(\bar{s}, s)$, traction $\mathbf{P}_{lj}^*(\bar{s}, s)$ and stress $\boldsymbol{\sigma}_l^*(\bar{s}, \boldsymbol{y})$ are presented explicitly in [3,4]. The functions $\dot{\boldsymbol{u}}_j(s)$, $\dot{\boldsymbol{p}}_j(s)$ and $\dot{\boldsymbol{\varepsilon}}^p(\boldsymbol{y})$ describe the distribution of displacements, tractions on the boundary and plastic strains in the domain, respectively. Both the boundary and the domain in PIES are defined in a parametric reference system using curves and surfaces. The boundary is composed of n segments represented by any curves Γ, whose beginning and end are determined by s_{j-1} and s_j. The domain is modeled by any surface Ω. Variables s and $\bar{s}$ are parameters in the mentioned parametric reference system, $J_j(s)$ is the Jacobian, $\mathbf{y} \in \Omega$ and $l, j = 1..n$.

3 The Plastic Zone Modeled by the Trimmed Surface

In elastoplastic problems, it is not enough to define the boundary itself, as it is not able to create a domain representing a plastic area. In FEM [1] to form such a domain, regardless of the problem, the whole body is divided into finite elements (Fig. 1a). In BEM [2] only the region with predicted plastic strains is modeled by cells (Fig. 1b). It is more effective, but still requires discretization of part of the domain and it can be troublesome if the region needs to be remodeled. PIES also requires defining only the yield region, but it is represented as a whole, without discretization, using a Bezier surface. In other words, the Bezier surface is like one global element as shown in Fig. 1c. What is important, the defined initially surface does not change when solving the problem (same as in BEM), and the plastic zone is determined by yielded interpolation points at each increment.

As can be seen in Fig. 1a,b,c, the strategy used in PIES is more effective than those applied in FEM and BEM. Posing a large number of elements requires declaring many nodes and modifying them in case of any shape changes. In PIES the whole expected plastic region is defined by the small number of control points of the surface [5]. Its modification is also very simple, it is enough to change the position of individual points. The only problem that arises is the complexity of the plastic zone, especially when it cannot be modeled with a single Bezier surface. The strategy proposed in this paper is to use the trimming technique.

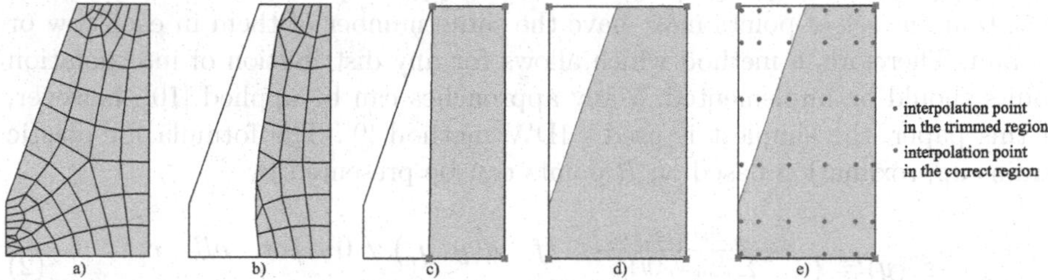

Fig. 1. Modeling in: a) FEM, b) BEM, c) PIES d) PIES with the trimmed surface and e) interpolation points arrangement in the correct and trimmed regions (Color figure online)

The main idea of the trimming technique is to use a bilinear Bezier surface, from which the expected domain is determined by trimming curves (Bezier curves in this paper). A certain orientation rule should be followed when defining them. It says that the correct region is to the left of the curve. Figure 1d presents the characteristic of the trimming technique using the geometry from Fig. 1c. As can be seen, the bilinear Bezier surface is modeled using four control points (blue squares) and the trimming curve (red line) designates the right area (grey region). Bezier curves of any degree can be used for trimming. In addition to the obvious advantage of being able to model complex domains, this strategy in many cases relieves the need to predict the shape of the plastic region, as the whole body area can be treated as potentially plastic.

4 Numerical Solving of PIES with the Trimmed Surface

To solve PIES (1), functions $\dot{\boldsymbol{u}}_j(s)$, $\dot{\boldsymbol{p}}_j(s)$ and $\dot{\boldsymbol{\varepsilon}}^p(\boldsymbol{y})$ should be found. They can be approximated by series with various base functions. In recent papers, the Lagrange polynomials were most often used for this purpose [3]. The number of expressions in the mentioned series depends on the number of assumed collocation (for displacements and tractions) or interpolation points (for plastic strains). The latter are arranged globally within the whole surface according to a predefined scheme e.g. uniformly or at places corresponding to roots of various kinds of polynomials. Using the trimming technique some of them can be outside the correct region (red points in Fig. 1e) and should not be used for the approximation of plastic strains. To determine them, a projection scheme is used [7].

For each interpolation point, the closest projection from it to the trimming curve is found. Then using the magnitude of the vector from the interpolation to the projection point and the tangential vectors of the trimming curve at the projection point the cross product is obtained. Taking into account the orientation rule, if its direction is coming out of the plane, the analyzed interpolation point is located outside of the correct region.

However, considering only some of interpolation points causes a problem with applying the Lagrange polynomials for approximation. It comes from the fact, that the set of points must have the same number of them in each row or column. Therefore, a method which allows for any distribution of interpolation points should be implemented. Many approaches can be applied [10], however, in this paper, the simplest is used - IDW method [9]. The formula for plastic strains approximation based on R points can be presented by

$$\dot{\boldsymbol{\varepsilon}}^{p}(\boldsymbol{y}) = \{ \begin{array}{ll} \frac{\sum_{r=0}^{R} \omega_r(\boldsymbol{y})\dot{\boldsymbol{\varepsilon}}^{p}(\boldsymbol{y}_r)}{\sum_{r=0}^{R} \omega_r(\boldsymbol{y})} & if \quad d(\boldsymbol{y},\boldsymbol{y}_r) \neq 0 \quad for \quad all \quad r, \\ \dot{\boldsymbol{\varepsilon}}^{p}(\boldsymbol{y}_r) \quad if \quad d(\boldsymbol{y},\boldsymbol{y}_r) = 0 \quad for \quad some \quad r, & \end{array} \tag{2}$$

where $\omega_r(\boldsymbol{y}) = \frac{1}{d(\boldsymbol{y},\boldsymbol{y}_r)^p}$ is a weighting function, d is a distance from the known $\boldsymbol{y}_r$ to the unknown $\boldsymbol{y}$ point and p is a power parameter. To predict a value for any unmeasured location, IDW uses the measured values from a neighborhood of influence. In this paper it is determined by a circle with a radius r and the center at the unmeasured point.

The next step in solving PIES is to substitute the expression (2) into the formula (1) and to calculate the integrals, before applying the collocation method. The strategy used for boundary integrals is the same as for elastic problems, and the domain integrals are calculated over the whole surface using a higher-order quadrature [3]. Unfortunately, this time the strategy has to be modified since a part of the domain is trimmed. In elastoplastic problems, some singularities in domain integrals appear. The last integral from (1) is weakly singular, but this singularity can be canceled by employing the transformation technique [2,3], in which the surface is divided into triangles at a point of singularity. The same technique is used for calculating strongly singular integrals in the stress integral identity [3]. It can be also, after some modifications, a direct solution to the problem of integration over the trimmed surface. Figure 2 presents the way of division into triangles in the original version of PIES and with trimmed surfaces depending on the location of the singular point.

Figure 2 shows that instead of using corner points of the surface, the real vertices of the considered body are used. The division takes place in the domain of the surface (unit square), therefore vertices should be recalculated to that parameter space. It can be done using formulas describing the surface by a point inversion algorithm, which in this paper is implemented only for the bilinear surface. For more complicated cases (e.g. trimming curves form a concave boundary), another strategy may be required. The initial geometry is divided into the smallest number of triangles using existing vertices. They are treated as separate surfaces and within them, the transformation technique is applied.

It should be emphasized, that the approximation formula (2) remains global over the trimmed surface. Moreover, if the trimming causes there are too few

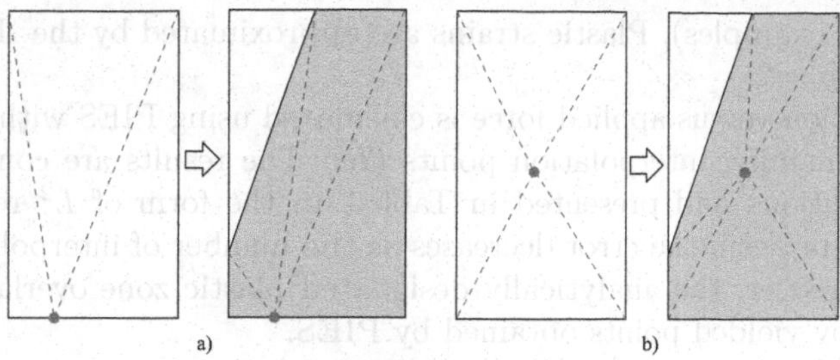

Fig. 2. The transformation technique with and without the trimmed surface: a) for collocation point, b) for interpolation point

interpolation points around the trimming curve, they can be easily generated according to the selected distribution along the given curve.

5 Examples

The first example is selected for initial verification as it has an analytical solution [11], although it could be solved without the trimming technique. The cantilever beam (Fig. 3a) is end-loaded, defined as plane stress with the material parameters: $E = 2 * 10^{11} Pa$, $\nu = 0.25$, $\sigma_0 = 20Pa$ and $H = 0$. The von Mises yield criterion is assumed (like in the remaining examples).

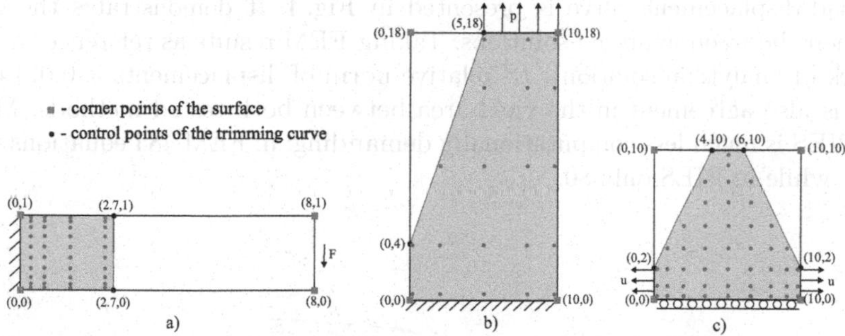

Fig. 3. The considered trimmed geometry for: a) first, b) second and c) third example

The whole considered domain is modeled by the bilinear surface and the assumed plastic zone (a part of that surface) is declared by the linear trimming curve. Initially, interpolation points (36,64,100) are arranged globally at the roots of Chebyshev polynomials of the first kind (this also applies to the remaining examples), but finally, the trimming curve and the projection scheme eliminate some of them (Fig. 3a). The area designated by the trimming curve is divided into triangles for calculating singular integrals (the same technique is used in the remaining examples). Plastic strains are approximated by the IDW method with $r = 0.4$, $p = 2$.

Tip deflection versus applied force is calculated using PIES with the various number of remaining interpolation points (R). The results are compared with analytical solutions and presented in Table 1 in the form of L^2 relative error norm. As can be seen, the error decreases as the number of interpolation points increases. Moreover, the analytically designated plastic zone overlaps with the area defined by yielded points obtained by PIES.

Table 1. L^2 error norm for tip deflection for various number of interpolation points

R	12	24	40
L^2	0.051991	0.045149	0.038378

The next considered geometry together with applied boundary conditions is presented in Fig. 3b. The domain of the problem is modeled by the bilinear surface and one linear trimming curve. The plane stress conditions with the following material properties are considered: $E = 70000\,\text{MPa}$, $\nu = 0.2$, $\sigma_0 = 150\,\text{MPa}$ and $H' = 0$. Within the initial surface, 25 interpolation points are arranged. Some of them are finally not inside the plastic region, therefore they are eliminated by the projection scheme. The distribution of the remaining 20 interpolation points is shown in Fig. 3b. IDW is used with $r = 5$ and $p = 2$.

The displacements at the point $(10, 18)$ are obtained by PIES and FEM. The load-displacement curve is presented in Fig. 4. It demonstrates the good agreement between analyzed solutions. Taking FEM results as reference (due to the lack of analytical solution), L^2 relative norm of displacements is 0.0214231. There is also agreement in the yield area between both tested methods. Moreover, PIES is much less computationally demanding: in FEM 484 equations were solved, while in PIES only 80.

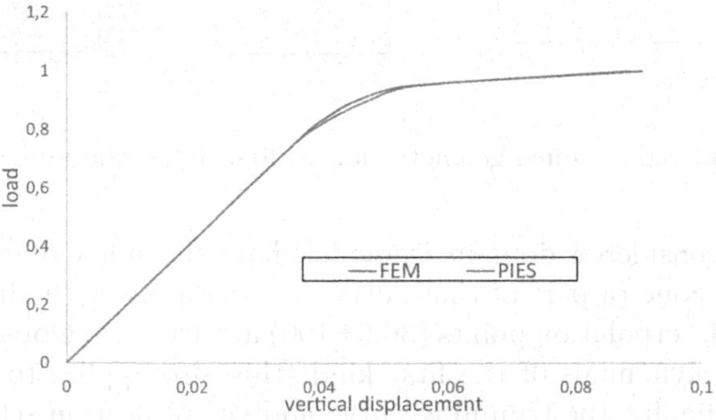

Fig. 4. The load-displacement curve

The last example concerns the geometry under the imposed displacement modeled using the surface trimmed by two curves (Fig. 3c). The perfect plasticity under plane stress state with $E = 1\,\text{MPa}$, $\nu = 0.3$, $\sigma_0 = 0.9\,\text{MPa}$ is assumed. 47 of the globally placed points remain in the correct area. The IDW parameters are $r = 5$, $p = 2$.

The von Mises equivalent stresses σ_{VM} along the cross-section $x = 5$ are obtained. Due to the lack of analytical solution, they are compared with FEM results and L^2 norm is calculated. It equals 0.0229775 which means that the results are very similar, but computational requirements are much smaller for PIES than for FEM (PIES 84, FEM 586 equations).

6 Conclusions

The paper presents PIES with trimmed surfaces for solving elastoplastic problems. The domain is modeled by a surface and its unnecessary part is trimmed by curves. The approximation of the plastic strains is performed by the IDW method, which allows for any arrangement of interpolation points. Finally, the domain integrals are calculated over the trimmed surface using the modified division technique that cancels out the singularity in the integrand. In more complicated cases, the earlier division into multiple surfaces may be required.

The proposed approach is tested on three examples with various numbers of trimming curves. The obtained results are in good agreement with analytical and FEM solutions. Further research is required, on more complex examples, e.g. with holes or a curved boundary.

References

1. Zienkiewicz, O.C.: The Finite Element Methods. McGraw-Hill, London (1977)
2. Aliabadi, M.H.: The Boundary Element Method. Applications in Solids and Structures, vol. 2. Wiley, Chichester (2002)
3. Bołtuć, A.: Elastoplastic boundary problems in PIES comparing to BEM and FEM. Comput. Math. Appl. **72**(9), 2343–2363 (2016)
4. Bołtuć, A.: 2D elastoplastic boundary problems solved by PIES without strongly singular surface integrals. Eur. J. Mech. A-Solid **65**, 233–242 (2017)
5. Salomon, D.: Curves and Surfaces for Computer Graphics. Springer, New York (2006). https://doi.org/10.1007/0-387-28452-4
6. Czarny, O., Huysmans, G.: Bézier surfaces and finite elements for MHD simulations. J. Comput. Phys. **227**(16), 7423–7445 (2008)
7. Hyun-Jung, K., Yu-Deok, S., Sung-Kie, Y.: Isogeometric analysis for trimmed CAD surfaces. Comput. Method Appl. Mech. Eng. **198**, 2982–2995 (2009)
8. Marussig, B., Hughes, T.J.R.: A review of trimming in isogeometric analysis: challenges, data exchange and simulation aspects. Arch. Comput. Method E **25**,1059–1127 (2018)
9. Shepard, D.: A two-dimensional interpolation function for irregularly-spaced data. In: Proceedings of the 1968 ACM National Conference, pp. 517–524. Association for Computing Machinery, USA (1968)
10. Bołtuć, A., Zieniuk, E.: PIES for 2D elastoplastic problems with singular plastic strain fields. Comput. Math. Appl. **103**, 53–64 (2021)
11. Lubliner, J.: Plasticity Theory. Macmillan Publishing Company, New York (1990)

Linear Computational Cost Implicit Variational Splitting Solver with Non-regular Material Data for Parabolic Problems

Paweł Maczuga[1], Maciej Paszyński[1(✉)], and Victor Calo[2]

[1] AGH University of Science and Technology, Kraków, Poland
maciej.paszynski@agh.edu.pl
[2] Curtin University, Perth, Australia

Abstract. We employ a variational splitting for the Crank-Nicolson method and Pennes bioheat equation modeling the heating of the human head as a result of the cellphone antenna radiation. The solution of the system of equations resulting from the 3D discretization of the implicit time integration scheme with the Crank-Nicolson method has $\mathcal{O}(N^2)$ complexity using direct solver, resulting in the exact solution. Iterative solvers (e.g., multi-grid solvers) deliver $\mathcal{O}(Nk)$ computational cost resulting in an approximate solution. The alternating direction implicit solver delivers $\mathcal{O}(N)$ complexity instead; it provides the exact solution (as the direct solver). Still, it requires a regular tensor product structure of the material data. In this paper, we propose a method for generalizing the linear computational cost alternating direction implicit solver using the Crank-Nicolson scheme into non-regular material data.

Keywords: Pennes problem · Variational splitting · Implicit method · Non-regular material data · Linear computational cost

1 Introduction

Splitting methods modify the original linear systems of equations seeking to reduce computation costs [11]. The methods have been originally proposed for finite differences [8,9] and later generalized for isogeometric analysis (IGA) [3]. We focus on parabolic equations discretized in space using tensor-product IGA grid. We adopt an implicit time integration scheme based on the Crank-Nicolson method. We employ the Kronecker product decomposition of the matrix $\mathcal{M} = \mathcal{M}_x \otimes \mathcal{M}_y \otimes \mathcal{M}_z$, that allows for linear cost factorization $\mathcal{M}^{-1} = \mathcal{M}_x^{-1} \otimes \mathcal{M}_y^{-1} \otimes \mathcal{M}_z^{-1}$. The results detailed in [1,5–7] show that the Kronecker product-based solvers result in a linear cost for every time step. They offer an attractive alternative for multi-frontal solver algorithms [2]. Splitting solvers require the regular tensor product structure of the material data. We propose a method for generalizing the solvers into non-regular material data, preserving

D. Groen et al. (Eds.): ICCS 2022, LNCS 13351, pp. 132–138, 2022.
https://doi.org/10.1007/978-3-031-08754-7_18

the linear cost. We present exemplary numerical results of the heating of the human head based on a non-regular MRI scan and Pennes equations [10], and the cellphone radiation data available from [4]. Our method can be employed to a broad spectrum of applications of the variational-splitting solvers [1,5,6].

2 Varying Coefficients in Alternating Directions Solver

Lemma 1. *It is possible to factorize in a linear $\mathcal{O}(N)$ computational cost a system of equations, resulting from the problem $\frac{\partial u(x,y;t)}{\partial t} - \nabla\left(\epsilon(x,y;t)\nabla u(x,y;t)\right) = f(x,y;t)$ discretized in space with B-spline basis functions and in time with Crank-Nicolson implicit scheme, having the non-regular material data $\epsilon(x,y;t)$.*

Proof. We discretize and test with B-splines. We however cut each one into N_qN_q pieces (see Fig. 1), so we have a total of $\{\mathcal{I}^x_k(x)\mathcal{I}^y_l\}_{k=1,\ldots,N_qN_x;l=1,\ldots,N_qN_y}$ test functions. In such a way we obtain a weak formulation. Each equation with $B^x_mB^y_n$ test function is replaced by N_qN_q equations with suitable test functions $\{\mathcal{I}^x_k(x)\mathcal{I}^y_l\}_{k=1,\ldots,N_q;l=1,\ldots,N_q}$. In the limit of N_q the integrals equations converge

$$\sum_{k=1,\ldots,N_q;l=1,\ldots,N_q} \int \mathcal{F}(x,y)\mathcal{I}^x_k\mathcal{I}^y_l dxdy \xrightarrow[N_q\to\infty]{} \int \mathcal{F}(x,y)B^x_mB^y_n dxdy \quad (1)$$

With our partitioned test functions, *we assume* that each equation can be approximated by one quadrature point. We generate a *rectangular* system of equations, with $N\times N_qN_q$ equations and N unknowns, repeated for N_qN_q sets of N equations (see Fig. 1). Later, we will replace N equations from NN_qN_q system of equations by a linear combination of N_qN_q equations to obtain quadratic $N\times N$ system. We will glue the test functions $B^x_mB^y_n$ back together from the pieces $\{\mathcal{I}^x_k\mathcal{I}^y_l\}_{k=1,\ldots,N_q,l=1,\ldots,N_q}$. The left-hand side is of the following form

$$LHS = \sum_{i,j}\Big(\int B^x_iB^y_j\mathcal{I}^x_k\mathcal{I}^y_l + \eta\int \epsilon(x,y)\partial_xB^x_iB^y_j\partial_x\mathcal{I}^x_k\mathcal{I}^y_l$$
$$+\eta\int \epsilon(x,y)B^x_i\partial_yB^y_j\mathcal{I}^x_k\partial_y\mathcal{I}^y_l\Big)\mathcal{U}_{i,j} \quad \forall k=1,\ldots,N_xN_q; l=1,\ldots,N_yN_q \quad (2)$$

and the right-hand side terms, one for each test function $\mathcal{I}^x_k\mathcal{I}^y_l$, are derived from the Crank-Nicolson scheme. We introduce quadrature points (ξ_k,ξ_l), one for each test function $\mathcal{I}^x_k(x)\mathcal{I}^y_l(y)$, and the Kronecker product approximation of the left-hand side, ignoring terms of higher order with respect to η

$$LHS = \sum_{i=1,\ldots,N_x;j=1,\ldots,N_y} w_kw_l\left(B^x_i(\xi_k)\mathcal{I}^x_k(\xi_k)B^y_j(\xi_l)\mathcal{I}^y_l(\xi_l)\right.$$
$$+\eta\epsilon(\xi_k,\xi_l)B^x_i(\xi_k)\mathcal{I}^x_k(\xi_k)\partial_yB^y_j(\xi_l)\partial_y\mathcal{I}^y_l(\xi_l)$$
$$\left.+\eta\epsilon(\xi_k,\xi_l)\partial_xB^x_i(\xi_k)\partial_x\mathcal{I}^x_k(\xi_k)B^y_j(\xi_l)\mathcal{I}^y_l(\xi_l)\right)\mathcal{U}_{i,j} \quad (3)$$

$$\approx \sum_{i=1,...,N_x;j=1,...,N_y} w_k w_l (B^x_i(\xi_k)\mathcal{I}^x_k(\xi_k) + \eta\epsilon_{k,l}\partial_x B^x_i(\xi_k)\partial_x \mathcal{I}^x_k(\xi_k))$$
$$(B^y_j(\xi_l)\mathcal{I}^y_l(\xi_l) + \eta\epsilon_{k,l}\partial_y B^y_j(\xi_l)\partial_y \mathcal{I}^y_l(\xi_l))\,\mathcal{U}_{i,j} = RHS$$
$$\forall k = 1, ..., N_x N_q; l = 1, ..., N_y N_q \quad (4)$$

Notice that $\epsilon_{k,l} = \epsilon(\xi_k, \xi_l)$ is a given constant value of material data at the quadrature point defined for a test function $\mathcal{I}^x_k \mathcal{I}^y_l$.

$$\hat{\mathcal{A}}_k = w_k \times$$
$$\begin{bmatrix} (B^x_1(\xi_1)\mathcal{I}^x_1(\xi_1) + \eta\epsilon_{1,k}\partial_x B^x_1(\xi_1)\partial_x \mathcal{I}^x_1(\xi_1)) \cdots \\ \cdots (B^x_{N_x}(\xi_1)\mathcal{I}^x_1(\xi_k) + \eta\epsilon_{1,1}\partial_x B^x_{N_x}(\xi_1)\partial_x \mathcal{I}^x_1(\xi_1)) \\ \cdots \\ \left(B^x_1(\xi_{N_xN_q})\mathcal{I}^x_{N_xN_q}(\xi_{N_xN_q}) + \eta\epsilon_{N_xN_q,k}\partial_x B^x_1(\xi_{N_xN_q})\partial_x \mathcal{I}^x_{N_xN_q}(\xi_{N_xN_q})\right) \cdots \\ \cdots \left(B^x_{N_x}(\xi_{N_xN_q})\mathcal{I}^x_{N_xN_q}(\xi_{N_xN_q}) + \eta\epsilon_{N_xN_q,k}\partial_x B^x_{N_x}(\xi_{N_xN_q})\partial_x \mathcal{I}^x_{N_xN_q}(\xi_{N_xN_q})\right) \end{bmatrix} \quad (5)$$

for each $k = 1, ..., N_y N_q$. Here, to save space, we have partitioned the first and the last row of the matrix into two consecutive rows. The size of $\mathcal{A}_k$ is $N_x N_q \times N_x$.

$$\hat{\mathcal{B}}_{l,j} = w_l \begin{bmatrix} (B^y_j(\xi_l)\mathcal{I}^y_l(\xi_l) + \eta\hat{\epsilon}_{1,l}\partial_y B^y_j(\xi_l)\partial_y \mathcal{I}^y_l(\xi_l)) & \cdots & 0 \\ & \cdots & \\ 0 & \cdots & (B^y_j(\xi_l)\mathcal{I}^y_l(\xi_l) + \eta\hat{\epsilon}_{N_x,l}\partial_y B^y_j(\xi_l)\partial_y \mathcal{I}^y_l(\xi_l)) \end{bmatrix} \quad (6)$$

for each $l = 1, ..., N_y N_q$, $j = 1, ..., N_y$. The size of each $\mathcal{B}_{l,j}$ block is $N_x \times N_x$. Using this notation, see Fig. 1, we can rewrite our system in the following form $\mathcal{ABU} = \mathcal{F}$

$$\begin{bmatrix} \hat{\mathcal{A}}_1 \cdots & 0 \\ \cdots\cdots & \cdots \\ 0 & \cdots \hat{\mathcal{A}}_{N_yN_q} \end{bmatrix} \begin{bmatrix} \hat{\mathcal{B}}_{1,1} & \cdots & \hat{\mathcal{B}}_{1,N_y} \\ \cdots & \cdots & \cdots \\ \hat{\mathcal{B}}_{N_yN_q,1} & \cdots & \hat{\mathcal{B}}_{N_yN_q,N_y} \end{bmatrix} \begin{bmatrix} \mathcal{U}_{1,1} \\ \cdots \\ \mathcal{U}_{N_x,N_y} \end{bmatrix} = \begin{bmatrix} \hat{\mathcal{F}}_{1,1} \\ \cdots \\ \hat{\mathcal{F}}_{N_xN_q,N_yN_q} \end{bmatrix} \quad (7)$$

where $\hat{\mathcal{F}}_{k,l} = \int \mathcal{RHS}_{k,l}(x,y)\mathcal{I}^x_k \mathcal{I}^y_j$ for $k = 1, ..., N_x N_q$, $l = 1, ..., N_y N_q$, and $\mathcal{RHS}_{k,l}$ is defined according to the Crank-Nicolson employed time integration scheme. $\mathcal{A}$ matrix has dimension $N_x N_q N_y N_q \times N_x N_y N_q$, $\mathcal{B}$ matrix has dimension $N_x N_q N_y \times N_x N_y$, $\mathcal{U}$ has dimension $N_x N_y$ and $\mathcal{F}$ have dimensions $N_x N_q N_y N_q$.

$$\begin{bmatrix} \hat{\mathcal{G}}_1 \\ \cdots \\ \hat{\mathcal{G}}_{N_x,N_yN_q} \end{bmatrix} = \begin{bmatrix} \hat{\mathcal{B}}_{1,1} & \cdots & \hat{\mathcal{B}}_{1,N_y} \\ \cdots & \cdots & \cdots \\ \hat{\mathcal{B}}_{NyNq,1} & \cdots & \hat{\mathcal{B}}_{N_yN_q,N_y} \end{bmatrix} \begin{bmatrix} \mathcal{U}_{1,1} \\ \cdots \\ \mathcal{U}_{N_x,N_y} \end{bmatrix} \quad (8)$$

$$\begin{bmatrix} \hat{\mathcal{A}}_1 \cdots & 0 \\ \cdots\cdots & \cdots \\ 0 & \cdots \hat{\mathcal{A}}_{N_yN_q} \end{bmatrix} \begin{bmatrix} \hat{\mathcal{G}}_1 \\ \cdots \\ \hat{\mathcal{G}}_{N_x,N_yN_q} \end{bmatrix} = \begin{bmatrix} \hat{\mathcal{F}}_1 \\ \cdots \\ \hat{\mathcal{F}}_{N_xN_q,N_yN_q} \end{bmatrix} \quad (9)$$

We sum up blocks of N_q rows to recover B-spline test functions along x

$$\begin{bmatrix} \mathcal{A}_1 \cdots & 0 \\ \ldots \ldots & \ldots \\ 0 & \cdots \mathcal{A}_{N_y} \end{bmatrix} \begin{bmatrix} \hat{\mathcal{G}}_1 \\ \ldots \\ \hat{\mathcal{G}}_{N_x,N_yN_q} \end{bmatrix} = \begin{bmatrix} \mathcal{F}_1 \\ \ldots \\ \mathcal{F}_{N_x,N_yN_q} \end{bmatrix} \tag{10}$$

$$\mathcal{A}_k = w_k \begin{bmatrix} (B^x_1(\xi_k)B^x_1(\xi_k) + \eta\hat{\epsilon}_{1,k}\partial_x B^x_1(\xi_k)\partial_x B^x_1(\xi_k)) \quad \cdots \\ \cdots \quad (B^x_{N_x}(\xi_k)B^x_1(\xi_k) + \eta\hat{\epsilon}_{1,k}\partial_x B^x_{N_x}(\xi_k)\partial_x B^x_1(\xi_k)) \\ \cdots \\ (B^x_1(\xi_k)B^x_{N_x}(\xi_k) + \eta\hat{\epsilon}_{N_x,k}\partial_x B^x_1(\xi_k)\partial_x B^x_{N_x}(\xi_k)) \quad \cdots \\ \cdots \quad (B^x_{N_x}(\xi_k)B^x_{N_x}(\xi_k) + \eta\hat{\epsilon}_{N_x,k}\partial_x B^x_{N_x}(\xi_k)\partial_x B^x_{N_x}(\xi_k)) \end{bmatrix} \tag{11}$$

for each $k = 1, ..., N_x$. Here, again, to save space, we have partitioned the first and the last row of the matrix into two consecutive rows. Here

$$\begin{gathered} (B^x_i(\xi_k)B^x_j(\xi_k) + \eta\hat{\epsilon}_{i,k}\partial_x B^x_i(\xi_k)\partial_x B^x_j(\xi_k)) \approx \\ \sum_{r=1,\ldots,N_q} (B^x_i(\xi_k)\mathcal{I}^x_r(\xi_k) + \eta\epsilon_{r,k}\partial_x B^x_i(\xi_k)\partial_x \mathcal{I}^x_r(\xi_k)) \end{gathered} \tag{12}$$

for $i = 1, ..., N_x$ and in particular $\hat{\epsilon}_{i,k} = \frac{\sum_{r=1,\ldots,N_q}\epsilon_{r,k}}{N_q}$ is selected in such a way that these sums are as close as possible. Moreover $\mathcal{F}_{k,l} = \int \mathcal{RHS}_{k,l}(x,y)B^x_k B^y_l$ for $k = 1, ..., N_x$, $l = 1, ..., N_yN_q$. Notice, that this summation of rows does not change the number of unknowns, which is related to the number of columns in matrix $\mathcal{A}$. We solve this system for $\left[\hat{\mathcal{G}}_1 \cdots \hat{\mathcal{G}}_{N_x,N_yN_q}\right]^T$. Next, we take

$$\begin{bmatrix} \hat{\mathcal{B}}_{1,1} & \cdots & \hat{\mathcal{B}}_{1,N_y} \\ \cdots & \cdots & \cdots \\ \hat{\mathcal{B}}_{NyNq,1} & \cdots & \hat{\mathcal{B}}_{N_yN_q,Ny} \end{bmatrix} \begin{bmatrix} \mathcal{U}_{1,1} \\ \cdots \\ \mathcal{U}_{N_x,N_y} \end{bmatrix} = \begin{bmatrix} \hat{\mathcal{G}}_1 \\ \cdots \\ \hat{\mathcal{G}}_{N_x,N_yN_q} \end{bmatrix} \tag{13}$$

and we sum-up N_q rows to recover the full B-splines along y

$$\begin{bmatrix} \mathcal{B}_{1,1} & \cdots & \mathcal{B}_{1,N_y} \\ \cdots & \cdots & \cdots & \cdots \\ \mathcal{B}_{N_y,1} & \cdots & \mathcal{B}_{N_y,N_y} \end{bmatrix} \begin{bmatrix} \mathcal{U}_{1,1} \\ \cdots \\ \mathcal{U}_{N_x,N_y} \end{bmatrix} = \begin{bmatrix} \mathcal{G}_1 \\ \cdots \\ \mathcal{G}_{N_x,N_y} \end{bmatrix} \tag{14}$$

$$\mathcal{B}_{l,j} = \begin{bmatrix} \left(\int_y B^y_1(\xi_l)B^y_l(\xi_l) + \int_y \eta\tilde{\epsilon}_{1,l}\partial_y B^y_j(\xi_l)\partial_y B^y_l(\xi_l)\right) & \cdots & 0 \\ \cdots \\ 0 \quad \cdots \quad \left(\int_y B^y_j(\xi_l)B^y_l(\xi_l) + \int_y \eta\tilde{\epsilon}_{N_x,l}\partial_y B^y_j(\xi_l)\partial_y B^y_l(\xi_l)\right) \end{bmatrix} \tag{15}$$

for each $l = 1, ..., N_y$, $j = 1, ..., N_y$. Here

$$\begin{gathered} (B^y_j(\xi_l)B^y_j(\xi_l) + \eta\tilde{\epsilon}_{j,l}\partial_y B^x_j(\xi_l)\partial_y B^y_j(\xi_l)) \approx \\ \sum_{r=1,\ldots,N_q} (B^y_j(\xi_l)\mathcal{I}^x_r(\xi_l) + \eta\hat{\epsilon}_{r,l}\partial_y B^y_j(\xi_l)\partial_y \mathcal{I}^x_r(\xi_l)) \end{gathered} \tag{16}$$

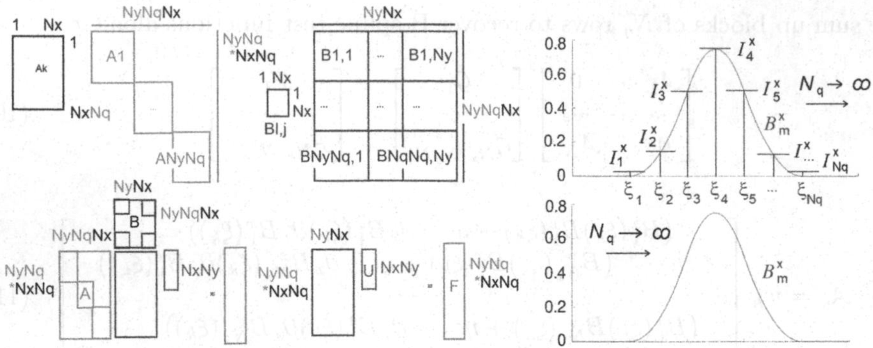

Fig. 1. Partitioning of a matrices. Partition of a test function B^x_m.

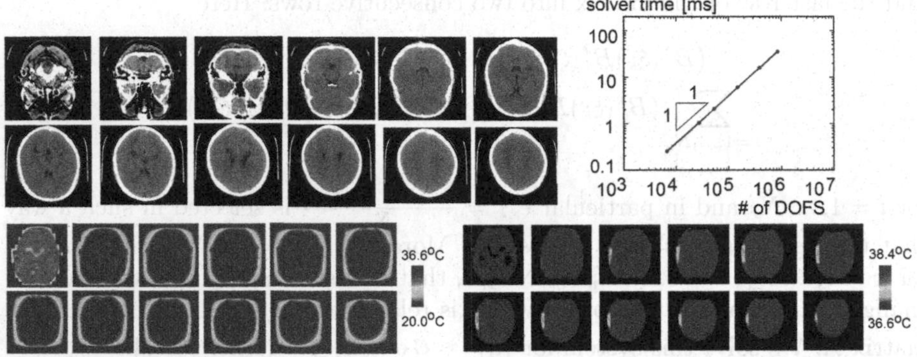

Fig. 2. (First row) Slices of the MRI scan of the human head. Linear cost of the solver. (Second row) The temperature of the head without the cell-phone antenna. The temperature of the head with the cell-phone antenna radiation.

Table 1. Material data used in simulation.

Material	Air	Brain	Skull	Material	Air	Brain	Skull
$\rho\ [kg/m^3]$	1.16	1039	1645	$q_m\ [W/m^3]$	0	7100	590
$c\ [J/kg^oC]$	1006	3700	1300	$W_b c_b\ [W/m^{3o}C]$	0	40000	3300
$K\ [W/m^oC]$	0.02	0.57	0.4	$u_{a0}\ [^oC]$	20	36.6	36.6

for $j = 1, ..., N_y$ and in particular $\tilde{e}_{j,l} = \frac{\sum_{r=1,...,N_q} \hat{e}_{r,l}}{N_q}$ is selected in such a way that these sums are as close as possible. Finally, we solve for $\mathcal{U}_1, ..., \mathcal{U}_{N_x,N_y}$

$$\begin{bmatrix} \mathcal{B}_1 \cdots \mathcal{B}_{N_y} \\ \cdots\cdots\ \cdots\ \cdots \\ \mathcal{B}_1 \cdots \mathcal{B}_{N_y} \end{bmatrix} \begin{bmatrix} \mathcal{U}_1 \\ \cdots \\ \mathcal{U}_{N_x N_y} \end{bmatrix} = \begin{bmatrix} \mathcal{G}_1 \\ \cdots \\ \mathcal{G}_{N_x N_y} \end{bmatrix} \tag{17}$$

Both systems (10) and (17) can be solved in a linear computational cost due to the banded structures of matrices build with one-dimensional B-splines.

3 Variational Splitting for Pennes Bioheat Equation

In this example, we start from the MRI scan of the head of Maciej Paszyński, transformed into a 3D bitmap. We formulate the Pennes bio-heat equation

$$\rho c\frac{\partial u}{\partial t} - \nabla\cdot(K\nabla u) = W_b c_b\,(u_{a0} - u) + q_m + q_{SAR} \text{ in } \Omega \tag{18}$$

where K represents the thermal conductivity, q_m is the metabolism, $W_b c_b$ stands for the perfusion, and q_{SAR} is the heat source from the cellphone. We employ the parameters from Table 1, following [4]. They are selected according to the MRI scan data. We employ the alternating-direction solver with varying material data. We derive the weak form with the Crank-Nicolson scheme; we discretize with B-splines, and we approximate the left-hand side with Kronecker product form ignoring the terms which are of the higher-order with respect to τ

$$\sum_{i,j,k}\left(\int_{\Omega_x} B_i^x B_m^x - \frac{1}{\rho c}\tau\frac{K}{2}\partial_x B_i^x \partial_x B_m^x\right)$$

$$\left(\int_{\Omega_y} B_j^y B_n^y - \frac{1}{\rho c}\tau\frac{K}{2}\partial_y B_j^y \partial_y B_n^y\right)\left(\int_{\Omega_z} B_k^z B_l^z - \frac{1}{\rho c}\tau\frac{K}{2}\partial_z B_k^z \partial_z B_l^z\right) u_{ijk}^{t+1}$$

$$= \sum_{i,j,k}\int_{\Omega} B_i^x B_j^y B_k^z B_m^x B_n^y B_l^z u_{ijk}^t + \frac{\tau}{\rho c}\int_{\Omega}\frac{K}{2}\sum_{i,j,k}\nabla B_i^x B_j^y B_k^z u_{ijk}^t\cdot\nabla(B_m^x B_n^y B_l^z)$$

$$+\frac{\tau}{\rho c}\int_{\Omega}(W_b[c_b u_{a0} - c_b\sum_{i,j,k} B_i^x B_j^y B_k^z u_{ijk}^t] + u_t + q_m + q_{SAR})B_m^x B_n^y B_l^z \tag{19}$$

This allows applying Lemma 1 generalized for 3D. We assume that the air is located where the intensity of the bitmap is ≤ 1, the skin or brain (tissue in general) where the intensity is in the range of $(1, 240)$ and the skull, where the intensity is ≥ 240. As the initial condition, we select the temperature of 36.6 C of the human head and 20.0 C of the air. We assume first no additional heat source and no presence of the cell phone antenna radiation, $q_{SAR} = 0$. The cross-sections of the 3D mesh after 10 min of the simulation are presented in Fig. 2. As denoted by the red color, the maximum temperature is 36.6 C. The blue color outside the head represents the air with a temperature of 20.0 C. Next, we assume the q_{SAR} as estimated in Fig. 6.13 [4]. The resulting heating of the human head after 10 min of the radiation is illustrated in Fig. 2. As denoted by the red color, the maximum temperature is 38.4°C. The blue color of the human head represents the temperature of 36.6 C. We illustrate the linear computational cost of the solver in Fig. 2. We use computational grids of size $16 \times 16 \times 16$, which using cubic B-splines results in $N = (16+3)^3 = 6,859$, then $24\times 24\times 24$, which results in $N = (24 + 3)^3 = 19,683$, then $(32+3)^3 = 42,875$, then $(48+3)^3 = 132,651$, then $(64 + 3)^3 = 300,763$ and finally $(96 + 3)^3 = 970,299$. The computations are performed on the Linux workstation with a 2.4 GHz processor with 64 GB of RAM. The computational burden related to distinguishing different material data in comparison to homogeneous material is negligibly small. Our solver is a direct solver, and it provides the exact solution.

Conclusion. We can vary material data in implicit variational splitting solvers, preserving the linear cost. We can vary the material data $\epsilon_{k,l}$ at quadrature points. In the solver, we average them along lines parallel to the axis of the coordinate system for each support of the test functions. We test the method on the Pennes bioheat equation, and we verify the linear cost of the solver.

Acknowledgement. National Science Centre, Poland grant no. 2017/26/M/ST1/00281. Research project partly supported by program "Excellence initiative - research university" for the University of Science and Technology. The research presented in this paper was partially supported by the funds of Polish Ministry of Education and Science assigned to AGH University of Science and Technology. The European Union's Horizon 2020 Research and Innovation Program of the Marie Skłodowska-Curie grant agreement No. 777778 provided additional support.

References

1. Behnoudfar, P., Calo, V.M., Deng, Q., Minev, P.D.: A variationally separable splitting for the generalized-α method for parabolic equations. Int. J. Numer. Meth. Eng. **121**(5), 828–841 (2020)
2. Calo, V., Collier, N., Pardo, D., Paszyński, M.: Computational complexity and memory usage for multi-frontal direct solvers used in p finite element analysis. Procedia Comput. Sci. **4**, 1854–1861 (2011)
3. Gao, L., Calo, V.M.: Fast isogeometric solvers for explicit dynamics. Comput. Methods Appl. Mech. Eng. **274**, 19–41 (2014)
4. Kyunogjoo, K.: Finite element modeling of the radiation and induced heat transfer in the human body. Ph.D. dissertation, The University of Texas at Austin (2013)
5. Łoś, M., Munoz-Matute, J., Muga, I., Paszyński, M.: Isogeometric residual minimization method (iGRM) with direction splitting for non-stationary advection-diffusion problems. Comput. Math. Appl. **79**(2), 213–229 (2020)
6. Łoś, M., Paszyński, M., Kłusek, A., Dzwinel, W.: Application of fast isogeometric l2 projection solver for tumor growth simulations. Comput. Methods Appl. Mech. Eng. **316**, 1257–1269 (2017)
7. Łoś, M., Woźniak, M., Paszyński, M., Dalcin, L., Calo, V.M.: Dynamics with matrices possessing Kronecker product structure. Procedia Comput. Sci. **51**, 286–295 (2015)
8. Peaceman, D.W., Rachford, H.H., Jr.: The numerical solution of parabolic and elliptic differential equations. J. Soc. Ind. Appl. Math. **3**(1), 28–41 (1955)
9. Samarskii, A.A.: The Theory of Difference Schemes, vol. 240. CRC Press (2001)
10. Schaefer, R., Los, M., Sieniek, M., Demkowicz, L.F., Paszyński, M.: Quasi-linear computational cost adaptive solvers for three dimensional modeling of heating of a human head induced by cell phone. J. Comput. Sci. **11**, 163–174 (2015)
11. Sportisse, B.: An analysis of operator splitting techniques in the stiff case. J. Comput. Phys. **161**(1), 140–168 (2000)

A Hadamard Matrix-Based Algorithm to Evaluate the Strength of Binary Sequences

Amparo Fúster-Sabater[1(✉)], Verónica Requena[2], and Sara D. Cardell[3]

[1] Instituto de Tecnologías Físicas y de la Información, C.S.I.C., Madrid, Spain
amparo@iec.csic.es

[2] Departament de Matemàtiques, Universitat d'Alacant, Alacant, Spain
vrequena@ua.es

[3] Centro de Matemática, Computação e Cognição, Universidade Federal do ABC (UFABC), Santo André, SP, Brazil
s.cardell@ufabc.edu.br

Abstract. Nowadays, a wide range of critical services relies on Internet of Things (IoT) devices. Nevertheless, they often lack proper security, becoming the gateway to attack the whole system. IoT security protocols are often based on stream ciphers, where pseudo-random number generators (PRNGs) are an essential part of them. In this work, we introduce a novel algorithm based on Hadamard matrices to evaluate the strength (unpredictability) of binary sequences, a key part of the IoT security stack. A comparative study with other algorithms that compute the same parameter is also presented.

Keywords: Hadamard matrix · Unpredictability · PRNG · IoT

1 Introduction

Nowadays, diverse critical services such as smart-grid, e-health, e-govern or industrial automation depend on an IoT infrastructure. At any rate, as the services around IoT grow dramatically so do the security risks [3]. Low-cost IoT devices, currently characterized by their resource constrains in processing power, memory, size and energy consumption, are also characterized by their minimum security. Combining lack of security with network dependability, they become the perfect gateway to compromise the whole network. This is the reason why 5G related research [8] or specific calls such as that of NIST for lightweight cryptography primitives [10], are addressing this concerning topic. In brief, lightweight cryptography as well as stream ciphers (binary sequence generators) are the key stones for designing security protocols.

This work was supported in part by the Spanish State Research Agency (AEI) of the Ministry of Science and Innovation (MICINN), project P2QProMeTe (PID2020-112586RB-I00/AEI/10.13039/501100011033), co-funded by the European Regional Development Fund (ERDF, EU).

D. Groen et al. (Eds.): ICCS 2022, LNCS 13351, pp. 139–145, 2022.
https://doi.org/10.1007/978-3-031-08754-7_19

In this work, we propose a novel algorithm based on the fractal structure of the Hadamard matrices that analyses the unpredictability (linear complexity) of binary sequences with application in cryptography. Finally, we discuss a comparison among different algorithms that measure the same feature.

2 Preliminaries

Basic notation and concepts are now introduced.

Binary Sequences: let $\{u_n\}_{n\geq 0} = \{u_0, u_1, u_2, \ldots\}$ be a binary sequence with $u_n \in \mathbb{F}_2$. Here we will just consider binary sequences with period a power of 2.

Linear Feedback Shift Register (LFSR): an LFSR is an electronic device with L interconnected memory cells (stages) with binary content. Maximum-length LFSRs generate PN-sequences with period $T = 2^L - 1$, see [5].

Linear Complexity: the LC of a sequence measures its unpredictability and is related with the amount of sequence we need in order to reconstruct the whole sequence. In cryptographic applications, LC must be as large as possible.

Generalized Sequences: they are a family of binary sequences $\{s_n\}_{n\geq 0}$ obtained by means of the self-decimation of a PN-sequence, see [6]. The period of any generalized sequence is a divisor of 2^{L-1} and the linear complexity satisfies $2^{L-2} < LC \leq 2^{L-1} - (L-2)$ as proved in [4].

Binomial Sequences: a binomial sequence $\left\{\binom{n}{k}\right\}_{n\geq 0}$, k being an integer, is a binary sequence

$$\left\{\binom{n}{k}\right\}_{n\geq 0} = \left\{\binom{0}{k}, \binom{1}{k}, \binom{2}{k}, \ldots\right\}_{mod\,2} \tag{1}$$

whose terms are the binomial numbers $\binom{n}{k}$ reduced mod 2. The sequence given in (1) is the ***k*-th binomial sequence**. Additional characteristic and properties of such sequences can be found in [1].

3 Binomial Representation of Binary Sequences

Every binary sequence $\{u_n\}$ whose period $T = 2^t$ is a power of 2 can be written as a linear combination of binomial sequences [1,2]:

$$\{u_n\} = \sum_{i=0}^{2^t-1} c_i \left\{\binom{n}{i}\right\}, \tag{2}$$

where t is a non-negative integer, $\left\{\binom{n}{i}\right\}$ is the i-th binomial sequence and the c_i are binary coefficients. The previous equation is the binomial representation of the sequence $\{u_n\}$. Moreover, $imax$ is an integer $(0 \leq imax \leq 2^t - 1)$ such that the coefficient c_{imax} of the binomial representation satisfies that $c_{imax} \neq 0$ while $c_i = 0$ for all index i in the range $(imax < i \leq 2^t - 1)$.

The coefficient c_{imax} and the binomial representation provide information about two fundamental parameters of the sequence: the period T of $\{u_n\}$ is that of the binomial sequence $\{\binom{n}{imax}\}$ [1, Proposition 3] while the linear complexity LC of $\{u_n\}$ is that of the binomial sequence $\{\binom{n}{imax}\}$ [1, Corollary 14], that is $LC = imax + 1$. The **binomial matrix** H_{2^t} is a binary Hadamard matrix [7] of size $2^t \times 2^t$ constructed as:

$$H_{2^t} = \begin{bmatrix} H_{2^{t-1}} & H_{2^{t-1}} \\ 0_{2^{t-1}} & H_{2^{t-1}} \end{bmatrix},$$

being $0_{2^{t-1}}$ the $2^{t-1} \times 2^{t-1}$ null-matrix. Indeed, H_{2^t} exhibits the typical structure of a Hadamard matrix: three identical blocks plus the null-block. In turn, each block $H_{2^{t-1}}$ is a Hadamard matrix too. In addition, any binomial matrix H_{2^t} can be easily constructed from the binomial sequences such as follows: (a) its rows correspond to the first 2^t bits of the first 2^t binomial sequences and (b) its columns correspond to shifted versions of the first 2^t binomial sequences starting each of them in its first 1.

We write the binomial matrix $H_{2^4} = H_{16}$ that will be a basic structure for the construction of other binomial matrices of higher dimensions.

$$H_{16} = H_{2^4} = \begin{bmatrix} H_{2^3} & H_{2^3} \\ 0_{2^3} & H_{2^3} \end{bmatrix},$$

where

$$H_8 = H_{2^3} = \begin{bmatrix} 1&1&1&1&1&1&1&1 \\ 0&1&0&1&0&1&0&1 \\ 0&0&1&1&0&0&1&1 \\ 0&0&0&1&0&0&0&1 \\ 0&0&0&0&1&1&1&1 \\ 0&0&0&0&0&1&0&1 \\ 0&0&0&0&0&0&1&1 \\ 0&0&0&0&0&0&0&1 \end{bmatrix}.$$

Due to the particular structure of the binomial sequences, we can reformulate the binomial representation of $\{u_n\}$ given in (2) and convert it into a matrix equation including the binomial matrix H_{2^t}. That is:

$$(c_0, c_1, \ldots, c_{2^t-1}) = (u_0, u_1, \ldots, u_{2^t-1}) \cdot H_{2^t} \bmod 2, \tag{3}$$

where $(u_0, u_1, \ldots, u_{2^t-1})$ corresponds to the 2^t successive terms of the sequence $\{u_n\}$ and $(c_0, c_1, \ldots, c_{2^t-1})$ are the binary coefficients of the Eq. (2).

4 An Algorithm to Compute the *LC* of Binary Sequences

The Eq. (3) is the core of a new algorithm that computes the LC of a sequence. In fact, if the binomial matrix is written in terms of its column vectors $H_{2^t} = (\boldsymbol{h_0}, \boldsymbol{h_1}, \ldots, \boldsymbol{h_{2^t-1}})$, then the coefficients c_i are easily calculated as:

$$c_i = (u_0, u_1, \ldots, u_{2^t-1}) \cdot \boldsymbol{h_i} \qquad (0 \leq i \leq 2^t - 1). \tag{4}$$

The computation starts with the coefficient c_{2^t-1} and proceeds in reverse order until the first coefficient $c_i \neq 0$ is reached. In that case, $imax = i$ and LC is easily computed as $LC = imax + 1$. Algorithm 1 illustrates such a computation. Now, two basic ideas can be drawn:

Algorithm 1: Computation of the LC of a given sequence

Input: *seq*: sequence of period 2^t, H_t: the $(2^t \times 2^t)$ binomial matrix.
$imax = 0$; $i = length(seq) - 1$;
while $i \geq 0$ **do**
  $c_i = (u_0, u_1, \ldots, u_{2^t-1}) \cdot \boldsymbol{h_i}$;
  if $c_i \neq 0$ **then**
    $imax = i$;
    Break;
  endif
  $i = i - 1$;
endwhile
Output: $LC = imax + 1$: Linear complexity of the sequence.

1. The computation of LC is reduced to products modulo 2 of binary vectors. Clearly, its computational complexity is lower than that of other algorithms found in the literature, see Sect. 4.2.
2. If the column $\boldsymbol{h_{imax}}$ has many $0's$ and only a few $1's$, then only a few terms of the sequence $\{u_n\}$ will be required to compute its LC.

The previous algorithm is particularly useful when we analyse sequences whose LC is upper bounded by a maximum value LC_{max}. In that case, the computation of coefficients is simplified as $c_i = 0$ for every coefficient in the range $(imax < i \leq 2^t - 1)$. Thus, the Algorithm 1 starts with the index $i = imax = LC_{max} - 1$ computing directly the coefficient c_{imax}.

4.1 Application of the Algorithm to Generalized Sequences

The generalized sequences $\{s_n\}_{n \geq 0}$ are ideal candidates for the application of Algorithm 1. In fact, their T is a power of 2 and their LC is upper bounded. Therefore, the Algorithm 1 starts with $i = imax = 2^{L-1} - (L-1)$ and computes the value of c_{imax}. If the coefficient $c_{imax} \neq 0$, then the complexity of the generalized sequence is $LC = 2^{L-1} - (L-2)$, otherwise LC will take a lower value. The column $\boldsymbol{h_{imax}}$ corresponds to the $(L-1)$-th column of the matrix $H_{2^{L-1}}$ read from right to left. As far as L increases by 1, the column $\boldsymbol{h_{imax}}$ is shifted one position to the left. Next, we will apply these Hadamard matrices to the computation of LC for different values of L.

For L in the Range $2 \leq L \leq 17$: we fix H_{16} as our reference matrix, called 16-box and depicted in Table 1. The successive matrices $H_{2^{L-1}}$ in this range are made up of 16-boxes. Then, we divide the period $T = 2^{L-1}$ of the generalized sequence by 16 to determine the number of 16-boxes included in its binomial matrix $H_{2^{L-1}}$. Next, we count the number of $1's$ in the column $\boldsymbol{h_{imax}}$ of the 16-box and, finally, we multiply this number by the number of 16-boxes in $H_{2^{L-1}}$ to get the total number of $1's$ in the general column $\boldsymbol{h_{imax}}$.

Table 1. 16-box to analyse generalized sequences with $2 \leq L \leq 17$

	H_8	H_8
	0_8	H_8
$L =$	**17 ... 14 13 ... 10**	**9 ... 6 5 ... 2**

For L in the Range $18 \leq L \leq 33$: we now use a 32-box as shown in Table 2, where H_{16} is the 16-box. Next, we divide the period T of $\{s_n\}$ by 32 and analyse the number of $1's$ in the successive columns $\boldsymbol{h_{imax}}$ of the 32-box.

Table 2. 32-box to analyse generalized sequences with $18 \leq L \leq 33$

	H_{16}	H_{16}
	0_{16}	H_{16}
$L =$	**33 ... 26 25 ... 18**	17 ... 10 9 ... 2

For L in the Range $34 \leq L \leq 65$: we now use a 64-box as shown in Table 3, where H_{32} is the 32-box. Now we divide the period T of the sequence by 64 and analyse the $1's$ of the successive columns of the 64-box.

For L in the Range $66 \leq L \leq 129$: the study is similar to that of the previous intervals but now we would use a 128-box made up of 64-boxes according to the typical structure of a Hadamard-matrix. We would divide the period T of the sequence by 128 and would analyse the number of $1's$ in the successive columns of the 128-box.

In all this analysis, the least suitable cases are generalized sequences whose columns $\boldsymbol{h_{imax}}$ in $H_{2^{L-1}}$ correspond to binomial sequences $\left\{\binom{n}{2^m}\right\}$ as half their digits are $1's$. Consequently, $T/2$ digits of $\{s_n\}$ are needed to compute its LC. Conversely, the most suitable cases are generalized sequences whose columns $\boldsymbol{h_{imax}}$ in $H_{2^{L-1}}$ correspond to binomial sequences $\left\{\binom{n}{2^m-1}\right\}$ where only $T/2^m$ of their digits are $1's$. Consequently, $T/2^m$ digits of $\{s_n\}$ are enough to compute its LC.

Table 3. 64-box to analyse generalized sequences with $34 \leq L \leq 65$

	H_{16}	H_{16}	H_{16}	H_{16}
	0_{16}	H_{16}	0_{16}	H_{16}
	0_{32}		H_{16}	H_{16}
			0_{16}	H_{16}
L=	**65 ...**	**50 49 ... 34**	33 ... 18	17 ... 2

Remark 1. *With only four boxes (16, 32, 64 and 128-boxes), we have easily got values of L in the range $L > 128$, which is the real cryptographic range.*

Moreover, we realize that Algorithm 1 will never require the knowledge of the whole sequence $\{s_n\}$ to compute its LC as there will always be at least a null-block in the binomial matrix. Consequently, the amount of sequence required is much less than that of other algorithms, see next sub-section.

4.2 Comparison with Other Algorithms

We compare the proposal here developed with other algorithms computing LC. The comparison is summarized in Table 4, where l is the sequence length, r the number of binomial sequences in the binomial decomposition and 5G denotes algorithm focussed on 5G Technologies.

Table 4. Comparison between proposed and existing algorithms to calculate LC

Authors	Merits	Demerits	O(.)	Length	5G
Berlekamp et al. [9] (1969)	For sequences of any length	High requirements in length	$O(l^2)$	$2 \times l$	N
Cardell et al. [1] (2019)	Improvement in sequence length requirements	Applicable to seq. With length a power of 2 Sequential	$O(r \times l)$	$l - log\, l$	N
Martin et al. [7] (2020)	Improvement in sequence length requirements It outperforms previous algorithms Concurrent	Applicable to seq. With length a power of 2	$O(l)$	$l - log\, l$	Y
Fúster et al. (This work) (2022)	Great improvement in sequence length requirements It outperforms all the previous algorithms Concurrent	Applicable to seq. With length a power of 2	$O(l/2)$	$l/2$	Y

5 Conclusions

In this work, a new algorithm, the Hadamard matrix-based algorithm to compute the linear complexity of binary sequences, has been introduced. It exhibits a better performance (lower computational complexity and sequence requirements) than that of similar algorithms computing LC. This is a big step in the study of binary sequences with period a power of 2 as well as it makes easier to detect flaws such as predictability in this kind of sequences. Moreover, the binomial decomposition as a way to extract information from a given sequence is an innovative tool and it is left for future work its application to other features, e.g. auto-correlation, balancedness, compression, of binary sequences.

References

1. Cardell, S.D., Fúster-Sabater, A.: Binomial representation of cryptographic binary sequences and its relation to cellular automata. Complexity **1**, 1–13 (2019). https://doi.org/10.1155/2019/2108014
2. Díaz Cardell, S., Fúster-Sabater, A.: Cryptography with Shrinking Generators. SM, Springer, Cham (2019). https://doi.org/10.1007/978-3-030-12850-0
3. Chin, W.L., Li, W., Chen, H.H.: Energy big data security threats in IoT-based smart grid communications. IEEE Commun. Mag. **2017**(55), 70–75 (2017)
4. Fúster-Sabater, A., Cardell, S.D.: Linear complexity of generalized sequences by comparison of PN-sequences. Revista de la Real Academia de Ciencias Exactas, Físicas y Naturales (RACSAM) **114**(4), 79–97 (2020)
5. Golomb, S.W.: Shift Register-Sequences, 2nd edn. Aegean Park Press, Laguna Hill (1982)
6. Yupu, H., Xiao, G.: Generalized self-shrinking generator. IEEE Trans. Inf. Theory **50**(4), 714–719 (2004)
7. Martin-Navarro, J.L., Fúster-Sabater, A.: Folding-BSD algorithm for binary sequence decomposition. Computers **9**(4), 100 (2020). https://doi.org/10.3390/computers9040100
8. Mavromoustakis, C.X., Mastorakis, G., Batalla, J.M. (eds.): Internet of Things (IoT) in 5G Mobile Technologies. MOST, vol. 8. Springer, Cham (2016). https://doi.org/10.1007/978-3-319-30913-2
9. Massey, J.L.: Shift-register synthesis and BCH decoding. IEEE Trans. Inf. Theory **15**(1), 122–127 (1969)
10. NIST. Lightweight Cryptography Project. https://csrc.nist.gov/Projects/Lightweight-Cryptography. Accessed 28 Mar 2022

Private and Public Opinions in a Model Based on the Total Dissonance Function: A Simulation Study

Michał Jarema[1(✉)] and Katarzyna Sznajd-Weron[2]

[1] Department of Operations Research and Business Intelligence, Wrocław University of Science and Technology, 50-370 Wrocław, Poland
michal.jarema@pwr.edu.pl
[2] Department of Theoretical Physics, Wrocław University of Science and Technology, 50-370 Wrocław, Poland
katarzyna.weron@pwr.edu.pl

Abstract. We study an agent-based model of opinion dynamics in which an agent's private opinion may differ significantly from that expressed publicly. The model is based on the so-called *total dissonance function*. The behavior of the system depends on the competition between the latter and social temperature. We focus on a special case of parental and peer influence on adolescents. In such a case, as the temperature rises, Monte Carlo simulations reveal a sharp transition between a state with and a state without private-public opinion discrepancy. This may have far-reaching consequences for developing marketing strategies.

Keywords: Agent-based model · Opinion dynamics · Private and public opinion · Expressed opinion · Dissonance · Marketing

1 Introduction

As early as the 1980s, it was noted how problematic public opinion polling can be due to *a genuine difference between people's private opinions and their public opinions* [14]. This problem was addressed also within agent-based models in several contexts, such as preference falsification [16], pluralistic ignorance [11,20], the emperor's dilemma [5,19], or hypocrisy [8]. Probably the first agent-based model in which agents differ in their beliefs and convictions was proposed in 2005 [5]. However, only recently increased interest in this topic has been observed [2,6,10,12,17,21].

To the best of our knowledge, all models dealing with private-public opinion discrepancies are algorithmic, i.e., based on precise rules. However, there is another possibility to model social systems, which is based on the so-called Hamiltonian, or in social terms – the *total dissonance function* [7]. The advantage of this approach, which minimizes a certain global function, is that there is

Supported by the National Science Center (NCN, Poland) through Grant No. 2019/35/B/HS6/02530.

D. Groen et al. (Eds.): ICCS 2022, LNCS 13351, pp. 146–153, 2022.
https://doi.org/10.1007/978-3-031-08754-7_20

a lot of freedom in choosing how the system is updated, in contrast to models based on dynamical rules [12]. Therefore, in this paper, we propose a model with *private* and *public* (also called *expressed*) opinions based on the Hamiltonian. We study the model on a one-dimensional lattice. We are aware that such a structure is not the best one to describe a social system. However, it is very convenient to visualize temporal–spatial evolution and thus we use it as a zero-level approach. In the future, we plan to investigate the model on various heterogeneous graphs.

2 The Model

The system is a chain of N agents, each of them has two binary opinions: public $S_i = \pm 1$ and private $\sigma_i = \pm 1$. We assume here only pairwise interactions, although more complex ones could also be included in the future. Therefore, the Hamiltonian of the model reads

$$\begin{aligned} H = &- J_1 \sum_{i=1}^{N} S_i S_{i+1} - J_2 \sum_{i=1}^{N} S_i S_{i+2} - K_1 \sum_{i=1}^{N} \sigma_i \sigma_{i+1} - K_2 \sum_{i=1}^{N} \sigma_i \sigma_{i+2} \\ &- R_1 \sum_{i=1}^{N} \sigma_i S_{i+1} - R_2 \sum_{i=1}^{N} \sigma_i S_{i+2} - M_0 \sum_{i=1}^{N} \sigma_i S_i, \end{aligned} \tag{1}$$

where periodic boundary conditions are used, i.e. $S_{N+1} \equiv S_1$, $S_{N+2} \equiv S_2$, $\sigma_{N+1} \equiv \sigma_1$, $\sigma_{N+2} \equiv \sigma_2$. The nearest neighbors can be interpreted as family and the next nearest as friends, as in the first and the second social circle [9]. We introduce several restrictions on the model's parameters:

1. $M_0 > 0$, due to *cognitive dissonance*, i.e., the mental conflict that occurs when an agent's public and private opinions do not align;
2. $J_1 < 0$, agents do not agree publicly with their nearest neighbors;
3. $J_2 > 0$, agents agree publicly with their next nearest neighbors;
4. $R_1 > 0$, agents agree privately with public opinion of their nearest neighbors;
5. $R_2 = 0$, private opinion of agents is not influenced by public opinion of their next nearest neighbors;
6. $K_1 = K_2 = 0$, neighbors' private opinions are not known to agents.

Assumptions 1, 5, and 6 are realistic independently of the studied problem. Assumptions 2, 3, and 4 can be treated just as a special case, representing the relation between parents and their teenage children [1,4].

3 Monte Carlo Simulations

In this paper, we use one of the most popular Monte Carlo (MC) simulation algorithms that is used to study Hamiltonian systems, so-called Metropolis algorithm [15]. Within such an algorithm, each MC step (MCS) consists of N updates of randomly selected agents (random sequential updating). The update of agent i consists of two MC trial moves: $S_i \rightarrow -S_i$ followed by $\sigma_i \rightarrow -\sigma_i$. However, several other updating schemes were used and they all led to the same stationary state, which is the main advantage of the presented approach. The Metropolis algorithm is as follows:

1. Prepare initial state of S_i and σ_i;
2. Until the requested number of MCS is performed, select randomly an agent and perform a trial move;
3. Calculate $\Delta E = E' - E$, where E', E are the values of H defined by Eq. (1) after and before the trial move respectively;
4. If $\Delta E \leq 0$ accept the trial move and go to step 2; otherwise:
5. Take a pseudorandom number $0 < r < 1$ from uniform distribution;
6. If $r < e^{-\Delta E/T}$, where T is the social temperature [3], accept the trial move and go to 2. Otherwise accept the old configuration again and go to 2.

To describe the limiting behavior of the system, we measure the following quantities: the correlation functions

$$g_S = \frac{1}{N}\sum_{i=1}^{N} S_i S_{i+1}, \quad g_\sigma = \frac{1}{N}\sum_{i=1}^{N} \sigma_i \sigma_{i+1}, \tag{2}$$

which measure the local order and identify the states: consensus (ferromagnetic state in physics) g_S, $g_\sigma = 1$, disagreement (antiferromagnetic state) g_S, $g_\sigma = -1$, and random (interchangeably referred to as noisy or disordered) state g_S, $g_\sigma = 0$.

Additionally, we measure the private-public opinion discrepancy defined as:

$$D = \frac{1}{2} - \frac{1}{2N}\sum_{i=1}^{N} S_i \sigma_i, \tag{3}$$

which takes values in the interval $[0, 1]$. When public and private opinions are the same, then the discrepancy $D = 0$. When they are opposite, then $D = 1$.

4 Results

In the lack of any noise, i.e., at $T = 0$ the system is blocked after a few iterations if $M_0 \neq R_1$. Only for $M_0 = R_1$ private opinions σ are free to change and therefore we focus here on the case $M_0 = R_1$. We introduce the parameter A

$$M_0 = R_1 = |J_1| + A, \tag{4}$$

which measures the competition between two forces – the strength of cognitive dissonance, described by M_0 (note that $M_0 > 0$, so the absolute value is not needed) and the strength of the interactions with the nearest neighbors, given by $|J_1|$. For $|J_1| \geq M_0$ (i.e. $A \leq 0$) the system evolves towards disagreement in the public opinion S and random state in the private opinion σ. For $|J_1| < M_0$ (i.e. $A > 0$) the system evolves towards consensus in both S and σ.

Since in every system there is noise, so the assumption $T = 0$ is not realistic. Therefore, from now we focus on $T > 0$. We present time–space evolution of the

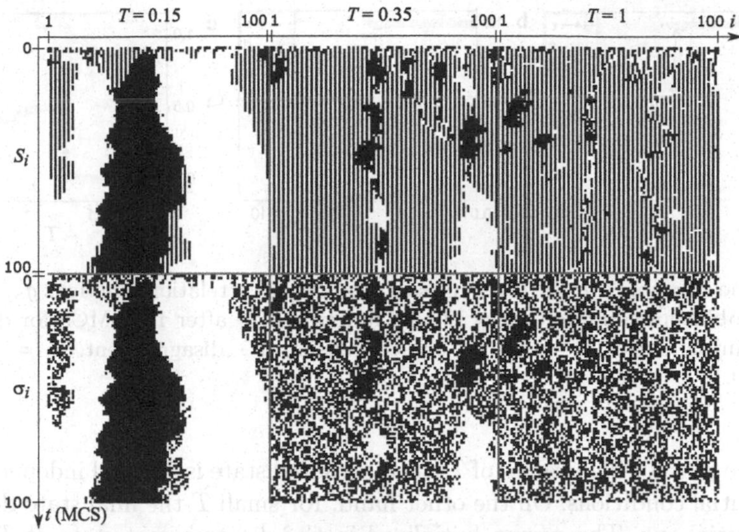

Fig. 1. Time evolution of public S_i (upper row) and private σ_i (bottom row) opinions for random initial conditions at three selected temperatures T indicated above the upper plots and $J_1 = -J_2 = -1$, $A = 0.1$.

system for several values of T in Fig. 1: public opinions are shown in the upper panels, whereas private opinions in the bottom ones. Each row in each plot corresponds to the state of the system at a given time t: white pixels correspond to positive opinions, whereas the black pixels to negative ones. Therefore, white domains represent the area of positive consensus, black domains correspond to negative consensus, alternating white with black to disagreement, and randomly mixed white and black to the noisy state.

We see that for low values of T (e.g., $T = 0.15$) the system tends to have large clusters of consensus on both public and private levels. In the finite system, eventually the full consensus is reached, i.e., all agents have the same opinions. However, the time needed to reach such a consensus increases with the system size N and for $N \to \infty$ the system consists of opposite domains forever. Interestingly, these domains are interspersed with areas of disagreement at the public level of opinion and with noisy (random) areas at the private level. For larger values of T ($T = 0.35$ in Fig. 1), the system evolves towards disagreement state in the public opinions S_i and random state in the public opinions σ_i. Only thin consensus inclusions are visible at the boundary between disagreement domains. For $T = 1$ more fluctuations appear and eventually, for very large values of T, the fluctuations destroy the disagreement on the public level.

To analyze the model more systematically and quantitatively, we calculate the average values of correlation functions g_S, g_σ, private-public opinion discrepancy D for different values of T and 3 different types of initial conditions: (r) random, (c) consensus, and (d) disagreement. They are ensemble averaged over 100 configurations, each taken after 1000 MCS, which this was sufficient to reach the stationary state. Corresponding results are shown in Fig. 2.

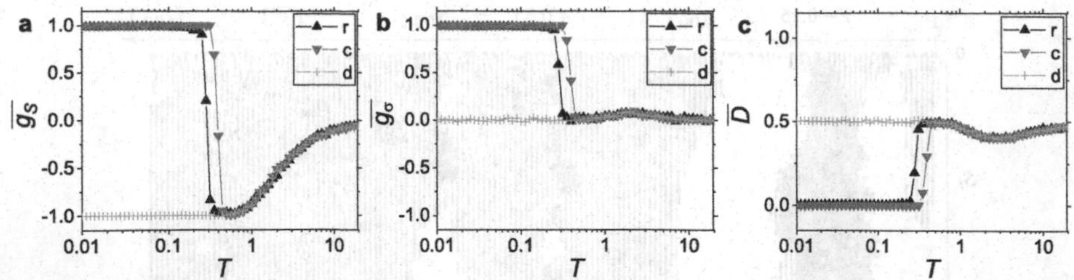

Fig. 2. The ensemble average (over 100 samples) of the correlation functions g_S, g_σ, and private–public opinion discrepancy D as a function of T after 1000 MCS for different types of initial states: r – random, c – consensus, d – disagreement. $J_1 = -J_2 = -1$, $A = 0.1$

We see that for high values of T the disordered state is reached independently on the initial conditions. On the other hand, for small T the final state depends on the initial one. The system initialized in the disagreement state at $T < 0.4$ reaches the state with $g_S = -1$, $g_\sigma = 0$ i.e. disagreement in public (S) and disordered in private (σ) opinions. Opinions on both levels are uncorrelated and therefore $D = 0.5$. For two other types of initial conditions (consensus and random), for $T < 0.3$ the system reaches consensus in both public and private opinions ($g_S = g_\sigma = 1$) with no discrepancy between them ($D = 0$). For $0.3 < T < 0.5$ the final state depends on the initial one – this phenomenon is called hysteresis. For $0.5 < T < 1$ we have $g_S = -1$, $g_\sigma = 0$ in the final state, i.e., disagreement is reached on the public level and a disordered state on the private one. To summarize, there are two transitions in the public opinion S: one sharp at $T = T_1$, between consensus and disagreement, and one smooth at $T = T_2$,

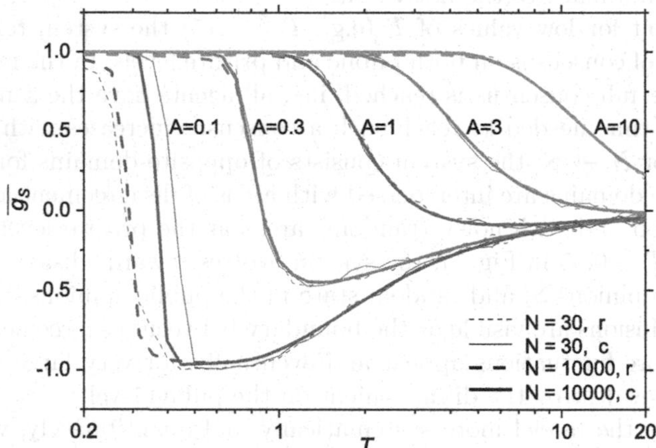

Fig. 3. Dependence of the $\overline{g_S}$ function on the size N of the system and on the type of initialization: r - random, c - consensus.

between disagreement and disorder. In σ there is only one sharp transition at T_1 between consensus and disorder.

Values of T_1, T_2 depends on parameter A, i.e., on the interplay between internal harmony and interactions with others, as shown in Fig. 3. There are two other phenomena shown in Fig. 3. Firstly, the hysteresis exists only for $A = 0.1$ (therefore results for this value are presented in Figs. 1 and 2) and its width decreases with increasing system size. Secondly, apart from hysteresis shrinkage, the results for small ($N = 30$) and large ($N = 10^4$) systems are the same.

5 Conclusions

In this short paper, we conduct Monte Carlo simulations of a relatively simple agent-based model to study opinion formation among parents and their teenage children. The huge spending power and significant influence on family purchase decisions makes adolescents a critical consumer segment. Parents, peers, and media are the three essential factors influencing adolescents' consumer attitudes and purchase behavior [18].

Although we focus on one specific case, the model itself is much more general and allows us to describe many phenomena in which discrepancies between private and public opinions may arise. The novelty with respect to the previous literature on private–public opinion discrepancy is that the model is built around the total dissonance function. This makes it insensitive to the updating scheme, as opposed to many other agent-based models [12].

More specifically, we show that as a result of the competition between different types of interactions (with parents, peers, own private–public opinions) and social temperature T, three interesting phenomena can be observed. Firstly, there is a critical temperature $T = T_1$, below which all agents have the same private and public opinion, so there are no private–public opinion discrepancies. For $T > T_1$ on average half of the agents have different public and private opinions.

Secondly, there is a threshold $T = T_2$, which concerns only the public opinion. For $T_1 < T < T_2$ a disagreement phase appears on the public level: each agent disagrees with the nearest neighbors (interpreted as members of the first social circle – the family), but agrees with the next nearest neighbors (interpreted as the second social circle – peers). Interestingly, for $T_1 < T < T_2$ there is no such phase on the private level – private opinions are just random, which can be interpreted as independent behaviour (not influenced by others). This leads to the third observation: on the private level agents are independent already for $T > T_1$, whereas on the public level the range of independence is smaller and starts at $T = T_2 > T_1$. Identifying and studying such regimes may help marketers formulate appropriate marketing communication strategies that can be effective in resolving parent–child purchase disagreements [13].

Of course, we realize that the one-dimensional case considered here is not very realistic from the social point of view. However, the model based on the total dissonance function can be easily studied on any other type of graph, including actual social networks. This, however, is left for future studies.

References

1. Allen, M., Donohue, W., Griffin, A., Ryan, D., Turner, M.: Comparing the influence of parents and peers on the choice to use drugs - a meta-analytic summary of the literature. Crim. Justice Behav. **30**(2), 163–186 (2003)
2. Anderson, B., Ye, M.: Recent advances in the modelling and analysis of opinion dynamics on influence networks. Int. J. Autom. Comput. **16**(2), 129–149 (2019)
3. Bahr, D., Passerini, E.: Statistical mechanics of opinion formation and collective behavior: micro-sociology. J. Math. Sociol. **23**(1), 1–27 (1998)
4. Biddle, B., Bank, B., Marlin, M.: Parental and peer influence on adolescents. Soc. Forces **58**(4), 1057–1079 (1980)
5. Centola, D., Willer, R., Macy, M.: The emperor's dilemma: a computational model of self-enforcing norms. Am. J. Sociol. **110**(4), 1009–1040 (2005)
6. Gaisbauer, F., Olbrich, E., Banisch, S.: Dynamics of opinion expression. Phys. Rev. E **102**(4), 042303 (2020)
7. Galesic, M., Olsson, H., Dalege, J., Van Der Does, T., Stein, D.: Integrating social and cognitive aspects of belief dynamics: towards a unifying framework. J. R. Soc. Interface **18**(176), 20200857 (2021)
8. Gastner, M., Takács, K., Gulyás, M., Szvetelszky, Z., Oborny, B.: The impact of hypocrisy on opinion formation: a dynamic model. PLoS ONE **14**(6), e0218729 (2018)
9. Hamill, L., Gilbert, N.: Social circles: a simple structure for agent-based social network models. JASSS **12**(2), 3 (2009)
10. Hou, J., Li, W., Jiang, M.: Opinion dynamics in modified expressed and private model with bounded confidence. Phys. A **574**, 125968 (2021)
11. Huang, C.Y., Wen, T.H.: A novel private attitude and public opinion dynamics model for simulating pluralistic ignorance and minority influence. JASSS **17**(3), 8 (2014)
12. Jędrzejewski, A., Marcjasz, G., Nail, P., Sznajd-Weron, K.: Think then act or act then think? PLoS ONE **13**(11), e0206166 (2018)
13. Kim, C., Lee, H., Han, S.L.: A study of parent-adolescent interaction: the impact of family communication patterns on adolescents' influence strategies and parents' response strategies. Eur. J. Marketing **52**(7–8), 1651–1678 (2018)
14. King, S.: Conflicts between public and private opinion. Long Range Plan. **14**(4), 90–105 (1981)
15. Landau, D.P., Binder, K.: A Guide to Monte Carlo Simulations in Statistical Physics. Cambridge University Press, Cambridge (2000)
16. Makowsky, M., Rubin, J.: An agent-based model of centralized institutions, social network technology, and revolution. PLoS ONE **8**(11), e80380 (2013)
17. Manfredi, R., Guazzini, A., Roos, C., Postmes, T., Koudenburg, N.: Private-public opinion discrepancy. PLoS ONE **15**(11), e0242148 (2020)
18. Mishra, A., Maity, M.: Influence of parents, peers, and media on adolescents' consumer knowledge, attitudes, and purchase behavior: a meta-analysis. J. Consum. Behav. **20**(6), 1675–1689 (2021)

19. Muhammad, A., Zia, K., Saini, D.: Population dynamics necessary to avert unpopular norms. LNAI **11352**, 64–75 (2019)
20. Seeme, F., Green, D., Kopp, C.: Pluralistic ignorance: a trade-off between group-conformity and cognitive dissonance. In: Gedeon, T., Wong, K.W., Lee, M. (eds.) ICONIP 2019. LNCS, vol. 11954, pp. 695–706. Springer, Cham (2019). https://doi.org/10.1007/978-3-030-36711-4_58
21. Ye, M., Qin, Y., Govaert, A., Anderson, B., Cao, M.: An influence network model to study discrepancies in expressed and private opinions. Automatica **107**, 371–381 (2019)

Analysis of Public Transport (in)accessibility and Land-Use Pattern in Different Areas in Singapore

Hoai Nguyen Huynh(✉)

Institute of High Performance Computing, Agency for Science, Technology and Research, Singapore, Singapore
huynhhn@ihpc.a-star.edu.sg

Abstract. As more and more people continue to live in highly urbanised areas across the globe, reliable accessibility to amenities and services plays a vital role in sustainable development. One of the challenges in addressing this issue is the consistent and equal provision of public services, including transport for residents across the urban system. In this study, using a novel computational method combining geometrical analysis and information-theoretic measures, we analyse the accessibility to public transport in terms of the spatial coverage of the transport nodes (stops) and the quality of service at these nodes across different areas. Furthermore, using a network clustering procedure, we also characterise the land-use pattern of those areas and relate that to their public transport accessibility. Using Singapore as a case study, we find that the commercial areas in its central business district expectedly have excellent accessibility and the residential areas also have good to very good accessibility. However, not every residential area is equally accessible. While the spatial coverage of stops in these areas is very good, the quality of service indicates substantial variation among different regions, with high contrast between the central and eastern areas compared to the others in the west and north of the city-state. We believe this kind of analysis could yield a good understanding of the current level of public transport services across the urban system, and their disparity will provide valuable and actionable insights into future development plans.

Keywords: Public transport · Land use · Spatial pattern · Entropy · GIS · Singapore

1 Introduction

With countries having pledged to reach net-zero emissions in the next few decades [1], plans are being put together by governments around the world to achieve the goal. Among the courses of action, a feasible way to accomplish this is to reduce the use of private vehicles and shift toward the more sustainable use of public transit. As part of the process, improving the quality and accessibility of existing public transport infrastructure is vital in achieving high ridership.

D. Groen et al. (Eds.): ICCS 2022, LNCS 13351, pp. 154–161, 2022.
https://doi.org/10.1007/978-3-031-08754-7_21

Furthermore, from a social perspective, equal public transport service provision also contributes to sustainable development in urban areas [12,14].

Public transport accessibility has received much attention from various communities. Among the measures of accessibility, the public transport accessibility level (PTAL) [15], which combines the walking distance to the transport nodes and the frequency of the services, has been frequently used for its simple calculation. However, the method's main disadvantage lies in using a heuristic distance threshold from a point of interest to its nearby transport nodes, which may produce artefacts in certain areas. More sophisticated methods have been proposed, such as analysing the walking distance and time to the nearest bus stop using the actual walking paths [8]. Yet, the approach may not be suitable for large-scale analysis where the amount of spatial data required would make the procedure computationally inefficient. Other authors have also looked at the optimal spacing of bus stops [13] so that the system's overall performance could be improved and the link with land use could be established [2,10].

While public transport accessibility is intuitively about the ability and ease of users to reach their destination, it could also be examined from the opposite perspective of inaccessibility, which could be imagined to hold the hidden information about the system. Along that line, in this study, we formulate the study of public transport accessibility as an inverse problem, that is, to view the spatial inaccessible area as "conjugate" of its accessible counterpart. This approach has received relatively little attention in the literature, but it can be shown to provide interesting insights into the spatial organisation of an urban system. Using this approach, we will explore the public transport accessibility of different areas in terms of the bus stops' spatial (non-)coverage and their associated quality of service. The methodology will be demonstrated in the city-state of Singapore, and the accessibility of its different areas will be analysed.

In the Singapore context, several studies have been performed to analyse the accessibility to the Mass Rapid Transit (MRT) stations [11,17,18] or the impact of the expansion of the MRT network on accessibility to employment [3]. Yet, few have been done for the bus network. Therefore, this work also contributes to discerning the accessibility of bus services in Singapore. While a public transport system typically contains train and bus services, it could be argued that analysing the public bus network alone is sufficient in terms of the spatial accessibility of different local areas. This is because bus stops are also well presented at the train stations, and the bus network has a high degree of penetration into the residential areas.

2 Data and Methods

Data. For this study, the **public transport** data in Singapore is obtained from the Land Transport Authority (LTA) [9] and includes the information on the bus stops and the services serving those bus stops. For every bus stop, we filter the data to select only the regular services, i.e. available throughout the day with the lowest dispatch frequency not longer than 30 min. On the other hand, the **land-use** data is obtained from the Urban Redevelopment Authority (URA) through the Singapore's open data portal [16], comprising 33 land-use types, ranging from commercial, residential to industrial use. We exclude the areas being used for ports or airports as they are generally not accessible to the public, leaving the remaining 32 types in the analysis. Similarly, the **planning area** data is also available from the same portal. There are 55 planning areas in Singapore and they will serve as the spatial unit of analysis in this study in terms of both public transport accessibility and land-use patterns.

Analysis of Land-Use Pattern. In this study, we characterise each planning area by the types of land use that make up its area. Specifically, the land-use profile of a planning area can be presented as a 32-dimensional vector $\boldsymbol{p} = (p_1, p_2, \ldots, p_{32})$ with p_i being the percentage of each land-use type within the planning area. After that, we employ cosine similarity to measure how close a pair of vectors are to one another or how similar the two corresponding planning areas are. The similarity score ranges between 0 (totally different) and 1 (perfectly identical). After measuring the similarity between the land-use configuration of the planning areas, a network can be constructed by treating each area as a node and adding a link between a pair of planning areas if their similarity score is above 0.7 (corresponding to an angle of 45°, which geometrically suggests some degree of similarity). The planning area network will be analysed using the clustering procedure as described in [7] to identify the groups of planning areas that are most similar to one another and classify them. The results of land-use classification will form the basis for selecting planning areas for the subsequent analysis of public transport accessibility (see Sect. 3).

Analysis of Public Transport Nodes. A place of interest is said to have good access to public transport if its distance to the nearest public transport node (bus stop) is reasonably short. Using this picture, we can identify the so-called "non-covered" area at a given distance ρ, or the area that is outside (all) the circles of radius ρ centred at the bus stops. For a suitable value of ρ, if the area covered by the buffer circles is considered the accessible area, we can call the area that is not covered by any of the circles the inaccessible one.

To determine the suitable value of ρ, we can look at the spatial structure of the patches formed in the process of identifying the non-covered area. As the value of ρ increases from 0, the non-covered area transits from a connected land mass to vanishing isolates when most of the area is covered at a large value of ρ. In between, we can observe that the non-covered area becomes highly fragmented at some intermediate value of ρ. Using the information-theoretic measure of entropy, we can quantify the degree of fragmentation of the non-covered patches. This is similar to the idea of measuring the complexity of a spatial point pattern [5] that has been applied to study the relative location of public transport nodes in the urban context [4,6]. The entropy measure peaks at some intermediate value of ρ, called the critical distance, at which the inaccessible space is most fragmented, marking the onset of accessibility. Subsequently, we

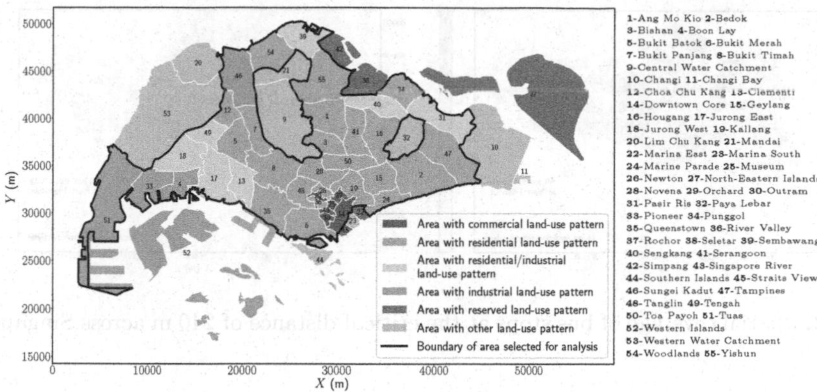

Fig. 1. Classification of planning areas based on composition of land-use types.

can use this critical distance to assess the spatial coverage of public transport nodes across different planning areas in Singapore.

3 Results and Discussion

Classification of Areas Based on Land-Use Pattern. While different urban areas typically contain complex land-use patterns, they can generally be clustered into a small number of groups using the procedure described in Sect. 2. The 55 planning areas (PAs) in Singapore can be grouped into six major categories (see Fig. 1), namely commercial, residential, residential mixed with industrial, industrial, reserved, and others, based on their composition of the 32 land-use types [16]. The first group comprises 6 PAs with a very high proportion of land for commercial and hotel usage. These areas are indeed in Singapore's Central Business District (CBD), and we label them as "commercial" land-use pattern. The second group contains 23 PAs with predominantly land use of residential type and, hence, is labelled as "residential". The third group includes 7 PAs, which have a high proportion of land for both residential use and business 2 (clean, light industry and general industry or warehouse). These areas suggest the co-location of residential development alongside the industrial facilities and are labelled as "residential/industrial". The fourth group contains 5 PAs with land chiefly for business 2 and, hence, is labelled as "industrial". The fifth group has 5 PAs with land use mainly of reserved type and, hence, is labelled as "reserved". The final group contains 9 PAs with very distinct land-use profiles from the rest of the areas and, thus, is labelled as "others".

For subsequent analysis, we decided to exclude the 5 areas with "reserved" land-use pattern as they are reserved by the government for future development and currently do not accommodate any urban activities. The 9 "other" areas are also excluded as they do not reflect the typical urban characteristics in Singapore. Furthermore, we also exclude the area of Western Islands (labelled 52 in Fig. 1) as

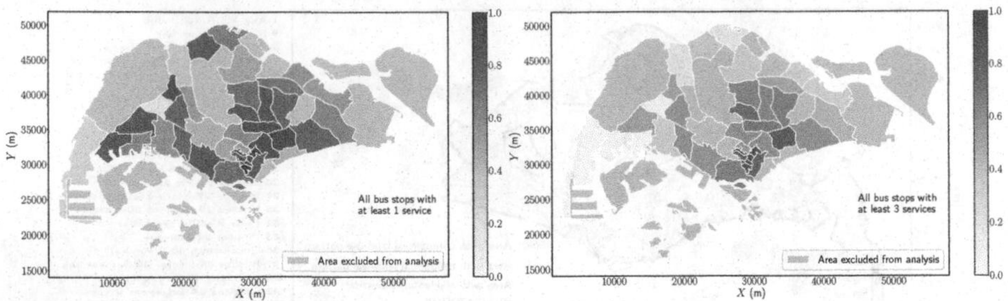

Fig. 2. Spatial coverage of bus stops at the critical distance of 240 m across Singapore.

they are not served by public transport in Singapore. Altogether, we only retain 40 areas for the analysis of public transport accessibility. The final selection of areas for analysis is shown with a thick black boundary in Fig. 1 (without the ports and airports, which have earlier been excluded).

Measure of Accessibility to Public Transport Nodes. The analysis of the fragmentation of the non-covered patches in Sect. 2 reveals a critical distance at which the inaccessible space is most fragmented, signifying a transition of state [5]. In Singapore, this critical value is found to be 240 m and agrees reasonably well with the perception of the typical distance that most people would find comfortable for walking (3 to 4 min). This might reflect the result of the transport planning that in most areas, people would walk no more than 240 m to reach the nearest bus stops. While this value reflects an overall good spatial coverage of bus stops across Singapore, we can also utilise it to analyse and assess the public transport accessibility of different areas in the city-state.

Using the buffer distance of 240 m, we identify the area covered by the bus stops served by at least a regular bus service in Singapore, or the "accessible area". We compute the total (union) accessible area within each PA as a fraction of its total area. The result of such measure over all selected PAs is shown as a choropleth map in Fig. 2. It could be observed that most of the PAs selected for analysis have very good spatial coverage of the public transport nodes. Notably, the CBD has excellent coverage when the PAs it contains are almost entirely covered by the bus stops at 240 m. Outside the city centre, the majority of the PAs of the residential type also have a high quality of bus stop coverage. The PAs with a mixture of residential and industrial land use generally have fair coverage. The remaining industrial-type areas typically have low spatial coverage.

As argued earlier, spatial coverage is only part of the measure of the accessibility of public transport in an area. Besides the spatial coverage, the quality of service at the bus stops also contributes significantly to accessibility. Bus stops served by more services are arguably more "accessible" as they provide travel options to more places. To probe the service quality, we look at the spatial coverage of bus stops served by at least 3 regular services, using the same buffer distance of 240 m. The map in Fig. 2 shows that the CBD still enjoys an excellent coverage of stops with at least 3 services. However, the quality of the coverage starts to decline outside the CBD. Yet, the decline takes different rates in different regions, with many areas having poorer coverage than others. Across Singapore, the central and eastern parts appear to have much better accessibility than the city-state's western and northern parts.

Discussion. It should be noted that the method of geometrical analysis described in this work can also shed light on the spatial structure of an area. For example, the high fraction of area covered by the bus stops at a certain buffer radius can be related to the compactness of an area, reflecting the high density of bus stops within a small area, like the case of the planning areas in the CBD of Singapore. On the other hand, the concept of the critical distance allows us to gain insight into the sparseness of the (location of the) bus stops, which in most cases reflects the level of development [6].

The result of the spatial coverage in Fig. 2 could highlight the difference in the quality of accessibility across different areas in the same city, which might otherwise not be apparent. In the Singapore context, the results suggest that the new towns of Punggol and Sembawang still need much improvement in transport planning, whereas middle-aged towns like Jurong East still have a lot of potential for further development. In contrast, the areas with a high density of private housing like Bukit Timah or Marine Parade, despite having only average spatial coverage of the bus stops, appear to have a good quality of the services provided. While it is commonly understood that the residents in those areas belong to the higher income group and have a high rate of ownership of private vehicles, improving public transport accessibility in terms of shorter walking distance to bus stops could nudge their transport behaviour toward a more sustainable one.

4 Conclusion

In this study, we develop a computational method to analyse the quality of public transport accessibility in relation to the pattern of land use in different urban areas and apply it to the case study of the city-state of Singapore. The method combines geometrical analysis and information-theoretic measures to quantify the area that is not within a certain buffer distance of the public transport nodes, called the non-covered or inaccessible area. It is argued that the spatial structure of such inaccessible area undergoes a phase transition with the entropy measure maximising at some critical value of the buffer distance. This critical value of the buffer distance is where the inaccessible area is most fragmented, marking the onset of accessibility within the system. In Singapore, this distance is about 240 m, indicating an overall high density of bus stops. However, analysis at the individual area level reveals that despite having good spatial coverage of the bus stops, the quality of the service at these stops varies across the country.

On the other hand, we also analyse the pattern of land use in these areas and relate it to the public transport accessibility, providing more context for its interpretation. Typically, the commercial and residential areas of the city-state are found with very good accessibility. However, residential areas in different parts of the country exhibit marked differences, with much better accessibility in the central and eastern regions than in the west and north of Singapore. The results obtained from this research can be useful for the relevant urban and transport planning authorities in further developing the public transport network. For example, the results could help identify areas where improvements are needed and devise policies to nudge people's behaviour toward more sustainable public transport usage.

Acknowledgement. The author would like to acknowledge the support of the Social Sciences & Technology (SST) Horizontal Technology Programme Office through the 2021 Social Sciences Innovation Seed Fund (ref C211618005).

References

1. United Nations Climate Change Conference. https://ukcop26.org/
2. Chen, J., Currie, G., Wang, W., Liu, Z., Li, Z.: Should optimal stop spacing vary by land use type?: New methodology. Transp. Res. Rec. **2543**(1), 34–44 (2016)
3. Conway, M.W., Byrd, A., van Eggermond, M.: Accounting for uncertainty and variation in accessibility metrics for public transport sketch planning. J. Transp. Land Use **11**(1), 541–558 (2018)
4. Huynh, H.N.: Continuum percolation and spatial point pattern in application to urban morphology. In: D'Acci, L. (ed.) The Mathematics of Urban Morphology. MSSET, pp. 411–429. Springer, Cham (2019). https://doi.org/10.1007/978-3-030-12381-9_18
5. Huynh, H.N.: Spatial point pattern and urban morphology: perspectives from entropy, complexity, and networks. Phys. Rev. E **100**(2), 022320 (2019)
6. Huynh, H.N., Makarov, E., Legara, E.F., Monterola, C., Chew, L.Y.: Characterisation and comparison of spatial patterns in urban systems: a case study of U.S. cities. J. Comput. Sci. **24**, 34–43 (2018)
7. Jiang, Z., Huynh, H.N.: Unveiling music genre structure through common-interest communities. Soc. Network Anal. Min. **12**, 35 (2022)
8. Kaszczyszyn, P., Sypion-Dutkowska, N.: Walking access to public transportation stops for city residents. A comparison of methods. Sustainability **11**(14), 3758 (2019)
9. Land Transport DataMall. https://datamall.lta.gov.sg/content/datamall/en.html
10. Lee, S., Hickman, M., Tong, D.: Development of a temporal and spatial linkage between transit demand and land-use patterns. J. Transp. Land Use **6**(2), 33–46 (2013)
11. Li, Z., et al.: Equality of public transit connectivity: the influence of mass rapid transit services on individual buildings for Singapore. Transportmetrica B Transp. Dyn. **7**(1), 576–595 (2019)
12. Ribeiro, J., Fontes, T., Soares, C., Borges, J.L.: Accessibility as an indicator to estimate social exclusion in public transport. Transp. Res. Procedia **52**, 740–747 (2021)
13. Sahu, P.K., Mehran, B., Mahapatra, S.P., Sharma, S.: Spatial data analysis approach for network-wide consolidation of bus stop locations. Public Transp. **13**(2), 375–394 (2021). https://doi.org/10.1007/s12469-021-00266-0

14. Scheurer, J., Curtis, C., McLeod, S.: Spatial accessibility of public transport in Australian cities: does it relieve or entrench social and economic inequality? J. Transp. Land Use **10**(1), 911–930 (2017)
15. Shah, J., Adhvaryu, B.: Public transport accessibility levels for Ahmedabad, India. J. Public Transp. **19**(3), 19–35 (2016)
16. Singapore's open data portal. https://data.gov.sg/
17. Wibowo, S.S., Olszewski, P.: Modeling walking accessibility to public transport terminals: case study of Singapore mass rapid transit. J. Eastern Asia Soc. Transp. Stud. **6**, 147–156 (2005)
18. Zhu, X., Liu, S.: Analysis of the impact of the MRT system on accessibility in Singapore using an integrated GIS tool. J. Transp. Geogr. **12**(2), 89–101 (2004)

Pseudo-Newton Method with Fractional Order Derivatives

Krzysztof Gdawiec[1](✉), Agnieszka Lisowska[1], and Wiesław Kotarski[2]

[1] Institute of Computer Science, University of Silesia, Będzińska 39, 41-200 Sosnowiec, Poland
{krzysztof.gdawiec,agnieszka.lisowska}@us.edu.pl
[2] Katowice, Poland
kotarski@ux2.math.us.edu.pl

Abstract. Recently, the pseudo-Newton method was proposed to solve the problem of finding the points for which the maximal modulus of a given polynomial over the unit disk is attained. In this paper, we propose a modification of this method, which relies on the use of fractional order derivatives. The proposed modification is evaluated twofold: visually via polynomiographs coloured according to the number of iterations, and numerically by using the convergence area index, the average number of iterations and generation time of polynomiographs. The experimental results show that the fractional pseudo-Newton method for some fractional orders behaves better in comparison to the standard algorithm.

Keywords: Pseudo-Newton method · Riemann–Liouville derivative · Caputo derivative · Dynamics

1 Introduction

Newton's method is one of the most famous and important algorithms in numerical analysis. It has a local quadratic convergence and is undefined for critical points. This simple algorithm has a long history and ample bibliography [13]. In recent years many modifications of Newton's method have been proposed.

An interesting modification of the Newton's method is the pseudo-Newton method [8]. That method effectively finds the local maximal values of the modulus of complex polynomials over the unit disc on the complex plane.

In recent years, various fractional derivatives have become an intensive field of study in root-finding area. The first method, in which the classical derivative was replaced by the fractional ones, was the Newton's method [5]. Then, the Newton-type method with convergence of order ν was proposed [1], where ν is the order of the fractional derivative. Next, the use of fractional derivatives and various iteration processes in the Newton's method was shown [6]. Then, a variant of Chebyshev's method [2] and a two-step iterative scheme with fractional derivatives [3] were introduced.

W. Kotarski—Independent Researcher.

D. Groen et al. (Eds.): ICCS 2022, LNCS 13351, pp. 162–168, 2022.
https://doi.org/10.1007/978-3-031-08754-7_22

In this paper, we replace the classic derivative with the fractional derivatives in the pseudo-Newton method. This leads to a new class of pseudo-Newton's methods – fractional pseudo-Newton methods. The performed numerical experiments suggest that for some values of ν the fractional pseudo-Newton method is better in comparison to the standard pseudo-Newton one.

The paper is organised as follows. In Sect. 2, the definitions of the Riemann–Liouville and Caputo derivatives, are presented. In Sect. 3, the pseudo-Newton method, introduced in [8], is described. In Sect. 4, the application of the fractional derivatives into the pseudo-Newton method is proposed. In Sect. 5, the experimental results are shown. Finally, Sect. 6 concludes this paper.

2 Fractional Derivatives

Integer order derivatives and integrals are commonly known and used. As a natural generalisation, the fractional derivative was introduced [10]. It can be defined in many ways. In this paper, we use Riemann–Liouville and Caputo derivatives as the most commonly used ones. We recall their definitions [10,11].

Let Γ be the well-known gamma function. The Riemann–Liouville derivative (RL-derivative) of order $\nu \in (n-1, n]$, $n \in \mathbb{N}$ is defined as:

$$D^{\nu}_{RL} f(t) := \begin{cases} \frac{1}{\Gamma(n-\nu)} \frac{d^n}{dt^n} \int_0^t \frac{f(\tau)}{(t-\tau)^{\nu+1-n}} d\tau, & \text{if } \nu \in (n-1, n), \\ \frac{d^n}{dt^n} f(t), & \text{if } \nu = n. \end{cases} \tag{1}$$

We also recall that the Caputo derivative (C-derivative) of order $\nu \in (n-1, n]$, $n \in \mathbb{N}$, of a real–valued function f is defined as:

$$D^{\nu}_{C} f(t) := \begin{cases} \frac{1}{\Gamma(n-\nu)} \int_0^t \frac{f^{(n)}(\tau)}{(t-\tau)^{\nu+1-n}} d\tau, & \text{if } \nu \in (n-1, n), \\ \frac{d^n}{dt^n} f(t), & \text{if } \nu = n. \end{cases} \tag{2}$$

Both of these fractional derivatives are linear.

In this paper, we calculate the fractional derivatives of polynomials. Thus, to determine them, we can only consider monomial t^m, thanks to linearity. So,

$$D^{\nu}_{RL} t^m = \frac{\Gamma(m+1)}{\Gamma(m-\nu+1)} t^{m-\nu}, \; D^{\nu}_{C} t^m = \begin{cases} \frac{\Gamma(m+1)}{\Gamma(m-\nu+1)} t^{m-\nu}, & \text{if } m > n-1, \\ 0, & \text{if } m \leq n-1, \end{cases} \tag{3}$$

where $\nu \in (n-1, n)$, $n \in \mathbb{N}$, $m \in \mathbb{R}$.

Let us note that for a constant function and $\nu \neq 1$ we obtain that $D^{\nu}_{RL} c \neq 0$ and $D^{\nu}_{C} c = 0$. So, these derivatives are not equal.

So far, we presented the fractional derivatives of functions defined on $\mathbb{R}$. But we are going to use them on $\mathbb{C}$. However, we cannot replace the real variable t by a complex variable z in the definitions in (1) and (2) because of the multi-valuedness of expressions that are present under integrals in both derivatives. Nevertheless, in the case of analytic functions the formulas for the RL- and C-derivative for monomial z^m are the same as in (3), but only for a complex variable z such that $z \in \mathbb{C} \setminus \{c \in \mathbb{C} : Im(c) = 0 \wedge Re(c) < 0\}$, and $m \neq -1, -2, -3, \ldots$. This additional assumption is related to the branch cut line that is needed to eliminate the multi-valuedness of z^m if $z \in \mathbb{C}$ and $m \in \mathbb{R}$.

3 The Pseudo-Newton Method

Let p be a non-constant complex polynomial over the unit disk $D = \{z \in \mathbb{C} : |z| \leq 1\}$. Let us then consider the problem of finding the points for which we attain the maximal modulus over D, i.e., $\|p\|_\infty = \max\{|p(z)| : z \in D\}$.

According to the Maximum Modulus Principle [7], $\|p\|_\infty$ is attained at the boundary of D [8]. Moreover, we have that a point $z_* \in D$ is a local maximum of $|p(z)|$ if and only if

$$z_* = \left(\frac{p(z_*)}{p'(z_*)}\right) \Big/ \left(\left|\frac{p(z_*)}{p'(z_*)}\right|\right). \tag{4}$$

Equation (4) is the test for checking if z_* is a local maximum of $|p(z)|$ over D. But instead of solving (4) one can solve the following equation [8,9]:

$$G(z) = p(z)|p'(z)| - zp'(z)|p(z)| = 0. \tag{5}$$

All the solutions of (5) are the fixed points of (4). To find them one can use the pseudo-Newton method [8]. This method has the following form:

$$z_{n+1} = z_n - \frac{G_n(z_n)}{G_n'(z_n)}, \quad n = 0, 1, 2, \ldots, \tag{6}$$

where $z_0 \in \mathbb{C}$ is a given starting point and $G_n(z) = p(z)|p'(z_n)| - zp'(z)|p(z_n)|$. Let us observe that the functions G_n are easily differentiable with respect to z because the modules in them are constant values.

The proof of convergence of the pseudo-Newton method can be found in [12], where this method is converted to some equivalent convergent Newton-like method. Moreover, the pseudo-Newton method can be easily generalised to higher-order methods [4].

4 The Pseudo-Newton Method with Fractional Derivatives

In recent years, many applications of fractional derivatives appeared in the literature. A good example is the use of Riemann–Liouville and Caputo derivatives in the classical Newton's method [1,5]. In this section, we present a similar combination of the pseudo-Newton method with the fractional order derivatives.

Let us denote by D_*^ν any of the two considered fractional derivatives, i.e., D_{RL}^ν or D_C^ν. By replacing the classical first derivative G_n' in (6) by D_*^ν, we get

$$z_{n+1} = z_n - \frac{G_n(z_n)}{D_*^\nu G_n(z_n)}, \quad n = 0, 1, 2, \ldots. \tag{7}$$

Such defined methods are called fractional pseudo-Newton methods.

In these methods, we need to calculate the derivative $D_*^\nu G_n(z)$. When we look closely at the form of function G_n, we can notice that for a fixed n the values of

$|p'(z_n)|$, $|p(z_n)|$ are constant. Therefore, the terms $p(z)|p'(z_n)|$ and $zp'(z)|p(z_n)|$ are polynomials of argument z. So, due to the linearity property of D^{ν}_{RL} and D^{ν}_{C}, the derivative $D^{\nu}_{*}G_n(z)$ has the following form

$$D^{\nu}_{*}G_n(z) = |p'(z_n)|D^{\nu}_{*}(p(z)) - |p(z_n)|D^{\nu}_{*}(zp'(z)). \tag{8}$$

5 Numerical Results

In this section, we present the numerical results of application of the proposed methods in practice. We start by presenting the polynomiographs that show the speed of convergence and the dynamics of the proposed method graphically. Then, we show the dependencies between some numerical measures and the order of the considered fractional derivatives.

To generate a polynomiograph in the given area, we take each point of this area as a starting point for (7). Then, we map the number of the performed iterations to a colour by using the colour map from Fig. 1. Basing on the polynomiograph, we compute the following numerical measures: the average number of iterations (ANI) in the considered area, the convergence area index (CAI, i.e., the ratio of the number of the points that converged to the number of all points in the considered area), and the generation time of the polynomiograph.

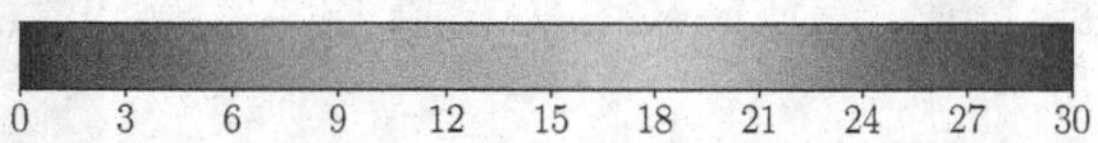

Fig. 1. The colour map used in the experiments.

The experiments were performed for a number of polynomials, but due to the lack of space, we present here the complete results (i.e., the polynomiographs and the plots of numerical measures) only for $p_4(z) = z^4 - 10z^2 + 9$. To generate the polynomiographs we used the following parameters: the area is fixed as $[-3, 3]^2$, the maximal number of iterations equals to 30, accuracy $\varepsilon = 0.001$, and image resolution is 800×800 pixels. The experiments were performed on the computer with: Intel i5-9600K (@ 3.70 GHz) processor, 32 GB DDR4 RAM, and Windows 10 (64-bit). The software was implemented in Processing.

We start with the polynomiographs for p_4. In Fig. 2, we see the polynomiograph generated by the pseudo-Newton method with the classical derivative. Next, in Figs. 3 and 4, we see the polynomiographs obtained with the fractional versions of the pseudo-Newton method. We have chosen only the most representative ones. The presented dynamics is related to the number of iterations needed to achieve the maximum modulus of polynomials via the investigated algorithms. In general, the blue colour in the polynomiographs means quick convergence, the green one means average convergence and the red colour denotes slow convergence. The polynomiographs from Figs. 3, 4 show less dynamic in comparison to the reference one from Fig. 2. Indeed, one can see that the larger or lower the value of ν related to $\nu = 1$ the slower the convergence of the polynomiographs (we can observe more red colour). Additionally, careful analysis of them suggests that there exist values of ν for which the fractional pseudo-Newton methods could be better (in the mean of higher CAI and lower ANI values) in comparison to the pseudo-Newton method with the classic derivative.

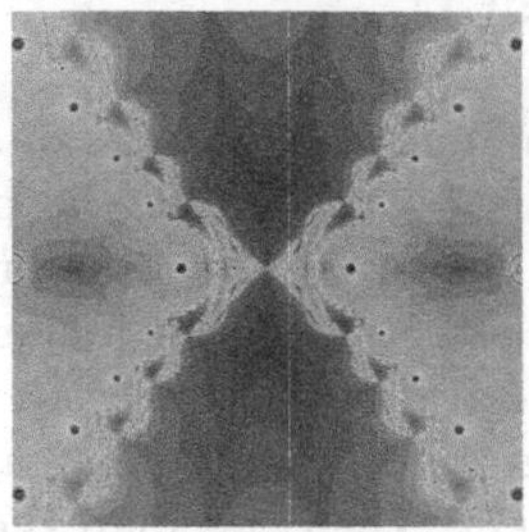

Fig. 2. The dynamics for the classical derivative ($\nu = 1$) for p_4.

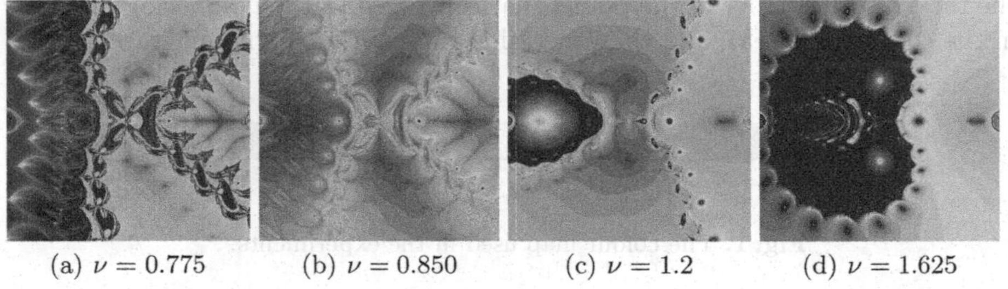

(a) $\nu = 0.775$ (b) $\nu = 0.850$ (c) $\nu = 1.2$ (d) $\nu = 1.625$

Fig. 3. Examples of dynamics for the RL-derivative for different values of ν for p_4.

The dependencies between the considered numerical measures (ANI, CAI, and generation time) and the order ν of the fractional derivatives for p_4 are presented in Fig. 5. The best values of the numerical measures are the following:

- classical derivative – ANI: 8.578, CAI: 0.998, time: 1.180 s,
- RL–derivative – min. ANI: 8.197 ($\nu = 0.93$), max. CAI: 0.999 ($\nu = 0.835$), min. time: 5.316 s ($\nu = 0.940$),
- C–derivative – min. ANI: 7.747 ($\nu = 0.850$), max. CAI: 0.999 ($\nu = 0.805$), min. time: 4.387 s ($\nu = 0.890$).

From the results presented above and the plots shown in Fig. 5, one can see that by using fractional derivatives one can decrease the value of ANI and improve the convergence (higher values of CAI). The decrease of ANI and the increase of CAI compared to the classical case can be observed for $\nu < 1$, but in the neighbourhood of 1. For $\nu > 1$, the results for the fractional case are worse

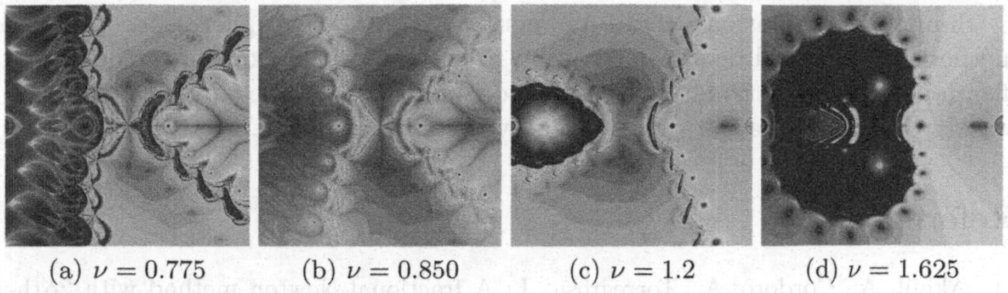

(a) $\nu = 0.775$ (b) $\nu = 0.850$ (c) $\nu = 1.2$ (d) $\nu = 1.625$

Fig. 4. Examples of dynamics for the C-derivative for different values of ν for p_4.

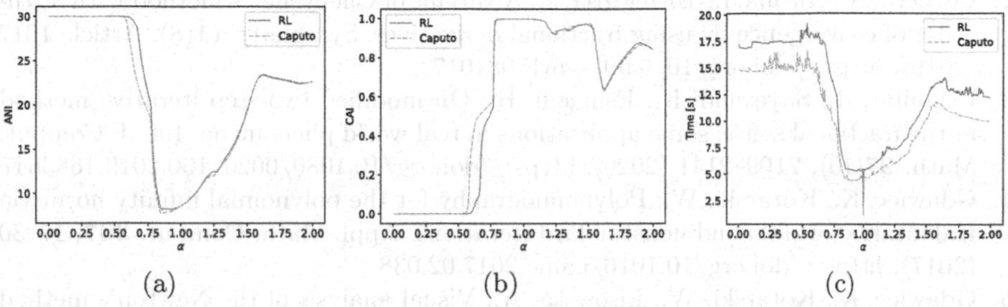

(a) (b) (c)

Fig. 5. The plots of (a) ANI, (b) CAI, and (c) time (in seconds), for polynomial p_4.

than for the classical derivative. Unfortunately, the generation time of the polynomiographs via the fractional pseudo-Newton method cannot be improved. In general, calculation cost is higher for fractional derivatives compared to classical derivatives. It is because in the case of the classical derivatives, we raise to a power with only an integer exponent, whereas in the fractional case we raise to real-valued exponents that is more computationally expensive. Additionally, one can observe that for C-derivative, the plots of ANI in some intervals below $\nu = 1$ are lying below those for RL-derivative. The same occurs for time plots. Moreover, for CAI plots it is conversely. This generally denotes that C-derivatives should be preferred over the RL-derivatives since the former ones converge faster.

6 Conclusions

In this paper, we proposed the use of the fractional derivatives instead of the classical one in the pseudo-Newton method. The experimental results showed that the proposed approach can improve the standard pseudo-Newton method in some aspects. Namely, for some values of order ν, the value of ANI is lower and the value of CAI is higher. Unfortunately, the generation time of polynomiographs is higher, which is clear because we must perform more computationally complex calculations for the fractional derivatives.

Similar investigations could be performed for the higher-order pseudo-methods [4]. It could be also interesting to check the behaviour of further modifications of the fractional pseudo-Newton methods obtained by replacing the standard Picard iteration with various types of iterations [4,6].

References

1. Akgül, A., Cordero, A., Torregrosa, J.: A fractional Newton method with 2αth-order of convergence and its stability. Appl. Math. Lett. **98**, 344–351 (2019). https://doi.org/10.1016/j.aml.2019.06.028
2. Cordero, A., Girona, I., Torregrosa, J.: A variant of Chebyshev's method with 3αth-order of convergence by using fractional derivatives. Symmetry **11**(8), Article 1017 (2019). https://doi.org/10.3390/sym11081017
3. Erfanifar, R., Sayevand, K., Esmaeili, H.: On modified two-step iterative method in the fractional sense: some applications in real world phenomena. Int. J. Comput. Math. **97**(10), 2109–2141 (2020). https://doi.org/10.1080/00207160.2019.1683547
4. Gdawiec, K., Kotarski, W.: Polynomiography for the polynomial infinity norm via Kalantari's formula and nonstandard iterations. Appl. Math. Comput. **307**, 17–30 (2017). https://doi.org/10.1016/j.amc.2017.02.038
5. Gdawiec, K., Kotarski, W., Lisowska, A.: Visual analysis of the Newton's method with fractional order derivatives. Symmetry **11**(9), Article ID 1143 (2019). https://doi.org/10.3390/sym11091143
6. Gdawiec, K., Kotarski, W., Lisowska, A.: Newton's method with fractional derivatives and various iteration processes via visual analysis. Numer. Algorithms **86**(3), 953–1010 (2020). https://doi.org/10.1007/s11075-020-00919-4
7. Kalantari, B.: A geometric modulus principle for polynomials. Am. Math. Mon. **118**(10), 931–935 (2011). https://doi.org/10.4169/amer.math.monthly.118.10.931
8. Kalantari, B.: A necessary and sufficient condition for local maxima of polynomial modulus over unit disc. arXiv:1605.00621 (2016)
9. Kalantari, B., Andreev, F., Lau, C.: Characterization of local optima of polynomial modulus over a disc. Numer. Algorithms **3**, 1–15 (2021). https://doi.org/10.1007/s11075-021-01208-4
10. Miller, K., Ross, B.: An Introduction to the Fractional Calculus and Fractional Differential Equations. John Wiley & Sons Inc., New York (1993)
11. Samko, S., Kilbas, A., Marichev, O.: Fractional Integrals and Derivatives: Theory and Applications. Gordon and Breach Science Publishers, Amsterdam (1993)
12. Usurelu, G., Bejenaru, A., Postolache, M.: Newton-like methods and polynomiographic visualization of modified Thakur process. Int. J. Comput. Math. **98**(5), 1049–1068 (2021). https://doi.org/10.1080/00207160.2020.1802017
13. Ypma, T.: Historical development of Newton-Raphson method. SIAM Rev. **37**(4), 531–551 (1995). https://doi.org/10.1137/1037125

An Energy Aware Clustering Scheme for 5G-Enabled Edge Computing Based IoMT Framework

Jitendra Kumar Samriya[1], Mohit Kumar[1], Maria Ganzha[2], Marcin Paprzycki[2], Marek Bolanowski[3](✉), and Andrzej Paszkiewicz[3]

[1] Dr. B. R. Ambedkar National Institute of Technology, Jalandhar, India
kumarmohit@nitj.ac.in

[2] Systems Research Institute Polish Academy of Sciences, Warszawa, Poland

[3] Rzeszow University of Technology, Rzeszów, Poland
marekb@prz.edu.pl

Abstract. 5G networks offer novel communication infrastructure for Internet of Things applications, especially for healthcare applications. There, edge computing enabled Internet of Medical Things provides online patient status monitoring. In this contribution, a Chicken Swarm Optimization algorithm, based on Energy Efficient Multi-objective clustering is applied in an IoMT system. An effective fitness function is designed for cluster head selection. In a simulated environment, performance of proposed scheme is evaluated.

Keywords: Energy efficiency · Network lifetime · Clustering · Cluster head selection · Delay · Chicken swarm optimization · Sensor networks · Adaptive networks

1 Introduction

Within Internet of Things (IoT), availability of 5G networks empowers raise of Internet of Everything [1, 2]. IoT materializes also in healthcare, and is often referred to as Internet of Medical Things (IoMT) [3, 4]. Typically, IoMT systems are linked with wireless body area networks (WBAN) connecting biosensor nodes [5], which act like a personal digital assistant ([6, 7]). However, if the energy in the biosensor is exhausted, the WBAN collapses [8, 9]. Note that biosensor replacement is very difficult, when it is placed inside the patient [10]. Here, energy-efficient clustering protocols are needed to achieve effective cluster head selection [11, 12]. However, existing energy-aware clustering and routing schemes suffer from network overhead [13, 14]. Separately, fuzzy control based energy efficient clustering protocol [15] still lacks in energy consumption. Moreover, heterogeneity based energy aware clustering protocols have been designed in [16, 17]. The key contribution of this work is to propose a clustering approach, which offers energy-aware communication in 5G enabled, edge-based ecosystems. Here, IoMT deployment consists of resource-limited wearable sensors (SNs), which transmit data through a 5G-enabled base station (BS). Transmission and reception of data takes more

D. Groen et al. (Eds.): ICCS 2022, LNCS 13351, pp. 169–176, 2022.
https://doi.org/10.1007/978-3-031-08754-7_23

power. Hence, to maximize lifespan of a system, a multi objective cluster head (CH) selection, based on Chicken Swarm Optimization (CSO), is used for cluster formation.

2 System Model and Assumptions

In this work, it is assumed that uniform level of energy is allocated to all wearable SNs and energy needed to perform intra-cluster communication is represented by an arbitrary value, within the pre-determined range (including "sleeping mode"). Network lifespan is reduced when SN battery is drained. Hence, energy-efficiency has to be taken into account when electing the CH, amid the accessible SNs. The model of the system illustrating the proposed clustering scheme is depicted in Fig. 1.

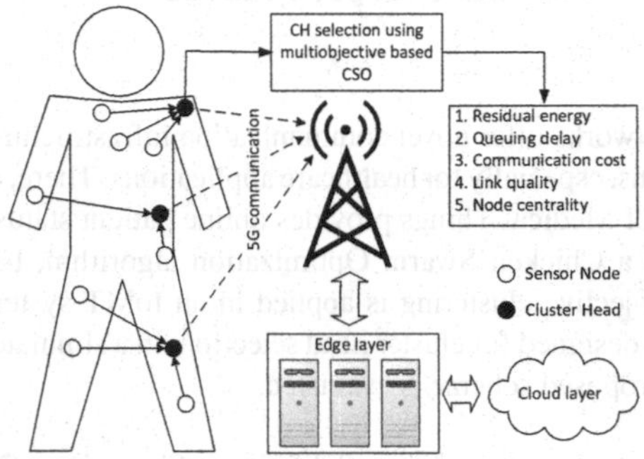

Fig. 1. Proposed clustering scheme system model framework

In what follows, the energy model, found in [18], has been selected. The equation for calculating energy consumption of data packet of size s bits for distance (d) is $E_{Trans}(d) = (TA_{FS}d^{\alpha} + E_D)s$. E_D denotes energy consumption of a device, TA_{FS} is the free space model amplifier of a transmitter, and α denotes the path loss exponents, with $2 \le \alpha \le 4$. Energy use to obtain data packet is represented by $E_{Rec}(d) = s \times E_D$. The cumulative energy use, of each wearable SN (to send or receive data), is based on distance d, and represented as $E_{Cum} = \{(TA)d^{\alpha} + 2(E_D)\}s$. Selection of cluster head relies on the objective function. Here, selection of energy efficient CH depends on residual energy, queuing delay, communication cost, link quality and node centrality.

Residual Energy: Initially, wearable SNs, deployed inside the IoMT, gather sensitive patient data and forward it to the CH. Energy consumption of CHs, during data gathering from SNs, is:
$E_{CH-SN} = D_B \times \left(E_{PB_F} + AE_{PB} \times \left(\sqrt{(a_{CH} - a_{SN})^2 + (b_{CH} - b_{SN})^2}\right)\right)$, where (a_{CH}, b_{CH}) is the position of CH and (a_{SN}, b_{SN}) is the position of SN; D_B is the number of bits in the data packet, E_{PB_F} is the energy needed, per bit, for data forwarding, and AE_{PB} is the amplification energy. Data forwarding from CH to BS can be computed as follows: $E_{BS-CH} = D_B \times \left(E_{PB_F}\left(\frac{N}{Y} - 1\right) + \left(E_{PB_G} \times \left(\frac{N}{Y}\right)\right) + E_{PB_F} + AE_{PB} \times \left(\sqrt{(a_{BS} - a_{CH})^2 + (b_{BS} - b_{CH})^2}\right)\right)$,

where $(a_{BS} - b_{BS})$ is the position of BS, E_{PB_F} is the energy used for data forwarding, N is the total number of *SN*s in the IoMT system, Y denotes the number of SNs in the cluster. Finally, the cumulative energy consumption of each cluster is computed as: $E_C = E_{BS-CH} + ((\frac{N}{Y}) - 1) \times E_{CH-SN}$.

Communication Cost: Commination cost is defined as the power needed for data forwarding: $Com_C = \frac{d_{avg}^2}{d_0^2}$, where d_{avg} denotes the average distance between given SN and its neighbor SNs, and d_0 represents the forwarding radius of an SN.

Queuing Delay: D_{Que}, depends on the rate of arrival of packets (to SN), and the outward link forwarding capacity. For A_R, the arrival rate of packets P_i to the SN and F_C the forwarding capacity, the queuing delay D_{Que}becomes: $D_{Que} = (A_R + F_C)/P_i$.

Link Quality: In IoMT, fading of a channel is highly irregular. If the receiver does not receive the complete signal, re-forwarding happens. This requires additional energy from the transmitter. Therefore, the link quality is estimated as:$LQ = \frac{LQ_i - LQ_{min}}{LQ_{max} - LQ_{min}}$, where LQ_{max} and LQ_{min} denote upper and lower range of re-forwarding; and LQ_i represents entire re-transmission cost among neighbors and given (*i-th*) SN.

Node Centrality: Node centrality measure *i* determines number of times a node acts as a link on the shortest paths among two nodes. It is computed as: $N_C = \sum_{m \neq r \neq n \in R} \frac{\lambda_{mn(i)}}{\lambda_{mn}}$, where λ_{mn} is the number of shortest paths between node *m* and *n*, and $\lambda_{mn(i)}$ is the number of paths via *i*. Here, every node follows the fitness function based on calculated objective function values, along with the weighted coefficients, as follows: $Fitness_{final} = w_1 \times E_C + w_2 \times \left(\frac{1}{Com_C}\right) + w_3 \times \left(\frac{1}{D_{Que}}\right) + w_4 \times LQ + w_5 \times N_C$. Here, $w_1 + w_2 + w_3 + w_4 + w_5 = 1$ and, $0 \leq w_i \leq 1, \forall i, 1 \leq i \leq 5$. The central goal is to: $Maximize \sum_{i=1}^{|CH|} Fitness_{final}$ such that $1 \leq i \leq |CH|$. Node, which fulfills all objectives will be selected as a CH. In every cluster, the selected CH is responsible for data gathering and forwarding to BS. Specifically, after CH selection, for each CH, route will be established for transferring collected data to BS.

The proposed approach is based on the chicken swarm optimization (CSO) introduced in [19] for CH selection. The most important aspects of CSO, in the considered problem, are as follows.

Chicken Movement: "best node" is the rooster, "worst node" is the chick, while the remaining nodes are hens. Let R_n be count of roosters, H_n count of hens, C_n count of chicks, and M_n count of mother hens; while B – be the number of iterations. Chicken positions can be denoted $cu_{i,j}^{t_a}$ where $i \in [1, 2, \ldots\ldots N]$ and $j \in [1, 2, \ldots\ldots D]$, for time t_a in D dimensional space. In the proposed approach, the rooster is the CH with optimal fitness value.

Rooster Movement: Following [19], movement of roosters is computed as:

$$cu_{i,j}^{t_a+1} = cu_{i,j}^{t_a} \times \left[1 + Randn\left(0, \sigma^2\right)\right] \tag{1}$$

$$\sigma^2 = \begin{cases} 1 iff_i \leq f_k \\ exp\left(\frac{f_k - f_i}{|f_i| + \varepsilon}\right) otherwise; k \in [1, N], k \neq i \end{cases}$$

where $cu_{i,j}^{t_a+1}$ depicts the movement of the rooster, $Randn\left(0, \sigma^2\right)$ denotes the Gaussian distribution, with mean value 0 and standard deviation σ^2, ε denotes a constant value added to avoid zero-division, k implies the index of the rooster, selected randomly from the group, and f_i denotes the value of fitness of rooster x_i.

Hen Movement: Following [19], hen movement is represented as:

$$cu_{i,j}^{t_a+1} = cu_{i,j}^{t_a} + S1 \times Rand \times \left(cu_{r1,j}^{t_a} - cu_{i,j}^{t_a}\right) + S2 \times Rand \times \left(cu_{r2,j}^{t_a} - cu_{i,j}^{t_a}\right) \tag{2}$$

where $S1 = exp\left(\frac{f_i - f_{r1}}{abs(f_i + \varepsilon)}\right)$, $S2 = exp(f_{r2} - f_i)$,$Rand$ is a random number in [0, 1], $r1 \in [1, 2, ...N]$ is the index of the mate of ith hen, $r2 \in [1, 2, ...N]$ is the index of randomly chosen rooster (or hen), $S1$ and $S2$ are the influence factors.

Chick Movement: Following [17], chick movement can be formulated as:

$$cu_{i,j}^{t_a+1} = cu_{i,j}^{t_a} + FL \times \left(cu_{m,j}^{t_a} - cu_{i,j}^{t_a}\right) \tag{3}$$

where $cu_{m,j}^{t_a}$ denotes the location of the mother of ith chick, for $m \in [1, 2, ...N]$, $FL \in [1, 2]$ denotes the randomly selected speed of the chick following the mother.

For selecting the CH, accessible SNs become chickens; nodes with best fitness values become roosters, with worst fitness are chicks, while the remaining nodes are hens. In each round, location of the rooster is updated using formula (1). Following the rooster, location of every hen is updated using formula (2). The chicks searching for food around their mother explore search spaces, which is captured in formula (3). Ranking of chickens maintains hierarchical order. Based on fitness values, chickens are ranked. After ranking, relationships between mothers and chicks are identified, to find differences between the chicks. ***Algorithm 1*** depicts the proposed algorithm for CH selection. The SN, selected by the CSO algorithm, becomes a CH, while remaining SNs form its cluster. After CH selection, patient data is sent to the CH, and can be removed.

Algorithm 1: Multi objective based CSO Algorithm for CH selection
Input: N number of CHs, CSO parameters; **Output:** Pareto Solution S indicating the nodes that act as CHs.

1. *Initialize all the parameters* $R_n, H_n, C_{n,}, M_n$ *and* B
2. *Initialize the chickens in the swarm randomly as* $C_{U_i}(i = 1,2,\ldots\ldots.y)$
3. *Initialize the total count of iterations as* Max_{itr}
4. *While* $T_r < Max_{itr}$ *do*
5. *If* $(T_r\%B = 0)$ *then*
6. *Establish the hierarchical order through ranking of chickens*
7. *Partition the swarm group and identify the mother-child relationship*
8. *End if*
9. *For* $(i = 1)$*do*
10. *If* $(i == rooster)$*do*
11. *Perform local update of the rooster's location using (1)*
12. *End if*
13. *If* $(i == hen)$*do*
14. *Perform local update of the hen's location using (2)*
15. *End if*
16. *If* $(i == chick)$ *do*
17. *Perform local update of the chick's location using (3)*
18. *End if*
19. *Estimate the fitness of the obtained solution using* $Fitness_{final}$
20. *If the solution outperforms the older one* → *update location*
21. *End for*
22. *Label the best solution as pareto optimal solution S*
23. *End while*
24. *Return S*

3 Experimental Results and Discussion

Performance of the proposed solution was measured using: *cluster formation time; energy consumption:* energy consumed by SNs (in mJ); *network lifetime:* for how many rounds, network remains operational; *throughput:* CHs-BS (Mb/s); *delay:* transmission time SN-BS via CH (ms). Proposed approach was compared to EO-μGA [20], ABCSA [21], BCO [22] and PSO [23]. Simulated network parameters were: Number of SNs: 1000; IoMT sensing area: 500m^2; BS position: (500,500); Packets Size:1500 bits; Max Throughput: 1 Mbps; Initial Node Energy: 2J; Electronics energy: 30 nJ/bit; Data aggregation energy: 3 nJ/bit/signal; Transmitting power: 9 mW; Max number of rounds: 500. The CSO algorithm parameters were: Population Size: 100; Number of rosters: 3; Number of hens: 5; Update time steps: 10; Maximum Iterations: 150.

As shown in Fig. 2a, CSO-based clustering has the lowest cluster formation time. Moreover, the proposed CSO minimizes the cost by 1.9%, 2.7%, 3.8% and 4.9% in comparison to EO-μGA, ABCSA, BCO and PSO, respectively (Fig. 2b). Next, when number of SNs varied from 50 to 1000 (Fig. 2c), the proposed scheme minimized energy consumption by 3.4% to 7.1%. It was also most optimal from the perspective of energy consumption, for transmission power between -25 *dBm* and -5 *dBm* (Fig. 2d). Proposed solution improved network lifetime (for 50 to 1000 SNs; Fig. 2e) by 3.2% to 17%. Network lifetime was also evaluated with respect to the number of clusters (from 3 to 10; Fig. 2f). Here, the gain was between 5.7% and 21.3%. The throughput was simulated for 50 to 1000 SN's (Fig. 2g). The performance gain was 0.1% to 39%. Throughput was also evaluated when varying transmission power (-25 *dBm* to -5 *dBm*; Fig. 2h) and the improvement was 6.8% to 48,2%. Finally, Fig. 2i depicts propagation delay for varying number of SN, from 50 to 1000. Results confirm that proposed SCO scheme reduces propagation delay by 0.05 ms to 0.56 ms.

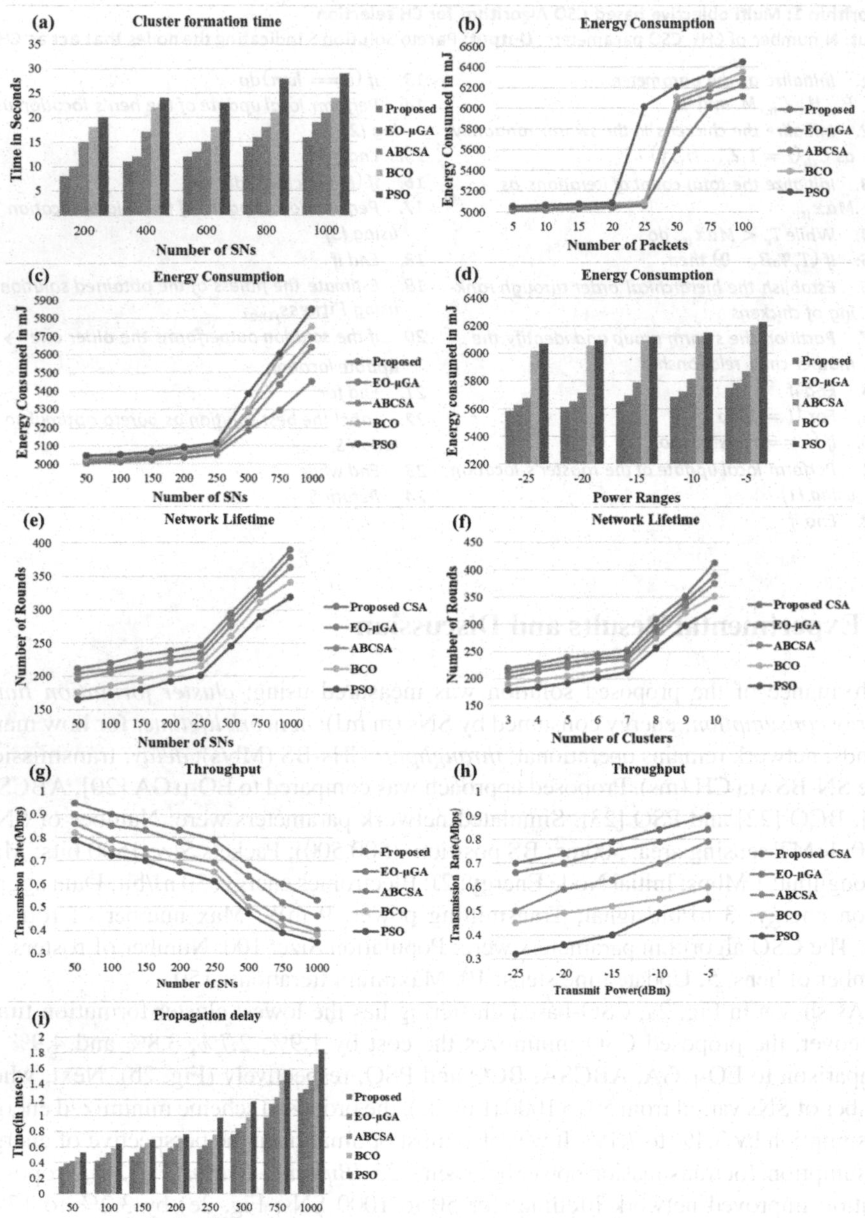

Fig. 2. (a) Cluster formation time; (b) Energy consumption (EC)/number of packets; (c) EC/number of SNs; (d) EC/transmission power ranges; (e) Network Lifetime (NL)/number of SNs; (f) NL/number of Clusters; (g) Throughput (T)/number of SNs; (h) T/transmission power ranges; (i) Propagation delay/number of SNs.

4 Concluding Remarks

In this work, an energy efficient CSO-based clustering scheme was proposed for IoMT ecosystems. The proposed scheme uses fitness function, based on residual energy, queuing delay, communication cost, link quality and node centrality. Additional details about the approach, including extensive literature review can be found in [24]. The performance of the proposed scheme was compared with EO-µGA, ABCSA, BCO and PSO approaches. CSO-based approach was more efficient in all categories, with reduction of energy consumption by 3–7%. In the future, the proposed scheme will be extended with respect to mobility of nodes, body actions, and cross layer optimization.

References

1. Li, S., Xu, L., Zhao, S.: 5G Internet of Things: a survey. J. Ind. Inf. Integr. **10**, 1–9 (2018)
2. Zahra, S.R., Chishti, M.A.: Assesing the services, security threaths, challenges and solutions in the Internet of Things. SCPE. **20**, 457–484 (2019)
3. Al-Turjman, F., Nawaz, M., Ulusar, U.: Intelligence in the internet of medical things era: a systematic review of current and future trends. Comp. Comm. **150**, 644–660 (2020)
4. Hattori, Y., Tanaka, T., Kajiwara, Y., Shimakawa, H.: Estimation of intimacy change in team using vital signs. In: Communication Papers of 2018 Federated Conference on Computer Science and Information Systems. https://doi.org/10.15439/2018F90, (2018)
5. Takabayashi, K., Tanaka, H., Sakakibara, K.: Toward an advanced human monitoring system based on a smart body area network for industry use. Electronics **10**(6), 688 (2021)
6. Jafer, E., Hussain, S., Fernando, X.: A wireless body area network for remote observation of physiological signals. IEEE Consumer Electronics Magazine **9**(2), 103–106 (2020)
7. Raj, S.: An efficient IoT-based platform for remote real-time cardiac activity monitoring. IEEE Trans. Consum. Electron. **66**(2), 106–114 (2020)
8. Kalaivaani, P., Krishnamoorthi, R.: Design and implementation of low power bio signal sensors for wireless body sensing network applications. Microprocess. Microsyst. **79**, 103271 (2020)
9. Bilandi, N., Verma, H., Dhir, R.: Energy-efficient relay node selection scheme for sustainable wireless body area networks. Sustainable Comput.: Inform. Syst. **30**, 100516 (2021)
10. Kumar, A., Sharma, K., Sharma, A.: Genetically optimized Fuzzy C-means data clustering of IoMT-based biomarkers for fast affective state recognition in intelligent edge analytics. Appl. Soft Comput. **109**, 107525 (2021)
11. Saleh, N., Kassem, A., Haidar, A.: Energy-efficient architecture for wireless sensor networks in healthcare applications. IEEE Access. **6**, 6478–6486 (2018)
12. Qureshi, K., Tayyab, M., Rehman, S., Jeon, G.: An interference aware energy efficient data transmission approach for smart cities healthcare systems. Sustain. Cities Soc. **62**, 102392 (2020)
13. Shukla, A., Tripathi, S.: A multi-tier based clustering framework for scalable and energy efficient WSN-assisted IoT network. Wireless Netw. **26**(5), 3471–3493 (2020). https://doi.org/10.1007/s11276-020-02277-4
14. Deebak, D., Al-Turjman, F.: A hybrid secure routing and monitoring mechanism in IoT-based wireless sensor networks. Ad Hoc Netw. **97**, 102022 (2020)
15. Preeth, S., Dhanalakshmi, R., Kumar, R., Shakeel, P.: An adaptive fuzzy rule based energy efficient clustering and immune-inspired routing protocol for WSN-assisted IoT system. Journal of Ambient Intelligence and Humanized Computing. (2018)

16. Gupta, V., Pandey, R.: An improved energy aware distributed unequal clustering protocol for heterogeneous wireless sensor networks. Eng. Sci. Technol., an Int. J. **19**(2), 1050–1058 (2016)
17. Singh, S.: An energy aware clustering and data gathering technique based on nature inspired optimization in WSNs. Peer-to-Peer Networking Appl. **13**(5), 1357–1374 (2020). https://doi.org/10.1007/s12083-020-00890-w
18. Heinzelman, W., Chandrakasan, A., Balakrishnan, H.: Energy-efficient communication protocol for wireless microsensor networks. In: Proceedings of the 33rd Annual Hawaii International Conference on System Sciences
19. Wu, D., Xu, S., Kong, F.: Convergence analysis and improvement of the chicken swarm optimization algorithm. IEEE Access. **4**, 9400–9412 (2016)
20. Majumdar, A., Debnath, T., Biswas, A., Sood, S.K., Baishnab, K.L.: An energy efficient e-healthcare framework supported by Novel EO-μGA (Extremal Optimization Tuned Micro-Genetic Algorithm). Inf. Syst. Front. **23**(4), 1039–1056 (2020). https://doi.org/10.1007/s10796-020-10016-5
21. Anguraj, D., Thirugnanasambandam, K.: Enriched cluster head selection using augmented bifold cuckoo search algorithm for edge-based internet of medical things. Int. J. Commun. Syst. 34(9), e4817 (2021)
22. El-shafeiy, E., Sallam, K., Chakrabortty, R., Abohany, A.: A clustering based swarm intelligence optimization technique for the internet of medical things. Expert Syst. Appl. **173**, 114648 (2021)
23. Bharathi, R., et al.: Energy efficient clustering with disease diagnosis model for IoT based sustainable healthcare systems. Sustainable Comput.: Inform. Syst. **28**, 100453 (2020)
24. https://www.researchgate.net/publication/359865819

A Framework for Network Self-evolving Based on Distributed Swarm Intelligence

Changbo Tian[1,2], Yongzheng Zhang[3], and Tao Yin[1,2(✉)]

[1] Institute of Information Engineering, Chinese Academy of Sciences, Beijing 100093, China
{tianchangbo,yintao}@iie.ac.cn

[2] School of Cyber Security, University of Chinese Academy of Sciences, Beijing 100049, China

[3] Chinese Asset Cybersecurity Technology Co., Ltd., Beijing 100041, China

Abstract. More and more users are attracted by P2P networks characterized by decentralization, autonomy and anonymity. The management and optimization of P2P networks have become the important research contents. This paper presents a framework for network self-evolving problem based on distributed swarm intelligence, which is achieved by the collaboration of different nodes. Each node, as an independent agent, only has the information of its local topology. Through the consensus method, each node searches for an evolving structure to evolve its local topology. The self-evolving of each node's local topology makes the whole topology converge to the optimal topology model. In the experiments, two simulated examples under different network topologies illustrate the feasibility of our approach.

Keywords: Self-evolving network · Swarm intelligence · Distributed optimization · Distributed computing · Distributed consensus

1 Introduction

Benefitted from the decentralization, scalability, autonomy and anonymity, P2P network has been extensively applied in many fields, such as file exchange, peer-to-peer computing, cooperative work, instant communication, search engine and so on. That applications and users based on P2P network rapidly increase has raised a new challenge to the management and optimization of P2P network. Many technologies can be used to achieve the optimization of P2P network, such as game theory [1,2], neural network [3,4], distributed computation [5–7] and so on. But in practical applications, many approaches confront with the problem about computational complexity, optimization effectiveness, computational convergence and so on.

Supported by the National Key Research and Development Program of China under Grant No. 2019YFB1005203.

D. Groen et al. (Eds.): ICCS 2022, LNCS 13351, pp. 177–183, 2022.
https://doi.org/10.1007/978-3-031-08754-7_24

In dynamic network environment, each node in P2P network is allowed to join or exit the network freely. In addition, in some special applications, such as anti-tracking network, anonymous network, etc., the topology structures of such networks need to be hidden for the purpose of security and privacy that brings a big challenge to the network self-optimization. To address this problem, we present a framework for network self-evolving based on distributed swarm intelligence. Each node acts as an independent agent to collaborate with other nodes and searches for the optimal local topology which is beneficial to all relevant nodes. Then, according to the consensus results, the relevant nodes evolve the local topology. Our approach also designs an optimal topology model to guide each node's local topology evolving to guarantee the convergence of their evolving processes.

2 Model of Self-evolving Network

Based on an unstructured P2P network, our approach achieves the topology self-evolving towards the optimal topology through the distributed swarm intelligence. We denote the self-evolving network as an undirected graph $G(V, E)$, in which V denotes the node collection and E denotes the edge collection. Assume graph G has N nodes, each node $v_i (1 \le i \le N)$ is regarded as a biological species having a fitness value f_i which directly reflects node v_i' local topology's status. Node v_i collaborates with its neighboring nodes and adjusts its local topology to search for a local optimal f_i.

The optimal topology model is defined as a kind of network topology which has the uniform distribution of all nodes' degrees. We set a degree threshold τ to limit the density of each node's degree. Formally, the optimal topology T_o can be defined as shown in Eq. 1 in which $D(v_i)$ denotes the degree of node v_i.

$$T_o = \{v_i \in V | D(v_i) \to \tau\} \tag{1}$$

For node v_i, its local clustering coefficient c_i can be calculated by Eq. 2 in which Z_i denotes the edge number in the group of node v_i and its neighboring nodes.

$$c_i = \frac{2 \times Z_i}{d_i(d_i - 1)} \tag{2}$$

Assume node v_i has the degree of τ, we consider two extreme cases, one is that any two neighboring nodes of node v_i has no links, the other is that node v_i and its neighboring nodes form a full-connected graph. Then, we can calculate the lowest and highest local clustering coefficients of node v_i with the degree τ, separately denoted as $c_i^l = \frac{2}{\tau - 1}$ and $c_i^h = 1$.

Combined with the node degree and local clustering coefficient, the fitness value f_i of each node v_i can be calculated by Eq. 3, in which $\bar{c_i}$ denotes the average local clustering coefficient of node v_i' neighboring nodes, α and β denote the normalized coefficients. The average local clustering coefficient $\bar{c_i}$ can be calculated by Eq. 4 in which n_i denotes the number of node v_i' neighboring nodes, N_i denotes the neighboring node collection of node v_i, c_v denotes the local clustering coefficient of the neighboring node v.

$$f_i = F_i(N_i) = \frac{|d_i - \tau|}{\alpha} + \frac{|c_i - \bar{c_i}|}{\beta} \tag{3}$$

$$\bar{c_i} = \frac{1}{n_i}\sum_{v \in N_i} c_v \tag{4}$$

3 Self-evolving Architecture

3.1 Distributed Swarm Intelligence Algorithm

Assume P2P network contains n nodes, $V = \{v_1, v_2, \cdots, v_n\}$. Each node $v_i(1 \leq i \leq n)$ independently executes the swarm intelligence algorithm, searches for its optimal local topology which minimizes the fitness value calculated by Eq. 3 according to node v_i' neighboring nodes set N_i. As shown in Eq. 5, the neighboring nodes set N_i^* of node v_i represents its optimal local topology. All nodes collaborate with its neighboring nodes to search for the global optimal topology. Based on the distributed swarm intelligence algorithm, the objective function F_G of the global topology can be shown as Eq. 6.

$$N_i^* = arg \min_{N_i \subseteq V} F_i(N_i) \tag{5}$$

$$F_G \triangleq \frac{1}{n}\sum_{i=1}^{n} F_i(N_i) \tag{6}$$

Since each node v_i observes only one component of the objective function F_G while exchanging the information with its neighboring nodes, each node v_i estimates a local optimal topology Lo_i with its own swarm intelligence algorithm separately. We define the evolving area of each node v_i as $D_i = \{u|u \in V \; and \; dst(u, v_i) \leq 2\}$, in which $dst(v, u)$ denotes the distance of node v and u. Each node v_i collaborates with all nodes in its evolving area D_i to evolve its local topology. Each node only has two atomic operations in the optimization of its local topology: (1) breaks connection with its neighboring nodes and (2) builds connection with other nodes.

$$\begin{cases} f_i^1 = F_i(N_i^1) \; , \; N_i^1 \leftarrow AO_1(n_i, u_1) \\ f_i^2 = F_i(N_i^2) \; , \; N_i^2 \leftarrow AO_2(n_i, u_2) \\ \quad \vdots \\ f_i^m = F_i(N_i^m) \, , \, N_i^m \leftarrow AO_m(n_i, u_2) \end{cases} \tag{7}$$

Assume that there are m nodes in the evolving area D_i of node v_i, which are labelled as $u_1, u_2, \cdots, u_m$. For each node $u_j(1 \leq j \leq m)$, node v_i evaluates the fitness value after it implements the atomic operation. Here, $AO(x, y)$ denotes the atomic operation between node x and y, N_i^j denotes node v_i's new neighboring nodes set after the implementation of $AO(v_i, u_j)$, f_i^0 denotes the fitness value without any atomic operations. Then, node v_i can get m fitness values, as shown in Eq. 7.

Node v_i minimizes the fitness value to search for its optimal local topology, but cannot be at the cost of the relevant nodes' local topology. So, node v_i firstly chooses the "better" fitness values and collaborates with relevant nodes to make a consensus. We define the candidate set which is consist of fitness value f_i^j, candidate node u_j and the relevant atomic operation AO_j as shown in Eq. 8. Each node $u_j (1 \leq j \leq m)$ in the evolving area D_i will also get its own Vec_j. Then, node v_i needs to search for an atomic operation which minimizes fitness values of both sides to make a consensus.

$$Vec_i = \{(f_i^j, C_i, AO_j(n_i, u_j)) | f_i^j > f_i^0, 1 \leq j \leq m\} \tag{8}$$

3.2 Consensus Search

For each node v_i and its candidate set Vec_i, node v_i needs to implement the following steps to make consensuses with the candidate nodes in Vec_i.

(1) *Target node search.* Node v_i searches for the target nodes of which the candidate sets also contains the atomic operation related with node v_i.
(2) *Atomic operation evaluation.* Node v_i evaluates the effects of the atomic operations on the relevant target nodes' fitness values and searches for the atomic operations which minimize the fitness values of both sides.
(3) *Local topology evolution.* Node v_i and its target node exchange the information with each other to make consensuses for local topology evolution.

At first, each node v_i iterates the candidate nodes in C, and requests that if node u_j's candidate nodes set also contains v_i. If so, it means that the atomic operation between node v_i and u_j benefits both parties. Then, node v_i chooses node u_j as target node for further atomic operation. At the same time, node u_j sends its fitness value about its current local topology to node v_i. After the step of *target node search*, node v_i will get two important sets: the target nodes set T and the corresponding fitness values set FV.

Secondly, node v_i evaluates the effect of the atomic operation with each target node and calculates the evaluation value ev for quantitative evaluation of the local topology evolution. The calculation of evaluation value ev is shown in Eq. 9, in which, α and β are weight coefficients, f_u and f_v separately denote the fitness values of the current local topology of node u_j and v_i, f_u^0 and f_v^0 separately denote the fitness values of the evolved local topology of node u_j and v_i after the corresponding atomic operation. In general, the values of α and β are 0.5.

$$ev = \alpha * (f_u^0 - f_u) + \beta * (f_v^0 - f_v) \tag{9}$$

The evaluation value ev measures the effect of an atomic operation on the change of the fitness values of the relevant nodes. Node v_i searches for the min evaluation value, and evolves its local topology based on the corresponding atomic operation to make sure that the topology evolution approaches to the optimal topology T_o defined in Eq. 1. After the step of *atomic operation evaluation*, node v_i get the evaluation values set EV.

Node v_i sorts the evaluation values set EV according to the evaluation value ev from the big to small. Node v_i chooses the biggest one and negotiates with the corresponding target node for the implementation of the atomic operation. If the target node makes a consensus with node v_i, they collaborate to evolve the local topology based on the corresponding atomic operation. If not, node v_i chooses the next target node in the sorted evaluation values set EV for the negotiation of the local topology evolution until node v_i succeeds in making a consensus with one target node, or the iteration of EV is completed.

3.3 Topology Self-evolving

In a n-nodes P2P network, a stable condition is introduced for each node to estimate necessity of consensus search and local topology evolving. For each node $v_i (1 \leq i \leq n)$, the criterion of the stable condition can be defined as σ, which can be calculated as Eq. 10. In Eq. 10, m denotes the node number in node v_i' evolving area, $EA(v_i)$ denotes the node set of node v_i' evolving area, f_{v_i} and f_u separately denote the fitness values of node v_i and u in the current local topology.

$$\sigma = \frac{1}{m} \sum_{u \in EA(v_i)} |f_{v_i} - f_u| \tag{10}$$

According to the optimal topology definition shown in Eq. 1, the optimal topology has the uniform distribution of all nodes' degree and each node's degree is close to the degree threshold τ. So, with the determined value τ, the fitness value of each node also approaches to a determined value. Because an absolute uniform distribution of nodes' degree can not be achieved, we set an optimal fitness value interval $<f_{down}, f_{up}>$ to conduct each node to evolve its local topology. If in node v_i's evolving area, every node's fitness value is in the interval, we believe node v_i is in stable condition. Every node in stable condition will not implement evolving calculation and topology optimizing except its local topology has been changed.

Based on the interval $<f_{down}, f_{up}>$ and Eq. 10, the max value of criterion value σ is $(f_{up} - f_{down})/m$, and the min value is 0. So, the stable condition can be defined as that each node v_i' criterion value σ is in the interval $<0, (f_{up} - f_{down})/m>$ for a given optimal fitness value interval $<f_{down}, f_{up}>$. In other words, the optimal fitness value interval $<f_{down}, f_{up}>$ determined the eventual status of network self-evolving. The fitness value interval can be calculated by the optimal topology' status.

For example, in our previous work [8], we conceived a topology model, named convex-polytope topology(CPT). When CPT reaches the maximum connectivity, the average degree of all nodes in CPT approaches to 6. We take CPT as an example, then $\tau = 6$. In the optimal structure of CPT, the degree interval is <5, 7>. According to the CPT's property and Eq. 2, the corresponding clustering coefficient interval is <2/3, 1>. Then, according to Eq. 3 in which we set $\alpha = 1$ and $\beta = 1$, we can calculate the optimal fitness value interval is <0, 4/3>.

4 Performance Evaluation

To evaluate the performance of network self-evolving of our proposal, we use ring topology and centralized topology to construct two networks with 100 nodes, and deploy our algorithm on each node for network self-evolving. We use d_{min}, d_{max} and d_{avg} to denote the minimum node degree, maximum node degree and average node degree respectively. We define the degree threshold τ of the optimal topology model T_o as 6 for our proposal to calculate the fitness value.

As illustrated in Fig. 1(a), the network is constructed in ring topology initially, in which each node has the degree of 2. So, the values of d_{max}, d_{min} and d_{avg} are same with 2 at the beginning. After each round of network self-evolving, d_{max} increases until its value reaches to 7. Because we set the degree threshold $\tau = 6$, when one node's degree is bigger than 6, its fitness value will be decreased. The value of d_{min} increases until it reaches to 5, because our proposal cannot achieve the absolute uniform distribution of nodes' degree. But, the value of d_{avg} finally reaches to 5.3 which is very close to the degree threshold.

As illustrated in Fig. 1(b), the network is constructed in centralized topology initially. The central node connects with all other nodes which has the degree of 99, and the other nodes only have one connection with central node which have the degree of 1. So, the d_{max} and d_{min} is 99 and 1 at the beginning. The value of d_{max} decreases until it approaches to 9, the value of d_{min} increases until it reaches to 5. After the network self-evolving, the node degree of this network inclines to balanced, and the degree of all nodes is in the interval <5, 9>. Experiments on different topologies show the effectiveness of our proposal on the network self-evolving. In both of centralized and ring topologies, our proposal can achieve the topology self-evolving towards the optimal topology model T_o.

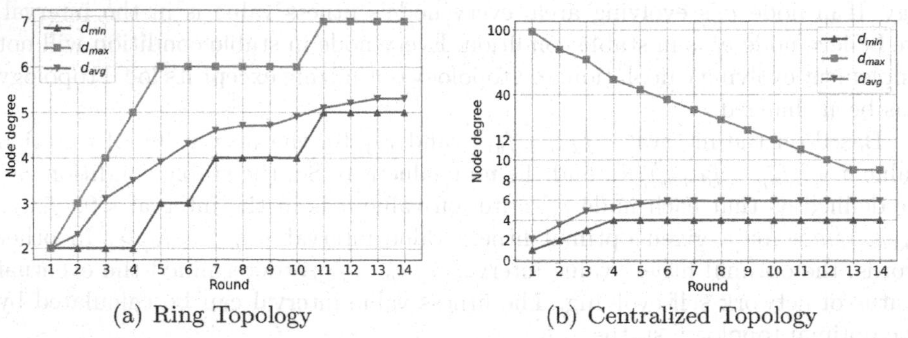

(a) Ring Topology (b) Centralized Topology

Fig. 1. The change of d_{min}, d_{max} and d_{avg} in different topologies.

5 Conclusion

In this paper, we present a framework for network self-evolving based on distributed swarm intelligence to solve the management and optimization problem of dynamic network topology without the global view of the whole P2P network. Each node, as an independent agent, evaluates the atomic operations with the other nodes in its evolving area to search for the optimal local topology. Then, each node negotiates with the relevant node for local topology adjustment to make the whole network converge to the optimal topology model. We also evaluate the feasibility of our approach on two simulated examples which are ring topology and centralized topology.

Acknowledgements. The authors would like to thank the anonymous reviewers for their insightful comments and suggestions on this paper. This work was supported in part by the National Key Research and Development Program of China under Grant No. 2019YFB1005203.

References

1. Rozario, F., Han, Z., Niyato, D.: Optimization of non-cooperative P2P network from the game theory point of view. In: 2011 IEEE Wireless Communications and Networking Conference, pp. 868–873. IEEE (2011)
2. Charilas, D.E., Panagopoulos, A.D.: A survey on game theory applications in wireless networks. Comput. Netw. **54**(18), 3421–3430 (2010)
3. Lee, S.W., Palmer-Brown, D., Roadknight, C.M.: Performance-guided neural network for rapidly self-organising active network management. Neurocomputing **61**, 5–20 (2004)
4. Auvinen, A., Keltanen, T., Vapa, M.: Topology management in unstructured P2P networks using neural networks. In: 2007 IEEE Congress on Evolutionary Computation, pp. 2358–2365. IEEE (2007)
5. Tian, C., Zhang, Y., Yin, T.: Topology self-optimization for anti-tracking network via nodes distributed computing. In: Gao, H., Wang, X. (eds.) CollaborateCom 2021. LNICST, vol. 406, pp. 405–419. Springer, Cham (2021). https://doi.org/10.1007/978-3-030-92635-9_24
6. Berahas, A.S., Bollapragada, R., Keskar, N.S., Wei, E.: Balancing communication and computation in distributed optimization. IEEE Trans. Autom. Control **64**(8), 3141–3155 (2018)
7. Wang, H., Liao, X., Huang, T., Li, C.: Cooperative distributed optimization in multiagent networks with delays. IEEE Trans. Syst. Man Cybern. Syst. **45**(2), 363–369 (2014)
8. Tian, C., Zhang, Y., Yin, T.: Modeling of anti-tracking network based on convex-polytope topology. In: Krzhizhanovskaya, V.V., Závodszky, G., Lees, M.H., Dongarra, J.J., Sloot, P.M.A., Brissos, S., Teixeira, J. (eds.) ICCS 2020. LNCS, vol. 12138, pp. 425–438. Springer, Cham (2020). https://doi.org/10.1007/978-3-030-50417-5_32

Investigating an Optimal Computational Strategy to Retrofit Buildings with Implementing Viscous Dampers

Farzin Kazemi[1(✉)], Neda Asgarkhani[2], Ahmed Manguri[1], and Robert Jankowski[1]

[1] Faculty of Civil and Environmental Engineering, Gdańsk University of Technology, ul. Narutowicza 11/12, 80-233 Gdansk, Poland
{farzin.kazemi,ahmed.manguri,jankowr}@pg.edu.pl

[2] Department of Civil Engineering, Faculty of Engineering and Technology, Imam Khomeini International University, PO Box 34149-16818, Qazvin, Iran
n.asgarkhani@edu.ikiu.ac.ir

Abstract. Civil engineering structures may seriously suffer from different damage states result of earthquakes. Nowadays, retrofitting the existing buildings is a serious need among designers. Two important factors of required performance level and cost of retrofitting play a crucial role in the retrofitting approach. In this study, a new optimal computational strategy to retrofit structures by implementing linear Viscous Dampers (VDs) is investigated to achieve a higher performance level with lower implementation cost. Regarding this goal, a Tcl programming code was developed with the capability of considering damaged structure due to earthquake-induced structural pounding. The code allows us to improve structural models to take into account the real condition of buildings using both MATLAB and Opensees software simultaneously. To present the capability of this strategy, the 3-, and 6-story colliding Steel Moment-Resisting Frames (SMRFs) were selected. Incremental Dynamic Analysis (IDA) was performed based on the interstory drift ratio of floor levels as engineering demand parameter, and $S_a(T_1)$ as intensity measure. Interstory median IDAs of floor levels of colliding SMRFs were plotted to find out the floor level prone to damage and to retrofit only this floor level instead of all stories. The results show that implementing only two linear VDs with a cost of two units can achieve a higher life safety performance level in the case of 3-, and 6-story SMRFs. Moreover, the proposed computational strategy can be used for any structure (with and without pounding conditions), and in all performance levels prescribed in FEMA 356 code.

Keywords: Optimal computational strategy · Opensees programming · Retrofitting of buildings · Viscous damper · Structural pounding · Earthquakes

1 Introduction

During severe earthquakes, buildings and bridge structures may suffer from different damage states, from local damage to the total collapse [1, 2]. In the case of buildings,

D. Groen et al. (Eds.): ICCS 2022, LNCS 13351, pp. 184–191, 2022.
https://doi.org/10.1007/978-3-031-08754-7_25

many researchers proposed procedures of retrofitting using structural elements or energy dissipation devices. Using additional structural elements can increase the stiffness of the whole structure and can dissipate lower energy than using energy dissipation devices, such as fluid Viscous Dampers (VDs), which would not alter the stiffness of the structure, and consequently, the frequency of vibration. These devices can perform in a wide range of temperatures (between -40 °C to 70 °C) and lower maintenance is required during longer life service. Therefore, this type of energy dissipation device was widely investigated and implemented in single structures, or between adjacent structures [3]. Kazemi et al. [4] investigated a seismic retrofitting procedure that uses diagonal VDs in all story levels of buildings. The results confirmed that using this strategy significantly influenced the seismic collapse capacity of the structures. In addition, Kazemi et al. [5] proposed using VDs between adjacent structures, which can prevent pounding during earthquakes. The distribution of VD within a structure is a critical decision due to its effects on the seismic response. While a large number of VD placements have been proposed, a limited comparison was conducted to investigate the effectiveness of these methods [6]. Predicting the seismic limit state or collapse capacity of a structure is a useful tool to determine the performance level of the structure during severe earthquake or due to impact forces induced by the pounding phenomenon [7, 8]. This performance level can help a designer to determine the damage state of a building [9–11]. Regarding this issue, the main purpose of this study is to investigate a new computational strategy to optimize the performance levels of structure and cost of VDs implementation. Regarding this goal, a Tcl programming code was developed with the capability of considering damaged structure due to earthquake-induced structural pounding.

2 Modeling Approach

The three and six story level (3-story and 6-story) Steel Moment-Resisting Frames (SMRFs), designed according to ASCE 7–10 [12], were used in this study (see also [4–6]). Figure 1 presents the documentation and structural elements of them.

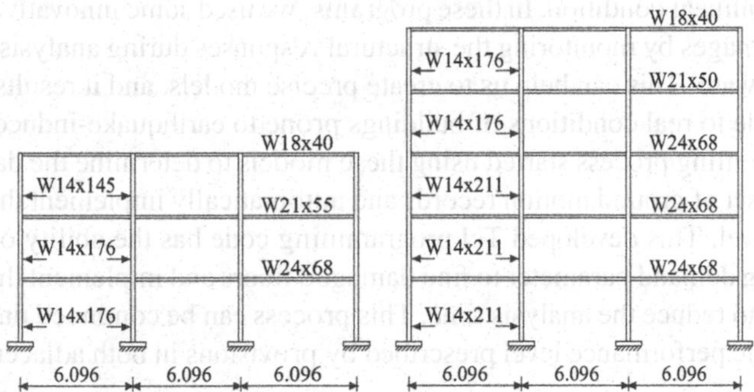

Fig. 1. Documentation and structural elements of the 3-, and 6-story SMRFs.

The plan presented in Fig. 2 was used to design the structures. According to this plan, there are four SMRFs in the three-dimensional building and one of them was modeled in this study. All columns were considered as leaning columns assuming the P-delta effect, except for those which belong to this single SMRF [13–15]. In addition, to model beams and columns, nonlinear rotational spring was used according to the Modified Ibarra–Krawinkler bilinear-hysteretic model [16].

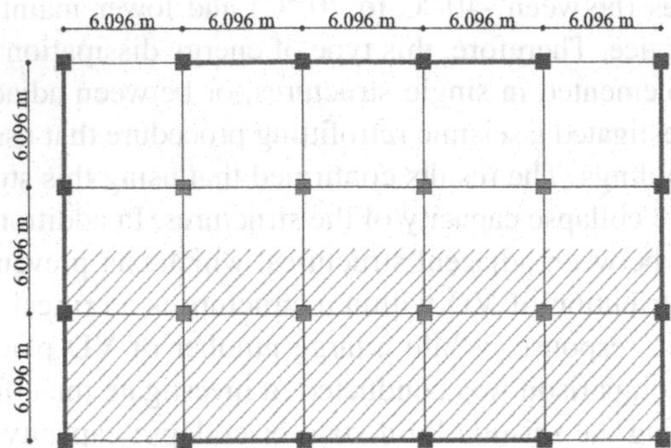

Fig. 2. Structural plan of the 3-, and 6-story SMRFs.

3 Computational Retrofitting Strategy

Previous researches have confirmed that using linear VDs show higher performance than using nonlinear VDs [4–6]. Therefore, in this paper, the effects of linear VDs are investigated. Kazemi et al. [4–6] used equations and procedures to model the pounding phenomenon, calculating an allowable clear distance between structures and implementing linear VDs. In order to simulate more accurately the real condition of structures exposed to pounding, the 3-story and 6-story SMRFs were modeled in MATLAB [17] and Opensees [18] softwares using a developed Tcl programming code with the capability of considering damages during analysis. Tcl program can analyze a model with high accuracy in nonlinear condition. In these programs, we used some innovative approaches to consider damages by monitoring the structural responses during analysis using MATLAB [17] software. This can help us to create precise models, and it results would have higher resemble to real conditions of buildings prone to earthquake-induced pounding. Then, the retrofitting process started using these models to determine the damaged floor level during a set of ground motion records and automatically implement the linear VDs in that floor level. This developed Tcl programming code has the ability of controlling the engineering demand parameter to find damaged floors and implement the linear VDs automatically to reduce the analysis time. This process can be continued until achieving a higher seismic performance level prescribed by provisions in both adjacent structures.

4 Results of Strategy

Incremental Dynamic Analysis (IDA) is a method to determine the total seismic collapse capacity. In this method, the potential levels of a ground motion, known as the intensity measure (e.g. $S_a(T_1)$), and an engineering demand parameter (e.g. Interstory Drift Ratio (IDR)) are used to control the structure condition [4–6, 19, 20]. In this study, to perform IDAs, Near-field Pulse-Like (NPL) ground motion records suggested by FEMA P695 [21] were used, and IDA curves were plotted for all floor levels of the adjacent 3-, and 6-story SMRFs using IDR for all story levels. To compare the IDA curves, the Interstory Median of IDA curves (IM-IDAs) were determined. Figure 3 presents IM-IDAs of all floor levels of the 3-, and 6-story SMRFs in pounding conditions subjected to NPL record subset assuming a clear distance of 0.298 m. To better illustrate the current state of a structure during severe earthquakes, the performance level was used. According to FEMA 356 [22], for primary structural elements, three performance levels of Immediate Occupancy (IO), Life Safety (LS), and Collapse Prevention (CP) were assumed regarding damages occurring in the structure. Therefore, the performance levels of IO, LS, and CP for SMRFs have the values of IDR of 0.7%, 2.5%, and 5.0%, respectively. In this study, the optimal retrofitting strategy using linear VDs is investigated to improve the performance level of LS. Therefore, IM-IDAs of all floor levels were compared in LS performance level (2.5%).

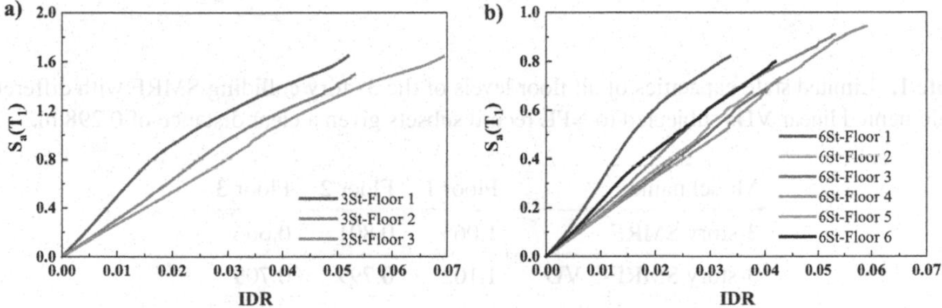

Fig. 3. IM-IDA curves of all floor levels of the 3-, and 6-story colliding SMRFs subjected to NPL record subset given clear distance of 0.298 m.

According to Fig. 3(a), the third floor of 3-story SMRFs has the lowest value of $S_a(T_1)$ in the LS performance level. Moreover, regarding Fig. 3(b), the fourth floor of 6-story SMRFs has the lowest value of $S_a(T_1)$ in the LS performance level. Therefore, these story levels were automatically selected for implementing one linear VD in the first retrofit analysis. Figure 4(a) presents IM-IDA curves of all floor levels of the 3-story colliding SMRF retrofitted with one linear VD in the third story level. It can be seen that using the linear VD in the third floor level increases IM-IDA curves of the 1st, 2nd, and 3rd floor levels by 3.47%, –0.24%, and 6.45%, respectively. In addition, the third floor level still has the lowest value of $S_a(T_1)$ in the LS performance level. Then this floor level was automatically selected for adding the second linear VD in the second retrofit computational analysis. The results of IM-IDA curves of all floor levels of the 3-story colliding SMRF retrofitted with two linear VDs in the third story level are presented in Fig. 4(b). According to this figure, IM-IDA curves of the 1st, 2nd, and 3rd floor levels have increased by 33.89%, 42.57%, and 97.29%, respectively. Therefore, optimal placement of linear VDs in 3-story colliding SMRF with higher performance level achieved by using only two linear VDs that can be implemented by a cost of two units (cost of each linear VD assumed as one unit). Table 1 presents the values of $S_a(T_1)$ in the LS performance level for the 3-story colliding SMRF.

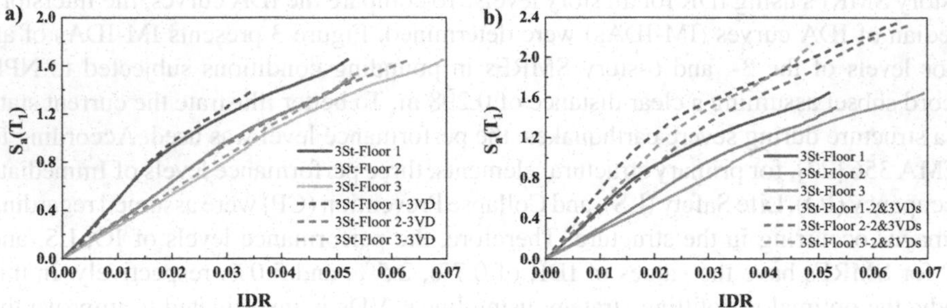

Fig. 4. Comparison between IM-IDA curves of all floor levels of the 3-story colliding SMRF retrofitted with, a) one linear VD in third story level, b) two linear VDs in second and third story levels, subjected to NPL record subset given clear distance of 0.298 m.

Table 1. Limited state capacities of all floor levels of the 3-story colliding SMRF with different implemented linear VDs subjected to NPL record subsets given a clear distance of 0.298 m.

Model name	Floor 1	Floor 2	Floor 3
3-story SMRF	1.065	0.801	0.666
3-story SMRF-3 VD	1.102	0.799	0.709
3-story SMRF-2 and 3 VDs	1.426	1.142	1.314

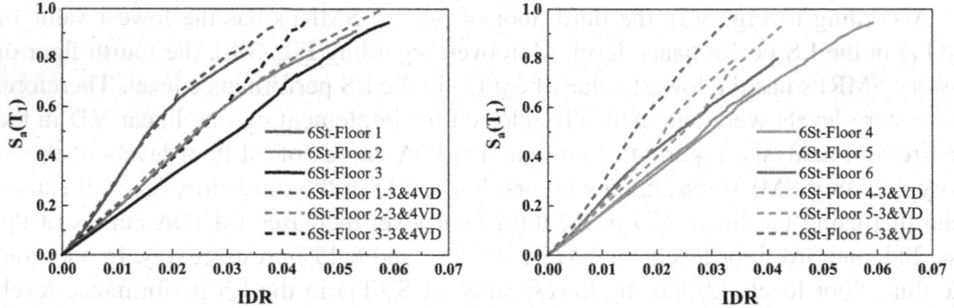

Fig. 5. Comparison between IM-IDA curves of all floor levels of the 6-story colliding SMRF retrofitted with two linear VDs in third and fourth story levels, subjected to NPL record subset given clear distance of 0.298 m.

Table 2. Limited state capacities of all floor levels of the 6-story colliding SMRF with different implemented linear VDs subjected to NPL record subsets given a clear distance of 0.298 m.

Model name	Floor 1	Floor 2	Floor 3	Floor 4	Floor 5	Floor 6
6-story SMRF	0.696	0.509	0.421	0.398	0.437	0.519
6-story SMRF-4 VD	0.669	0.502	0.449	0.466	0.561	0.677
6-story SMRF-3 and 4 VDs	0.719	0.521	0.506	0.504	0.620	0.767

In the next, linear VD added on the fourth floor level of the 6-story SMRF and the third floor level has the lowest value of $S_a(T_1)$ and selected for retrofitting. Figure 5 presents the results of all floor IM-IDA curves of the 6-story colliding SMRF in the second retrofit computational analysis. It is shown that IM-IDA curves of the 1st, 2nd, 3rd, 4th, 5th, and 6th floor levels have increased by 3.3%, 2.35%, 27.13%, 19.71%, 41.87%, and 47.78%, respectively. Therefore, optimal placement of linear VDs in 6-story colliding SMRF with higher performance level has been achieved by using only two linear VDs that can be implemented by a cost of two units. Table 2 presents the values of $S_a(T_1)$ in the LS performance level for the 6-story colliding SMRF.

5 Conclusion

This study investigated a computational strategy that uses a developed Tcl programming code with the capability of considering damaged structure due to earthquake-induced structural pounding. The developed code allows us to improve structural models to take into account the real condition of buildings using both MATLAB [17] and Opensees [18] softwares simultaneously. Implementing linear VDs on the adjacent structures can be assumed as a retrofitting strategy, while the cost of implementation is an important factor. In the research, IM-IDAs based on IDRs were determined performing IDA analysis subjected to NPL record subset, and these curves were used as the main purpose of retrofitting based on the LS performance level. Regarding this issue, the lower value of $S_a(T_1)$ in each floor level in the LS performance level was selected to implement linear LVDs, and the results were compared to the previous state. This type of retrofitting strategy can help designers to find out the floor level prone to damage, and retrofit this particular floor level instead of all stories. It should be noted that this strategy could be used for any structure with and without pounding conditions and in all performance levels prescribed in FEMA 356 [22].

References

1. Mohebi, B., Asadi, N., Kazemi, F.: Effects of using gusset plate stiffeners on the seismic performance of concentrically braced frame. Int. J. Civ. Environ. Eng. **13**(12), 723–729 (2019)
2. Rezaei, H., Moayyedi, S.A., Jankowski, R.: Probabilistic seismic assessment of RC box-girder highway bridges with unequal-height piers subjected to earthquake-induced pounding. Bull. Earthq. Eng. **18**(4), 1547–1578 (2020). https://doi.org/10.1007/s10518-019-00764-4

3. Lu, Z., Wang, Z., Zhou, Y., Lu, X.: Nonlinear dissipative devices in structural vibration control: A review. J. Sound Vib. **423**, 18–49 (2018)
4. Kazemi, F., Mohebi, B., Yakhchalian, M.: Enhancing the seismic performance of adjacent pounding structures using viscous dampers. In: Proceedings of the 16th European Conference on Earthquake Engineering (16ECEE), June 2018, pp. 18–21. Thessaloniki, Greece (2018)
5. Kazemi, F., Miari, M., Jankowski, R.: Investigating the effects of structural pounding on the seismic performance of adjacent RC and steel MRFs. Bull. Earthq. Eng. **19**(1), 317–343 (2021). https://doi.org/10.1007/s10518-020-00985-y
6. Kazemi, F., Mohebi, B., Jankowski, R.: Predicting the seismic collapse capacity of adjacent SMRFs retrofitted with fluid viscous dampers in pounding conditions. Mech. Syst. Signal Process. **161**, 107939 (2021)
7. Elwardany, H., Seleemah, A., Jankowski, R., El-Khoriby, S.: Influence of soil-structure interaction on seismic pounding between steel frame buildings considering the effect of infill panels. Bull. Earthq. Eng. **17**(11), 6165–6202 (2019)
8. Sołtysik, B., Falborski, T., Jankowski, R.: Preventing of earthquake-induced pounding between steel structures by using polymer elements – experimental study. Procedia Eng. **199**, 278–283 (2017)
9. Kazemi, F., Mohebi, B., Yakhchalian, M.: Predicting the seismic collapse capacity of adjacent structures prone to pounding. Can. J. Civ. Eng. **47**(6), 663–677 (2020)
10. Mohebi, B., Yazdanpanah, O., Kazemi, F., Formisano, A.: Seismic damage diagnosis in adjacent steel and RC MRFs considering pounding effects through improved wavelet-based damage-sensitive feature. J. Build. Eng. **33**, 101847 (2021)
11. Yazdanpanah, O., Mohebi, B., Kazemi, F., Mansouri, I., Jankowski, R.: Development of fragility curves in adjacent steel moment-resisting frames considering pounding effects through improved wavelet-based refined damage-sensitive feature. Mech. Syst. Signal Process. **173**, 109038 (2022)
12. Minimum Design Loads for Buildings and Other Structures (ASCE/SEI 7–10), first, second, and third printings. Minimum Design Loads for Buildings and Other Structures (2010)
13. Mohebi, B., Kazemi, F., Yakhchalian, M.: Investigating the P-Delta effects on the seismic collapse capacity of adjacent structures. In: Proceedings of the 16th European Conference on Earthquake Engineering (16ECEE), June 2018, pp. 18–21. Thessaloniki, Greece (2018)
14. Kazemi, F., Mohebi, B., Yakhchalian, M.: Evaluation of the P-delta effect on collapse capacity of adjacent structures subjected to far-field ground motions. Civ. Eng. J. **4**(5), 1066–1073 (2018)
15. Asgarkhani, N., Yakhchalian, M., Mohebi, B.: Evaluation of approximate methods for estimating residual drift demands in BRBFs. Eng. Struct. **224**, 110849 (2020)
16. Lignos, D.G., Krawinkler, H.: Deterioration modeling of steel components in support of collapse prediction of steel moment frames under earthquake loading. J. Struc. Eng. **137**(11), 1291–1302 (2011)
17. MATLAB/Simulink as a Technical Computing Language.: Engineering Computations and Modeling in MATLAB (2020)
18. McKenna, F., Fenves, G.L., Filippou, F.C., Scott, M.H.: Open system for earthquake engineering simulation (OpenSees). Pacific Earthquake Engineering Research Center, University of California, Berkeley (2016). http://OpenSees.berkeley.edu
19. Yakhchalian, M., Asgarkhani, N., Yakhchalian, M.: Evaluation of deflection amplification factor for steel buckling restrained braced frames. J. Build. Eng. **30**, 101228 (2020)
20. Yakhchalian, M., Yakhchalian, M., Asgarkhani, N.: An advanced intensity measure for residual drift assessment of steel BRB frames. Bull. Earthq. Eng. **19**(4), 1931–1955 (2021). https://doi.org/10.1007/s10518-021-01051-x

21. Applied Technology Council and United States. Federal Emergency Management Agency. Quantification of building seismic performance factors. US Department of Homeland Security, FEMA (2009)
22. FEMA-356: Prestandard and Commentary for the Seismic Rehabilitation of Buildings. Federal Emergency Management Agency, Washington (2000)

ARIMA Feature-Based Approach to Time Series Classification

Agnieszka Jastrzebska[1], Wladyslaw Homenda[1,2(✉)], and Witold Pedrycz[3,4]

[1] The Faculty of Mathematics and Information Science, Warsaw University of Technology, Warsaw, Poland
{A.Jastrzebska,homenda}@mini.pw.edu.pl

[2] The University of Information Technology and Management in Rzeszow, Rzeszow, Poland

[3] The University of Alberta, Edmonton, Canada
wpedrycz@ualberta.ca

[4] The Systems Research Institute, Polish Academy of Sciences, Warsaw, Poland

Abstract. Time series classification is a supervised learning problem that aims at labelling time series according to their class belongingness. Time series can be of variable length. Many algorithms have been proposed, among which feature-based approaches play a key role, but not all of them are able to deal with time series of unequal lengths. In this paper, a new feature-based approach to time series classification is proposed. It is based on ARIMA models constructed for each time series to be classified. In particular, it uses ARIMA coefficients to form a classification model together with sampled time series data points. The proposed method was tested on a suite of benchmark data sets and obtained results are compared with those provided by the state-of-the-art approaches.

Keywords: Time series · ARIMA · Classification · SVMs · Random forest

1 Introduction

Time series classification emerged as a vital area of study in machine learning. A popular group of algorithms are feature-based methods that convert time series into a collection of attributes, which are then subjected to classification. Feature-based approaches reduce problem's dimensionality, as the number of features extracted for each time series is usually much smaller than its length, i.e., the number of data points of the time series. Feature-based methods typically do not need time series to be of equal length.

The project was funded by POB Research Centre Cybersecurity and Data Science of Warsaw University of Technology within the Excellence Initiative Program - Research University (ID-UB).

D. Groen et al. (Eds.): ICCS 2022, LNCS 13351, pp. 192–199, 2022.
https://doi.org/10.1007/978-3-031-08754-7_26

In this paper, a new approach to time series classification is proposed. The idea is to process time series models instead of the time series themselves. It has several advantages. First of all, time series models are defined as a set of parameters, which quantity is much smaller than the length of the original time series. Secondly, the approach is suitable to process sets of time series of differing lengths. Finally, operating on model parameters instead of temporal sequences allows employing general-purpose classifiers. The above properties make our approach a feature-based method, where model parameters can be seen as time series features.

The quality of the new method is evaluated in a series of experiments and compared with the quality of several state-of-the-art approaches. Results show that this technique provides satisfying results.

The particular novel contribution of this paper, not present in previous studies on feature-based approaches to time series classification, is the usage of ARIMA models describing time series and viewing their parameters as patterns to be classified. In this light, we may note that in this paper, we introduce a fusion of several approaches: baseline (plain), feature-based, and model-based.

2 Literature Review

The baseline approach to time series classification would treat each time series data point as a single attribute. Thus, we may apply any standard classifier, for example, random forest, to a data frame in which one row corresponds to one time series and one column corresponds to one moment in time. We would have to make sure that each time series is of the same length and starts at the same moment in time. Surveys show that this baseline approach achieves surprisingly satisfying results [2].

There are numerous algorithms dedicated specifically to the time series classification problem. They fall into two main categories: distance-based and feature-based. The distance-based approaches aim at computing distances between time series. The vast research volume in this area was devoted to studies on various distance measures [1]. In feature-based classification methods, time series are transformed into feature vectors and classified using a conventional classifier such as a neural network or a decision tree. Many approaches fall into this group. For instance, spectral methods, such as discrete Fourier transform [11] or discrete wavelet transform [9] are used to provide features of the frequency domain, which are the basis for classifier construction. We shall also distinguish feature-based methods that are called Bag of Patterns. These are dictionary approaches, in which one extracts specific attributes describing time series and uses those attributes to classify them. The literature offers a wide range of such approaches. For example, we can name methods that transform time series into strings and, at the same time, reduce the length of an input sequence. A particular example of such an approach is Symbolic Aggregate Approximation (SAX) [10]. This method discretizes time series that were previously normalized. Another example is the SAX and Vector Space Model (SAXVSM), which joins the idea of time series to string conversion via discretization, but it adds token (character) weighting [12]. SAX was also fused with DTW in the method named DTW Features. It uses distances computed with the DTW with SAX histograms [8]. It is worth mentioning a method called Bag-of-Features [3]. It produces random subsequences of a given time series from which features are extracted. Fulcher and Jones presented a technique that is the most relevant in the context of the approach introduced in this paper [5]. It extracts time series summaries in terms of correlations, distributions, entropy, stationarity, and scaling. A classification model is constructed using such features.

3 The Method

Feature-based approaches rely on extracting attributes from a raw time series. In this paper, we propose to use the parameters of a time series theoretical model as attributes. We use the AutoRegressive Integrated Moving Average model (ARIMA). The ARIMA model consists of three elements: autoregressive model (AR), moving average model (MA), and model integration (I).

An autoregressive model of order p, denoted as AR(p), uses p previous values to describe (predict) the current value. Its general form is given as:

$$z_t^{'} = c + b_1 z_{t-1} + b_2 z_{t-2} + \ldots + b_p z_{t-p} \tag{1}$$

where $z_t^{'}$ is a description (prediction) of the current value z_t, c is an intercept, describing the drift. The autoregressive model is analogous to the multiple regression model, but with lagged values of time series data points, instead of standard attributes used as predictors. Parameters $b_1, b_2, \ldots, b_p$ describe how strong is a relationship between history and a current value. Autoregressive models are typically applied to stationary data. Thus, time series with a trend or seasonal regularities need to be preprocessed. The description of the current value is an approximation of the real value z_t with the error ε_t:

$$z_t = c + b_1 z_{t-1} + b_2 z_{t-2} + \ldots + b_p z_{t-p} + \varepsilon_t \tag{2}$$

A moving average model, another component of ARIMA, uses past forecast errors in a regression-like model:

$$a_{t-1}\varepsilon_{t-1} + a_{t-2}\varepsilon_{t-2} + \ldots + a_{t-q}\varepsilon_{t-q} \tag{3}$$

q denotes the order of the moving average model; we denote it as MA(q). $a_1, a_2, \ldots, a_q$ are discovered coefficients. While the autoregressive model uses past values, the moving average uses past distortions to model a time series.

The third component of the ARIMA model is integration (I). Integration, in this context, is the action opposite to differentiating. If we join the three components together, we obtain ARIMA(p, d, q) model, where d is the degree of first differentiating applied. The model can be written as:

$$\begin{aligned} \text{ARIMA}(p, d, q) = c + b_1 z'_{t-1} + b_2 z'_{t-2} + \ldots + b_p z'_{t-p} \\ + a_{t-1}\varepsilon_{t-1} + a_{t-2}\varepsilon_{t-2} + \ldots + a_{t-q}\varepsilon_{t-q} + \varepsilon_t \end{aligned} \tag{4}$$

To automatically detect the structure of the model, that is p, d and q, one may use the Hyndman-Khandakar algorithm [6]. It combines unit root tests, minimization of the Akaike information criterion, and Maximum Likelihood Estimation to obtain an ARIMA model.

Parameters of the ARIMA model describe the properties of the time series. In particular, they describe the intensity of the influence of historical values and distortions of the time series on the current state of the process. We postulate to use ARIMA model parameters to distinguish between time series belonging to different classes. We fit an ARIMA model for each time series separately and use computed parameters as attributes: d (the differentiating degree), c (the intercept), $b_1, \ldots, b_p$, (coefficients associated with consecutive values in the autoregressive model), and $a_1, \ldots, a_q$ (coefficients from the moving average model). The following coefficients are generated for a particular training set:

- $1 + 1$ (for d and c)
- p_j (the highest discovered order of autoregressive model in time series in a given training set, j is the number of this time series $j = 1, \ldots, K$))
- q_r (the highest discovered order of moving average model in time series in a given training set, r is the number of this time series $r = 1, \ldots, K$)

If for a given time series, we have obtained an autoregressive model's order lower than p_j or a moving average model's order lower than q_r, then irrelevant coefficients are set to 0. ARIMA coefficients make the elementary data frame that can be subjected to classification using a standard classifier.

The usage of ARIMA parameters as indirect features in pattern mining in time series is already present in the literature. Kalpakis et al. [7] and Wang et al. [13] used it to cluster time series.

To reinforce the efficacy of classification, apart from considering ARIMA parameters, we may include randomly chosen samples from time series. We may append an arbitrary number of such attributes. The simplest way would be to select indexes of time series observations randomly to be appended. The randomization of indexes must be from the range $[1, M_f]$, where M_f is the length of the shortest time series in the training set.

Extracted attributes are subject to a standard classification, which may be preceded by removing correlated columns or thinning a training set.

4 Results

We have conducted experiments using a collection of publicly available time series from http://timeseriesclassification.com/ and compared our results with other methods. The data sets on the web page were already standardized and split into train and test sets. We implemented the method introduced in this paper in R language with the use of e1071, randomForest, and forecast packages.

First experiments were conducted for the scenario when only ARIMA parameters were used as attributes. Next, apart from ARIMA parameters, we added 1%, 5%, 10%, 20%, and 30% of randomly selected time series data points. We

Table 1. Comparison of accuracy (in %) achieved with our method and two groups of other methods: plain and other feature-based approaches. We apply colors to improve visibility: the greener the color the better our model was in the comparison.

data set name	best accuracy feature	plain	ARIMA	differences: ARIMA – feature	– plain	ARIMA model
Beef	80.00%	93.33%	80.00%	0.00%	-13.33%	RF, 20% (94)
BeetleFly	90.00%	90.00%	90.00%	0.00%	0.00%	RF, 5% (26)
BirdChicken	100.00%	85.00%	85.00%	-15.00%	0.00%	RF, 30% (154)
ChlorineConcentration	71.98%	92.42%	78.05%	6.07%	-14.38%	SVM, 30% (50)
Coffee	100.00%	100.00%	100.00%	0.00%	0.00%	RF,SVM, 5% (14)
CricketY	75.38%	61.03%	61.03%	-14.36%	0.00%	RF, 20% (60)
DiatomSizeReduction	93.14%	96.41%	93.14%	0.00%	-3.27%	RF, 20% (69)
DistalPhalanxOutlnAgeG.	78.26%	82.25%	80.67%	2.41%	-1.58%	RF, 30% (24)
DistalPhalanxOutlineCor.	84.17%	80.58%	84.50%	0.33%	3.92%	RF, 10% (8)
DistalPhalanxTW	69.78%	70.50%	78.75%	8.97%	8.25%	RF, 30% (24)
Earthquakes	74.82%	74.82%	81.99%	7.17%	7.17%	SVM, 0
FordA	92.95%	84.47%	90.92%	-2.04%	6.45%	RF, 0
FordB	75.06%	77.16%	86.52%	11.46%	9.36%	SVM, 10% (50)
GunPoint	100.00%	94.00%	95.33%	-4.67%	1.33%	RF, 20% (30)
Herring	64.06%	65.63%	68.75%	4.69%	3.13%	RF, 1% (5)
InlineSkate	51.64%	37.09%	48.18%	-3.45%	11.09%	RF, 0
InsectWingbeatSound	63.28%	65.61%	64.85%	1.57%	-0.76%	RF, 30% (77)
ItalyPowerDemand	96.02%	97.28%	96.89%	0.87%	-0.39%	RF, 10% (2)
Lightning2	85.25%	75.41%	77.05%	-8.20%	1.64%	RF, 10% (64)
Lightning7	75.34%	72.60%	78.08%	2.74%	5.48%	RF, 20% (64)
MiddlePhalanxOutlnCor.	57.79%	61.69%	77.50%	19.71%	15.81%	SVM, 30% (24)
MiddlePhalanxTW	59.74%	62.99%	65.16%	5.42%	2.18%	RF, 20% (16)
MoteStrain	90.34%	88.90%	89.46%	-0.88%	0.56%	RF, 20% (17)
Plane	100.00%	99.05%	99.05%	-0.95%	0.00%	RF, 10% (14)
ProximalPhalanxOutlnCor.	84.88%	86.83%	86.34%	1.46%	-0.49%	RF, 30% (24)
ProximalPhalanxTW	81.46%	82.44%	80.50%	-0.96%	-1.94%	RF, 30% (24)
ScreenType	51.20%	44.80%	46.67%	-4.53%	1.87%	SVM, 0
Strawberry	97.57%	97.30%	94.45%	-3.11%	-2.84%	RF, 10% (24)
SwedishLeaf	92.16%	88.16%	87.04%	-5.12%	-1.12%	RF, 30% (38)
Trace	100.00%	93.00%	95.00%	-5.00%	2.00%	SVM, 10% (28)

used two classifiers: Support Vector Machine (SVM) and random forest (RF). Those two classifiers are very popular. In all experiments, we removed columns with variance lower than 0.01. All SVMs were using Gaussian kernels. We individually tuned parameters with a simple search procedure. We used 500 trees in RF, which is the recommended default value in the used library. All models were trained using train sets. We only used test sets for quality evaluation.

In the designed experiment, we decided to compare the results achieved with our method with two kinds of algorithms: (i) plain classifiers, run directly on time series and (ii) other feature-based algorithms.

The first group of methods is assumed to provide a bottom-line efficiency in time series classification. We considered the following algorithms: Naive Bayes, C4.5 decision tree, SVM with linear (SVML) and quadratic kernel (SVMQ), Bayesian Network, RF with 500 trees, rotation forest with 50 trees, and multi-layer perceptron.

The second group is made of feature-based algorithms. Those are the competitors belonging to the same group as the method addressed in the paper. These were: Bag of Patterns [10], Symbolic Aggregate Approximation – Vector Space Model (SAXVSM) [12], Bag of SFA Symbols (BOSS) [4], Time Series Forest (TSF) [11], Time Series Bag of Features (TSBF) [3].

Table 1 provides aggregated results. They concern test sets. We give the best accuracy achieved by a classifier from each group: plain, feature-based, and the proposed ARIMA feature-based. In the last column in Table 1, we outline which configuration produced our model: a percentage of time series data points added as attributes, the name of a classifier that gave this particular result, and, in brackets, a specific number of data points that was added.

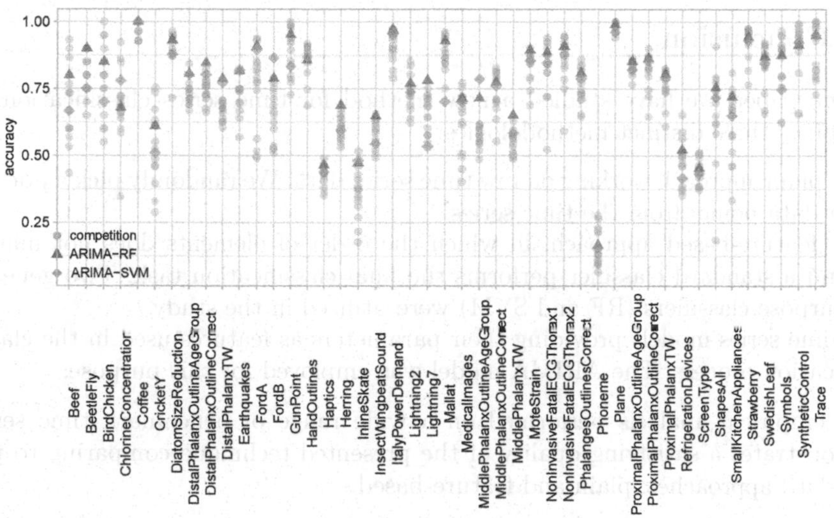

Fig. 1. The accuracy achieved using our method with SVM (light green diamonds) and RF (dark green triangles) as classifiers contrasted with particular results achieved by our 13 competitors (semi-transparent yellow circles). Results concern all 30 data sets. (Color figure online)

In 15 out of 30 data sets, the proposed method outperformed plain classifiers. In the further 5 cases, the best plain classifier produced the same accuracy as our method. In 14 out of 30 data sets, the new method outperformed other feature-based approaches to time series classification. In many cases, the advantage of the proposed method is very high. In 4 cases, the best feature-based competitor gave the same accuracy as the method introduced in this paper. When we compare the ARIMA feature-based method with plain classifiers, in two cases, our technique achieved accuracy at least 10% greater than this group of approaches. If we compare our method with other feature-based approaches, the results are slightly worse. Still, in six cases, the ARIMA feature-based method outperformed its best-performing competitor by at least 5%. RF turned out to be frequently outperforming SVM-based models.

In Fig. 1, we illustrate accuracy achieved using our method with SVM and RF as classifiers contrasted with particular results achieved by our 13 competitors. The plot shows that in several cases, there are noticeable differences between accuracy obtained using SVM and RF. We also see that the quality of classification highly depends on a data set. There are cases, such as Coffee, HandOutlines, Haptics, and others, where all algorithms reached a very similar accuracy. In contrast, for sets such as Beef, BirdChicken, and ChlorineConcentration the differences were substantial. For example, in the Beef data set, SAXVSM achieved accuracy equal to 0.43 while SVMQ achieved accuracy equal to 0.93. The figure demonstrates that the proposed method performs well.

5 Conclusion

In the paper, we have studied a new method for time series classification. It combines three distinct methodologies:

- A plain approach working on raw time series data. We randomly pick a portion of data points from the time series.
- A feature-based approach, in which the order of elements does not matter and a standard classifier performs the final classification task. Two general-purpose classifiers (RF and SVM) were utilized in the study.
- Time series models providing their parameters as features used in the classification process. The ARIMA model was employed for this purpose.

The empirical analysis performed on a wide range of benchmark time series demonstrates a satisfying quality of the presented technique comparing to two standard approaches: plain and feature-based.

References

1. Abanda, A., Mori, U., Lozano, J.A.: A review on distance based time series classification. Data Min. Knowl. Disc. **33**(2), 378–412 (2018). https://doi.org/10.1007/s10618-018-0596-4
2. Bagnall, A., Lines, J., Bostrom, A., Large, J., Keogh, E.: The great time series classification bake off: a review and experimental evaluation of recent algorithmic advances. Data Min. Knowl. Disc. **31**(3), 606–660 (2016). https://doi.org/10.1007/s10618-016-0483-9

3. Baydogan, M.G., Runger, G., Tuv, E.: A bag-of-features framework to classify time series. IEEE Trans. Pattern Anal. Mach. Intel. **35**(11), 2796–2802 (2013)
4. Deng, H., Runger, G., Tuv, E., Vladimir, M.: A time series forest for classification and feature extraction. Inf. Sci. **239**, 142–153 (2013)
5. Fulcher, B.D., Jones, N.S.: Highly comparative feature-based time-series classification. IEEE Trans. Knowl. Data Eng. **26**(12), 3026–3037 (2014)
6. Hyndman, R., Khandakar, Y.: Automatic time series forecasting: the forecast package for R. J. Stat. Softw. **27**(3), 1–22 (2008)
7. Kalpakis, K., Gada, D., Puttagunta, V.: Distance measures for effective clustering of ARIMA time-series. In: Proceedings 2001 IEEE International Conference on Data Mining, pp. 273–280 (2001)
8. Kate, R.J.: Using dynamic time warping distances as features for improved time series classification. Data Min. Knowl. Disc. **30**(2), 283–312 (2015). https://doi.org/10.1007/s10618-015-0418-x
9. Li, D., Bissyande, T.F., Klein, J., Traon, Y.L.: Time series classification with discrete wavelet transformed data. Int. J. Softw. Eng. Knowl. Eng. **26**(09n10), 1361–1377 (2016)
10. Lin, J., Khade, R., Li, Y.: Rotation-invariant similarity in time series using bag-of-patterns representation. J. Intell. Inf. Syst. **39**(2), 287–315 (2012)
11. Schäfer, P.: The BOSS is concerned with time series classification in the presence of noise. Data Min. Knowl. Disc. **29**(6), 1505–1530 (2014). https://doi.org/10.1007/s10618-014-0377-7
12. Senin, P., Malinchik, S.: SAX-VSM: interpretable time series classification using sax and vector space model. In: 2013 IEEE 13th International Conference on Data Mining, pp. 1175–1180 (2013)
13. Wang, X., Smith, K., Hyndman, R.: Characteristic-based clustering for time series data. Data Min. Knowl. Disc. **13**(3), 335–364 (2006)

A Note on Adjoint Linear Algebra

Uwe Naumann(✉)

Department of Computer Science, RWTH Aachen University,
52056 Aachen, Germany
naumann@stce.rwth-aachen.de
http://www.stce.rwth-aachen.de

Abstract. A new proof for adjoint systems of linear equations is presented. The argument is built on the principles of Algorithmic Differentiation. Application to scalar multiplication sets the base line. Generalization yields adjoint inner vector, matrix-vector, and matrix-matrix products leading to an alternative proof for first- as well as higher-order adjoint linear systems.

Keywords: Algorithmic differentiation · Adjoint · Linear algebra

1 Motivation

Algorithmic Differentiation [3,5] of numerical programs builds on a set of elemental functions with known partial derivatives with respect to their arguments at the given point of evaluation. The propagation of adjoint derivatives relies on the associativity of the chain rule of differential calculus. Differentiable combinations of elemental functions yield higher-level elementals. Efficient implementation of AD requires the highest possible level of elemental functions.

Basic AD assumes the set of elemental functions to be formed by the arithmetic operators and intrinsic functions built into the given programming language. While its application to linear algebra methods turns out to be straight forward basic AD is certainly not the method of choice from the point of view of computational efficiency. Elementals of the highest possible level should be used. Their derivatives should be formulated as functions of high-level elementals in order to exploit benefits of corresponding optimized implementations.

Following this rationale this note presents a new way to derive adjoint systems of linear equations based on adjoint Basic Linear Algebra Subprograms (BLAS) [4]. It is well known (see [2] and references therein) that for systems $A \cdot \mathbf{x} = \mathbf{b}$ of n linear equations with invertible A and *primal* solution $\mathbf{x} = A^{-1} \cdot \mathbf{b}$ first-order *adjoints* $A_{(1)}$ of A (both $\in \mathbb{R}^{n \times n}$ with $\mathbb{R}$ denoting the real numbers) and $\mathbf{b}_{(1)}$ of $\mathbf{b}$ (both $\in \mathbb{R}^n$) can be evaluated at the primal solution $\mathbf{x} \in \mathbb{R}^n$ as

$$\begin{pmatrix} \mathbf{b}_{(1)} = A^{-T} \cdot \mathbf{x}_{(1)} \\ A_{(1)} = -\mathbf{b}_{(1)} \cdot \mathbf{x}^T \end{pmatrix}. \tag{1}$$

D. Groen et al. (Eds.): ICCS 2022, LNCS 13351, pp. 200–206, 2022.
https://doi.org/10.1007/978-3-031-08754-7_27

The main contribution of this note is an alternative proof for Eq. (1) that builds naturally on the adjoint BLAS used in the context of state of the art AD. For consistency with related work we follow the notation in [5], that is, $v^{(1)} \in \mathbb{R}$ denotes the value of the first-order directional derivative (or tangent) associated with a variable $v \in \mathbb{R}$ and $v_{(1)} \in \mathbb{R}$ denotes the value of its adjoint.

2 Prerequisites

The Jacobian $\nabla F = \nabla F(\mathbf{x}) \equiv \frac{dF}{d\mathbf{x}}(\mathbf{x}) \in \mathbb{R}^{m \times n}$ of a differentiable implementation of $\mathbf{y} = F(\mathbf{x}) : \mathbb{R}^n \to \mathbb{R}^m$ as a computer program induces a linear mapping $\mathbf{y}^{(1)} = \nabla F \cdot \mathbf{x}^{(1)} : \mathbb{R}^n \to \mathbb{R}^m$ implementing the tangent of F. The corresponding *adjoint* operator $\nabla F^* = \nabla F^*(\mathbf{x})$ is formally defined via the inner vector product identity

$$\langle \nabla F \cdot \mathbf{x}^{(1)}, \mathbf{y}_{(1)} \rangle = \langle \mathbf{x}^{(1)}, \nabla F^* \cdot \mathbf{y}_{(1)} \rangle \tag{2}$$

yielding $\nabla F^* = \nabla F^T$ [1]. In the following all (program) variables are assumed to be alias- and context-free, that is, distinct variables do not overlap in memory and F is assumed to be not embedded in an enclosing computation. We distinguish between *active* and *passive* variables. Derivatives of all active outputs of the given program are computed with respect to all active inputs. We are not interested in derivatives of passive outputs nor are we computing derivatives with respect to passive inputs.

3 BLAS Revisited

In its basic form AD builds on known tangents and adjoints of the arithmetic functions and operators built into programming languages. Tangents and adjoints are propagated along the flow of data according to the chain rule of differential calculus. We enumerate entries of vectors $\mathbf{v} \in \mathbb{R}^n$ staring from zero as $v_0, \ldots, v_{n-1}$.

From the perspective of AD adjoint versions of higher-level BLAS are derived as adjoints of lower-level BLAS. Optimization of the result aims for implementation using the highest possible level of BLAS. For example, adjoint matrix-matrix multiplication (level-3 BLAS) is derived from adjoint matrix-vector multiplication (level-2 BLAS) yielding efficient evaluation as two matrix-matrix products (level-3 BLAS) as shown in Lemma 4. Rigorous derivation of this result requires bottom-up investigation of the BLAS hierarchy. We start with basic scalar multiplication (Lemma 1) followed by the inner vector (Lemma 2) and matrix-vector (Lemma 3) products as prerequisites for the matrix-matrix product.

Lemma 1. *The adjoint of scalar multiplication $y = a \cdot x$ with active $a, x, y \in \mathbb{R}$ is computed as*

$$\begin{aligned} a_{(1)} &= x \cdot y_{(1)} \\ x_{(1)} &= a \cdot y_{(1)} \end{aligned} \tag{3}$$

for $y_{(1)} \in \mathbb{R}$ yielding $a_{(1)}, x_{(1)} \in \mathbb{R}$.

Proof. Differentiation of $y = a \cdot x$ with respect to a and x yields the tangent

$$y^{(1)} = \left\langle \begin{pmatrix} a^{(1)} \\ x^{(1)} \end{pmatrix}, \begin{pmatrix} x \\ a \end{pmatrix} \right\rangle$$

for $y^{(1)}, a^{(1)}, x^{(1)} \in \mathbb{R}$. Equation (2) implies

$$\langle y^{(1)}, y_{(1)} \rangle = y^{(1)} \cdot y_{(1)} = \left\langle \begin{pmatrix} a^{(1)} \\ x^{(1)} \end{pmatrix}, \begin{pmatrix} a_{(1)} \\ x_{(1)} \end{pmatrix} \right\rangle = \left\langle \begin{pmatrix} a^{(1)} \\ x^{(1)} \end{pmatrix}, \begin{pmatrix} x \\ a \end{pmatrix} \right\rangle \cdot y_{(1)}$$

yielding

$$\begin{pmatrix} a_{(1)} \\ x_{(1)} \end{pmatrix} = \begin{pmatrix} x \\ a \end{pmatrix} \cdot y_{(1)}$$

and hence Eq. (3).

Lemma 2. *The adjoint of an inner vector product*

$$y = \langle \mathbf{a}, \mathbf{x} \rangle \equiv \mathbf{a}^T \cdot \mathbf{x} = \sum_{i=0}^{n-1} a_i \cdot x_i$$

with active inputs $\mathbf{a} \in \mathbb{R}^n$ *and* $\mathbf{x} \in \mathbb{R}^n$ *yielding the active output* $y \in \mathbb{R}$ *is computed as*

$$\begin{aligned} \mathbf{a}_{(1)} &= \mathbf{x} \cdot y_{(1)} \\ \mathbf{x}_{(1)} &= \mathbf{a} \cdot y_{(1)} \end{aligned} \tag{4}$$

for $y_{(1)} \in \mathbb{R}$ *yielding* $\mathbf{a}_{(1)} \in \mathbb{R}^n$ *and* $\mathbf{x}_{(1)} \in \mathbb{R}^n$.

Proof. Differentiation of $y = \mathbf{a}^T \cdot \mathbf{x}$, for $\mathbf{a} = (a_i)_{i=0,\ldots,n-1}$ and $\mathbf{x} = (x_i)_{i=0,\ldots,n-1}$, with respect to $\mathbf{a}$ and $\mathbf{x}$ yields the tangent

$$\begin{aligned} y^{(1)} &= \sum_{i=0}^{n-1} (x_i \; a_i) \cdot \begin{pmatrix} a_i^{(1)} \\ x_i^{(1)} \end{pmatrix} = \sum_{i=0}^{n-1} \left(x_i \cdot a_i^{(1)} + a_i \cdot x_i^{(1)} \right) \\ &= \sum_{i=0}^{n-1} x_i \cdot a_i^{(1)} + \sum_{i=0}^{n-1} x_i^{(1)} \cdot a_i = \mathbf{x}^T \cdot \mathbf{a}^{(1)} + \mathbf{a}^T \cdot \mathbf{x}^{(1)} = (\mathbf{x}^T \; \mathbf{a}^T) \cdot \begin{pmatrix} \mathbf{a}^{(1)} \\ \mathbf{x}^{(1)} \end{pmatrix}. \end{aligned}$$

Equation (2) implies

$$y_{(1)} \cdot y^{(1)} = (\mathbf{a}_{(1)}^T \; \mathbf{x}_{(1)}^T) \cdot \begin{pmatrix} \mathbf{a}^{(1)} \\ \mathbf{x}^{(1)} \end{pmatrix} = y_{(1)} \cdot (\mathbf{x}^T \; \mathbf{a}^T) \cdot \begin{pmatrix} \mathbf{a}^{(1)} \\ \mathbf{x}^{(1)} \end{pmatrix}$$

yielding $(\mathbf{a}_{(1)}^T \; \mathbf{x}_{(1)}^T) = y_{(1)} \cdot (\mathbf{x}^T \; \mathbf{a}^T)$ and hence Eq. (4).

The following derivation of adjoint matrix-vector and matrix-matrix products relies on serialization of matrices. Individual rows of a matrix $A \in \mathbb{R}^{m \times n}$ are denoted as $\mathbf{a}_i \in \mathbb{R}^{1 \times n}$ for $i = 0, \ldots, m-1$; columns are denoted as $\mathbf{a}^j \in \mathbb{R}^m$ for $i = 0, \ldots, n-1$. (Row) Vectors in $\mathbb{R}^{1 \times n}$ are denoted as $(v_j)^{j=0,\ldots,n-1}$; (column) vectors in $\mathbb{R}^m$ are denoted as $(v_i)_{i=0,\ldots,m-1}$; Consequently, a row-major serialization of A is given by $\left(\mathbf{a}_i^T\right)_{i=0,\ldots,m-1}$. A column-major serialization of A is given by $\left(\mathbf{a}^j\right)_{j=0,\ldots,n-1}$. Tangents and adjoints of the individual entries of A define

$$A^{(1)} = (\mathbf{a}_i^{(1)})_{i=0,\ldots,m-1} = (a_{i,j}^{(1)})_{i=0,\ldots,m-1}^{j=0,\ldots,n-1}$$

and

$$A_{(1)} = (\mathbf{a}_{(1)i})_{i=0,\ldots,m-1} = (a_{(1)i,j})_{i=0,\ldots,m-1}^{j=0,\ldots,n-1},$$

respectively.

Lemma 3. *The adjoint of a matrix-vector product*

$$\mathbf{y} = A \cdot \mathbf{x} \equiv (\mathbf{a}_i \cdot \mathbf{x})_{i=0,\ldots,m-1}$$

with active inputs $A \in \mathbb{R}^{m\times n}$ and $\mathbf{x} \in \mathbb{R}^n$ yielding the active output $\mathbf{y} \in \mathbb{R}^m$ is computed as

$$\begin{aligned} \mathbf{x}_{(1)} &= A^T \cdot \mathbf{y}_{(1)} \\ A_{(1)} &= \mathbf{y}_{(1)} \cdot \mathbf{x}^T \end{aligned} \tag{5}$$

for $\mathbf{y}_{(1)} \in \mathbb{R}^m$ yielding $\mathbf{x}_{(1)} \in \mathbb{R}^n$ and $A_{(1)} \in \mathbb{R}^{m\times n}$.

Proof. Differentiation of $\mathbf{y} = A\cdot\mathbf{x}$, where $A = (\mathbf{a}_i)_{i=0,\ldots,m-1}$, $\mathbf{x} = (x_j)_{j=0,\ldots,n-1}$ and $\mathbf{y} = (y_i)_{i=0,\ldots,m-1}$, with respect to A and $\mathbf{x}$ yields the tangent

$$\begin{aligned} \mathbf{y}^{(1)} &= \left(\left\langle \begin{pmatrix} \mathbf{x} \\ \mathbf{a}_i^T \end{pmatrix}, \begin{pmatrix} \mathbf{a}_i^{(1)^T} \\ \mathbf{x}^{(1)} \end{pmatrix} \right\rangle\right)_{i=0,\ldots,m-1} = \left(\mathbf{x}^T \cdot \mathbf{a}_i^{(1)^T} + \mathbf{a}_i \cdot \mathbf{x}^{(1)}\right)_{i=0,\ldots,m-1} \\ &= \left(\mathbf{x}^T \cdot \mathbf{a}_i^{(1)^T}\right)_{i=0,\ldots,m-1} + \left(\mathbf{a}_i \cdot \mathbf{x}^{(1)}\right)_{i=0,\ldots,m-1} \\ &= \left(\mathbf{a}_i^{(1)} \cdot \mathbf{x}\right)_{i=0,\ldots,m-1} + \left(\mathbf{a}_i \cdot \mathbf{x}^{(1)}\right)_{i=0,\ldots,m-1} \\ &= \left(\mathbf{a}_i^{(1)}\right)_{i=0,\ldots,m-1} \cdot \mathbf{x} + (\mathbf{a}_i)_{i=0,\ldots,m-1} \cdot \mathbf{x}^{(1)} = A^{(1)} \cdot \mathbf{x} + A \cdot \mathbf{x}^{(1)}. \end{aligned}$$

Equation (2) implies

$$\begin{aligned} \left\langle \mathbf{y}_{(1)}, \mathbf{y}^{(1)} \right\rangle &= \left\langle \begin{pmatrix} \left(\mathbf{a}_{(1)i}^T\right)_{i=0,\ldots,m-1} \\ \mathbf{x}_{(1)} \end{pmatrix}, \begin{pmatrix} \left(\mathbf{a}_i^{(1)T}\right)_{i=0,\ldots,m-1} \\ \mathbf{x}^{(1)} \end{pmatrix} \right\rangle \\ &= \left(\mathbf{a}_{(1)i}^T\right)_{i=0,\ldots,m-1}^T \cdot \left(\mathbf{a}_i^{(1)T}\right)_{i=0,\ldots,m-1} + \mathbf{x}_{(1)}^T \cdot \mathbf{x}^{(1)} \\ &= \mathbf{y}_{(1)}^T \cdot \left(A^{(1)} \cdot \mathbf{x} + A \cdot \mathbf{x}^{(1)}\right) = \mathbf{y}_{(1)}^T \cdot A^{(1)} \cdot \mathbf{x} + \mathbf{y}_{(1)}^T \cdot A \cdot \mathbf{x}^{(1)} \\ &= \left(\left(y_{(1)i} \cdot \mathbf{a}_i^{(1)T}\right)_{i=0,\ldots,m-1}\right)^T \cdot (\mathbf{x})_{i=0,\ldots,m-1} + \mathbf{y}_{(1)}^T \cdot A \cdot \mathbf{x}^{(1)} \\ &= \underbrace{\left(\left(y_{(1)i} \cdot \mathbf{x}\right)_{i=0,\ldots,m-1}\right)^T}_{=\left(\left(\mathbf{a}_{(1)i}^T\right)_{i=0,\ldots,n-1}\right)^T} \cdot \left(\mathbf{a}_i^{(1)T}\right)_{i=0,\ldots,m-1} + \underbrace{\mathbf{y}_{(1)}^T \cdot A}_{=\mathbf{x}_{(1)}^T} \cdot \mathbf{x}^{(1)}, \end{aligned}$$

where $(\mathbf{x})_{i=0,\ldots,m-1} \in \mathbb{R}^{m\cdot n}$ denotes a concatenation of m copies of $\mathbf{x} \in \mathbb{R}^n$ as a column vector. Equation (5) follows immediately.

Lemma 4. *The adjoint of a matrix-matrix product $Y = A \cdot X$ with active inputs $A \in \mathbb{R}^{m\times p}$, $X \in \mathbb{R}^{p\times n}$ yielding the active output $Y \in \mathbb{R}^{m\times n}$ is computed as*

$$\begin{aligned} A_{(1)} &= Y_{(1)} \cdot X^T \\ X_{(1)} &= A^T \cdot Y_{(1)} \end{aligned} \tag{6}$$

for $Y_{(1)} \in \mathbb{R}^{m\times n}$ yielding $A_{(1)} \in \mathbb{R}^{m\times p}$ and $X_{(1)} \in \mathbb{R}^{p\times n}$.

Proof. Differentiation of $Y = A \cdot X$, where $A = (\mathbf{a}_i)_{i=0,\ldots,m-1}$, $X = (\mathbf{x}^k)^{k=0,\ldots,p-1}$ and $Y = (\mathbf{y}^k)^{k=0,\ldots,p-1}$, with respect to A and X yields tangents

$$\mathbf{y}^{(1)k} = \left(\left\langle \begin{pmatrix} \mathbf{x}^k \\ \mathbf{a}_i^T \end{pmatrix}, \begin{pmatrix} \mathbf{a}_i^{(1)^T} \\ \mathbf{x}^{(1)k} \end{pmatrix} \right\rangle\right)_{i=0,\ldots,m-1} = A^{(1)} \cdot \mathbf{x}^k + A \cdot \mathbf{x}^{(1)k}.$$

for $k = 0, \ldots, p-1$ and hence

$$Y^{(1)} = A^{(1)} \cdot X + A \cdot X^{(1)}.$$

Equation (2) implies

$$\begin{aligned} \left\langle \mathbf{y}^k_{(1)}, \mathbf{y}^{(1)k} \right\rangle &= \left\langle \begin{pmatrix} \left(\mathbf{a}^T_{(1)i}\right)_{i=0,\ldots,m-1} \\ \mathbf{x}_{(1)k} \end{pmatrix}, \begin{pmatrix} \left(\mathbf{a}_i^{(1)T}\right)_{i=0,\ldots,m-1} \\ \mathbf{x}^{(1)k} \end{pmatrix} \right\rangle \\ &= \underbrace{\left(\left(y^k_{(1)i} \cdot \mathbf{x}^k\right)_{i=0,\ldots,m-1}\right)^T}_{=\left((\mathbf{a}^T_{(1)i})_{i=0,\ldots,m-1}\right)^T} \cdot \left(\mathbf{a}_i^{(1)T}\right)_{i=0,\ldots,m-1} + \underbrace{\mathbf{y}^{k^T}_{(1)} \cdot A}_{=\mathbf{x}^{k^T}_{(1)}} \cdot \mathbf{x}^{(1)k} \end{aligned}$$

for $k = 0, \ldots, p-1$ and hence the Eq. (6).

4 Systems of Linear Equations Revisited

Lemmas 5 and 6 form the basis for the new proof of Eq. (1).

Lemma 5. *The tangent*

$$Y^{(1)} = A \cdot X^{(1)} \cdot B$$

of $Y = A \cdot X \cdot B$ for active $X \in \mathbb{R}^{n\times q}$, $Y \in \mathbb{R}^{m\times p}$ and passive $A \in \mathbb{R}^{m\times n}$, $B \in \mathbb{R}^{q\times p}$ implies the adjoint

$$X_{(1)} = A^T \cdot Y_{(1)} \cdot B^T.$$

Proof.

$$Y^{(1)} = Z^{(1)} \cdot B \quad \Rightarrow \quad Z_{(1)} = Y_{(1)} \cdot B^T$$

follows from application of Lemma 4 to $Y = Z \cdot B$ with passive B.

$$Z^{(1)} = A \cdot X^{(1)} \quad \Rightarrow \quad X_{(1)} = A^T \cdot Z_{(1)}$$

follows from application of Lemma 4 to $Z = A \cdot X$ with passive A. Substitution of $Z^{(1)}$ and $Z_{(1)}$ yields Lemma 5.

Lemma 6. *The tangent*

$$Y^{(1)} = \sum_{i=0}^{k-1} A_i \cdot X_i^{(1)} \cdot B_i$$

of $Y = \sum_{i=0}^{k-1} A_i \cdot X_i \cdot B_i = \sum_{i=0}^{k-1} Y_i$ *with active* $X_i \in \mathbb{R}^{n_i \times q_i}$, $Y \in \mathbb{R}^{m \times p}$ *and with passive* $A_i \in \mathbb{R}^{m \times n_i}$, $B_i \in \mathbb{R}^{q_i \times p}$ *implies the adjoint*

$$X_{i(1)} = A_i^T \cdot Y_{(1)} \cdot B_i^T$$

for $i = 0, \ldots, k-1$

Proof. From

$$Y_i^{(1)} = A_i \cdot X_i^{(1)} \cdot B_i$$

follows with Lemma 5

$$X_{i(1)} = A_i^T \cdot Y_{i(1)} \cdot B_i^T$$

for $i = 0, \ldots, k-1$. Moreover, $Y^{(1)} = \sum_{i=0}^{k-1} Y_i^{(1)}$ implies $Y_{i(1)} = Y_{(1)}$ due to identity Jacobians of Y with respect to Y_i for $i = 0, \ldots, k-1$ and hence Lemma 6.

Theorem 1. *Adjoints of systems* $A \cdot \mathbf{x} = \mathbf{b}$ *of* n *linear equations with invertible* $A \in \mathbb{R}^{n \times n}$ *and right-hand side* $\mathbf{b} \in \mathbb{R}^n$ *are evaluated at the primal solution* $\mathbf{x} = A^{-1} \cdot \mathbf{b} \in \mathbb{R}^n$ *by Eq. (1).*

Proof. Differentiation of $A \cdot \mathbf{x} = \mathbf{b}$ with respect to A and $\mathbf{b}$ yields the tangent system

$$A^{(1)} \cdot \mathbf{x} + A \cdot \mathbf{x}^{(1)} = \mathbf{b}^{(1)}$$

which implies

$$\mathbf{x}^{(1)} = A^{-1} \cdot \mathbf{b}^{(1)} \cdot I_n - A^{-1} \cdot A^{(1)} \cdot \mathbf{x}$$

with identity $I_n \in \mathbb{R}^{n \times n}$. Lemma 6 yields

$$\begin{aligned} \mathbf{b}_{(1)} &= A^{-T} \cdot \mathbf{x}_{(1)} \cdot I_n^T \\ A_{(1)} &= -\underbrace{A^{-T} \cdot \mathbf{x}_{(1)}}_{=\mathbf{b}_{(1)}} \cdot \mathbf{x}^T \end{aligned}$$

and hence Eq. (1).

5 Conclusion

As observed previously by various authors a possibly available factorization of A can be reused both for the tangent ($A \cdot \mathbf{x}^{(1)} = \mathbf{b}^{(1)} - A^{(1)} \cdot \mathbf{x}$) and the adjoint ($A^T \cdot \mathbf{b}_{(1)} = \mathbf{x}_{(1)}$) systems. The additional worst case computational cost of $O(n^3)$ can thus be reduced to $O(n^2)$. Higher-order tangents [adjoints] of linear systems amount to repeated solutions of linear systems with the same [transposed] system matrix combined with tangent [adjoint] BLAS.

References

1. Dunford, N., Schwartz, J.: Linear Operators. I. General Theory. With the assistance of W. G. Bade and R. G. Bartle. Pure and Applied Mathematics, vol. 7. Interscience Publishers, Inc., New York (1958)
2. Giles, M.B.: Collected matrix derivative results for forward and reverse mode algorithmic differentiation. In: Bischof, C., Bücker, M., Hovland, P., Naumann, U., Utke, J. (eds.) Advances in Automatic Differentiation. 64, pp. 35–44. Springer, Heidelberg (2008). https://doi.org/10.1007/978-3-540-68942-3_4
3. Griewank, A., Walther, A.: Evaluating Derivatives. Principles and Techniques of Algorithmic Differentiation. Other Titles in Applied Mathematics, vol. OT105, 2nd edn. SIAM, Philadelphia (2008)
4. Lawson, C., Hanson, R., Kincaid, D., Krogh, F.: Basic linear algebra subprograms for Fortran usage. ACM Trans. Math. Softw. **5**, 308–323 (1979)
5. Naumann, U.: The Art of Differentiating Computer Programs. An Introduction to Algorithmic Differentiation. Software, Environments, and Tools, vol. SE24. SIAM, Philadelphia (2012)

Approximate Function Classification

Martin Lukac[1(✉)], Krzysztof Podlaski[2], and Michitaka Kameyama[3]

[1] Nazarbayev University, Nur-sultan, Kazakhstan
martin.lukac@nu.edu.kz
[2] University of Lodz, Lodz, Poland
krzysztof.podlaski@uni.lodz.pl
[3] Emeritus Professor Tohoku University, Sendai, Japan
kameyama@ecei.tohoku.ac.jp

Abstract. Classification of Boolean functions requires specific software or circuits to determine the class of a function or even to distinguish between two different classes. In order to provide a less costly solution, we study the approximation of the NPN function classification by a artificial neural network (ANN), and shown that there are configurations of ANN that can perfectly classify four-bit Boolean functions. Additionally, we look at the possibility of learning the classification of four-bit Boolean functions using a set of three-bit Boolean neural classifiers, and determine the scalability. Finally we also learn a discriminator that can distinguish between two functions and determine their similarity or difference in their NPN classes. As a result we show that the approximate neural function classification is a convenient approach to implement an efficient classifier and class discriminator directly from the data.

Keywords: Function classification · Approximate learning · Neural networks

1 Introduction

The classification of functions has several applications in both industry and research. Knowing the properties of groups of functions allows one to study solutions to a problem from a group-like perspective. This includes the cost of circuit estimation or group functions implementations. It also helps the industry to design circuit from an approximate cost: functions grouped into specific groups can have similar realization properties and thus similar and predictable cost [1, 10,12]. Finally, function classification can be used to discover structures in the groups of functions that are not completely understood [2,4,5].

The classification can be performed by a standard look-up or by a functional approach. However, such a direct approach is not well scalable because it requires the definition of all possible input value combinations. Therefore alternative methods for generating classes of functions using approximate methods are desired. The classification of functions using machine learning has been previously explored [6–9,11]. However, most of the available works focus on learning Boolean functions using machine learning.

D. Groen et al. (Eds.): ICCS 2022, LNCS 13351, pp. 207–213, 2022.
https://doi.org/10.1007/978-3-031-08754-7_28

The function classification becomes function approximation when the classification is a many to one mapping. In particular, when the classification results in group properties such that the cost of any function within the group is the same, then a logic design process can be simplified by searching for equivalence groups rather than for single functions. The most recognized classifications of Boolean functions are P-equivalent, NP-equivalent, NPN-equivalent [5].

Two Boolean functions are NPN-*equivalent* if they can be transformed into each other by one or more of the following transformations: the negation of any input variable, permutation of input variables, and negation of the function.

In the paper, we consider the possibility of classification of functions using off-the-shelf machine learning approaches [6]. The target of such classification will show if the NPN classification can be efficiently learned. Therefore, if the groups' properties that represent linear transformations can also be easily represented in the embedded feature space. In addition, we are also interested to see if we can learn the classification of NPN classes on a subset of functions and then generalize it to other functions not used for learning. The expected result will show if such an approximate learning and NPN classes representation can be used to learn classifications of functions where there are too many samples to enumerate. Finally, we consider the proposed method as a general template for classification: the NPN classification presented in this paper is only a starting point for a more general classification framework.

The machine learning approach to classification requires learning data. For functions with more than four input bits, the number of NPN classes and the number of functions is too large to be used for learning. Therefore, we look at the problem from the point of view of learning n-bit NPN classes by composing the machine learning classifier from $n-1$-bit functions. We train accurate $n-1$ NPN function classifiers, and by combining several of them, we train the n bit classifier on partial data. We show under which conditions such an approach is possible and what amount of data is required for accurate prediction. We also investigate the discrimination of NPN classes by a trainable neural network. We show that a single network is quite capable to distinguish between functions of different classes without intermediary determination of their respective NPN classes. Therefore, using the neural methods for distinguishing NPN classes provides a considerable advantage when compared to classical, circuit-based methods.

2 Experiments

2.1 Datasets

For the experiments, we have created two datasets. The first dataset is the set of all four-bit irreversible functions, and the mapping to be learned is $\mathcal{M} : \{0,1\}^{2^4} \rightarrow \mathbb{L}_4$, with $\mathbb{L}_4$ being the set of all NPN classes for four-variable irreversible functions. There are 222 NPN classes for four-variable irreversible functions, and the number of data samples in the dataset is 65536.

The second set of datasets contains NPN labels and functions for one, two, three, and four input bits. The target of this dataset is to use it to learn and to determine the scalability hardness of the mapping from a set of functions of $n-1$-variables to n-NPN labels: $\mathcal{N} : \{0,1\}^{n-1} \times \ldots \times \{0,1\}^{n-1} \rightarrow \mathbb{L}_n$. The $\mathbb{L}_n$ represents the labels of n-bit functions.

2.2 Four-Variable Irreversible Logic Functions

In preliminary experiments, we evaluated various machine learning algorithms: multi-layer perceptron (MLP), support vector machine (SVM), decision tree (DT), and random forests (RF), but only the MLP showed good enough learning convergence and final classification accuracy. Therefore, we decided to analyze if it's possible to obtain a perfect classifier using MLP and what architecture is required. For the experiment, we took 4-bit Boolean functions. The training set contains 65% of all functions in the dataset, and a test set contains the rest. We discovered that by learning the default network for up to 300 epochs, we could not obtain 100% positive answers on the test set with one fully connected hidden layer MLP with up to 120 neurons. Therefore we have added a second hidden layer and performed a grid search.

Table 1. Accuracy of MLP classifier with two hidden layers, l_1, l_2 denote sizes of first and second hidden layer respectively.

$l_1 \backslash l_2$	10	15	20	25	30	35	40	45
10	0.128	0.171	0.250	0.234	0.450	0.295	0.267	0.240
30	0.173	0.436	0.718	0.791	0.804	0.823	0.829	0.829
40	0.159	0.700	0.767	0.814	0.809	0.821	0.895	0.865
60	0.341	0.547	0.787	0.922	0.889	0.963	0.968	0.950
70	0.211	0.444	0.760	0.919	0.971	0.945	0.987	0.995

$l_1 \backslash l_2$	50	55	60	70	75	80	85	90
80	0.997	0.995	**1.000**	0.999	0.999	**1.000**	0.997	0.998
90	0.992	0.991	**1.000**	0.996	**1.000**	**1.000**	0.998	**1.000**
100	0.998	0.996	0.996	0.999	0.999	0.999	0.999	**1.000**
110	0.998	0.994	0.999	0.999	**1.000**	0.997	**1.000**	0.996
120	0.996	0.999	0.996	**1.000**	0.999	**1.000**	**1.000**	**1.000**

The parameters of the network and the obtained accuracy on the training set after learning 300 epochs are presented in Table 1. The rows of Table 1 show the number of neurons in the first hidden layer, and the columns show the number of neurons in the second hidden layer. As can be seen, the full precision of the NPN classification was achieved when the first layer had at least 80 neurons and the second one at least 60. It agrees with the observation that the MLP is able to approximate the multivariate function only with two hidden layers [3].

2.3 From $n-1$ to n Classification

Classifying functions with more than four bits is more difficult due to very large number of functions and therefore does not allow efficient learning. Consequently, we evaluate alternative models to classify NPN labels for functions of n bits as a function of various compositions of functions of $n-1$ bits.

The main idea of this approach is shown in Fig. 1: a pre-trained neural network for a $n-1$ NPN classification is used as a module (component neural network) for predicting $n-1$ NPN labels from a n-variable subfunction extracted from the target n-variable function. Each component network has as inputs the bits representing a $n-1$-variable function and outputs an NPN class. The results of these component networks are then used to train a n bit NPN classifier. The main concept is to extract a set of $n-1$-variable NPN function labels from the n-variable function and then use these labels to predict the n-variable NPN label. Note that in Fig. 1, each component network is the same, and only the last layer (denoted n to NPN) of the whole network is trained on the n-variable function dataset.

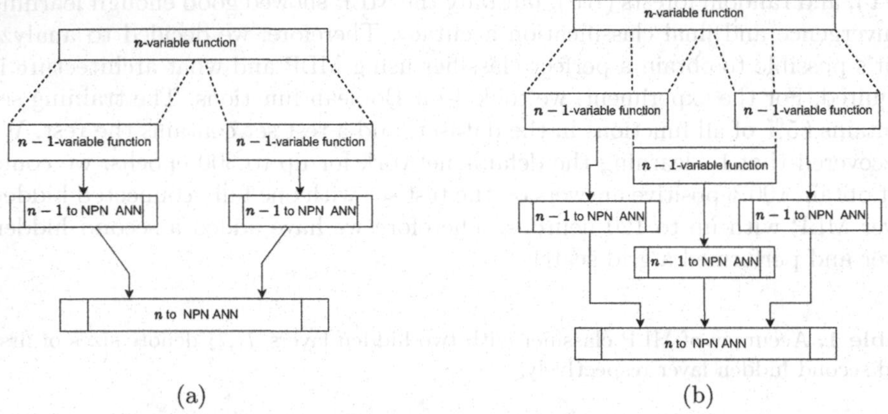

Fig. 1. Schematic diagram of estimating the NPN classes of a four-variable functions using pretrained classifiers for the NPN classes of a three bit logic function.

There are in total $\binom{2^n}{2^{n-1}}$ possible selection of inputs to each component network. However, we are mainly interested in selecting 2^{n-1} out of 2^n values in a continuous manner. Therefore we tested three very simple configurations that contain: two, three, and five-component neural networks.

The input to the network is represented by 2^n values and the 2^{n-1} values for the components networks were extracted using positional indexes. For the two network models, each component network accepts 2^{n-1} bits at positions 0 to $2^n/2-1$ and $2^n/2$ to 2^n-1 respectively. For three-component networks, the first two-component networks take bits in a similar fashion to the two-component network model. The third-component network uses bits starting from the $2^n/4$ and ending at $2^n/4+2^n/2$ bit. For the five-component networks model, the fourth and fifth-component neural networks are simply added to the three component neural network model. The fourth and fifths components ANN had inputs fed from the $2^n/8$ to $2^n/8+2^n/2$ bit and from the $2^n/2-2^n/8$ to $2^n-2^n/8$ bit respectively. As an example, Fig. 1 shows the three component model classifier: the blocks labeled $n-1$-bit function represents the functions of $n-1$-bits extracted from n-variable input function and the blocks labeled $n-1$ to NPN represents the pre-trained NPN component network classifiers for $n-1$-variable functions.

The component networks were trained independently and prior to the final experiment. In the presented case the component networks were trained on three-variable Boolean functions and their NPN labels. In this case the mapping is $\mathbb{K} : \{0,1\}^3 \rightarrow \mathcal{J}_3$. The three-variable NPN classification networks were trained up to 100% accuracy. Each component network had two hidden layers with 50 and 20 neurons, respectively, and the n-bit NPN classifier network had the same structure as well.

Table 2. Results of learning NPN classification for four-variable Boolean functions using three-bit neural NPN classifiers.

Model	3 bit NPN ANN specs	Accuracy	4 bit NPN ANN specs	Accuracy
2 subnets	50/20	1	50/20	0.679 @ 0.7
3 subnets	50/20	1	50/20	0.769 @ 0.65
5 subnets	50/20	1	50/20	0.87 @ 0.26

The results of these experiments are shown in Table 2. Each row shows the result of a particular configuration and its experimental accuracy. The second and fourth columns show the model's configuration: column two shows the component network configuration, while column four shows the whole model configuration. Column three shows the component network accuracy; it is always one because the networks were pre-trained to the highest accuracy. Column five shows the accuracy of the whole model. The results are reported as accuracy@train-set size. The first observation is that the accuracy increases when more component networks are used. It is natural as there are only thirteen NPN labels for functions of three bits. Therefore, using two components networks does not generate enough output combinations to reach the required 222 NPN classes of four-variable functions. Therefore the most accurate result is obtained with five components networks. The second observation is that less data is required when more component networks are provided. As can be seen with two-component networks the accuracy of 67% was reached at $r_e = \frac{0.3}{0.7}$ but the highest recorded accuracy was 87% for five-component networks at $r_e = \frac{0.74}{0.26}$. Therefore the observation seems to lead to a conclusion that knowledge from functions on less bits can be efficiently used to learn the classification of larger functions with a fraction of training data.

3 NPN Classes Discrimination

The final set of experiments is aimed at finding out at what cost a machine learning approach can be used to distinguish two functions from different or the same NPN class. The most natural approach is to classify two functions directly using a binary classifier. The inputs to such a model are two 2^n long binary vectors representing two n-variable functions. The problem that we are thus facing is a solution to a mapping $\mathcal{J}: \{0,1\}^{2^n} \times \{0,1\}^{2^n} \rightarrow \{0,1\}$.

Remembering that the number of functions grows rapidly with the number of bits, already for four-variable functions, the number of samples is $(2^{16})^2$. We implemented a pseudo-random method, allowing us to quickly get a large number of samples. Not all combinations can be generated. Therefore, four different datasets were generated: each dataset being a multiple of the total number of functions of n-variables. For each dataset a same size test dataset was generated so that the train and evaluation minimizes the amount of overlap between the training and testing data. The experiment was run with a multi-layer perceptron with 33 neurons in the input layer, 100 neurons in the hidden layer and two neurons in the output layer. Each experiment was run for 2000 learning epochs.

The first experiment was performed only with the training dataset. Thus, the learning and the evaluation were performed using the same dataset. The reason for this was to determine the learning difficulty. The summary of this first learning experiment for comparing four-variable NPN functions is shown in Table 3. As can be seen, the accuracy of the discrimination remains constant with the increasing size of the dataset. Each dataset contains more instances of similar and dissimilar pairs of functions. Interestingly the learning occurred at only a very high learning rate $\lambda = 0.1$. Such high λ is unusual considering that standard experiments in machine learning use much lower values for the learning rate. Also, it is worth noting that the learning accuracy grows with the increasing epochs. With the current model, the accuracy of distinguishing four-variable NPN classes was determined to be maximal at 97.8% when 10000 epochs were performed.

Table 3. Results of the discrimination learning

	Data size				
	×1	×2	×4	×6	×8
Training set accuracy	95.1%	95.3%	94.6%	93.9%	93.3%
Test set accuracy	89.7%	89.2%	85.0%	84.9%	77.6%

Next, we evaluated the networks on test datasets different from the train sets as much as possible. As was expected, the average accuracy on the test set is lower than the discrimination accuracy on the training dataset. We observed that in both cases, the False-Negatives are, in general, two to three times more numerous than the False-Positives. A similar pattern occurs for the smaller dataset. Certain groups of functions are more learnable than others.

4 Conclusion

In this paper, we presented the study of approximate Boolean functions classification using machine learning. We demonstrated that the NPN classification could be perfect when the parameter space is searched systematically. An interesting observation is that one can generate NPN classification for n-variable function from $n-1$-variable NPN function classes. Finally, we also showed that one can train a classifier for function discrimination without explicit NPN class computation. Some of the classes are easier to train, some tougher. It depends on the size of the class as well as on their representatives. In future works, we will test if these tougher classes also have a more complicated circuit representation.

The usefulness of this approach can seen in the possible application to logic synthesis. In particular, the learnability of the NPN classes could be used for estimating the cost of logic circuits before synthesis by learning the function to be synthesised. The learning result can be then used to determine the synthesis method. Therefore the future work is to looking closer at the individual classes learned and determine the difficulty to learn specific categories such as linear, symmetric, bent, etc.

References

1. Debnath, D., Sasao, T.: Efficient computation of canonical form under variable permutation and negation for Boolean matching in large libraries. IEICE Trans. Fundam. Electron. Commun. Comput. Sci. **E89–A**(12), 3443–3450 (2006)
2. Edwards, C.R.: The application of the Rademacher-Walsh transform to Boolean function classification and threshold logic synthesis. IEEE Trans. Comput. **100**(1), 48–62 (1975)
3. Guliyev, N.J., Ismailov, V.E.: Approximation capability of two hidden layer feedforward neural networks with fixed weights. CoRR abs/2101.09181 (2021)
4. Harrison, M.A.: On the classification of Boolean functions by the general linear and affine groups. J. Soc. Ind. Appl. Math. **12**(2), 285–299 (1964)
5. Hurst, S.L.: The Logical Processing of Digital Signals. Crane Russak & Company Inc., Edward Arnold, New York, London (1978)
6. Lukac, M., Moraga, C., Kameyama, M.: Properties of bent functions in the truth domain. In: 2019 International Conference on Information and Digital Technologies (IDT), pp. 304–310 (2019)
7. Mhaskar, H., Liao, Q., Poggio, T.A.: Learning real and Boolean functions: when is deep better than shallow. CoRR abs/1603.00988 (2016)
8. Oliveira, A.L., Sangiovanni-Vincentelli, A.: Learning complex Boolean functions: algorithms and applications. In: Proceedings of the 6th International Conference on Neural Information Processing Systems (NIPS 1993), p. 911–918. Morgan Kaufmann Publishers Inc., San Francisco (1993)
9. Sadohara, K.: Learning of Boolean functions using support vector machines. In: Abe, N., Khardon, R., Zeugmann, T. (eds.) ALT 2001. LNCS, vol. 2225, pp. 106–118. Springer, Heidelberg (2001). https://doi.org/10.1007/3-540-45583-3_10
10. Sasao, T.: Switching Theory For Logic Synthesis. Cluver Academic Publishers, Boston/London/Dordrecht (1999)
11. Tavares, A.R., Avelar, P., Flach, J.M., Nicolau, M., Lamb, L.C., Vardi, M.: Understanding Boolean function learnability on deep neural networks (2020)
12. Tsai, C.C., Marek-Sadowska, M.: Boolean functions classification via fixed polarity reed-muller forms. IEEE Tran. Comput. **46**(2), 173–186 (1997)

Acceleration of Optimized Coarse-Grid Operators by Spatial Redistribution for Multigrid Reduction in Time

Ryo Yoda[1(✉)], Matthias Bolten[2], Kengo Nakajima[1], and Akihiro Fujii[3]

[1] The University of Tokyo, Tokyo, Japan
ryo_yoda@mist.i.u-tokyo.ac.jp
[2] Wuppertal University, Wuppertal, Germany
[3] Kogakuin University, Tokyo, Japan

Abstract. The multigrid reduction in time (MGRIT) method is one of the parallel-in-time approaches for time-dependent PDEs and typically uses rediscretized coarse-grid operators. As their convergence struggle with hyperbolic problems, an optimization method for coarse-grid operators has been proposed to deal with these problems. This method improves convergence using coarse-grid operators with a slightly increased number of nonzero elements. However, it is more desirable for coarse-grid operators to be cheaper than fine-grid operators, and there is room for improvement in terms of parallel implementation. This work combines the spatial redistribution technique for MGRIT, which accelerates coarse-grid solvers using agglomerated idle processors, with the above optimization method. This combination attempts to achieve better scaling performance while maintaining good convergence. Numerical experiments demonstrate a 23% runtime reduction at most among the various assignments tried with specific amount of parallelism.

Keywords: Parallel-in-time approaches · Multigrid methods · Coarse-grid optimization · Spatial redistribution

1 Introduction

This paper considers parallel numerical solvers for time-dependent partial differential equations (PDEs). In modern computing systems with increasing number of cores, spatial parallelism obtained by domain decomposition methods is exhausted because of over-decomposition. Therefore, attempts to extract temporal parallelism, called parallel-in-time approaches [4,6], have attracted much attention recently. Some examples of the most powerful parallel-in-time solvers are space-time multigrid, Parareal, and multigrid reduction in time (MGRIT) [3]. This paper focuses on MGRIT. While the application of MGRIT has been successful for various parabolic problems, the convergence of MGRIT struggles with hyperbolic problems [1,5]. This failure can occur even with implicit time discretization or spatial coarsening to stabilize the coarse-grid problem.

D. Groen et al. (Eds.): ICCS 2022, LNCS 13351, pp. 214–221, 2022.
https://doi.org/10.1007/978-3-031-08754-7_29

De Sterck et al. firstly identified that the convergence deterioration was due to rediscretized coarse-grid operators and proposed an optimization method for coarse-grid operators for linear advection problems [1]. This optimization method constructs coarse-grid operators with a realistic number of nonzero elements that minimizes the spectral difference from the ideal operator, dramatically improving the convergence of MGRIT. However, it should be noted that the cost of operators increases from fine- to coarse- level, it is not negligible as mentioned below in terms of parallel performance; because when running at high temporal parallelism, the percentage of coarse-level operations increases, and the overhead is not negligible. Thus, there is room for improvement of the parallel implementation here.

The aim of this paper is to accelerate MGRIT with optimized coarse-grid operators using the spatial redistribution technique [7]. This technique accelerates the coarse-level solvers by assigning temporal agglomerated idle processors to redistributed spatial domains and reduces the overhead. Hence, the novel combination of these existing methods is expected to offset the increased cost of operators optimized for improved convergence.

2 Multigrid Reduction in Time

First, we consider sequential time-stepping for linear time-dependent PDEs. Let N_x be the number of spatial grid points and N_t be the number of time steps. We assume that the spatial and temporal discretized governing equations have the relation: $\mathbf{u}^{i+1} = \Phi\mathbf{u}^i + \mathbf{g}^{i+1}$, where $\mathbf{u}$ and $\mathbf{g}$ denote the unknown and force vector, respectively, and the temporal index i ranges from 0 to $N_t - 1$. The time-stepping proceeds sequentially N_t times according to the above relation.

Multigrid reduction in time (MGRIT) [3] is an all-at-once approach that extracts time parallelism by solving all time steps at once. It yields the linear space-time system on the temporal fine-grid based on the time-stepping method:

$$A\mathbf{u} = \begin{bmatrix} I & & & \\ -\Phi & I & & \\ & \ddots & \ddots & \\ & & -\Phi & I \end{bmatrix} \begin{bmatrix} \mathbf{u}^0 \\ \mathbf{u}^1 \\ \vdots \\ \mathbf{u}^{N_t-1} \end{bmatrix} = \begin{bmatrix} \mathbf{g}^0 \\ \mathbf{g}^1 \\ \vdots \\ \mathbf{g}^{N_t-1} \end{bmatrix} = \mathbf{g}, \tag{1}$$

where Φ is called a fine-grid operator. MGRIT uses relaxations in parallel by delimiting the time dependency based on C-points and F-points. C-points correspond to each m-th time-step, m is called a coarsening factor, and the others are labeled F-points. F- or C-relaxation perform the time-stepping method at F-points or C-points only, respectively. FCF-relaxation also performs F-, C-, and F-relaxations in that order.

MGRIT also constructs a coarse-grid system similar to Eq. 1 with N_t/m time steps and a coarse-grid operator Ψ. The restriction and prolongation operators, which transfer between fine and coarse grids, are defined as injections on C-points. In general, we obtain Ψ with a rediscretization approach that enlarges the

Table 1. Convergence rates of MGRIT with various numbers of nonzero elements for one-dimensional advection problem with $N_x = 2^8$, $N_t = 2^9$ and $\mathrm{nnz}(\phi) = 10$.

	$m=2$	$m=4$	$m=8$	$m=16$	$m=32$
$\mathrm{nnz}(\tilde{\phi}^m)$	19	36	60	89	105
$\nu = 10$	0.092	0.119	0.190	0.469	1.233
$\nu = 11$		0.088	0.166	0.362	0.610
$\nu = 12$			0.157	0.227	0.388
$\nu = 13$			0.085	0.226	0.362
$\nu = 14$				0.183	0.307
$\nu = 15$				0.105	0.311
$\nu = 16$				0.073	0.232

time-step width Δt by a m factor. On the other hand, Ψ can also be obtained by an optimization method [1], which minimizes the spectral difference from optimal operators. Here we briefly it introduce according to the notation of [1]. Let λ be the eigenvalues of fine-grid operators Φ and let μ be the eigenvalues of coarse-grid operators. Assuming periodic boundary conditions for spatial domains, using the first columns of each operator and the DFT matrix $\mathcal{F} \in \mathbb{C}^{N_x \times N_x}$, we can compute the eigenvalues $\lambda^m = \mathcal{F}\tilde{\phi}^m$ and $\mu = \mathcal{F}\psi$ for Φ^m and Ψ, respectively. Based on this computation, the optimization problem for coarse-grid operators is formulated by

$$\psi := \underset{\hat{\psi} \in \mathbb{R}^\nu}{\mathrm{argmin}} \left\| W_\lambda^{1/2} \mathcal{F} \left(\tilde{\phi}^m - \mathcal{R}^T \hat{\psi} \right) \right\|_2^2, \tag{2}$$

where $W_\lambda = \mathrm{diag}(w(|\lambda_k|))$ is a weighting matrix, and $k = 0, \ldots, N_x - 1$ denotes the spatial index. We adopt $w(z) = 1/(1 - z + \epsilon)^2$ as weight function, where $\epsilon = 10^{-6}$. The operator $\mathcal{R} \in \mathbb{R}^{\nu \times N_x}$ constrains sparsity to obtain practical coarse-grid operators with $\nu \ll N_x$. By solving the normal equation of Eq. 2, we obtain the solution ψ and construct the optimized coarse-grid operator Ψ.

Finally, we briefly review the convergence improvement of the optimization approach for a one-dimensional linear advection problem based on the experiment in Fig. 6 in [1]. Table 1 shows the convergence rates of MGRIT, derived by the two-level reduction analysis [2], for the above problem with each m and ν. We confirm that this approach provides good convergence even for hyperbolic problems by slightly increasing ν for m. While this increase is acceptable and practical, there is room for improvement in terms of parallel performance, which will be addressed in the next section.

3 Spatial Redistribution Technique

The spatial redistribution technique [7] for MGRIT assigns temporally agglomerated idle processes to spatially redistributed domains, accelerating the coarse-grid operators. Therefore, this technique decreases temporal parallelism P_t and increases spatial parallelism P_x to reduce the coarse-level spatial solver on the coarse level. See [7] for more details.

In this paper, we use this technique to reduce the cost of optimized coarse grid operators. Moreover, it does not affect convergence; we improve scaling while maintaining good convergence. Also, it offers flexibility in how processes are agglomerated and reassigned, and there are no restrictions on application problems. Therefore, in later numerical experiments, after constructing the coarse-grid operators via the optimization approach, this solver is parallelized in the spatial direction by row-wise one-dimensional block partitioning according to spatial parallelism.

4 Numerical Experiments

This section investigates the effectiveness of MGRIT with optimized coarse-grid operators and spatial redistribution. Numerical experiments are conducted for the one- or two-dimensional advection problems on structured grids with periodic boundary conditions: $\mathbf{u}_t - \alpha \mathbf{u}_x = 0$, where $(\mathbf{x}, t) \in [0,1]^d \times [0,T]$ and $\alpha = 1$ in our experiments. The initial conditions are $\sin(\pi x)^4$ or $\sin(\pi x)^4 \cdot \sin(\pi y)^4$, respectively. We discretize them with third-order upwind discretization for space and explicit third-order Runge-Kutta for time. The CFL number on the finest level is set to 0.85 times the CFL limit, where the rediscretization approach does not work well. The respective coefficients are based on the setup in [1].

To simplify the comparison of solvers, we use the following abbreviations: "M1" corresponds to the MGRIT with optimized coarse-grid operators. "M2" denotes a solver that combines the above with spatial redistribution. The convergence tolerance for both solvers isthat the relative residual 2-norm becomes less than 10^{-10}. We evaluate each solver on the Wisteria/BDEC-01 Odyssey supercomputer system equipped with 2.2GHz Fujitsu A64FX, which has 12×4 cores. We implement flat MPI and MPI/OpenMP modes for each solver using Fujitsu compiler and MPI v4.7.0. In the latter mode, the number of threads T_x is fixed to 12, considering A64FX architecture. The runtime is the minimum value of three measurements.

The first experiment is a one-dimensional problem with $N_x = 2^{15}$, $N_t = 2^{14}$, $T = 5.528$, and flat MPI mode. Both MGRITs use $L = 6$ levels with the coarsening factor $m = 4$. Figure 1 shows the strong scaling results. Both figures include the same results of the sequential time-stepping solver, and we can see it reduces the runtime up to $P_x = 1,024$. However, it stagnates at $P_x = 4,096$ and indicates the exhaustion of spatial parallelism. Next, we move on to M1 on the left and M2 on the right. Both MGRITs converged after only four iterations for this problem. These optimization approaches showed good convergence while the rediscretization approach diverged (not shown in the figure). The scaling tests fix the spatial parallelism P_x from 1 to 256 and increase the temporal parallelism P_t. We can see from the left figure that while M1 scales well at small temporal parallelism, the improvement degrades as the temporal parallelism approaches 4,096, the maximum. The reason is that as temporal parallelism increases, the

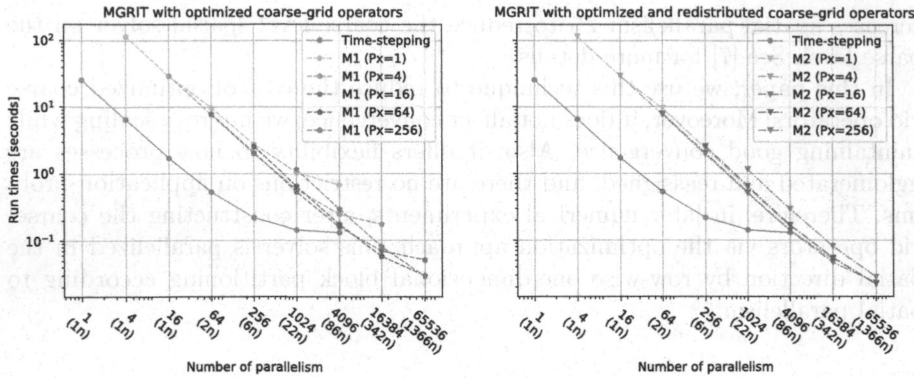

Fig. 1. Strong scaling experiment for one-dimensional advection problem with $N_x = 2^{15}$, $N_t = 2^{14}$, $L = 6$, and $M = 6$. Each color depends on the spatial parallelism specified at the finest level. "M1": MGRIT with optimized coarse-grid operators and "M2" with these and spatial redistribution.

coarse levels occupy a larger proportion of the total runtime, and these levels use optimized coarse-grid operators that are slightly more expensive than the finest grid. In contrast, M2 scales well without stagnation, even at high parallelism. This good scaling is achieved by accelerating the optimized coarse-grid operators. In order to see the improvement of M2 in more detail, we use a runtime breakdown of the fastest case with $P_x = 64$ and $P_t = 1,024$ in Fig. 2. This figure compares the runtime of M1 and M2 at each level, decomposing it into four representative operations: sparse matrix-vector (SpMV), vector operations, communication in space, and communication in time. In the SpMV part, we can see that the runtime increases as the level increases, except for $L = 0$, which contains the convergence check part. This increase directly corresponds to an increase in the number of nonzero elements in optimized coarse-grid operators. Since there is no temporal agglomeration and spatial redistribution on $L = 0$ and 1 for P_t and N_t in this problem, there is no significant difference in the runtime of the two solvers on these levels. After $L = 2$, M2 reduces the SpMV and vector operation parts due to the spatial redistribution. This solver also increases communication to some extent on coarse levels due to the increased spatial parallelism, but it is negligibly small.

Finally, we confirm the different parallelism assignments of each fastest case at maximum nodes 1,366 nodes in Fig. 1: M1 with $P_x = 256$ and $P_t = 256$ and M2 with $P_x = 64$ and $P_t = 1,024$. Comparing the two, M2 achieves a 23% runtime reduction over M1. The M2 parallelism assignment decreases the spatial parallelism and increases the temporal parallelism, compared to the fastest case of M1. A similar trend can be observed for smaller nodes. We believe this is because we can simultaneously benefit from the high temporal parallelism of MGRIT and the coarse-level overhead reduction due to spatial redistribution.

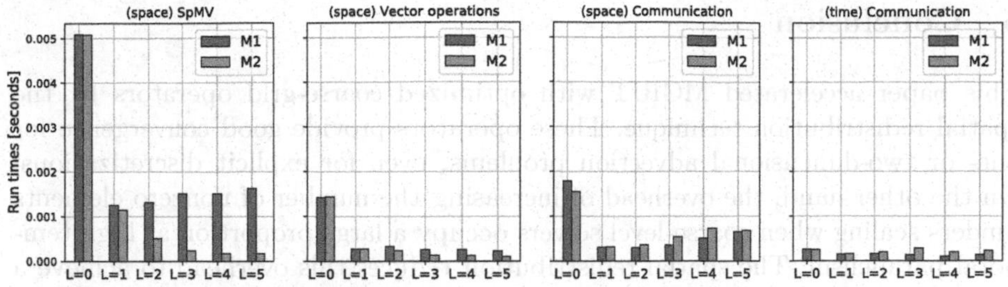

Fig. 2. Runtime breakdown with $P_x = 64$ and $P_t = 1,024$ at 1366 nodes in Fig. 1.

The second experiment is the two-dimensional problem with $N_x^2 = (2^9)^2$, $N_t = 2^{10}$, $T = 2.764$, and hybrid mode. The spatial direction is parallelized using a hybrid MPI/OpenMP approach with a fixed number of threads $T_x = 12$, and the temporal direction is parallelized using MPI only. We set $L = 4$ and $m = 4$ for both MGRITs. In the two-dimensional case, we obtained similar results as in the previous section. Figure 3 shows the stagnation of time-stepping and M1 at high parallelism and the well-scaling of M2. Focusing on the results at 1,024 nodes, we compare both MGRITs' performance. M2 with $P_x = 16$, $P_t = 256$, $T_x = 12$ achieves a 17% improvement compared to M1 with $P_x = 64$, $P_t = 64$, $T_x = 12$.

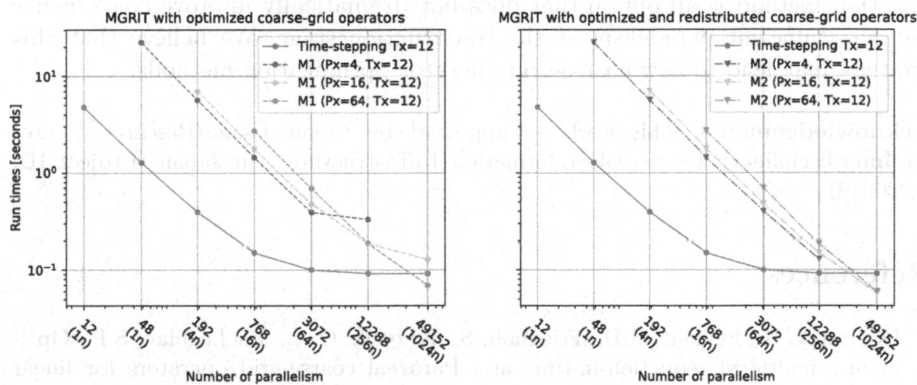

Fig. 3. Strong scaling experiment for two-dimensional advection problem with $N_x^2 = (2^9)^2$, $N_t = 2^{10}$, $L = 4$, $M = 4$, and $T_x = 12$. "M1": MGRIT with optimized coarse-grid operators and "M2" with these and spatial redistribution.

5 Conclusion

This paper accelerated MGRIT with optimized coarse-grid operators by the spatial redistribution technique. These operators provide good convergence for one- or two-dimensional advection problems, even for explicit discretizations. On the other hand, the overhead of increasing the number of nonzero elements hinders scaling when coarse-level solvers occupy a large proportion at high temporal parallelism. The spatial redistribution reduces this overhead to achieve a good convergence and a scalable solver, using agglomerated idle processes on coarse levels. Our numerical experiments show a 23% improvement for the one-dimensional problem and a 17% improvement for the two-dimensional problem compared with the fastest parallelism assignment.

In numerical experiments on a two-dimensional problem, we evaluate a hybrid MPI/OpenMP implementation with a fixed number of threads. However, given that the cost of coarse-grid operators varies with the optimization approach and spatial parallelism changes with the spatial redistribution, the optimal configuration of the number of processes and threads will be different. Future work will investigate them as they vary on the coarse level.

Our approach can tolerate some increase in the cost of optimized coarse-level operators. Ideally, if coarse-grid operators are accelerated by m times, a coarsening factor of MGRIT, then an increase in the cost of m times is acceptable for an optimization method. The present optimization process is performed sequentially and constructs coarse-grid operators. However, for example, an optimization method is an option that does not dramatically improve convergence but has sufficient parallelism at the construction stage. We believe that this strategy may lead to new coarse-grid operator optimization methods.

Acknowledgements. This work is supported by "Joint Usage/Research Center for Interdisciplinary Large-scale Information Infrastructures" in Japan (Project ID: jh220049).

References

1. De Sterck, H., Falgout, R.D., Friedhoff, S., Krzysik, O.A., MacLachlan, S.P.: Optimizing multigrid reduction-in-time and Parareal coarse-grid operators for linear advection. Numer. Linear Algebra Appl. **28**(4), e2367 (2021)
2. Dobrev, V., Kolev, T., Petersson, N., Schroder, J.: Two-level convergence theory for multigrid reduction in time (MGRIT). SIAM J. Sci. Comput. **39**(5), S501–S527 (2017)
3. Falgout, R.D., Friedhoff, S., Kolev, T.V., MacLachlan, S.P., Schroder, J.B.: Parallel time integration with multigrid. SIAM J. Sci. Comput. **36**(6), C635–C661 (2014)
4. Gander, M.J.: 50 years of time parallel time integration. In: Carraro, T., Geiger, M., Körkel, S., Rannacher, R. (eds.) Multiple Shooting and Time Domain Decomposition Methods. CMCS, vol. 9, pp. 69–113. Springer, Cham (2015). https://doi.org/10.1007/978-3-319-23321-5_3

5. Howse, A., Sterck, H., Falgout, R., MacLachlan, S., Schroder, J.: Parallel-in-time multigrid with adaptive spatial coarsening for the linear advection and inviscid burgers equations. SIAM J. Sci. Comput. **41**(1), A538–A565 (2019)
6. Ong, B., Schroder, J.: Applications of time parallelization. Comput. Visual. Sci. **23**(11), 1–15 (2020)
7. Yoda, R., Bolten, M., Nakajima, K., Fujii, A.: Assignment of idle processors to spatial redistributed domains on coarse levels in multigrid reduction in time. In: HPCAsia2022, pp. 41–51 (2022)

Networks Clustering-Based Approach for Search of Reservoirs-Analogues

Andrey Bezborodov and Irina Deeva(✉)

ITMO University, Saint Petersburg, Russia
iriny.deeva@gmail.com

Abstract. This article presents a new look at the problem of finding reservoirs-analogues, representing reservoirs as a network and solving the problem of finding reservoirs-analogues as a problem of finding communities in the network. The proposed network approach allows us to effectively search for a cluster of reservoirs-analogues and restore missing parameters in the target reservoir based on the found clusters of reservoirs-analogues. Also, the network approach was compared with the baseline approach and showed greater efficiency. Three approaches were also compared to restore gaps in the target reservoir using clusters of reservoirs-analogues.

Keywords: Networks · Community detection algorithms · Distance metrics · Oil and gas reservoirs · Reservoirs-analogues

1 Introduction

In the early stages of exploration and development of reservoirs, two problems usually arise - high uncertainty in the data and the lack of these data. The decisions that are made at these stages can have a severe impact on the economic development of the entire project. A common practice for solving the above problems is to use information about already investigated reservoirs similar to our target reservoir. Such reservoirs are called reservoirs-analogues. Expert knowledge-based approaches are the simplest method for finding reservoirs-analogues in time and effort. Voskresenskiy et al. proposed searching for reservoirs-analogues using several filters - parameters [8]. The most critical parameter is the distance to the current area with a target reservoir. It is also possible to determine reservoirs-analogues using various machine learning technologies. Various clustering algorithms can be used to obtain groups of reservoirs-analogues with similar properties [3,6].

The methods listed above have their drawbacks, for example, expert knowledge may not always be available, and clustering algorithms are susceptible to dimensionality growth. Therefore, it was decided to look at the problem of finding reservoirs-analogues as the problem of finding communities in a graph structure since all reservoirs can be represented as a network. A wide range of existing applications is in social network analysis [1,9].

D. Groen et al. (Eds.): ICCS 2022, LNCS 13351, pp. 222–228, 2022.
https://doi.org/10.1007/978-3-031-08754-7_30

Suppose we assume that each reservoir is an object in a multidimensional space of features. All objects are interconnected by edges, which are proportional to some distance metric. In that case, the analogous reservoirs are some clusters in this network that are tightly connected. This formulation of the problem requires solving several problems: (1) how exactly to build networks of reservoirs, (2) how to cluster the network to search for reservoirs-analogues, (3) how to use the resulting markup to search for a cluster of reservoirs-analogues for the target reservoir.

The purpose of this article is to solve the above problems and demonstrate the advantage of a network approach for searching for reservoirs-analogues, as well as to present a network approach for recovering gaps in reservoirs using the found reservoirs-analogues clusters Fig. 1.

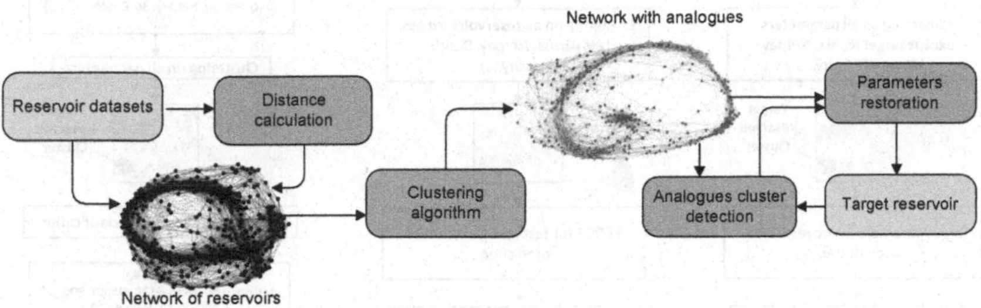

Fig. 1. The pipeline of the proposed network approach to the oil and gas reservoirs-analogues searching.

2 Algorithms and Methods

2.1 Networks Clustering Algorithms

For building a network of reservoirs, an approach was proposed in which a particular distance metric was measured between the parameters of the reservoirs. An edge is drawn between the reservoirs if the metric value is greater than a specific threshold value. As possible distance metrics, we propose cosine distance, Hamming distance, Gower distance and several modified Gower distances - HEOM and HVDM [10]. It is possible to use various community detection approaches in a formed graph. In our study, three algorithms will be compared - Louvain algorithm [2], Leiden algorithm, Newman algorithm [4].

2.2 Search for a Cluster of Reservoirs-Analogues for the Target Reservoir

When reservoirs are already labelled with clusters of reservoirs-analogues, how to use the resulting labelling for a new target reservoir that has gaps in the data, that is, we are simulating a situation where a specialist has a certain reservoir with missing parameters and wants to find a suitable cluster of reservoirs-analogues for this reservoir and use it to restore the gaps. Three approaches are possible to solve this problem (Fig. 2). The first approach is that the missing parameters are excluded from the parameters for building the network. The target reservoir is completely excluded from the network construction in the second approach. Its membership in any cluster is made based on the prediction of FEDOT pipeline [5]. In the third approach, the missing values are first pre-filled based on the values of the entire database with FEDOT pipelines.

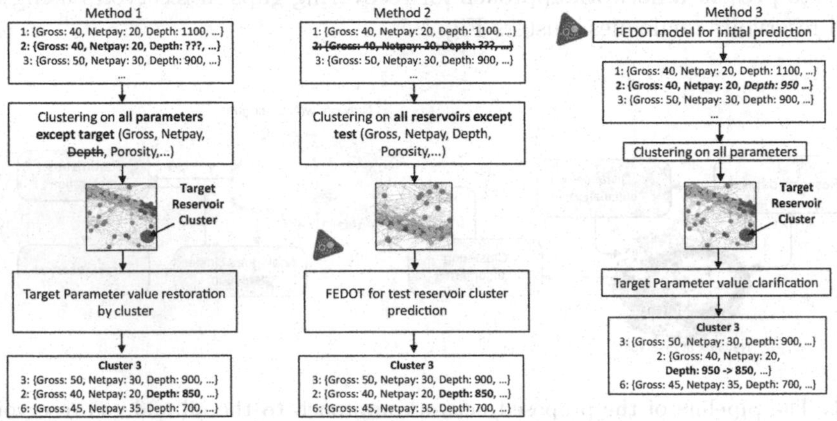

Fig. 2. Three methods of using a network of reservoirs-analogues to restore gaps in the target reservoir.

3 Experiments and Results

3.1 Using Different Distance Metrics

The analysis and testing of the proposed approaches were carried out on a production database containing 442 reservoirs without missing parameters and 72 reservoirs with gaps in various data. The following experiments were carried out to find out which distance metric is the best for building a network of reservoirs:

1. A network is built based on the distance metric;
2. Clusters are searched in the network by the Leiden algorithm;
3. A reservoir is selected from the database, and it is determined to which cluster it belongs;
4. The values in the reservoir are sequentially deleted and restored based on the values in the cluster. The average restores continuous parameters; the most frequent category restores discrete parameters;

After filling the gaps in the parameters for all reservoirs, we propose to use $1 - the\_average\_accuracy$ over the cluster for categorical parameters. The Root Mean Squared Error (RMSE) divided by the parameter's range is used for continuous parameters. Figure 3 shows visualizations of graphs constructed using various metrics with nodes-reservoirs in the color of clusters for Leiden algorithm. The results obtained for these metrics with the further restoration of gaps in the parameters are presented in Table 1. It can be seen from the table that for most parameters, the cosine distance metric shows the best results.

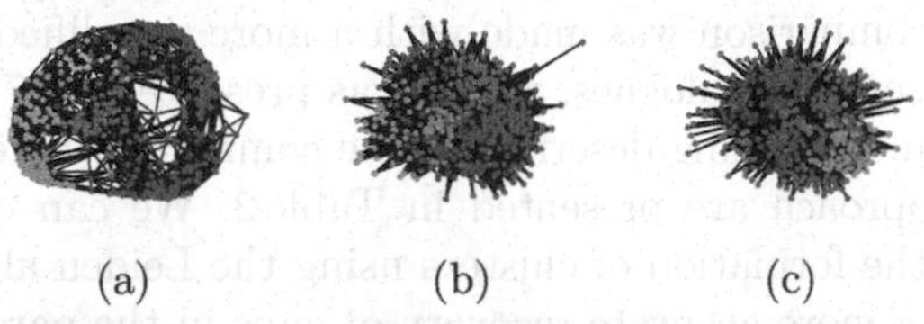

(a) (b) (c)

Fig. 3. Clustered reservoir networks built on different distance metrics. (a) - cosine distance, (b) - Gower distance, (c) - Hamming distance.

Table 1. The results of restoring parameter values based on the found clusters of analogs for different distance metrics (less is better).

Parameter	Cosine	Gower	Hamming	HEOM	HVDM
Depth	**0.098**	0.184	0.182	0.174	0.184
Gross	**0.061**	0.118	0.119	0.117	0.118
Netpay	**0.076**	0.092	0.098	0.094	0.093
Permeability	**0.093**	0.122	0.119	0.126	0.123
Porosity	**0.124**	0.139	0.135	0.141	0.138
Tectonic regime	**0.5**	**0.5**	0.45	0.52	0.52
Hydrocarbon type	**0.21**	0.24	0.25	0.29	0.25
Structural setting	0.67	0.65	**0.63**	0.72	0.68
Period	**0.69**	0.79	0.73	0.72	0.7
Lithology	0.46	0.44	**0.42**	0.43	0.36
Depositional system	0.65	0.62	0.61	0.59	**0.58**
Trapping mechanism	**0.51**	**0.51**	0.54	0.52	0.54

3.2 Using Different Clusters Detection Algorithms

The structure of experiments is the same as in the previous section (Sect. 3.1). Figure 4 shows the results of clustering the reservoirs network by three different algorithms.

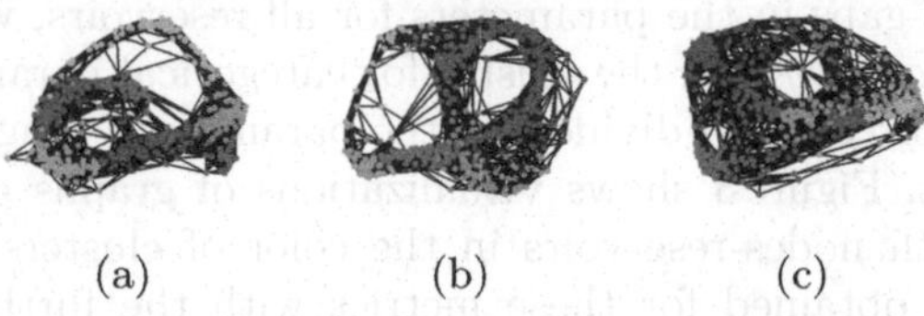

Fig. 4. The results of clustering networks of reservoirs by three algorithms - (a) - Louvain algorithm, (b) - Leiden algorithm, (c) - Newman algorithm.

An additional comparison was made with a more simplified approach (baseline) for finding reservoir-analogues, which was presented in [7]. The gap recovery results in parameters using described three community detection algorithms with a baseline approach are presented in Table 2. We can conclude that for most parameters, the formation of clusters using the Leiden algorithm makes it possible to obtain a more accurate recovery of gaps in the parameters.

Table 2. The results of restoring parameter values based on the found clusters of analogs for different clusters detection algorithms (less is better).

Parameter	Leiden Algorithm	Louvain Algorithm	Newman Eigenvector Algorithm	Baseline Approach
Depth	**0.098**	0.103	0.099	0.155
Gross	**0.061**	0.065	0.068	0.076
Netpay	**0.076**	0.078	0.081	0.081
Permeability	**0.093**	0.101	0.097	0.118
Porosity	**0.124**	**0.124**	0.125	0.127
Tectonic regime	**0.500**	0.580	0.560	0.570
Hydrocarbon type	**0.210**	0.260	0.240	0.250
Structural setting	**0.670**	0.790	0.690	0.810
Period	**0.690**	0.730	0.740	0.790
Lithology	**0.460**	0.490	**0.460**	0.650
Depositional system	**0.650**	**0.650**	0.670	0.720
Trapping mechanism	**0.510**	0.520	0.550	0.520

3.3 Methods for Parameters Recovery

We explored three approaches (Sect. 2.2) using the Leiden algorithm to find clusters with reservoirs-analogues. For the experiment, ten target reservoirs were selected with four random missing parameters; in total, 30 runs for each method were performed. The results obtained are presented in Table 3. For the prediction task, FEDOT [5] was used, since it allows us to find the best model for classification and regression problems. The Fig. 5 shows the most common pipelines for classification (cluster prediction) and regression (gap prefill). For most parameters, the third method, showed the best results.

Table 3. The results of restoring parameter values based on cluster values using three approaches of working with the network and the target reservoir (less is better).

Parameter	Method 1	Method 2	Method 3
Depth	0.238	0.278	**0.223**
Gross	0.228	0.216	**0.205**
Netpay	0.161	0.145	**0.132**
Permeability	0.219	0.266	**0.197**
Porosity	0.238	0.229	**0.212**
Tectonic regime	**0.47**	0.54	0.53
Hydrocarbon type	**0.17**	0.25	0.19
Structural setting	0.74	0.73	**0.64**
Period	0.71	0.76	**0.67**
Lithology	**0.44**	0.49	0.51
Depositional system	0.67	0.64	**0.62**
Trapping mechanism	0.59	0.56	**0.41**

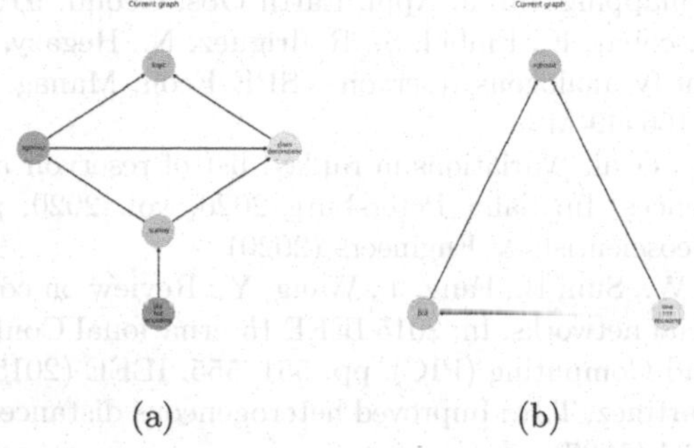

Fig. 5. The most frequent FEDOT pipelines for classification and regression problems.

4 Conclusion

This article presented an approach for searching for analogous reservoirs based on a network representation of reservoirs and searching for clusters in such a network. The reservoirs-analogues obtained by using the Leiden algorithm turned out to be the most interconnected compared to other studied approaches. Furthermore, all proposed clustering algorithms turned out to be more efficient than the baseline approach. Also, the most efficient method was obtained for processing gaps in the target parameters for further clustering. In the future, it would be interesting to create a kind of hybrid approach that considers expert knowledge and the results of the community search algorithm.

Acknowledgement. This research is financially supported by the Ministry of Science and Higher Education, agreement FSER-2021-0012.

References

1. Atay, Y., Koc, I., Babaoglu, I., Kodaz, H.: Community detection from biological and social networks: a comparative analysis of metaheuristic algorithms. Appl. Soft Comput. **50**, 194–211 (2017)
2. Blondel, V.D., Guillaume, J.L., Lambiotte, R., Lefebvre, E.: Fast unfolding of communities in large networks. J. Stat. Mech: Theory Exp. **2008**(10), P10008 (2008)
3. Dumont, M., Reninger, P.A., Pryet, A., Martelet, G., Aunay, B., Join, J.L.: Agglomerative hierarchical clustering of airborne electromagnetic data for multi-scale geological studies. J. Appl. Geophys. **157**, 1–9 (2018)
4. Newman, M.E.: Finding community structure in networks using the eigenvectors of matrices. Phys. Rev. E **74**(3), 036104 (2006)
5. Nikitin, N.O., et al.: Automated evolutionary approach for the design of composite machine learning pipelines. Futur. Gener. Comput. Syst. **127**, 109–125 (2022)
6. Ren, Z., Sun, L., Zhai, Q.: Improved k-means and spectral matching for hyperspectral mineral mapping. Int. J. Appl. Earth Obs. Geoinf. **91**, 102154 (2020)
7. Rodríguez, H., Escobar, E., Embid, S., Rodriguez, N., Hegazy, M., Lake, L.: New approach to identify analogous reservoirs. SPE Econ. Manag. **6** (2013). https://doi.org/10.2118/166449-MS
8. Voskresenskiy, A., et al.: Variations in ranked list of reservoir analogs as an effect of search preferences. In: Saint Petersburg 2020, vol. 2020, pp. 1–5. European Association of Geoscientists & Engineers (2020)
9. Wang, C., Tang, W., Sun, B., Fang, J., Wang, Y.: Review on community detection algorithms in social networks. In: 2015 IEEE International Conference on Progress in Informatics and Computing (PIC), pp. 551–555. IEEE (2015)
10. Wilson, D.R., Martinez, T.R.: Improved heterogeneous distance functions. J. Artif. Intell. Res. **6**, 1–34 (1997)

KP01 Solved by an n-Dimensional Sampling and Clustering Heuristic

Maria Harita(✉), Alvaro Wong, Dolores Rexachs, and Emilio Luque

Universidad Autonoma de Barcelona, 08193 Barcelona, Spain
MariaDeLosAngeles.Harita@autonoma.cat,
{alvaro.wong,dolores.rexachs,emilio.luque}@uab.es

Abstract. In the field of optimization, NP-Hard problems play an important role concerning its real-world applications, such as resource allocation, scheduling, planning, logistics, etc. In this paper, we propose a heuristic search algorithm based on Montecarlo along with a clustering strategy that analyzes density and performs k-means partitions to solve the classic binary Knapsack Problem (KP01). Our heuristic method, which was designed to solve combinatorial optimization problems, has evolved and can adapt to other optimization problems, such as the KP01 that can be organized in an n-Dimensional search space. Regarding the methodology, we substantially reduced the search space while the areas of interest were located in the clustering stage, which brings us closer to the best solutions. After the experiments, we obtained a high-quality solution, which resulted in an average success rate of above 90%.

Keywords: Heuristic method · Clustering · 01 Knapsack Problem

1 Introduction

In optimization, we set the objectives when looking for the best possible configuration of a set of variables. We can deal with these problems by discretizing the variables or taking the real values. Thus, each optimization problem is specified by defining the possible solutions (or states) and a general objective. The classical optimization paradigm is understood as to how the solution identifies by enumeration or differential calculus, the existence of an assumed (unique) solution, or the convergence of classical optimization methods for the solution of the corresponding first-order conditions.

Probabilistic distribution methods such as Montecarlo offer flexible forms of approximation, with some advantages regarding cost. In this sense, metaheuristics are high-level algorithms that are capable of determining a sufficiently

This research has been supported by the Agencia Estatal de Investigacion (AEI), Spain and the Fondo Europeo de Desarrollo Regional (FEDER) UE, under contract PID2020-112496GB-I00 and partially funded by the Fundacion Escuelas Universitarias Gimbernat (EUG).

D. Groen et al. (Eds.): ICCS 2022, LNCS 13351, pp. 229–236, 2022.
https://doi.org/10.1007/978-3-031-08754-7_31

satisfactory (almost optimal) solution for an approximate optimization problem, while other approaches use similarity.

We propose a heuristic algorithm that works in conjunction with clustering techniques based on [1]. The Montecarlo K-Means (MCKM) method was improved and evaluated with the optimization benchmark functions [4], which are high-Dimensional problems in a large search space. In all the benchmark cases analyzed, we came up with the improved heuristic Montecarlo Density K-Means (MCDKM), obtaining promising results comparable with those in the literature.

For this paper, our proposal goes a step further in solving different types of combinatorial optimization problems where search processes are involved. We have selected the Knapsack Problem (KP), which seeks to maximize the profits provided by a selection of elements to enter a Knapsack. The objective function would be defined as the total profit of the selected objects. KP is a classic combinatorial optimization problem with various applications in different industries, such as resource allocation, tasks, resource management/scheduling, or energy allocation management. KP is also part of a historical list of NP-Complete problems elaborated by Richard Karp [5].

Martello [8], on the background of this classic NP-Hard problem, offers an extensive review of the many effective heuristics to solve the KP. For example, a population-based approach such as incremental learning algorithm based on a greedy strategy in [7]. In addition, other different heuristic techniques, such as [3,6,9] are proposed to provide an acceptable solution to this problem.

Regarding the contribution of our heuristic method MCDKM, it has been validated that it can find feasible solutions. In the methodology, we will see how the partitions in n-dimensions of the search space may reduce the ordering times of the axes. The results section compares the solution quality found when varying the Dimensional arrays that organize the data, ranging from bi-dimensional to more than 50-dimension arrangement.

The paper organizes as follows. Section 2 describes the MCDKM heuristic method. Section 3 introduces the Knapsack problem into the MCDKM method, Sect. 4 details the implementation and the results, and finally, Sect. 5 presents the conclusions and future work.

2 Description of the MCDKM Heuristic Method

MCKM heuristic method [1] was initially developed to solve a particular combinatorial optimization problem. Our proposal goes a step further by adding a previous step. This is, we prepare the problem by parameterizing it and organizing the search in an n-dimensional space. Furthermore, we also perform a density analysis of the generated map. Bringing the problem into our parameters will contribute to the solution's efficiency since, first, the search space can be significantly reduced. Second, the ordering times of the axes are improved, even when sorting them in a more significant number of dimensions. All this finally translates into a general improvement that contributes to increasing the

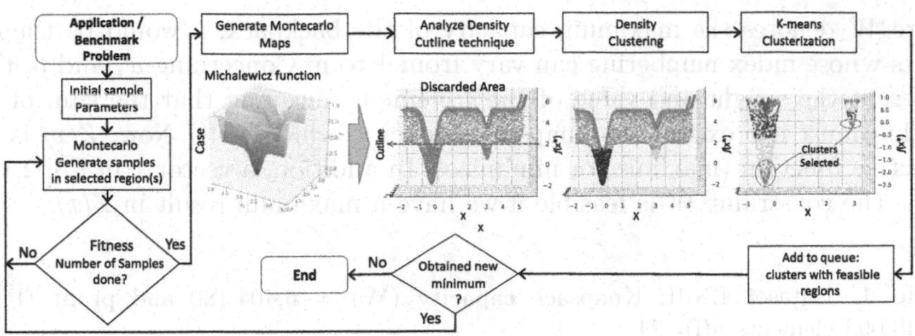

Fig. 1. Flowchart showing the redesigned MCKM heuristic algorithm, highlighting its stages and showing benchmarks as a solution example.

quality of the solution and a good ratio between the size of the problem and the analyzed sample.

The procedure is shown in detail in Fig. 1. First, with Montecarlo, we generated a uniform sample that formed a map that is distributed through the whole search space. Then, the sample is reduced through the density analysis (for which we use the DBScan [2] algorithm), sweeping data to discard the sample that is not worth analyzing while keeping the remaining areas to perform another clustering process. Such clustering is made with k-means, which is now able to correctly partition the sample to group the feasible solutions into clusters for its selection.

The now more robust MCDKM heuristic method was able to find the best solution values, which turned out to be very similar and comparable to the optimum solution of the benchmark functions. The success rate was between 97% and 99%, as seen in the results from [4].

3 The Knapsack Problem into the MCDKM Method

We are bringing the Knapsack Problem KP01 into our heuristic method to find a comparable efficient solution. This combinatorial optimization problem is sometimes exemplified as: the objective is for a supposed person to choose the elements that will allow him to maximize the benefit without exceeding the allowed capacity of a Knapsack. The problem can be stated as a problem of vectors $X = (x_1, x_2, x_3, ...x_n)$, which have components of zeros and ones, shown in Eq. 1, and have at most the restriction of the objective function *Z(x)*, as seen in Eq. 2. Mathematically, we have the following:

$$W(x) = \sum_{i=1}^{n} x_i w_i \leq W \tag{1}$$

$$Z(x) = \sum_{i=1}^{n} x_i p_i \tag{2}$$

where W denotes the maximum capacity of the backpack, x would be the elements whose index numbering can vary from 1 to n. Concerning w_i and p_i they represent the weight and value of the i element, meaning that the sum of the weights must not exceed the knapsack capacity, which is W. Now, $Z(x)$ is the objective function (maximize or minimize). In addition, a vector x in Eq. 2 that meets the constraint W is feasible if we have a maximum result in $Z(x)$.

Table 1. Dataset EX01: Knapsack capacity (W) = 6,404,180 and profit (P) = 13,549,094 elements n(i)=24

Dimensional array for the Knapsack EX01 Problem												
ID element (n)	1	2	3	4	5	6	7	8	9	10	11	12
weight (i)	382,745	799,601	909,247	729,069	467,902	44,328	34,610	698,150	823,460	903,959	853,665	551,830
profit (i)	825,594	1,677,009	1,676,628	1,523,970	943,972	97,426	69,666	1,296,457	1,679,693	1,902,996	1,844,992	1,049,289
2-Dimensional	1st dim											
3-Dimensional	1st dim								2nd dim			
4-Dimensional	1st dim						2nd dim					
6-Dimensional	1st dim				2nd dim				3rd dim			
8-Dimensional	1st dim			2nd dim			3rd dim			4th dim		
12-Dimensional	1st dim		2nd dim		3rd dim		4th dim		5th dim		6th dim	
24-Dimensional	1st dim	2nd dim	3rd dim	4th dim	5th dim	6th dim	7th dim	8th dim	9th dim	10th dim	11th dim	12th dim
ID element (n)	13	14	15	16	17	18	19	20	21	22	23	24
weight (i)	610,856	670,702	488,960	951,111	323,046	446,298	931,161	31,385	496,951	264,724	224,916	169,684
profit (i)	1,252,836	1,319,836	953,277	2,067,538	675,367	853,655	1,826,027	65,731	901,489	577,243	466,257	369,261
2-Dimensional	2nd dim											
3-Dimensional	2nd dim				3rd dim							
4-Dimensional	3rd dim						4th dim					
6-Dimensional	4th dim				5th dim				6th dim			
8-Dimensional	5th dim			6th dim			7th dim			8th dim		
12-Dimensional	7th dim		8th dim		9th dim		10th dim		11th dim		12th dim	
24-Dimensional	13th dim	14th dim	15th dim	16th dim	17th dim	18th dim	19th dim	20th dim	21th dim	22th dim	23th dim	24th dim

Analyzing these statements is one of the essential steps in the methodology to adapt the parameters. First, we find the relationship between the variables to sort the data into a Montecarlo map.

Since this KP modality is a decision problem, our strategy considers the elements as belonging to different dimensions. This is, we divide the search space into multiple dimensions distributing the elements among them, and subsequently ordering according to all the possible combinations between them and identifying them with an ID.

We selected two available open-source datasets, EX01 from Kreher and Stinson which was checked with a branch and bound (B&B) algorithm. EX02 is a dataset available in the ORLibrary. EX03 dataset was based on EX01.

Table 1 shows a Dimensional array we designed for EX01. As seen, for a 2-D search space, 12 elements will be placed in each dimension, and the ordering is made out of all the possible combinations between these 12 elements. In the case of a 24-D search space, each element belongs to a dimension. Ordering the elements should be more efficient, which could be taken as an advantage in terms of resource use.

Table 2. Detail of the possible combinations (C) in the 1st-dim. of an 8-dim array.

Combinations (Cn)			Elements weight				
Id	Binary representation	Element	1	2	3	Weight combination	Profit combination
C0	000	–	–	–	–	–	0
C1	001	1	382,745	–	–	382,745	825,594
C2	010	2	–	799,601	–	799,601	1,677,009
C3	011	1,2	382,745	799,601	–	1,182,346	2,502,603
C4	100	3	909,247	–	–	909,247	1,676,628
C5	101	1,3	382,745	–	909,247	1,291,992	2,502,222
C6	110	2,3	–	799,601	909,247	1,708,848	3,353,637
C7	111	1,2,3	382,745	799,601	909,247	2,091,593	4,179,231

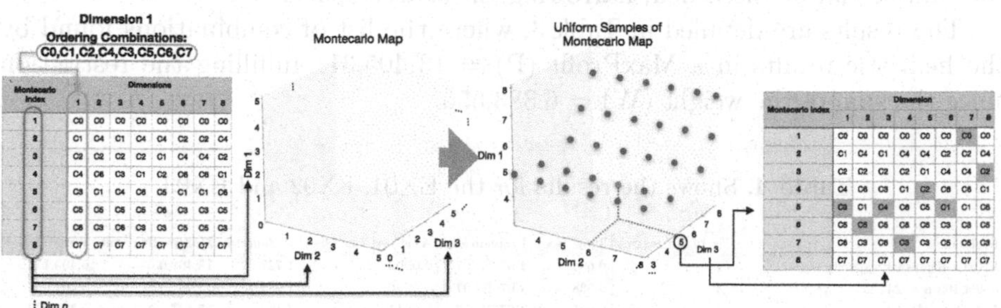

Fig. 2. Establishing a Montecarlo index for each combination in every dimension. Every sample fulfills the restriction in an 8-Dimensional search space.

Table 3. Solution found by the heuristic method for a 8-Dimensional search space.

	Combination			Elements weight				
Dimension	Id	Binary representation	Elements	1	2	3	Weight combination	Profit combination
1	C3	011	1,2	382,745	799,601	–	1,182,346	2,502,603
2	C5	101	4,6	729,069	–	44,328	773,397	1,621,396
3	C4	001	9	823,460	–	–	823,460	1,679,693
4	C3	011	10,11	903,959	853,665	–	1,757,624	3,747,988
5	C2	010	14	–	670,702	–	670,702	1,319,836
6	C1	001	16	951,111	–	–	951,111	2,067,538
7	C0	000	-	–	–	–	0	0
8	C2	100	23	–	–	224,916	224,916	466,257
					Total Weight		6,383,556	
							Total Profit	13,405,311

The Montecarlo map generates by crossing data between the combinations of the elements, which must not exceed the restriction (Eq. 1) or the value is rejected. The uniform map is formed with the samples that meet the weight constraint. The binary code serves as a coordinate representation of such combinations since there are only two possible states. Each digit position represents whether an element will be packed or not, so we know what elements are within each coordinate. Afterward, ordering the axes is carried out in ascending order of best profit. In Table 2 we show how ordering a dimension takes place.

In Fig. 2, we illustrate how we define the map by establishing a Montecarlo index and referring to the position of each combination of elements in each dimension, which returns an integer when generating random samples.

After creating the Montecarlo map, we seek to reduce the search space. Now, the density clustering algorithm can be applied directly to this map, but it has to be adjusted to work efficiently.

We make such adjustment by cutting the map along the objective function axis. Cutting the map helps improve the clustering, and we can avoid analyzing the complete map. Sweeping of the data must occur, and then the stopping criterion is activated when more than 1 cluster is found. We aim for the largest number of clusters since it interprets as many feasible areas. Now, the rest of the sample can be discarded, narrowing the search space.

The results are detailed in Table 3, where the list of combinations found by the heuristic results in a MaxProfit (P) = 13,405,311 fulfilling the restriction since the sum of the weight (W) = 6,383,556.

Table 4. Shows the results for the EX01, EX02 and EX03.

Problem	# of dimensions	Time (sec)	Problem size	Axis sort tme (sec)	# Samples	Max profit result	Weight result
EX01 - Knapsack capacity =	2	0.694	1.676E+07	0.0992	1,727	13,463,657	6,403,149
6,404,163 n = 24	4	0.538	1.676E+07	0.0030	1,644	13,461,329	6,402,653
Fitness = 1E−01	6	0.995	1.676E+07	0.0011	3,147	13,437,475	6,375,863
Optimal profit (P) = 13,549,094	8	2.322	1.676E+07	0.0009	7,004	13,436,707	6,399,256
with a weight (w) = 6,402,560	12	2.048	1.676E+07	0.0006	4,429	13,385,615	6,389,719
Time required = 1.2 s	24	0.432	1.676E+07	0.0010	146	13,300,418	6,392,566
EX02 - Knapsack capacity = 850 n = 50	2	1,277.291	1.125E+15	1276.6851	2,084	6,242	805
Fitness = 1E−01	5	2.043	1.125E+15	0.0489	1,417	6,643	818
Optimal profit (P) = 7,534	10	2.616	1.125E+15	0.0023	2,856	6,962	810
with a weight (w) = 850	25	3.903	1.125E+15	0.0016	5,044	6,233	849
Time required = 0.9 s	50	0.822	1.125E+15	0.0011	524	6,445	772
EX03- Synthetic Knapsack capacity = 64,041,630 n = 240	20	3.302	1.766E+72	1.1231	557	130,279,429	63,791,741
Fitness = 1E−01	30	2.127	1.766E+72	0.0868	1,622	130,589,231	63,981,629
Optimal profit = 135,789,553	40	1.088	1.766E+72	0.0226	899	130,448,449	63,941,709
with a weight (w) = 64,041,800	48	1.873	1.766E+72	0.0162	1,148	130,824,332	63,923,428
Time required = 13,057.79 s	60	5.010	1.766E+72	0.0089	3,170	130,805,991	63,987,106
	80	0.469	1.766E+72	0.0081	172	130,011,304	63,741,250

4 Implementation and Empirical Results

The headers of Table 4 detail the best solutions found for each data-set of the Knapsack Problem, such as the capacity *(W)*, the number of elements *(n)*, the fitness used, and the Optimal Profit *(P)*, obtained with a branch and bound (B&B) algorithm. The rest of the Table represents the best results obtained by the MCDKM heuristic out of 30 executions for each space partition.

As expected, as the number of dimensions increased, the Axis sort time decreased when it required fewer elements, thus decreasing the total time until finding a solution. Therefore, it can be considered an advantage against other types of search algorithms capable of finding the optimal but entailing a high computational cost (like B&B). Also, as seen in Table 4, we emphasize the ratio between the number of samples and the size problem. As a result, the total samples needed to find an efficient solution was less than 1% in all cases.

Regarding EX03, we increased the complexity to compare us with other algorithms. In this case, we used branch and bound. We verified that, as the complexity of the problem increases, the execution time might also increase, influencing the prediction quality. Nevertheless, the solution maintained a prediction quality above 95% in less than 1 s (when partitioned into 80 dimensions). In contrast, the execution time was considerably longer even though the branch and bound algorithm obtained the optimal solution.

Therefore, heuristic search methods, especially ours that perform stochastic optimization, provide a good way to solve large combinatorial problems and offer a high-quality solution.

5 Conclusions and Future Work

The Knapsack Problem was entirely adapted and brought into our parameters. As a result, we found an efficient solution and achieved better results in the execution time when we increased the no. of partitions of the search space. Furthermore, we significantly reduced the number of samples needed to reach the best solutions compared to the full sample (which is the problem size) in all cases. Space ordering and sorting is an essential step in the methodology.

As we present in this paper, the MCDKM heuristic search algorithm proved some of its advantages, such as its applicability in combinatorial optimization problems. The MCDKM method has the potential to solve any KP01 using an n-Dimensional search space approach obtaining an efficient, high-quality solution with a success rate above 90% average. Solving the KP01, grants MCDKM the capability to deal with a range of other combinatorial optimization problems, such as resource allocation, production scheduling, distribution processes, etc.

Our heuristic model is constantly improving and considering more factors, for example, the relationship among multiple items. In addition, our model is expected to solve multi-objective and multi-Dimensional optimization problems. We believe that parallelization of the method would provide a fast convergence with the improvement in the execution times.

References

1. Cabrera, E., Taboada, M., Iglesias, M.L., Epelde, F., Luque, E.: Optimization of healthcare emergency departments by agent-based simulation. Procedia Comput. Sci. **4**, 1880–1889 (2011)
2. Ester, M., Kriegel, H.P., Sander, J., Xu, X.: A density-based algorithm for discovering clusters in large spatial databases with noise. In: Proceedings of the Second International Conference on Knowledge Discovery and Data Mining, KDD 1996, pp. 226–231. AAAI Press (1996)
3. Gao, Y., Zhang, F., Zhao, Y., Li, C.: Quantum-inspired wolf pack algorithm to solve the 0–1 knapsack problem. Math. Probl. Eng. **2018**, 1–10 (2018)
4. Harita, M., Wong, A., Rexachs, D., Luque, E.: Evaluation of the quality of the "Montecarlo plus k-means" heuristics using benchmark functions. In: Short Papers of the 8th Conference on Cloud Computing Conference, Big Data & Emerging Topics (JCC-BD&ET), pp. 36–39 (2020)

5. Karp, R.M.: Reducibility among combinatorial problems. In: Complexity of Computer Computations, pp. 85–103. Springer, Cham (1972). https://doi.org/10.1007/978-1-4684-2001-2_9
6. Li, Z., Li, N.: A novel multi-mutation binary particle swarm optimization for 0/1 knapsack problem. In: 2009 Chinese Control and Decision Conference, pp. 3042–3047 (2009)
7. Liu, L.: Solving 0–1 knapsack problems by greedy method and dynamic programming method. Adv. Mater. Res. **282–283**, 570–573 (2011)
8. Martello, S., Pisinger, D., Toth, P.: New trends in exact algorithms for the 0–1 knapsack problem. Eur. J. Oper. Res. **123**, 325–332 (2000)
9. Zhao, J., Huang, T., Pang, F., Liu, Y.: Genetic algorithm based on greedy strategy in the 0–1 knapsack problem. In: 2009 Third International Conference on Genetic and Evolutionary Computing, pp. 105–107 (2009)

A Deep Neural Network as a TABU Support in Solving LABS Problem

Dominik Żurek(✉), Marcin Pietroń, Kamil Piętak, and Marek Kisiel-Dorohinicki

AGH University of Science and Technology, al. Adama Mickiewicza 30, 30-059 Krakow, Poland
{dzurek,pietron,kpietak,doroh}@agh.edu.pl

Abstract. One of the leading approaches for solving various hard discrete problems is designing advanced solvers based on local search heuristics. This observation is also relevant to the low autocorrelation binary sequence (LABS) – an open hard optimisation problem that has many applications. There are a lot of dedicated heuristics such as the steepest-descent local search algorithm (SDLS), Tabu search or xLostovka algorithms. This paper introduce a new concept of combining well-known solvers with neural networks that improve the solvers' parameters based on the local context. The contribution proposes the extension of Tabu search (one of the well-known optimisation heuristics) with the LSTM neural network to optimise the number of iterations for which particular bits are blocked. Regarding the presented results, it should be concluded that the proposed approach is a very promising direction for developing highly efficient heuristics for LABS problem.

Keywords: LABS · TABU · Neural network · LSTM

1 Introduction

This paper concentrates on solving hard discrete problems using a combination of local optimisation heuristics and neural networks that improves the chosen parameters of heuristics based on the current computation context. The concept is verified on the low autocorrelation binary sequence problem (LABS) and the Tabu search algorithm.

LABS [10] consists of finding a binary sequence $S = \{s_0, s_1, ..., s_{L-1}\}$ with length L, where $s_i \in \{-1, 1\}$, which minimises energy function $E(S)$:

$$C_k(S) = \sum_{i=0}^{L-k-1} s_i s_{i+k} \quad \text{and} \quad E(S) = \sum_{k=1}^{L-1} C_k^2(S) \tag{1}$$

There may varied techniques that try to solve the problem. The simplest method of solving LABS is the application of exhaustive enumeration; this provides the best results, but can be applied only to small values of L. There are

D. Groen et al. (Eds.): ICCS 2022, LNCS 13351, pp. 237–243, 2022.
https://doi.org/10.1007/978-3-031-08754-7_32

also a lot of various heuristic algorithms that use some plausible rules to locate good sequences more quickly. A well-known method for such techniques is *steepest descend local search* (SDLS) [1] or Tabu search [8]. In recent years, a few modern solvers based on the 'self-avoiding walk' concept have been proposed. The most promising solvers are *lssOrel* [3] and *xLostavka* [4], which are successfully used for finding skew-symmetric sequences of lengths between 301 and 401 [5]. These techniques can also be parallelised utilising GPGPU architectures as has been shown in the literature [12,15].

A different direction of research is the application of agent-based biologically-inspired computational systems. One of such meta-heuristics approach is the concept of an evolutionary multi-agent system successfully applied for solving complex continues and discrete optimisation problems such as LABS [12,15].

In the presented work, a deep learning network is incorporated for estimating optimal Tabu search parameters. The LSTM architecture was chosen as a base model in different optimisation domains such as sequence-to-sequence learning [13] or as a foundation of the pointer networks which are used to resolve the TSP problem. In order to resolve this NP-hard problem, *Vinyals et al.* [14] proposed training the recurrent neural network in a supervised manner to anticipate the order of visited cities. This approach was extended in research [2] in which the optimisation of the recurrent neural network which is resolving the TSP problem was gained by using a policy gradient method. Those researches have become the motivation for us to introduce the recurrent neural network as a support for Tabu search algorithm in resolving LABS problem.

The paper is organised as follows. In the next section, the Tabu search is presented as a starting point for further extensions. In the third section, a new concept of Tabu search with LSTM extensions (for predicting the M parameter) are described in detail. Then, the experimental results are presented and conclusions are drawn in the following sections. The paper is summarised in the last section where future work is also suggested.

2 Tabu Search for LABS Problem

The Tabu search is a well-known heuristic [9], adjusted to the LABS problem [6, 7]. The Tabu method can be seen as an extension of steepest descent local search with so-called banned states. In a similar manner to SDLS, Tabu explores the neighbourhood of a given sequence (i.e. all sequences with one bit changed) and also introduces a Tabu array, with the same length as a LABS sequence, which contains blocked indices. When a new better solution is found in an iteration, the index which led to this sequence is placed in the Tabu array and is therefore banned from further changes (in M following iterations). The details of the Tabu search are presented in Algorithm 1.

In contrast to the SDLS algorithm, applying the Tabu mechanism with blocked states helps to escape the attraction basins of a local minimum. The consecutive iterations don't have to produce sequences with lower energy if the Tabu array contains the best sequences from the current neighbourhood. This fact leads to proper directions in LABS solutions space exploration [11].

Algorithm 1. The Tabu search algorithm for the LABS problem (based on [7]). Symbols: S – input sequence; L – length of the sequence; $maxIters$ – number of iterations; E – sequence evaluation; $tabu$ – integer vector that describes how long sequence elements are blocked; $minTabu$, $extraTabu$ – auxiliary numbers used to set values in vector $tabu$; M – number of iterations for which a chosen bit should be blocked; $changedBit$ – an index of element (a bit) that has been changed.

```
function TabuSearch(S, maxIters)
    int[] tabu                                   ▷ integer vector initialised with zeros
    minTabu = maxIters/10
    extraTabu = maxIters/50
    S_start = S_best = S
    E_best = EVALUATE(S_best)
    for i = 0 to maxIters − 1 do
        E_i = ∞
        for j = 0 to L − 1 do
            S_tmp = S_start
            S_tmp[j] = −1 * S_tmp[j]
            E_tmp = EVALUATE(S_tmp)
            if i ≥ tabu[j] or E_tmp < E_best then
                if E_tmp < E_i then
                    S_i = S_tmp, E_i = E_tmp
                    changedBit = j
        if S_i ≠ null then                       ▷ if better, non-blocked sequence is found
            S_start = S_i                        ▷ set the current sequence as the starting point
            M = minTabu + RAND(0, extraTabu)
            tabu[changedBit] = i + M
        if E_i > E_best then
            S_best = S_i, E_best = E_i
    return S_best
```

3 Improving Tabu Search with LSTM

LSTM is an example of artificial recurrent neural network architecture. Unlike standard feedforward neural networks, LSTM has feedback connections. It can process not only single data points, but also entire sequences of data.

The main goal of this paper is to device a more effective technique to determine the value of the M parameter, the use of which, enables finding a more optimal value of the energy. For this purpose we introduce the LSTM neural network which is able to predict the most effective value of this parameter for a particular input sequence S with the length L and corresponding to this sequence energy E (see next subsection). For each different value of the parameter L there is a separate trained model.

Prepare Training Data. In order for it to be possible to train the neural network, for each value of the L parameter used in this paper, 1048576 (2^{20})

sequences were randomly generated and evaluated. The number of generated sequences represents X%, Y%, Z% percent of the total solution space, respectively for lengths 128, 192, and 256 (based on the formula $\frac{2^{20}}{2^L}$). Each generated sequence has received its own random value of the M parameter from the range [2;18] (adopted based on max iteration, see Sect. 4). Each tuple of sequence S and parameter M are put as the input to the module which is searching optimal sequence *S'* which means with a minimum value of energy E. This searching is realised through the use of very effective GPGPU implementation of Tabu search which was proposed in previous work [12]. Consequently, there is 2^{20} *S*, *M*, *S'* and energy E which was calculated based on sequence *S'* and its value of parameter M. Finally, this value is the corresponding energy for the sequence S. In order to teach the neural network to predict the optimal value of parameter M for any sequence, as an input to neural network there is placed a pair sequence S and energy E (the *S'* sequence is not used in the next phases).

Finding Optimal Parameters. The optimal value of parameters which were used in the training phase and to build neural network architecture, were determined empirically. Through the use of a very small part of the generated data, we trained a separate model for each L with a different number of hidden units (#HU), LSTM layers (#LL) and the value of learning-rate parameter (lr). The best results were gained for lr equal to 0.0001, #HU equals 50 and #LL equals 2. The model was trained over seventy epochs with use of the MSE as loss function and the Adam optimiser. For the best gained results, we built the optimal architecture of LSTM which then was used in real training and during the test phase.

Training Process. During the training phase (Fig. 1), as an input to the neural network, the sequence with the length L and normalised value of energy calculated on the basis of this sequence is given (energies are normalised to the range: [–1;1], through the usage the *MinMaxScaler* from *scikit-learn* library[1]). In order to enhance the meaning of the energy, this value is copied to the input vector L times (without this repetition, the energy value would be imperceptible by LSTM model). During the training, the neural network is taught to extract dependencies between the arranging binary data in the input sequences S, corresponding to this sequence energy E and the value of the M parameter which is returned as an output of the neural network. The model is trained through 70 epochs to reduce the difference between the output value of the LSTM *M'* and the original value of this parameter M, which was used to calculate the value of the corresponding energy in preparing the training data phase. The training process was performed on the Nvidia Tesla V100-SXM2-32GB[2]. This process was run once for each value of the L parameter and this took around ten hours.

[1] https://scikit-learn.org/.
[2] https://www.nvidia.com/en-us/data-center/v100/.

Test Process. The optimal solution of the LABS problem should have the minimum possible value of energy. Thereby, during the test phase, for each randomly generated test sequence, in place of real value of the energy (as occurred in the training phase), we put the smallest possible value of the energy that can be achieved, which after normalisation is equal to –1. As was the case in the training phase, this value of energy is copied L times to the input vector. As a result, the neural network returns value of the M', which allows obtaining the smallest possible value of the energy for an given input sequence. The input sequence with predicted M' value is given as a input for the Tabu calculation module and the final energy is returned (Fig. 1).

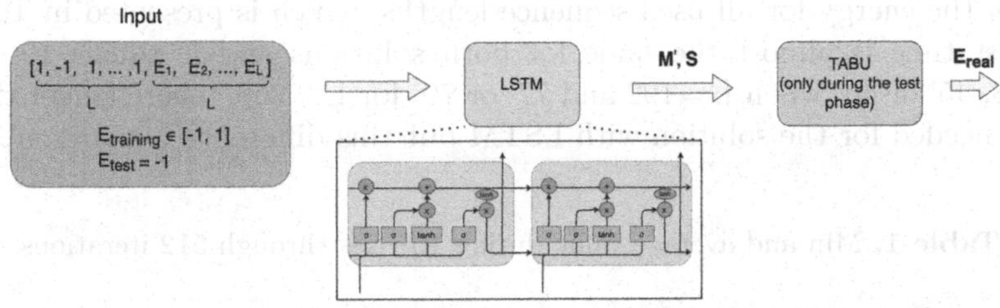

Fig. 1. Training and test phase of TABU with LSTM

4 Effectiveness of the Proposed Algorithms

In order to measure the effectiveness of the proposed solution, energies from the version of Tabu with a trained LSTM neural network, are compared with the basic version of the Tabu search which was implemented in the Python language. In the original Tabu, the number of banned steps M, is determined as a sum of two factors: *minTabu* and *extraTabu* (Algorithm 1, line 19). In our experiments, the value of the *maxItcr* parameter was set based on promising results which was achieved in our previous work [12] and it is set to 128. Consequently, the value of the M parameters in the reference version of Tabu search is between [13] [16].

Each algorithm was seeking the optimal solution for three different input lengths (128, 192, 256). In a single test, 512 random sequences are generated for which the Tabu calculations are run in both scenarios: i) base Tabu search, ii) tabu search with support of LSTM, where M parameter is prompted though the use of the neural network. The entire process was run 10 times, so consequently, 512×10 random sequences were tested for each value of parameter L.

Figure 2 demonstrates the mean value of energies from 10 independent runs after each iteration. As could be observed for each size of the problem, the best solution was obtained by the Tabu which was supported by the neural network. Moreover, with this solution for all test data, it was possible to find the lowest

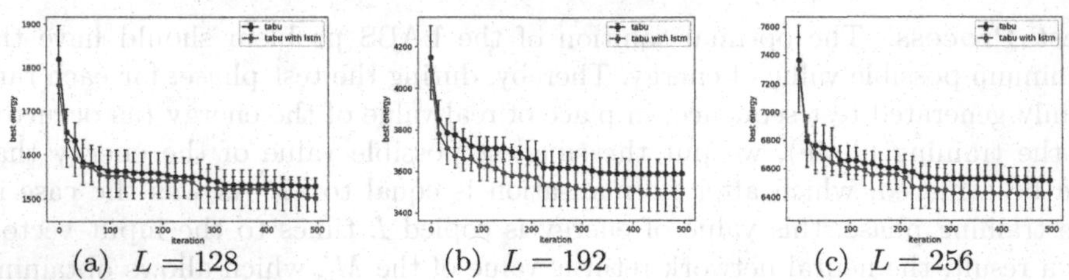

(a) $L = 128$ (b) $L = 192$ (c) $L = 256$

Fig. 2. Energies achieved by basic Tabu search and Tabu search with the LSTM neural network for the different value of L parameter

value of the energy for all used sequence lengths, which is presented in Table 1. The test time is almost the same for both solutions and it equals 19' when L = 128, 45' or 46' when L = 192 and 77' or 79' for L = 256, where longer time is always needed for the solution with LSTM but this difference is marginal.

Table 1. Min and average value during 10 runs through 512 iterations

Sequence lenght	Method	Min value	Average value	Sequence lenght	Method	Min value	Average value
128	**TABU**	1484	1555.48	**128**	**TABU with LSTM**	1444	1539.80
192	**TABU**	3432	3567.54	**192**	**TABU with LSTM**	3284	3567.54
256	**TABU**	6372	6595.80	**256**	**TABU with LSTM**	6180	6518.98

5 Conclusions and Further Work

The presented paper is the latest of our research related to finding an efficient approach to resolving the LABS problem. The presented work, as our first step of combining a neural network with local optimisation heuristics for hard discrete problems, shows that incorporating deep learning models for predicting the internal parameters of a Tabu search significantly improves the final results from our attempt to solve the LABS problem. It is also worth mentioning that the computational overhead caused by the additional step of the prediction of the M parameter is almost negligible. The presented method seems to be very promising and allows us to identify a few further lines of research such as improvement of the LSTM architecture and its topological impact on parameters relating to prediction efficiency, applying the concept to other local search techniques and the optimisation of the developed algorithms with GPGPU.

Acknowledgments. The research presented in this paper was realised with funds of Polish Ministry of Science and Higher Education assigned to AGH University of Science and Technology and it was supported in part by PLGrid Infrastructure.

References

1. Bartholomew-Biggs, M.: The Steepest Descent Method, pp. 1–8. Springer, US, Boston, MA (2008). https://doi.org/10.1007/978-0-387-78723-7_7
2. Bello, I., Pham, H.: Workshop track -iclr 2017 neural combinatorial optimization with reinforcement learning (2017)
3. Bošković, B., Brglez, F., Brest, J.: Low-Autocorrelation Binary Sequences: On Improved Merit Factors and Runtime Predictions to Achieve Them. arXiv e-prints arXiv:1406.5301 (June 2014)
4. Brest, J., Bošković, B.: A heuristic algorithm for a low autocorrelation binary sequence problem with odd length and high merit factor. IEEE Access **6**, 4127–4134 (2018). https://doi.org/10.1109/ACCESS.2018.2789916
5. Brest, J., Bošković, B.: In searching of long skew-symmetric binary sequences with high merit factors (2020)
6. Dotú, I., Van Hentenryck, P.: A note on low autocorrelation binary sequences. In: Benhamou, F. (ed.) CP 2006. LNCS, vol. 4204, pp. 685–689. Springer, Heidelberg (2006). https://doi.org/10.1007/11889205_51
7. Gallardo, J.E., Cotta, C., Fernandez, A.J.: A memetic algorithm for the low autocorrelation binary sequence problem. In: Proceedings of the 9th Annual Conference on Genetic and Evolutionary Computation, pp. 1226–1233. GECCO 2007, ACM, New York, NY, USA (2007)
8. Gallardo, J.E., Cotta, C., Fernández, A.J.: Finding low autocorrelation binary sequences with memetic algorithms. Appl. Soft Comput. **9**(4), 1252–1262 (2009)
9. Glover, F., Laguna, M.: Tabu Search. Kluwer Academic Publishers, Norwell, MA, USA (1997)
10. Golay, M.: Sieves for low autocorrelation binary sequences. IEEE Trans. Inf. Theory **23**(1), 43–51 (1977)
11. Naick, B.S., Kumar, P.R.: Detection of low auto correlation binary sequences using meta heuristic approach. Int. J. Comput. Appl. **106**(10) (2014)
12. Piętak, K., Żurek, D., Pietroń, M., Dymara, A., Kisiel-Dorohinicki, M.: Striving for performance of discrete optimisation via memetic agent-based systems in a hybrid CPU/GPU environment. J. Comput. Sci. **31**, 151–162 (2019)
13. Sutskever, I., Vinyals, O., Le, Q.V.: Sequence to sequence learning with neural networks. In: Proceedings of the 27th International Conference on Neural Information Processing Systems, vol. 2, pp. 3104–3112. NIPS'14, MIT Press, Cambridge, MA, USA (2014)
14. Vinyals, O., Fortunato, M., Jaitly, N.: Pointer networks. In: Cortes, C., Lawrence, N., Lee, D., Sugiyama, M., Garnett, R. (eds.) Advances in Neural Information Processing Systems, vol. 28. Curran Associates, Inc. (2015)
15. Żurek, D., Piętak, K., Pietroń, M., Kisiel-Dorohinicki, M.: Toward hybrid platform for evolutionary computations of hard discrete problems. Procedia Comput. Sci. **108**, 877–886 (2017)

Classification and Generation of Derivational Morpho-semantic Relations for Polish Language

Wiktor Walentynowicz(✉), Maciej Piasecki, and Mateusz Gniewkowski

Faculty of Information and Communication Technology,
Wrocław University of Science and Technology, Wrocław, Poland
wiktor.walentynowicz@pwr.edu.pl

Abstract. In this article, we take a new look on automated analysis and recognition of morpho-semantic relations in Polish. We present a combination of two methods for join exploration on word-form information – generating new forms and classifying pairs of words in derivational relations. As a method of generation, we used the Transformer architecture in the seq-2-seq task. Classification is performed using a neural network and using the fastText representation method. At the very end, we discussed the results obtained in the experiments.

Keywords: Derivational morphology · Morpho-semantic relations · Word formation · Relation classification

1 Introduction

Word formation processes can be observed in the vast majority of natural languages: *derivatives* are formed from *derivational bases* by means of language specific derivational mechanisms, e.g. *a teacher* from *to teach*, *a duchess* from *a duke* or Polish *domeczek* ≈'a nice, little house' from *dom* 'a house', *białość* ≈'a state of being white' from *biały* ≈'white'. For some natural languages, especially for the inflectional Slavic languages, such mechanisms constitute a very productive system. That is why native speakers can recognise a new derivative as a language unit and identify its derivational base with high precision. What is more derivational relations, in contrast to morpho-syntactic word formation processes signal a meaning change between a basis and the derivative, also predictive to a very large extent, e.g. *palarnia* ≈ 'a place for smoking' derived from *palić* 'to smoke'. Due to this, they are called *morphosemantic relations* [5].

Morphosemantic relations combine two transformations: between word forms and, in parallel, between lexical meanings, that are tightly coupled: different types of word form transformations are characteristic for semantic derivations, e.g. *kierowniczka* ≈ 'a female head or manager' derived from *kierownik* 'a head or manager' primarily by the suffix *-ka*. Derivation rules can be described to some extent by a combination of suffixes, prefixes and inside stem alternations. However such word form level rules are semantically, e.g., the suffix *-ka* mostly

D. Groen et al. (Eds.): ICCS 2022, LNCS 13351, pp. 244–251, 2022.
https://doi.org/10.1007/978-3-031-08754-7_33

signals: `+Male` → `+Female`, but it appears in tool name derivation, too: *wiercić* 'to drill' → *wiertarka* 'a driller', and can be also misleading: *pierwiastka* 'a woman giving birth for the first time' is not a female form of *pierwiastek* 'root', in spite of '*ka*'. Thus proper recognition and interpretation of derivational requires taking into account both types of transformations: morphological and semantic.

The goal of our work is to developed a mechanism for recognition and interpretation of derivatives in a way combining morphological and lexico-semantic level. For a given word, a potential derivative, we want to recognise a set of words with which it is in a certain lexico-semantic relation, and also its derivational basis. We propose machine learning means for both levels: word form and semantic. The unique feature of our approach is a combination of transformer-based neural architecture for modelling derivational patterns tightly coupled with recognition of lexico-semantic relations based on non-contextual word embeddings as semantic representation. We focus on the Polish language for which a large and rich model of morphosemantic relations is included in plWordNet [3]. Contrary to many other wordnets and derivational dictionaries, the plWordNet morphosemantic relations link particular senses of two words, not the word forms. In addition, these relations are always directed according to the derivational processes in Polish: from a derivational basis to the derivative.

2 Related Work

Derivational relations are often described in morphological dictionaries as links between lemmas, e.g. [6,19] or a very large morphological and derivational network DeriNet [16], only later automatically classified to 5 very coarse-grained semantic classes [18]. In [18] the training data were pairs of words (not senses) and classification was based on morphological features of word forms. Semantic annotation of word pairs was adopted for wordnets (lexico-semantic networks), e.g. RoWordNet [10], BulNet [2,10] or CroWN [20]. However, in wordnets, links between lemmas are additionally labelled with semantic relations, i.e. mapped onto morphosemantic relations. plWordNet [3] showed that such an approach is simplification and prone to errors, as different morphosemantic relations may be valid only for selected senses of lemmas. Thus, we focus on morphosemantic relations as linking senses, but signalled by derivational associations.

In [14] two character-level transducers extracted from training data and combined with internal stem alternations were proposed. Relations suggested by transducers were next filtered by grammatical patterns, corpus frequency and semantic classifiers for word pairs. trained a combination of features describing word distributions in a large corpus. The best results were reported for the set of 9 most populated relations: 36.84 (the young being relation) up to 97.19 (femininity) of F1. However, it should be emphasised that in this case wordnet-internal knowledge about assignment of lemmas to WordNet domains [4] was utilised. We do not use such knowledge in our approach. In a similar approach [8], but much more supported by hand-crafted knowledge F1 = 0.682 was achieved for verb and noun synset pairs in BulNet. A sequential pattern mining technique based on regular expressions as features for ML was proposed in [9] and tested on Polish and Spanish. It was trained on "1500 pairs of base words with their derivatives". However, the annotation guidelines are unknown, semantics of the links was not taken into account, as well as the direction of derivation. Finally, the accuracy of 82.33% was achieved with "53.5 thousand links in the network".

Word embeddings (word2vec and neural language models) were investigated in [11] for the Czech coarse-grained derivational relations. Neural character encoder-decoder was applied to predict a derivative from a derivational base in [17]. It used occurrence context too, but was limited to deverbal nouns.

Our main contribution is a combined method for transforming a word form into its derivational basis, both perceived as a lexical units, and recognising the morphosemantic relation. For transformation, we applied Transformer [15] to characters. Information about the relation linking the input and output words is delivered to the decoder, so the encoder is relation-independent, and can be used as a source of embeddings vector for characters containing derivational information. The semantic aspect is represented by a classifier filtering the transformer results as lexical units from the morphosemantic relation perspective. The combination of these two allows for detection and suggestion of derivational relations between lexical units in any wordnet as our method does not require any language-specific knowledge resource, except a training set of relation instances.

3 Data

Table 1. Relationships found in plWordNet at different granularities. *HP* - Hidden Predicate, *CCS* - Cross-Categorial Synonymy

Coarse-grained	Fine-grained	Cardinality	Coarse-grained	Fine-grained	Cardinality
aspectuality	pure aspectuality	31030	role ADJ-V	agent	1694
	secondary aspectuality	7457		time	167
characteristic	characteristic	5366		location	937
markedness	diminutives	4184		instrument	322
	augmentatives	886		patient	306
	young being	83		product	85
markedness-intensity	markedness-intensity	996		cause	427
state/feature bearer	state/feature bearer	1410	role material	material	1315
similarity	similarity	2171	state/feature	state/feature	1410
predisposition	habituality	120	*CCS*	ADJ-N	4507
	quantification	15		ADV-ADJ	11355
	appreciation	21		N-ADJ	4506
	potential	334		N-V	30262
role	agent	153		V-N	30262
	time	36		for relational	17069
	location	25	role inclusion	agent inclusion	124
	instrument	299		time inclusion	38
	patient	1039		location inclusion	46
	product	1521		instrument inclusion	515
	agent of *HP*	10		patient inclusion	234
	location of *HP*	250		product inclusion	786
	product of *HP*	3762	femininity	femininity	3789

The applied dataset from plWordNet (the 25.02.2020 dump) consists of triples: a derivational base, a derivational relation and a derivative; 111,955 triples, of which 19,441 triples with multi-words. The data were divided into training and test: 9 to 1. The division was done in two ways. The first one is a split balanced in terms of the number of relations. The second is a split in which the same derivational bases do not occur in both sets. For the relation classification task, we considered only relations with at least 300 instances. For our classification experiments, we selected relation instances consisting of only single words. plWordNet derivational relations can be divided according to two levels of granularity – sparse (coarse grained) and dense (fine grained), see Table 1 and [14].

4 Proposed Approach and Experiments

The tasks is: for a triple (a derivational base lemma, a derivative lemma and the derivational relation), given two of these three pieces of information, we want to obtain the value of the unknown information. This results in three tasks: generation of a derivational base lemma (*derivational analysis*), generation of a derivative lemma (*derivational generation*), a *classification of a derivational relation*. The first two require transforming a character sequence into another character sequence, i.e., *sequence-to-sequence method*. The third task is classification. The *Derivator* system proposes a combination of these two types.

The derivational form generation is done by a sequence-to-sequence neural network based on Transformer [15]. Our Transformer model has a character-based input and is not pre-trained. In our approach, the decoder receives as an additional, special first token – relation tag for which it performs the transformation. The Transformer decoder, performs non-autoregressive decoding, so for our purpose it was enhanced with an autoregressive decoding module. This was necessary, because the length of the generated sequence is not known beforehand. Token selection, in each decoding step, is done using a greedy step-by-step method. In addition, we prepared a derivational analyser: a derivative to its base lemma, also taking into account the derivational relation. The analyser has the same architecture as the generator, the only difference is the transformation direction to be learned. The implementation is based on *PyTorch* [12].

For the derivation generation task, we performed experiments on the two defined splits of the datasets with and without multi-word lexical units (MWE): 75.35% (stratified, without MWE), 73.88% (grouped, without MWE), 76.07% (stratified, with MWE) and 77.14% (grouped, with MWE). In the derivation analysis task we obtained: 82.65% (stratified, without MWE), 74.42% (grouped, without MWE), 83.20% (stratified, with MWE) and 77.20% (grouped, with MWE). The given measure is the sequence identity accuracy.

The derivation relation classifier was based on Multilayer Perceptrons trained as one-vs-rest, that triggered a problem arises of selecting negative examples. We tested different approaches. The first was to use examples from the other relations as negative ones. However, it resulted in a highly unbalanced training set. For training our classifiers, we also chose examples from other relations as negative examples only that include lexical units of the same Part-of-Speech (PoS) pairs. This is well suited, given that our method of filtering candidate derivation pairs uses PoS information, so a classifier to classify a given relation, will never encounter a pair that does not match. Finally, we also tried samples for which the correspondence occurred at the PoS level of the word base as negative examples. The experiments were performed with *scikit-learn* [13].

We divided the classification problem into three experiments according to the input representation. In the first, we used difference of the derivative and base fastText vectors [1] from KGR10 [7]. In the second the difference vector was concatenated with the base and derivative vectors. In the last the difference vector was expanded with the difference of the vectors of pseudoaffixes. Pseudoaffixes were obtained by finding the longest common sequence between the lemmas of a base and a derivative, next separating prefixes and suffixes and building fastText vectors for them. In the presentation, we use the following notations: 1D – word vector difference; 1DNG – word vector difference vector only using fastText n-grams alone; 1DAF – concatenation of a word difference vector and pseudoaffixes difference vectors; 2DAF – concatenation of pseudoaffixes difference vectors; 3D – concatenation of word difference vector and word vectors. Table 2 presents classification results per relation. For the other two experiments, the differences in results were about 1–2 percentage points more for the weighted average. The implementation and experimental results are available in a repository at: https://gitlab.clarin-pl.eu/morphology/derywator.

Table 2. F1-Score measure values for all relations classifier.

	General relation	Detailed relation	1D	1DNG	1DAF	2DAF	3D	Support
one-vs-rest classifiers	aspectuality	pure aspectuality	0,92	0,91	0,95	0,94	0,93	939
	aspectuality	secondary aspectuality	0,50	0,00	0,67	0,68	0,61	170
	characteristic	characteristic	0,65	0,66	0,70	0,66	0,62	411
	cross-categorial synonymy	ADJ-N	0,93	0,93	0,94	0,94	0,92	328
	cross-categorial synonymy	ADV-ADJ	0,99	0,98	0,99	0,99	0,98	578
	cross-categorial synonymy	for relational	0,84	0,83	0,86	0,85	0,86	1386
	cross-categorial synonymy	N-ADJ	0,95	0,95	0,96	0,94	0,96	328
	cross-categorial synonymy	N-V	0,98	0,98	0,98	0,98	0,98	1126
	cross-categorial synonymy	V-N	1,00	1,00	1,00	1,00	1,00	1126
	femininity	femininity	0,97	0,97	0,97	0,90	0,96	324
	markedness	augmentatives	0,00	0,00	0,83	0,78	0,64	74
	markedness	diminutives	0,86	0,86	0,94	0,86	0,93	306
	role	agent	0,79	0,80	0,81	0,77	0,81	122
	role	agent of hidden predicate	0,92	0,92	0,94	0,91	0,92	325
	role	instrument	0,66	0,64	0,70	0,38	0,64	78
	role	location of hidden predicate	0,86	0,76	0,87	0,78	0,84	39
	role	product	0,67	0,64	0,68	0,64	0,67	125
	role ADJ-V	agent	0,88	0,92	0,91	0,95	0,91	66
	role inclusion	instrument inclusion	0,65	0,26	0,71	0,47	0,70	44
	role inclusion	product inclusion	0,00	0,04	0,70	0,27	0,73	55
	role material	material	0,05	0,00	0,08	0,00	0,20	117
	similarity	similarity	0,34	0,35	0,40	0,31	0,38	184
	state/feature	state/feature	0,47	0,46	0,55	0,43	0,57	125
	state/feature bearer	state/feature bearer	0,60	0,63	0,69	0,61	0,61	125
	Mean	*uniform*	*0,69*	*0,65*	*0,78*	*0,71*	*0,76*	*8501*
	Mean	*weighted*	*0,85*	*0,83*	*0,88*	*0,86*	*0,87*	*8501*

5 Discussion

The character Transformer model handles the derivative analysis task more efficiently than the generation one, which is expected because the analysis task has a smaller target domain. The addition of multi-word lemmas to the training and evaluation data had positive effect. We believe that data that increases the variation of the training samples, even if they are more complex samples – increasing the sequence fragments that do not change – improves generalisation of the model. Considering the averages, the best classifier uses the combined information of the difference vectors for: words, prefixes and endings. It achieved the highest weighted average values in all types of environment – including all relationships, filtering based on base PoS and filtering based on whole relationship PoS. The fact that the classifier is not based on vectors of specific words – like the 3D classifier – is also a good signal for generalisation. Comparing the results of the 1DAF classifier with those from [14] for selected relations we observe improvement 6 to 51% points.

Explainability is important for a classification task, as it provides additional information on the quality and usability. The features that are most important to the classifier do not always make sense. In order to verify our model predictions, we try to estimate an importance of a word's n-grams. We do that by recalculating prediction probability of a previously trained model for every missing n-gram. If the probability of a sample not being in a given relation grows significantly, then the feature is relevant. The importance score of an n-gram is a difference between prediction probabilities (of a negative class) of the sample and its neighbour. As fastText generates a word embedding as the average of all the n-grams and the word representation, we remove each subsequent components of the sum. We do that only for derivatives. The input vector for a classifier is then calculated as described before – by subtracting an unchanged base vector from a newly created derivative embedding. We observed that for most relations, the key element in the classification is the ending, which agrees with linguistic research. It is also common for word endings to be considered as elements 2–3 characters long. For significant n-grams, the vast majority are 5–6 character n-grams.

6 Conclusion and Future Work

In this paper, we present a combined method for the generation of derivational word forms taking into account a potential morphosemantic relation, using a sequence-to-sequence technique based on Transformer and merged with semantic relation classifiers. We identified important features for decision-making by classifiers and compared them with linguistic research. Our system can be used to detect potential relations by generating potential derivational bases (derivatives) for different relations using Transformer and next verifying the results with the semantic classifier. The method can also be used to generate derivatives from the base lemmas and next to increase the coverage of a lexico-semantic network.

The obtained results are better or at least comparable to those of the previous methods, but with more fine grained semantic classification. Our method works on lexical units, contrary to the previous methods focused on words alone. Definitely, there is still plenty of space for improvement. One of the future directions is tighter integration of relation classification with generation of new words.

Acknowledgements. Financed by the European Regional Development Fund as a part of the 2014–2020 Smart Growth Operational Programme, CLARIN - Common Language Resources and Technology Infrastructure, project no. POIR.04.02.00-00C002/19.

References

1. Bojanowski, P., Grave, E., Joulin, A., Mikolov, T.: Enriching word vectors with subword information. arXiv preprint arXiv:1607.04606 (2016)
2. Dimitrova, T., Tarpomanova, E., Rizov, B.: Coping with derivation in the Bulgarian WordNet. In: Proceedings of the Seventh Global Wordnet Conference (2014)
3. Dziob, A., Piasecki, M., Rudnicka, E.: plWordNet 4.1 - a linguistically motivated, corpus-based bilingual resource. In: Proceedings of the 10th Global Wordnet Conference (2019)
4. Fellbaum, C. (ed.): WordNet - An Electronic Lexical Database. The MIT Press, Cambridge (1998)
5. Fellbaum, C., Osherson, A., Clark, P.E.: Putting semantics into WordNet's "morphosemantic" links. In: Vetulani, Z., Uszkoreit, H. (eds.) LTC 2007. LNCS (LNAI), vol. 5603, pp. 350–358. Springer, Heidelberg (2009). https://doi.org/10.1007/978-3-642-04235-5_30
6. Kanuparthi, N., Inumella, A., Sharma, D.M.: Hindi derivational morphological analyzer. In: Proceedings of the Twelfth Meeting of the Special Interest Group on Computational Morphology and Phonology (2012)
7. Kocoń, J., Gawor, M.: Evaluating KGR10 Polish word embeddings in the recognition of temporal expressions using BiLSTM-CRF. Schedae Informaticae 27 (2018)
8. Koeva, S., Leseva, S., Stoyanova, I., Dimitrova, T., Todorova, M.: Automatic prediction of morphosemantic relations. In: Proceedings of the 8th Global WordNet Conference (GWC) (2016)
9. Lango, M., Ševčíková, M., Žabokrtský, Z.: Semi-automatic construction of word-formation networks (for Polish and Spanish). In: Proceedings of the Eleventh International Conference on Language Resources and Evaluation (LREC 2018) (2018)
10. Mititelu, V.B.: Adding morpho-semantic relations to the Romanian WordNet. In: Proceedings of the Eighth International Conference LREC 2012 (2012)
11. Musil, T., Vidra, J., Mareček, D.: Derivational morphological relations in word embeddings. In: Proceedings of the 2019 ACL Workshop BlackboxNLP: Analyzing and Interpreting Neural Networks for NLP (2019)
12. Paszke, A., et al.: Pytorch: an imperative style, high-performance deep learning library. In: Advances in Neural Information Processing Systems, vol. 32 (2019)
13. Pedregosa, F., et al.: Scikit-learn: Machine learning in Python (2011)
14. Piasecki, M., Ramocki, R., Maziarz, M.: Recognition of polish derivational relations based on supervised learning scheme. In: Proceedings of the Eight International Conference on Language Resources and Evaluation (LREC 2012) (2012)

15. Vaswani, A., et al.: Attention is all you need. In: Advances in Neural Information Processing Systems (2017)
16. Vidra, J., Žabokrtský, Z., Ševčíková, M., Kyjánek, L.: DeriNet 2.0: towards an all-in-one word-formation resource. In: Proceedings of the Second International Workshop on Resources and Tools for Derivational Morphology (2019)
17. Vylomova, E., Cotterell, R., Baldwin, T., Cohn, T.: Context-aware prediction of derivational word-forms. In: Proceedings of the 15th Conference of the European Chapter of the ACL (2017)
18. Ševčíková, M., Kyjánek, L.: Introducing semantic labels into the DeriNet network. J. Linguist./Jazykovedný casopis **70**(2), 412–423 (2019)
19. Šnajder, J.: DerivBase.hr: a high-coverage derivational morphology resource for Croatian. In: Proceedings of the Ninth International Conference LREC 2014 (2014)
20. Šojat, K., Srebačić, M.: Morphosemantic relations between verbs in Croatian WordNet. In: Proceedings of the Seventh Global Wordnet Conference (2014)

On the Explanation of AI-Based Student Success Prediction

Farzana Afrin(✉), Margaret Hamilton, and Charles Thevathyan

School of Computing Technologies, RMIT University, Melbourne, VIC 3000, Australia
s3862196@student.rmit.edu.au, {margaret.hamilton,charles.thevathyan}@rmit.edu.au

Abstract. Student success prediction is one of the many applications of artificial intelligence (AI) which helps educators identify the students requiring tailored support. The intelligent algorithms used for this task consider various factors to make accurate decisions. However, the decisions produced by these models often become ineffective due to lack of explainability and trust. To fill this gap, this paper employs several machine learning models on a real-world dataset to predict students' learning outcomes from their social media usage. By leveraging the SHapley Additive exPlanations (SHAP) to investigate the model outcomes, we conduct a critical analysis of the model outcomes. We found several sensitive features were considered important by these models which can lead to questions of trust and fairness regarding the use of such features. Our findings were further evaluated by a real-world user study.

Keywords: Student success prediction · Machine learning explainability · User study

1 Introduction and Background

The recent developments of AI in education have shown potential to make informed decisions. Student success is one of the important metrics applied to the performance of education service providers. Moreover, by predicting student success, the educators can come up with actionable plans ahead of time. Eventually, this could contribute to improving the overall student experience. Machine learning algorithms have been successful in predicting student success. In general, machine learning algorithms incorporate data from various sources to model student success at different stages of their academic journey [1,5,7,9,15].

With the proliferation of AI-based technologies, concerns have been raised about the incorporation of sensitive predictors in the automated decision-making process. This may influence making unfair decisions regarding student success [2, 10]. The lack of explainability of the deployed models is one of the main reasons for this challenge, which eventually results in trust issues among the system users. Therefore, the AI systems need to be transparent and the users must

D. Groen et al. (Eds.): ICCS 2022, LNCS 13351, pp. 252–258, 2022.
https://doi.org/10.1007/978-3-031-08754-7_34

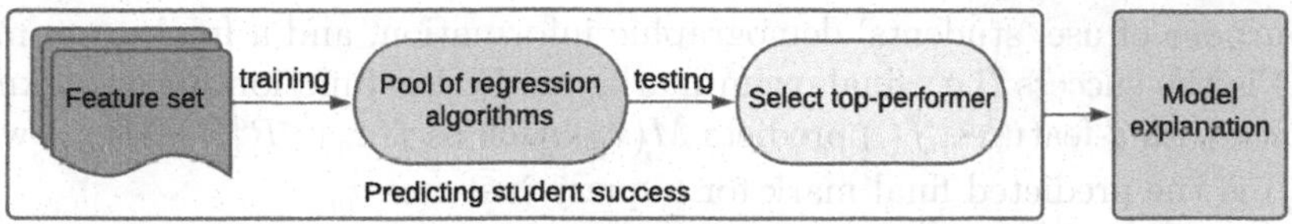

Fig. 1. An overarching view of the methodology

understand the extent to which they can trust the outcomes of the underlying machine learning technologies [12].

The need for explainable AI is gaining momentum in many industries [4,11]. There are many directions of research targeting explainable AI. Some research is directed towards finding effective ways to explain the model outcomes such as local, global and counterfactual [3,13,16]. However, there is limited research on their applicability in educational settings. In this paper, we investigate this issue by formulating a student success prediction problem by utilising data related to students' social media usage and their demographics. We also investigate how such a model evaluates the features to be considered for making predictions. In particular, the contributions of this paper are as follows:

- Prediction of student success from their social media usage. In addition we provide a SHAP-based explanation of the prediction outcomes.
- Critical analysis of explainability for AI-based student success prediction. We also evaluate our findings through a real-world user study.

We discuss our methodological approach in Sect. 2 which is followed by the experiments and discussion of results in Sects. 3 and 4 respectively. Finally, the paper concludes in Sect. 5 leaving some directions for future studies.

2 Methodology

Our approach contains two main segments: i) Student success predictions and ii) Explaining the predictions. An overview of the methodology is illustrated in Fig. 1.

2.1 Student Success Prediction

For student success prediction, we train a collection of state-of-the-art regression algorithms to predict the student success (i.e., final exam marks). For training, we utilise a set of features representing information about students' social media usage times, demographic and background. We formalize our student success prediction problem as follows:

Let, $M_{fe} = \{M_{s1}, M_{s2}, ..., M_{sn}\}$ be the set of final marks of n students. In the final dataset, each instance x is described by a d-dimensional vector of attributes R^d related to the usage time of different social media by students, segregated by the purpose of use, students' demographic information, and a final exam mark. If $f(.)$ is the success (i.e., final exam mark) prediction function for an unknown instance with d-features, $f(.)$ predicts $\hat{M}(x_q)$ such as $f(x_q) : R^d \rightarrow \hat{M}(x_q)$ where $\hat{M}(x_q)$ is the predicted final mark for a query instance x_q.

2.2 Explaining the Predictions

In this step, we consider the top-performing regression algorithm based on produced error, and then investigate the model outputs in terms of global feature-importance along with the relationship between each attribute and the target variable (i.e., final exam mark). A detailed description is given the in following subsections.

3 Experimental Setup

3.1 The Dataset

We utilise a dataset published by [14] which contains information about social media usage and final marks obtained of 505 students (221 males and 284 females) in a course. The dataset was collected from a large metropolitan Australian university, across three teaching session in 2017–2018. The subject, considered to build this dataset, is compulsory for students enrolled across business, commerce, law, engineering, science and information technology disciplines.

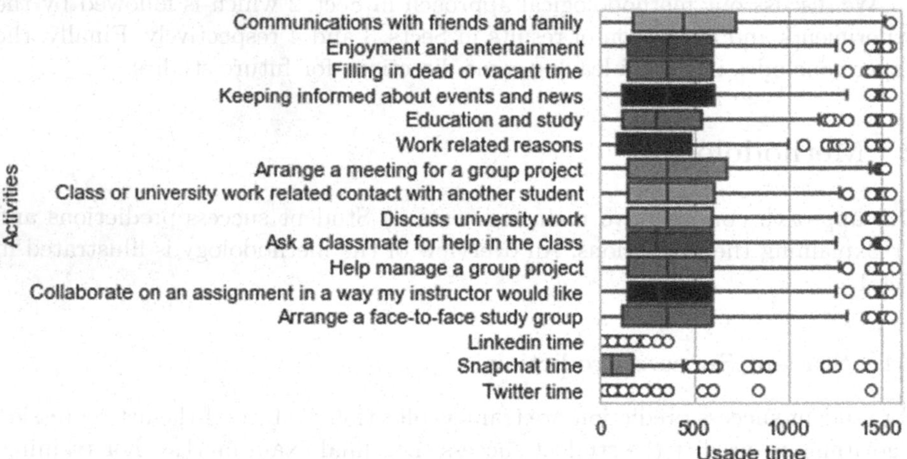

Fig. 2. Time distributions of LinkedIn, Snapchat, Twitter and different types of Facebook usage. Note that the breakdown of Facebook usage time was originally derived by multiplying the total usage time per day (in minutes) with the extent and likelihood students indicate for different reasons for Facebook usage.

In particular, the dataset contains the usage times of Facebook, LinkedIn, Snapchat, and Twitter are logged in this dataset. The Facebook usage times are further decomposed into several purpose of usage. In addition demographic and background information including 'Age', 'Gender', 'WAM'(weighted average mark which is similar to grade point average) of the participant students are also provided. A box-plot illustrating the time distribution of LinkedIn, Snapchat, Twitter and different types of Facebook usage is given in Fig. 2. We can see some flier points go past the upper whiskers. We also find a few instances of LinkedIn and Twitter usage times which are scattered in nature.

3.2 Prediction Results

We setup the experiment as per the formulation discussed in Sect. 2.1. We employ a pool of regression algorithms implemented in scikit-learn [8] including Linear, Random Forest, GBM, LGBM, and XGBM. We randomly split 80% of the data for traing these models and the rest are used for testing. The prediction outcomes are evaluated against two metrics: mean squared error (MSE) and root mean squared error (RMSE). As shown in Table 1, the performance of all the regressors is similar, however, the Random Forest produces slightly less errors than others in the pool.

Table 1. Prediction error by different models

Regression model	MSE	RMSE
GBM	0.025	0.012
LGBM	0.026	0.013
XGBM	0.026	0.013
Linear	0.025	0.013
Random Forest	**0.024**	**0.011**

3.3 Prediction Explanation

To investigate the model outcomes, we employ SHapley Additive exPlanations (SHAP)[1] which is a game theoretic approach to explain the output of any machine learning model. Figure 3 shows the global feature-importance (i.e., overall contribution of each feature in the model outcome) in terms of mean SHAP value. We can see that WAM, which is equivalent to grade point average, has highest predictive power which is followed by Age, Gender and Snapchat time.

We further investigate the relationships (i.e., positive and negative) between all the predictors and the target variable. We plot these relationships in Fig. 4, where red and blue dots indicate a higher and lower features values respectively. We found that a higher value of WAM, age, gender (1-male, 0-female) results

[1] https://shap.readthedocs.io/en/latest/index.html.

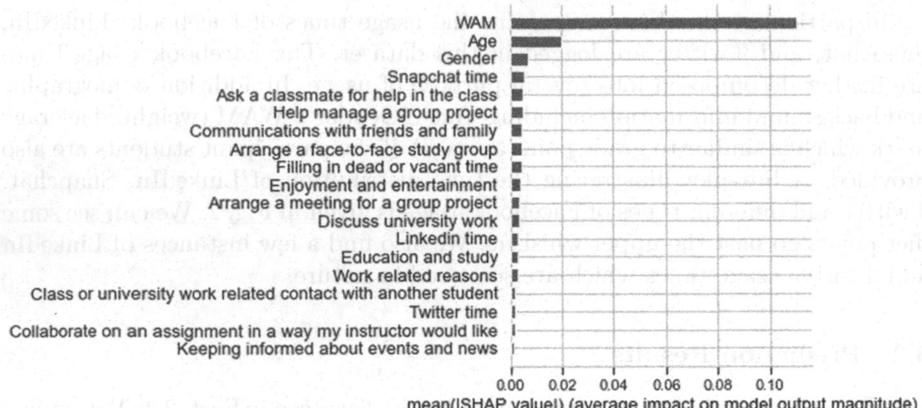

Fig. 3. Global feature-importance

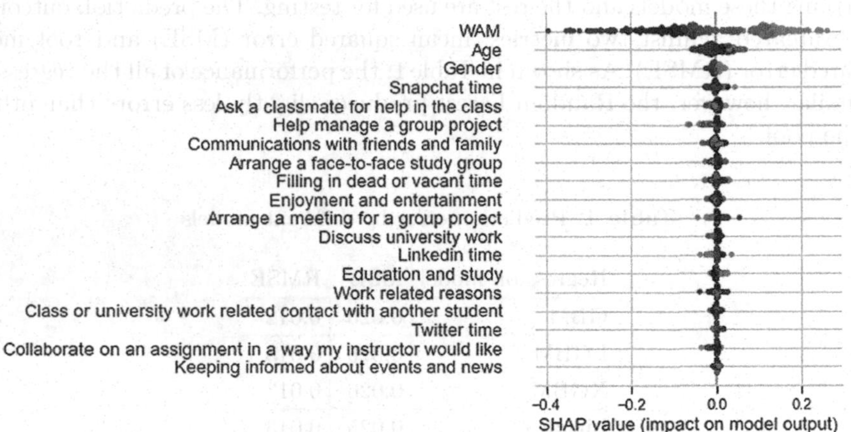

Fig. 4. Relationship between predictors and the target variable. Note: red and blue colors indicated high and low feature values respectively (Color figure online)

in a higher SHAP value, which influences the prediction outcome positively. Next, we investigate how students perceive this proposition as previous research highlighted the sensitiveness of demographics and background related features.

4 Discussion and Analysis

In practice, the consideration of WAM, age, and gender in the modelling of student success may cause trust issues as these values are meant to be private and influence making an unfair decision. We further examine the perceived fairness through a real-world user study as discussed below.

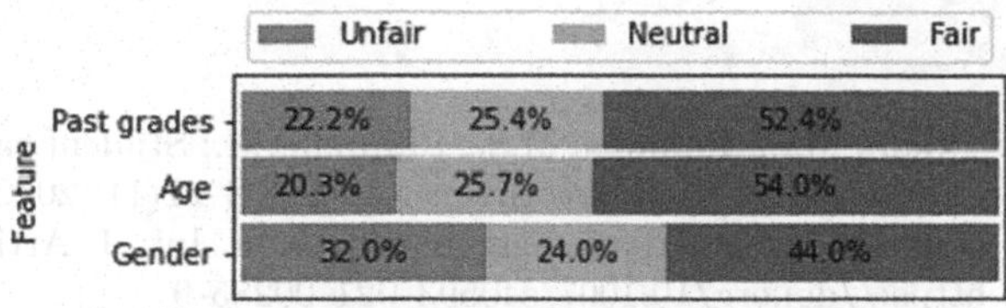

Fig. 5. Proportion of user perceptions regarding gender, age and past grades.

4.1 Investigating User Perceptions

We investigate user perceptions about the incorporation of various features in machine learning algorithms to predict student success by adapting a framework presented in [6]. The framework implements a survey where the participant students were given a set of features (predictors) related to demographics, internal evaluation, psychometric and course content, as highlighted in the recent literature that are commonly used to predict student success. Then we ask students whether they perceive a specific feature as fair or not when used to generate a model outcome. We collected and analyzed 1658 completed responses contributed by university students from Australia, Bangladesh and Saudi Arabia. We observed that the fairness perception of a specific features differs across cohorts of participants. Figure 5, illustrates the proportion of perceived responses for three selected features out of a large pool.

5 Conclusions and Future Work

This paper investigates the explanation of model outcomes in predicting student success in terms of final exam score. We found that the Random Forest regressor was the best performing model in our pool, which considers WAM (previous grade point average), age, gender, snapchat usage time and time spent on asking friends for help as the top-5 features that influence the model outcome. The consideration of sensitive features (age, gender and past results) are an obvious reason that questions the fairness and trustworthiness of the deployed model as it is also verified by our user study consisting of 1658 tertiary students from three different countries.

Future research may consider ways to enhance model outcomes without considering these sensitive features, enabling trust and fairness in automated decision making. Future research also may include the use of counterfactuals which are capable of describing a causal situation (e.g. what values of a subset of features make a specific student perform better). Another direction could be to investigate user perception under various scenarios.

Acknowledgements. Farzana was supported through an Australian Government's RTP Scholarship.

References

1. Afrin, F., Rahaman, M.S., Rahman, M.S., Rahman, M.: Student satisfaction mining in a typical core course of computer science. AJSE **11**(1) (2012)
2. Baker, R.S., Hawn, A.: Algorithmic bias in education. Int. J. Artif. Intell. Educ. 1–41 (2021). https://doi.org/10.1007/s40593-021-00285-9
3. Briz-Ponce, L., Juanes-Méndez, J.A., García-Peñalvo, F.J., Pereira, A.: Effects of mobile learning in medical education: a counterfactual evaluation. J. Med. Syst. **40**(6), 1–6 (2016)
4. Gade, K., Geyik, S.C., Kenthapadi, K., Mithal, V., Taly, A.: Explainable AI in industry. In: Proceedings of the 25th ACM SIGKDD, pp. 3203–3204 (2019)
5. Giunchiglia, F., Zeni, M., Gobbi, E., Bignotti, E., Bison, I.: Mobile social media usage and academic performance. Comput. Hum. Behav. **82**, 177–185 (2018)
6. Grgic-Hlaca, N., Redmiles, E.M., Gummadi, K.P., Weller, A.: Human perceptions of fairness in algorithmic decision making: a case study of criminal risk prediction. In: Proceedings of the 2018 World Wide Web Conference, pp. 903–912 (2018)
7. Liu, Z.: A practical guide to robust multimodal machine learning and its application in education. In: Proceedings of the Fifteenth WSDM, p. 1646. New York (2022)
8. Pedregosa, F., et al.: Scikit-learn: machine learning in Python. J. Mach. Learn. Res. **12**, 2825–2830 (2011)
9. Rodrigo, M.M.T., et al.: Affective and behavioral predictors of novice programmer achievement. In: Proceedings of the 14th ITiCSE, pp. 156–160 (2009)
10. Sha, L., et al.: Assessing algorithmic fairness in automatic classifiers of educational forum posts. In: AIED, pp. 381–394 (2021)
11. Slijepcevic, D., et al.: Explaining machine learning models for clinical gait analysis. ACM TCH **3**(2), 1–27 (2021)
12. Toreini, E., Aitken, M., Coopamootoo, K., Elliott, K., Zelaya, C.G., van Moorsel, A.: The relationship between trust in AI and trustworthy machine learning technologies. In: Proceedings of the 2020 Conference on Fairness, Accountability, and Transparency, pp. 272–283. FAT* 2020, New York (2020)
13. Verma, M., Ganguly, D.: LIRME: locally interpretable ranking model explanation. In: SIGIR, pp. 1281–1284. New York (2019)
14. Wakefield, J., Frawley, J.K.: How does students' general academic achievement moderate the implications of social networking on specific levels of learning performance? Comput. Educ. **144**, 103694 (2020)
15. Yu, R., Li, Q., Fischer, C., Doroudi, S., Xu, D.: Towards accurate and fair prediction of college success: evaluating different sources of student data. International Educational Data Mining Society (2020)
16. Zytek, A., Liu, D., Vaithianathan, R., Veeramachaneni, K.: Sibyl: explaining machine learning models for high-stakes decision making. In: Extended Abstracts of the 2021 CHI Conference on Human Factors in Computing Systems. New York (2021)

The Importance of Scaling for an Agent Based Model: An Illustrative Case Study with COVID-19 in Zimbabwe

Sarah Wise[1(✉)], Sveta Milusheva[2], and Sophie Ayling[1]

[1] Centre for Advanced Spatial Analysis (CASA), University College London, London, UK
{s.wise,sophie.ayling.10}@ucl.ac.uk

[2] World Bank Group, Washington, D.C., USA
smilusheva@worldbank.org

Abstract. Agent-based models frequently make use of scaling techniques to render the simulated samples of population more tractable. The degree to which this scaling has implications for model forecasts, however, has yet to be explored; in particular, no research on the spatial implications of this has been done. This work presents a simulation of the spread of Covid-19 among districts in Zimbabwe and assesses the extent to which results vary relative to the samples upon which they are based. It is determined that in particular, different geographical dynamics of the spread of disease are associated with varying population sizes, with implications for others seeking to use scaled populations in their research.

Keywords: Agent-based modelling · Scaling · Synthetic population · Agent-based modeling · Simulation

1 Introduction

Agent Based Models (ABMs) are often designed and built to model complex behaviours among large populations. However, the combination of complex behavior and a large population can require extensive memory or CPU usage and quickly become too computationally intensive to run efficiently in terms of speed or memory use. Thus, different methodologies have arisen to deal with the challenge of large-scale simulations in ABM literature (see Bithell and Parry [9] for a few of the most common). This is a point of particular interest as researchers have sought to lend a hand to the time-sensitive problem of disease forecasting (see for example [4,6], or [5]).

Briefly, some models approach this problem by reducing the level of complexity of the model or the number of agents in order to enable it to run [10]. Others revert to equation-based modelling or hybrid approaches to reduce some of the burden in terms of computational intensity [5]. Still others have the option to simply increase computational power through either computer hardware or

D. Groen et al. (Eds.): ICCS 2022, LNCS 13351, pp. 259–265, 2022.
https://doi.org/10.1007/978-3-031-08754-7_35

parallelisation (see, for example, [8]). Some researchers restructure their model to enable each "super individual" to represent multiple agents, which risks the dynamics of a larger population not being reflected beyond a certain point - both spatially and temporally [1].

None of these approaches are without drawbacks, and it is important to understand the implications of the trade-off decisions researchers are making. As such, this work provides an illustrative example of the impact of using different sizes of population in an ABM simulating the spread of SARS-CoV-2. The model explores the spread of disease through a representative sampled population in Zimbabwe during the first wave of the pandemic, which began in March 2020.

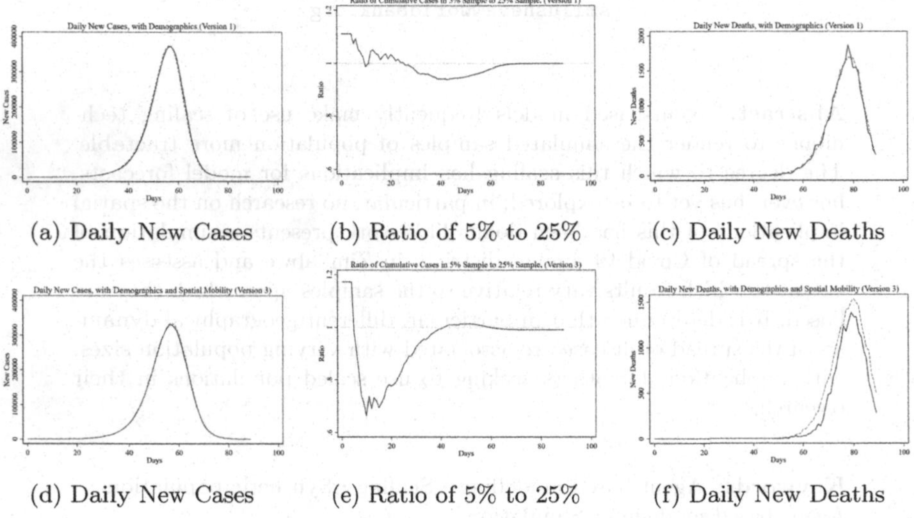

(a) Daily New Cases (b) Ratio of 5% to 25% (c) Daily New Deaths

(d) Daily New Cases (e) Ratio of 5% to 25% (f) Daily New Deaths

Fig. 1. Comparison of 5% (orange dotted) and 25% (green solid) Sample for New Symptomatic Cases and Deaths Across Two Models. Note: Panels a-c use a model where individuals all live in one administrative unit. Panels d-f use a model where individuals can move between the different districts in Zimbabwe. (Color figure online)

2 Methodology

In this paper we use the example of an agent based model simulating the spread of the COVID-19 pandemic in Zimbabwe to illustrate the role of sample size in the model outputs. We will present a very brief overview of the model and its functioning using the ODD Protocol [3].

2.1 Overview

The **purpose** of the model is to forecast the spread of an infectious disease throughout a population of spatially distributed humans.

The **entities** in the model are either individual humans or infections. The human agents, called **Persons**, are characterised by their age and sex. Persons are assigned to households - physical locations where a consistent, designated group of Persons gather in the evenings. They move around their communities, sometimes travelling to other nearby districts, interacting with other people and potentially transmitting infections to one another. **Infections** represent a case of the given disease. An Infection must be assigned with a host Person, and may progress over time based on the age of the host.

The model **environment** represents space. Persons can be located within either household or community locations, both of which are situated within districts (larger spatial units which together make up the country of Zimbabwe). A Person who is visiting a district outside of their own home district must be out in the "community" - in their own home district, they may either be out in the community or else in their own household. Persons make decisions every 4 h and interact with others based on their location.

The **processes** represented in this model include movement and infectious behaviours. Individual Persons choose whether to go out into the community every day; they may visit the community of their own district or to travel to another district and visit the community there. If they have an Infection, they will potentially prompt the Persons with which they interact to generate their own Infections. Infections develop over time, developing from being exposed all the way to either the recovery or death of the host. In advanced stages, the Infection may render the host immobile, disallowing them from moving.

2.2 Design

The design of the model allows for the emergence of local outbreaks and hotspots within target districts. Interactions between agents give rise to the spread of disease, and the movement of Persons between districts allows for the disease to spread between otherwise relatively closed communities.

2.3 Details

The **initialisation** of the model is significant because after the populations have been generated in their target households and districts, a set number of Infections are generated in hosts in the target districts. The hosts are randomly chosen for each instantiation of the simulation.

The input data, being of particular interest in this paper, has been specified in its own section, Sect. 3. The submodels are simply the movement module and the infection module. In the former, Persons choose whether to leave the house with some probability; if they choose to leave, they will select a target destination based on the movement matrix described in Section . If they move to a community, they will interact with some set number of individuals present in the same community, potentially prompting an Infection in them. At the end of 8 h in the community, they will return home and interact with those in their household.

The infection behaviour sees the Infection transitioning from the state of being exposed to, potentially, either symptomatic or asymptomatic. In the latter case, after a stochastic period of time, the Infection will resolve into recovery; in the former case, the Infection may either resolve into recovery or transition to a mild case. It may continue along this path, at each stage either recovering or worsening into a severe case, a critical case, and ultimately death. An individual Person who has recovered becomes susceptible to new Infections. If the Infection progresses beyond the stage of being mild, the Person is set to be immobile and restricted to their household.

3 Data

The model makes use of a number of different forms of data to motivate and contextualise the simulation, including:

- ***Census*** The Zimbabwe Census data from 2012 was taken from IPUMS International. The data that is available is a 5% sample of the original census of 15 million individuals. This data contains information on the age, sex, economic status, household ID, and district of origin of every agent in the model.
- ***Mobility Data*** Mobility data was calculated from approximately 8.1 billion Call Data Records (CDR) and reflect the levels of mobility as monitored from and to each of Zimbabwe's 60 districts. The detailed data were aggregated by a telecom operator into daily origin and destination matrices using code developed by the research team (see [7]). Only the aggregated, fully anonymised output was shared by the telecom company with the research team. The Origin Destination Matrix shows trips between two districts, relative to day of week. This variable is a combination of two trip types.
- ***Epidemiological Data*** A series of parameters to define the characteristics of infection were also input into the model to establish the infection dynamics. These included the susceptibility by age to transmission, hospitalization rates by age, critical rates by age, fatality rates by age when non critical and death rates by age. The characteristics of SARS-CoV-2 such as the incubation period, infectious period, and recovery times were also included and taken from the Covasim model which in turn were taken from Ferguson et al. (2020) [2] and Verity et al. (2020) [6,11].
- ***Case data*** The aggregate district level case numbers from March 2020 provided by the Ministry of Health of Zimbabwe to the World Bank representing the number of cases. These are used to inform seeding of cases in the districts in model version 3 (V3) presented here.

3.1 Synthetic Populations

In order to assess how the complexity of the model interacts with the size of the sample, we generate two synthetic populations for the model:

- V1 - a sample of the population in which individuals have representative ages and live in households of an appropriate size. Everyone is in the same geographic location, so when in the community, every person in the sample can interact with every other person.
- V3 - as in V1, individuals have ages and household sizes drawn from real data. In V3, households are further assigned to individual districts to create a spatially reasonable distribution of population across the country.

For each version, we create a 5% sample and a 25% sample. For the 5% sample we use directly the 5% census sample that we have for the population, using the characteristics provided. To get the 25% sample, we expand the original 5% sample by generating identical replicas of each household. Therefore, differences that arise between the two versions can be more easily attributed to the increased size of the sample. Obviously, it generates a somewhat contrived population - it should be understood as a mechanism for testing rather than a realistic census distribution.

4 Results

Each of the two model versions is run with the each of the synthetic populations ten times. The outputs are scaled up directly to the full population (for the 5% sample we use a factor of 20 and for the 25% sample we use a factor of 4) to facilitate comparison.

Looking across the trajectory of the disease, the numbers of new cases track fairly closely between samples (Fig. 1a and b). However, it seems that adding geographic variation and mobility across districts leads to larger differences between the scaled 5% and 25% samples when it comes to other metrics (see Fig. 1d–f). Note the death rates - while deaths are rarer events and their curves less smooth, the inclusion of mobility sees the number of new deaths drop consistently lower in the 5% sample than in the 25% sample.

In considering the difference between the scaled district cases across the two samples as a proportion of the scaled cases in the 25% sample, the differences are clear (Fig. 2). In the map, the areas in blue and green demonstrate districts where there are more cases registered in the 25% sample, as compared to orange areas where more cases were generated in the 5% sample. The 25% sample shows obvious cases where an individual has arrived in remote districts which are not reached by anyone in the 5% scenario (blue districts on the west side of the country and blue district in the north of Zimbabwe). These districts, which might be judged to be not at risk by the smaller sample, show up clearly in the 25% version of the simulation.

Additionally, if we focus on when districts have their initial case, we see that this happens much quicker in the 25% sample, with many districts getting their first case before the 5% sample would predict (Fig. 2). In particular, while we seed initial cases in the same four districts for both the 5% and the 25% samples, we see that in the 5% sample, the disease does not spread beyond these

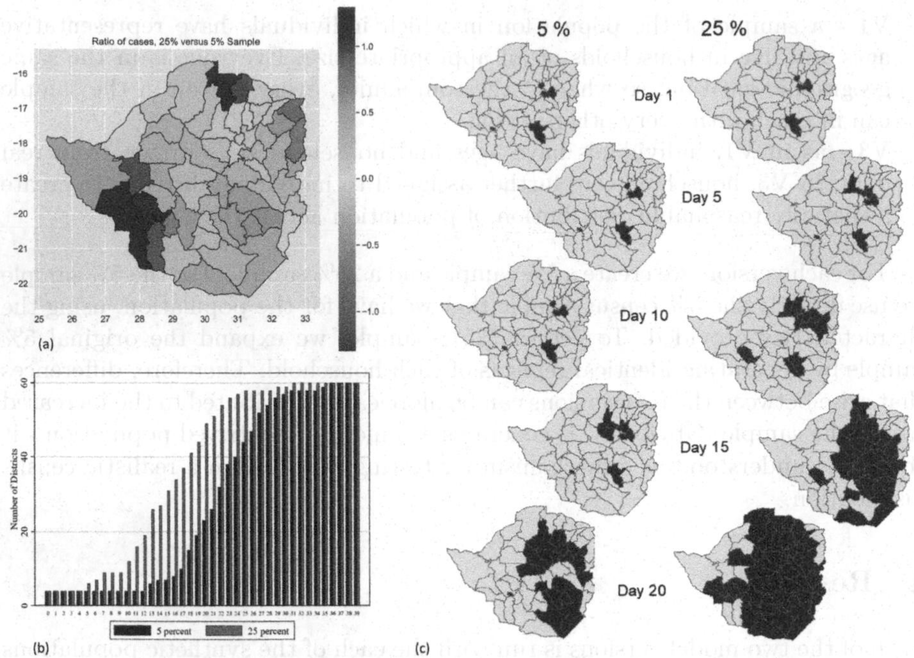

Fig. 2. Spatial differences when using different initial samples. *(a) Difference between 25% and 5% median daily symptomatic case counts normalized by district cases for the second month of the V3 case (beta 0.3). (b) Number of districts with at least one cumulative case by day of outbreak. (c) Districts with at least one cumulative case by days 1,5,10 15, and 20 of the pandemic.*

four districts until day 14. In contrast, with the 25% sample, the disease spreads to two new districts by day 6. This is visualized in Fig. 2, which also illustrates that some of the first districts to which the disease spreads are different across the two samples.

5 Discussion and Conclusion

We demonstrate that researchers must take care in selecting the scale of the population sample in their models, particularly if there is interest in understanding the initial phases of a pandemic when case counts and death numbers are low. Importantly, when there is a spatial component to the model, the bias generated when using a small sample becomes much larger. The smaller sample leads to both an underestimate in the cumulative cases early on in the disease cycle, and an underestimate of deaths across the entire time period.

Small samples also obviously lead to much higher uncertainty in these models. This is not only in the early stages, but for the entire curve. Given that there is already a high level of uncertainty in these epidemiological models due to the large number of assumptions that are made, the added uncertainty generated by the small sample size may reduce the reliability of such a model for policy planning.

From a policy perspective, there is interest in understanding when a disease might spread to a new geographic area. This geographic analysis can be one of the important advantages of an agent based model, which allows for the simulation of how different agents might move between areas. Yet, we see that if the sample used is very small, it may not accurately portray the timing of when a disease will spread to a new area. This is important from a policy perspective since identifying when a disease might first enter an area is important for mitigation strategies and planning.

Overall, these trade-offs are increasingly relevant to researchers and we hope this work can contribute to both discussion and awareness of them.

References

1. Ben-Dor, G., Ben-Elia, E., Benenson, I.: Population downscaling in multi-agent transportation simulations: a review and case study. Simul. Model. Pract. Theory **108**, 102233 (2021)
2. Ferguson, N.M., et al.: Report 9: impact of non-pharmaceutical interventions (NPIs) to reduce COVID19 mortality and healthcare demand. Imperial Coll. Lond. **10**, 491–497 (2020)
3. Grimm, V., et al.: The ODD protocol for describing agent-based and other simulation models: a second update to improve clarity, replication, and structural realism. J. Artif. Societies Social Simul. **23**(2), 1–20 (2020)
4. Hoertel, N., et al.: A stochastic agent-based model of the SARS-CoV-2 epidemic in France. Nat. Med. **26**(9), 1417–1421 (2020)
5. Hunter, E., Mac Namee, B., Kelleher, J.: A hybrid agent-based and equation based model for the spread of infectious diseases. J. Artif. Soc. Social Simul. **23**(4), 14 (2020)
6. Kerr, C.C., et al.: Covasim: an agent-based model of COVID-19 dynamics and interventions. PLOS Comput. Biol. **17**(7), e1009149 (2021)
7. Milusheva, S., et al.: Challenges and opportunities in accessing mobile phone data for COVID-19 response in developing countries. Data Policy **3**, e20 (2021)
8. Minson, R., Theodoropoulos, G.K.: Distributing RePast agent-based simulations with HLA. Concurr. Comput. Pract. Exp. **20**(10), 1225–1256 (2008)
9. Parry, H.R., Bithell, M.: Large scale agent-based modelling: a review and guidelines for model scaling. Agent-Based Models Geograph. Syst. (2012)
10. Perez, L., Dragicevic, S.: An agent-based approach for modeling dynamics of contagious disease spread. Int. J. Health Geograph. **8**(1), 1–17 (2009)
11. Verity, R., et al.: Estimates of the severity of coronavirus disease 2019: a model-based analysis. Lancet Infect. Dis. **20**(6), 669–677 (2020)

Phase-Field Modelling of Brittle Fracture Using Time-Series Forecasting

Minh Ngoc Dinh[1(✉)], Chien Trung Vo[2], Cuong Tan Nguyen[1], and Ngoc Minh La[1]

[1] School of Science, Engineering and Technology, RMIT University, Ho Chi Minh City, Vietnam
minh.dinh4@rmit.edu.vn

[2] Mechanical Engineering and Materials Science, University of Pittsburgh, Pittsburgh, USA

Abstract. The crack propagation behavior can be considered a time-series forecasting problem and can be observed based on the changes of the Phase-field variable. In this work, we study the behavior of the Isotropic Brittle Fracture Model (BFM), and propose a hybrid computational technique that involves a time-series forecasting method for finding results faster when solving variational equations with a fine-grained. We use this case study to compare and contrast two different time-series forecasting approaches: ARIMA, a statistical method, and LSTM, a neural network learning-based method. The study shows both methods come with different strengths and limitations. However, ARIMA method stands out due to its robustness and flexibility, especially when training data is limited because it can exploit a priori knowledge.

Keywords: Brittle fracture model · Phase-field · ARIMA · LSTM

1 Introduction

Predicting the propagation of cracks that occur on a target material is an important problem in continuum damage mechanics. The BFM [1] is one of the typical models, which bases on the concept of critical energy release rate in order to evaluate and quantify crack formation and propagation. Importantly, the performance of modelling crack formation and propagation is mesh size-dependent, and many have tried to address this issue [2]. Recently, collecting experimental/computational data to predict the crack propagation process using machine learning techniques becomes a mainstream approach.

There are two main approaches in the field of machine learning: statistical methods and artificial neural network (ANN) architectures. Statistical methods such as the Holt-Winters [3] or Autoregressive Intergrated Moving Average (ARIMA) have shown to be robust and flexible, especially when fewer training data is available because they exploit a priori knowledge. However, these

Supported by RMIT University, Vietnam, Internal Research Grant 2, 2020.

D. Groen et al. (Eds.): ICCS 2022, LNCS 13351, pp. 266–274, 2022.
https://doi.org/10.1007/978-3-031-08754-7_36

techniques often assume the linearity in the data. Besides, ANNs such as Long Short Term Memory (LSTM) [4] can approximate almost any function, given enough training data. In this work, we study the behaviors of the Isotropic BFM. We propose a hybrid framework where we utilize data collected from computational brittle fracture simulations to enhance the time-series forecasting capacity of machine learning techniques, especially when working with fine-grained mesh. We use this case study to compare two different time-series forecasting approaches: ARIMA and LSTM. Our main contributions are (1) a hybrid framework involves a time-series forecasting method for predicting crack propagation behaviors, and (2) comparing the performance of traditional forecasting techniques to deep learning-based algorithms.

Section 2 presents the Phase-field variable model. In Sect. 3, we present the framework using ARIMA and LSTM as time-series forecasting techniques. We describe our case study in Sect. 4. We show how time-series forecasting can predict crack propagation behaviors, and evaluate both LSTM and ARIMA models in terms of accuracy and compute performance in Sect. 5. Section 6 presents the conclusion and future work.

2 Isotropic Brittle Fracture Model

2.1 Phase-Field Approximation

The Phase-field variable $\phi(x)$ is an approximated exponential function depending on the spatial variable x, which is used to describe the damage state of materials at each position in a continuum domain. The value of $\phi(x)$ varies in $[0,1]$ as x varies from $(-\infty,0]$ and $[0,+\infty)$ such that for $\phi(0)=1$ denotes the cracked material and $\phi(\pm\infty)=0$ represents the intact one. The diffusion of the phase-field variable depends on the length scale parameter l_c, and $\phi(x)$ is the solution of the homogeneous differential equation, also called the Euler equation of the variational principle: $\phi(x)-l_c^2\Delta\phi(x)=0\ in\ \Omega$.

Following the variational approach, the total potential Ψ forms the crack is:

$$\Psi(\phi,\mathbf{u})=\int_{\Omega}\left[\left(1-\phi^2\right)+\kappa\right]\psi_0(\varepsilon)\,d\Omega+\int_{\Omega}\frac{\mathcal{G}_c}{2}\left[l_c\nabla\phi\cdot\nabla\phi+\frac{1}{l_c}\phi^2\right]d\Omega \quad (1)$$

On the other hand, the total potential energy is equivalent to the internal energy of the system. The variation of the internal energy increment $\delta \mathrm{W_{int}}$ and the external work increment $\delta \mathrm{W_{ext}}$ are similarly expanded as follows:

$$\delta\mathrm{W_{int}}=\delta\Psi=\frac{\partial\Psi}{\partial\varepsilon_{\mathrm{ij}}}\delta\varepsilon_{\mathrm{ij}}+\frac{\partial\Psi}{\partial\phi}\delta\phi;\qquad \delta\mathrm{W_{ext}}=\int_{\Omega}b_j\delta u_j\,d\Omega+\int_{\partial\Omega}h_j\delta u_j\,d\partial\Omega \quad (2)$$

The residual $\delta\mathrm{W_{int}}-\delta\mathrm{W_{ext}}$ is a system of equations containing displacement variables and phase-field computed by the principle of virtual work.

2.2 Treating the Crack Propagation Process as a Time-Series

The crack diffusion of brittle materials, as described by the Phase-field variable, could be fixed in an unchanged-dimensional matrix. Importantly, the brittle fracture model is quasi-static, in which the load steps could be considered as time steps in the process of simulation. Msekh et al. [5] describe the crack propagation as temporal-spatial dependent process. As every mesh node on an arbitrary position has a temporal attribute, we investigate how crack propagation can be predicted using time-series modelling techniques.

3 Time-Series Forecasting

Time-series forecasting analyses observations to (1) uncover potential structure such as autocorrelation, trend or seasonal variation, and (2) produce monitoring and forecasting capacity. There are two main approaches in time-series forecasting: statistics-based method such as ARIMA and the deep learning-based architectures such as LSTM [4]. This section demonstrates how both techniques can be applied to forecast the crack propagation process.

3.1 Statistics Based Time-Series Forecasting with ARIMA

Assuming time is a discrete variable and X_t denotes the observation at time t, and ϵ_t denotes the zero-mean random noise term at time t. The autoregressive (AR) process, and moving average (MA) can be combined as:

$$X_t = \sum_{i=1}^{k} \alpha_i X_{t-i} + \sum_{i=1}^{q} \beta_i \epsilon_{t-i} + \epsilon_t \tag{3}$$

The combination of the two schemes [6] takes both previous observations and the progressive noise term into account: the "autoregressive" term k: presents the lags of the series in the forecast, and the "moving average" term q: presents the lags of the forecast errors. The ARIMA framework introduces an extra parameter d, and performs time series forecasting for nonstationary series. An ARIMA (k, d, q) forecasting model acts as a "signal filter" and a trend "filter" that uses d^{th} order differences to induce past observations into future forecasts.

$$\nabla^d X_t = \epsilon_t + \sum_{i=1}^{k} \alpha_i \nabla^d X_{t-i} + \sum_{i=1}^{q} \beta_i \epsilon_{t-i}; \qquad \tilde{X}_t = \nabla^d \tilde{X}_t + \sum_{i=1}^{d-1} \nabla^i X_{t-i} \tag{4}$$

We studied an application of ARIMA forecasting technique in our previous work, where soft-errors from a running scientific simulation can be identified through time-series analysis [7]. Because computational simulations such as the Brittle Fracture Model do not incur noise, term q can be set to zero ($q = 0$). We focus on terms d and k in below sections.

Determine Differencing Term '*d*'. We apply the Augmented Dickey-Fuller Test (ADF) [7] on our time-series. ADF's p-value was 0.689 (higher than the

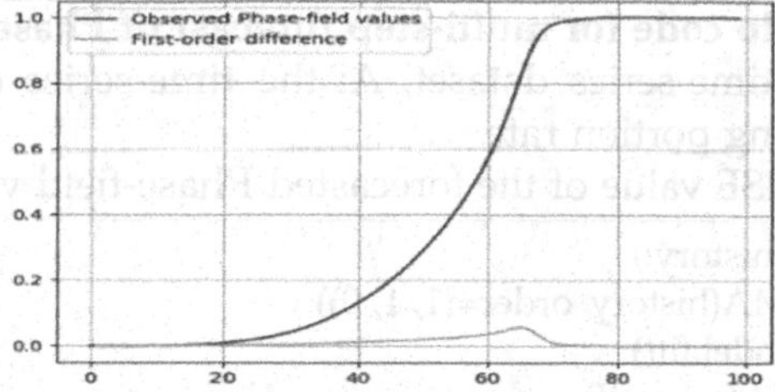

Fig. 1. Observed values per 100 timesteps (y/x-axis) with 1st order differencing.

significant level 0.05), indicating the non-stationarity. Figure 1 shows that a 1st order of differencing (d = 1) is appropriate for our time series.

The Autoregressive Term '*k*'. Because our forecasting approach handles data observations arriving sequentially and updates the models simultaneously, which is more natural for many time-step based applications, $k = 1$ (i.e. AR(1)) is applicable for ARIMA model. As a result, we has the ARIMA(1, 1, 0) model which is coined the "differenced first-order autoregressive model".

Applying ARIMA to Phase-Field Value Prediction. Because ARIMA is a linear regression-based approach, we develop an algorithm to perform multi-step out-of-sample forecast with re-estimation (i.e., the model is re-fitted each time the model makes a prediction) [8]. Accordingly, the Algorithm 1 below takes a set of Phase-field values obtained from a BFM simulation with a parser grid, builds a forecast model, performs the forecast, and reports the RMSE of the predictions. We describe how train and test sets are configured (i.e., the training portion rate R) along with the performance metrics in Sect. 4 below.

3.2 Deep-Learning Based Time-Series Modelling with LSTM

Recurrent Neural Network (RNN) uses sequential observations and learns from the earlier stages to forecast future trends. Especially, Long Short-Term Memory neural networks (LSTMs) captures and processes past and present data items in a form of a transport line. Each cell is equipped with "gates" to allow data items to be either disposed, filtered, or added to the next cells. Each sigmoid layer produces outputs between [0, 1]. LSTM network consists of:

- **Forget Gate**: outputs 0 when the data item should be completely ignored.
- **Memory Gate**: selects a new data item to be stored in the cell.
- **Output Gate**: decides what a cell will output.

Algorithm 1 : Pseudo code for multi-step forecast of Phase-field value.
Input: A_{init}: initial time-series dataset, A: the time-series dataset to be predicted, and R: training portion rate
Output: average RMSE value of the forecasted Phase-field values

```
procedure ARIMA(history:)
    model ← ARIMA(history, order=(1, 1, 0))
    model fit ← model.fit()
    return model_fit.forecast()
#Split data into train set and test set
size ← length(A); train ← A[0..size] * R; test ← A_init[length(train)..size]
observed ← train; predicts ← (); t ← length(train) + 1
while t < size then
    #build ARIMA time series model and deliver forecast
    predict.append(arima(observed))
    observed.append(test[t]); t ← t+1
end while
return sqrt(mean square error(predicts,observed[length(train)..size)])
```

With LSTM, the process of training a subsequence in a continuous, non-ending time-series might break down. Therefore, the internal state values of the LSTM network must be reset by a forget gate at an appropriate time. We enhanced the LSTM architecture [9] with a forget gate embedded to reset the internal state value for each subsequent prediction (Fig. 2). Specifically, the internal connections of the LSTM consist of 4 cells in each layer. The forget gate f_t acts as the filter to remove unnecessary information. When $f_t = 0$, the information is discarded, and when $f_t = 1$ the necessary information is retained. Also, we assume that the data generated by the BFM process bias-free. Thus, the performance of LSTM cells does not depend on the bias term.

Applying LSTM for Phase-Field Variable Prediction. The overall process to apply LSTM for Phase-field value prediction is different from the ARIMA algorithm, where we use a set of Phase-field values obtained from a BFM simulation with a parser grid for training the LSTM model. Similar to the ARIMA algorithm, we output the Root Mean Square Error (RMSE) of the predictions.

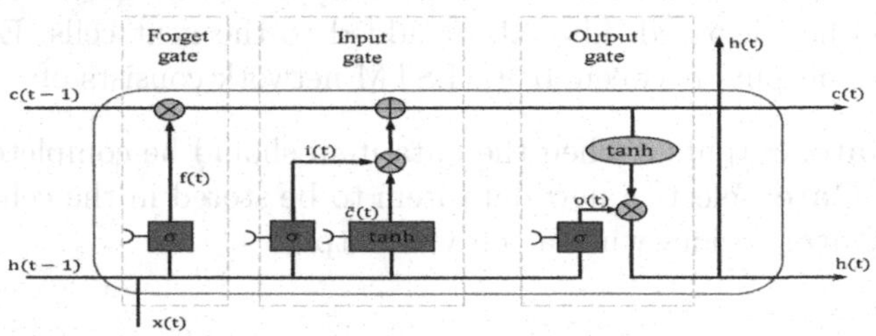

Fig. 2. LSTM cell with a forget gate.

4 Case Study - Single Edge Notched Tensile

We consider a two-dimensional plane strain plate with size 1×1mm with a crack described as a line at the left of the plate and positioned as (Fig. 3 (a)). The material parameters are specified as: the Young's modulus $E = 210\,\mathrm{kN/mm^2}$, the Poisson's ratio $\nu = 0.3$, the critical energy release rate is $\mathcal{G}_c = 5\,\mathrm{N/mm}$ and the length scale parameter $l_c = 2\,\mathrm{mm}$. The plate is fixed at the bottom edge and the load-displacement u at the top edge consists of 100 iterations with the increment of displacement $\Delta u = 5e{-}7$ mm. In this example, the specimen with four mesh sizes and gradually increases the number of elements as 40×40, 50×50, 60×60, and 70×70. In this fracture model, the Phase-field variable is significant to describe the formation of cracks. With mesh size 70×70, the values of the phase-field variable in the vicinity of the crack will be affected by l_c (Fig. 3 (b)), which determines the width and the diffusion of cracks. In this case study we keep $l_c = 8h$, with h being the size of each rectangular element. After the phase-field variable of the sample changes, the displacement-reaction force relations of mesh sizes are similar in the trend of behaviors.

The data of Phase-field variables at various mesh sizes are collected. We vary the size of the Train set from 30%, 40%, and 50% of the data. We simulate 100 timesteps in our fracture simulation as 30% means 30 historical time-steps. We observe how increasing the amount of training data affects the performance of different time-series forecasting methods. We use the remaining data values (e.g., the Test set) to measure the accuracy of using RMSE.

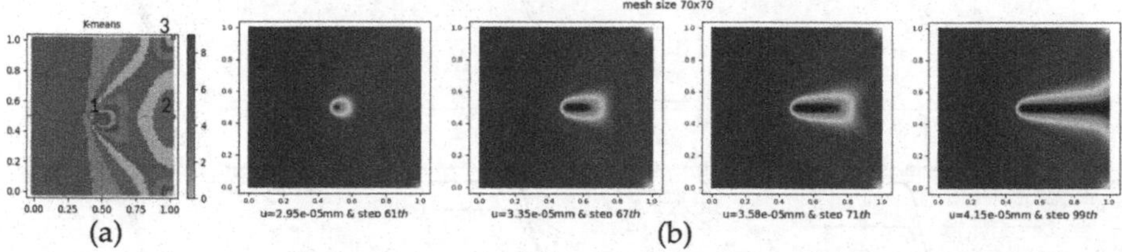

Fig. 3. (a) Observed points in the plate, (b) The propagation of the crack.

5 Results

In this experiment, some critical points are observed by using K-means clustering to validate the utility and advantages of LSTM and ARIMA in the simulation. However, the point in center of the plate is picked because it is the initial position of crack propagation where the phase-field value varies from 0 to 1 over the period of simulation. Once a time-series forecasting model is trained (using either 30% to 50% of the recorded Phase-field values), we keep predicting Phase-field values across 100 time-steps using either the LSTM model or the ARIMA model. All prediction values are stored and later used to compute the average RMSE value against the recorded Phase-field values (observations).

5.1 Prediction Accuracy Rate

The time-series data of the phase-field variable at three point (1), (2), (3) at Fig. 3(a) are shown with respect to Fig. 4(a), (b), (c) and Fig. 4(d), (e), (f). Overall, the results indicate that the statistical method (captured by red-dashed lines) outperforms the deep-neural network approach (captured by blue-dashed line) in all configurations (mesh size combining with different amounts of training data). ARIMA's core principle is to describe the autocorrelation of a time-series, thus it performs better given sudden changes in the changing rate of a time series (for example, around time-step 65). Figure 4 depicts the ARIMA one-step-ahead efficient forecast for this sort of behavior in our Phase-field variable. Finally, the figures as well as the RMSE results, show that ARIMA forecasting performs consistently, especially when training data is limited.

Deep learning methods such as LSTM requires more training data. However, our LSTM model can over-fit when the extrapolating time series presenting a trend. Consider the mesh size 40×40 (Fig. 4(a), (b), (c)). Our LSTM model performs quite well as the LSTM curves track closely comparing to the observation curve. However, as we increase the mesh size, our LSTM model fails to predict the plateau in Phase-field values from time-steps 65 and beyond. As Fig. 4(d), (e), (f) show this over-fitting behavior as predictions shoot above or below the expected Phase-field values.

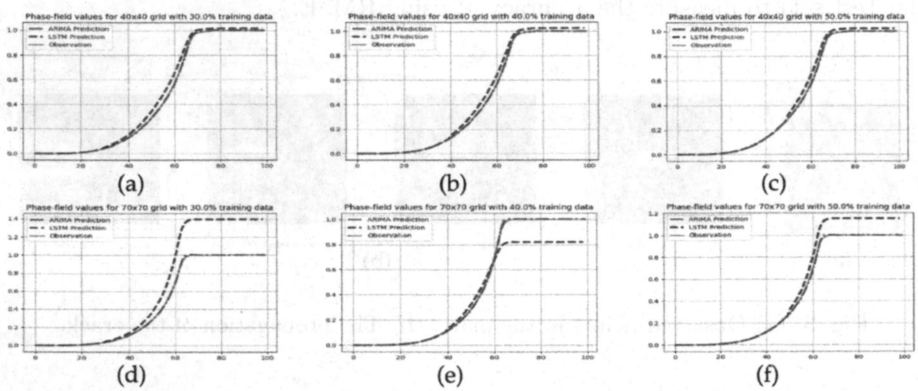

Fig. 4. Forecasting Phase-field values (y-axis) at different steps (x-axis) for mesh size 40×40 at (a), (b), (c) and mesh size 70×70 at (d), (e), (f). (Color figure online)

5.2 Runtime Performance of the Prediction Models

We observed that the computational costs for both machine learning models depend on two factors: the size of the mesh and the Phase-field value data used for training. Figure 5 shows that increasing the amount of training data (30% to 50%) increases the runtime for both models. ARIMA's runtime increases around 2000 milliseconds between 30% and 50%, while it was only an increase of 800 milliseconds for the LSTM model. More importantly, with LSTM, as we increase the mesh size, the training time remains relatively constant. ARIMA model, on the other hand, shows a decrease in training time as the mesh size increases from 40×40 to 50×50 and 60×60. Overall, ARIMA's runtime is still less than LSTM's runtime regardless of the size of data. That shows that statistical approaches for time-series forecasting are still more favorable at large scales.

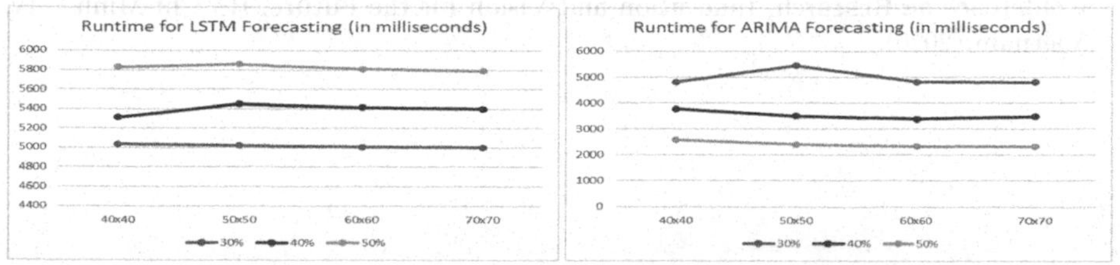

Fig. 5. Runtime (in milliseconds) for time-series forecasting models.

6 Conclusion and Future Work

The use of the phase-field variable in the illustration of crack propagation is one of the state-of-art methods. The properties of a brittle fracture model could be considered as a time-series for an online machine learning analysis and prediction. In this paper, we described how forecasting approaches were used to track the changes of the Phase-field value evaluated the performance of techniques. Our results show ARIMA delivers higher precision and accuracy and is cheaper to train and faster to fit in predicting Phase-field values, compared to LSTM. However, this work is limited to predicting Phase-field values from running brittle fracture simulations with varying mesh sizes. For future work, we also apply machine learning techniques to identify regions with the concentration of high values of the Phase-field variable.

References

1. Griffith, A.A.: VI. The phenomena of rupture and flow in solids. Philos. Trans. R. Soc. London Ser. A **221**(582–593), 163–198 (1921)
2. Moës, N., Dolbow, J., Belytschko, T.: A finite element method for crack growth without remeshing. Int. J. Numer. Meth. Eng. **46**(1), 131–150 (1999)
3. Gardner, E.S., Jr.: Exponential smoothing: the state of the art. J. Forecast. **4**, 1–28 (1985)
4. Yu, Y., Si, X., Hu, C., Zhang, J.: A review of recurrent neural networks: LSTM cells and network architectures. Neural Comput. **31**, 1235–1270 (2019)

5. Molnár, G., Gravouil, A.: 2D and 3D Abaqus implementation of a robust staggered phase-field solution for modeling brittle fracture. Finite Elem. Anal. Des. **130**, 27–38 (2017)
6. Hamilton, J.D.: Time Series Analysis. Princeton University Press, Princeton (1994)
7. Dinh, M.N., Vo, C.T., Abramson, D.: Tracking scientific simulation using online time-series modelling. In: 20th IEEE/ACM International Symposium on Cluster, Cloud and Internet Computing, Melbourne, Australia (2020)
8. Hyndman, R.J., Athanasopoulos, G.: Forecasting: Principles and Practice (2018)
9. Vo, C.T., Dinh, M.N., Dimla, E.: Predicting phase-field behavior of brittle fracture model based on LSTM time series forecasting model. In: IEEE International Conference on Research, Innovation and Vision for the Future, Ho Chi Minh City, Vietnam (2020)

Advances in High-Performance Computational Earth Sciences: Applications and Frameworks

Calculation of Cross-correlation Function Accelerated by Tensor Cores with TensorFloat-32 Precision on Ampere GPU

Yuma Kikuchi[1(✉)], Kohei Fujita[1,2(✉)], Tsuyoshi Ichimura[1,2,3(✉)], Muneo Hori[4], and Lalith Maddegedara[1(✉)]

[1] Earthquake Research Institute and Department of Civil Engineering, The University of Tokyo, Bunkyo, Tokyo, Japan
{kikuchi-y,fujita,ichimura,lalith}@eri.u-tokyo.ac.jp
[2] Center for Computational Science, RIKEN, Kobe, Japan
[3] Center for Advanced Intelligence Project, RIKEN, Tokyo, Japan
[4] Japan Agency for Marine-Earth Science and Technology, Research Institute for Value-Added-Information Generation, Yokohama, Japan
horimune@jamstec.go.jp

Abstract. The cross-correlation function appears in many fields with time-series data, and speeding up the computation is essential given the recent accumulation of significant amounts of data. The cross-correlation function can be calculated as a matrix-matrix product, and a significant speed-up can be expected utilizing Tensor Core, which is a matrix-matrix product acceleration unit of the latest NVIDIA Graphics Processing Units (GPUs). In this research, we target a new precision data type called the TensorFloat-32, which is available in the Ampere architecture. We develop a fast calculation method considering the characteristics of the cross-correlation function and TensorCore. Our method achieved a very high performance of 53.56 TFLOPS in the performance measurement assuming seismic interferometry using actual data, which is 5.97 times faster than cuBLAS, a widely used linear algebra library on NVIDIA GPUs. In addition, the accuracy of the calculation result is sufficiently high compared to the 64-bit floating-point calculation, indicating the applicability of Tensor Core operations using TensorFloat-32 for scientific calculations. Our proposed method is expected to make it possible to utilize a large amount of data more effectively in many fields.

Keywords: Cross-correlation function · GPU computing · Tensor core

1 Introduction

The cross-correlation function expresses the similarity or difference between two time-series data. This calculation is widely used in many fields dealing with

D. Groen et al. (Eds.): ICCS 2022, LNCS 13351, pp. 277–290, 2022.
https://doi.org/10.1007/978-3-031-08754-7_37

time-series data, such as radar detection [2], the discovery of gravity waves in physics [3], and detecting earthquakes and volcanic events by matched filtering [4,5]. A large amount of data has been amassed with the advancement of observation technology, and it is crucial to reduce the computational cost of the cross-correlation function to utilize these data more effectively.

In recent years, many computers equipped with Graphics Processing Units (GPUs) have appeared, and many scientific calculations have been accelerated utilizing GPUs [6,7]. A GPU-based approach has been proposed to calculate cross-correlation functions [8], but further speed-up can be expected by developing methods based on the latest computer architecture.

NVIDIA's Volta architecture [9] and later GPUs have not only the usual arithmetic units but also matrix-matrix product acceleration units called Tensor Core [10]. Tensor Core can perform matrix-matrix products as a hardware function and maintains exceptionally high theoretical performance, which is one of the main features of the latest NVIDIA GPUs.

A study on accelerating the cross-correlation function calculation targeting the Tensor Core with the Volta architecture was performed by Yamaguchi et al. [11]. The Tensor Core with the Volta architecture supports only 16-bit floating-point arithmetic (FP16), making it challenging to use for scientific calculations that require high precision and a wide dynamic range. Yamaguchi et al. solved this problem by introducing local normalization considering the characteristics of the cross-correlation function and Tensor Core; they achieved a 4.47 fold speed-up while maintaining accuracy compared to the matrix-matrix product function of cuBLAS [12], a matrix arithmetic library on GPUs that also utilizes Tensor Core.

Many systems equipped with NVIDIA Ampere architecture GPU [13] have been established in the last few years. In addition to FP16, the Tensor Core with the Ampere architecture can handle various data precision types such as 8-bit integer (INT8), Brain Floating-Point (bfloat16), and TensorFloat-32 (TF32). Further computation speed-up can be expected utilizing these precision data types. TF32 is a new type of floating-point that supports the same range of values as the 32-bit floating-point (FP32) with the same precision as FP16, and in many cases, calculations that require scaling to avoid over/underflow in FP16 can be performed without scaling.

The calculation using TF32 with Tensor Core can be executed through cuBLAS, and some speed-up can be easily obtained. However, since the computation speed of Tensor Core is much faster than the memory access bandwidth of the GPU, the performance of Tensor Core may not be sufficiently high due to the memory access bandwidth limitation. This problem can be avoided by understanding the behavior of Tensor Core and implementing it in accordance with the characteristics of the cross-correlation function calculation, and further acceleration can be expected.

In this research, we develop a method to accelerate the calculation of the cross-correlation function using a TF32 Tensor Core with Ampere architecture based on the method of Yamaguchi et al. Since Ampere is the first architecture equipped with Tensor Core that can handle TF32 precision, we initially investigate its behavior, and then develop a method for calculating cross-correlation functions that can exploit the high performance of the Ampere architecture. Our proposed method enables faster calculation of the cross-correlation function compared to the matrix-matrix product function of cuBLAS, which can also use Tensor Core with TF32.

Since the calculation of the cross-correlation function is mathematically a one-dimensional convolution operation, the method in this study is expected to be effective in many fields.

The rest of this paper is organized as follows. Section 2 describes how to use Tensor Core with TF32 and implement it to achieve higher performance. Afterward, we describe the specific calculation method of the cross-correlation function. In Sect. 3, we apply our proposed method to the cross-correlation function calculation using real observed waveforms in the seismic interferometry and discuss its performance and accuracy. Finally, Sect. 4 summarizes the paper.

The code we developed can be available on GitHub repository [1].

2 Methodology

2.1 Usage of Tensor Cores with TF32

Tensor Core is a matrix-matrix product acceleration unit introduced in the Volta architecture, and can execute Fused Multiple-Add instruction $\mathbf{C} \leftarrow \mathbf{AB} + \mathbf{C}$ for small matrices as a hardware function. TF32 is a new precision data type available in the Tensor Core with Ampere architecture. It consists of a 1-bit sign part, an 8-bit exponent, and a 10-bit mantissa, for a total of 19 bits, and has the same dynamic range and precision as FP32 and FP16, respectively. Tensor Core operations specify the size of the matrices that can be executed; in the case of $\mathbf{A}$ and $\mathbf{B}$ being TF32 and $\mathbf{C}$ being FP32, the matrix sizes $\mathbf{A}^{m\times k}, \mathbf{B}^{k\times n}$, and $\mathbf{C}^{\ m\times n}$ must be $(m, n, k) = (16, 16, 8)$. Hereinafter, we use the notation $\mathbf{A}, \mathbf{B}, \mathbf{C}$ for the whole matrix, not the small submatrix. In CUDA C++, Tensor Core operations are available through Warp Matrix Multiply-Accumulate (WMMA) API. Figure 2 shows the basic usage of WMMA API. The execution flow is as follows. Here, 32 threads (1 warp) work together to compute the matrix-matrix product of $(m, n, k) = (16, 16, 8)$.

(1) Define a fragment (i.e., a variable for each thread required for execution in Tensor Core) (`wmma::fragment`)
(2) Load data from shared memory to the fragment (`wmma::load_matrix_sync`)
(3) Perform matrix-matrix product using the fragment (`wmma::mma_sync`)
(4) Store the result in the fragment of the shared memory (`wmma::store_matrix_sync`)

In the case of a matrix-matrix product on a GPU, the shared memory can be used as a buffer for the global memory to perform the calculations more efficiently. When the kernel is executed, data required for computation is stored in the global memory. Global memory can be accessed by all threads and has a large capacity, but it requires long cycles for memory access. Therefore, hiding

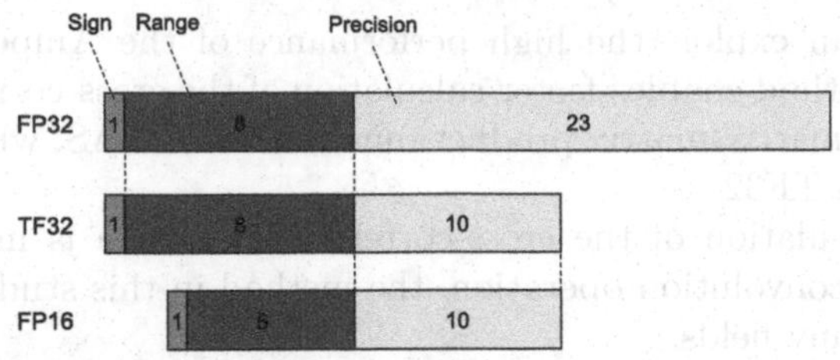

Fig. 1. Comparison of FP32, TF32, and FP16 format

```
// a_shmem, b_shmem and c_shmem are shared memory
__device__ void matmul_tensor_16_16_8 (float *a_shmem, float *b_shmem, float *c_shmem)
{
    wmma::fragment<wmma::matrix_a, 16, 16, 8, wmma::precision::tf32, wmma::row_major> a_frag;
    wmma::fragment<wmma::matrix_b, 16, 16, 8, wmma::precision::tf32, wmma::col_major> b_frag;
    wmma::fragment<wmma::accumulator, 16, 16, 8, float> c_frag;

    // Load values from shared memory to fragment
    wmma::load_matrix_sync(a_frag, a_shmem, 8);
    wmma::load_matrix_sync(b_frag, b_shmem, 8);

    wmma::fill_fragment(c_frag, 0.0f);

    // convert to TF32
    for (int i = 0; i < a_frag.num_elements; ++i) a_frag.x[i] = wmma::__float_to_tf32(a_frag.x[i]);
    for (int i = 0; i < b_frag.num_elements; ++i) b_frag.x[i] = wmma::__float_to_tf32(b_frag.x[i]);

    // execute matrix Fused Multiply-Add
    wmma::mma_sync(c_frag, a_frag, b_frag, c_frag);

    // store values from fragment to shared memory
    wmma::store_matrix_sync(c_shmem, c_frag, wmma::col_major);
}
```

Fig. 2. A basic usage of Tensor Core operations with TF32 by calling WMMA API in CUDA C++

the latency becomes difficult when access to global memory occurs frequently. On the other hand, shared memory can only be accessed from within the same thread block and has limited capacity, but the number of cycles associated with access is shorter than that of global memory, so performance improvement can be expected if it is well utilized (Fig. 1).

As mentioned above, 32 threads (1 warp) performs matrix-matrix product of $(m, n, k) = (16, 16, 8)$ in the operation using Tensor Core with TF32. In this case, each matrix component must be distributed and stored in each of the 32 threads in the warp, which is called a fragment. The result of the matrix-matrix product is also distributed and stored in the fragment of each thread. The mapping pattern between this matrix and the fragment is complex. Therefore, in CUDA C++, `wmma::{load,store}_matrix_sync` is provided as an API to facilitate this mapping. These functions do not require us to consider complicated fragment mapping patterns, but they need shared memory for mapping to and from fragments. Therefore, when the functions are called frequently, the amount of access between the shared memory as a buffer and for this purpose increases. Given the extremely high computation speed of Tensor Core, the performance is easily constrained by the shared memory access bandwidth. The fragment is internally represented as a set of registers, and the mapping pattern of the Volta architecture with FP16 has been analyzed in the previous study [14]. Therefore, by taking the same approach in this research, it is possible to directly map the corresponding values from the shared memory as a buffer to the registers without using `wmma::{load,store}_matrix_sync`. However, since TF32 is the first type introduced in Tensor Core with Ampere architecture and the supported matrix size is different from FP16, we cannot correctly calculate it by the mapping pattern described in previous work. Here, we explore the mapping pattern of the fragment with TF32 in the Ampere architecture and implement it in a way suitable for the Ampere architecture.

The fragment is a CUDA C++ structure that holds the number of elements it owns as `num_elements` and the values of the `num_elements` fragments as member variables `x[i]`. Thus, we can figure out the mapping pattern by actually outputting the thread number and the value of the fragment in the warp, as shown in Fig. 4, First, we present fragments of the matrix **A**, **B** (called matrix_A and matrix_B in the WMMA API). These mapping patterns can be determined by generating a matrix with unique non-overlapping elements (e.g., 0,1, $\cdots$, 127) executing `wmma::load_matrix_sync`, and then outputting the values of the fragment of each thread and comparing it with the original matrix. The mapping pattern of matrix **C** (called accumulator in the WMMA API) can be obtained by setting matrix **A**, **B** so that the result of the matrix-matrix product is unique and without duplication, and then outputting the fragment after calling `wmma::mma_sync` in the same way as matrix_A and matrix_B.

Figures 5 and 6 display the mapping patterns of matrix_A (row_major) and matrix_B (col_major), and the accumulator stored in col_major format, respectively. Here, {col, row}_major specifies whether the original two-dimensional matrix is column-first or row-first when it is stored in memory in one dimension. The number in the matrix signifies the element order, and the number in parentheses denotes the storage order in the register of each thread. For example, in matrix_A of row_major and matrix_B of col_major, each thread has four elements out of 16×8 (8×16) distributed elements, and the thread in which `threadIdx.x` % 32 is 0 has 0th, 64th, 4th, 68th, and 0th, 4th, 64th, 68th elements in the 0-index, respectively. In the accumulator stored as col_major, each thread has eight elements out of 16×16 distributed elements, and the thread in which `threadIdx.x` % 32 is 0 has the 0th, 1st, 128th, 129th, 8th, 9th, 136th, and 137th elements in the 0-index. These mappings allow the values to be stored directly in the registers without calling `wmma::load_matrix_sync`, but the matrix-matrix product execution function of Tensor Core `wmma::mma_sync` can only pass the values through the fragment and cannot execute using registers. We overcame this problem using CUDA PTX inline assembly similar to Yamaguchi et al. PTX is a pseudo-assembly language in CUDA, that allows us to write low-level code with higher flexibility than the standard API. Figure 7 shows the PTX inline assembly of `wmma::mma_sync` for TF32. Based on these mapping patterns, together with the inline PTX assembly, we can realize fast memory access and

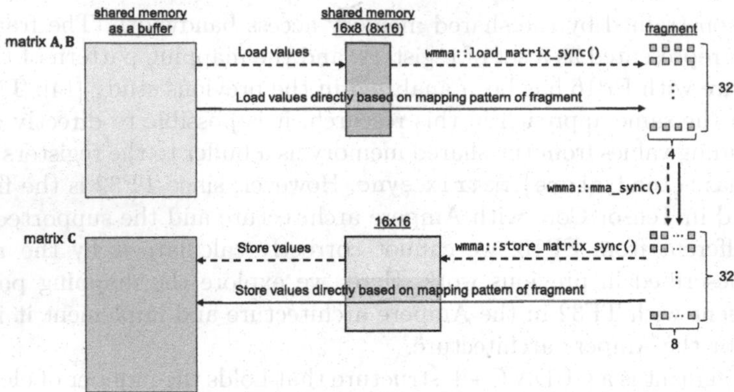

Fig. 3. A simplified data flow of the Tensor Core execution with TF32. Dashed line: use `wmma::load_matrix_sync` API to load/store values, Solid line: directly load/store values based on mapping patterns without using the API (proposed method).

```
...
int thid = threadIdx.x%32;
for (i = 0; i < a_frag.num_elements; ++i) printf("%d %f\n", thid, a_frag.x[i]);
for (i = 0; i < b_frag.num_elements; ++i) printf("%d %f\n", thid, b_frag.x[i]));
...
wmma::mma_sync(c_frag,a_frag,  b_frag, c_frag);
for (i = 0; i < c_frag.num_elements; ++i) printf("%d %f\n", thid, c_frag.x[i]);
```

Fig. 4. A way to examine the mapping pattern from shared memory to Tensor Core fragments and from Tensor Core fragments to shared memory with TF32

execution without `wmma::load_matrix_sync` even when TF32 is used in the Ampere architecture (Fig. 3).

2.2 Calculation of Cross-correlation Function Using Tensor Cores

The cross-correlation is a function calculated to check the similarity or deviation between two time-series data. The cross-correlation function between two waveforms X_i and X_j of length T can be expressed as a function of the time shift τ, as in Eq. (1).

$$CC_{i,j}(\tau) = \sum_{t=1}^{T} X_i(t)X_j(t+\tau) \tag{1}$$

The cross-correlation function can also be calculated in the frequency domain, but we only deal with it in the time domain. Focusing on a single τ, the calculation of the cross-correlation function is the inner product of two waveforms, which can be calculated as a matrix-matrix product operation for multiple waveform pairs with multiple time shifts simultaneously, as presented in Eq. (2). We apply the Tensor Core operation to this matrix-matrix product calculation.

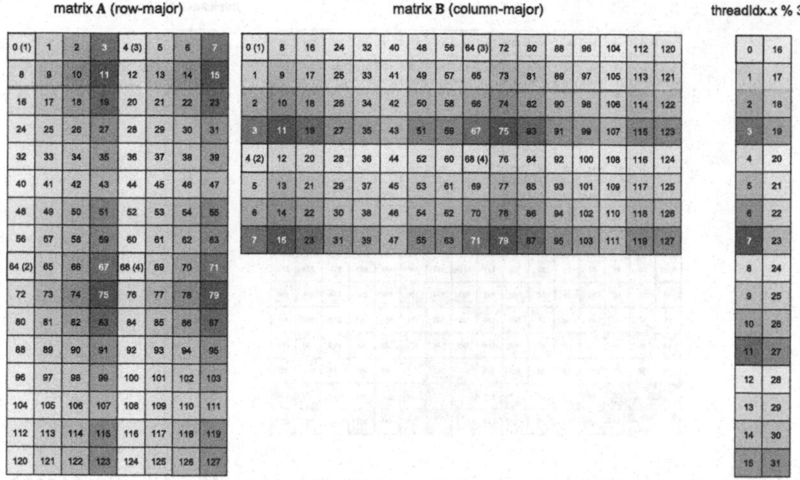

Fig. 5. Mappings of `wmma:fragment<wmma::matrix_a, 16, 16, 8, wmma::precision::tf32, wmma::row_major>` and `wmma:fragment<wmma::matrix_b, 16, 16, 8, wmma::precision::tf32, wmma::col_major>` with Ampere architecture. The number in parentheses shows the order of each value in a fragment.

$$\begin{pmatrix} CC_{1,1}(0) & CC_{1,1}(1) & CC_{1,1}(2) & CC_{1,1}(3) \\ CC_{2,1}(0) & CC_{2,1}(1) & CC_{2,1}(2) & CC_{2,1}(3) \\ CC_{3,1}(0) & CC_{3,1}(1) & CC_{3,1}(2) & CC_{3,1}(3) \\ CC_{4,1}(0) & CC_{4,1}(1) & CC_{4,1}(2) & CC_{4,1}(3) \end{pmatrix} = \begin{pmatrix} X_1(1) & X_1(2) & X_1(3) & X_1(4) \\ X_2(1) & X_2(2) & X_2(3) & X_2(4) \\ X_3(1) & X_3(2) & X_3(3) & X_3(4) \\ X_4(1) & X_4(2) & X_4(3) & X_4(4) \end{pmatrix} \begin{pmatrix} X_1(1) & X_1(2) & X_1(3) & X_1(4) \\ X_1(2) & X_1(3) & X_1(4) & X_1(5) \\ X_1(3) & X_1(4) & X_1(5) & X_1(6) \\ X_1(4) & X_1(5) & X_1(6) & X_1(7) \end{pmatrix} \quad (2)$$

In Yamaguchi et al.'s method, cross-correlations are simultaneously computed for time shifts of N time steps for 16 waveforms of length 256 per 1 warp (32 threads). Thus, in each warp, the product of matrix $\mathbf{A}$ of size 16×256 and matrix $\mathbf{B}$ of size $256 \times N$ is calculated by executing the matrix-matrix product of size 16×16 and 16×16 multiple times. In this study, we target matrix-matrix product of these sizes, but the size of matrices supported by Tensor Core vary between FP16 and TF32. Thus, we calculate the matrix-matrix product of size 16×16 by performing the matrix product of size 16×8 and 8×16 twice.

As explained above, the matrix-matrix product calculation on the GPU can be performed more efficiently utilizing shared memory. In this research, shared memory is also used as a buffer for global memory so that the access to global memory with long latency is minimized.

In the case of calculating the cross-correlation function, the matrix $\mathbf{B}$ can be constructed by sequentially shifting single time-series data, as shown on the right-hand side of Eq. (2). Since the size of matrix $\mathbf{B}$ is $256 \times N$, a typical calculation requires a total of $256 \times N$ elements to be loaded from the global

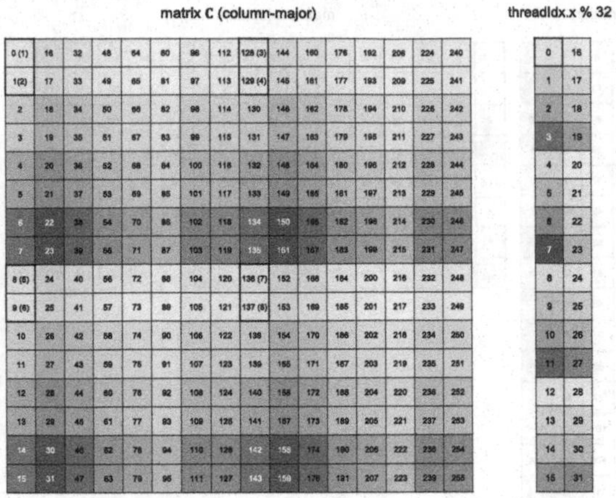

matrix C (column-major)

0 (1)	16	32	48	64	80	96	112	128 (3)	144	160	176	192	208	224	240
1(2)	17	33	49	65	81	97	113	129 (4)	145	161	177	193	209	225	241
2	18	34	50	66	82	98	114	130	146	162	178	194	210	226	242
3	19	35	51	67	83	99	115	131	147	163	179	195	211	227	243
4	20	36	52	68	84	100	116	132	148	164	180	196	212	228	244
5	21	37	53	69	85	101	117	133	149	165	181	197	213	229	245
6	22	38	54	70	86	102	118	134	150	166	182	198	214	230	246
7	23	39	55	71	87	103	119	135	151	167	183	199	215	231	247
8 (5)	24	40	56	72	88	104	120	136 (7)	152	168	184	200	216	232	248
9 (6)	25	41	57	73	89	105	121	137 (8)	153	169	185	201	217	233	249
10	26	42	58	74	90	106	122	138	154	170	186	202	218	234	250
11	27	43	59	75	91	107	123	139	155	171	187	203	219	235	251
12	28	44	60	76	92	108	124	140	156	172	188	204	220	236	252
13	29	45	61	77	93	109	125	141	157	173	189	205	221	237	253
14	30	46	62	78	94	110	126	142	158	174	190	206	222	238	254
15	31	47	63	79	95	111	127	143	159	175	191	207	223	239	255

threadIdx.x % 32

0	16
1	17
2	18
3	19
4	20
5	21
6	22
7	23
8	24
9	25
10	26
11	27
12	28
13	29
14	30
15	31

Fig. 6. A mapping of `wmma:fragment<wmma::accumulator, 16, 16, 8, float>` with Ampere architecture. The number in parentheses shows the order of each value in a fragment.

```
asm("{¥n¥t"
    "wmma.mma.sync.aligned.row.col.m16n16k8.f32.tf32.tf32.f32 {%0,%1,%2,%3,%4,%5,%6,%7},
     {%8,%9,%10,%11},{%12,%13,%14,%15},{%16,%17,%18,%19,%20,%21,%22,%23};¥n¥t"
    "}"
    : "=f"(cv[0]),"=f"(cv[1]),"=f"(cv[2]),"=f"(cv[3]),"=f"(cv[4]),"=f"(cv[5]),"=f"(cv[6]),"=f"(cv[7])
    : "r"(av[0]),"r"(av[1]),"r"(av[2]),"r"(av[3])
      "r"(bv[0]),"r"(bv[1]),"r"(bv[2]),"r"(bv[3])
      "f"(cv[0]),"f"(cv[1]),"f"(cv[2]),"f"(cv[3]),"f"(cv[4]),"f"(cv[5]),"f"(cv[6]),"f"(cv[7]));
```

Fig. 7. PTX inline assembly of Tensor Core operation with TF32

memory. Conversely, by employing the characteristics of the cross-correlation function calculation, we only need to read a total of $256 + N - 1$ data, significantly reducing memory access cost and usage. As N increases, the amount of data read decreases in proportion to the amount of computation, which is expected to improve the performance. However, as memory usage increases, hiding the latency associated with operations becomes more difficult, resulting in performance degradation. Yamaguchi et al. stated that $N = 96$ is the equilibrium point in the performance measurement, and the same tendency was observed in our experiment, so the calculation is performed with $N = 96$ in the next section.

For the Tensor Core with Volta architecture, matrices **A** and **B** are supported only for FP16. FP16 has a much narrower dynamic range than the FP32/64-bit floating-point (FP64), which is generally used in scientific computing and may cause over/underflow during matrix-matrix product operation. In Yamaguchi et al.'s method, local scaling is applied to matrices **A**, **B** to prevent over/underflow during the computation. Although the calculation of the scaling value was quickly performed via the shuffle instructions in the warp, it was necessary to multiply the scaling value locally after every Tensor Core execution. Therefore, another set of registers was allocated and used as buffers for the calculation results, and the scaling values were multiplied when adding them to the buffers. In short, the matrix **C** was assigned zero for every calculation, and buffer **D** was assigned **C** multiplied by the local scaling value after performing $\mathbf{C} \leftarrow \mathbf{AB}+\mathbf{C}$. In contrast, TF32 can handle the same range of values as FP32, so the calculation can be performed correctly without scaling in many cases. Consequently, the scaling restriction mentioned above is eliminated, and computation can be performed only using one set of registers, meaning that the matrix **C** only needs to be initialized to zero once at the beginning. Afterward, the register containing the calculation result can be used directly for the Tensor Core execution in the subsequent corresponding execution, as in $\mathbf{C} \leftarrow \mathbf{AB}+\mathbf{C} \leftarrow \mathbf{AB}+\mathbf{C} \leftarrow \cdots$. This reduces the number of registers required for execution and enables more efficient computation.

TF32 is currently supported only for operations on Tensor Core and cannot be used for normal calculations on a CPU and GPU. Therefore, both matrices **A** and **B** are transferred to the GPU as FP32 at the kernel execution time and converted to TF32 by `wmma::__float_to_tf32` when they are stored in shared memory as buffers. The input and output of `wmma::__float_to_tf32` is FP32, but the output is numerically TF32. If the FP32 type, which is numerically TF32, is mixed with the usual FP32 type operations, the precision and range of the results are undefined [15], but this is not problematic in this time because there is no need to perform scaling or other operations on matrices **A** or **B** on the GPU.

An overview of the above computation procedure per thread block is shown in Fig. 8.

3 Application and Performance Measurement

3.1 Application Example

An example dealing with the calculation of cross-correlation functions in seismology is the seismic interferometry method [16]. In the seismic interferometry, the cross-correlation function of the waveforms observed at two different stations is calculated and stacked over a long period. In this way, a pseudo-response waveform (Green's function) in which one station is regarded as the source and the other as the observation point can be synthesized [17] (Fig. 9). The earth is constantly vibrating due to natural earthquakes and other microtremors such as pulsations caused by ocean waves and constant microtremors due to human activities. By applying the seismic interferometry to these wavefields, it is expected that Green's function can be obtained without the use of artificial seismic sources [18].

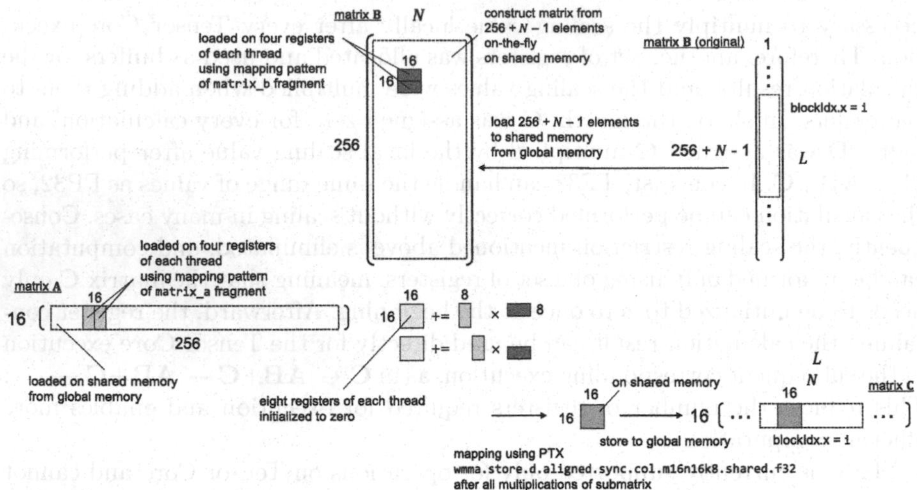

Fig. 8. An overview of the cross-correlation function calculation using Tensor Core with TF32 per thread block. L is the total time shift of the cross-correlation function.

In recent years, many seismic observation networks, such as the MOWLAS [19], have been operating in Japan, and a large amount of data has been accumulated. By continuously applying seismic interferometry to these observation data, it is expected to lead to the realization of the continuous monitoring of underground structures without the need for artificial seismic sources [20,21]. However, since the number of calculations of the cross-correlation function increases proportionately to the square of the number of observation points, we are currently able to handle only a portion of the data (e.g., by narrowing down the number of observation points).

In this measurement, we calculate the cross-correlation function for 256 time steps of 16 channels and 1 channel for a total of 4.32×10^6 time shifts, as in Yamaguchi et al. In other words, we calculate the cross-correlation function as a matrix-matrix product of size 16×256 and size $256 \times (4.32 \times 10^6)$ and measure the performance of our proposed method. The matrices are constructed using the actual observed data of K-net [22] which is one of the MOWLAS.

3.2 Performance Measurement

At this time, we measure the elapsed time and accuracy of three types of kernels, including the proposed method. The first is the kernel of Yamaguchi et al. that performs computations on FP16 with local scaling and runs on a NVIDIA V100 GPU with the Volta architecture. The other is a kernel using `cublasGemmEx`, which is a dense matrix-dense matrix product function in cuBLAS, a linear algebra library provided by NVIDIA, and runs on a NVIDIA A100 GPU with the Ampere architecture. In `cublasGemmEx`, we can select the data precision type and execution mode through arguments and options. We set the calculation to

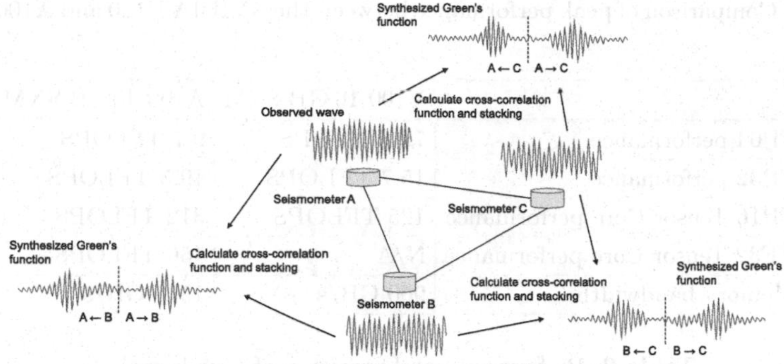

Fig. 9. A basic concept of seismic interferometry. Green's function can be synthesized by calculating the cross-correlation function of the observed waves at two observation points and stacking them over a long period of time.

be performed using Tensor Core with TF32 precision. Unlike the other kernels, matrix **B** is constructed explicitly in advance, and the computation is performed. Lastly, we analyze our proposed method using TF32 with Tensor Core, executed on A100 GPU.

Table 1 displays the peak hardware performance of the NVIDIA V100 and A100 GPUs used in this measurement. We use nvcc V11.2.67 in nvhpc 21.2 as the compiler and `--generate-code arch=compute_{70,80},code=sm_{70,80} -O3 --use_fast_math` as the compile option for V100 and A100, respectively. The elapsed time is measured using nsys nvprof, and the FLOP is counted manually.

We also evaluate the calculation result accuracy of each kernel in comparison with the calculation results of FP64 on the CPU. The accuracy is calculated using Eq. (3), which is defined as the largest absolute error between the result of each kernel and FP64. In order to compute the cross-correlation function on the CPU, we use `cblas_dgemm`, a dense matrix-dense matrix product function in the Intel Math Kernel Library (MKL). We use the Xeon Platinum 8360Y (36 cores), and the peak FP64 performance of this CPU is 2.765 TFLOPS. Additionally, We use icc 19.1.3.304 as the compiler and set `-O3 -mkl=parallel` as the compile option and `KMP_AFFINITY=compact` as the environment variable. For reference, we also measure the MKL kernel performance.

$$ERR = \max_{i,j} \left| CC_{i,j} - CC_{i,j}^{FP64} \right| \qquad (3)$$

Table 2 lists the elapsed time, execution performance, and ERR of each kernel. First, Yamaguchi et al.'s kernel using FP16 demonstrates a performance of 25.90 TFLOPS. This is higher than the theoretical performance ratio of FP32 in V100, which is 20.72% of the theoretical performance of FP16 with Tensor Core in V100. In addition, local scaling keeps the error as small as 5.6×10^{-4}.

Table 1. Comparison of peak performance between the NVIDIA V100 and A100 GPUs.

	V100 16 GB SXM	A100 40 GB SXM
FP64 performance	7.8 TFLOPS	9.7 TFLOPS
FP32 performance	15.7 TFLOPS	19.5 TFLOPS
FP16 Tensor Core performance	125 TFLOPS	312 TFLOPS
TF32 Tensor Core performance	N/A	156 TFLOPS
Memory bandwidth	900 GB/s	1555 GB/s

Table 2. Performance and precision of each kernel.

Kernel	Elapsed time	TFLOPS (ratio to peak)	*ERR*
Yamaguchi (2019) FP16 (V100)	1.366 ms	25.90 (20.72%)	5.6×10^{-4}
cuBLAS TF32 (A100)	3.947 ms	8.97 (5.75%)	2.0×10^{-4}
Proposed TF32 (A100)	0.661 ms	53.56 (34.34%)	2.0×10^{-4}
MKL FP64 (Xeon Platinum)	84.91 ms	0.42 (15.07%)	0.0

On the other hand, our method using TF32 with Tensor Core achieves a very high execution performance of 53.56 TFLOPS, which is 34.34% of the theoretical performance of TF32 with Tensor Core on A100 and higher than the peak performance ratio of Yamaguchi et al.'s kernel. While the theoretical performance of A100's TF32 with Tensor Core is 1.25 times higher than that of V100's FP16 with Tensor Core, the obtained execution performance is 2.07 times higher, likely because the wide dynamic range of TF32 eliminates the need for scaling and allows for more efficient use of Tensor Core. In terms of accuracy, the maximum error is 2.03×10^{-4}, which is approximately half that of the FP16 calculation. In addition, compared with the result on cuBLAS with TF32 with Tensor Core, our proposed method achieves a speed-up of 5.97 times, indicating the proposed method's effectiveness. Finally, we compare the elapsed time on the CPU for reference. In MKL, for matrices of the sizes targeted in this measurement, the overhead of copying the matrices to the buffer becomes relatively large, and the execution performance is as low as 0.42 TFLOPS (15.07% of the peak performance). On the other hand, the proposed method achieves high performance even for matrices of such sizes, resulting in a speed-up of 128.46 times when the theoretical performance ratio is about 56.42 times. By implementing the calculation accounting for the characteristics of Tensor Core and the cross-correlation function, we developed a method that can take advantage of the high performance of Tensor Core with Ampere architecture.

4 Closing Remarks

In this study, we developed a fast computation method for cross-correlation functions in the time domain utilizing Tensor Core with a NVIDIA Ampere architecture GPU. Tensor Core with the Ampere architecture supports various precision data types. We can expect a significant speed-up in many computations involving the matrix-matrix product by utilizing these data types. Tensor Core has very high theoretical performance, but its computational performance is easily limited by the global and shared memory bandwidth when ordinary APIs are used because of its extremely high speed. We first investigated the mapping pattern between shared memory and the fragment for Tensor Core operation using TensorFloat-32 precision on the Ampere architecture. By combining this with a low-level description using PTX inline assembly, we have achieved an implementation that does not require a standard API, and our method is less constrained by memory bandwidth. In the performance measurement of the cross-correlation function calculation in seismic interferometry using actual seismic data, our proposed method achieved a significant performance level of 53.56 TFLOPS, which is 34.34% of the theoretical performance of TF32 with Tensor Core. Moreover, our proposed method is 5.97 times faster than cuBLAS using TF32 with Tensor, a linear algebra library commonly used on NVIDIA GPUs. The cross-correlation function appears in many fields that deal with time-series data, and speeding up the computation has become an important issue in light of the accumulation of observation data, especially in recent years. It is expected that our proposed method can enable analysis that makes fuller and more effective use of big data.

Acknowledgment. We acknowledge support from the Japan Society for the Promotion of Science (18H05239).

References

1. TC-enhanced Cross-correlation Function. https://github.com/nlnxfkl/TC-enhanced_Cross-correlation_Function. Accessed 10 Apr 2022
2. Venu, D., Rao, N.V.K.: A cross-correlation approach to determine target range in passive radar using FM broadcast signals (2016). http://dx.doi.org/10.1109/WiSPNET.2016.7566190
3. Alonso, D., Cusin, G., Ferreira, P.G., Pitrou, C.: Detecting the anisotropic astrophysical gravitational wave background in the presence of shot noise through cross-correlations (2020). https://doi.org/10.1103/PhysRevD.102.023002
4. Shearer, P.M.: Global seismic event detection using a matched filter on long-period seismograms (1994). https://doi.org/10.1029/94JB00498
5. Aso, N., Ohta, K., Ide, S.: Volcanic-like low-frequency earthquakes beneath Osaka Bay in the absence of a volcano (2011). https://doi.org/10.1029/2011GL046935
6. Norman, M.R., et al.: Unprecedented cloud resolution in a GPU-enabled full-physics atmospheric climate simulation on OLCF's summit supercomputer (2021). https://doi.org/10.1177/10943420211027539
7. Ichimura, T., et al.: A fast scalable implicit solver for nonlinear time-evolution earthquake city problem on low-ordered unstructured finite elements with artificial intelligence and transprecision computing (2018). https://doi.org/10.1109/SC.2018.00052
8. Beaucé, E., Frank, W.B., Romanenko, A.: Fast Matched Filter (FMF): an efficient seismic matched-filter search for both CPU and GPU architectures (2017). https://doi.org/10.1785/0220170181

9. NVIDIA TESLA V100 GPU ARCHITECTURE. https://images.nvidia.com/content/volta-architecture/pdf/volta-architecture-whitepaper.pdf. Accessed 04 Feb 2022
10. Markidis, S., Chien, S.W.D., Laure, E., Peng, I.B., Vetter, J.S.: NVIDIA tensor core programmability, performance & precision (2018). https://doi.org/10.1109/IPDPSW.2018.00091
11. Yamaguchi, T., Ichimura, T., Fujita, K., Kato, A., Nakagawa, S.: Matched filtering accelerated by tensor cores on Volta GPUs With improved accuracy using half-precision variables (2019). https://doi.org/10.1109/LSP.2019.2951305
12. cuBLAS. https://docs.nvidia.com/cuda/cublas/index.html. Accessed 04 Feb 2022
13. NVIDIA A100 Tensor Core GPU Architecture. https://images.nvidia.com/aem-dam/en-zz/Solutions/data-center/nvidia-ampere-architecture-whitepaper.pdf. Accessed 17 Jan 2022
14. Raihan, M.A., Goli, N., Aamodt, T.M.: Modeling deep learning accelerator enabled GPUs (2019). https://doi.org/10.1109/ISPASS.2019.00016
15. CUDA C++ Programming Guide. https://docs.nvidia.com/cuda/cuda-c-programming-guide/index.html. Accessed 04 Feb 2022
16. Curtis, A., Gerstoft, P., Sato, H., Snieder, R., Wapenaar, K.: Seismic interferometry-turning noise into signal (2006). https://doi.org/10.1190/1.2349814
17. Wapenaar, K., Fokkema, J.: Green's function representations for seismic interferometry (2006). https://doi.org/10.1190/1.2213955
18. Chen, Y., Saygin, E.: Empirical Green's function retrieval using ambient noise source-receiver interferometry (2020). https://doi.org/10.1029/2019JB018261
19. National Research Institute For Earth Science And Disaster Resilience: NIED MOWLAS (Monitoring of Waves on Land and Seafloor) (2019). https://nied-ir.bosai.go.jp/?action=repository_uri&item_id=2151&lang=english, https://doi.org/10.17598/NIED.0009
20. Dales, P., Audet, P., Olivier, G.: Seismic interferometry using persistent noise sources for temporal subsurface monitoring (2017). https://doi.org/10.1002/2017GL075342
21. Voisin, C., Guzmán, M.A.R., Réfloch, A., Taruselli, M., Garambois, S.: Groundwater monitoring with passive seismic interferometry (2017). https://doi.org/10.4236/jwarp.2017.912091
22. National Research Institute For Earth Science And Disaster Resilience: NIED K-NET, KiK-net (2019). https://nied-ir.bosai.go.jp/?action=repository_uri&item_id=2146&lang=english, https://doi.org/10.17598/NIED.0004

Developing an ELM Ecosystem Dynamics Model on GPU with OpenACC

Peter Schwartz, Dali Wang(✉), Fengming Yuan, and Peter Thornton

Environmental Sciences Division, Oak Ridge National Laboratory,
Oak Ridge, TN 37830, USA
{schwartzpd,wangd,yuanf,thorntonpe}@ornl.gov

Abstract. Porting a complex scientific code, such as the E3SM land model (ELM), onto a new computing architecture is challenging. The paper presents design strategies and technical approaches to develop an ELM ecosystem dynamics model with compiler directives (OpenACC) on NVIDIA GPUs. The code has been refactored with advanced OpenACC features (such as deepcopy and routine directives) to reduce memory consumption and to increase the levels of parallelism through parallel loop reconstruction and new data structures. As a result, the optimized parallel implementation achieved more than a 140-time speedup (50 ms vs 7600 ms), compared to a naive implementation that uses OpenACC routine directive and parallelizes the code across existing loops on a single NVIDIA V100. On a fully loaded computing node with 44 CPUs and 6 GPUs, the code achieved over a 3.0-times speedup, compared to the original code on the CPU. Furthermore, the memory footprint of the optimized parallel implementation is 300 MB, which is around 15% of the 2.15 GB of memory consumed by a naive implementation. This study is the first effort to develop the ELM component on GPUs efficiently to support ultra-high-resolution land simulations at continental scales.

Keywords: Earth system models · Exascale energy earth system model · E3SM land model · OpenAcc · Functional unit testing · Ecosystem dynamics

1 Introduction

The Exascale Energy Earth System Model (E3SM) is a fully coupled Earth system model that uses code optimized for the Department of Energy (DOE)'s advanced computers to address the most critical scientific questions facing the

This research was supported as part of the Energy Exascale Earth System Model (E3SM) project, funded by the U.S. Department of Energy, Office of Science, Office of Biological and Environmental Research. This research used resources of the Oak Ridge Leadership Computing Facility and Experimental Computing Laboratory at the Oak Ridge National Laboratory, which are supported by the Office of Science of the U.S. Department of Energy under Contract No. DE-AC05-00OR22725.

D. Groen et al. (Eds.): ICCS 2022, LNCS 13351, pp. 291–303, 2022.
https://doi.org/10.1007/978-3-031-08754-7_38

US and the society [5]. The E3SM contains several community models to simulate major Earth system components: atmosphere, ocean, land, sea ice, and glaciers. Within the E3SM framework, the E3SM Land Model (ELM) is designed to simulate how the changes in terrestrial land surfaces interact with other Earth system components and has been used to understand hydrologic cycles, biogeophysics, and ecosystem dynamics [3]. The ELM software has several distinguishing computational features: 1) all the biophysical and biochemical processes are simulated on individual land surface units (i.e., gridcell) independently; 2) highly customized globally accessible, hierarchical data structures are used to represent the heterogeneity of Earth's landscape; 3) none of the subroutines are computationally intensive [11].

Most current high-end supercomputers use heterogeneous hardware with accelerators [2]. The E3SM, consisting of millions of lines of code developed for traditional homogeneous multicore processors, cannot automatically benefit from the advancement of these supercomputers. Refactoring and optimizing the E3SM models for new architectures with accelerators is challenging but inevitable.

Rewriting a large-scale legacy code in a new programming language (such as CUDA) is not practical, two general approaches (compiler directives and the use of GPU-ready libraries) have been adapted to develop E3SM code for computing systems with accelerators. For example, the Kokkos libraries have been used to increase the performance on the E3SM atmosphere model on NVIDIA GPUs in the Summit supercomputer [1] with a performance similar to that of the CPU code. The OpenACC and Athread have been used to develop the community atmosphere model and parallel ocean program on many-core 64-bit RISC processors in the Taihu Light supercomputer [10] with performance improvements of up-to 8 times. However, the performance improvement in this effort mainly came from the extensive programming using Athread. The OpenACC is mainly used as a pre-preparation for code porting onto the RISC system. The OpenACC has been used to accelerate the Model for Prediction Across Scales (MPAS) microphysics WSM6 (WRF Single Motion) model on a single NVIDIA GPU [6] with a performance improvement of up to 2.4 times.

With the availability of high-resolution atmospheric forcing (such as temperature, precipitation, shortwave radiation, and vapor pressure) [7] and land surface proprieties (such as vegetation and soil properties maps), it is desirable to conduct high-fidelity land simulations with ELM at 1km-resolution to deliver a "gold standard" set of results describing the surface weather and climate, as well as the energy, water, carbon, and biogeochemistry processes at a continental scale. The ultra-high-resolution ELM simulation over North America covers a landscape of 24 million gridcells, 350 million columns, and 700 million vegetation patches, is only feasible with highly efficient use of the accelerators within high-end supercomputers. Furthermore, for the reason that there are over one thousand subroutines in ELM, the majority of which are computationally non-intensive, using compiler directives is the appropriate approach to accelerate ELM onto GPU systems. The paper reports development of an ecosystem dynamics model within ELM on NVIDIA GPU using OpenACC.

1.1 Computational Platform and Software Environment

The computational platform used in the study is the Summit leadership computing system at the Oak Ridge National Laboratory. Summit has 4,608 computing nodes, most of which contain two 22-core IBM POWER9 CPUs, six 16-GB NVIDIA Volta GPUs, and 512 GB of shared memory. Technically, this study uses 42 CPU cores (2 CPU cores are reserved for system functions) and all the non-tensor cores in GPUs. The software environment used in our study include NVIDIA HPC 20.11 and several libraries: OpenMPI (spectrum-mpi/10.4.0.3-20210112), NetCDF (netcdf-c/4.8.0, netcdf-fortran/4.4.5), pnetcdf(1.12.2), HDF (1.10.7), and CUDA (11.1).

2 Method

The ecosystem dynamics model simulates the biogeochemical cycles of the ecosystem, including carbon, nitrogen, and phosphorus. It contains many function groups, such as nitrogen deposition/fixation, maintenance and growth respiration, phosphorus deposition, and soil litter decomposition. The ecosystem dynamics model is the most sophisticated model within ELM that contains over 90 subroutines and accesses over 2000 globally accessible variables, many of them 3D arrays. All the ELM routines can access these hierarchical global data structures during the simulation.

We have developed a Functional Unit Testing (FUT) framework to generate standalone ELM models to accelerate code porting and performance tuning. The FUT is a python toolkit built upon the previous software system designed to facilitate scientific software testing [8,9]. We first generate a standalone ecosystem dynamics model for code porting and performance evaluation. To quickly assess the performance of ELM, we also create synthesized data using the observational data from AmeriFlux (ameriflux.lbl.gov) as the forcing data set for the code development. The forcing data and the global variables for 6000 gridcells take 11 GB GPU memory (approximate 1.8 MB data per gridcell). Each NVIDIA V100 contains 16 GB of memory, so 5 GB of shared GPU memory is available for other globally accessible variables, history data, external forcing data, and ELM kernels. The ecosystem dynamics model takes an hourly timestep. An ELM spin-up simulation[1] generally covers a period of 800 to 1000 years, and an ELM transit simulation[2] runs over a period of 100–200 years. Therefore, the ecosystem dynamics model usually is executed around 8–10 million times on each gridcell.

2.1 ELM Data Structure and Ecosystem Dynamics Model

The ELM uses highly customized, hierarchical data structures (Fig.Z 1) to represent the heterogeneity of Earth's landscape. Due to historical reasons, these

[1] The spin-up simulation is used for the ELM to reach a state of statistical equilibrium under the applied climatic forcing.

[2] The transit simulation is referred to the ELM simulation of post industry revolution period (1850-present) with rising CO2, greenhouse emission, dynamical vegetation, and land use and land change, including urbanization.

data structures are declared as globally accessible entities that can be referenced and modified by individual ELM functions during the simulation. Specifically, ELM contains 5 landscape datatypes (gridcell, topographic unit, land cover, soil column, and vegetation patch) representing several aspects of the land surface. Each gridcell represents a small region on the Earth's surface, and is a customized datatype that is derived from a standalone geospatially explicit data library and has dozens of global variables. Each gridcell contains 9 landcover units, 11 column units (each soil column can have up to 20 soil layers and 5 bedrock layers), and 24 vegetation units. Furthermore, The ELM has another 25+ customized model-related datatypes (such as the state and flux datatype for carbon, nitrogen, temperature, and soil). Altogether, these derived datatypes (around 90) contain over 2000 global variables, many of which are multidimensional arrays.

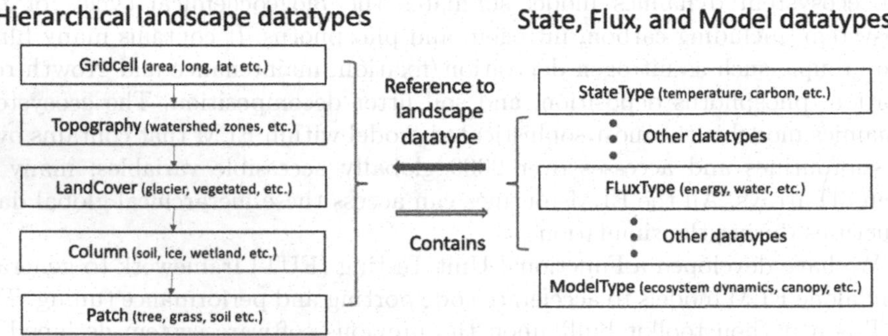

Fig. 1. Highly customized, hierarchical ELM data structure

The ecosystem dynamics model is the most complex submodel within ELM, containing four major subroutine groups (initialization, leaching, noleaching1, and noleaching2). In total, these subroutine groups contain over 90 subroutines and access around 2000 global variables. From the function perspective, the ecosystem dynamics model can be grouped into several modules, such as carbon, nitrogen, and phosphorus dynamics, phenology, soil litter, gap mortality, and fire (Fig. 2). For a better illustration, many secondary and supporting functions (such as IO, timing, and other utility functions) are not shown in the picture.

2.2 Performance Comparison and a Naive OpenACC Implementation

This study adopts a node-level performance comparison. A similar full workload is placed onto all the CPUs or GPUs within a single node. The execution times of the original CPU code and GPU implementation at a single hourly simulation timestep are collected for performance evaluation. Specifically, on a single Summit node, we assign 36000 gridcells on GPUs (6000 gridcells on each GPU) and 36036 gridcells on CPUs (858 gridcells on each CPU core). With the workload

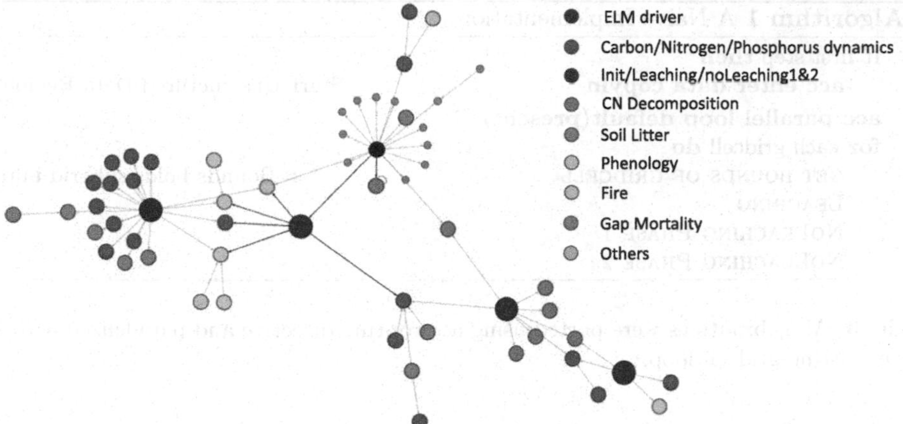

Fig. 2. The ecosystem dynamics model contains several function modules. There are 129 nodes, each represents a subroutine. The four blue node represent the entrance to subroutine groups: initialization, leaching, noleaching phase 1, noleaching phase 2.

of 858 gridcell, the original CPU-based ecosystem dynamics model takes 150 milliseconds (ms) to finish a single timestep (hourly) simulation. Note that the node-level comparison with a similar workload has also been adopted by other scientific code porting [1,6].

The original code organizes gridcells into clumps on each CPU core, and the ecosystem dynamics model runs over these clumps of gridcells. Therefore one of the most straightforward implementations is to instrument OpenACC directives into the original code and onto these existing loops. Specifically, this implementation contains four major steps: 1) use the OpenACC data directive to create a data region and copy all the input data into the data region once at the beginning of simulation; 2) use the routine directives to generate GPU kernels of the majority of ecosystem dynamics subroutines, including three subroutine groups, Leaching, NoLeaching Phase 1, and NoLeaching Phase 2, and many subroutines inside them; 3) place parallel loop constructs to loop over the gridcells, then 4) launch these GPU subroutine kernels in parallel (Fig. 3). This naive implementation does work but comes with abysmal performance (over 7 s) and consumes a large amount of memory (2.15 GB) for just the kernel.

2.3 Code Optimization

Several optimizations were developed to improve the code performance, such as reducing the memory consumption, restructuring parallel loops, deploying reduction clauses, and increasing parallelism over independent elements.

Algorithm 1 A Naive Implementation

```
if first step then
    acc enter data copyin                  ▷ Start Unstructured Data Region
acc parallel loop default(present)
for each gridcell do
    GET BOUNDS OF GRIDCELL                 ▷ Bounds holds subgrid info
    LEACHING
    NOLEACHING PHASE 1
    NOLEACHING PHASE 2
```

Fig. 3. All subroutines were ported using acc routine directive and parallelized across the existing gridcell loop.

Reducing the Memory Allocation of Local Variables: Each NVIDIA V100 has 16 GB of memory to contain all the data and ELM kernels. Memory allocation operations on GPU are more expensive than those on the host, so we need to reduce the memory consumption of each kernel. In this study, we deployed many methods to reduce the size of these local variables, such as converting arrays into scalars and compressing the sparse arrays into dense arrays. For example: In SoilLittVertTransp, 9 local arrays would be allocated with a size of 238 doubles for each gridcell (that is 17136 (9*238*8) bytes in total). After refactorization, these arrays were replaced with a couple of 64-bit scalar and 4 dense arrays (less than 550 bytes in total). The memory reduction decreases the total execution time of the ecosystem dynamics model to 150 ms from 7600 ms. This memory saving is also necessary since we want to put up to 6000 gridcells on a single GPU.

Restructure Parallel Loops: The routine directive provides the ability to test functions on the GPU quickly, but for subroutines with internal loop structures and nested function calls, performance degradation is expected. In the ecosystem dynamics model, routines that loop over the same subgrid element (Column or Patch) tend to be clustered together. For example, Fig. 4 shows a group of functions labeled as SetValues, where the ELM subroutines compute over the Patch and Column arrays, respectively (Algorithm 2). A reliable optimization strategy is to refactor these routines to remove their internal loops and change the external gridcell based loops to the relevant subgrid element, which allows them to be grouped under different parallel loop constructs (Algorithm 3). In this case, each subroutine is actually completely independent of the others so that the SetValues group of functions can utilize asynchronous kernel launch. The loop re-factorization decreases the execution time from 30 ms (CPU code) to 16 ms (GPU code) and proves that the routine directive can provide a significant speedup over the CPU implementation.

Algorithm 2 Parallelize Gridcell	**Algorithm 3** Parallelize Patch

acc parallel loop
for each Gridcell **do**
    VEGCFSETVALUES(array)
    VEGNFSETVALUES(array)
    VEGPFSETVALUES(array)
	acc parallel loop async
	for each Patch **do**
	    VEGCFSETVALUES(Patch_index)
	acc parallel loop async
	for each Patch **do**
	    VEGNFSETVALUES(Patch_index)
	acc parallel loop async
	for each Patch **do**
	    VEGPFSETVALUES(Patch_index)

Fig. 4. The SetValues routines were refactored to remove the internal loops. In the original code (Algorithm 2), the SetValues routines take global Patch arrays as arguments and calculations are performed on the elements of these arrays. After the refactorization, the outer loop is over Patches with only the Patch index passed as an argument, and the multiple kernels can be launched asynchronously. (Algorithm 3).

Accelerate Internal Loops: For subroutines that must have their internal loops, we forego the routine directive and deploy different parallel techniques specific to each loop. One good example is the fire module that includes two large subroutines (FireFluxes and FireArea), each also contains many internal loops. When deploying the OpenACC routine directive within the fire module, a single step execution on GPU takes around 50 ms. After the internal loop acceleration, the execution time becomes 4.84 ms (Table 1).

Algorithm 4 CPU Code	**Algorithm 5** GPU Code

for each active Patch **do**
    for each SoilLayer **do**
        Get: Column from Patch
        Sum: Patch vars to Column
	acc parallel loop collapse gang worker
	for each SoilLayer **do**
	    for each active Column **do**
	        Init: sums
	        acc vector reduction(+:sums)
	        for each Patch in Column **do**
	            if Patch is active **then**
	                Reduce: Patch vars to Column

Fig. 5. CPU Code(left) only uses two loops, the GPU code(right) required an additional loop to prevent race conditions during reduction ,but this requires looping over all Patches rather than only the active Patches. Despite the increase in loop size, the achieved speedup is over 2x for the entire FireMod section (Table 1)

The ecosystem dynamics model uses hierarchical data structures, and the lower-level variables (i.e. Patch variables) are aggregated into the corresponding higher-level variables (i.e. Column variables). Figure 5 shows an example of aggregating variables from Patch to Column in the FireFluxes subroutine. The code is optimized by using gangs and workers to parallelize the *Soil Layer* and *Column* (collapsed) loops with an inner vector loop performing the reduction. The reduction operation is very efficient in our case because there are a maximum of 33 patches in each column and each warp of the NVIDIA V100 has 32 threads. With the reduction, we can finish the operations in 0.3 ms. After the optimizations, the execution time of the entire GPU-based fire module is reduced to 4 ms, which includes 20 kernels in total.

Parallelize over Nutrients or Output Variables: To further improve the performance of the ecosystem dynamics model, we investigate the task parallelism of algorithms that don't distinguish among nutrients, such as carbon, nitrogen, and phosphorus (CNP), aside from the input and output variables (i.e. sources and sinks of CNP). A good example are the transport calculations inside the SoilLittVertTransp subroutine (Fig. 6), where we take advantage of asynchronous compute to pipeline the creation of local arrays for tridiagonal coefficients. A new array of derived types is created with each element pointing to a set of 3D arrays corresponding to a CNP nutrient input or output variable. This data structure is initialized and moved to the GPU only at the start of a run via deepcopy [4]. We can then either collapse the nutrient loop into the parallel loop construct or use the nutrient loop to asynchronously launch kernels to improve the code performance.

Algorithm 6 Init Pointer List

```
Loop over ntypes (C, N, P)
for each ntype do
    list[ntype]%conc → ntype%conc
    list[ntype]%src → ntype%src
    list[ntype]%trcr → ntype%trcr
acc copyin(list)
```

Algorithm 7 Parallelize Over ntypes

```
for each ntype do
    acc parallel loop async(ntype)
    for each decomp source do
        for each soil levels do
            for each soil column do
                Set outputs using list
                Solve TRANSPORT
```

Fig. 6. (Left) An array of a derived type is created whose fields point to an output field of a CNP global variable and copied to device at start of run. (Right) The CPU nutrient type loop launches the transport kernels asynchronously so that independence of CNP provides a fourth level of parallelism to occupy the GPU.

The left panel of Fig. 6 illustrates the creation of the new derived type, and the right panel shows the new asynchronous loop structure of the transport algorithm inside the SoilLittVertTransp module. The vertical transport among soil levels in each soil column is calculated using a tridiagonal solver. The solving algorithm does not have enough dimensions (vertical soil layers) to fully utilize the GPU resources, so we take advantage of the independence between CNP inputs and outputs to occupy the device. The asynchronous launching of these kernels allows the whole SoilLittVertTransp module to finish in 5 ms, compared to over 40 ms using the routine directive after removing excess memory allocations. This method of using arrays of pointers is also used to reorganize the code structure of other models within ELM, so we can use OpenACC directive to explore the parallelism of these models efficiently. A good example is the ELM output model that generates a history buffer containing over 500 aggregated and averaged variables to be discussed in the future work on the full ELM simulation.

3 Results

3.1 Overall Performance Improvement

We gather the subroutines in the ecosystem dynamics model into many groups and collect the individual execution time sequentially (Table 1). The GPU-based ecosystem dynamics model achieves a 3.0 times speedup over the original code on the CPU when the Summit node is fully loaded. The timing data of the majority (10 out of 15) function groups show good speedups (ranging from 2.3 to 8.3 times). Especially, gap mortality (GapMortality) contains many global-variable operations (similar to SetValues) that have been refactored into parallel do loop and are launched asynchronously on GPU. Respiration contains two simple loops that have been refactored into parallel do loops with a reduction clause. The execution times of the other 5 groups are relatively short (less than 2 ms). The slow down of these groups is mainly because the overhead associated with GPU kernels overshadowed the small computational part of these subroutines. After optimization, the entire ecosystem dynamics model (GPU kernels with all the nested subroutines) requires around 300 MB of memory.

3.2 Profiling Details

To determine the limitations of OpenACC's routine directive and better understand performance improvements, we used NVIDIA's Nsight Compute[3] and Nsight Systems[4] to collect GPU metrics and traces for SoilLittVertTransp, FireMod, and SetValues subroutines at various stages of optimization. Table 2 shows the metrics of the kernels that illustrate different shortcomings of prior versions of the GPU kernels. Several metrics, such as wallclock time, kernel launch overhead, and total instructions issued, are measured to frame performance discussion in this work. The overheads listed in Table 2 are the percentage of wallclock time spent on launching the kernel, allocating memory, and creating device streams (if applicable). Nsight Systems drastically increases initial kernel launch times but agrees with Nsight Compute in reporting the kernel compute time, so to consistently calculate overhead, we subtract the kernel compute time from the total wallclock time of non-profiling runs and take the percentage.

[3] AnNVIDIAinteractivekernelprofiler, https://developer.nvidia.com/nsight-compute.
[4] An NVIDIA performance analysis tool, https://developer.nvidia.com/nsight-systems.

Table 1. Comparison of average execution time between GPU and CPU versions of the ecosystem dynamics model (single timestep) on a fully loaded single Summit node. Each node has the the same size grid divided between 6 GPUs or 42 CPUs

Function group	GPU(ms)	CPU(ms)	Speedup
SetValues	15.49	46.16	2.98
NDeposition/Fixation	0.09	0.02	0.24
Respiration/PDeposition	0.33	2.77	8.30
Decomp. rate	1.31	3.08	2.35
Vertical decomp.	2.74	7.77	2.84
Alloc Phase 1	1.51	0.93	0.62
SoilLittDecomp 1	6.12	23.30	3.81
SoilLittDecomp 2	1.18	1.89	1.60
Phenology	2.26	0.58	0.26
Growth and root respiration	1.03	0.43	0.42
StateUpdate 1	2.79	12.21	4.39
SoilLittVertTransp	5.94	17.35	2.92
GapMortality	2.56	17.54	6.85
StateUpdate 2	1.88	4.95	2.64
FireMod	4.84	11.19	2.31
Total	50.05	150.2	3.00

SoilLittVertTransp: The initial OpenACC implementation of SoilLittVertTransp had the slowest performance of the ecosystem dynamics model and even the full E3SM Land Model runs. The reliance of the CPU version on dynamically allocated arrays can be distilled to the over-inflated number of instructions reported on the GPU, which is nearly three orders of magnitude larger than that with the arrays refactored out (Table 2). The memory workload and warp state metrics further support that this memory is the bottleneck by showing very high L1 and L2 hit rates and related stalls (not listed in the paper).

After the initial refactoring (Mem Opt. in Table 2), the performance is still 2.6 times slower than a single CPU core given a similar workload. While the kernel only uses half the threads in a warp (on average), the profilers report the kernel compute time as around 10 ms. The overheads associated with launching the kernel contribute the most to the total time, and the subroutine is simply too big and complicated to be ported using OpenACC routine directives. The metrics for the final OpenACC implementation illustrate much more efficient utilization of the GPU's warps and pipelines due to smaller kernel size and techniques described in Sect. 2. Note that the massive over 35 billion instructions in the original code is caused by the excessive memory allocations.

FireFluxes: The initial naive implementation of OpenACC worked better for the FireFluxes subroutine, which consists exclusively of computations on global data types. However, the overheads are significant because many nested loop structures and the hundreds of global variables have to be passed as arguments by the kernel launcher. The final optimizations resulted in a great increase in the required instructions due to the changes shown in Fig. 5. With the asynchronous kernel launches and high-efficient utilization of the GPUs resources, the model performance on the GPU delivers a 2.3 speedup over the performance on the CPU.

Table 2. High-level metrics for three kernels showing differences between optimization methods. The routine directive can result in very high overheads, and poor memory allocations result in large excess in instructions issued. For FireFluxes, the excess instructions are due to extra loops but is alleviated by better saturation of GPU.

Kernel	Time(ms)	Overhead(%)	Instruction (millions)
SoilLittVertTransp			
Orig	10,000	49.8	34,500
Mem Opt	44.0	70.0	48.7
Final Opt	6.95	11.8	40.9
FireFluxes			
Orig.	45.0	92	9.83
Opt.	3.67	27.9	197
SetValues			
Orig.	29.4	70.2	15.3
Opt.	15.4	11.6	14.5

SetValues: The naive OpenACC implementation of the SetValues functional group had a speedup of around 1.5 times over the CPU code, but the total time taken was significant relative to the other modules. The SetValues subroutines involved initializing hundreds of arrays of three derived types at the Patch and Column level, which can all be done independently. While the OpenACC routine directive allows for multiple levels of parallelism, compiling as a sequential routine is the most reliable and free of compilation errors for our use case. The profiling data identified that kernel launch overhead was the bottleneck, and so the routines were simplified as described in Sect. 2.3. The overhead is greatly reduced and mostly due to CUDA API calls to allocate memory on the host and device and create streams with a final speedup of 3.0.

4 Conclusion and Future Work

The study reports design and optimization strategies for developing an ELM ecosystem dynamics model using compiler directives (OpenACC) on NVIDIA

GPUs. We have restructured the code to reduce the memory footprint and to increase the parallelism, so that the code can be programmed with OpenACC directives efficiently. The routine directive and deepcopy capabilities of OpenACC provided a robust method for accelerating very complex ELM modules, with certain functions receiving immediate speedup.

After code analyses and refactoring, the parallel GPU implementation (with a small memory footprint of 300 MB) achieved a 3.0-times speedup over the original CPU code on a fully loaded Summit computing node. Computationally, ELM doesn't have a single submodel that dominates run time but around a dozen complex components that contribute more or less equally. Accelerating the full code base requires a flexible approach that can handle different algorithms, and this work demonstrates methods for reworking algorithms and data structures in tandem with compiler directives to tackle exascale climate simulations. Although this study is not intended to improve the ELM performance on the CPUs, this study revealed several limitations of the original CPU code that was implemented without prudent considerations of memory usage and allocation. Some techniques mentioned in the study (such as reducing the memory footprint) also have benefits for the CPU-based ELM code. Future work will focus on the parallelization of other models within ELM and further integrated performance tuning. For the ultra-high resolution ELM simulation over North America, if we assign 36000 gridcells to each node, the 24 million gridcells of North America require 680 nodes, which is around 15% of the total capability of the 4608-node Summit. With an estimated 3.0 times overall speedup, a 100-year simulation over North America takes around 2 days.

References

1. Bertagna, L., et al.: A performance-portable nonhydrostatic atmospheric dycore for the energy exascale earth system model running at cloud-resolving resolutions. In: SC20: International Conference for High Performance Computing, Networking, Storage and Analysis, pp. 1–14. IEEE (2020)
2. Bourzac, K.: Supercomputing poised for a massive speed boost. Nature **551**(7680), 554–557 (2017)
3. Burrows, S., et al.: The doe e3sm v1. 1 biogeochemistry configuration: Description and simulated ecosystem-climate responses to historical changes in forcing. J. Adv. Model. Earth Syst. **12**(9), e2019MS001766 (2020)
4. Ghane, M., Chandrasekaran, S., Cheung, M.S.: Assessing performance implications of deep copy operations via microbenchmarking. CoRR abs/1906.01128 (2019). http://arxiv.org/abs/1906.01128
5. Golaz, J.C., et al.: The doe e3sm coupled model version 1: overview and evaluation at standard resolution. J. Adv. Model. Earth Syst. **11**(7), 2089–2129 (2019)
6. Kim, J.Y., Kang, J.S., Joh, M.: Gpu acceleration of mpas microphysics wsm6 using openacc directives: performance and verification. Comput. Geosci. **146**, 104627 (2021)
7. Thornton, P.E., Shrestha, R., Thornton, M., Kao, S.C., Wei, Y., Wilson, B.E.: Gridded daily weather data for North America with comprehensive uncertainty quantification. Sci. Data **8**(1), 1–17, 104627 (2021)

8. Wang, D., et al.: Scientific functional testing platform for environmental models: an application to community land model. In: International Workshop on Software Engineering for High Performance Computing in Science, 37th International Conference on Software Engineering (2015)
9. Wang, D.: A functional test platform for the community land model. Environ. Model. Softw. **55**, 25–31 (2014)
10. Zhang, S.: Optimizing high-resolution community earth system model on a heterogeneous many-core supercomputing platform. Geosci. Model Dev. **13**(10), 4809–4829 (2020)
11. Zheng, W., Wang, D., Song, F.: XScan: an integrated tool for understanding open source community-based scientific code. In: Rodrigues, J.M.F., et al. (eds.) ICCS 2019. LNCS, vol. 11536, pp. 226–237. Springer, Cham (2019). https://doi.org/10.1007/978-3-030-22734-0_17

Artificial Intelligence and High-Performance Computing for Advanced Simulations

Optimization-Free Inverse Design of High-Dimensional Nanoparticle Electrocatalysts Using Multi-target Machine Learning

Sichao Li, Jonathan Y. C. Ting, and Amanda S. Barnard(✉)

Australian National University, Acton, ACT 2601, Australia
amanda.s.barnard@anu.edu.au

Abstract. Inverse design that directly predicts multiple structural characteristics of nanomaterials based on a set of desirable properties is essential for translating computational predictions into laboratory experiments, and eventually into products. This is challenging due to the high-dimensionality of nanomaterials data which causes an imbalance in the mapping problem, where too few properties are available to predict too many features. In this paper we use multi-target machine learning to directly map the structural features and property labels, without the need for exhaustive data sets or external optimization, and explore the impact of more aggressive feature selection to manage the mapping function. We find that systematically reducing the dimensionality of the feature set improves the accuracy and generalizability of inverse models when interpretable importance profiles from the corresponding forward predictions are used to prioritize inclusion. This allows for a balance between accuracy and efficiency to be established on a case-by-case basis, but raises new questions about the role of domain knowledge and pragmatic preferences in feature prioritization strategies.

Keywords: Inverse design · Machine learning · Catalysis

1 Introduction

Chemical reactions are essential to maintaining sustainable production of fuels, medicines and materials, and the majority of industrially important reactions are facilitated by catalysts, which are additives that speed up chemical reactions without being consumed in the process [1]. Catalysts enable a shorter reaction time without affecting the yield, and considerable research has been directed toward finding the right catalyst for a given reaction and industrial objective. Applications include energy generation [2–6] using metallic nanoparticles (typically 1–100 nm), which include single crystals, polycrystals and clusters in a variety of sizes and shapes [7,8]. Metallic nanoparticles have high surface-area-to-volume ratio, enabling high catalytic activity to be achieved since more active sites are available for reactant adsorption compared to larger catalysts, and offer

D. Groen et al. (Eds.): ICCS 2022, LNCS 13351, pp. 307–318, 2022.
https://doi.org/10.1007/978-3-031-08754-7_39

a large number of engineering degrees of freedom [9]. This presents both an opportunity and a challenge.

Nanoinformatics [10] has recently emerged as an complementary approach to typical trial-and-error nanomaterials design, ideal for situations where the design space is large; employing both computational simulations and data-driven methods such as machine learning in conjunction with standard experimental and computational tools [11] to infer relationships between the structural characteristics (features, that can be control) and the properties (target labels, that are required) [12,13]. Past studies have focused on forward prediction [14–17], but an alternative approach is to use inverse predictions, that provide a 'recipe' for experimentalists to follow [18–20]. This is difficult however, due to the highly imbalanced mapping functions. We have many more structural features we need to predict than properties we have to base them on.

Recently an entirely machine learning approach to inverse design has been reported which directly maps multiple structural features to multiple property labels, simultaneously, without the need for additional global searching or optimization [21]. This method has considerable advantages over conventional inverse design that requires exhaustive data sets, external optimization algorithms, and can potentially predict multiple candidates with no means for discrimination. In this method the mapping imbalance problem is alleviated by using a matching forward model for features selection, but has so far only been demonstrated for low-dimensional cases or where the mapping can be reduced to 1-to-1, which is unlikely to be true for complex materials such as electrocatalysts. In this study we explore the use of the optimization-free inverse design method for a high-dimensional set of platinum nanoparticle catalysts and show that, while restricting the models to features identified using importance profiles and recursive feature elimination, a systematic improvement in model performance can be achieved by balancing the mapping function, even if it is imbalanced in the corresponding forward model.

2 Methods

2.1 Data Set

The raw data used in this work are generated through molecular dynamics simulations of the sintering and coarsening of platinum nanoparticles under different temperatures and atomistic deposition rates, which exist in the form of three dimensional spatial coordinates [22], and can be obtained from Reference [23]. The entire data set contains 1300 instances, over a range of temperatures, sizes (54 to 15837 atoms), growth rates and shapes. A complete list of the manually extracted features and labels is provided with the meta data in the online repository [23]. The data set is labelled by the proportion of surface atoms with particular ranges of coordination number, or surface coordination numbers as an indicator of catalytic activity of face-centred cubic (fcc) metallic nanoparticles for carbon monoxide oxidation reactions, oxygen reduction reactions, and hydrogen oxidation and evolution reactions [24–26].

2.2 Multi-target Random Forest Regression

Previous works have demonstrated that random forest (RF) methods can be used to fit high-dimensional target labels simultaneously with high accuracy [21]. Random forest is an ensemble technique based on decision trees with bootstrap aggregation (bagging), that has been shown to perform well for predicting the properties of metallic nanoparticle compared with other regressors [27]. A decision tree predicts the value of a label following decision rules inferred from the features and bagging randomly sampled from the training set with replacement, reducing the variance.

For each tree in the forest a bootstrap sample, i is selected from S, giving $S^{(i)}$. A decision-tree is learnt such that instead of examining all possible feature-splits at each node of the tree, a randomly subset of the features $f \subseteq F$ is used, where F is a set containing all features. Each node splits on the best feature in f, which is smaller than F, making it more efficient, and allows for the accumulation of importance profiles based on how many times a decision is split on a given feature. The pseudocode is illustrated in Algorithm 1.

Algorithm 1. Random forest

```
Require: A training set S := (x1, y1), . . . ,(xn, yn), features F, and number of trees
  in forest B.
  function RandomForest(S, F)
      H ← 0
      for i ∈ 1, ..., B  do
          S^(i) ← A bootstrap sample from S
          h_i ← RandomizedTreeLearn(S^(i), F)
          H ← H ∪ {h_i}
      end for
      return H
  end function
  function RandomizedTreeLearn(S, F)
      for Each node: do
          f ← very small subset of F
          Split on best feature in f
      end for
      return The learned tree
  end function
```

Constructing a large number of decision trees with random feature selection grows into a random forest, in which the decision of the individual tree is counted to output the mode of the classes for classification or average prediction for regression [28,29]. Features are ranked during training increasing the diversity and avoiding over-fitting. Feature importance is calculated as the decrease in node impurity weighted by the probability of reaching that node. The node probability can be calculated by the number of samples that reached the node, divided by the total number of samples. The feature selection criterion for internal nodes is the Gini impurity or information gain for classification and variance reduction for regression. This machine learning method can intrinsically handle multi-task problems since the leaf nodes can refer to any collection of relevant labels. To extend the traditional single-target ensemble predictor to solve multi-target RF learning problems, users can simply substitute the typical univariate trees for multivariate trees, where leaf nodes refer to multiple classes or target labels [30].

2.3 Workflow

The workflow for multi-target machine learning-based forward and inverse design requires data preprocessing (outlier removal and feature engineering) and splitting the data for model training and validation. We used Tukey's method [31] to detect outliers based on the quartiles of the data, and the results were optimal when the threshold was set to 4, leaving 1114 instances. Constant features and strongly correlated features (with $> 95\%$ correlation) were omitted to avoid introducing bias during learning and each label was stratified to reduce the impact of imbalanced distributions, resulting in 37 features retained in the feature space. All data are standardized and normalized (both features and labels). Essential hyper-parameter tuning of all models was undertaken here using a random grid search with 5-fold cross validation and evaluated using the mean squared error (MSE) and the mean absolute error (MAE). 10-fold cross validation was used during training, and all models used stratified 25/75 test/train splits with the same random seed (both forward and inverse, during tuning and training).

When applying this approach to inverse design, a preliminary multi-target forward model is often necessary to reduce the number of features and focus the inverse model on attributes that contribute to a strong relationship. Robust feature selection in a forward ML workflow is typically data-driven, such that final feature set used to describe the raw data is selected by computational algorithms [32]. This reduces model complexity, improves performance, and indicates which features are likely to be most influential. When using RF regression, this involves using the feature importance profile and recursive feature elimination (RFE) to extract the subset of important features that are sufficient to simultaneously predict the target labels without significant loss of performance. A multi-target inverse model can be re-tuned and then re-trained to simultaneously predict the reduced (important) set of structural features (referred to as "meta-labels") using the set of labels (referred to as "meta-features"), as reported elsewhere [21].

The entire workflow is provided in Fig. 1.

3 Discussion of Results

After data cleaning 37 features were retained to predict the 4 target labels. This was done both separately (single target RF regression) and simultaneously

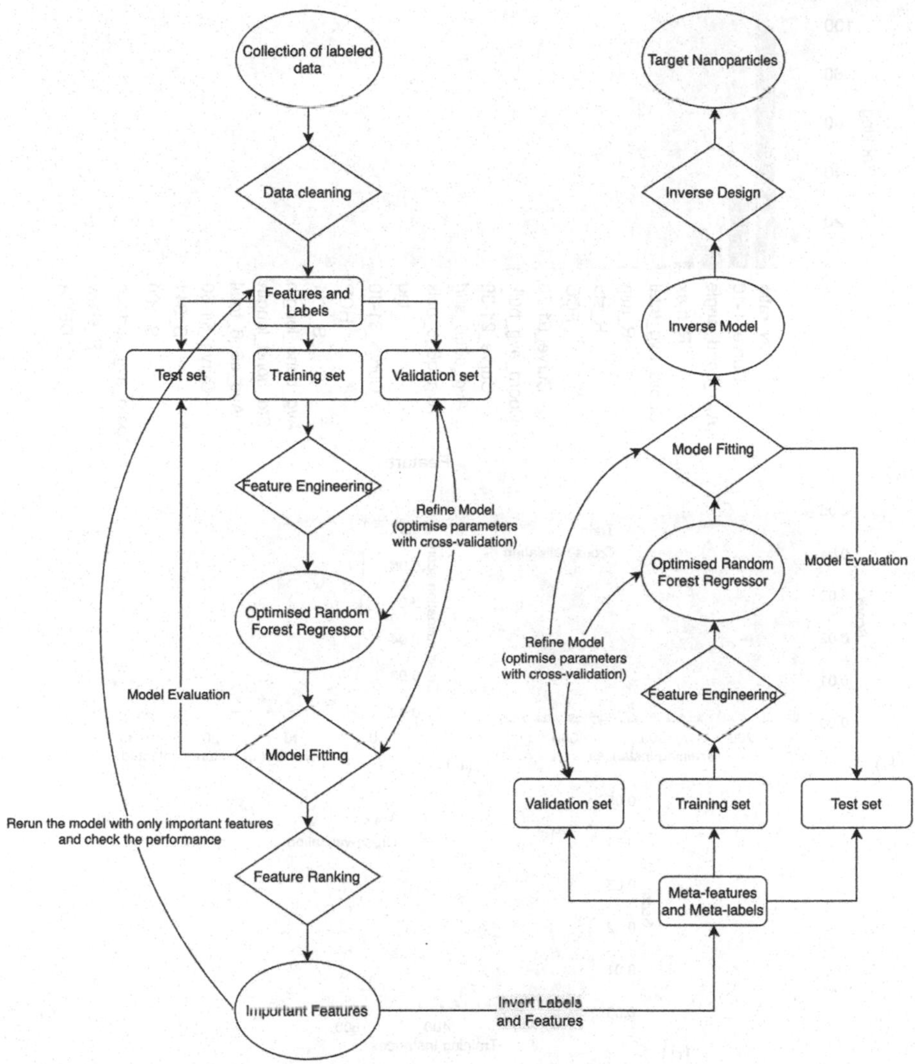

Fig. 1. Workflow for the inverse design methodology, based on multi-target regression and feature selection using a forward model to reduce the mapping function. Reproduced with permission from Reference [21].

(multi-target RF regression). The results for each of the four single-target models (which provide the baselines) are compared with the multi-target prediction in Table 1 and presented in the Fig. 2. We can see from Fig. 2 that there is no bias error (under-fitting) and less than 0.5% variance error (over-fitting) when we use all of the 37 features retained after cleaning, indicating the forward model can achieve high accuracy and generalizability. The relative importance (magnitude) rapidly decays after the top three features in the feature importance profile. The

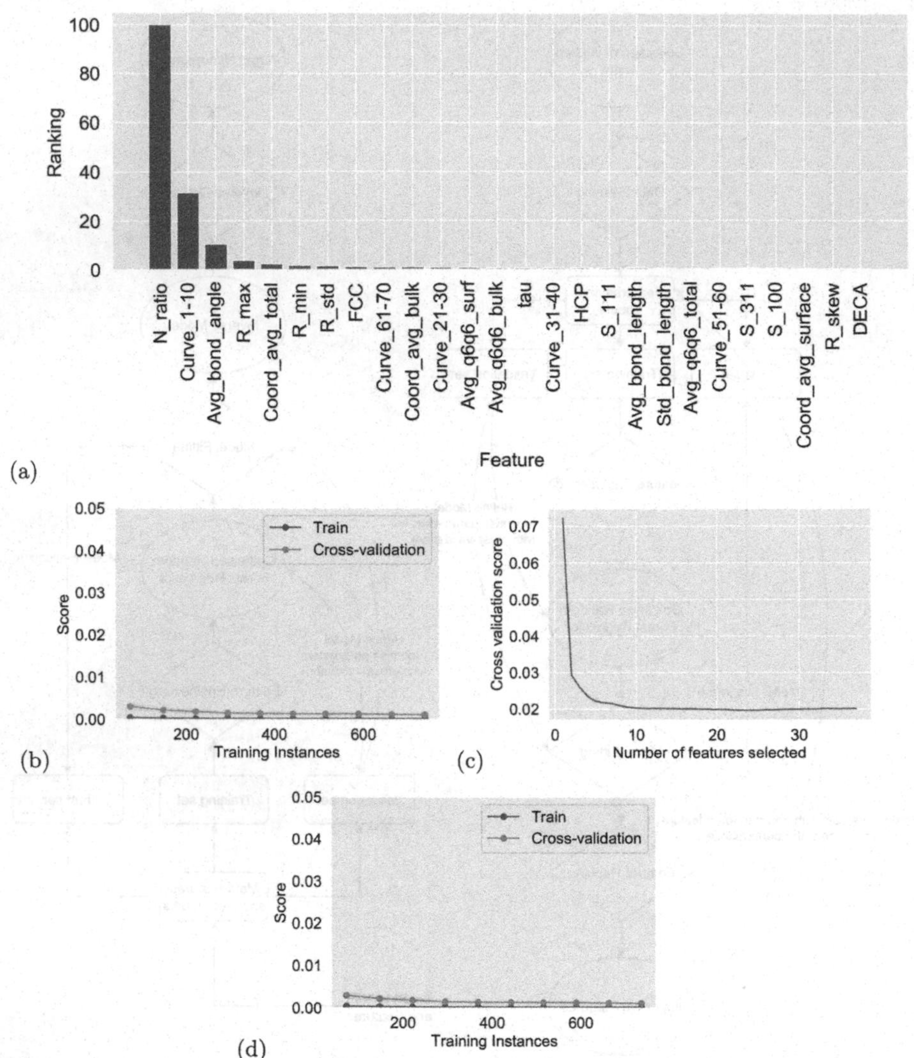

Fig. 2. Results for forward multi-target regression simultaneously predicting all four target properties of the Pt catalysts trained on 37 features retained after data cleaning, including (a) the feature importance profile rankings, (b) the learning curve trained on the 37 retained features, (c) the recursive feature elimination selecting the top 24 features, and (d) the learning curve trained on the 24 most important features, showing the accuracy and generalizability. Scores are the mean squared error.

forward features selection using RFE indicates that the model may be reduced to the top 24 features, which exhibits an increase in performance due to the significant reduction in model complexity and the removal of nuisance variables, as measured by the MAE and MSE in Table 1. The training process for the final

Table 1. Forward predictions using the multi-target regression model of Pt catalyst nanoparticles, with all of the features retained after data cleaning, the top 24 important features identified using recursive feature elimination to predict single property individually, or all 4 properties simultaneously; and inverse predictions using the 24 most important features or the 16, 8 or 4 top features as ranked in the forward feature importance profile. The results are evaluated using the mean absolute error (MAE) and mean squared error (MSE).

Prediction	Feature set (Number)	Target (Number)	MAE	MSE
Forward	Retained (37)	`Surface_Defects` (1)	0.0205	0.0016
Forward	Retained (37)	`Surface_Microstructures` (1)	0.0155	0.0007
Forward	Retained (37)	`Surface_Facets` (1)	0.0181	0.0007
Forward	Retained (37)	`Formation_E` (1)	0.0129	0.0003
Forward	Retained (37)	All Properties (4)	0.0211	0.0012
Forward	Important (24)	All Properties (4)	0.0208	0.0012
Inverse	All Properties (4)	Important (24)	0.0540	0.0085
Inverse	All Properties (4)	Top (16)	0.0499	0.0071
Inverse	All Properties (4)	Top (8)	0.0410	0.0048
Inverse	All Properties (4)	Top (4)	0.0284	0.0022

multi-target forward model is visualised in Fig. 2(d). Overall the multi-target model exhibited performance comparable to the single-target models, and all models converge well with almost no loss of accuracy or generalizability.

To develop the inverse model the RF regressor was re-tuned and retrained on the same data set, but the properties become "meta-features" and the structural characteristics become "meta-labels". The models are evaluated using the same strategy, and the results are presented in Table 1. All training processes are visualized using learning curves in Fig. 3(a), and the comparison of the predicted values of an instance from the testing set with its corresponding ground truth in Fig. 3(b), for the set containing the top 24 meta-labels. The MAE for the inverse model is slightly more than double the MAE for the forward model, due to the significant imbalance between the number of meta-features and meta-labels, but is still less than 6%. As we reduce the feature set below the optimal list identified using RFE in the forward model, a systematic improvement in inverse model performance can be achieved by balancing the mapping function, even if it is imbalanced in the corresponding forward model. This is shown in the decrease in the MAE and MSE in Table 1 and in Fig. 3(c,d) for 16 meta-labels, Fig. 3(e,f) for 8 meta-labels, and Fig. 3(g,h) for 4 meta-labels (representing a balanced function).

Conventional inverse design approaches are based on screening of a set of (forward) structure/property relationships [33–38] to determine the structures with the right properties, which eventually casts it as an optimization problem.

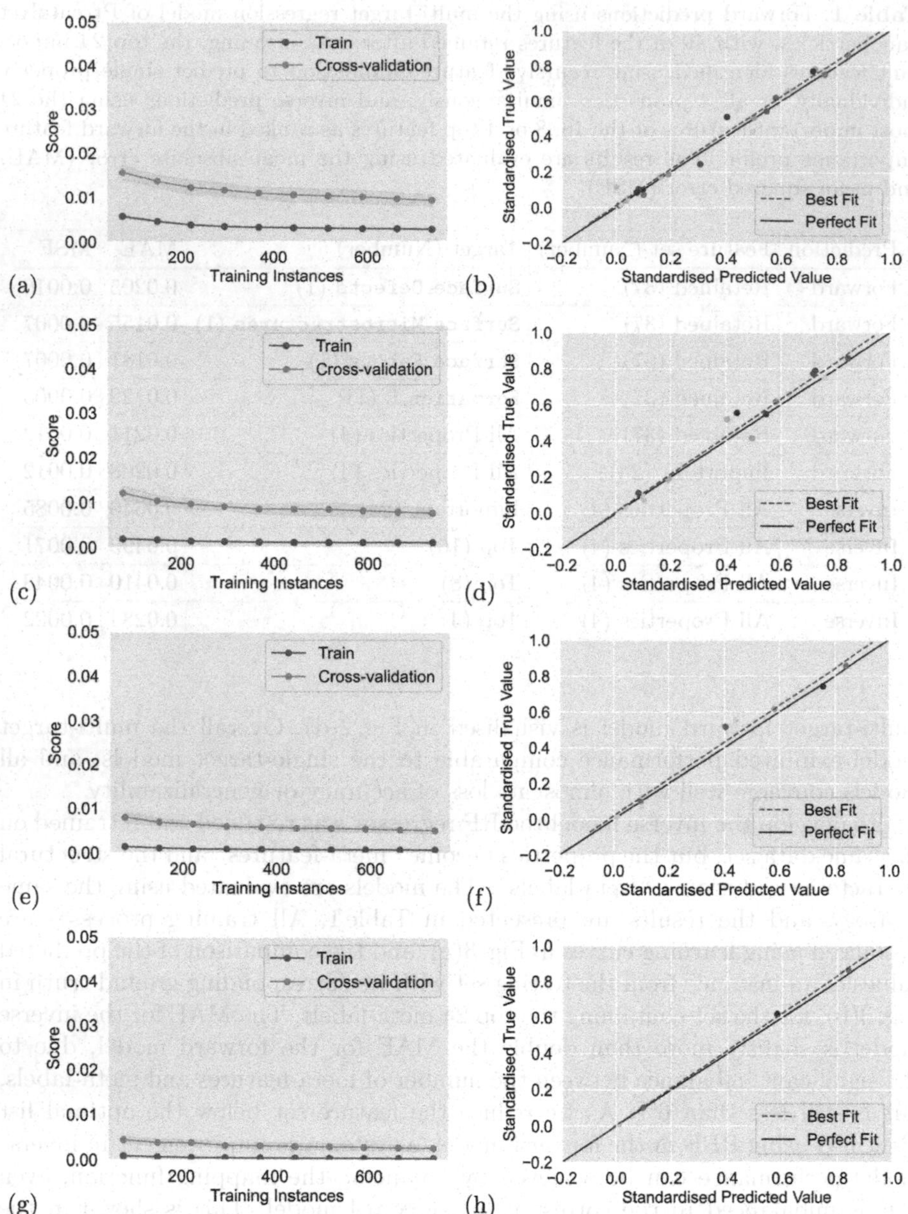

Fig. 3. Results for inverse multi-target regression simultaneously predicting (a,b) 24 meta-labels (structural characteristics), (c,d) 16 meta-labels, (e,f) 8 meta-labels, and (g,h) 4 meta-labels, for the Pt catalysts trained on all 4 target meta-features (nanoparticle stability and catalytic property indicators). The learning curves are shown to the left (a,c,f,g) and the 45 °C plot comparing the prediction for an instance in the testing set with the ground truth to the right (b,d,f,h). Scores are the mean squared error, and the colours in (b,d,f,h) are the features in the profile shown in Fig. 2(a).

Several approaches have been proposed [39,40]. Tominaga *et al.* designed a procedure that used genetic algorithms and Yang *et al.* developed IM^{2}ODE (Inverse design of Materials by Multi-Objective Differential Evolution) based on multi-objective differential evolution, both capable of global searching of large databases. Zunger *et al.* also identified materials with specific functionalities using an inverse design framework based on a global searching task combined with high-throughput density functional theory [41–43]. These approaches suffer from introducing more overhead (beyond the model training), more uncertainty (as optimization is imperfect), and can only identify candidate materials that are already in the set. They also become computationally intensive and impractical in nanomaterials design, where the design space is larger, and the probability of predicting implausible configurations increases [44]. By eliminating the need for the optimization step the workflow is accelerated, with the added advance of increasing specificity and relaxing the need for big data.

Given a profile of desired properties for a catalyst, the well-trained inverse models can output the most appropriate values of each structural feature for a given set of industrial requirements. Property indicators for the oxygen reduction reaction could be suppressed in favour of indicators for hydrogen evolution reactions, for example, or a particle could be designed that can be simultaneously be used for both reactions. A combination of structural feature could be predicted for the specified properties without requiring additional optimization or global searching. A question that arises, however, is how flexible the feature (meta-label) selection process can be? Some of the top 24, 16, 8 or even top 4 features cannot be directly controlled in the lab. Including domain knowledge and selecting features based on practical considerations could accelerate translations of predictions in to the lab, and eventually the factory, but would represent a departure from an entirely data-driven, evidence based approach. The impact of combining, or prioritising, user preference over data-driven feature selection is the topic of future work, to explore how tolerant inverse models (such as these) will be to human intervention and whether inverse models can be tailored to experimental or industrial needs.

4 Conclusions

In the present study we have demonstrated the use of forward and inverse design models to predict the structure/property and property/structure relationship of Pt nanoparticle catalysts, respectively. These models included performance indicators relevant to the nanoparticle stability, carbon monoxide oxidation reactions, oxygen reduction reactions, and hydrogen oxidation and evolution reactions. We have used interpretable multi-target random forest regression to predict multiple property indicators and structural characteristics simultaneously, which better accounts for the fact that material properties can be correlated and that certain structural features can drive more than one type of reaction. The models generally show low errors, less than 2%, with no bias error and minimal variance error. The absolute prediction error for the inverse model is more than twice the error in the corresponding forward model, likely due (at least in part) to the significant imbalance in the mapping function; predicting 24 meta-labels with only 4 meta-features.

To improve the performance of the inverse model, we explored the impact of artificially reducing the set of meta-labels to better balance the mapping function, while retaining the most important features in the forward model at each stage. We find that systematically reducing the number of meta-labels lowers the model error, and when the number of meta-features and meta-labels are balanced a similar performance to the forward model can be achieved. Reducing the meta-label imbalance also simplifies the inverse model and can offer some advantages when using it in practice. This raises questions as to whether human intervention in the feature selection process to accommodate practical consideration, will be detrimental to performance once a fully data-driven approach is abandoned. This is an interesting topic of ongoing work and will be reported elsewhere.

Acknowledgement. Computational resources for this project were supplied by the National Computing Infrastructure national facility under grant p00.

References

1. Laidler, K.J.: A glossary of terms used in chemical kinetics, including reaction dynamics (IUPAC Recommendations 1996). Pure Appl. Chem. **68**(1), 149–192 (1996)
2. Shokrlu, Y.H., Babadagli, T.: Viscosity reduction of heavy oil/bitumen using micro- and nano-metal particles during aqueous and non-aqueous thermal applications. J. Petrol. Sci. Eng. **119**, 210–220 (2014)
3. González-Gil, R., Herrera, C., Larrubia, M.A., Mariño, F., Laborde, M., Alemany, L.J.: Hydrogen production by ethanol steam reforming over multimetallic RhCeNi/Al2O3 structured catalyst: pilot-scale study. Int. J. Hydrogen Energy **41**(38), 16786–16796 (2016)
4. Guo, W., Vlachos, D.G.: Patched bimetallic surfaces are active catalysts for ammonia decomposition. Nature Commun. **6**(1), 8619 (2015)
5. Wu, Z.P., et al.: Alloying-realloying enabled high durability for Pt-Pd-3d-transition metal nanoparticle fuel cell catalysts. Nature Commun. **12**(1), 1–14 (2021)
6. Sheng, W., et al.: Non-precious metal electrocatalysts with high activity for hydrogen oxidation reaction in alkaline electrolytes. Energy Environ. Sci. **7**(5), 1719–1724 (2014)
7. Wu, Z., Yang, S., Wu, W.: Shape control of inorganic nanoparticles from solution. Nanoscale **8**(3), 1237–1259 (2016)
8. Baig, N., Kammakakam, I., Falath, W.: Nanomaterials: a review of synthesis methods, properties, recent progress, and challenges. Mater. Adv. **2**(6), 1821–1871 (2021)
9. Rodrigues, T.S., da Silva, A.G.M., Camargo, P.H.C.: Nanocatalysis by noble metal nanoparticles: controlled synthesis for the optimization and understanding of activities. J. Mater. Chem. A **7**(11), 5857–5874 (2019)
10. Barnard, A.S., Motevalli, B., Parker, A.J., Fischer, J.M., Feigl, C.A., Opletal, G.: Nanoinformatics, and the big challenges for the science of small things. Nanoscale **11**(41), 19190–19201 (2019)

11. Rajan, K.: Materials informatics. Mater. Today **8**(10), 38–45 (2005)
12. Kotsiantis, S.B., Zaharakis, I., Pintelas, P.: Supervised machine learning: a review of classification techniques. Emerg. Artif. Intell. Appl. Comput. Eng. **160**(1), 3–24 (2007)
13. Sammut, C., Webb, G.I. (eds.): Supervised Learning, p. 941. Springer, Boston (2010)
14. Parker, A.J., Opletal, G., Barnard, A.S.: Classification of platinum nanoparticle catalysts using machine learning. J. Appl. Phys. **128**(1), 1–11 (2020)
15. Sun, B., Fernandez, M., Barnard, A.S.: Machine learning for silver nanoparticle electron transfer property prediction. J. Chem. Inf. Model. **57**(10), 2413–2423 (2017)
16. Janet, J.P., Kulik, H.J.: Predicting electronic structure properties of transition metal complexes with neural networks. Chem. Sci. **8**(7), 5137–5152 (2017)
17. Takigawa, I., Shimizu, K.I., Tsuda, K., Takakusagi, S.: Machine-learning prediction of the d-band center for metals and bimetals. RSC Adv. **6**(58), 52587–52595 (2016)
18. Christiansen, R.E., Michon, J., Benzaouia, M., Sigmund, O., Johnson, S.G.: Inverse design of nanoparticles for enhanced Raman scattering. Opt. Exp. **28**(4), 4444–4462 (2020)
19. Lee, J.W., Park, W.B., Do Lee, B., Kim, S., Goo, N.H., Sohn, K.S.: Dirty engineering data-driven inverse prediction machine learning model. Sci. Rep. **10**(1), 20443 (2020)
20. Hassan, S.A.: Artificial neural networks for the inverse design of nanoparticles with preferential nano-bio behaviors. J. Chem. Phys. **153**(5), 54102 (2020)
21. Li, S., Barnard, A.S.: Inverse design of nanoparticles using multi-target machine learning. Adv. Theory Simul. **5**(2), 2100414 (2022)
22. Barron, H., Opletal, G., Tilley, R.D., Barnard, A.S.: Dynamic evolution of specific catalytic sites on Pt nanoparticles. Catal. Sci. Technol. **6**(1), 144–151 (2016)
23. Barnard, A., Opletal, G.: Platinum nanoparticle data set, v1. CSIRO Data Collection (2019). https://doi.org/10.25919/5d3958d9bf5f7
24. Zhao, Z., Chen, Z., Zhang, X., Lu, G.: Generalized surface coordination number as an activity descriptor for CO2 reduction on Cu surfaces. J. Phys. Chem. C **120**(49), 28125–28130 (2016)
25. Sun, B., Barron, H., Wells, B., Opletal, G., Barnard, A.S.: Correlating anisotropy and disorder with the surface structure of platinum nanoparticles. Nanoscale **10**(43), 20393–20404 (2018)
26. Parker, A.J., Barnard, A.S.: Machine learning reveals multiple classes of diamond nanoparticles. Nanoscale Horiz. **5**(10), 1394–1399 (2020)
27. Barnard, A.S., Opletal, G.: Selecting machine learning models for metallic nanoparticles. Nano Futures **4**(3), 035003 (2020)
28. Breiman, L.: Random forests. Mach. Learn. **45**(1), 5–32 (2001)
29. Breiman, L.: Bagging predictors. Mach. Learn. **24**, 123–140 (1996)
30. Kamiński, B., Jakubczyk, M., Szufel, P.: A framework for sensitivity analysis of decision trees. Cent. Eur. J. Oper. Res. **26**(1), 135–159 (2018)
31. Hoaglin, D.C., Iglewicz, B., Tukey, J.W.: Performance of some resistant rules for outlier labeling. J. Am. Stat. Assoc. **81**(396), 991–999 (1986)
32. Liu, T., Barnard, A.S.: Fast derivation of shapley based feature importances through feature extraction methods for nanoinformatics. Mach. Learn. Sci. Technol. **2**(3), 035034 (2021)
33. Zunger, A.: Inverse design in search of materials with target functionalities. Nat. Rev. Chem. **2**, 0121 (2018)

34. Sanchez-Lengeling, B., Aspuru-Guzik, A.: Inverse molecular design using machine learning: generative models for matter engineering. Science **361**, 360 (2018)
35. Jørgensen, P.B., Schmidt, M.N., Winther, O.: Deep generative models for molecular science. Mol. Inf. **37**, 1700133 (2018)
36. Hanakata, P.Z., Cubuk, E.D., Campbell, D.K., Park, H.S.: Accelerated search and design of stretchable graphene Kirigami using machine learning. Phys. Rev. Lett. **121**, 255304 (2018)
37. Wan, J., Jiang, J.-W., Park, H.S.: Thermal conductivity versus the density of holes for porous graphene at room temperature. Carbon **157**, 262 (2020)
38. Ma, C., et al.: Accelerated design and characterization of non-uniform cellular materials via a machine-learning based framework. npj Comput. Mater. **6**, 40 (2020)
39. Tominaga, D., Koga, N., Okamoto M.: In: Proceedings of the 2nd Annual Conference on Genetic and Evolutionary Computation, pp. 251–258. ACM Press, New York (2000)
40. Zhang, Y.-Y., Gao, W., Chen, S., Xiang, H., Gong, X.-G.: Inverse design of materials by multi-objective differential evolution. Comput. Mater. Sci. **98**, 51–55 (2015)
41. Dudiy, S., Zunger, A.: Searching for alloy configurations with target physical properties: impurity design via a genetic algorithm inverse band structure approach. Phys. Rev. Lett. **97**, 046401 (2006)
42. Yu, L., Kokenyesi, R.S., Keszler, D.A., Zunger, A.: Inverse design of high absorption thin-film photovoltaic materials. Adv. Energy Mater. **3**, 43 (2013)
43. Yang, D., et al.: Functionality-directed screening of Pb-free hybrid organic-inorganic perovskites with desired ontrinsic photovoltaic functionalities. Chem. Mater. **29**, 524 (2017)
44. Zunger, A.: Beware of plausible predictions of fantasy materials. Nature **566**, 447 (2019)

CNNs with Compact Activation Function

Jindong Wang[1], Jinchao Xu[2(✉)], and Jianqing Zhu[3]

[1] School of Mathematical Sciences, Peking University, 100871 Beijing, China
jdwang@pku.edu.cn

[2] Department of Mathematics, Pennsylvania State University, 16802 University Park, PA, USA
xu@math.psu.edu

[3] Faculty of Science, Beijing University of Technology, 100124 Beijing, China
jqzhu@emails.bjut.edu.cn

http://www.personal.psu.edu/jxx1/

Abstract. Activation function plays an important role in neural networks. We propose to use hat activation function, namely the first order B-spline, as activation function for CNNs including MgNet and ResNet. Different from commonly used activation functions like ReLU, the hat function has a compact support and no obvious spectral bias. Although spectral bias is thought to be beneficial for generalization, we show that MgNet and ResNet with hat function still exhibit a slightly better generalization performance than CNNs with ReLU function by our experiments of classification on MNIST, CIFAR10/100 and ImageNet datasets. This indicates that CNNs without spectral bias can have a good generalization capability. We also illustrate that although hat function has a small activation area which is more likely to induce vanishing gradient problem, hat CNNs with various initialization methods still works well.

Keywords: Spectral bias · Convolutional neural network · Activation function

1 Introduction

Activation function is an important part of neural networks. The most popular activation function is the Rectified Linear Unit (ReLU) [14]. The ReLU activation function can speed up the learning process with less computational complexity as observed in [3,11,12]. Although many other activation functions have been proposed in the last few years [6,9,13,17,21], ReLU is still the most commonly used activation function for CNNs in image classification due to its simplicity and the fact that other activation functions such as ELU [2] and GELU [9] have no significant advantage over ReLU.

Despite being heavily over-parameterized, deep neural networks have been shown to be remarkably good at generalizing to natural data. There is a phenomenon known as the spectral bias [16] or frequency principle [1,23,24] which claims that activation functions such as ReLU make the networks prioritize learning the low frequency modes and the lower frequency components of trained

D. Groen et al. (Eds.): ICCS 2022, LNCS 13351, pp. 319–327, 2022.
https://doi.org/10.1007/978-3-031-08754-7_40

networks are more robust to random parameter perturbations. This has raised the important problem of understanding the implicit regularization effect of deep neural networks and is one main reason for good generalization accuracy [15,16,20,24]. Recently, a theoretical explanation for the spectral bias of ReLU neural networks is provided in [10] by leveraging connections with the theory of finite element method and hat activation function for neural networks that different frequency components of error for hat neural networks decay at roughly the same rate and so hat function does not have the same spectral bias that has been observed for networks based on ReLU, Tanh and other activation functions.

In this paper, we consider CNN with the hat activation function which is defined as follows

$$\text{Hat(x)} = \begin{cases} x, & x \in [0,1], \\ 2-x, & x \in [1,2], \\ 0, & \text{otherwise}, \end{cases} \tag{1}$$

and is actually the B-Spline of first order. Different from ReLU function, hat function has a compact set. We use the hat activation function for CNNs including MgNet [4] and ResNet [7].

MgNet [4] is strongly connected to multigrid method and a systematic numerical study on MgNet in [5] shows its success in image classification problems and its advantages over established networks. We use MgNet in the experiment since it relates to multiscale structure that can handle the frequency variation in different resolution data. Note that hat function has a compact support which is more likely to result in the vanishing gradient problem and no spectral bias, it still obtains comparable generalization accuracy and can even perform slightly better than CNNs with ReLU activation function on MNIST, CIFAR10/100 and ImageNet datasets. This also questions whether the spectral bias is truly significant for regularization. We illustrate that the scale of hat activation function in different resolution layer is of importance and should be set properly for MgNet to adapt the frequency variation in the network. Furthermore, considering the performance of a neural network also depends on how its parameters are initialized, we also test several initialization methods for neural networks including Xavier initialization [3] and Kaiming initialization [6], the results show that all these initialization methods work well for these CNNs with hat activation function.

2 Hat Function for MgNet

Different from ReLU function, the hat function has a compact support of $[0,2]$. Neural networks with hat function also have the universal approximation property [18,19,22]. Hat function is closely related to finite element method and we can adjust the compact support of this function to change its frequency. Thus, we define the following scaled hat activation function with parameter M such that

$$\text{Hat}(x;M) = \begin{cases} x, & x \in [0, \frac{M}{2}], \\ M-x, & x \in [\frac{M}{2}, M], \\ 0, & \text{otherwise}. \end{cases} \tag{2}$$

It has the advantage that its frequency can vary by changing the parameter M. It is shown in [10] that different frequency components of error for hat neural networks decay at roughly the same rate and thus hat function does not have the same spectral bias that has been observed for ReLU networks. To make full good use of this property, we introduce to use MgNet with hat activation function.

MgNet [4] is a convolutional neural network that strongly connects to multi-grid methods and also has a good performance in comparison with existing CNN models [5]. The network consists of several iterative blocks shown in Fig. 1 in both data space and feature space.

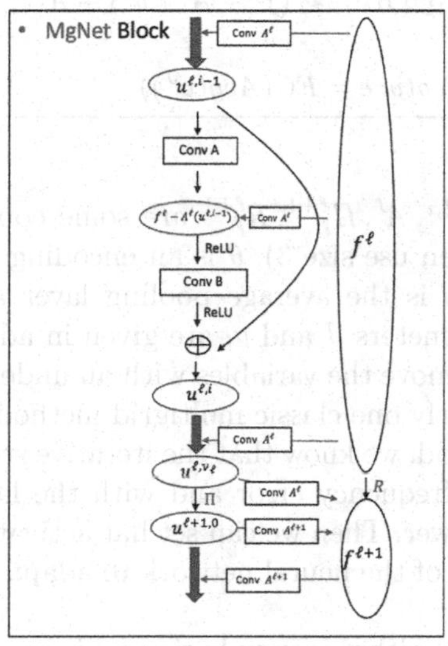

Fig. 1. MgNet Iterative block.

This block is related to the following residual correction step in multigrid method

$$u^i = u^{i-1} + B^i * (f - A^i * u^{i-1}). \tag{3}$$

where B^i, A^i are the convolution operators, f and u denote the source term (image) and solution (feature) respectively. Besides, downscaling the primary image f into coarse resolution requires the following iterative step that projects the high resolution data to the low resolution data

$$u^{\ell+1} = \Pi_\ell^{\ell+1} *_2 u^\ell, \tag{4}$$

$$f^{\ell+1} = R_\ell^{\ell+1} *_2 (f^\ell - A^\ell * u^\ell) + A^{\ell+1} * u^{\ell+1}, \tag{5}$$

where $\Pi_\ell^{\ell+1}, R_\ell^{\ell+1} *_2$ represent the convolution with stride 2. MgNet imposes some nonlinear activation function in the iterative steps above to exact the feature from image as shown in the Algorithm 1.

Algorithm 1. $Feature = \text{MgNet}(f; J, \nu_1, \cdots, \nu_J)$

Initialization: $f^1 = \underline{\sigma} \circ \underline{\theta}(f)$, $u^{1,0} = 0$
for $\ell = 1 : J$ **do**
  for $i = 1 : \nu_\ell$ **do**

$$u^{\ell,i} = u^{\ell,i-1} + \underline{\sigma} \circ B^{\ell,i} * \underline{\sigma}(f^\ell - A^\ell * u^{\ell,i-1}). \tag{6}$$

  end for
  Note $u^\ell = u^{\ell,\nu_\ell}$

$$u^{\ell+1,0} = \underline{\sigma} \circ \Pi_\ell^{\ell+1} *_2 u^\ell \tag{7}$$

$$f^{\ell+1} = \underline{\sigma} \circ R_\ell^{\ell+1} *_2 (f^\ell - A^\ell * u^\ell) + A^{\ell+1} * u^{\ell+1,0}. \tag{8}$$

end for
Fully connected layer: $Feature = FC(Avg(u^J))$

In this algorithm, $B^{\ell,i}, A^\ell, \Pi_\ell^{\ell+1}, R_\ell^{\ell+1}$ are some convolution operators of a given kernel size (we often use size 3), θ is an encoding layer that increases the number of channel, *Avg* is the average pooling layer and σ is the activation function. The hyperparameters J and ν_i are given in advance.

We note that if we remove the variables with an underline, namely $\underline{\sigma}$ and $\underline{\theta}$ in Algorithm 1, we get exactly one classic multigrid method. From the convergence theory of multigrid method, we know that the iterative step (6) is associated with the elimination of high frequency error and with the layer getting deeper, the frequency of data gets lower. Then we can set hat activation functions with various M in different layers of the neural network to adapt this frequency variation in the network.

The MgNet model algorithm is very basic and it can be generalized in many different ways. It can also be used as a guidance to modify and extend many existing CNN models [5]. The following result shows Algorithm 1, admits the following identities

$$r^{\ell,i} = r^{\ell,i-1} - A^\ell \circ \sigma \circ B^{\ell,i} \circ \sigma(r^{\ell,i-1}), \quad i = 1 : \nu_\ell \tag{9}$$

where

$$r^{\ell,i} = f^l - A^\ell * u^{\ell,i}. \tag{10}$$

and (9) represents pre-act ResNet [8].

3 Experiments

Since MgNet has strongly connection with ResNet [7], we evaluate the performance of hat activation function on image classification for MgNet and ResNet compared with ReLU neural networks. For MgNet, we consider $J = 4$ and $\nu_1 = \nu_2 = \nu_3 = \nu_4 = 2$ as stated in Algorithm 1, thus there are four different resolution layers.

The following benchmark datasets are used: (i) MNIST, (ii) CIFAR10, (iii) CIFAR100, and (iv) ImageNet.

We consider using SGD method with the batchsize of 256 for 150 epochs. The initialization method of parameters is Kaiming's uniform intialization [6]. The initial learning rate is 0.1 with a decay rate of 0.1 per 50 epochs. The following results are the average of 3 runs. In the Table 1, the numbers [5,10,15,20] denote the scaling of hat activation in different resolution layers since the size of data have four different resolution levels in MgNet and ResNet18.

Table 1 shows that hat activation function has slightly better generalization capability than ReLU activation function for both MgNet and ResNet. To illustrate the argument of MgNet, we evaluate the performance of MgNet with different scale settings of hat activation function. As is shown in Table 2, it is better to use the hat function with larger support in the coarser resolution data which is consistent of the frequency variation of MgNet.

To exclude the influence of training process, we train the CNNs with more epochs of 300. As is shown in Table 3, the test accuracy increases both for hat MgNet and ReLU MgNet, and hat activation function still maintains slightly better generalization capability than ReLU activation function which indicates that hat activation function is truly powerful.

Since the scale of hat function is fixed which can be a potential disadvantage, we also regard these scale numbers as parameters in the network. Table 4 gives the results of trainable scale hat MgNet on CIFAR10/100 datasets and we also record the scale numbers of the model. Though using two different settings of initial scale, the results all demonstrate that it is better to use the hat function with larger support in the coarser resolution level and the support intend to be getting small during the training. The results show that the generalization accuracy of MgNet is still competitive with a much smaller support in the first few layers without adding any neurons. We also note that it is available for a combination of hat function and ReLU function for CNNs with trainable scale hat function and we can replace the activation function of the encoding layer with ReLU function.

Kaiming's initialization has been shown to work well for ReLU networks, the experiments show that hat CNNs also work well with this initialization method. Furthermore, we also consider Xavier's uniform initialization [3] for hat CNNs on CIFAR10/100 datasets. The results in Table 5 and Fig. 2 show that the initialization methods make almost no difference on test accuracy but for the CIFAR100 dataset the loss of Kaiming's initialization converges slightly fast.

Table 1. Comparison of hat CNNs and ReLU CNNs for image classification.

Dataset	Model	Activation function	Test accuracy
MNIST	MgNet	ReLU	99.66
		hat-[5,10,15,20]	**99.68**
CIFAR10	MgNet	ReLU	93.13
		hat-[5,10,15,20]	**93.23**
	ResNet	ReLU	94.64
		hat-[20,15,10,5]	**94.79**
CIFAR100	MgNet	ReLU	70.85
		hat-[5,10,15,20]	**70.96**
	ResNet	ReLU	76.21
		hat-[5,10,15,20]	**76.47**
ImageNet	MgNet	ReLU	72.36
		hat-[10,20,30,40]	**72.69**

Table 2. Comparison of different scale setting of hat function for MgNet.

Dataset	Activation function	Test accuracy
MNIST	hat-[5,10,15,20]	**99.68**
	hat-[20,15,10,5]	99.64
CIFAR10	hat-[5,10,15,20]	**93.23**
	hat-[20,15,10,5]	92.87
CIFAR100	hat-[5,10,15,20]	**70.96**
	hat-[20,15,10,5]	70.56
ImageNet	hat-[10,20,30,40]	**72.69**
	hat-[40,30,20,10]	71.87

Table 3. MgNet results of 300 epochs.

Dataset	Activation function	Test accuracy
CIFAR10	hat-[5,10,15,20]	**93.97**
	hat-[20,15,10,5]	93.80
	ReLU	93.79
CIFAR100	hat-[5,10,15,20]	**72.06**
	hat-[20,15,10,5]	71.68
	ReLU	71.73

Table 4. MgNet with hat function of trainable scale (300 epochs).

Dataset	Test accuracy	Initial scale	Final scale
CIFAR10	94.15	[5,10,15,20]	[1.0561,1.6974,1.7712,3.2502]
	93.99	[20,15,10,5]	[1.2643,1.7570,2.8338,3.2682]
CIFAR100	72.18	[5,10,15,20]	[1.4278,2.5441,2.7431,9.6464]
	72.24	[20,15,10,5]	[2.0015,2.4351,2.7342,9.8020]

Table 5. Comparison of different initialization methods for hat-CNNs.

Dataset	Model	Activation function	Test accuracy(Kaiming)	Test accuracy(Xavier)
CIFAR10	MgNet	hat-[20,15,10,5]	93.182	93.20
		hat-[5,10,15,20]	93.233	93.26
	ResNet	hat-[20,15,10,5]	94.786	94.77
		hat-[5,10,15,20]	94.453	94.47
CIFAR100	MgNet	hat-[20,15,10,5]	70.56	70.15
		hat-[5,10,15,20]	70.963	70.89
	ResNet	hat-[20,15,10,5]	76.186	76.15
		hat-[5,10,15,20]	75.996	76.22

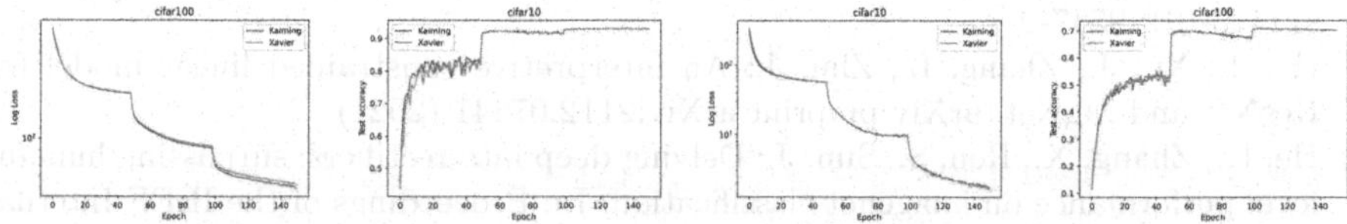

Fig. 2. Comparison of loss curve and test accuracy curve of MgNet for CIFAR10 and CIFAR100 datasets versus initialization methods.

4 Conclusion

We introduce the hat activation function, a compact function for CNNs and evaluate its performance on several datasets in this paper. The results show that hat activation function which has a compact activation area still has competitive performance in comparison with ReLU activation function for MgNet and ResNet although it does not have those properties of ReLU which are deemed to be important. Besides, activation function with a small compact set can cause gradient vanishing easily but this has no influence on the performances of CNNs with hat function. Specifically, from the experiments we note that the scale setting of hat activation function also influences the performance, which is related to the frequency variation in the network. Furthermore, commonly used initialization methods are also shown to be viable for hat CNNs. These numerous experiments show that hat function is indeed a viable choice of activation functions for CNNs and indicate the spectral bias is not significant for generalization accuracy.

Acknowledgment. The work of Jinchao Xu is supported in part by the National Science Foundation (Grant No. DMS-2111387). The work of Jianqing Zhu is supported in part by Beijing Natural Science Foundation (Grant No. Z200002). The work of Jindong Wang is supported in part by High Performance Computing Platform of Peking University.

References

1. Basri, R., Galun, M., Geifman, A., Jacobs, D., Kasten, Y., Kritchman, S.: Frequency bias in neural networks for input of non-uniform density. In: International Conference on Machine Learning, pp. 685–694. PMLR (2020)
2. Clevert, D.A., Unterthiner, T., Hochreiter, S.: Fast and accurate deep network learning by exponential linear units (elus). arXiv preprint arXiv:1511.07289 (2015)
3. Glorot, X., Bengio, Y.: Understanding the difficulty of training deep feedforward neural networks. In: Proceedings of the Thirteenth International Conference on Artificial Intelligence and Statistics, pp. 249–256. JMLR Workshop and Conference Proceedings (2010)
4. He, J., Xu, J.: MgNet: a unified framework of multigrid and convolutional neural network. Sci. China Math. **62**(7), 1331–1354 (2019). https://doi.org/10.1007/s11425-019-9547-2
5. He, J., Xu, J., Zhang, L., Zhu, J.: An interpretive constrained linear model for ResNet and MgNet. arXiv preprint arXiv:2112.07441 (2021)
6. He, K., Zhang, X., Ren, S., Sun, J.: Delving deep into rectifiers: surpassing human-level performance on imagenet classification. In: Proceedings of the IEEE International Conference on Computer Vision, pp. 1026–1034 (2015)
7. He, K., Zhang, X., Ren, S., Sun, J.: Deep residual learning for image recognition. In: Proceedings of the IEEE Conference on Computer Vision and Pattern Recognition, pp. 770–778 (2016)
8. He, K., Zhang, X., Ren, S., Sun, J.: Identity mappings in deep residual networks. In: Leibe, B., Matas, J., Sebe, N., Welling, M. (eds.) ECCV 2016. LNCS, vol. 9908, pp. 630–645. Springer, Cham (2016). https://doi.org/10.1007/978-3-319-46493-0_38
9. Hendrycks, D., Gimpel, K.: Gaussian error linear units (gelus). arXiv preprint arXiv:1606.08415 (2016)
10. Hong, Q., Siegel, J., Tan, Q., Xu, J.: On the activation function dependence of the spectral bias of neural networks. preprint (2022)
11. Klambauer, G., Unterthiner, T., Mayr, A., Hochreiter, S.: Self-normalizing neural networks. Adv. Neural Inf. Process. Syst. **30**, 1–10 (2017)
12. Krizhevsky, A., Sutskever, I., Hinton, G.E.: Imagenet classification with deep convolutional neural networks. Adv. Neural Inf. Process. Syst. **25**, 1–9 (2012)
13. Maas, A.L., Hannun, A.Y., Ng, A.Y., et al.: Rectifier nonlinearities improve neural network acoustic models. In: Proceedings of ICML, vol. 30, p. 3. Citeseer (2013)
14. Nair, V., Hinton, G.E.: Rectified linear units improve restricted boltzmann machines. In: ICML (2010)
15. Poggio, T., et al.: Theory of deep learning iii: the non-overfitting puzzle. CBMM Memo **73**, 1–38 (2018)
16. Rahaman, N., et al.: On the spectral bias of neural networks. In: International Conference on Machine Learning, pp. 5301–5310. PMLR (2019)
17. Ramachandran, P., Zoph, B., Le, Q.V.: Searching for activation functions. arXiv preprint arXiv:1710.05941 (2017)
18. Siegel, J.W., Xu, J.: Characterization of the variation spaces corresponding to shallow neural networks. arXiv preprint arXiv:2106.15002 (2021)
19. Siegel, J.W., Xu, J.: Improved approximation properties of dictionaries and applications to neural networks. arXiv preprint arXiv:2101.12365 (2021)
20. Soudry, D., Hoffer, E., Nacson, M.S., Gunasekar, S., Srebro, N.: The implicit bias of gradient descent on separable data. J. Mach. Learn. Res. **19**(1), 2822–2878 (2018)

21. Trottier, L., Giguere, P., Chaib-Draa, B.: Parametric exponential linear unit for deep convolutional neural networks. In: 2017 16th IEEE International Conference on Machine Learning and Applications (ICMLA), pp. 207–214. IEEE (2017)
22. Xu, J.: The finite neuron method and convergence analysis. arXiv preprint arXiv:2010.01458 (2020)
23. Xu, Z.Q.J., Zhang, Y., Luo, T., Xiao, Y., Ma, Z.: Frequency principle: fourier analysis sheds light on deep neural networks. arXiv preprint arXiv:1901.06523 (2019)
24. Xu, Z.J.: Understanding training and generalization in deep learning by fourier analysis. arXiv preprint arXiv:1808.04295 (2018)

Deep Neural Networks and Smooth Approximation of PDEs

Kamil Doległo[1], Maciej Paszyński[1(✉)], and Leszek Demkowicz[2]

[1] AGH University of Science and Technology, Kraków, Poland
maciej.paszynski@agh.edu.pl
[2] Oden Institute, The University of Texas in Austin, Austin, USA
leszek@oden.utexas.edu

Abstract. We focus on Isogeometric Analysis (IGA) approximations of Partial Differential Equations (PDEs) solutions. We consider linear combinations of high-order and continuity base functions utilized by IGA. Instead of using the Deep Neural Network (DNN), which is the concatenation of linear operators and activation functions, to approximate the solutions of PDEs, we employ the linear combination of higher-order and continuity base functions, as employed by IGA. In this paper, we compare two methods. The first method trains different DNN for each coefficient of the linear computations. The second method trains one DNN for all coefficients of the linear combination. We show on model L-shape domain problem that training several small DNNs learning how to span B-splines coefficients is more efficient.

Keywords: Partial differential equations · Isogeometric analysis · Physics informed neural networks · Stochastic gradient descent

1 Introduction

Isogeometric Analysis (IGA) [1] employs smooth high-order and continuity base functions for approximation of solutions of Partial Differential Equations (PDEs). Physics Informed Neural Networks (PINN) [2] approximate the solution of a given PDE with Deep Neural Network (DNN) being the concatenation of several linear operators and non-linear activation functions. The Stochastic Gradient Descent (SGD) [3] is used to find the coefficients of the DNN approximating a given PDEs. In this presentation, we consider how smooth linear combinations of higher-order and continuity base functions used by IGA can be employed by DNNs and SGD method to approximate solutions of PDEs. In [4], we described how IGA could be used to approximate the coefficients of a linear combination of B-splines, employed for the solution of a family of PDEs depending on the right-hand side and boundary condition functions. In this presentation, we compare two methods. The first method, following [4] is to set up and train different DNNs for different coefficients of linear combinations. The second method is to

D. Groen et al. (Eds.): ICCS 2022, LNCS 13351, pp. 328–332, 2022.
https://doi.org/10.1007/978-3-031-08754-7_41

set up and train one DNNs for all coefficients of linear combination. We consider two problems. First, the simple family of one-dimensional problem

$$u''(x) = f(x) = n^2\pi^2 sin(n\pi x), u(0) = 0, u'(1) = g(n) = n\pi cos(n\pi) \quad (1)$$

for $x \in (0,1)$, used to explain our strategy. The family of solutions $u_n(x) = sin(n\pi x)$ depends on the parameter n defining the forcing and boundary condition. Second, we focus on model two-dimensional L-shape domain problem. The DNNs proposed in this paper learn how to span the smooth IGA combinations of B-splines approximating the solutions of PDEs.

Several Deep Neural Network Approximating Coefficients of IGA Base Functions. First, we assume that we approximate the solution of PDE by a linear combination of B-splines $u(x;n) \approx u_n(x) = \sum_i u_i(n)B^x_{i,p}$. Following [4] we approximate coefficients of linear combinations by several DNNs, one $u_i(n) \approx DNN_i(n)$ for each i-th coefficient

$$u(x;n) \approx u_n(x) = \sum_i u_i(n)B_{i,p}(x) \approx \sum_i DNN_i(n)B_{i,p}(x) \quad (2)$$

$$DNN_i(n) = c_i\sigma\left(a_in + b_i\right) + d_i = \frac{c_i}{1 + exp(-a_in - b_i)} + d_i \quad (3)$$

We define the error function

$$error_i(n) = 0.5\left(DNN_i(n) - u_i(n)\right)^2 \quad (4)$$

with $u_i(n)$ computed using IGA method. The training procedure is employed independently for each $DNN_i(n)$ approximating different coefficients of the linear combination as a function of forcing and boundary conditions depending on n parameter. These DNNs learn an operator of different forcing and boundary conditions (parameterized by n) into coefficients of B-spline base functions approximating solutions of PDE (depending on forcing and boundary conditions).

One Deep Neural Network Approximating All Coefficients of IGA Base Functions. The second method consists in approximating of coefficients of linear combinations by one DNN, namely $DNN(n) = (u_1, ..., u_N)$ computing all the coefficients

$$u(x;n) = \sum_i u_i(n)B_{i,p}(x) = \sum_i \left(DNN(n)\right)_i B_{i,p}(x) \quad (5)$$

$$DNN(n) = C\sigma\left(An + B\right) + D = \frac{C}{1 + exp(An - B)} + D \quad (6)$$

where A, B, C, D are vectors of size $N \times 1$. We define

$$error(x) = \frac{\left(\sum_i \left(DNN(n)\right)_i - u_i(n)\right)^2}{2} \quad (7)$$

with $u_i(n)$ computed using IGA method. Here we have a single DNN learning an operator from forcing and boundary conditions (parameterized by n) into coefficients of B-spline basis functions approximating the solution of PDE (depending on forcing and boundary conditions parameterized by n).

Physics Informed Neural Network. Finally, we compare to PINN [2] approach, which approximates a single solution of (1) for a fixed n, with concatenation of linear operators and non-linear activation functions

$$PINN(x) = c\sigma(ax + b) + d = \frac{c}{1 + exp(-ax - b)} + d \tag{8}$$

where $\sigma(x) = \frac{1}{1+exp(-x)}$ is e.g. the sigmoid activation function. We define the error functions for the approximation of PDE and b.c.

$$\begin{aligned} error_1(x) = 0.5\left(PINN''(x)\right)^2, error_2 = 0.5\left(PINN'(1) - g(n)\right)^2, \\ error_3 = 0.5\left(PINN(0)\right)^2. \end{aligned} \tag{9}$$

Here, DNN learns how to approximate a single solution of the PDE directly.

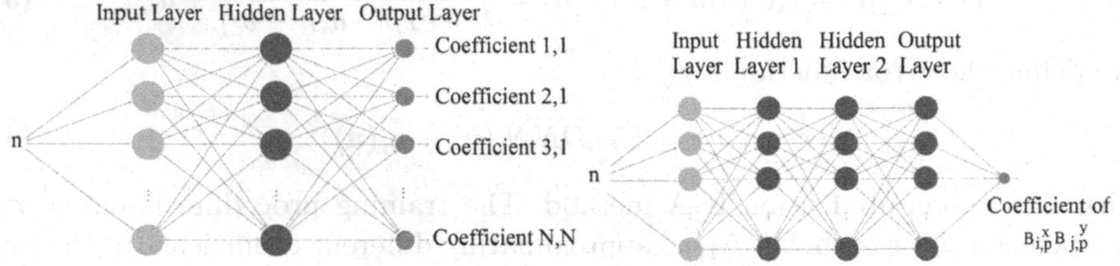

Fig. 1. Visualization of the first and the second DNN architectures.

Experimental Verification. We verify the methods on model 2D L-shape domain problem $\Delta u = 0$ with zero Dirichlet b.c., and the Neumann b.c. $\frac{\partial u}{\partial n} = g(x, y)$. We embed L-shape in a square and set 1/4 of the domain to 0. We assume the exact solution of this PDE of the form $u_{exact}(x, y) = sin(2\pi n \cdot x) \cdot sin(2\pi n \cdot y)$. using the manufactured solution with $g = \frac{\partial u_{exact}}{\partial n}$. We discretize with B-spline basis functions, seeking $u_h(x, y) = \sum_{i=1,\dots,N;j=1,\dots n} u_{ij} B^x_{i,p} B^y_{j,p}$. In our experiment, we have 42 × 42 two-dimensional cubic B-splines of C^2 continuity, with $N = 42 + 3 = 45$, the total of 2025 coefficients. In general the DNN is given by

$$\begin{aligned} l^{(0)} = x, l^{(1)} = \sigma^{(1)}\left(W^{(1)} l^{(0)} + b^{(1)}\right), \cdots \\ \cdots, l^{(n)} = \sigma^{(n)}\left(W^{(n)} l^{(n-1)} + b^{(n)}\right), l^{(out)} = W^{(out)} l^{(n)} + b^{(out)}. \end{aligned} \tag{10}$$

We employ two methods, see Fig. 1. The first uses one DNN for each coefficient of the linear combination. Each DNN has input with n parameter, 600 neurons in the input layer, 600 neurons in two hidden layers, and 600 neurons in the output layers, sigmoid activation function, and one output value of the B-spline coefficient. So we have 2025 DNNs approximating 2025 coefficients of B-splines. The second one uses one DNN to compute all the linear combination coefficients. This DNN has input with n parameter, 1000 neurons in the input layer, 1000 neurons in the hidden layer, and 1000 neurons in the output layer, ReLU activation function, and $N \times N$ output B-spline coefficients (2025 output values in our case). Figure 2 shows that training with 100 samples is cheaper and more accurate if we train several small DNNs, learning how to span B-splines.

Conclusions. We show that DNN can be used to approximate coefficients of linear combinations of higher-order and continuity base functions employed by IGA to approximate solutions of PDEs. The DNN learnt how to span the linear combinations of smooth B-splines. We obtained the approximation of the solution of a family of PDEs that is of higher-order and continuity, smooth and easily differentiate. In the future work we may employ the idea proposed by the BSDE method [5], approximating the gradients of the solution of PDE.

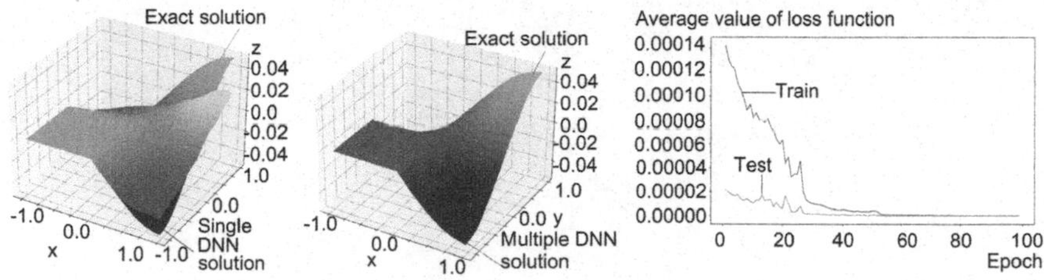

Fig. 2. Several DNNs, one for each coefficients work well as opposed to one DNN for all coefficients. Exact solution (blue). **Left panel:** Solution from one DNN for all coefficients (red), MSE = 1.41e–4. **Middle panel:** Solution from several DNNs one for each coefficients. MSE = 8.42e–7. **Right panel:** Training and testing with 100 samples of one DNN approximating a single B-spline coefficient. (Color figure online)

Acknowledgement. The European Union's Horizon 2020 Research and Innovation Program of the Marie Skłodowska-Curie grant agreement No. 777778.

References

1. Cottrell, J.A., Hughes, T.J.R., Bazilevs, Y.: Isogeometric Analysis: Towards Unification of Computer Aided Design and Finite Element Analysis. John Wiley and Sons, Hoboken (2009)
2. Maziar, R., Perdikaris, P., Karniadakis, G.E.: Physics-informed neural networks: a deep learning framework for solving forward and inverse problems involving nonlinear partial differential equations. J. Comput. Phys. **378**, 686–707 (2019)
3. Léon, B.: Online Algorithms and Stochastic Approximations, Online Learning and Neural Networks. Cambridge University Press, Cambridge (1998)

4. Doległo, K., Paszyńska, A., Paszyński, M., Demkowicz, L.: Deep neural networks for smooth approximation of physics with higher order and continuity B-spline base functions, pp. 1–44 (2022) arXiv:2201.00904
5. Wang, Y., Ni, Y.H.: Deep BSDE-ML learning and its application to model-free optimal control, pp. 1–20 (2022) arxiv:2201.01318

Isogeometric Analysis of Bound States of a Quantum Three-Body Problem in 1D

Quanling Deng(✉)

School of Computing, Australian National University, Canberra, ACT 2601, Australia
Quanling.Deng@anu.edu.au

Abstract. In this paper, we initiate the study of isogeometric analysis (IGA) of a quantum three-body problem that has been well-known to be difficult to solve. In the IGA setting, we represent the wavefunctions by linear combinations of B-spline basis functions and solve the problem as a matrix eigenvalue problem. The eigenvalue gives the eigenstate energy while the eigenvector gives the coefficients of the B-splines that lead to the eigenstate. The major difficulty of isogeometric or other finite-element-method-based analyses lies in the lack of boundary conditions and a large number of degrees of freedom required for accuracy. For a typical many-body problem with attractive interaction, there are bound and scattering states where bound states have negative eigenvalues. We focus on bound states and start with the analysis for a two-body problem. We demonstrate through various numerical experiments that IGA provides a promising technique to solve the three-body problem.

Keywords: Isogeometric analysis · Three-body problem · Bound state

1 Introduction

While there are still unsolved questions in the classical three-body problem [5,26], the quantum mechanical three-body problem also has unanswered questions [15, 20,34]. Accurate and rigorous solutions are highly desirable both for answering these open questions as well as for studies of three-body correlations within various many-body systems. The two-body problem is generally considered as "solved" due to the momentum conservation that leads to a second-order ordinary differential equation which can be solved effectively. For three-body problem, the space is six-dimensional in the center of mass system. The total angular momentum conservation leads to three coupled second-order nonlinear differential equations in classical mechanics [27]. In quantum mechanics, we have a more complicated system that admits no analytic solution in general and existing numerical methods are not satisfactory in the sense of robustness, efficiency, and reliability.

In quantum chemistry and molecular physics, the Born–Oppenheimer (BO) approximation has been the most well-known and widely-used mathematical approximation since the early days of quantum mechanics [3,8,28,32]. The method is based on the assumption that the nuclei are much heavier than the

D. Groen et al. (Eds.): ICCS 2022, LNCS 13351, pp. 333–346, 2022.
https://doi.org/10.1007/978-3-031-08754-7_42

electrons which consequently leads to that the wave functions of atomic nuclei and electrons in a molecule can be treated separately. For instance, BO was used in [16,18] to study the Efimov effect in few-body systems. BO was adopted in [21,22] recently to establish the universality in a one-dimensional three-body system. A more efficient numerical method based on BO was developed using the tensor-product structure [35]. BO has been the standard method to describe the interaction between electrons and nuclei but it can fail whenever the assumption fails, for example, in graphene [30]. When the mass ratios of the interacting bodies are close to one, the assumption fails and BO is generally inaccurate. Other methods such as the pseudospectral method based on Fourier analysis [4] and Skorniakov and Ter-Martirosian (STM) method based on exact integral equations [33] have been developed to obtain the three-body bound states for arbitrary mass ratios.

In this paper, we develop a general numerical method to solve one-dimensional quantum two- and three-body problems with arbitrary mass ratios and any interaction potentials that lead to bound states. With this goal in mind, we initiate the study of finite element analysis (FEA) based methods to find the bound states of three-body systems. In particular, we adopt the more advanced method isogeometric analysis (IGA) for this purpose. IGA, first developed in [11,23], has been widely-used as a numerical analysis tool for various simulations that are governed by partial differential equations (PDEs). IGA adopts the framework of classic Galerkin FEA and uses B-splines or non-uniform rational basis splines (NURBS) instead of the Lagrange polynomials as its basis functions. These basis functions have higher-order continuity (smoothness) which consequently improves the accuracy of the FEA numerical approximations. The work [10] applied IGA to study a structural vibration problem that is modeled as a Laplacian eigenvalue problem. It has been shown that IGA improved the accuracy of the spectral approximation significantly compared with FEA [25]. Further advantages of IGA over FEA on spectral accuracy have been studied in [24,31]. With the advantages in mind, we adopt IGA to solve the quantum three-body problem as a second-order differential eigenvalue problem.

The rest of this paper is organized as follows. Section 2 presents the two- and three-body problems under consideration. We then unify these two problems as a single differential eigenvalue problem in one or two dimensions where 1D refers to the two-body problem and 2D refers to the three-body problem. We show an example of solutions to a two-body problem for both bound and scattering states, which serves as a motivation of the proposed method that solves only the bound states over an approximate finite domain. We then present the IGA discretization method in Sect. 3 to solve the unified problem for the bound states. Section 4 collects and discusses various numerical tests to demonstrate the performance of the proposed method. We also perform the numerical study of the impact of domain size on the approximation accuracy of the bound states. Concluding remarks are presented in Sect. 5.

2 The Two- And Three-Body Problems

In this section, we first present the heavy-light two-body and heavy-heavy-light three-body problems that are modeled as the dimensionless stationary Schrödinger equations recently studied in [21,22]. We then generalize the problems for any mass ratios and unify them as a single differential eigenvalue problem. A numerical example is followed to show the bound and scattering states of a two-body problem. The shape of bound states gives a motivation to pose the differential eigenvalue problem on a finite domain with a size to be specified depending on the differential operator and accuracy tolerance.

The heavy-light quantum two-body system with an attractive interaction via a potential of finite range, after eliminating the center-of-mass motion, is modeled as a dimensionless stationary Schrödinger equation

$$\Big[-\frac{1}{2}\frac{\partial^2}{\partial x^2} - v(x) \Big] \psi^{(2)} = E^{(2)} \psi^{(2)}, \tag{1}$$

where $E^{(2)}$ is the binding energy and $\psi^{(2)}$ is the two-body wave function. The corresponding three-body system is modeled as

$$\Big[-\frac{\alpha_x}{2}\frac{\partial^2}{\partial x^2} - \frac{\alpha_y}{2}\frac{\partial^2}{\partial y^2} - v(x+y/2) - v(x-y/2) \Big] \psi = E\psi, \tag{2}$$

where E is the eigenenergy and $\psi = \psi(x,y)$ is the three-body wave function describing the relative motions. The coefficients

$$\alpha_x = \frac{1/2 + m_h/m_l}{1 + m_h/m_l}, \qquad \alpha_y = \frac{2}{1 + m_h/m_l}, \tag{3}$$

where m_h denotes the mass of two heavy particles and m_l denotes the mass of the light particle. The potential

$$v(\xi) = \beta f(\xi) \tag{4}$$

with $\beta > 0$ denoting a magnitude and f denoting the shape of the interaction potential. We assume that the f is symmetric and describes a short-range interaction, that is, $|\xi|^2 f(|\xi|) \to 0$ as $|\xi| \to \infty$.

2.1 The Unified Problem

The two-body problem (1) is posed on an infinite domain $\Omega = \mathbb{R}$ while the three-body problem (2) is posed on $\Omega = \mathbb{R}^2$. Mathematically, problems (1) and (2) are differential eigenvalue problems where the differential operator is a Hamiltonian. Moreover, we observe that (1) is of one variable while (2) is of two variables which can be regarded as 1D and 2D spatial variables, respectively. With this in mind, we unify problems (1) and (2) to obtain a differential eigenvalue problem

$$-\nabla \cdot (\kappa \nabla u) - \gamma u = \lambda u \qquad \forall\, \boldsymbol{x} \in \Omega, \tag{5}$$

where ∇ is the gradient operator, $\nabla\cdot$ is the divergence operator. $\lambda = E^{(2)}, \gamma = v(x), \kappa = \frac{1}{2}$ in 1D while $\lambda = E, \gamma = v(x+y/2)+v(x-y/2)$ and $\kappa = (\frac{\alpha_x}{2}, 0; 0, \frac{\alpha_y}{2})$ being a diagonal matrix in 2D. Herein, u denotes an eigenstate.

From now on, we focus on the unified problem (5). There are three major difficulties in solving this problem using a Galerkin FEA-based discretization method.

- (a) The attractive interaction may lead to negative eigenvalues. Consequently, the discretization of the differential operator $\mathcal{L} = -\nabla \cdot (\kappa\nabla) - \gamma$ leads to a stiffness matrix that is not necessarily positive-definite. This in return brings a potential issue when solving the resulting linear algebra problem.
- (b) The domain Ω is infinite. This makes it impossible to discretize the domain with a finite number of elements with each element being of finite size.
- (c) There are no boundary conditions provided. A Galerkin FEA-based discretization method requires setting appropriate boundary conditions for the resulting linear algebra system to be non-singular.

For (a), an eigenvalue shift will resolve the issue. That is, we rewrite $-\nabla \cdot (\kappa\nabla u) - \gamma u = \lambda u$ by adding a positive scale to obtain $-\nabla \cdot (\kappa\nabla u) - (\gamma - \gamma_0)u = (\lambda + \gamma_0)u$ where $\gamma_0 > 0$ is a constant such that $\gamma - \gamma_0 < 0$ for all $\boldsymbol{x} \in \Omega$. With a slight abuse of notation, the problem (5) can be rewritten as

$$-\nabla \cdot (\kappa\nabla u) + \gamma u = \lambda u \qquad \forall\, \boldsymbol{x} \in \Omega. \tag{6}$$

To overcome the difficulties (b) and (c), we first present an example of a solution to the two-body problem in the next subsection.

2.2 A Solution Example of the Two-Body Problem

For attractive interaction, the eigenenergies for certain eigenstates can be negative. For a potential vanishing at $\pm\infty$, a negative eigenvalue implies a bound state while a positive eigenvalue implies a scattering state [19].

Figure 1 shows an example of state solutions to the two-body problem (1) with $\kappa = 1/2$, potential $f(\xi) = e^{-\xi^2}$ and $\beta = 1$ for the left plot while $\beta = 2$ for the right plot. The bound states eigenenergies are marked in the figure. Herein, we apply IGA with 5000 elements and C^6 septic B-spline basis functions. We present the details of the IGA method in the next section. We observe that there is one bound state for $\beta = 1$ and two bound states for $\beta = 2$. All other states are scattering states. When $x \to \pm\infty$, the wavefunctions go to zeros exponentially fast for bound states while they do not go to zeros for scattering states. A theoretical explanation can be found in [1].

In this paper, our goal is to find the eigenenergies and eigenstates for bound states. In the case of $\beta = 1$, the bound state solution decays to zero approximately at $x = \pm 10$ with an error of 5.9×10^{-5}. We observe similar behaviour for the two bound states of the case $\beta = 2$. This decaying behaviour provides an insight to overcome the difficulties (b) and (c) listed in Sect. 2.1. The idea is

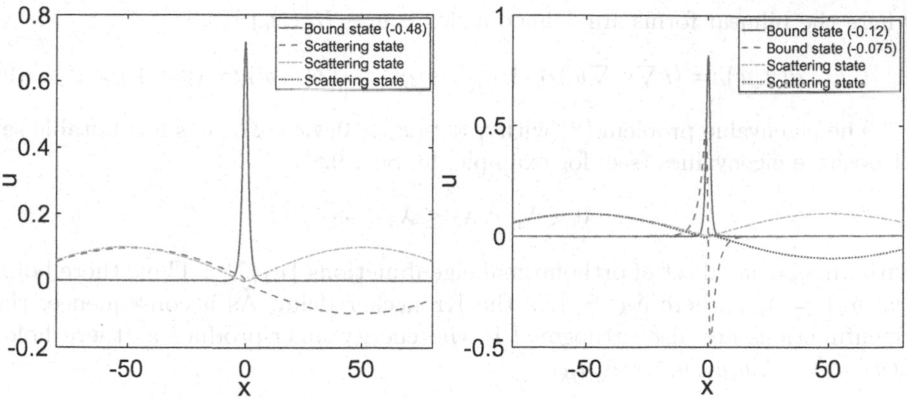

Fig. 1. Bound and scattering states of two-body problems.

that for a given tolerance $\epsilon > 0$, we propose to solve the problem (6) on a finite domain $\Omega_\epsilon = [-\boldsymbol{x}_\epsilon, \boldsymbol{x}_\epsilon]^d, d = 1, 2$ with homogeneous boundary condition

$$u = 0, \qquad \forall\, \boldsymbol{x} \in \partial\Omega_\epsilon. \tag{7}$$

Remark 1. For smaller error tolerance, one expects to apply a larger finite domain. A detailed study is presented in Sect. 4.2. We also point out that a simple transformation such as $\hat{x} = \tanh(x)$ that transfers the infinite domain Ω to a finite domain $\hat{\Omega} = (-1, 1)^d, d = 1, 2$ can not avoid the difficulties of (b) and (c) listed in Sect. 2.1.

3 Isogeometric Analysis

In this section, we present the IGA method for the unified problem (6) on Ω_ϵ supplied with the boundary condition (7). We also give an a priori error estimate for the bound states and their eigenenergies.

3.1 Continuous Level

Let $\Omega_\epsilon = [-\boldsymbol{x}_\epsilon, \boldsymbol{x}_\epsilon]^d \subset \mathbb{R}^d, d = 1, 2$ be a bounded domain with Lipschitz boundary $\partial\Omega_\epsilon$. We adopt the standard notation for the Hilbert and Sobolev spaces. For a measurable subset $S \subseteq \Omega_\epsilon$, we denote by $(\cdot,\cdot)_S$ and $\|\cdot\|_S$ the L^2-inner product and its norm, respectively. We omit the subscripts when clear. For an integer $m \geq 1$, we denote the H^m-norm and H^m-seminorm as $\|\cdot\|_{H^m(S)}$ and $|\cdot|_{H^m(S)}$, respectively. In particular, we denote by $H_0^1(\Omega_\epsilon)$ the Sobolev space with functions in $H^1(\Omega_\epsilon)$ that are vanishing on the boundaries.

The variational formulation of (6) is to find eigenvalue $\lambda \in \mathbb{R}^+$ and eigenfunction $u \in H_0^1(\Omega_\epsilon)$ with $\|u\|_{\Omega_\epsilon} = 1$ such that

$$a(w, u) = \lambda b(w, u), \quad \forall\, w \in H_0^1(\Omega_\epsilon), \tag{8}$$

where the bilinear forms are defined as for $v, w \in H_0^1(\Omega_\epsilon)$

$$a(v,w) := (\kappa \nabla v, \nabla w)_{\Omega_\epsilon} + (\gamma v, w)_{\Omega_\epsilon}, \qquad b(v,w) := (v,w)_{\Omega_\epsilon}. \tag{9}$$

The eigenvalue problem (8) with $\gamma = \gamma(\boldsymbol{x}) > 0, \forall \boldsymbol{x} \in \Omega_\epsilon$, has a countable set of positive eigenvalues (see, for example, [6, Sec. 9.8])

$$0 < \lambda_1 < \lambda_2 \leq \lambda_3 \leq \cdots$$

with an associated set of orthonormal eigenfunctions $\{u_j\}_{j=1}^\infty$. Thus, there holds $(u_j, u_k) = \delta_{jk}$, where $\delta_{jk} = 1$ is the Kronecker delta. As a consequence, the eigenfunctions are also orthogonal in the energy inner-product as there holds $a(u_j, u_k) = \lambda_j b(u_j, u_k) = \lambda_j \delta_{jk}$.

3.2 IGA Discretized Level

At the discretized level, we first discretize the domain Ω_ϵ with a uniform tensor-product mesh. We denote a general element as τ and its collection as $\mathcal{T}_h$ such that $\overline{\Omega}_\epsilon = \cup_{\tau \in \mathcal{T}_h} \tau$. Let $h = \max_{\tau \in \mathcal{T}_h} \text{diameter}(\tau)$. In the IGA setting, for simplicity, we use the B-splines. The B-spline basis functions in 1D are given as the Cox-de Boor recursion formula; we refer to [12, 29] for details. Let $X = \{x_0, x_1, \cdots, x_m\}$ be a knot vector with a nondecreasing sequential knots x_j. The j-th B-spline basis function of degree p, denoted as $\phi_p^j(x)$, is defined recursively as

$$\begin{aligned} \phi_0^j(x) &= \begin{cases} 1, & \text{if } x_j \leq x < x_{j+1}, \\ 0, & \text{otherwise,} \end{cases} \\ \phi_p^j(x) &= \frac{x - x_j}{x_{j+p} - x_j} \phi_{p-1}^j(x) + \frac{x_{j+p+1} - x}{x_{j+p+1} - x_{j+1}} \phi_{p-1}^{j+1}(x). \end{aligned} \tag{10}$$

A tensor-product of these 1D B-splines produces the B-spline basis functions in multiple dimensions. We define the multi-dimensional approximation space as $V_p^h \subset H_0^1(\Omega_\epsilon)$ with:

$$V_p^h = \text{span}\{\phi_p^j\}_{j=1}^{N_h} = \begin{cases} \text{span}\{\phi_{p_x}^{j_x}(x)\}_{j_x=1}^{N_x}, & \text{in 1D,} \\ \text{span}\{\phi_{p_x}^{j_x}(x)\phi_{p_y}^{j_y}(y)\}_{j_x,j_y=1}^{N_x,N_y}, & \text{in 2D,} \end{cases}$$

where p_x, p_y specify the approximation order in each dimension. N_x, N_y is the total number of basis functions in each dimension and N_h is the total number of degrees of freedom. The isogeometric analysis of (6) in variational formulation seeks $\lambda^h \in \mathbb{R}$ and $u^h \in V_p^h$ with $\|u^h\|_{\Omega_\epsilon} = 1$ such that

$$a(w^h, u^h) = \lambda^h b(w^h, u^h), \quad \forall\, w^h \in V_p^h. \tag{11}$$

3.3 Algebraic Level

At the algebraic level, we approximate the eigenfunctions as a linear combination of the B-spline basis functions, i.e.,

$$u^h = \sum_{j=1}^{N_h} \nu_j \phi_p^j,$$

where $\nu_j, j = 1, \cdots, N_h$ are the coefficients. We then substitute all the B-spline basis functions for w^h in (11). This leads to the generalized matrix eigenvalue problem

$$\mathbf{KU} = \lambda^h \mathbf{MU}, \tag{12}$$

where $\mathbf{K}_{kl} = a(\phi_p^k, \phi_p^l), \mathbf{M}_{kl} = b(\phi_p^k, \phi_p^l)$, and $\mathbf{U}$ is the corresponding representation of the eigenvector as the coefficients of the B-spline basis functions. The homogeneous Dirichlet boundary condition (7) can be set by removing the rows and columns corresponding to the degrees of freedom associated with the nodes at the boundary. This matrix eigenvalue problem is to be solved in a computing program.

3.4 A Priori Error Estimates

IGA is a Galerkin finite element discretization method. On a rectangular domain with tensor-product grids, the only difference of IGA from the classical FEA is the basis functions. FEA adopts C^0 polynomials as basis function while IGA adopts $C^k, k \geq 1$ polynomials. We observe that C^0 is a larger space, i.e., $C^k(\Omega_\epsilon) \subset C^0(\Omega_\epsilon)$. In general, for an a priori error estimate that is established in the Galerkin FEA framework, the estimate also holds for IGA. Thus, we expect optimal convergence rates for the eigenvalues and eigenfunctions as in FEA [2,9,17]. We present the following estimate without a theoretical proof. Instead, we show numerical validation in Sect. 4.

Given the mesh configuration and IGA setting described above, let $(\lambda_j, u_j) \in \mathbb{R}^+ \times H_0^1(\Omega)$ solve (8) for bound states and let $(\lambda_j^h, u_j^h) \in \mathbb{R}^+ \times V_p^h$ solve (11) for bound states with the normalizations $\|u_j\|_{\Omega_\epsilon} = 1$ and $\|u_j^h\|_{\Omega_\epsilon} = 1$. Assuming elliptic regularity on the operator $\mathcal{L} = -\nabla \cdot (\kappa \nabla) + \gamma$ and high-order smoothness of the eigenfunctions u_j on Ω_ϵ, there holds:

$$|\lambda_j^h - \lambda_j| \leq Ch^{2p}, \qquad |u_j - u_j^h|_{H^1(\Omega)} \leq Ch^p, \tag{13}$$

where C is a positive constant independent of the mesh-size h. We remark that these estimates only hold for bound states and do not necessarily hold for scattering states.

4 Numerical Experiments

In this section, we present various numerical examples to demonstrate the performance of IGA. We first show the IGA approximation optimal convergence accuracy with a domain Ω_ϵ of large size. Then we study the impact of the domain size on accuracy and give an approximate formula that determines the size of the domain given a certain accuracy tolerance.

4.1 IGA Discretization Accuracy

We focus on the two- and three-body problems with a potential with polynomial decay

$$f(\xi) = \frac{1}{(1+\xi^2)^3} \tag{14}$$

of the cube of a Lorentzian and one with exponential decay

$$f(\xi) = e^{-\xi^2} \tag{15}$$

of a Gaussian. For these potentials, finding the exact analytical solutions is impossible. For the purpose of characterizing the errors, we use, as a reference solution to the exact one, the solution of IGA with a septic C^6 B-spline basis functions and a fine mesh. We focus on the eigenvalue error that is defined as

$$e_j = |\lambda_j^h - \hat{\lambda}_j|, \tag{16}$$

where λ_j^h is an IGA eigenvalue and $\hat{\lambda}_j$ is a reference eigenvalue that is of high accuracy approximating the exact one λ_j.

Figure 2 shows the eigenvalue error convergence rates for IGA of the two-body problem with $\kappa = \frac{1}{2}$ in 1D. We consider C^0 linear, C^1 quadratic, and C^2 cubic IGA elements. We study the problem with both potentials (14) and (15) and a fixed magnitude $\beta = 1$. For both potentials (14) and (15), there is one bound state. The state reference eigenvalue is $\hat{\lambda}_1 = -0.31658012845$ for (14) and $\hat{\lambda}_1 = -0.47738997738$ for (15), respectively. We solve the problem for the bound state using C^6 septic IGA with 5000 uniform elements over the domain $\Omega_\epsilon = [-20, 20]$. We observe optimal error convergence rates in all the scenarios. This confirms the theoretical prediction (13) in Sect. 3.4.

Now we consider a case where there are two bound states in the two-body problem. Let $\beta = 5$ and we apply a potential of polynomial decay (14). Figure 3 shows the two bound states solutions and their eigenenergies are -2.9149185630 and -0.25417134380. Herein, the numerical eigenstates are computed using C^6 septic IGA with 5000 uniform elements over the domain $\Omega_\epsilon = [-20, 20]$. The plot shows the eigenstate over $[-10, 10]$ for better focus while the problem is solved over the larger domain $\Omega_\epsilon = [-20, 20]$ for high accuracy.

Figure 4 shows the eigenenergy error convergence rates of the problem described above (also shown in Fig. 3). Again, we observe optimal error convergence rates that verify the theoretical prediction. Moreover, for IGA with higher-order elements, the eigenvalue errors reach small errors faster with coarser meshes. This validates that the reference solutions obtained by using C^6 septic IGA with 5000 elements are of high accuracy and can be used as highly accurate approximations to the exact solutions.

4.2 A Study on Domain Size

For bound states of the two-body problems as discussed in Sect. 2.2, the state values approach zero exponentially fast. The IGA discretization requires a finite

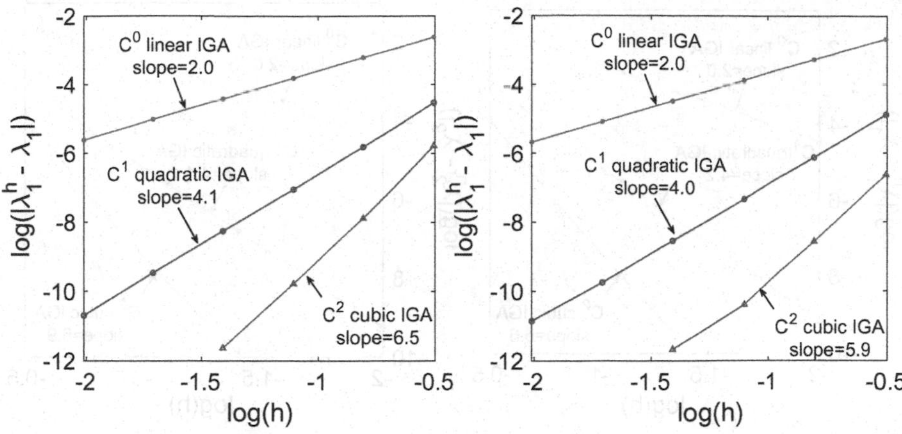

Fig. 2. Eigenvalue error convergence rates of IGA for the two-body problem with $\beta = 1$ on a domain $\Omega_\epsilon = [-20, 20]$. The potential has polynomial decay (14) for the left plot while exponential decay (15) for the right plot.

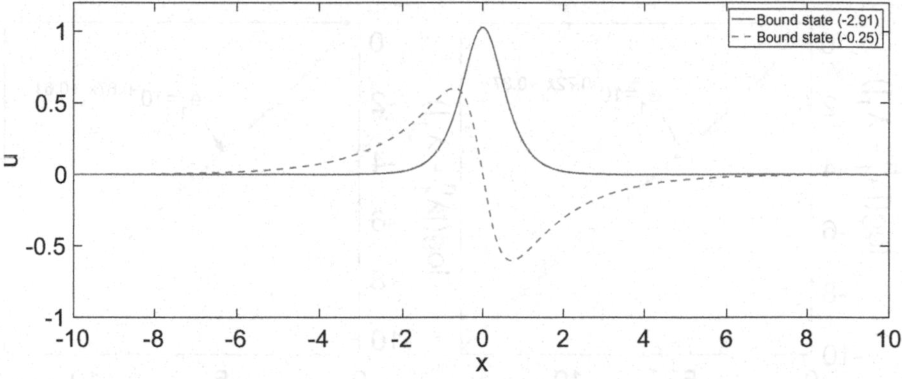

Fig. 3. The bound states of the two-body problem with $\beta = 5$ and a potential of polynomial decay (14).

domain with homogeneous boundary condition (7). The accuracy depends on the domain size x_ϵ for $\Omega_\epsilon = [-x_\epsilon, x_\epsilon]$.

To study the impact of the domain size x_ϵ on the accuracy, we apply the high-accuracy IGA method with C^6 septic B-spline elements. We apply uniform mesh grids with a fixed grid size $h = 0.01$. This setting of using a high-order element with fine grid size is to guarantee that the errors are dominated by the choice of the domain size. Figure 5 shows how the eigenvalue errors decrease when the domain size x_ϵ increases. We set $\kappa = 1/2$ in (6) and the potential magnitude $\beta = 1$. The left plot of Fig. 5 shows that the eigenvalue error decays exponentially when x_ϵ increases for the potential (14) while the right plot of Fig. 5 shows that of the potential (15). The fitted functions that establish the

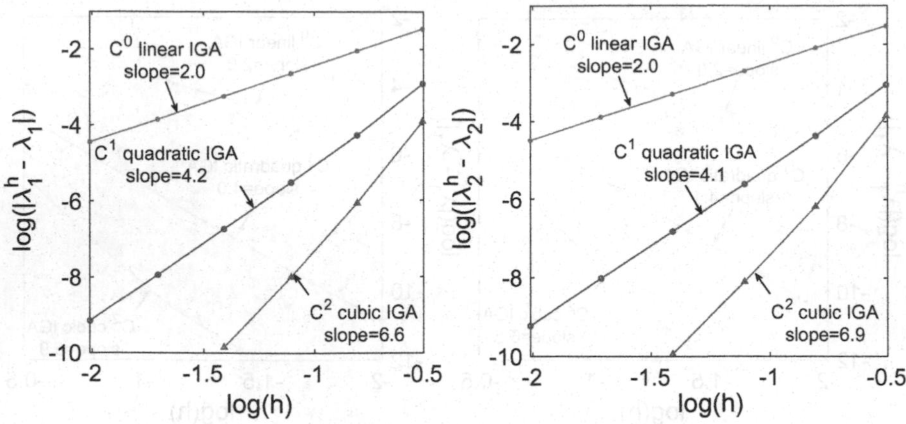

Fig. 4. Eigenvalue error convergence rates of IGA for the two-body problem with $\beta = 5$ and a polynomially decaying potential (14) on a domain $\Omega_\epsilon = [-20, 20]$. The left plot shows the eigenenergy errors of first bound state while the right plot shows these of the second bound state.

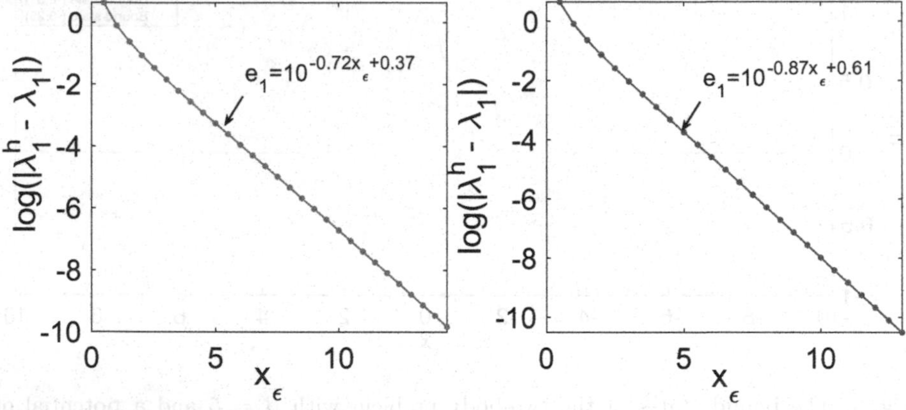

Fig. 5. Eigenvalue errors versus domain size x_ϵ for the two-body problem with polynomially decaying potential (14) (left plot) and exponentially decaying potential (15) (right plot).

relation between the error and the domain size are $e_1 = 10^{-0.72x_\epsilon+0.37}$ and $e_1 = 10^{-0.87x_\epsilon+0.61}$ for (14) and (15), respectively. Figure 6 shows the case where there are two bound states. Therein, potential (15) is used with $\beta = 5$. We observe a similar behaviour. The errors of the ground state reach an order of 10^{-12} when $x_\epsilon \geq 6.5$. This is due to that the IGA discretization error dominates the overall error (from discretization and approximation of the domain). The fitted functions give guidance for choosing the domain Ω_ϵ appropriately. For example, for the two-body problem with a potential (15) and $\beta = 5$, the fitted function for the second bound state (a larger domain is required to compute this mode) is $e_2 = 10^{-0.68x_\epsilon+0.69}$. Thus, to achieve an accuracy of error 10^{-15}, we set $10^{-15} = 10^{-0.68x_\epsilon+0.69}$ and solve for x_ϵ to get the domain $\Omega_\epsilon = [-x_\epsilon, x_\epsilon] = [-23.1, 23.1]$. This means that we require to solve the problem with a minimal domain size $x_\epsilon = 23.1$ to get an accuracy of order 10^{-15}.

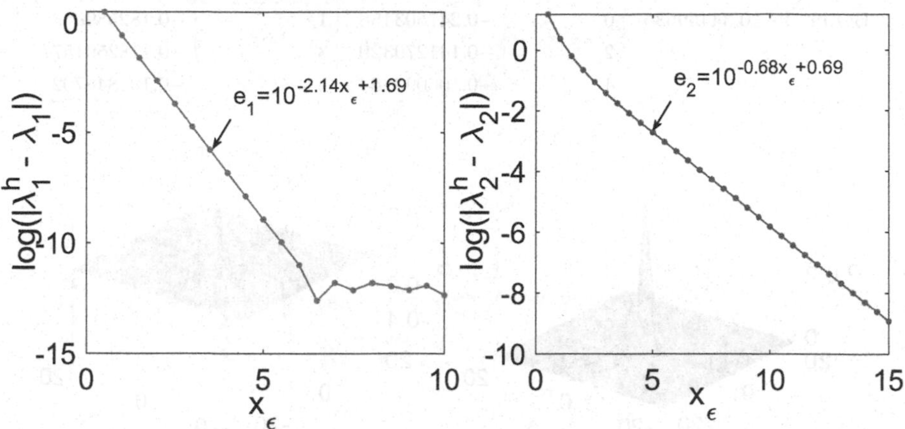

Fig. 6. Eigenvalue errors versus domain size x_ϵ for the two bound states of the two-body problem with exponentially decaying potential (15).

4.3 Three-Body Problem

Now, we consider the three-body problem with a heavy-light body ratio $m_h/m_l = 20$ that is studied in [21,22,35]. With such a mass ratio, κ in the unified problem (6) is a matrix with entries $\kappa = (41/84, 0; 0, 2/21)$. Our goal is to approximate the eigenenergies and eigenstates obtained by using the classical BO approximation in these papers. This preliminary numerical study demonstrates that the proposed IGA method is a promising alternative to the classical BO approximation method that a strong assumption is posed on the mass ratio.

To solve the three-body problem, we apply the highly accurate IGA method with C^6 septic B-spline elements. We set the domain as $\Omega_\epsilon = [-20, 20]$ and apply a non-uniform grid with 80×80 elements. Table 1 shows the eigenvalues of the bound states of the three-body problem with the exponentially decaying potential (15). The potential magnitude is $\beta = 0.344595351$. The IGA eigenvalues are close to the ones (scaled) shown in Table 1 of [35]. Figure 7 shows the first four bound state eigenfunctions. The eigenstate solution shapes match well with the ones obtained using the BO approximation in Fig. 4 of [21]. Moreover, we observe a similar universality behaviour as in [21] and we will present a detailed study in future work. In conclusion, the IGA method with a small mesh grid has the ability to approximate well both eigenenergies and eigenfunctions of the bound states of the three-body problem.

Table 1. Eigenenergies of the three-body problem with a mass ratio 20 and an exponentially decaying potential (15) when using IGA and BO approximation in [35].

Method	β	j (Bosons)	λ_j^h	j (Fermions)	λ_j^h
IGA	0.344595351	0	-0.2476034576	1	−0.1825896533
		2	−0.1412793292	3	−0.1182591543
		4	−0.1060931444	5	−0.1005294105
BO in [35]	0.34459535	0	−0.247603458	1	−0.182589653
		2	−0.141279329	3	−0.118259157
		4	−0.106093864	5	−0.102845702

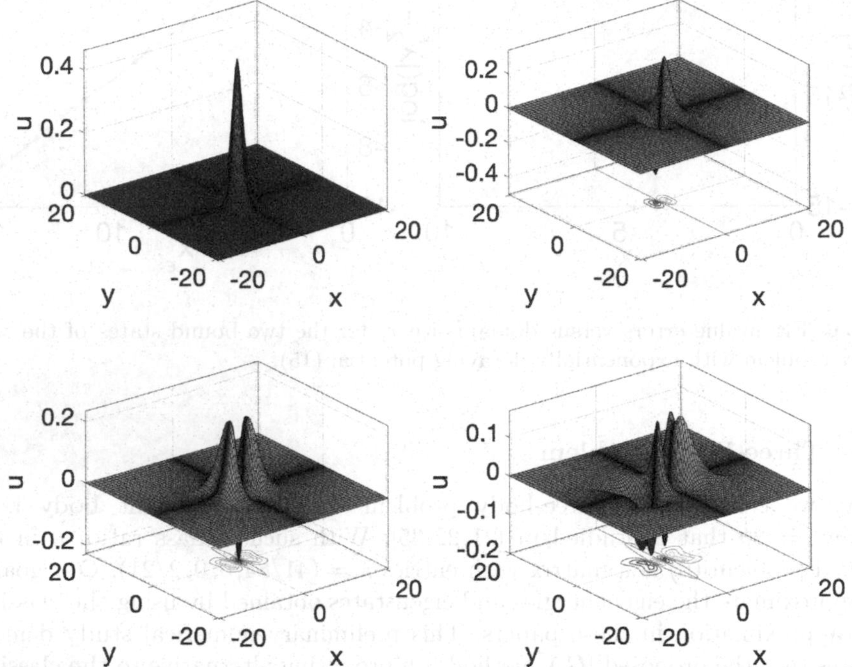

Fig. 7. Eigenstate wavefunctions of the three-body problem with a mass ratio 20, potential magnitude $\beta = 0.344595351$, and the exponentially decaying potential (15).

5 Concluding Remarks

In this paper, we initiated IGA of the quantum two- and three-body problems. IGA is developed based on the classical Galerkin FEA that has the advantages of a mature theoretical understanding of the error estimates, stabilities, and robustness. IGA is successfully applied to solve the bound states of the two- and three-body problems in 1D with arbitrary mass ratios and potential shapes.

As for future work, the first possible direction is a generalization to the two- and three-body problems in multiple dimensions. Tensor-product structures may be applied to reduce the computational costs [35]. Another direction of future work is that one may use the recently developed softFEM [14] and dispersion-minimized blending quadratures [7,13] to solve the three-body problem with higher accuracy and efficiency.

References

1. Agmon, S.: Lectures on Exponential Decay of Solutions of Second-Order Elliptic Equations. Princeton University Press, Princeton (2014)
2. Babuška, I., Osborn, J.: Eigenvalue problems. In: Handbook of Numerical Analysis, vol. II, pp. 641–787. North-Holland, Amsterdam (1991)
3. Baer, M.: Beyond Born-Oppenheimer: Electronic Nonadiabatic Coupling Terms and Conical Intersections. Wiley, Hoboken (2006)
4. Boyd, J.P.: Chebyshev and Fourier Spectral Methods. Courier Corporation, Chelmsford (2001)
5. Breen, P.G., Foley, C.N., Boekholt, T., Zwart, S.P.: Newton versus the machine: solving the chaotic three-body problem using deep neural networks. Mon. Notices Royal Astron. Soc. **494**(2), 2465–2470 (2020)
6. Brezis, H.: Functional Analysis, Sobolev Spaces and Partial Differential Equations. Universitext, Springer, New York (2011). https://doi.org/10.1007/978-0-387-70914-7
7. Calo, V., Deng, Q., Puzyrev, V.: Dispersion optimized quadratures for isogeometric analysis. J. Comput. Appl. Math. **355**, 283–300 (2019)
8. Cederbaum, L.S.: Born-Oppenheimer approximation and beyond for time-dependent electronic processes. J. Chem. Phys. **128**(12), 124101 (2008)
9. Ciarlet, P.G.: Finite Element Method for Elliptic Problems. Society for Industrial and Applied Mathematics, Philadelphia (2002)
10. Cottrell, J.A., Reali, A., Bazilevs, Y., Hughes, T.J.R.: Isogeometric analysis of structural vibrations. Comput. Meth. Appl. Mech. Eng. **195**(41–43), 5257–5296 (2006)
11. Cottrell, J.A., Hughes, T.J.R., Bazilevs, Y.: Isogeometric Analysis: Toward Integration of CAD and FEA. Wiley, Hoboken (2009)
12. De Boor, C.: A Practical Guide to Splines, vol. 27. Springer, New York (1978). https://doi.org/10.1007/978-1-4612-6333-3
13. Deng, Q., Calo, V.: Dispersion-minimized mass for isogeometric analysis. Comput. Meth. Appl. Mech. Eng. **341**, 71–92 (2018)
14. Deng, Q., Ern, A.: SoftFEM: revisiting the spectral finite element approximation of second-order elliptic operators. Comput. Math. Appl. **101**, 119–133 (2021)
15. Drut, J.E., McKenney, J.R., Daza, W.S., Lin, C.L., Ordóñez, C.R.: Quantum anomaly and thermodynamics of one-dimensional fermions with three-body interactions. Phys. Rev. Lett. **120**(24), 243002 (2018)
16. Efremov, M.A., Plimak, L., Berg, B., Ivanov, M.Y., Schleich, W.P.: Efimov states in atom-molecule collisions. Phys. Rev. A **80**(2), 022714 (2009)
17. Ern, A., Guermond, J.L.: Finite Elements II: Galerkin Approximation, Elliptic and Mixed PDEs. Springer, New York (2020). https://doi.org/10.1007/978-3-030-56923-5
18. Fonseca, A.C., Redish, E.F., Shanley, P.: Efimov effect in an analytically solvable model. Nucl. Phys. A **320**(2), 273–288 (1979)

19. Griffiths, D.J., Schroeter, D.F.: Introduction to Quantum Mechanics. Cambridge University Press, Cambridge (2018)
20. Guo, P., Gasparian, V.: Numerical approach for finite volume three-body interaction. Phys. Rev. D **97**(1), 014504 (2018)
21. Happ, L., Zimmermann, M., Betelu, S.I., Schleich, W.P., Efremov, M.A.: Universality in a one-dimensional three-body system. Phys. Rev. A **100**(1), 012709 (2019)
22. Happ, L., Zimmermann, M., Efremov, M.A.: Universality of excited three-body bound states in one dimension. J. Phys. B At. Mol. Opt. Phys. **55**(1), 015301 (2022)
23. Hughes, T.J.R., Cottrell, J.A., Bazilevs, Y.: Isogeometric analysis: CAD, finite elements, NURBS, exact geometry and mesh refinement. Comput. Meth. Appl. Mech. Eng. **194**(39), 4135–4195 (2005)
24. Hughes, T.J.R., Evans, J.A., Reali, A.: Finite element and NURBS approximations of eigenvalue, boundary-value, and initial-value problems. Comput. Meth. Appl. Mech. Eng. **272**, 290–320 (2014)
25. Hughes, T.J.R., Reali, A., Sangalli, G.: Duality and unified analysis of discrete approximations in structural dynamics and wave propagation: comparison of p-method finite elements with k-method NURBS. Comput. Meth. Appl. Mech. Eng. **197**(49–50), 4104–4124 (2008)
26. Letellier, C.: Chaos in Nature, vol. 94. World Scientific, Singapore (2019)
27. Nielsen, E., Fedorov, D.V., Jensen, A.S., Garrido, E.: The three-body problem with short-range interactions. Phys. Rep. **347**(5), 373–459 (2001)
28. Panati, G., Spohn, H., Teufel, S.: The time-dependent Born-Oppenheimer approximation. ESAIM: Math. Model. Numer. Anal. **41**(2), 297–314 (2007)
29. Piegl, L., Tiller, W.: The NURBS Book. Springer, Cham (1997). https://doi.org/10.1007/978-3-642-97385-7
30. Pisana, S., et al.: Breakdown of the adiabatic Born-Oppenheimer approximation in graphene. Nat. Mater. **6**(3), 198–201 (2007)
31. Puzyrev, V., Deng, Q., Calo, V.M.: Dispersion-optimized quadrature rules for isogeometric analysis: modified inner products, their dispersion properties, and optimally blended schemes. Comput. Meth. Appl. Mech. Eng. **320**, 421–443 (2017)
32. Scherrer, A., Agostini, F., Sebastiani, D., Gross, E., Vuilleumier, R.: On the mass of atoms in molecules: beyond the Born-Oppenheimer approximation. Phys. Rev. X **7**(3), 031035 (2017)
33. Skorniakov, G., Ter-Martirosian, K.: Three body problem for short range forces. I. Scattering of low energy neutrons by deuterons. Soviet Phys. JETP **4** (1957)
34. Sukhareva, O.M., Grigorenko, L.V., Kostyleva, D.A., Zhukov, M.V.: Validity of quasi-classical approaches to true three-body decays. In: Orr, N.A., Ploszajczak, M., Marqués, F.M., Carbonell, J. (eds.) FB22 2018. SPP, vol. 238, pp. 283–286. Springer, Cham (2020). https://doi.org/10.1007/978-3-030-32357-8_50
35. Thies, J., Hof, M.T., Zimmermann, M., Efremov, M.: Exploiting tensor structure for computing bound states of the quantum mechanical three-body problem. arXiv preprint arXiv:2111.02534 (2021)

1D Painless Multi-level Automatic Goal-Oriented *h* and *p* Adaptive Strategies Using a Pseudo-Dual Operator

Felipe Vinicio Caro[1,2](✉), Vincent Darrigrand[3], Julen Alvarez-Aramberri[2], Elisabete Alberdi Celaya[2], and David Pardo[1,2,4]

[1] Basque Center for Applied Mathematics (BCAM), Bilbao, Spain
[2] University of the Basque Country (UPV-EHU), Leioa, Spain
fcaro001@ikasle.ehu.eus
[3] CNRS-IRIT, Toulouse, France
[4] IKERBASQUE, Basque Foundation for Science, Bilbao, Spain

Abstract. The main idea of our Goal-Oriented Adaptive (GOA) strategy is based on performing global and uniform *h*- or *p*-refinements (for *h*- and *p*-adaptivity, respectively) followed by a coarsening step, where some basis functions are removed according to their estimated importance. Many Goal-Oriented Adaptive strategies represent the error in a Quantity of Interest (QoI) in terms of the bilinear form and the solution of the direct and adjoint problems. However, this is unfeasible when solving indefinite or non-symmetric problems since symmetric and positive definite forms are needed to define the inner product that guides the refinements. In this work, we provide a Goal-Oriented Adaptive (*h*- or *p*-) strategy whose error in the QoI is represented in another bilinear symmetric positive definite form than the one given by the adjoint problem. For that purpose, our Finite Element implementation employs a multi-level hierarchical data structure that imposes Dirichlet homogeneous nodes to avoid the so-called hanging nodes. We illustrate the convergence of the proposed approach for 1D Helmholtz and convection-dominated problems.

Keywords: Goal-oriented adaptivity · Pseudo-dual operator · Unrefinements · Finite element method · Multi-level

1 Introduction

One of the main challenges of Finite Element Methods (FEM) is to obtain accurate solutions with low memory requirements. Realistic models are often geometrically complex, and they usually exhibit inhomogeneities. Energy-norm-based adaptive techniques are often employed to model these complex problems. However, many engineering applications demand accurate solutions only in specific domain areas, for example, when the objective is to simulate some measurements

D. Groen et al. (Eds.): ICCS 2022, LNCS 13351, pp. 347–357, 2022.
https://doi.org/10.1007/978-3-031-08754-7_43

at particular receivers. In these scenarios, GOA strategies have shown success for more than twenty years [2,17].

The objective of goal-oriented adaptivity is to build an optimal finite-element grid that minimizes the size of the problem needed to achieve certain tolerance errors for some practical QoI, which is expressed in terms of a linear functional. It has been widely used across different areas of knowledge, including electromagnetic (EM) applications [1,12,16], Cahn-Hilliard-Navier-Stokes systems [10], visco-elasticity [19], and fluid-structure interactions [11]. Traditional approaches represent the error in the QoI by using the direct and adjoint solutions and the global bilinear form of the problem and dividing it in terms of local and computable quantities that are used to guide local refinements (see e.g. [15]).

Here, we follow a different approach. Based on Darrigrand et al. [4], we define an alternative pseudo-dual operator to represent the residual error of the adjoint problem. This new representation, which exhibits better properties than the original bilinear form (e.g., positive definiteness), has proved successful [5,13] and allows to compute the error in the QoI in a way similar to the classical approaches.

The present work combines the energy-based approach introduced in [3] (which uses the data structure proposed by Zander et al. [20–22]) and an alternative pseudo-dual operator for representation of the error in the QoI [4]. By doing so, we extend the Darrigrand et al. approach [3] to the context of h- and p-GOA algorithms.

This document is organized as follows: Sect. 2 describes the GOA strategy and the employed error estimators. Section 3 is devoted to the numerical results, and Sect. 4 summarizes the main conclusions.

2 Proposed Goal-Oriented Adaptive Algorithms

The h- and p-adaptive algorithms proposed in this work follow the next refinement pattern: first, we perform a global and uniform h- or p-refinement (for the h- and p-adaptive versions, respectively). Then, we perform a coarsening step, where some basis functions are removed. This procedure is illustrated in Algorithm 1, and it was already introduced in [3] in the context of energy-norm adaptivity. The critical part is the coarsening step that we describe next.

Algorithm 1: Adaptive process

```
Input: An initial mesh
Output: A final adapted mesh
while error > tolerance do
    Perform a global and uniform (h or p) refinement;
    Update the error;
    Execute the coarsening step (Algorithm 2) to the mesh;
end
```

Optimal unrefinements are performed following an element-by-element approach. Using the multi-level data structures proposed in Zander et al. [20–22], we compute element-wise error indicators of all *active* elements, i.e., those elements that do not have sons, or if they do, all new nodes of the sons have homogeneous Dirichlet boundary conditions. These element-based error indicators are one number per element for the h-adaptive version and d numbers for the p-adaptive version, where d is the dimension. The coarsening step procedure is depicted in Algorithm 2. The critical step here is the computation the element-wise error indicators, which we describe in the following subsection.

Algorithm 2: Coarsening processs

Input: A given mesh M.
Output: An unrefined adapted mesh, also denoted as M.
do
  Solve the problem on M;
  Compute the element-wise error indicators for the *active* elements;
  Mark the elements whose indicators are relatively small;
  Update M by unrefining the marked elements;
  If nothing has been marked, escape;
end ;

2.1 Error Indicators

We first introduce our boundary value problem in variational form using an abstract formulation:

Find $u \in V$ such that

$$b(u, v) = f(v), \quad \forall v \in V. \tag{1}$$

Here, f is a linear form, b represents a bilinear form and the space V is assumed to be $V = V(\Omega) := \{u \in H^1(\Omega) : u = 0 \text{ on } \Gamma_D,\ u' = 0 \text{ on } \Gamma_N\}$, where Ω is a one-dimensional (1D) computational domain, and Γ_D and Γ_N are the parts of the boundary where we impose homogeneous *Dirichlet* and *Neumann* boundary conditions.

The above *forward* problem has an associated *adjoint* (or *dual*) operator, whose formulation is given by:

Find $w \in V$ such that

$$b(v, w) = l(v), \quad \forall v \in V. \tag{2}$$

Here, l is a linear functional that represents the QoI.

The adjoint problem is often employed in the literature to guide goal-oriented refinements (see, e.g., [14]). However, for the case of indefinite or non-symmetric problems, we further need to introduce an inner product (symmetric and positive definite form) to guide the refinements.

To overcome this issue, we first define $\widetilde{w}$ as a projection of the dual solution w into a given subset of basis functions by simply removing the remaining basis functions' degrees of freedom (DoFs). For the p-adaptive case, the subset of basis functions results from a global p-unrefinement from the current mesh. Thus, $\widetilde{w}$ consists of taking w and replacing the DoF of the highest-order basis functions of each element with zero. In the h-adaptive case, the subset of basis functions results from a global h-unrefinement of the given mesh. Such projections can be trivially implemented in the context of the multi-level data structures proposed in Zander et al. [20–22]; but not when using traditional data structures like those described in [7–9]. Then, we introduce a *pseudo-dual* bilinear form $\hat{b}$, in our case, defined by the 1D Laplace operator (although it is possible to select other symmetric positive definite bilinear forms). Finally, we solve the following residual-based *pseudo-dual* problem:

Find ε such that

$$\hat{b}(v,\varepsilon) = l(v) - b(v,\widetilde{w}), \quad \forall v \in V. \tag{3}$$

The idea of using an elliptic error representation was already introduced by Romkes et al. [18] and applied by Darrigrand et al. [4] in the context of traditional data structures. However, it required dealing with two grids (fine and coarse) and projection based interpolation operators [6–8], which highly complicated its implementation and mathematical analysis. In here, we define problem (3) using simply $\widetilde{w}$ as the projection of w.

Thus, we define E_K as the error indicator associated with element K:

$$E_K := \left|\hat{b}(\bar{e}_K,\varepsilon)\right|, \tag{4}$$

where $\bar{e}_K$ is defined for the h-adaptive version as the DoF of u multiplied by the basis function whose support is within the father of the active element K. For the p-version, $\bar{e}_K$ is the DoF of u multiplied by the highest-order basis function whose support is contained within the active element K.

If we assume quasi-orthogonality of our multi-level basis functions, that is, $b(\bar{e}_K,\widetilde{w}) \simeq 0$, then:

$$E_K := \left|\hat{b}(\bar{e}_K,\varepsilon)\right| \simeq \left|b(\bar{e}_K,\widetilde{w}) + \hat{b}(\bar{e}_K,\varepsilon)\right| \simeq \left|b(e_K,\widetilde{w}) + \hat{b}(e_K,\varepsilon)\right| = |l(e_K)|. \tag{5}$$

In the above, we have used Eq. (3) and defined e_K as the error due to unrefining (in h or p) the element K.

The above error indicators can be extended to 2D and 3D. To account for the possibility of having multiple basis functions in the definition of $\bar{e}_K$, we divide the error indicators by the number of DoFs in $\bar{e}_K$. In 1D, this number is simply one.

3 Numerical Results

To illustrate the performance of our adaptive strategies, we consider two problems, governed by Helmholtz and convection-dominated equations. We provide the evolution of the relative error in the QoI for h- and p-adaptivity and for different values of the PDE parameters. To define the relative error in the QoI, we compute $l(u)$ on a globally refined mesh. Then, we describe the relative error as follows:

$$e_{\text{rel}} = \frac{|l(u) - l(u_c)|}{|l(u)|} \cdot 100, \tag{6}$$

where $l(u_c)$ is the QoI associated with the adapted mesh.

3.1 Helmholtz Goal-Oriented Problem

Let us consider the following wave propagation problem:

Find u such that,

$$-u'' - k^2 u = \mathbb{1}_{(0,\frac{2}{5})} \text{ in } (0,1), \tag{7}$$

$$u(0) = 0, \tag{8}$$

$$u'(1) = 0. \tag{9}$$

We define the QoI as $l(u) = 5 \cdot \int_{\frac{3}{5}}^{\frac{4}{5}} u\, dx$. Figures 1 and 2 show the evolution of the relative error in the QoI by using h- and p-adaptivity, respectively. Note that the larger the number of DoFs per wavelength, the faster the decrease of the relative error in the QoI. For example, in Fig. 1, for $k = 7 \cdot 2\pi$, 10 DoFs per wavelength are sufficient to enter into the so called asymptotic regime. In contrast, for $k = 28 \cdot 2\pi$, we need to consider at least 40 DoFs per wavelength. In Fig. 2, we select the initial mesh size such that the number of DoFs per wavelength is at least 3. This way, we satisfy the Nyquist rate. Both Figs. 1 and 2 show optimal convergence rates in both h- and p-adaptivity. As a curiosity, we observe that the curves in Fig. 1 are parallel, while the ones in Fig. 2 coincide. This occurs due to dispersion (pollution) error, which quickly disappears with the p-method.

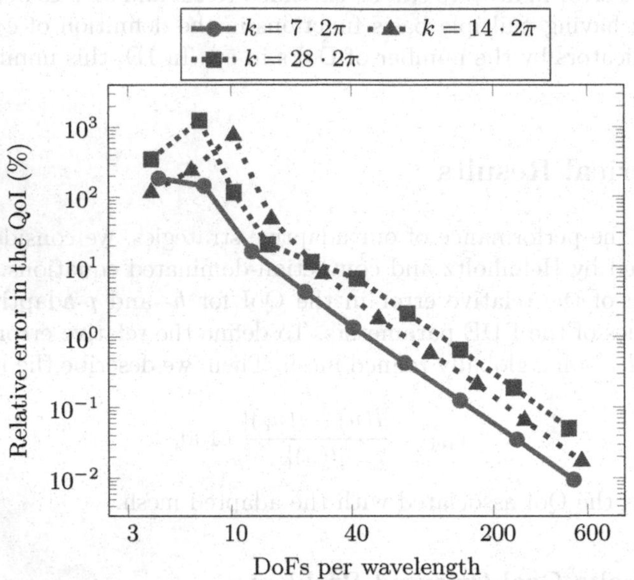

Fig. 1. Evolution of the relative error in the QoI by using h-adaptivity. Initial mesh size $h = \frac{1}{30}$ and uniform $p = 1$.

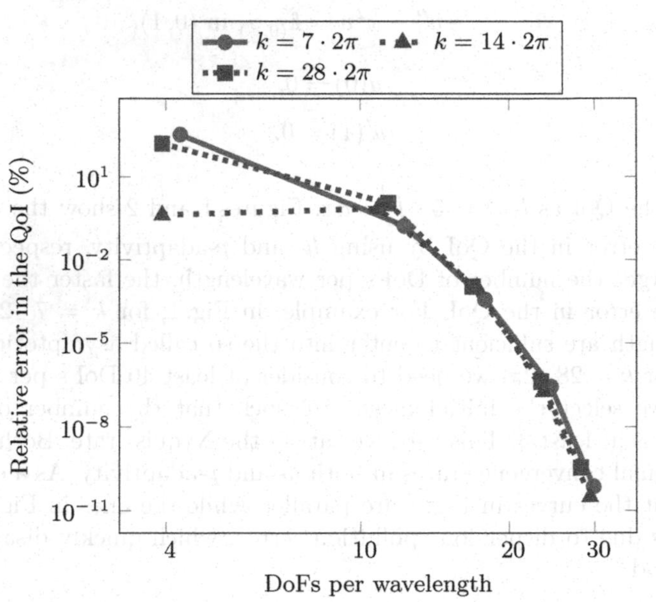

Fig. 2. Evolution of the relative error in the QoI by using p-adaptivity. Uniform mesh size $h = \frac{1}{30}$.

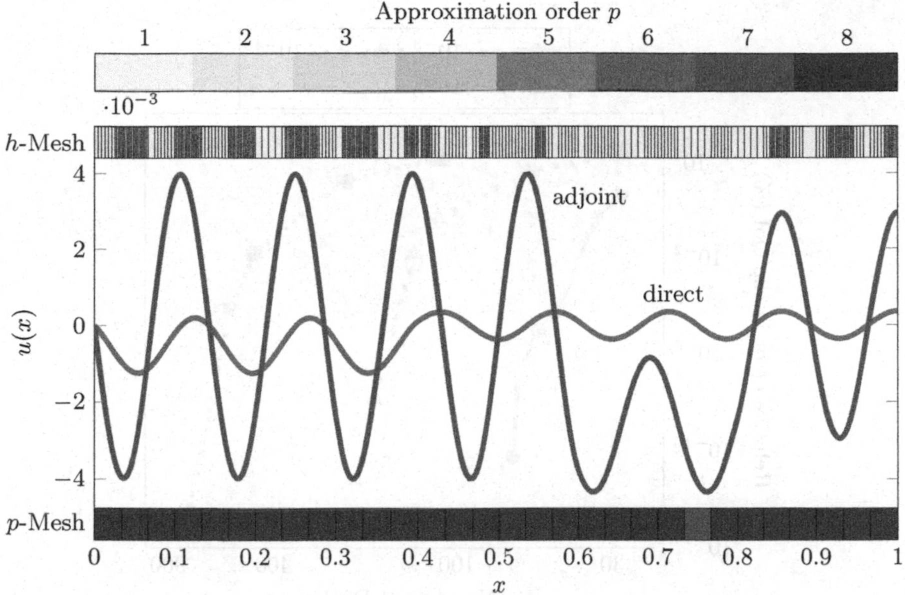

Fig. 3. Solutions with $k = 7 \cdot 2\pi$ problem as given after the h-adaptive process.

Figure 3 shows the solutions for the case $k = 7 \cdot 2\pi$. We also provide the corresponding h- and p-adaptive meshes. For the p-adaptive mesh, we show the mesh obtained in the 6th iteration, containing high approximation orders. To visualize the h-adaptive mesh, we show the mesh obtained in the 5th iteration. Finally, we show the solutions corresponding to the 6th iteration, which contains small localized values of the mesh size h.

3.2 Convection-Dominated Goal-Oriented Problem

Let us consider the boundary value problem associated with steady convective-diffusive transport:

Find u such that,

$$\begin{aligned} -\varepsilon u'' + \sigma \cdot u' &= \mathbb{1}_{(0,1)} \text{ in } (0,1), \\ u(0) = u(1) &= 0, \end{aligned} \tag{10}$$

with $\sigma = 1$, and $0 < \varepsilon \ll 1$ the diffusive coefficient. We define the QoI as $l(u) = 5 \cdot \int_{\frac{4}{5}}^{1} \nabla u \, dx$.

In Figs. 4 and 5, we represent the evolution of the relative error in the QoI by using h- and p-adaptivity, respectively. We observe optimal convergence rates. We note that the smaller the diffusive coefficient ε, the larger the number of DoFs to reach the convergence rates.

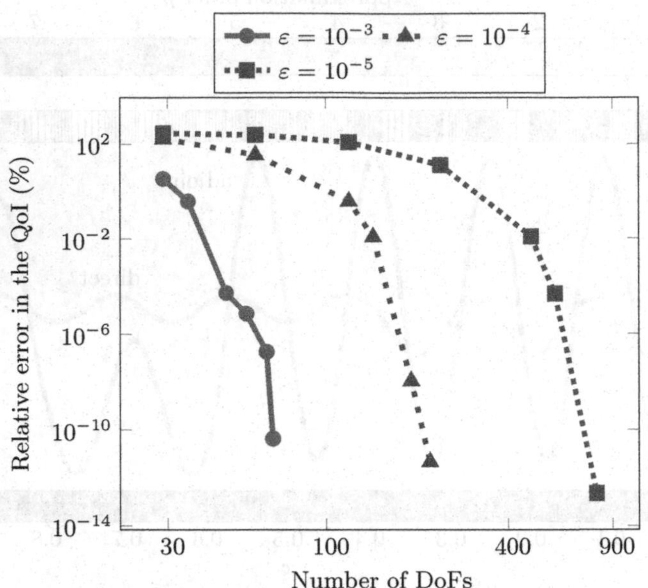

Fig. 4. Evolution of the relative error in the QoI by using h-adaptivity. Initial mesh size $h = \frac{1}{30}$ and uniform $p = 1$.

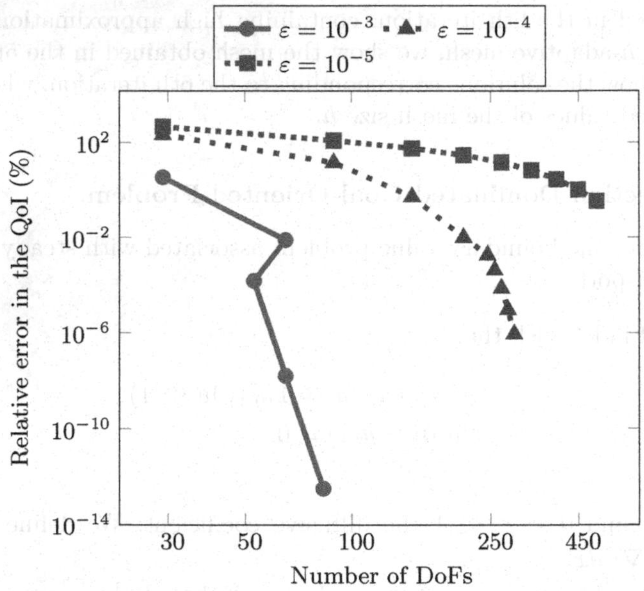

Fig. 5. Evolution of the relative error in the QoI by using p-adaptivity. Uniform mesh size $h = \frac{1}{30}$.

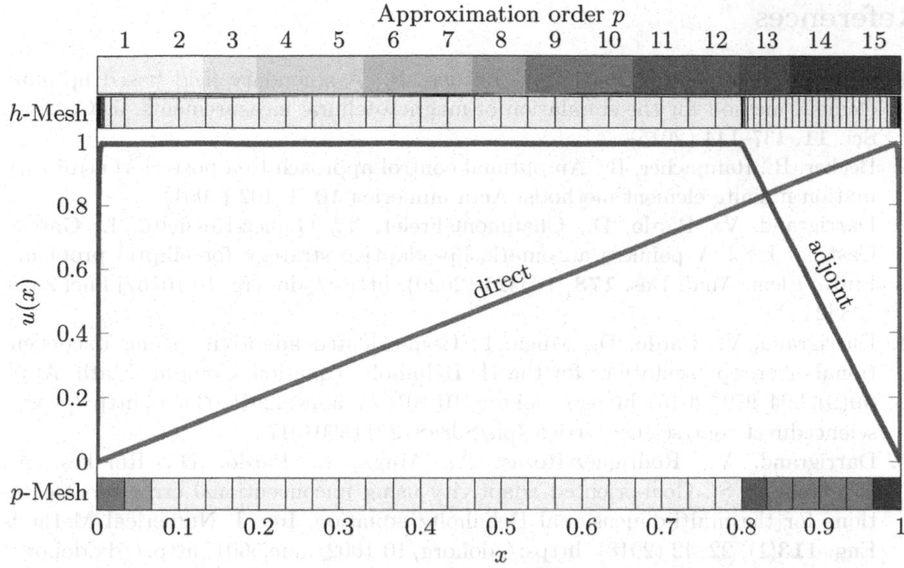

Fig. 6. Solutions with $\varepsilon = 10^{-3}$ problem as given after the h-adaptive process.

Figure 6 shows the solutions for the case $\varepsilon = 10^{-3}$. We also provide the meshes and the solutions corresponding to the last iteration in both h and p cases.

4 Conclusions

We propose h- and p-GOA strategies for possibly non-elliptic problems. These adaptive algorithms are simple-to-implement because they take advantage of multi-level data structures with hierarchical basis functions that avoid the problem of hanging nodes altogether.

The main idea of this approach consists of performing first a global and uniform refinement followed by a coarsening step, where some basis functions are removed. To select which basis functions are eliminated, we employ a representation of the error in the QoI that uses an unconventional symmetric and positive definite bilinear form.

1D numerical results show a proper convergence for Helmholtz and convection-dominated GOA problems when using the Laplace operator's pseudo-dual problem. This adaptive strategy can be easily extended to 2D and 3D problems, and it can be further exploited in other indefinite and/or non-symmetric problems.

References

1. Alvarez-Aramberri, J., Pardo, D., Barucq, H.: A secondary field based hp-finite element method for the simulation of magnetotelluric measurements. J. Comput. Sci. **11**, 137–144 (2015)
2. Becker, R., Rannacher, R.: An optimal control approach to a posteriori error estimation in finite element methods. Acta numerica **10**, 1–102 (2001)
3. Darrigrand, V., Pardo, D., Chaumont-Frelet, T., Gómez-Revuelto, I., Garcia-Castillo, L.E.: A painless automatic hp-adaptive strategy for elliptic problems. Finite Elem. Anal. Des. **178**, 103424 (2020). https://doi.org/10.1016/j.finel.2020.103424
4. Darrigrand, V., Pardo, D., Muga, I.: Goal-oriented adaptivity using unconventional error representations for the 1D Helmholtz equation. Comput. Math. Appl. **69**(9), 964–979 (2015). https://doi.org/10.1016/j.camwa.2015.03.006, http://www.sciencedirect.com/science/article/pii/S0898122115001017
5. Darrigrand, V., Rodríguez-Rozas, Á., Muga, I., Pardo, D., Romkes, A., Prudhomme, S.: Goal-oriented adaptivity using unconventional error representations for the multi-dimensional Helmholtz equation. Int. J. Numerical Methods Eng. **113**(1), 22–42 (2018). https://doi.org/10.1002/nme.5601, http://dx.doi.org/10.1002/nme.5601, nme.5601
6. Demkowicz, L., Rachowicz, W., Devloo, P.: A fully automatic *hp*-adaptivity. In: Proceedings of the Fifth International Conference on Spectral and High Order Methods (ICOSAHOM-01) (Uppsala), vol. 17, pp. 117–142 (2002). https://doi.org/10.1023/A:1015192312705,http://dx.doi.org/10.1023/A:1015192312705
7. Demkowicz, L.: Computing with *hp*-adaptive finite elements. One and two dimensional elliptic and Maxwell problems. Applied Mathematics and Nonlinear Science Series, vol. 1. Chapman & Hall/CRC, Boca Raton (2007). https://doi.org/10.1201/9781420011692, http://dx.doi.org/10.1201/9781420011692
8. Demkowicz, L., Kurtz, J., Pardo, D., Paszyński, M., Rachowicz, W., Zdunek, A.: Computing with *hp*-adaptive finite elements. Frontiers: Three Dimensional Elliptic and Maxwell Problems with Applications. Applied Mathematics and Nonlinear Science Series, vol. 2. Chapman & Hall/CRC, Boca Raton (2008)
9. Demkowicz, L., Oden, J.T., Rachowicz, W., Hardy, O.: Toward a universal hp adaptive finite element strategy, part 1. Constrained approximation and data structure. Comput. Methods Appl. Mech. Eng. **77**(1–2), 79–112 (1989)
10. Hintermüller, M., Hinze, M., Kahle, C., Keil, T.: A goal-oriented dual-weighted adaptive finite element approach for the optimal control of a nonsmooth Cahn-Hilliard-Navier-stokes system. Optim. Eng. **19**(3), 629–662 (2018)
11. Jhurani, C., Demkowicz, L.: Multiscale modeling using goal-oriented adaptivity and numerical homogenization. Part I: mathematical formulation and numerical results. Comput. Methods Appl. Mech. Eng. **213**, 399–417 (2012)
12. Key, K.: Mare2dem: a 2-D inversion code for controlled-source electromagnetic and magnetotelluric data. Geophys. J. Int. **207**(1), 571–588 (2016)
13. Muñoz-Matute, J., Alberdi, E., Pardo, D., Calo, V.M.: Time-domain goal-oriented adaptivity using pseudo-dual error representations. Comput. Methods Appl. Mech. Eng. **325**, 395–415 (2017)
14. Oden, J.T., Prudhomme, S.: Goal-oriented error estimation and adaptivity for the finite element method. Comput. Math. Appl. **41**(5-6), 735–756 (2001). https://doi.org/10.1016/S0898-1221(00)00317-5, http://dx.doi.org/10.1016/S0898-1221(00)00317-5

15. Ovall, J.S.: Asymptotically exact functional error estimators based on superconvergent gradient recovery. Numerische Mathematik **102**(3), 543–558 (2006)
16. Pardo, D., Demkowicz, L., Torres-Verdín, C., Paszynski, M.: Two-dimensional high-accuracy simulation of resistivity logging-while-drilling (LWD) measurements using a self-adaptive goal-oriented *hp* finite element method. SIAM J. Appl. Math. **66**(6), 2085–2106 (2006). https://doi.org/10.1137/050631732, http://dx.doi.org/10.1137/050631732
17. Prudhomme, S., Oden, J.T.: On goal-oriented error estimation for elliptic problems: application to the control of pointwise errors. Comput. Methods Appl. Mech. Eng. **176**(1–4), 313–331 (1999). https://doi.org/10.1016/S0045-7825(98)00343-0, http://dx.doi.org/10.1016/S0045-7825(98)00343-0
18. Romkes, A., Oden, J.T.: Adaptive modeling of wave propagation in heterogeneous elastic solids. Comput. Methods Appl. Mech. Eng. **193**(6–8), 539–559 (2004)
19. Van Der Zee, K.G., Tinsley Oden, J., Prudhomme, S., Hawkins-Daarud, A.: Goal-oriented error estimation for Cahn-Hilliard models of binary phase transition. Numer. Methods Partial Differ. Eqn. **27**(1), 160–196 (2011)
20. Zander, N., Bog, T., Elhaddad, M., Frischmann, F., Kollmannsberger, S., Rank, E.: The multi-level *hp*-method for three-dimensional problems: dynamically changing high-order mesh refinement with arbitrary hanging nodes. Comput. Methods Appl. Mech. Eng. **310**, 252–277 (2016). https://doi.org/10.1016/j.cma.2016.07.007, http://www.sciencedirect.com/science/article/pii/S0045782516307289
21. Zander, N., Bog, T., Kollmannsberger, S., Schillinger, D., Rank, E.: Multi-level *hp*-adaptivity: high-order mesh adaptivity without the difficulties of constraining hanging nodes. Comput. Mech. **55**(3), 499–517 (2015). https://doi.org/10.1007/s00466-014-1118-x
22. Zander, N.D.: Multi-level *hp*-FEM: dynamically changing high-order mesh refinement with arbitrary hanging nodes. Ph.D. thesis, Technische Universität München (2017)

Transfer Learning Approach to Prediction of Rate of Penetration in Drilling

Felix James Pacis[1(✉)], Sergey Alyaev[2], Adrian Ambrus[2], and Tomasz Wiktorski[1]

[1] University of Stavanger, Stavanger, Norway
{felix.j.pacis,tomasz.wiktorski}@uis.no

[2] NORCE Norwegian Research Centre, PB 22 Nygårdstangen, 5838 Bergen, Norway
{saly,aamb}@norceresearch.no

https://stavanger.ai/, https://www.norceresearch.no/

Abstract. The rate of penetration (ROP) is a key performance indicator in the oil and gas drilling industry as it directly translates to cost savings and emission reductions. A prerequisite for a drilling optimization algorithm is a predictive model that provides expected ROP values in response to surface drilling parameters and formation properties. The high predictive capability of current machine-learning models comes at the cost of excessive data requirements, poor generalization, and extensive computation requirements. These practical issues hinder ROP models for field deployment. Here we address these issues through transfer learning. Simulated and real data from the Volve field were used to pretrain models. Subsequently, these models were fine-tuned with varying retraining data percentages from other Volve wells and Marcellus Shale wells.

Four out of the five test cases indicate that retraining the base model would always produce a model with lower mean absolute error than training an entirely new model or using the base model without retraining. One was on par with the traditional approach. Transfer learning allowed to reduce the training data requirement from a typical 70% down to just 10%. In addition, transfer learning reduced computational costs and training time. Finally, results showed that simulated data could be used in the absence of real data or in combination with real data to train a model without trading off model's predictive capability.

Keywords: Rate of penetration model · Transfer learning · Deep learning

---

This work is part of the Center for Research-based Innovation DigiWells: Digital Well Center for Value Creation, Competitiveness and Minimum Environmental Footprint (NFR SFI project no. 309589, DigiWells.no). The center is a cooperation of NORCE Norwegian Research Centre, the University of Stavanger, the Norwegian University of Science and Technology (NTNU), and the University of Bergen, and funded by the Research Council of Norway, Aker BP, ConocoPhillips, Equinor, Lundin, TotalEnergies, and Wintershall Dea.

D. Groen et al. (Eds.): ICCS 2022, LNCS 13351, pp. 358–371, 2022.
https://doi.org/10.1007/978-3-031-08754-7_44

1 Introduction

According to a 2016 study by EIA [2], drilling constitutes 30–60% of the average cost per well, which varies from $4.9 MM to $8.3 MM for onshore wells and $120 MM to $230 MM for offshore wells. Thus, a modest improvement in the duration of drilling a well results in significant monetary savings. Among other factors such as preventing a non-productive time due to equipment failure or poor weather conditions, choosing the optimal drilling parameters to increase ROP is essential in reducing drilling duration.

Many attempts have been made on predicting the ROP. Although with some success [32], traditional physics-based models require frequent recalibration depending on the auxiliary data such as facies types, bit design, and mud properties [5,16,23,24]. This is challenging since facies types, in particular, are often unknown prior to drilling and would require correlation to data from nearby (offset) wells, if such wells exist.

Machine learning (ML) models try to address these challenges by using data to find correlations among many drilling variables. A study by Hegde et al. [17] showed an improvement in ROP prediction in accuracy from 0.46 to 0.84 when using random forest. Elkatatny et al. [11] also showed an improvement from 0.72 to 0.94 using an Artificial Neural Network (ANN).

Despite significant improvements in recent years, no ML approach has been widely used for ROP optimization to date [25]. The potential reason could be that the existing ML models are impractical for real-time ROP prediction tasks. Developing an ML ROP model is a multidimensional problem that does not revolve solely around prediction accuracy. Higher predictive capability comes at the cost of substantial data requirements, computational constraints, and generalization capability. From a practical perspective, tackling these constraints would be desirable for several reasons.

First, the need for large datasets for training a model for every well would limit the value creation. ANN training, such as Elkatatny et al. [11] and Abbas et al. [3], would require 70% of data for training; rendering these methods essentially not applicable in real scenarios since only a fraction of a well can benefit from such approach, see Fig. 1.

Second, ML models presented by O'Leary et al. [25], Mantha et al. [21], and Hegde et al. [17] require a priori knowledge on the formations being drilled. However, this information is rarely available prior to drilling the hole. This is problematic for wells drilled in new areas where offset wells do not exist yet.

Third, ROP prediction is a real-time regression problem. Unlike physics-based models that only require pre-identification of parameters, the ML requires training before deployment. Hence, one should consider the online computation requirements.

Fortunately, ROP ML models' issues are not foreign in other domains. Deep Learning models, in general, suffer from overfitting due to insufficient training data [33]. Transfer learning (TL) is an active research field in Deep Learning that deals with reusing a model trained from a more general task, termed base model or pre-trained model, to another specific tasks, termed target model. TL techniques

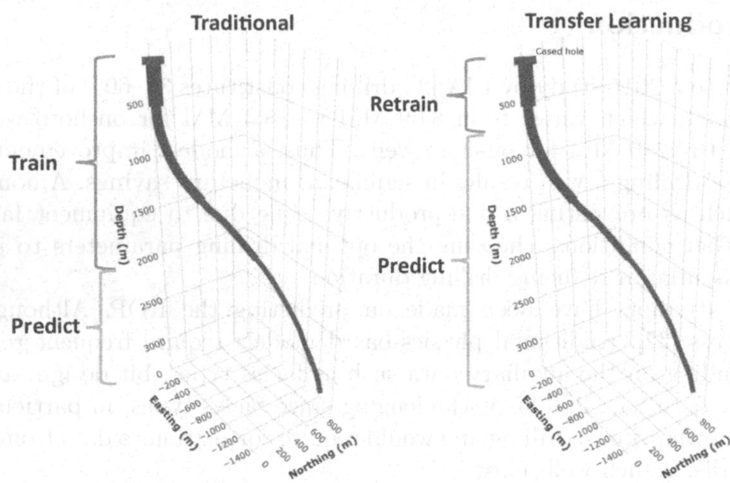

Fig. 1. Data utilization for traditional vs transfer learning approaches for well data.

have been proven successful in many domains such as computer vision and natural language processing [26].

In this paper, we present the application of TL to ROP prediction. To our knowledge, this is the first application of TL in the context of drilling. We train base models using real, simulated, and combined data from previously drilled wells. Then, we reconfigure each model by freezing some model parameters in order to limit the number of trainable parameters. Each reconfigured base model is retrained using a small fraction of target-well data, yielding a target model. This way a high quality target model is available already from the early stage of drilling operation, see Fig. 1. The performance of our TL models is compared to both the base models and models trained only for the data from the new well.

The paper is organized as follows. In Sect. 2, we briefly discuss the concept of TL. In Sect. 3, we describe the datasets and then proceed with the experimental setup, including the model architecture, input data, and the method for training and retraining. We also provide an end-to-end sample application of TL approach. Section 4 presents the results and lays out recommendations based on these. Section 5 concludes the paper.

2 Transfer Learning

Following the notations by Pan and Yang [26], Transfer Learning mainly involves a domain D and Task T. The domain, denoted by $D = \{X, P^X\}$, includes two components: a feature space X and a marginal probability distribution P^X, where each input instance is denoted by $x \in X$. On the other hand, the task, denoted by $T = \{Y, f(\cdot)\}$, includes all possible labels Y and a predictive function $f(\cdot)$ that predicts a corresponding label using unseen instances $\{x*\}s$. For a two domain scenario, given a source domain D_s and learning task T_s, a target domain D_t and learning task T_t, where $D_s \neq D_t$, or $T_s \neq T_t$, TL leverages learned knowledge from T_s to improve the T_t predictive function. Subscripts s and t here corresponds to source and target, respectively.

The most common TL technique is fine-tuning [29]. In the context of ANN, fine-tuning involves reusing the whole network or freezing certain hidden layers before updating the network weights during retraining for the target task. Fine-tuning works based on the premise that Deep Learning models learn different features at different layers. Thus, reusing a pre-trained model for a target task allows better performance with less training time by starting from "near truth" parameters than training a new model with randomly initialized parameters.

TL has been widely used both in computer vision and Natural Language Processing [26,37]. This is apparent from the proliferation of pre-trained networks e.g.,VCG-16 [31], XLNet [38], GPT-3 [7] using large datasets e.g., ImageNet[1], Giga5[2], and Common Crawl Dataset[3], and reused in domains where data is expensive or hard to obtain. For example in medical imaging, Shin et al. [30] fine-tuned AlexNet [20] - a pre-trained network using ImageNet dataset [9] with more than 14 million images belonging to around 20 thousand categories. They successfully achieved 85% sensitivity at 3 false positive per patient in thoraco-abdominal lymph node (LN) detection and interstitial lung disease (ILD) classification. Another successful application, Bird et al. [6] used a simulated scene from a computer game to train a model and resulted in an improvement for the real-world scene classification task.

Pre-trained networks also catalyzed the recent advances in Natural Language Processing (NLP). For example, Devlin et al. [10] introduced Bidirectional Encoder Representations from Transformers (BERT), which can be fine-tuned with adding an output layer to create state-of-the-art models for a wide range of tasks. Successful applications of BERT include text summarizing [10], modeling clinical notes and predicting hospital readmission [18], and machine reading comprehension [10].

The success of TL is apparent from its ubiquitous applications. This motivated websites, such as Hugging Face[4] and Model Zoo[5], which provide a platform to access many open-sourced pre-trained networks with ease.

TL has yet to be explored and applied broadly in the oil and gas domain. Since well-annotated datasets are expensive and difficult to obtain in the oil and gas industry, TL can be used to make rapid progress in this domain [15].

[1] https://www.image-net.org.
[2] https://catalog.ldc.upenn.edu/LDC2011T07.
[3] https://commoncrawl.org/the-data/.
[4] https://huggingface.co.
[5] https://modelzoo.co/.

3 Experimental Setup and Data

3.1 Methodology

TL requires the base model to be trained from several unique wells to improve their generalization capability. We freeze selected layers in these base models to keep the original weights and allow some to be trainable. These reconfigured layers are then fine-tuned using a pre-determined percentage of data from target wells. Hyperparameters during fine-tuning are carefully chosen to prevent vanishing or exploding gradients. This happens when the distribution of retraining data is entirely different, and the learning rate is too high; this impairs the base model's performance. In addition, a new model is also trained using the same retraining data. All these models are then tested using the remaining data from the target well.

We performed all computations using 2.3 GHz Dual-Core Intel Core i5.

3.2 Datasets

We used well data from three sources namely, Volve field data, Marcellus shale field data, and synthetic data. Table 1 summarizes the datasets. The well name, hole size, hole depth range, and the data source type for each dataset are provided for reference.

In general, when drilling an oil and gas well, bigger holes are drilled first, followed by smaller holes until they reach the predefined target. This is done to maintain well integrity, particularly when transitioning to a new geologic formation. Drillers use different drill bits, bit designs, and drilling fluid properties at each new hole size. This is similar to drilling an entirely new well from an engineering perspective. Thus, we produce independent datasets by segregating each data according to hole size from each well. These datasets contain recorded real-time drilling parameters such as hookload, stand pipe pressure, hole depth, weight on bit, mud weight, and rotations per minute. These measurements' frequency varies for every well depending on the equipment used.

In 2018 Equinor publicly shared raw real-time drilling data from 20 wells found in the Volve field in the North Sea [12], together with well logging data, surveying data, drilling reports, and other auxiliary information. Pre-processed Volve drilling logs can be found in a public data repository [35]. For this paper, we selected drilling data from 7 wells and separated them according to the hole size. In total, we compiled 12 independent datasets for the experiment. Volve data has an average sampling frequency of 0.4 Hz, corresponding to a time step of 2.5 s.

Marcellus shale is the most prolific natural gas-producing formation from the Appalachian basin in the United States [34]. A site owned and operated by Northeast Natural Energy, LLC provides several horizontal wells drilled in the Marcellus shale [1]. A specific long horizontal well spanning 2431 m, with an average measurement frequency of 0.176 Hz or 5.67 s time step, was chosen for the current study. This well data allows testing the models' generalization and re-usability outside Volve data.

To provide additional training data and investigate the feasibility of using simulated training data for the TL application, we generated eight synthetic datasets using a state-of-the-art drilling simulator which includes advanced hydraulics, mechanics, and heat transfer models [13]. The well architecture, trajectory, drilling mud properties, drill string configuration, and formation properties were based on the drilling reports extracted from the Volve public database [12]. The drilling set points (top-drive rotary speed, weight on bit, and flow rate) used as input to the simulations were based on the values from the Volve recorded drilling logs compiled by Tunkiel [35]. The simulation outputs were stored as time-series with a time step of 1 s.

Table 1. Description of Datasets.

Well name	Hole size (in)	Depth range (m)	Dataset type	Dataset source type	Test Case #
F-1 A	8.5	2602–3682	Train & Val.	Sim. & Real	
F-1 B	12.25	2603–3097	Train & Val.	Sim. & Real	
F-1 B	8.5	3097–3465	Retrain & Test	Real	4
F-1 C	12.25	2662–3056	Retrain & Test	Real	2
F-1 C	8.5	3067–4094	Train & Val.	Sim. & Real	
F-11 A	8.5	2616–3762	Train & Val.	Sim. & Real	
F-11 B	12.25	2566–3197	Train & Val.	Sim. & Real	
F-11 B	8.5	3200–4771	Retrain & Test	Real	3
F-15 A	17.5	1326–2591	Retrain & Test	Real	1
F-15 A	8.5	2656–4095	Train & Val.	Sim. & Real	
F-9 A	12.25	489–996	Train & Val.	Sim. & Real	
F-9 A	8.5	1000–1202	Train & Val.	Sim. & Real	
Marcellus Shale	8.75	1974–4405	Retrain & Test	Real	5

3.3 Setup of Experiments

Our model starts with an input layer, which receives four input parameters, followed by three successive pairs of dense and batch normalization layers. By embedding normalization as part of the model architecture, this prevents internal covariate shift [19] and causes a more predictable and stable behavior of the gradients [28], allowing higher learning rates without the risk of divergence [19, 28]. In addition, batch normalization eliminates the need for Dropout [33] for regularization [19]. We use rectified linear unit [14] as activation function. Finally, the output layer is a single-output dense layer with mean squared error as the loss function. A complete and detailed network structure is shown in Fig. 2.

To predict ROP, we used stand pipe pressure, weight on bit, mud weight, and rotations per minute (RPM). We based these inputs from the setup described

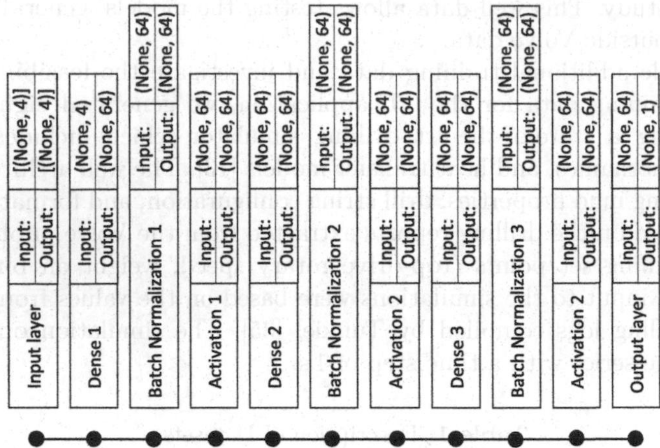

Fig. 2. Model architecture.

in Ambrus et al.'s work [4]. In general, when choosing our input parameters, two considerations were in place: first, despite using an ANN, the choice of input parameters should still reflect the physics of the drilling process. Second, selected inputs must always be available. During drilling, hundreds of parameters and metadata are recorded in real-time [36]. The inclusion of many drilling parameters as inputs to the ANN could be helpful but at the same time dangerous when one or more of these parameters are missing for the current well due to sensor failure or they were not necessarily recorded during the operation. Although one might infer the missing values, this would increase the model's prediction uncertainty when there are many inferred values.

Eight datasets were selected to build the base models out of the 12 available well sections from Volve. These were carefully selected to ensure that they contain values of the upper and lower boundaries of each input and output parameter. For example, the dataset with the highest ROP and lowest ROP values should be among the chosen eight.

To avoid overfitting the model, the first 80% of each well section is concatenated into the training dataset, whereas the remaining 20% are used for validation. This was done to all the three data source types - real, simulated, and combined. The shapes of concatenated training and validation data are shown in Table 2.

Data were scaled using a MinMaxScaler from Scikit-Learn [27] before passing to the model. This removes the harmful effects of having different value ranges for every input variable by scaling all of them to a (0,1) range.

Three separate runs for each data source type were conducted to build base models while keeping the model's hyperparameters the same. In particular, batch size was chosen to be 10000. This is relatively small, around 1% to 3% of each training dataset, to increase variation in batch statistics, thereby enabling better

Table 2. Traning and Validation Data Shapes.

Data Source Type	TrainX	TrainY	ValX	ValY
Real	(333400,4)	(333400,1)	(83350,4)	(83350,1)
Simulated	(649503,4)	(649503,1)	(162376,4)	(162376,1)
Combined	(982903,4)	(982903,1)	(245726,4)	(245726,1)

model generalization during the retraining process [22]. An early stopping Keras callback [8] was placed to cease training when the validation loss stops improving after 100 epochs. This allows us to generate two distinct base models for each data source type: one base model with the best validation loss and another based on the training loss. Altogether, we train six base models.

The four remaining well sections from real Volve data and the Marcellus shale horizontal well are used for retrain and test data. A sensitivity analysis was done by creating independent datasets with different retrain: test data ratios. These vary from 30:70, 20:80, 10:90, and 5:95, where the smaller partition corresponds to retraining data. Similar to the data preprocessing used for building the base model, each dataset is split sequentially and the values are scaled to a (0,1) range.

During retraining, we kept a similar model architecture to the base models, except that some layers were frozen. This allows us to retrain the model using smaller training datasets since fewer parameters are retrainable, and at the same time, model parameters are not initialized randomly. In addition, since models are pre-trained, a low learning rate is needed to reach the global minima. In our case, we used a learning rate of 0.0001 for all instances, with the exception of test case 4 that used 10^{-9}. Maximum epochs are set at 150000 for tests. Similar to training the base model, we set up an early stopping at 50 epochs based on the training loss.

These base models are reconfigured in three ways: freezing the first dense layer, first and second dense layers, and keeping all dense layers unfrozen. This gives us 18 reconfigured base models for retraining. All batch normalization layers were frozen in all these configurations to prevent the risk of vanishing or exploding gradients. In this context, freezing a layer means keeping the parameters learned during the initial training stage. Table 3 shows the number of trainable and non-trainable parameters for each configuration.

Table 3. Number of trainable and non trainable parameters.

Model configuration	Trainable parameters	Non-trainable parameters
Base Model	8897	384
Zero Frozen Layers	8513	768
One Frozen Layer	8257	1024
Two Frozen Layers	4161	5120

A randomly initialized new model with similar model architecture and hyper-parameters was trained for every retraining data configuration. This is to compare the performance of fine-tuning a pre-trained model with that of a new model trained from scratch on the same dataset.

3.4 Model Evaluation

We have six unique base models from previous sections, wherein each was reconfigured in three configurations based on the number of frozen layers. This gives us 18 unique models on top of the base models plus an entirely new trained model. In total, for every retraining data configuration, e.g., one test well, with unique retrain:test ratio, we tested 25 different models.

Model performance is evaluated by computing the mean absolute error ($MAE := L_1$) and root-mean-square error ($RMSE := L_2$) for every test data configuration:

$$L_k = \left(\frac{1}{N} \sum_{i=1}^{N} |y_i - \hat{y}_i|^k \right)^{\frac{1}{k}} \tag{1}$$

where N is the number for data points, $\hat{y}_i$ is the true value, and y_i is the predicted value. In addition, we kept a summary of a moving window MAE by dividing the test data into ten equal windows with the exact count of data points and computing MAE at each window. This enables us to measure prediction quality for various test data sizes. It is also important to emphasize that every well section data has a varying frequency of measurements and size, e.g., 30% of the well F-1C 12.25 in. section contains fewer data points than the F-11B 8.5 in. section.

3.5 Example of Usage

As discussed in Sect. 3.3, we train six base models then reconfigure by freezing layers. Subsequently, we derive four different datasets from well F-15A 17.5 in data. Each dataset differs on the retrain and test ratio as described previously. Each of the 18 reconfigured base models is then fine-tuned using retraining data from every dataset. In addition, we train an entirely new model using similar retraining data. From here, we have a total of 25 distinct models - 6 base models, 18 reconfigured base models, and one new model. These 25 models are then used for predicting ROP on the test datasets. Having 4 data split ratios from well F-15A 17.5 gives us a total of 100 test runs. For every test case, MAE is recorded.

4 Results

In Sect. 4.1, we analyze the results on well F-15A 17.5 in and compare the best models from several methods, which includes fine-tuning, training of an entirely new model, and direct use of the base model. Data from the other wells are not presented due to space constraints of this paper. This well was selected because results obtained on it were representative of both Volve and Marcellus shale test cases. In Sect. 4.2, we provide recommendations based on the results of five test cases. In Sect. 4.3, we provide results on the generalization capability of the approach.

4.1 3-way-comparison

After testing 100 models, we plot the predicted vs. expected ROP values plus a moving MAE window. In each plot, X-axis represents the hole depth, Y-axis to the left is the ROP with m/hr unit, and Y-axis to the right is the MAE. A, B, and C plots in Fig. 3 show the best model among fine-tuned base models, base models, and new models, respectively. Model configurations and metadata of these models can be found in Table 4.

Table 4. Test Case 1: F-15A 17.5 in

Frozen layers	BM loss type	Source type	Retrain (%)	Train (%)	MAE	RMSE
0	Validation	Simulated	30	N/A	3.674	4.886
2	Validation	Simulated	5	N/A	4.104	5.238
1	Validation	Combination	30	N/A	4.165	5.656
2	Validation	Simulated	30	N/A	4.234	5.562
1	Validation	Simulated	30	N/A	4.268	5.87
*New Model	Training	Real	N/A	30	5.061	5.845
*Base Model	Validation	Simulated	N/A	80	9.091	10.767

Retraining the base model reduces the MAE by 59.6% and 27.4% vs. using the base model and training an entirely new model, respectively. A relatively close RMSE to MAE also indicates that the ROP error disperses equitably across the data. Despite the base model not being trained with the same 17.5-inch hole size, it outperforms other models by tuning with the current well data. This is also on top of the fact that model A has fewer trainable parameters. Although not seen on the plot, the second-best model overall has an MAE of 4.104, despite only using 5% retraining data and 55% fewer trainable parameters compared to training a new model. Furthermore, both of the best two models were pre-trained using simulated data.

4.2 Recommendations Based on All Test Cases

Training Data Source Type. Four out of the five test cases suggest that training with simulated data provides better result than training with real data in terms of MAE. One explanation for this could be that predictions are less noisy since the simulated data is deterministic; thus, it produces better results when re-trained on a small section of the test set.

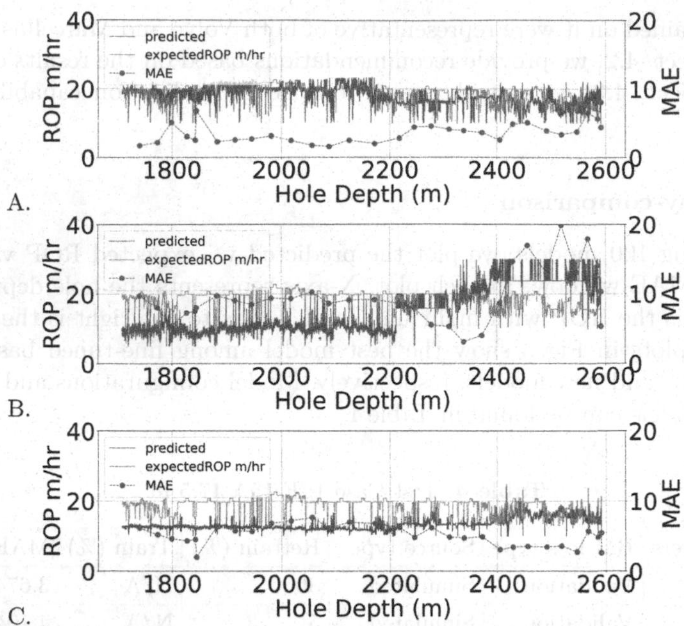

Fig. 3. ROP predicted by different models for well F-15A 17.5 in. A. Fine tuned model with TL. B. Base (pre-trained model) without fine-tuning. C. Newly trained model only using the data from the current well. Orange and blue lines refer to expected and predicted ROP values, respectively. Red markers are the computed MAE moving window e.g., one red marker is the MAE of the previous 2500 observations. All data are plotted against hole depth on the X-axis. Fine-tuned model performed best among other models with an MAE of 3.674.

Base Model Loss Type. Four out of the five test cases suggest that the best model should be based on the best validation loss rather than training loss. This is expected since the early stopping based on the validation loss helps reduce overfitting on the training data. Although the best retrained model in test case 3 was obtained using the training loss criterion, the base model using the validation loss criterion does not come far behind when considering the MAE.

Number of Frozen Layers. Four out of the five test cases suggest that fine-tuned models always perform the best. Case 4 performed just as well as the base model. Although, there was no clear relation between the number of frozen layers and the MAE. Paradoxically, increasing the number of frozen layers also increased the retraining time by 43 up to 247%. Thus, from retraining time perspective the ROP prediction problem benefits more from a pre-trained network without frozen layers. Another observation is that models with zero, one, and two frozen layers took an average retraining time of 7, 12, and 15 min, respectively, versus base models' 22 min.

Retraining Data Percent. As mentioned previously, every test well has a different length; therefore, even having the same retraining data percentage, the number of data points would still vary. There is no clear correlation between the number of data points and retraining data percentage for the best fine-tuned model based on the five test cases, although one could say that there could be a slight trade-off between the accuracy of the model and the length of the well to be predicted.

During our experiments we also observed that TL was sensitive to the choice of base model training data and learning rate. However the detailed analysis is out of the scope of this paper.

4.3 Test Outside Volve Data

We tested the approach on the Marcellus shale dataset to evaluate the generalization and re-usability of the TL approach. This well is entirely distinct from Volve data in terms of well profile (horizontal), type of formation (shale), location (onshore well), and equipment used. This is analogous to recognizing between breeds of dogs and breeds of cats for the computer vision domain. Clearly, the best retrained model reduced the MAE by 29% and 19% when compared with the newly trained model and base model, respectively. Relative to other test cases from Volve data, computed MAE is higher because of noise and lower measurement frequency in the Marcellus data. On top of improving the MAE, the retrained model only used 20% of retraining data while decreasing the trainable parameters by 10%. Furthermore, this result was achieved by training the base model with simulated data. This demonstrates the potential in using synthetic data generated with a high-fidelity drilling simulator for training the ANN ROP model that can be reconfigured for real operations with minimal amount of re-training.

5 Conclusions

We presented the application of TL for ROP prediction in oil and gas drilling. We trained, retrained, and tested a total of 100 models for each of the five test wells. Based on MAE evaluation, the TL approach for four out of five test wells outperforms both the newly trained model and the non-fine-tuned base model. For the fifth well the TL was on par with the traditional approach.

We explored the best model configurations based on the five test cases. In most cases the best results were obtained with the base models trained on the simulated data. Moreover the validation loss seems to be a good indicator of the model's performance on the new well. During fine-tuning, pre-trained models with zero frozen layers converge faster, although there was no clear relation between the MAE and the number of frozen layers. Despite uncertainty in the optimal number of frozen layers and retraining data percentage, results indicate that transfer learning is a valuable element in developing an adaptable, reusable, and more general ROP prediction model.

After successfully addressing the initial bottlenecks, new practical issues were identified. We noticed cases of negative transfer where some retrained models performed worse than their base model. The approach was also sensitive to the learning rate. Further work will focus on model optimization towards stability and improved accuracy while considering all practical bottlenecks.

References

1. Marcellus Shale Energy and Environment Laboratory. http://mseel.org. Accessed 11 Jan 2022
2. Trends in U.S. Oil and Natural Gas Upstream Costs. https://www.eia.gov/analysis/studies/drilling/pdf/upstream.pdf. Accessed 13 Jan 2022
3. Abbas, A.K., Rushdi, S., Alsaba, M.: Modeling rate of penetration for deviated wells using artificial neural network. In: Abu Dhabi International Petroleum Exhibition & Conference. OnePetro (2018)
4. Ambrus, A., Alyaev, S., Jahani, N., Wiktorski, T., Pacis, F.J.: Rate of penetration prediction using quantile regression deep neural networks (2022)
5. Bingham, G.: A new approach to interpreting rock drillability. Technical Manual Reprint. Oil Gas J. **1965**, 93 (1965)
6. Bird, J.J., Faria, D.R., Ekárt, A., Ayrosa, P.P.: From simulation to reality: CNN transfer learning for scene classification. In: 2020 IEEE 10th International Conference on Intelligent Systems (IS), pp. 619–625. IEEE (2020)
7. Brown, T.B., et al.: Language models are few-shot learners. arXiv preprint arXiv:2005.14165 (2020)
8. Chollet, F., et al.: Keras (2015). https://github.com/fchollet/keras
9. Deng, J., Dong, W., Socher, R., Li, L.J., Li, K., Fei-Fei, L.: Imagenet: a large-scale hierarchical image database. In: 2009 IEEE Conference on Computer Vision and Pattern Recognition, pp. 248–255. IEEE (2009)
10. Devlin, J., Chang, M.W., Lee, K., Toutanova, K.: Bert: pre-training of deep bidirectional transformers for language understanding. arXiv preprint arXiv:1810.04805 (2018)
11. Elkatatny, S., Al-AbdulJabbar, A., Abdelgawad, K.: A new model for predicting rate of penetration using an artificial neural network. Sensors **20**(7), 2058 (2020)
12. Equinor. Volve field data (CC BY-NC-SA 4.0) (2018). https://www.equinor.com/en/news/14jun2018-disclosing-volve-data.html
13. Gravdal, J.E., Ewald, R., Saadallah, N., Moi, S., Sui, D., Shor, R.: A new approach to development and validation of artificial intelligence systems for drilling. In: 2020 15th IEEE Conference on Industrial Electronics and Applications (ICIEA), pp. 302–307. IEEE (2020)
14. Hahnloser, R.H., Sarpeshkar, R., Mahowald, M.A., Douglas, R.J., Seung, H.S.: Digital selection and analogue amplification coexist in a cortex-inspired silicon circuit. Nature **405**(6789), 947–951 (2000)
15. Hajizadeh, Y.: Machine learning in oil and gas; a SWOT analysis approach. J. Petroleum Sci. Eng. **176**, 661–663 (2019)
16. Hareland, G., Rampersad, P.: Drag-bit model including wear. In: SPE Latin America/Caribbean Petroleum Engineering Conference. OnePetro (1994)
17. Hegde, C., Daigle, H., Millwater, H., Gray, K.: Analysis of rate of penetration (ROP) prediction in drilling using physics-based and data-driven models. J. Petroleum Sci. Eng. **159**, 295–306 (2017)

18. Huang, K., Altosaar, J., Ranganath, R.: Clinicalbert: modeling clinical notes and predicting hospital readmission. arXiv preprint arXiv:1904.05342 (2019)
19. Ioffe, S., Szegedy, C.: Batch normalization: accelerating deep network training by reducing internal covariate shift. In: International Conference on Machine Learning, pp. 448–456. PMLR (2015)
20. Krizhevsky, A., Sutskever, I., Hinton, G.E.: Imagenet classification with deep convolutional neural networks. Adv. Neural Inf. Process. Syst. **25**, 1097–1105 (2012)
21. Mantha, B., Samuel, R.: Rop optimization using artificial intelligence techniques with statistical regression coupling. In: SPE Annual Technical Conference and Exhibition. OnePetro (2016)
22. Masters, D., Luschi, C.: Revisiting small batch training for deep neural networks. arXiv preprint arXiv:1804.07612 (2018)
23. Maurer, W.: The perfect-cleaning theory of rotary drilling. J. Petroleum Technol. **14**(11), 1270–1274 (1962)
24. Motahhari, H.R., Hareland, G., James, J.: Improved drilling efficiency technique using integrated PDM and PDC bit parameters. J. Canadian Petroleum Technol. **49**(10), 45–52 (2010)
25. O'Leary, D., Polak, D., Popat, R., Eatough, O., Brian, T.: First use of machine learning for penetration rate optimisation on Elgin Franklin. In: SPE Offshore Europe Conference & Exhibition. OnePetro (2021)
26. Pan, S.J., Yang, Q.: A survey on transfer learning. IEEE Trans. Knowl. Data Eng. **22**(10), 1345–1359 (2010)
27. Pedregosa, F., et al.: Scikit-learn: machine learning in python. J. Mach. Learn. Res. **12**, 2825–2830 (2011)
28. Santurkar, S., Tsipras, D., Ilyas, A., Madry, A.: How does batch normalization help optimization? In: Proceedings of the 32nd International Conference on Neural Information Processing Systems, pp. 2488–2498 (2018)
29. Sarkar, D., Bali, R., Ghosh, T.: Hands-on Transfer Learning with Python: Implement Advanced Deep Learning and Neural Network Models Using TensorFlow and Keras. Packt Publishing Ltd (2018)
30. Shin, H.C., et al.: Deep convolutional neural networks for computer-aided detection: CNN architectures, dataset characteristics and transfer learning. IEEE Trans. Med. Imaging **35**(5), 1285–1298 (2016)
31. Simonyan, K., Zisserman, A.: Very deep convolutional networks for large-scale image recognition. arXiv preprint arXiv:1409.1556 (2014)
32. Soares, C., Daigle, H., Gray, K.: Evaluation of PDC bit ROP models and the effect of rock strength on model coefficients. J. Nat. Gas Sci. Eng. **34**, 1225–1236 (2016)
33. Srivastava, N., Hinton, G., Krizhevsky, A., Sutskever, I., Salakhutdinov, R.: Dropout: a simple way to prevent neural networks from overfitting. J. Mach. Learn. Res. **15**(1), 1929–1958 (2014)
34. Statistics, I., Analysis, U.E.I.A.: Marcellus shale play: geology review (2017)
35. Tunkiel, A.: Selected work repository. https://www.ux.uis.no/~atunkiel/ (2020)
36. Tunkiel, A.T., Wiktorski, T., Sui, D.: Drilling dataset exploration, processing and interpretation using volve field data. In: International Conference on Offshore Mechanics and Arctic Engineering, vol. 84430, p. V011T11A076. American Society of Mechanical Engineers (2020)
37. Weiss, K., Khoshgoftaar, T.M., Wang, D.D.: A survey of transfer learning. J. Big Data **3**(1), 1–40 (2016). https://doi.org/10.1186/s40537-016-0043-6
38. Yang, Z., Dai, Z., Yang, Y., Carbonell, J., Salakhutdinov, R.R., Le, Q.V.: Xlnet: generalized autoregressive pretraining for language understanding. Adv. Neural Inf. Process. Syst. **32** (2019)

Physics Informed RNN-DCT Networks for Time-Dependent Partial Differential Equations

Benjamin Wu[1,2], Oliver Hennigh[1], Jan Kautz[1], Sanjay Choudhry[1], and Wonmin Byeon[1(✉)]

[1] NVIDIA, Santa Clara, CA 95051, USA
{ohennigh,jkautz,schoudhry,wbyeon}@nvidia.com
[2] National Astronomical Observatory of Japan, Mitaka Tokyo 181-8588, Japan

Abstract. Physics-informed neural networks allow models to be trained by physical laws described by general nonlinear partial differential equations. However, traditional architectures of this approach struggle to solve more challenging time-dependent problems. In this work, we present a novel physics-informed framework for solving time-dependent partial differential equations. Using only the governing differential equations and problem initial and boundary conditions, we generate a latent representation of the problem's spatio-temporal dynamics. Our model utilizes discrete cosine transforms to encode spatial frequencies and re-current neural networks to process the time evolution. This efficiently and flexibly produces a compressed representation which is used for additional conditioning of physics-informed models. We show experimental results on the Taylor-Green vortex solution to the Navier-Stokes equations. Our proposed model achieves state-of-the-art performance on the Taylor-Green vortex relative to other physics-informed baseline models.

Keywords: Physics-informed neural networks · RNN · DCT · Numerical simulation · PDEs · Taylor-Green vortex

1 Introduction

Numerical simulations have become an indispensable tool for modeling physical systems, which in turn drive advancements in engineering and scientific discovery. However, as the physical complexity or spatio-temporal resolution of a simulation increases, the computational resources and run times required to solve the governing partial differential equations (PDEs) often grow drastically.

ML-driven Solvers. Recently, machine learning approaches have been applied to the domain of physical simulation to ameliorate these issues by approximating traditional solvers with faster, less resource-intensive ones. These methods

B. Wu and W. Byeon—Equal contribution.

B. Wu—Work done during NVIDIA AI Research Residency.

D. Groen et al. (Eds.): ICCS 2022, LNCS 13351, pp. 372–379, 2022.
https://doi.org/10.1007/978-3-031-08754-7_45

generally fall into two main paradigms: data-driven supervision [2,5,7,12,20] or physics-informed neural networks (PINNs) [1,15,16,19]. *Data-driven approaches* excel in cases for which reliable training data is available and if the underlying physical equations are unknown. These generally utilize deep neural networks to parameterize the solution operator. *PINN-based solvers* parameterize the solution function directly as a neural network. This is typically done by passing a set of query points through a multilayer perceptron (MLP) and minimizing a loss function based on the governing PDEs, initial conditions (ICs) and boundary conditions (BCs). The simulation becomes constrained by physics alone and does not require any training data. However, the accuracy of traditional PINN-based approaches is limited to simpler problems in low dimensions and time-independent physics [10].

Learning Time-dependent Problems with PINNs. For time-dependent problems, traditional PINN-based models use a continuous time approach which treats the temporal and spatial dimensions in the same manner. To avoid increasingly poor performance as the simulation evolves in time, methods have been developed to split the domain into many short-time problems and solve each step using continuous-time PINNs [13,14]. However, this results in additional model complexity and computational overhead. In principle, a well-constructed latent context grid allows the PINN to learn more easily while still relying on physics-constrained losses. In this work, we design a novel physics-informed MLP architecture by adding a new latent context generation process to effectively learn spatial-temporal physics problems.

Efficient Learning in Time and Space. Typical feedforward neural networks lack notions of temporal relationships. Recurrent neural networks (RNNs) form graphs directed along a temporal sequence, allowing learning of time-dependent dynamics. Long Short-Term Memory (LSTM) [9] and Gated Recurrent Units (GRUs) [3] provide a gating mechanism to solve the problem of vanishing gradients and have become popular choices for RNNs, exhibiting high performance and efficiency. For spatial features, Xu et al. [18] demonstrate that compressing large images in the frequency domain using digital signal processing improved model accuracy while greatly reducing input size. This allows for high model efficiency while maintaining compatibility with standard spatial CNN architectures.

Contributions. Although PINN solvers provide a well-principled, machine learning approach that could enhance the capabilities of numerical simulations, their current constraints to problems with simple geometries and short times severely limits their real-world impact. We address these shortcomings by introducing novel design choices that improve the simulation accuracy and efficiency of PINN solvers on more challenging problems, particularly in the regime of long time evolution where current PINNs severely struggle.

Our key contributions are as follows: **(1)** We propose a new approach for latent context generation that requires no additional data and enables PINNs to learn complex time-dependent physics problems. **(2)** To the best of our knowl-

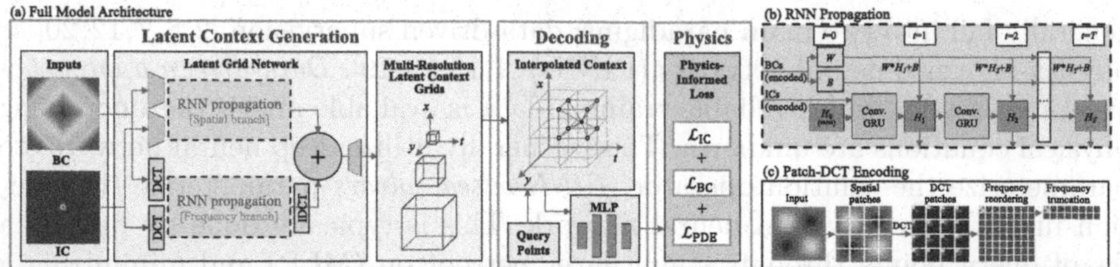

Fig. 1. (a) Full model architecture. (b) RNN propagation. (c) Patch-DCT encoding.

edge, our work is the first to directly address space-time-dependent physics using PINNs. This is achieved by utilizing convolutional GRUs for learning the spatio-temporal dynamics of simulations. **(3)** We separate the spatial and frequency domains, adding flexibility for the network to learn more diverse physical problems. **(4)** We test the new model against other architectures on benchmark transient simulation problems and demonstrate quantitative improvements in both accuracy and speed.

2 Methods

In this paper, we propose a new model that enables PINN-based neural solvers to learn temporal dynamics in both the spatial and frequency domains. Using no additional data, our architecture can generate a latent context grid that efficiently represents more challenging spatio-temporal physical problems.

Our full architecture is shown in Fig. 1(a). It consists of three primary parts, which are explained in more detail below: (1) latent context generation, (2) decoding, and (3) physics-informing. The latent context generation stage takes as input the problem ICs and BCs and outputs spatio-temporal latent context grids. For the decoding stage, spatio-temporal query points along with additional vectors interpolated from the latent context grid are used as input. For each set of points, the MLP predicts corresponding output values, on which the physics-constrained losses are applied. Upon minimization of these losses, the MLP approximates the function governed by the underlying PDEs.

Latent Grid Network. The primary novelty of our method is the latent grid network that can generate context grids which efficiently represent the entire spatio-temporal domain of a physical problem without requiring additional data.

This network requires two inputs for the problem-specific constraints: ICs and BCs. The ICs are defined as $u_0 = u(x_{1,..,N}, t = 0)$ for each PDE solution function u over N spatial dimensions. The BCs are defined based on the geometry of the problem for each spatial dimension. An additional spatial weighting by signed distance functions (SDFs) can also be applied to avoid discontinuities at, e.g., physical boundaries, but would not be necessary for, e.g., periodic BCs. Each tensor undergoes an encoding step in either the frequency or spatial domain.

The *frequency branch* transforms the spatial inputs to frequencies via the discrete cosine transform (DCT), motivated by [18]. Figure 1 (c) illustrates our patch-wise DCT encoding step. First, the ICs and BCs are separately split into spatial patches of size $p \times p$. DCTs are performed on each patch to yield the corresponding frequency coefficient array. The tensor is then reshaped such that the same coefficient across all patches forms each channel, and the channels are reordered by increasing coefficient (i.e., decreasing energy). After the reordering, the channels are truncated by $n\%$, so the lowest $n\%$ of frequency coefficients (largest energies) are kept. This outputs highly compressed representations for the ICs and BCs, which are used as inputs for an RNN propagation branch.

The *spatial branch* follows a traditional ResNet [6] architecture, in which the ICs and BCs each pass through separate convolutional encoders comprising sets of convolutional blocks with residual connections. The inputs are downsampled with strided convolutions before entering the spatial RNN propagation branch.

RNN Propagation. After compression, the representations enter the RNN propagation stage (Fig. 1 (b)), in which the BCs are split into an additive (B^{bc}) and multiplicative (W^{bc}) component and combined with an IC-informed state matrix (H_t). The final output at each timestep is computed as $S_t = W^{\text{bc}} H_t + B^{\text{bc}}$. This method offers flexibility and efficiency in learning the dynamics of compressed simulations [7]. To predict the simulation state at each successive timestep, the previous hidden state H_{t-1} is passed through a convolutional GRU (ConvGRU) along with the previous output S_{t-1}; for timestep 0, the initial state H_0 set to zero and ICs are used as inputs. This occurs in a recurrent manner until the final time T. Thus, for each timestep, the RNN propagation stage outputs S_t which is then sent to a decoding step corresponding to the original frequency or spatial encoding. $S_0 = u_0$, $H_0 = \mathbf{0}$, $H_t = \text{ConvGRU}(S_{t-1}, H_{t-1})$, $S_t = W^{\text{bc}} H_t + B^{\text{bc}}$, $t \in \{1, \ldots, T\}$. The RNN propagation stage is duplicated across both frequency and spatial branches.

Latent Grid Generation. After RNN propagation, the outputs are combined to form the latent grid. In the frequency branch, the output state at each timestep from the RNN is converted back into the spatial domain: 1) reshaping the frequencies from coefficients to patches 2) performing IDCTs, and 3) merging the patches to reconstruct the spatial domain. The output of the frequency branch is denoted as O_t^f. The representation in the spatial domain O_t^s is then added with learnable weights W_t^o. Thus, the final output is computed as: $O_t = W_t^o O_t^s + O_t^f$. These combined outputs O_t for each timestep are used to form the spatio-temporal latent context grids. Finally, the multiple resolutions of grids are generated by upsampling the outputs O_t using transpose convolutional blocks.

Decoding Step. The multi-resolution latent context grids generated from the previous step are then used to query points input to the MLP. This decoding step follows the same principles as [4]. Given a random query point $\mathbf{x} := (x, y, t)$, k neighboring vertices of $\mathbf{x}$ at each dimension are selected. Using these neighboring vertices, the final values of the context vector are then Gaussian-interpolated.

This process is repeated for each of the multi-resolution grids allowing the PINN framework to learn spatio-temporal quantities at multiple resolutions.

Physics-informed Loss. The MLP outputs predictions that are then subject to the loss function determined by the ICs, BCs, and the PDEs. The losses are backpropagated through the entire combined decoding and latent grid network and minimized via stochastic gradient descent. This end-to-end training allows our two-branch convGRU model to learn accurate time-evolution of the spatial and frequency domains in complex physical problems.

3 Experiments

We compare our model (**RNN-SpDCT**) against several other neural network solver architectures using the 2D Taylor-Green vortex problems. This problem is commonly used to test and validate spatial and temporal accuracy of both traditional and ML-based fluid solvers. We compare against PINN-based models and use the ICs, BCs, and PDE constraints for all comparing models. We used a single Tesla V100 16G or 32G for all experiments.

Taylor-Green Vortex. The Taylor-Green vortex describes a decaying vortex flow which follows a special case of the Navier-Stokes equations [17]. The incompressible Navier-Stokes equations in 2D are $\partial_x u + \partial_y v = 0$, $\partial_t u + u\partial_x u + v\partial_y u = -\partial_x \rho/\rho + \nu(\partial_{xx} u + \partial_{yy} u)$, $\partial_t v + u\partial_x v + v\partial_y v = -\partial_y \rho/\rho + \nu(\partial_{xx} v + \partial_{yy} v)$, where u and v are the x- and y-velocities, respectively, $\nu \in \mathbb{R}_+$ is the kinematic viscosity, and ρ is the density.

The exact closed form solution for the Taylor-Green vortex over the domain $x \times y \times t \in [0, 2\pi] \times [0, 2\pi] \times [0, T]$ is $u = \cos x \sin y F(t)$, $v = -\sin x \cos y F(t)$, $p = \frac{-\rho}{4}(\cos 2x + \cos 2y)F^2(t)$, where $F(t) = e^{-2\nu t}$ and p is the pressure.

Dataset. We used 2π seconds for both training and testing. During training, the input size is set to 32×32. The first timestep is used as an initial condition, and x and y are used as boundary conditions. For testing, the size of x and y are set to 128×128, and 10 time steps are uniformly sampled between 0 and 2π.

Experimental Setup. The number of interpolation points k used in the decoding step is 3, and the truncation ratio n is fixed to 25%. All models are trained with an Adam optimizer [11]. We found that patch-wise models need lower initial learning rate 1e-4 and the others with higher initial learning rate 4e-4 with 0.95 decay rate and different decay steps: 8000 steps for patch-based models, 2000 steps for other models.

Network Architecture. The patch size p in the patch-based DCT models is set to 8. The number of (encoding) residual blocks are 2 for the spatial branch and 1 for the frequency branch. In the spatial branch, 4 additional residual blocks are used with stride 2 for downsampling . There is no downsampling layer for the frequency branch. All convolutional layers have a filter size of 3×3, and there are two RNN propagation layers. The number of upsampling layers is searched between 1 and 4, and the reported numbers are with 4 layers.

Baseline Models. We compare our proposed model against several PINN-based approaches: MLP-PINN, RNN-S, RNN-pDCT, and RNN-SfDCT. All comparing models contain the RNN-propagation and decoding steps except for MLP-PINN and all use physics informed loss explained in Sect. 2. All use the same hyper-parameters as our model except for learning rate and decay steps. **MLP-PINN**: a traditional MLP-based PINN solver used as a default model from `SimNet` [8]. **RNN-S**: a PINN solver with a latent grid network consisting of a single spatial branch (ResNet). **RNN-pDCT**: a PINN solver with a latent grid network consisting of a single frequency branch (DCT). **RNN-SfDCT**:a PINN solver with a latent grid network consisting of both spatial and frequency branches. The frequency branch in this model applies DCT/IDCT to the full input, foregoing the patching, coefficient channel reordering, and truncation steps.

Table 1. Quantitative comparisons. The averaged mean squared error (MSE) over 10 uniformly sampled time steps for 2π sec. is reported. ν is the kinematic viscosity of the fluid. F and S indicate frequency and spatial branches. FullDCT applies DCT to the entire input. The numbers are in the magnitude of 10^{-2}.

Model name	Branch	DCT type	Taylor-Green Vortex					
			$\nu = 1.0$		$\nu = 0.1$		$\nu = 0.01$	
			Velocity	Pressure	Velocity	Pressure	Velocity	Pressure
MLP-PINN	–	–	0.033	5.910	1.769	0.782	0.824	0.522
RNN-S	S	–	6.683e-8	0.075	2.975e-7	0.138	2.527e-7	0.020
RNN-pDCT	F	Patch	1.979e-6	0.172	5.957e-7	1.383	8.804e-7	0.508
RNN-SfDCT	S+F	Full	9.171e-8	1.177	2.961e-7	0.301	7.015e-6	0.018
RNN-SpDCT	S+F	Patch	1.408e-7	**0.044**	3.107e-7	**0.101**	1.328e-6	**0.012**

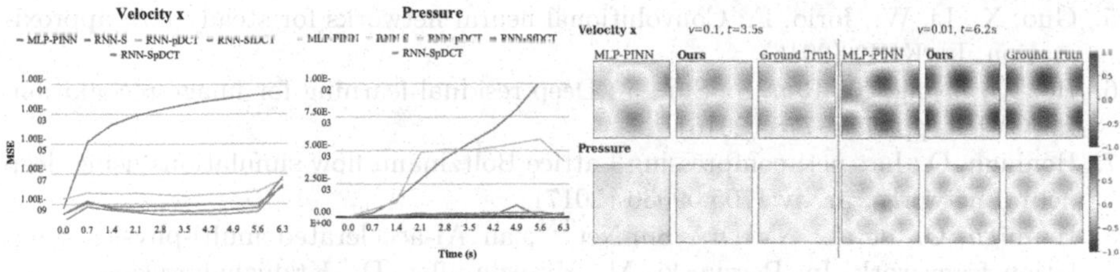

Fig. 2. Top: MSE comparisons over time. The MSE of the five models with respect to ground truth are shown over time for the Taylor-Green vortex problem. **Bottom: Visualization of the predictions** with the viscosity $\nu = 0.1$ at around 3.5 s. (left) and $\nu = 0.01$ at around 6 s. (right).

Results. Table 1 summarizes the performance of our model compared to the other PINN baselines. RNN-SpDCT achieves the best performance for all values of vorticity. All RNN models achieve extremely accurate velocities compared to

MLP-PINN. Figure 2 (top) shows the performance comparisons over time. Overall, for both x-velocity and pressure, RNN-based model produces much lower error throughout long time evolution compared to the baseline MLP-PINN, and RNN-SpDCT achieves the best overall performance. Figure 2 (bottom) visualizes the predictions and compares with the analytical solution. The model produces much more accurate predictions for longer time steps (up to 2π s) compared to MLP-based PINNs.

4 Conclusion

We presented a novel extension to the PINN framework designed especially for time-dependent PDEs. Our model utilizes RNNs and DCTs to generate a multi-resolution latent context grid to condition the traditional MLP PINN architecture. We demonstrated that our model can accurately predict the solution functions in Taylor-Green vortex simulations (especially for pressures) and achieve state-of-the-art results. Future directions include experiments on more complex problems, higher dimensions, and longer time evolution.

References

1. Bar, L., Sochen, N.: Unsupervised deep learning algorithm for PDE-based forward and inverse problems. arXiv:1904.05417 (2019)
2. Bhatnagar, S., Afshar, Y., Pan, S., Duraisamy, K., Kaushik, S.: Prediction of aerodynamic flow fields using convolutional neural networks. Comput. Mech. **64**(2), 525–545 (2019). https://doi.org/10.1007/s00466-019-01740-0
3. Cho, K., et al.: Learning phrase representations using RNN encoder-decoder for statistical machine translation. arXiv:1406.1078 (2014)
4. Esmaeilzadeh, S., et al.: MeshfreeFlowNet: a physics-constrained deep continuous space-time super-resolution framework. In: SC20, pp. 1–15. IEEE (2020)
5. Guo, X., Li, W., Iorio, F.: Convolutional neural networks for steady flow approximation. In: KDD (2016)
6. He, K., Zhang, X., Ren, S., Sun, J.: Deep residual learning for image recognition. In: CVPR, pp. 770–778 (2016)
7. Hennigh, O.: Lat-net: compressing Lattice Boltzmann flow simulations using deep neural networks. arXiv:1705.09036 (2017)
8. Hennigh, O., et al.: NVIDIA SimNet^{TM}: an AI-accelerated multi-physics simulation framework. In: Paszynski, M., Kranzlmüller, D., Krzhizhanovskaya, V.V., Dongarra, J.J., Sloot, P.M.A. (eds.) ICCS 2021. LNCS, vol. 12746, pp. 447–461. Springer, Cham (2021). https://doi.org/10.1007/978-3-030-77977-1_36
9. Hochreiter, S., Schmidhuber, J.: Long short-term memory. Neural Comput. **9**(8), 1735–1780 (1997)
10. Kashefi, A., Mukerji, T.: Physics-informed PointNet: a deep learning solver for steady-state incompressible flows and thermal fields on multiple sets of irregular geometries. arXiv:2202.05476 (2022)
11. Kingma, D.P., Ba, J.: Adam: a method for stochastic optimization. arXiv:1412.6980 (2014)

12. Li, Z., et al.: Fourier neural operator for parametric partial differential equations (2020)
13. Mattey, R., Ghosh, S.: A physics informed neural network for time-dependent nonlinear and higher order partial differential equations. arXiv:2106.07606 (2021)
14. Meng, X., Li, Z., Zhang, D., Karniadakis, G.E.: PPINN: parareal physics-informed neural network for time-dependent PDEs. Comput. Methods Appl. Mech. Eng. **370**, 113250 (2020)
15. Raissi, M., Perdikaris, P., Karniadakis, G.E.: Physics-informed neural networks: a deep learning framework for solving forward and inverse problems involving non-linear partial differential equations. J. Comput. Phys. **378**, 686–707 (2019)
16. Smith, J.D., Azizzadenesheli, K., Ross, Z.E.: EikoNet: solving the eikonal equation with deep neural networks. IEEE Trans. Geosci. Remote Sens. **59**(12), 10685–10696 (2020). IEEE
17. Taylor, G.I., Green, A.E.: Mechanism of the production of small eddies from large ones. Proc. R. Soc. Lond. Ser. A-Math. Phys. Sci. **158**(895), 499–521 (1937). The Royal Society London
18. Xu, K., Qin, M., Sun, F., Wang, Y., Chen, Y.K., Ren, F.: Learning in the frequency domain. In: CVPR, pp. 1740–1749 (2020)
19. Yu, B., et al.: The deep ritz method: a deep learning-based numerical algorithm for solving variational problems. arXiv:1710.00211 (2017)
20. Zhu, Y., Zabaras, N.: Bayesian deep convolutional encoder-decoder networks for surrogate modeling and uncertainty quantification. J. Comput. Phys. **366**, 415–447 (2018)

Recursive Singular Value Decomposition Compression of Refined Isogeometric Analysis Matrices as a Tool to Speedup Iterative Solvers Performance

Mateusz Dobija and Anna Paszynska(✉)

Faculty of Physics, Astronomy and Applied Computer Science, Jagiellonian University, Kraków, Poland
anna.paszynska@uj.edu.pl

Abstract. The isogeometric analysis (IGA) uses higher-order and continuity basis functions as compared to the traditional finite element method. IGA has many applications in simulations of time-dependent problems. These simulations are often performed using an explicit time-integration scheme, which requires the solution of a system of linear equations with the mass matrix, constructed with high-order and continuity basis functions. The iterative solvers are most commonly applied for large problems simulated over complex geometry. This paper focuses on recursive decomposition of the mass matrix using the Singular Value Decomposition algorithm (SVD). We build a recursive tree, where submatrices are expressed as multi-columns multiplied by multi-rows. When we keep the mass matrix compressed in such a way, the multiplication of a matrix by a vector, as performed by an iterative solver, can be performed in $O(Nr)$ instead of $O(N^2)$ computational cost, where N is the number of rows of input matrix, r is the number of singular values bigger than given value. Next, we focus on refined isogeometric analysis (rIGA). We introduce the C0 separators into IGA submatrices and analyze the SVD recursive compression and computational cost of an iterative solver when increasing the patch size and the order of B-spline basis functions.

Keywords: Refined isogeometric analysis · Hierarchically compressed matrix · Matrix-vector multiplication · Iterative solvers

1 Introduction

Isogeometric analysis (IGA) [1–3] is a generalization of the traditional finite element method into higher-order and continuity basis functions. It has many applications in the simulation of time-dependent problems, from wind turbine simulations [4], drug transport [5], to tumor growth simulations [6,7]. The refined isogeometric analysis [8–10] has been proposed to reduce the computational cost of IGA while keeping the high accuracy of IGA solutions. The time-dependent IGA solvers often rely on explicit dynamics formulations [6,7,11]. In the explicit

D. Groen et al. (Eds.): ICCS 2022, LNCS 13351, pp. 380–393, 2022.
https://doi.org/10.1007/978-3-031-08754-7_46

dynamics solvers, the system of linear equations with mass matrix constructed from IGA basis functions is solved in every time step. For large problems, iterative solvers are often employed [12]. In this paper, we discuss an algorithm for recursive compression of matrices. As an example, we consider the IGA mass matrix as applied in time-dependent simulations. Our algorithm employs a low-rank approximation of blocks, as performed by the Singular Value Decomposition algorithm [13]. This method follows the idea of hierarchical matrix compression as proposed by Hackbush [14]. The generation of isogeometric analysis mass matrices speeds up with low-rank approximation has been discussed in [15]. In our paper, we focus on the application of compression to speed up the iterative solver. We consider the refined isogeometric analysis and discuss the influence on the computational patch's different orders of approximations and dimensions.

2 Matrices of Refined Isogeometric Analysis

B-spline functions are commonly used in computer design and simulations thanks to the growing popularity of so-called isogeometric analysis popularized by prof. T. J. R. Hughes [1]. The formulas for B-splines are defined using Cox-de-Boor rule [3] in the following way:

$$B_{i,0}(\xi) = 1 \text{ for } \xi_i \leq \xi \leq \xi_{i+1} \text{ and 0 in the other case}$$
$$B_{i,p}(\xi) = \frac{\xi - \xi_i}{\xi_{i+p} - \xi_i} B_{i,p-1}(\xi) + \frac{\xi_{i+p+1} - \xi}{\xi_{i+p+1} - \xi_{i+1}} B_{i+1,p-1}(\xi) \tag{1}$$

Both linear B-splines and higher order B-splines can be described using the notation of the so-called knot vector. A knot vector will be a sequence of non-decreasing coordinates of points. Let's, for simplicity, assume points with integer coordinates. The degree of the basis functions is equal to the number of times the first (or last) point in the knot vector is repeated, minus one. In exemplary knot vector [0 0 0 0 1 2 3 4 5 5 5 6 7 8 9 10 10 10 10] the first and last points are repeated four times (0 0 0 0 and 10 10 10 10), so we introduce cubic basis functions of second continuity. There is a repeated knot in the vector (5 5 5), and the repetition of knot reduces the continuity at the point, so we have inserted the C0 separator at the center. In rIGA the C0 separators are inserted every l intervals. The two-dimensional B-splines are created by the tensor product of one-dimensional B-splines [3]. For example, the tensor product of the knot vectors [0 0 0 0 1 2 3 4 5 5 5 6 7 8 9 10 10 10 10] and [0 0 0 0 1 2 3 4 5 5 5 6 7 8 9 10 10 10 10] describes B-splines presented in Fig. 1, merging together four patches of 5×5 elements with cubic B-splines. The mass matrix is defined as the multiplication of two two-dimensional B-splines. The first B-splines are called trials and are used for the approximation of the solution. The former are called tests and are employed to generate different equations within the system (and for local approximation of the equation in the strong form).

$$M_{ij,k,l;p} = \int_{\Omega} B_{ij;p}(x,y) B_{kl;p}(x,y) dxdy = \int_{\Omega} B_{i;p}(x) B_{j;p}(y) B_{k;p}(x) B_{l;p}(y) dxdy \tag{2}$$

The mass matrix can be factorized in a linear cost only if the computational domain has a regular tensor product form. For the simulations on non-regular geometries in 2D, the direct solvers deliver $O(N^{1.5}p^2)$ computational cost, and the iterative solvers deliver $O(Nk)$ computational cost. Here N is the number of B-splines, p is the order of B-splines, and k is the number of iterations (depending on the geometry of the domain and the problem solved). The matrix-vector computations can be performed without forming a global matrix when we focus on refined isogeometric analysis and iterative solvers. This is called a matrix-free iterative solver algorithm [16]. The solution vector is obtained by assembling local elemental matrices multiplied by local portions of the right-hand side. In the context of refined isogeometric analysis, this can be generalized to patches of elements. In this paper, we focus on different patches of elements and different orders of B-splines, and we perform recursive SVD compression of the resulting mass matrix. Later, we compare the gain in terms of the number of floating-point operations when performing matrix-vector multiplication over the patch of elements, using the patches of rIGA.

Fig. 1. Two dimensional B-splines described as tensor product of two one-dimensional B-splines defined by knot vector [0 0 0 0 1 2 3 4 5 5 5 6 7 8 9 10 10 10 10].

3 Compression Algorithm

The main idea of the compression algorithm is the recursive dividing of the matrix into four smaller submatrices and performing the approximate singular value decomposition algorithm (SVD) for selected submatrices. The approximate SVD decomposes any matrix A into A = UDV, where D is s diagonal matrix with singular values sorted in a decreasing manner, U is the matrix of columns, an V is the matrix of rows. In the approximate SVD, we remove from D singular values smaller than prescribed epsilon. Thus, we can only keep r columns of U and r rows of V, where r is the number of singular values left. Additionally, in order to save the memory, the coefficients of the calculated matrix V are multiplied by corresponding singular values of the diagonal matrix D. In other words, instead of storing UDV we store UV′ where V′ is DV. The gain from the SVD decomposition is that it enables matrix-vector multiplication in a computational cost of $O(Nr)$ instead of $O(N^2)$, if we perform the multiplications in order U * (V′ * B) instead of A * B or (U * V′) * B, see Fig. 2. For simplicity we call these matrices U and V (remembering that V is multiplied by D).

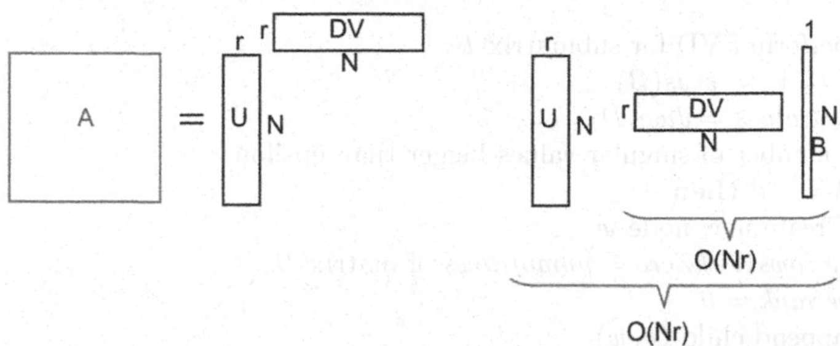

Fig. 2. The approximate SVD of matrix A, where we save r singular values, so we have N columns and N rows. The matrix-vector multiplication UDVB for the SVD compressed matrix A = UDV results in $O(Nr)$ computational cost.

The whole compression algorithm of a matrix can be seen as the recursive dividing of the matrix into four smaller submatrices and can be described in the following steps. If the submatrix consists of zeros, we remember its rank (zero) and its number of rows. If the submatrix has some nonzero values, the SVD algorithm is used. If the number of singular values found by SVD, which are bigger than desired epsilon, is zero, we also remember only the submatrix rank equal to zero and the number of its rows. If the number of singular values found by SVD, which are bigger than desired epsilon, is nonzero (denote this value by k), and additionally, (k <= r and k < numRows/2) or k == 1, where numRows is the number of rows in the matrix and r is the arbitrary boundary for rank, than we remember only the first k rows of matrix V and the first k column of matrix U found by the SVD algorithms. The algorithm, during recursively dividing of the matrix, creates the tree. Each node of the tree can be a leaf or it can have four sons representing four submatrices. In each node, some attributes are stored, like the rank of the corresponding matrix, the number of rows, or vector U and V found by the SVD algorithm. The input of the algorithm is the matrix A in a sparse form, the parameter epsilon, and the boundary rank value r. The algorithm of compression of the matrix is presented in Algorithm 1.

Algorithm 1 Compress matrix(A, epsilon,r)

```
Create new node v
Divide Matrix A into 4 submatrices (quarters): A1, A2, A3, A4
for each submatrix B = A1, A2, A3, A4 do
  if number on nonzero elements in B is equal to zero then
    Create new node w
    w.rowsWithZero = numofRows of matrix B
    w.rank = 0
    append child (v, w)
  else
    //perform SVD for submatrix B
    [U, D, V] = svds(B)
    eigenvalues = diag(D)
    k =number of singular values bigger than epsilon
    if k == 0 then
      Create new node w
      w.rowsWithZero = numofRows of matrix B
      w.rank = 0
      append child (v, w)
    else if (k <= r) AND (k < (numRows/2) OR (k == 1)) then
      Create new node w
      w.rank = k
      w.Ucolumns = U(:, 1 : k)
      w.Vrows = V'(1 : k, :)
      for i = 1 : k do
        w.Vrows(i, :) = w.Vrows(i, :) * node.eigenvalues(i)
      end for
      append child (v, w)
    else
      w =Compress Matrix(B, epsilon, r)
      append child (v, w)
    end if
  end if
end for
```

4 Multiplication Algorithm

The input for the algorithm for multiplication of the compressed matrix by a vector is the tree representing compressed matrix (the root node of the tree or its subtree) and the vector. The idea of the algorithm is the following. If the input node has no children, it means, that it represents a block of zeros or a block remembered as matrices *Ucolumns* and *Vrows* found by the SVD algorithm. In the first case, the resulting vector is a vector of zeros. In the second case, the result is the result of multiplication $node.Ucolumns * (node.Vrows * x)$. It must be underlined that the order of performing multiplication is critical, because it influences the computational cost of obtaining the results. In the last case, the input node has children. In this case, the partial multiplication for each submatrix (each child of the input node) by the corresponding part of the input vector has to be performed by the recursive call of the algorithm, followed by calculating the final result of the multiplication.

Algorithm 2 MultiplyMatrixByVector(node, x)

```
if node.noofchildren == 0 then
    if node.rank == 0 then
        result =vector consisting of node.rowsWithZero zeros
    else
        result = node.Ucolumns * (node.Vrows * x)
    end if
else
    numRows =number of rows of vector x
    x1 = v(1 : floor(numRows/2), :)//first part of vector x
    x2 = v(floor(numRows/2 + 1) : numRows, :)//second part of vector x
    //calculate the partial multiplication for each submatrix
    res1 = MultiplyMatrixByVector(node.children(1), x1)
    res2 = MultiplyMatrixByVector(node.children(2), x2)
    res3 = MultiplyMatrixByVector(node.children(3), x1)
    res4 = MultiplyMatrixByVector(node.children(4), x2)
    //calculate the final result of multiplication
    if res1 consist of zeros then
        res1res2 = res2
    else if res2 consist of zeros then
        res1res2 = res1
    else
        res1res2 = res1 + res2
    end if
    if res3 consist of zeros then
        res3res4 = res4
    else if res4 consist of zeros then
        res3res4 = res2
    else
        res3res4 = res3 + res4
    end if
    result = [res1res2; res3res4]
end if
```

The main idea of the recursive multiplication algorithm is the following. Let consider the compressed matrix consisting of four blocks denoted by C, D, E, and F, where each block is compressed by the SVD algorithm into matrices $C1$ and $C2$, $D1$ and $D2$, $E1$ and $E2$, $F1$ and $F2$, respectively, as presented in Fig. 3. The result of multiplication of this matrix by vector X ($X = [X1, X2]$) is a vector: $[C1*(C2*X1)+D1*(D2*X2), E1*(E2*X1)+F1*(F2*X2)]$. Let assume that the matrices $C2$, $D2$, $E2$, $F2$ are of size $r*N$ and the matrices $C1$, $D1$, $E1$, $F1$ are of size $N*r$. The vectors $X1$ and $X2$ have size $N*1$. The cost of performing multiplication of our input matrix by vector can be summarized step by step, as follows:

- The cost of multiplication C2 * X1 is N * r, and the result has size r * 1,
- The cost of multiplication C1 * (C2 * X1) is N * r * r, and the result has size N * 1,
- The cost of multiplication D2 * X2 is N * r, and the result has size r * 1,
- The cost of multiplication D1 * (D2 * X2) is N * r * r, the result has size N * 1,
- The cost of summing the results is N * 1,
- The cost of multiplication E2 * X1 is N * r, and the result has size r * 1,
- The cost of multiplication E1 * (E2 * X1) is N * r * r, and the result has size N * 1,
- The cost of multiplication F2 * X2 is N * r, and the result has size r * 1,
- The cost of multiplication F1 * (F2 * X1) is N * r * r, and the result has size N * 1,
- The cost of summing the results is N * 1.

Summing up, the cost of the multiplication our input compressed matrix by a vector is $N * r$ in the contrary to the classic multiplication of matrix by vector with the cost N^2.

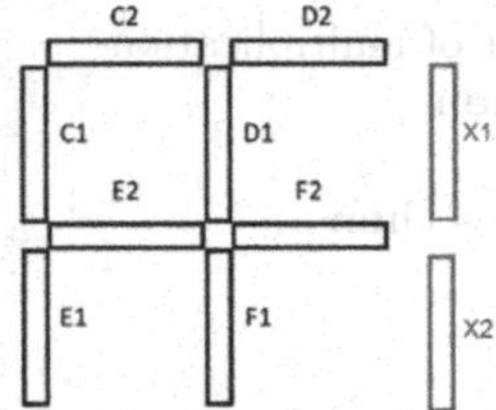

Fig. 3. Matrix-vector multiplication for the matrix recursively decomposed into four sub-matrices. The multiplications are performed $C1 * (C2 * X1) + D1 * (D2 * X2)$ and $E1 * (E2 * X1) + F1 * (F2 * X2)$ so the resulting computational cost is $O(Nr)$, where r is the number of rows and columns in the compressed matrices.

5 Results

The tests were performed comparing the number of floating point operations and accuracy for matrix by vector multiplication, for matrix compressed by the algorithm, and sparse matrix. The input matrix was the mass matrix in the compressed or sparse form. The tested mass matrices are generated for a different number of intervals, equal to 1, 2, 4, 8, 16, and 32, corresponding to different dimensions of patches in rIGA, and for different polynomial orders of approximation, equal 2, 3, 4 and 5. The compression algorithm was tested for epsilon equal to 0.00001, 0.0001, 0.001, 0.01 and 0.1 and r equal to 1, 2, 3, 4, 5, 6. For big values of *epsilon* (*epsilon* equal to 0.1) almost all singular values were smaller than epsilon and were omitted during the compression process thus the results of the multiplication of the compressed matrix by vector have low accuracy. Obtained results for r equal to 1 and 6 and small *epsilon* values

(0.00001) are summarized in this section. Figure 4 presents the number of floating point operations performed in the matrix by vector multiplication algorithm for compressed matrix and matrix in sparse form. The mass matrix was generated for quintic B-splines ($p = 5$), and the number of intervals was equal to 1, 2, 4, 8, 16, 32. The compression was performed for epsilon equal to 0.00001 and r = 1 (blue line) and r = 6 (grey line). The maximal error in coefficient in obtained vector (the result of matrix by vector multiplication) is presented in Table 1 and Table 2. The presented results show that for epsilon = 0.00001 and a smaller number of intervals (1, 2, 4) the compression performed for r = 1 gives bigger number of floating point operations for matrix by vector multiplication than classic approach. For r = 1 and number of intervals equal to 8 the number of floating point operations is almost the same, for 16 and 32 intervals, the number of floating point operations for multiplication of a compressed matrix by vector is smaller than for classic multiplication for sparse matrix. For r = 6 and epsilon = 0.00001, the number of floating point operations for compressed matrix by vector multiplication algorithm is up to five times smaller than the number of floating point operations performed by a sparse matrix by vector multiplication algorithm. The accuracy of a vector obtained by the compressed matrix by vector multiplication algorithm is two orders of magnitude better for r = 1 than for r = 6. However, the accuracy obtained for r = 6 is also satisfactory. In Fig. 5, the compressed mass matrices for quintic B-splines (p = 5) and the number of intervals equal to 2 (first and second panel) and 32 (third and fourth panel) are shown. The compression was performed for epsilon equal to 0.00001, r = 1 (first and third panel) and r = 6 (second and fourth). The white color in the matrix denotes blocks of zeros. For the case of performing the SVD algorithm for block and remembering only k-rows and k-columns, the block is represented as k rows and k columns denoted by black color and white color in other places of the block. It can be seen that for r = 1 the mass matrices were less compressed. For r = 6 the matrices were compressed into four big blocks, for the case of the mass matrix for 2 and 32 intervals.

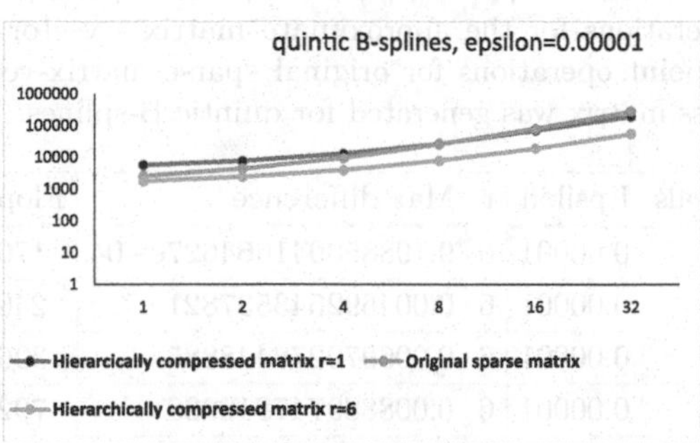

Fig. 4. The number of floating point operations in matrix by vector multiplication. The mass matrix generated for quintic B-splines, and the number of intervals equal to 1, 2, 4, 8, 16, 32. The compression performed for epsilon = 0.00001, r = 1 and r = 6.

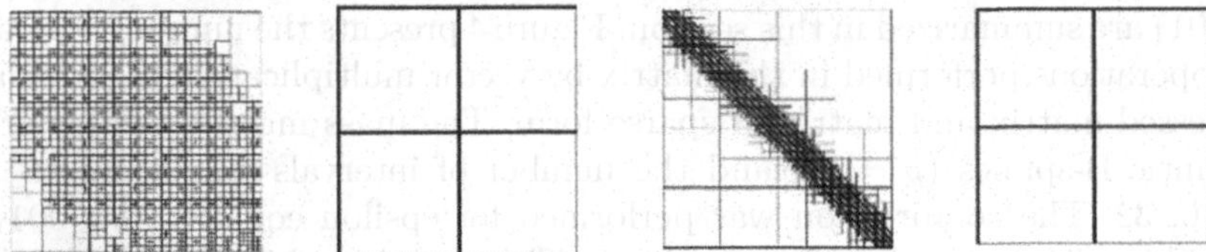

Fig. 5. The compressed mass matrix for quintic B-splines (p = 5) and the number of intervals equal to 2 (first and second panel) and 32 (third and fourth panel). The compression performed for epsilon equal to 0.00001, r = 1 (first and third panel) and r = 6 (second and fourth panel).

Table 1. The number of elements in a patch in a single direction, the epsilon used in the approximate svd compression, the maximal number of compressed rows and columns r = 1, the error of the approximate matrix-vector multiplication, the number of floating point operations for the approximate matrix - vector multiplication, the number of floating point operations for original sparse matrix-vector multiplication. The compressed mass matrix was generated for quintic B-splines.

Number of intervals	Epsilon	r	Max difference	Flops	Orginal flops
1	0.00001	1	7,04E−06	5674	2592
2	0.00001	1	1,49E−05	7657	4418
4	0.00001	1	2,26E−05	12971	9522
8	0.00001	1	2,41E−05	26826	25538
16	0.00001	1	2,34E−05	70412	80802
32	0.00001	1	3,01E−05	196064	284258

Table 2. The number of elements in a patch in a single direction, the epsilon used in the approximate svd compression, the maximal number of compressed rows and columns r = 6, the error of the approximate matrix-vector multiplication, the number of floating point operations for the approximate matrix - vector multiplication, the number of floating point operations for original sparse matrix-vector multiplication. The compressed mass matrix was generated for quintic B-splines.

Number of intervals	Epsilon	r	Max difference	Flops	Orginal flops
1	0.00001	6	9.198856941664627e−04	1764	2592
2	0.00001	6	0.004692543537821	2401	4418
4	0.00001	6	0.006970278448885	3969	9522
8	0.00001	6	0.008889747812938	7921	25538
16	0.00001	6	0.004687370741346	18945	80802
32	0.00001	6	0.001832486	55201	284258

Figure 6 presents the number of floating point operations performed in the matrix by vector multiplication algorithm for compressed matrix and matrix in sparse form. The mass matrix was generated for quartic B-splines (p = 4), and the number of intervals was equal to 1, 2, 4, 8, 16, 32. The compression was performed for epsilon equal to 0.00001 and r = 1 (blue line) and r = 6 (grey line). In Fig. 7, the compressed mass matrices for quartic B-splines (p = 4) and the number of intervals equal to 2 (first and second panel) and 32 (third and fourth panel) are shown. The compression was performed for epsilon equal to 0.00001, r = 1 (first and third panel) and r = 6 (second and fourth).

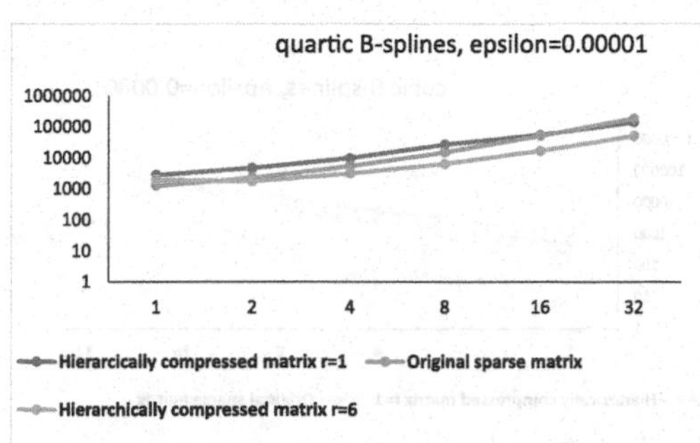

Fig. 6. The number of floating point operations in matrix by vector multiplication. The mass matrix generated for quartic B-splines, and the number of intervals equal to 1, 2, 4, 8, 16, 32. The compression performed for epsilon = 0.00001, r = 1 and r = 6.

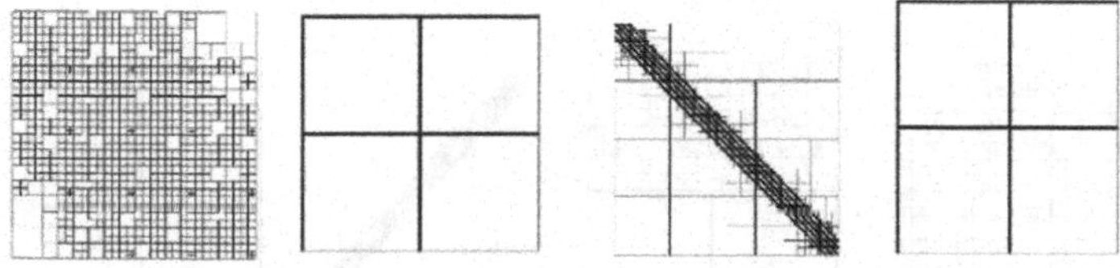

Fig. 7. The compressed mass matrix for quartic B-splines (p = 4) and the number of intervals equal to 2 (first and second panel) and 32 (third and fourth panel). The compression performed for epsilon equal to 0.00001, r = 1 (first and third panel) and r = 6 (second and fourth panel).

Figure 8 presents the number of floating point operations performed in the matrix by vector multiplication algorithm for compressed matrix and matrix in sparse form. The mass matrix was generated for cubic B-splines (p = 3), and the number of intervals was equal to 1, 2, 4, 8, 16, 32. The compression was performed for epsilon equal to 0.00001 and r = 1 (blue line) and r = 6 (grey line). The obtained results show that for the mass matrix generated for one interval or two intervals and cubic B-splines, the number of floating point operations for matrix by vector multiplication algorithm for matrix in compressed form for r equal to 1 and 6 is bigger than for the matrix in sparse form. For bigger number of intervals, for r = 6 and epsilon = 0.00001 the number of floating point operations for compressed matrix by vector multiplication algorithm is smaller than the number of floating point operations performed by sparse matrix by vector multiplication algorithm. In Fig. 9, the compressed mass matrices for cubic B-splines (p = 3) and the number of intervals equal to 2 (first and second panel) and 32 (third and fourth panel) are shown. The compression was performed for epsilon equal to 0.00001, r = 1 (first and third panel) and r = 6 (second and fourth).

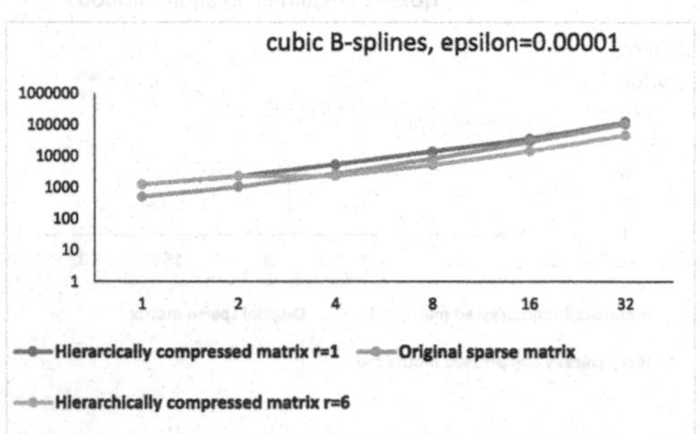

Fig. 8. The number of floating point operations in matrix by vector multiplication. The mass matrix generated for cubic B-splines, and the number of intervals equal to 1, 2, 4, 8, 16, 32. The compression performed for epsilon = 0.00001, r = 1 and r = 6.

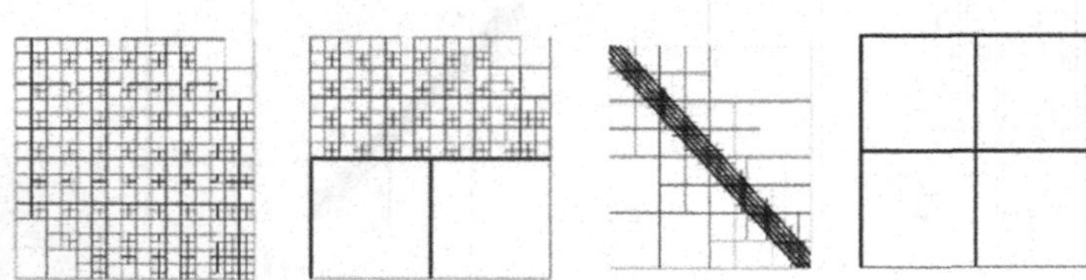

Fig. 9. The compressed mass matrix for cubic B-splines (p = 3) and the number of intervals equal to 2 (first and second panel) and 32 (third and fourth panel). The compression performed for epsilon equal to 0.00001, r = 1 (first and third panel) and r = 6 (second and fourth panel).

Figure 10 presents the number of floating point operations performed in the matrix by vector multiplication algorithm for compressed matrix and matrix in sparse form. The mass matrix was generated for qadratic B-splines (p = 2), and the number of intervals was equal to 1, 2, 4, 8, 16, 32. The compression was performed for epsilon equal to 0.00001 and r = 1 (blue line) and r = 6 (grey line). The obtained results show that for the mass matrix generated for one interval or two intervals and quadratic B-splines, the number of floating point operations for matrix by vector multiplication algorithm for matrix in compressed form for r equal to 1 and 6 is bigger than for the matrix in sparse form. For bigger number of intervals, for r = 6 and epsilon = 0.00001 the number of floating point operations for compressed matrix by vector multiplication algorithm is slightly smaller than the number of floating point operations performed by sparse matrix by vector multiplication algorithm. It can be seen that for r = 1 the mass matrices for 2 and 32 intervals as well as mass matrix for 2 intervals for r = 6 were less compressed. For r = 6 the mass matrix generated for 32 intervals was compressed into four big blocks. In Fig. 11, the compressed mass matrices for quadratic B-splines (p = 2) and the number of intervals equal to 2 (first and second panel) and 32 (third and fourth panel) are shown. The compression was performed for epsilon equal to 0.00001, r = 1 (first and third panel) and r = 6 (second and fourth). Similar results, with better compression but slightly lower accuracy were obtained for compression with epsilon equal to 0.0001.

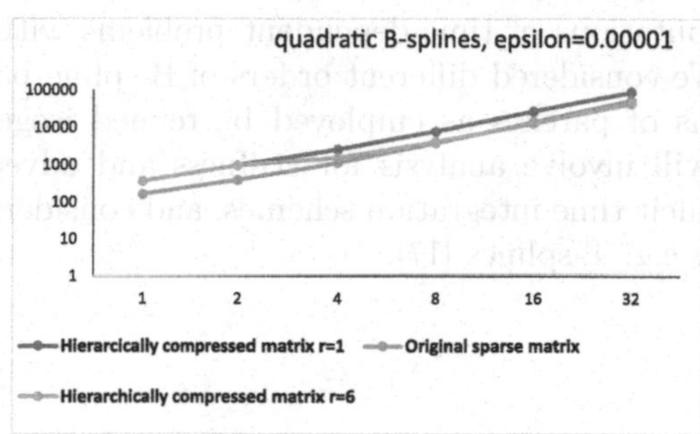

Fig. 10. The number of floating point operations in matrix by vector multiplication. The mass matrix generated for quadratic B-splines, and the number of intervals equal to 1, 2, 4, 8, 16, 32. The compression performed for epsilon = 0.00001, r = 1 and r = 6.

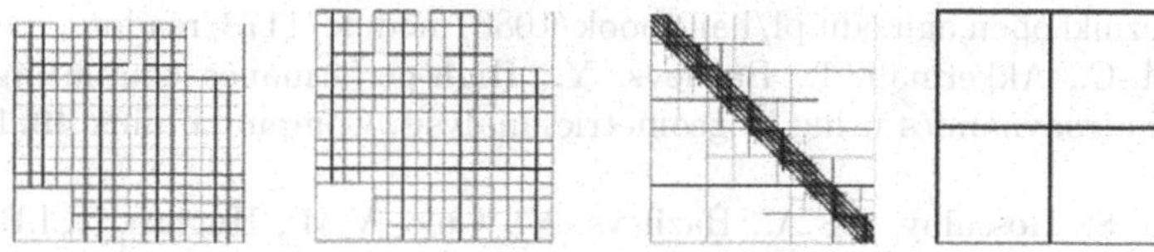

Fig. 11. The compressed mass matrix for quadratic B-splines (p = 2) and the number of intervals equal to 2 (first and second panel) and 32 (third and fourth panel). The compression performed for epsilon equal to 0.00001, r = 1 (first and third panel) and r = 6 (second and fourth panel).

6 Summary of the Results

The performed tests show that recursive compression of the matrix by SVD algorithm can speed up the process of matrix-vector multiplication. The best results were obtained for r = 6 and epsilon = 0.00001 for p = 5 (quintic B-splines). Especially, the number of floating point operations for compressed matrix - vector multiplication algorithm (55201 floating point operations) is up to five times smaller than the number of floating point operations performed by a sparse matrix - vector multiplication algorithm (284258 floating point operations) -see Table 1. In general, increasing the order of B-splines results in better compression and faster matrix-vector multiplication.

7 Conclusions and Future Work

In this paper, we focused on recursive compression of isogeometric analysis mass matrices. We showed that having the matrix recursively compressed, we can speed up the computations of time-dependent problems with iterative solvers up to five times. We considered different orders of B-spline basis functions and different dimensions of patches as employed by refined isogeometric analysis. The future work will involve analysis for stiffness and advection matrices to implement the implicit time integration schemes, and considering adaptive non-regular grids, using e.g. T-splines [17].

References

1. Austin Cottrell, J., Hughes, T.J.R., Bazilevs, Y.: Isogeometric Analysis: Toward Integration of CAD and FEA. Computational and Numerical Methods, Wiley, Hoboken (2009)
2. Hughes, T.J.R., Cottrell, J.A., Bazilevs, Y.: Isogeometric analysis: CAD, finite elements, NURBS, exact geometry and mesh refinement. Comput. Methods Appl. Mech. Eng. **194**(39), 4135–4195 (2005)
3. Paszyński, M.: Classical and isogeometric finite element method. https://epodreczniki.open.agh.edu.pl/handbook/1088/module/1173/reader
4. Hsu, M.-C., Akkerman, I., Bazilevs, Y.: High-performance computing of wind turbine aerodynamics using isogeometric analysis. Comput. Fluids **49**(1), 93–100 (2011)
5. Hossain, S., Hossainy, S.F.A., Bazilevs, Y., Calo, V.M., Hughes, T.J.R.: Mathematical modeling of coupled drug and drug-encapsulated nanoparticle transport in patient-speciffic coronary artery walls. Comput. Mech. **2**, 213–242 (2011)
6. Łoś, M., Kłusek, A., Hassaan, M.A., Pingali, K., Dzwinel, W., Paszyński, M.: Parallel fast isogeometric L2 projection solver with GALOIS system for 3D tumor growth simulations. Comput. Methods Appl. Mech. Eng. **343**, 1–22 (2019)
7. Łoś, M., Paszyński, M., Kłusek, K., Dzwinel, W.: Application of fast isogeometric L2 projection solver for tumor growth simulations. Comput. Methods Appl. Mech. Eng. **316**, 1257–1269 (2017)

8. Garcia, D., Pardo, D., Dalcin, L., Paszynski, M., Collier, N., Calo, V.M.: The value of continuity: refined isogeometric analysis and fast direct solvers. Comput. Methods Appl. Mech. Eng. **316**, 586–605 (2017)
9. Garcia, D., Pardo, D., Dalcin, L., Calo, V.M.: Refined isogeometric analysis for a preconditioned conjugate gradient solver. Comput. Methods Appl. Mech. Eng. **335**, 490–509 (2018)
10. Garcia, D., Pardo, M., Calo, V.M.: Refined isogeometric analysis for fluid mechanics and electromagnetics. Comput. Methods Appl. Mech. Eng. **356**, 598–628 (2019)
11. Łoś, M., Woz, M., Paszyński, M., Dalcin, L., Calo, V.M.: Dynamics with matrices possessing Kronecker product structure. Procedia Comput. Sci. **51**, 286–295 (2015)
12. Saad, J.: Iterative Methods for Sparse Linear Systems. Society for Industrial and Applied Mathematics (2003)
13. Golub, G.H., Van Loan, C.: Matrix Computations, 3rd edn. John Hopkins University Press, Baltimore (1996)
14. Hackbusch, W.: Hierarchical Matrices: Algorithms and Analysis. Springer, Heidelberg (2015). https://doi.org/10.1007/978-3-662-47324-5
15. Mantzaflaris, A., Juttler, B., Khoromskij, B., Langer, U.: Matrix generation in isogeometric analysis by low rank tensor approximation. In: International Conference on Curves and Surfaces, pp. 321–340 (2015)
16. Langville, A.N., Meyer, C.D.: Google's PageRank and Beyond: The Science of Search Engine Rankings. Princeton University Press, Princeton (2006)
17. Bazilevs, Y., et al.: Isogeometric analysis using T-splines. Comput. Methods Appl. Mech. Eng. **199**, 229–263 (2010)

Neural-Network Based Adaptation of Variation Operators' Parameters for Metaheuristics

Tymoteusz Dobrzański, Aleksandra Urbańczyk, Tomasz Peech-Pilichowski, Marek Kisiel-Dorohinicki, and Aleksander Byrski(✉)

AGH University of Science and Technology, Kraków, Poland
tdobrzanski@student.agh.edu.pl, {aurbanczyk,tomek,doroh,olekb}@agh.edu.pl

Abstract. The paper presents an idea of training an artificial neural network a relation between different parameters observed for a population in a metaheuristic algorithm. Then such trained network may be used for controlling other algorithms (if the network is trained in such way, that the knowledge gathered by it becomes agnostic regarding the problem). The paper focuses on showing the idea and also provides selected experimental results obtained after applying the proposed algorithm for solving popular benchmark problems in different dimensions.

Keywords: Adaptation of variation operators' parameters · Evolutionary algorithms · Neural networks

1 Introduction

Maintaining the balance between exploitation and exploration in metaheuristic is an important task in order to avoid premature loss of convergence and getting stuck in a local extremum by a population processed. The exploitation and exploration are rather volatile notions, but can be approximated by measuring the diversity, thus based on diversity one might propose way for controlling the search of a metaheuristic algorithm in order to avoid pitfalls (like already mentioned premature convergence). Different approaches were proposed in order to adapt the parameters of variation operators (based on dedicated rules, or e.g. modifying the parameters of mutation by the same algorithm which was used for solving the problem), but the intuition points out that the actual relation between different parameters observed in the population and the features of exploitation/exploration is complex.

The research presented in this paper has been financially supported by: Polish National Science Center Grant no. 2019/35/O/ST6/00570 "Socio-cognitive inspirations in classic metaheuristics." (A.U.) and Polish Ministry of Education and Science funds assigned to AGH University of Science and Technology (T.P-P., M.K-D., A.B.).

D. Groen et al. (Eds.): ICCS 2022, LNCS 13351, pp. 394–407, 2022.
https://doi.org/10.1007/978-3-031-08754-7_47

Why don't we use ANNs (Artificial Neural Networks) to grasp this complex relation and provide the metaheuristic with the knowledge required for reasonable (based on many experiments, not arbitrarily chosen rules) adaptation of the variation operators parameters? This is actually the idea proposed in this paper: to find a way for train an artificial neural network, gathering necessary knowledge and being able to reuse this knowledge in a similarly-defined (but not the same) problems.

In the course of this paper, after referring to the state of the art regarding neural networks and adaptation of the variation operators of metaheuristics, we discuss the proposed algorithm clearly, then show the experimental setting and provide the insight into first efficacious experiment results, finally we conclude our paper and point out the future work directions.

2 Neural Networks in Prediction and Control

For prediction and control purposes a number of methods and approaches can be exploited. The goal is to perform reliable data processing based on input data, initially preprocessed. An application of a predictor is based on adjusting its parameters (in a fixed or adaptive way). One of prediction approaches is to use neural networks instead of numerical formulas. Considering general architecture of neural networks, a neuron processes a set of input data, according to fixed weights and a bias. Supervised or unsupervised training is a process of obtaining network parameters. For example, simple time series one-step-ahead prediction with a neural network containing one input layer is comparable to a linear regression, nevertheless, prediction can be performed based on results of data classification.

A number of neural models applied for prediction purposes have been proposed. The presented approaches differ in architectures, objective functions or a paradigm used [6]. An application of a model relies on the prediction attributes, time-horizon and input data quality (sampling, consistency, statistical and frequency characteristics). For a given time horizon, predictions - as one step or multistep - can be computed as point forecasts (results in point estimates of values for a fixed horizon) or probabilistic ones (the distribution of future data is produced). For prediction purposes Multilayer Perceptron (MLP) networks, Convolutional Neural Networks (CNN) and Recurrent Neural Networks (RNN) can be used [8,29]. MLPs are a class of backpropagation neural networks used for prediction and regression of structured input data. A key feature is an ability of mapping from input datasets to output variables. MLPs can be used i.a. for time series classification and prediction - as mapping from input data to output data [27].

Neural networks (NNs) are applicable for dynamic process control (in particular, in industrial applications) [22,30] where processes are complex and they have to be operated with a small number of operators (computer systems are widely used for a process control and supervision) and automatic adjusting values of processes is required. Process control is based on data processing (i.a.

set points, estimated states) aimed at obtaining present and future values, for example for adaptive control purposes. Prediction of a system future behavior is essential, and a set of constraints is satisfied (in particular, in real-time systems [23]). Model Predictive Control (MPC) is focused on finding (computing) control by solving optimization problem in a loop according to predicted output (the behavior of the system) and fixed time-horizon. Neural networks can be exploited to adjust parameters of controllers of applied algorithms based on analytical calculations due to their limitations arising from a large number of variables, parameters and control loops as well as a need of performing computation in a real time systems according to specified time limits. NNs can be developed for solving optimal control problems and for tasks requiring the use of control loop, in particular - as NN-MPC (Neural Network based Model Predictive Control).

3 Adaptation of Evolutionary Search Parameters

In the group of evolutionary algorithms (Genetic Algorithms, Evolution Strategies and Evolutionary Programming and many hybridised versions of them) the key to success is good parameters setting. In the classic versions of them certain values for parameters (selection, mutation and crossover) were set on the beginning, very often by trial and error. Many researchers realised that setting them without the knowledge about the structure of the problem lead to sub-optimal solutions and result in another trial (known as "parameter tuning problem" [28]). Besides other solutions [13], the idea of parameter control grew on popularity. [17]. It states that instead of tuning the parameters in advance, these can be observed and adjusted during the run. This shift had many advantages, from releasing researchers from duty of initial parameters' setting, to broadening the spectrum of problems that can be solved by EAs (Evolutionary Algorithms).

The overview and classification of all possibly controlled parameters is extensively described in [17]. One of the axis of the classification is whether the method of control is dependent on the parameters itself or it is rather generic one.

There are numerous control methods intended for specific parameters. They can be based on one or few parameters combined. One of the most explored method related to one parameter is controlling and adaptation of mutation step size (σ), originally employed in ES (Evolution Strategies). The early ideas circled around modifying the step size proportionally to the distance to the optimum [3,26] or increasing/decreasing it by 0.2 accordingly to the success rate (*one-fifth rule*) [5,25]. The creation of the self-adaptive mutation was the next great leap in adaptive ESs: the mutation parameters (both the step-size and the covariance matrix) are tied to each individual and are also subject to mutation [9]. However, there is an algorithm that outperforms all of them and it is based on the adaptation of the full covariance matrix [14,16]. The core idea behind CMA-ES (Covariance Matrix Adaptation ES) is to use the algorithm's route to deterministically update the various mutation parameters. If the algorithm has taken a sequence of steps in the same direction, the step size should be increased to allow for greater steps and faster processing. The covariance matrix update is affected by similar ideas. Aside from single-parameter techniques, there have been numerous attempts to improve EAs by creating ways for controlling multiple parameters at the same time (e.g. [4,19]. However, they usually change components other than just variation parameters, therefore a detailed description is beyond the scope of this article.

The group of generic parameter control methods is the one with which we associate our work. In the recent extensive review of those methods by Gomes Pereira De Lacerda et al. [15] there are two main groups of general parameter controllers - one based on reinforcement learning and the other on prediction of probability of success for parameters at each point of time. The other attempts to design generic parameter control were based on Dynamic Programming under Markov assumption [2], fuzzy logic [18] and Bayesian networks [10]. We found no evidence of attempting to incorporate ANN as a generic controller for EA in the review or through our own research.

4 ANN-Based Adaptation of Evolutionary Search

The idea of adapting the parameters of metaheuristics' variation operators is not new, and we have cited several seminal papers in Sect. 3 to give the background on this. Observing the state-of-the-art algorithms we stated, that those approaches usually utilize the information perceived in the population, based on fixed rules. A notable approach is the one applied in evolution strategies, where the parameters of the probability distribution of mutation undergoes similar evolution as the genetic information contained in the individuals. Not degrading any of the approaches, we would like to express our deep conviction, that apparently, a relation between the observations of certain parameters of the population and the current (desired) parameters of the variation operators (which would help in attaining balance between exploration and exploitation) is complex. Thus, instead of assuming certain fixed rules, we propose to teach a dedicated neural network in a way agnostic to the actual problem, in order to use the knowledge gathered in the network to control the variation operator's parameters for similar problems.

The idea of the proposed approach is shown in Fig. 1. The idea consists in training a dedicated neural network, to learn the complex relation between a number of different population-related parameters and the parameters of the variation operators. The presented approach assumes that the neural network is trained in a supervised way. The trained data are the selected parameters of the population (e.g. diversity-related measures), and they are paired with the desired parameters of the variation operators (e.g. probability of applying crossover and mutation, mutation range).

In this paper we have focused on supervised training (in the future we will also consider unsupervised approaches). Now, in order to train the network, which should grasp the relation between the observed parameters and the parameters of the variation operators, we assume that those parameters (in the discussed

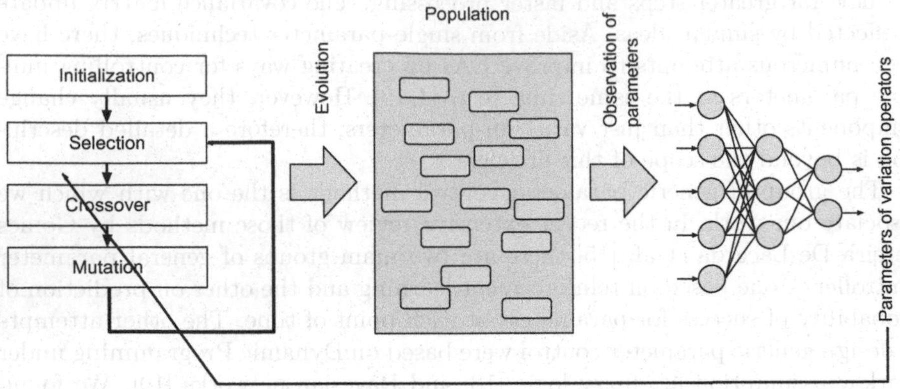

Fig. 1. Neural adaptation of variation operators' parameters

case) should have certain fixed relation with the exploration/exploitation phenomena. We assume, that an arbitrarily chosen function should, in the training phase, describe this relation (as a reasonable one) and become the template which will be later generalized by the network, applied to different problems. The function used in the presented approach is actually sigmoid-like (the details are given in Sect. 5) starting with high values, and gradually descending to lower values (yet not zero), assuming that a certain loss of diversity must be observed (as the population gradually ceases to show a "monte-carlo like" behavior and focuses on exploitation, still having certain exploration capabilities). The idea of training the network is shown in Fig. 2. Thus the network is trained on several selected benchmark problems, different chosen functions can be used for showing the progress of the parameters in training, and all the training related parameters are prepared in a way as agnostic as possible to a considered problem, the actual shape and dimension of the search space etc. so the knowledge gathered

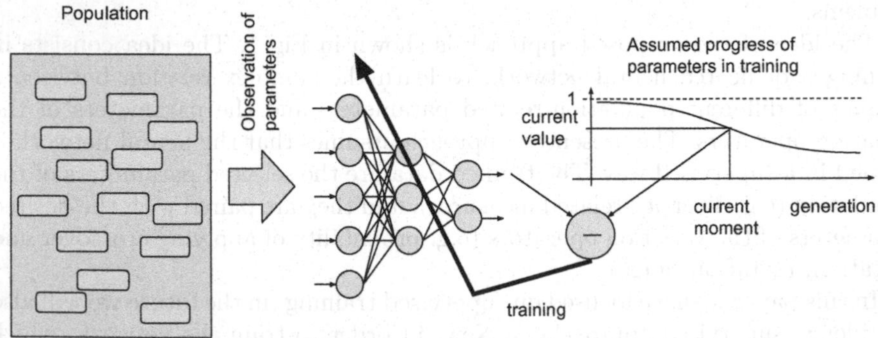

Fig. 2. Training the network

in the network (regarding the relation between the parameters and the variation operators' parameters) can be reused.

5 Experimental Results

In our experiments we have used as our metaheuristic algorithm a classic version of evolution strategy (μ, λ) algorithm applied to solving real-value optimization problems [20]. We have used apopulation of 100 individuals and as a stopping condition – reaching 10000 evaluations of the fitness function. We have repeated each experiment 1000 times, except results shown in Fig. 3 and 4, where we show result from single algorithm run.

Artificial Neural Networks (feed-forward fully connected Neural Network, with Rectified Linear Unit activation in hidden layers, and sigmoid activation in output layer) used in our work were built and trained using TensorFlow framework [1]. Evolution Strategy was implemented based on jMetalPy framework [7]. The benchmarks and ANN training process were evaluated on Windows based machine with AMD Ryzen 5 3600 (3.59 GHz) CPU, Nvidia GeForce GTX 1650 GPU and 16 GB of RAM memory.

The observed parameters of the population (after each generation) were:

- Current evaluation number.
- Standard deviation diversity (stddev) [31] - standard deviation of fitness values across population (P). Computed as:

$$stddev(P) = \sqrt{\frac{\sum_{i=1}^{n}(f_i - \overline{f})}{n-1}}$$

  Where: P – population, n – population size, f_i – fitness value of i individual, $\overline{f}$ – mean fitness value.
- Phenotype diversity (ptype) [31] - number of unique values (U) of fitness across population. Computed as:

$$ptype(P) = \frac{U-1}{n-1}$$

  Where: U – number of unique fitness value, n – population size This value was not used as input for ANN.
- Distance diversity (distance) - mean distance between all pairs of individuals. Computed as:

$$distance(P) = \frac{2}{(n-1)n}\sum_{i=1}^{n-1}\sum_{k=i+1}^{n} dist(P_i, P_k)$$

  Where: $dist$ – distance metric, P_i – genotype of i individual, n – population size. We can use different distance metric (e.g. Hamming for binary genotype) to compute this type of diversity measure.

- Moment of Inertia diversity (MoI) [21] - calculating system moment of inertia to define diversity. Computed as:
  Centroid (central point of system) defined as vector $C = (c_1, c_2, \ldots, c_d)$:

$$c_i = \frac{\sum_j^n x_{ij}}{n}$$

  Where: x_{ij} – value of i gen in j individual, n – population size
  Moment of Inertia:

$$MoI(P) = \sum_i^d \sum_j^n (x_{ij} - c_i)$$

  Where: x_{ij} – value of i gen in j individual, n – population size, d - problem dimensionality. In Artificial Neural Network as an input value, we used:

$$Input = MoI_c - MoI_r$$

  Where: MoI_c – MoI in current population, MoI_r – desired MoI diversity, calculated from sigmoid function multiplied by starting value of MoI.
- Current Mutation probability.
- Current Mutation range – determines range in which parameter can mutate
- Current Crossover probability.
- Current Crossover distribution index – parameter used in Simulated binary (SBX) crossover [11], to determine how far from parents, children's will be placed. Higher value means closer placement.
- Percentage loss of MoI diversity from start of algorithm runtime.
- Percentage loss of MoI diversity from last generation.

It is important to notice that almost all mentioned parameters are highly dependent on selected problem. Size of fitness function space or values range will have impact on computed parameters as well as prepared model. In order to make trained Neural Networks problem agnostic, diversity measurements have to be normalized. In our work, selected values were normalized by starting value of that parameter. This solution allow values to exceed 1, but it's not considered as a problem for trained ANN.

The biggest challenge in our work, was to create problem agnostic Artificial Neural Network, that can be used on any optimization function in any dimensionality, despite fact that selected parameters are size and dimension depended. For example MoI diversity in Ackley problem reaches 110 units, but in Griewank functions, that values can exceed 1400 units (in 100 dimensions). To achieve agnostic model we used normalization by starting value, that may not be optimal solution, but currently it works fine. In further work we will try to propose another, more efficient, normalization algorithm. Taking this information into consideration all size dependent parameters (e.g. Distance diversity, Standard deviation diversity) were normalize to create input in range [0, 1], other values that does not depend on problem (e.g. variation operators, percentage loss of MoI diversity) were used without preprocessing.

We have focused on popular benchmark functions: Ackley, Griewank, Rastrigin, Rosenbrock, Schwefel and De Jong [12]. The Ackley function (in 100 dimensions) was first used for training the network. All the problems were set in 100 dimensions except Rosenbrock which was set in 10, 20, 50, 100, 200 and 500 dimensions.

The parameters of the sigmoid function $sigm(x) = \frac{1}{1+e^{c_1*x-c_2}}$ describing the progress of the variation operators' parameters were: $c_1 = \frac{10}{m_e}$, m_e - number of maximum evaluations, $c_2 = 5$. Values range of this variant of sigmoid function is [0.99, 0.01]. That should demonstrate balance between exploration and exploitation phase of algorithm.

We have started our experiments with training the selected feed-forward networks having 10 input neurons observing the above-mentioned parameters, 2 output neurons producing the probabilities of crossover and mutation and

- Variant 1 - Hidden Layers size: 24-64-92-24-56-32-24-12
- Variant 2 - Hidden Layers size: 36-64-128-64-18-32-24-12
- Variant 3 - Hidden Layers size: 36-64-128-64-18-32-24-12 It has same hidden size that Variant 2, but training dataset was prepared in different way. Normally variations operators were adapted by sigmoid function, in this case they were generated randomly after each algorithm generation.
- Variant 4 - Hidden Layers size: 36-64-128-64-6-32-24-12.

We have used Ackley function for training. Hidden size was set up using trial and error method. Training dataset was prepared by evaluating 2000 runs of algorithm with adaptive mutation and crossover probability parameters and saving those mentioned above parameters.

Artificial Neural Network training parameters:

- Dataset size: 200000 records
- Validation dataset size: 20% of dataset
- Epochs: 4000
- Batch size: 512
- Learning rate: 0.001
- Optimizer: Adam
- Loss function: Mean Square Error (mse) [24]. Computed as:

$$mse = \frac{\sum_{i=1}^{n}(y_i - \lambda(x_i))^2}{n}$$

Where: n - number of outputs, y_i - true value, $\lambda(x_i)$ - model output.

The trained network was evaluated on the Ackley benchmark (applied to adapt the parameters of the variation operators) and we provide the insight into the process of optimization and the final values obtained at the end of computing.

In Fig. 3 the progress of the best fitness observed for different versions of neural network adapting the parameters of the variation operators is shown. Apparently the idea of ANN-controlled adaptation taught based on previously given sigmoid progress function works – the baseline algorithm (without the adaptation) does not get so close as the algorithms using adaptation. Moreover, we have also tested the algorithm with adaptation of variation parameters without the neural network used (so it was actually the algorithm used for teaching the ANNs – the results were similar as in the case of the baseline algorithm (see Fig. 5a, Adaptive ES version of the algorithm). Thus we claim the idea makes sense and delve into more detailed experiments. Let us check, what is actually happening in the course of computation, how (and if at all) does the ANN work (see Fig. 4). One can see that the actual relation between the observed parameters and the output (probabilities of mutation and crossover) is complex (as expected). Mutation and crossovers observed at the input are similar to the ones observed at the output (as these values are in fact controlled by the network). One can observe the increasing mutation probability in the beginning of computation, then for some time the crossover becomes the more frequently used operator, however finally the crossover is apparently discarded (at the moment when the measurement of MoI-From-Last becomes stable, later no peaks are

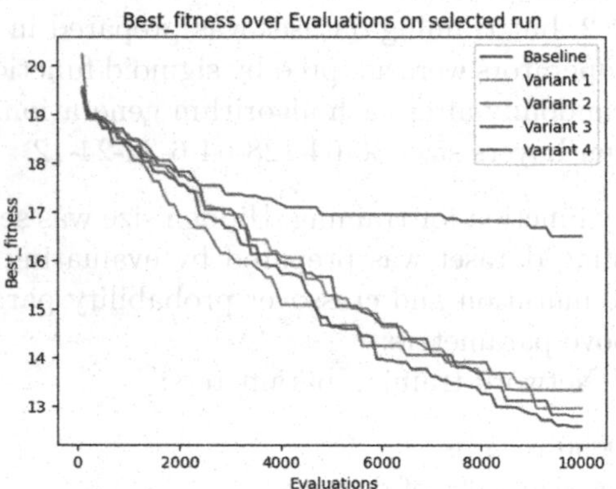

Fig. 3. Best fitness dependent on the number of evaluation, Ackley, 100 dimensions

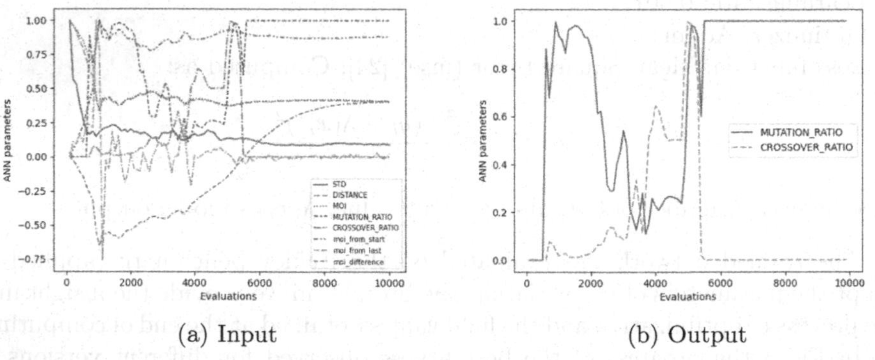

Fig. 4. Monitoring input and output parameters for the ANN used, Ackley 100 dimensions

observed in the graph showing this measure). This situation seems to be normal, after some time in this experiment the diversity of the search is lost, and because of that the mutation is used for escaping the local minimum (probably attained), in other words, to escape the situation when most (or all) of individuals are very similar.

In Fig. 5a we can observe the final values obtained at the end of computing for Ackley problem. It is easy to see, that the neural versions of the algorithm are significantly better than the version without the neural adaptation. At the same time, in Fig. 5b we can observe the final results obtained for Griewank problem. This time there is one neural network which turned out to be the best in adaptation. This is completely normal and highly possible situation, that only one network prevailed. Usually choosing the optimal parameters of neural network is realized in a tedious process of trial and error. Apparently the most complex problem is Schwefel (see Fig. 6a), as the results are far from optimum, however still one network prevails in this case. This might be an apparent case for more difficult problems, that only one network will prevail. Actually this observation is confirmed when observing the final results obtained for Sphere function (Fig. 6b), which is the easiest problem considered (this is a convex function). This time again the ANN-adapted versions of the algorithm prevail and it is difficult to tell, which structure of the network was the best.

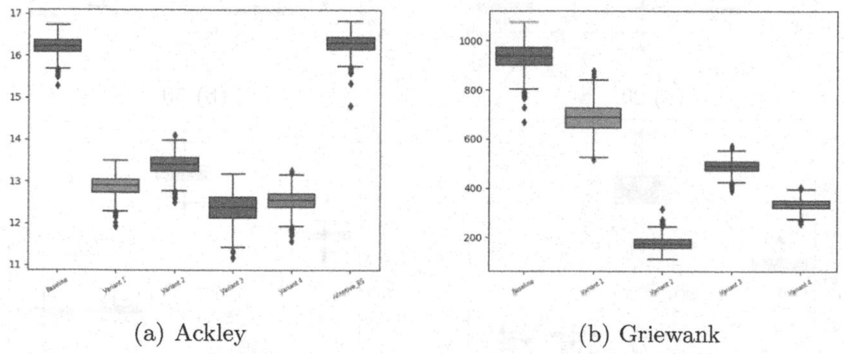

(a) Ackley (b) Griewank

Fig. 5. Final fitnesses for Ackley and Griewank problems, 100 dimensions

Finally let us check what happens if we change the dimensionality of the problem. Thus let us focus on Rosenbrock problem in different dimensions (see Fig. 7). Now it is easy to see that again, for simpler problems (20 and 50 dimensions) the algorithms show similar efficacy, while for more difficult problems (200 and 500 dimensions) one or several algorithms prevail. Moreover – it seems that the adaptation helped a lot in the case of those more difficult problems (the results are much closer to the global optimum than in the case of 20 and 50), at the same time acting mediocre (or being more or less useless) for simpler problems.

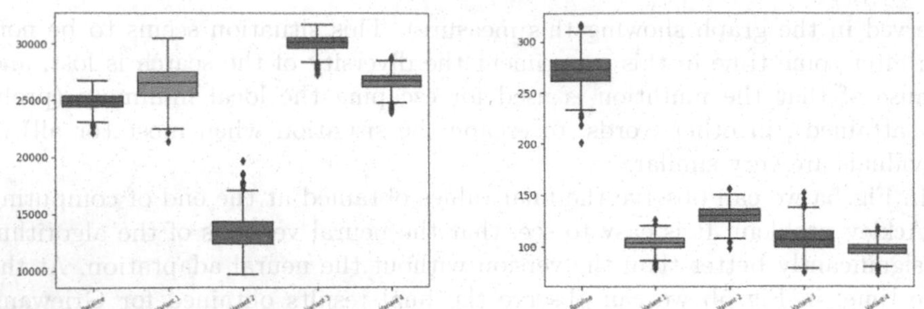

(a) Final fitness for different ANNs, Schwefel, 100 dimensions (b) Final fitness for different ANNs, Sphere, 100 dimensions

Fig. 6. Final fitnesses for Schwefel and Sphere problems, 100 dimensions

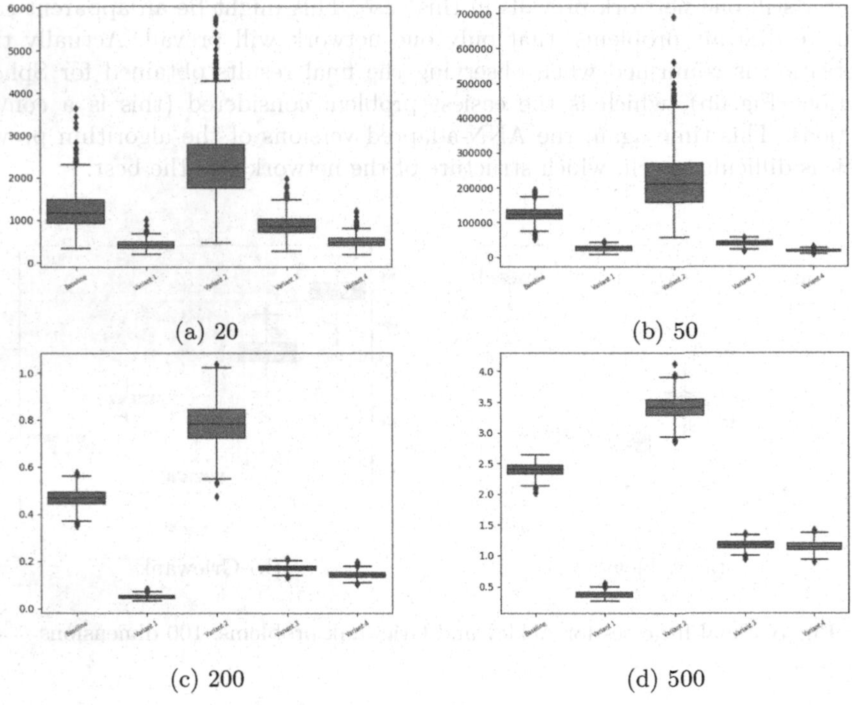

(a) 20 (b) 50

(c) 200 (d) 500

Fig. 7. Final fitnesses for Rosenbrock for different dimensions (20, 50, 200, 500)

6 Conclusion

We have presented a novel approach for adaptation of variation operators' parameters in metaheuristic algorithms, based on artificial neural networks. The idea consists of training the network on dedicated benchmark problems, working on the parameters of the search (and of the network) to make the knowledge gathered by the network agnostic to the problem, search space etc. and applying such network to adapt the parameters in other problems. We believe that the results presented are interesting, but we are sure that much more is to be done to further tune the algorithm and generalize it. Our plans for the future research are as follows:

- We have used a sigmoid function as a function showing the rational progress of the variation parameters – we will try experimenting also with other functions, to further explore our assumptions related the reasonable control over exploration and exploitation.
- We have followed the supervised learning paradigm, in the future we will also try to use unsupervised learning (e.g. hebbian learning) in order to avoid assuming concrete parameter progress functions.
- We will try to explore the relation between the network and the number of evaluation of the fitness function, perhaps this relation might be also generalized or omitted.
- In the presented research we have used one ANN for both parameters of variation operators, this was a simplification – we are sure we should apply different ANNs for each parameter controlled.
- We will aim at attaining agnostic knowledge gathered in the network in order to be able to easily reuse the trained network.

References

1. Abadi, M., et al.: TensorFlow: large-scale machine learning on heterogeneous systems (2015). https://www.tensorflow.org/, software available from tensorflow.org
2. Aine, S., Kumar, R., Chakrabarti, P.P.: Adaptive parameter control of evolutionary algorithms under time constraints. In: Tiwari, A., Roy, R., Knowles, J., Avineri, E., Dahal, K. (eds.) Applications of Soft Computing. AISC, vol. 36, pp. 373–382. Springer, Heidelberg (2006). https://doi.org/10.1007/978-3-540-36266-1_36
3. Auger, A., Le Bris, C., Schoenauer, M.: Dimension-independent convergence rate for non-isotropic $(1, \lambda)$ — *ES*. In: Cantú-Paz, E., et al. (eds.) GECCO 2003, Part I. LNCS, vol. 2723, pp. 512–524. Springer, Heidelberg (2003). https://doi.org/10.1007/3-540-45105-6_64
4. Bäck, T., Eiben, A.E., van der Vaart, N.A.L.: An emperical study on GAs "without parameters". In: Schoenauer, M., et al. (eds.) PPSN 2000. LNCS, vol. 1917, pp. 315–324. Springer, Heidelberg (2000). https://doi.org/10.1007/3-540-45356-3_31
5. Bassin, A., Buzdalov, M.: The 1/5-th rule with rollbacks. In: Proceedings of the Genetic and Evolutionary Computation Conference Companion, July 2019. https://doi.org/10.1145/3319619.3322067. http://dx.doi.org/10.1145/3319619.3322067
6. Benidis, K., et al.: Neural forecasting: introduction and literature overview. https://arxiv.org/abs/2004.10240 (2020)
7. Benítez-Hidalgo, A., Nebro, A.J., García-Nieto, J., Oregi, I., Del Ser, J.: jMetalPY: a python framework for multi-objective optimization with metaheuristics. Swarm Evol. Comput. **51**, 100598 (2019). https://doi.org/10.1016/j.swevo.2019.100598. https://www.sciencedirect.com/science/article/pii/S2210650219301397

8. Botalb, A., Moinuddin, M., Al-Saggaf, U.M., Ali, S.S.A.: Contrasting Convolutional Neural Network (CNN) with Multi-Layer Perceptron (MLP) for big data analysis. In: 2018 International Conference on Intelligent and Advanced System (ICIAS), pp. 1–5 (2018). https://doi.org/10.1109/ICIAS.2018.8540626
9. Bäck, T.: Self-adaptation in genetic algorithms. In: Proceedings of the First European Conference on Artificial Life, pp. 263–271. MIT Press (1992)
10. Corriveau, G., Guilbault, R., Tahan, A., Sabourin, R.: Bayesian network as an adaptive parameter setting approach for genetic algorithms. Complex Intell. Syst. **2**(1), 1–22 (2016). https://doi.org/10.1007/s40747-016-0010-z
11. Deb, K., Agrawal, R.B., et al.: Simulated binary crossover for continuous search space. Complex Syst. **9**(2), 115–148 (1995)
12. Digalakis, J., Margaritis, K.: An experimental study of benchmarking functions for evolutionary algorithms. Int. J. Comput. Math. **79**, 403–416 (2002)
13. Eiben, A., Smit, S.: Parameter tuning for configuring and analyzing evolutionary algorithms. Swarm Evol. Comput. **1**, 19–31 (2011). https://doi.org/10.1016/j.swevo.2011.02.001
14. Eiben, A.E., Smith, J.E., Michalewicz, Z., Schoenauer, M., Smith, J.E.: Parameter control in evolutionary algorithms. Stud. Comput. Intell. **54**, 19–46 (2007)
15. Gomes Pereira De Lacerda, M., Filipe De Araujo Pessoa, L., Buarque De Lima Neto, F., Ludermir, T.B., Kuchen, H.: A systematic literature review on general parameter control for evolutionary and swarm-based algorithms. Swarm Evol. Comput. **60**, 100777 (2021). https://doi.org/10.1016/j.swevo.2020.100777. www.elsevier.com/locate/swevo
16. Hansen, N., Ostermeier, A.: Adapting arbitrary normal mutation distributions in evolution strategies: the covariance matrix adaptation. In: Proceedings of IEEE International Conference on Evolutionary Computation, pp. 312–317 (1996). https://doi.org/10.1109/ICEC.1996.542381
17. Karafotias, G., Hoogendoorn, M., Eiben, A.E.: Parameter control in evolutionary algorithms: trends and challenges. IEEE Trans. Evol. Comput. **19**(2), 167–187 (2015). https://doi.org/10.1109/TEVC.2014.2308294
18. Maturana, J., Saubion, F.: On the design of adaptive control strategies for evolutionary algorithms. In: Monmarché, N., Talbi, E.-G., Collet, P., Schoenauer, M., Lutton, E. (eds.) EA 2007. LNCS, vol. 4926, pp. 303–315. Springer, Heidelberg (2008). https://doi.org/10.1007/978-3-540-79305-2_26
19. McGinley, B., Maher, J., O'Riordan, C., Morgan, F.: Maintaining healthy population diversity using adaptive crossover, mutation, and selection. IEEE Trans. Evol. Comput. **15**(5), 692–714 (2011)
20. Michalewicz, Z.: Genetic Algorithms + Data Structures = Evolution Programs. Springer, Heidelberg (1996). https://doi.org/10.1007/978-3-662-03315-9
21. Morrison, R., De Jong, K.: Measurement of population diversity, vol. 2310, pp. 31–41, October 2001
22. Narendra, K.S., Parthasarathy, K.: Neural networks and dynamical systems. Int. J. Approx. Reason. **6**(2), 109–131 (1992). https://doi.org/10.1016/0888-613X(92)90014-Q
23. Paternain, S., Morari, M., Ribeiro, A.: Real-time model predictive control based on prediction-correction algorithms. In: 2019 IEEE 58th Conference on Decision and Control (CDC), pp. 5285–5291. IEEE (2019). https://doi.org/10.1109/CDC40024.2019.9029408
24. Sammut, C., Webb, G.I. (eds.): Mean Squared Error, p. 653. Springer, Boston (2010). https://doi.org/10.1007/978-0-387-30164-8

25. Schumer, M., Steiglitz, K.: Adaptive step size random search. Autom. Contr. IEEE Trans. **AC13**, 270–276 (1968). https://doi.org/10.1109/TAC.1968.1098903
26. Schwefel, H.P.: Numerical Optimization of Computer Models. Wiley, New York (1981)
27. Shiblee, M., Kalra, P.K., Chandra, B.: Time series prediction with multilayer perceptron (MLP): a new generalized error based approach. In: Köppen, M., Kasabov, N., Coghill, G. (eds.) ICONIP 2008. LNCS, vol. 5507, pp. 37–44. Springer, Heidelberg (2009). https://doi.org/10.1007/978-3-642-03040-6_5
28. Smit, S.K., Eiben, A.E.: Comparing parameter tuning methods for evolutionary algorithms. In: 2009 IEEE Congress on Evolutionary Computation, pp. 399–406. IEEE (2009)
29. Wang, J., Li, X., Li, J., Sun, Q., Wang, H.: NGCU: a new RNN model for time-series data prediction. Big Data Res. **27**, 100296 (2022). https://doi.org/10.1016/j.bdr.2021.100296
30. Werbos, P.J.: Consistency of HDP applied to a simple reinforcement learning problem. Neural Netw. **3**(2), 179–189 (1990). https://doi.org/10.1016/0893-6080(90)90088-3
31. Zhu, K.Q., Liu, Z.: Population diversity in permutation-based genetic algorithm. In: Boulicaut, J.-F., Esposito, F., Giannotti, F., Pedreschi, D. (eds.) ECML 2004. LNCS (LNAI), vol. 3201, pp. 537–547. Springer, Heidelberg (2004). https://doi.org/10.1007/978-3-540-30115-8_49

Performance of Computing Hash-Codes with Chaotically-Trained Artificial Neural Networks

Jacek Tchórzewski[1](✉) and Aleksander Byrski[2]

[1] Cracow University of Technology, Kraków, Poland
jacek.tchorzewski@pk.edu.pl

[2] AGH University of Science and Technology, Kraków, Poland
olekb@agh.edu.pl

Abstract. The main goal of the research presented in this paper was to estimate the performance of applying neural networks trained with the usage of a chaotic model, that may serve as hashing functions. The Lorenz Attractor chaotic model was used for training data preparation, and Scaled Conjugate Gradient was used as a training algorithm. Networks consisted of two layers: a hidden layer with sigmoid neurons and an output layer with linear neurons. The method of bonding the input message with chaotic formula is presented. Created networks could return 256 or 512 bits of hash, however, this parameter can be easily adjusted before the training process. The performance analysis of networks is discussed (that is the time of hash computation) in comparison with popular standards SHA-256 and SHA-512 under the MATLAB environment. Further research may include analysis of networks' training parameters (like mean squared error or gradient) or analysis of results of the statistical tests performed on networks output. The presented solution may be used as a security algorithm complementary to a certificated one (for example for additional data integrity checking).

Keywords: Hashing algorithm · Artificial Neural Networks · Scalable cryptography algorithm · Hashing efficiency

1 Introduction

Hashing functions return a fixed-length bit string from an input bit string, [22]. This functionality may be utilized, for example, in passwords storage, data integrity checking, or digital signatures preparation. Three main features of hashing algorithms are:

1. the process of hash calculation may involve more than one usage of the hashing algorithm,

Supported by the funds assigned by The Polish Ministry of Education and Science to Cracow University of Technology (J.T.) and to AGH University of Science and Technology (A.B.).

D. Groen et al. (Eds.): ICCS 2022, LNCS 13351, pp. 408–421, 2022.
https://doi.org/10.1007/978-3-031-08754-7_48

2. it should be impossible to retrieve the content of the original message from its hash,
3. probability of returning the same hash value from two different messages should be minimal.

Currently, the most popular hashing standards, which are also certificated, are called SHA-2 and SHA-3 [14,15]. Those functions compute a fixed-length hash from a message. Available hashes lengths vary from 224 bits to 512 bits and the length determines the algorithm – e.g. SHA-256 will always return 256 bits of the hash. Certain certificated standards offer possibilities of choosing two different hashes lengths, for example, a hash equal to 224 bits is created from the truncation of the SHA-256 digest. However, the user has no more possibilities to adjust the hash length. It can be concluded that it is either one fixed length that depends on the algorithm or two possible lengths where the second one comes from the truncation of the hash of size equal to the first one. Hashing algorithms whose outputs are smaller, for example, vary from 80 bits to 160 bits, are called light cryptography hash functions [5].

Since late 1970, researchers proposed many different approaches to hash function construction. Most of the ideas are described in [19], for example, hashing function based on the block ciphers, cellular automata, discrete logarithm problem, or knapsack problem. Testing the strength and effectiveness of this kind of algorithms is still a problem. Existing test suits, like for example SHAVS presented in [13], are dedicated to the tested algorithm. Secondly, the National Institute of Standards and Technology (NIST) in the presented report state that [13]: '*The SHAVS is designed to test conformance to SHA rather than provide a measure of a product's security...*'.

In this paper, an idea of hashing neural networks proposed in [19–21] is further developed, showing the performance of the ANNs used. The discussed networks have two layers: a hidden layer with sigmoid neurons, and an output layer consisting of linear neurons. Two hash lengths were considered, that is 256 and 512 bits (to compare results with certificated standards). For each tested hash length seven networks were generated. Generated networks differed in the number of neurons in the hidden layer. The Lorenz Attractor, which appears to have chaotic behavior under appropriate conditions, was utilized for training data preparation. The length of the returned hash could be potentially set with the precision of one bit before the training process. Furthermore, the performance of the proposed networks is tested in comparison with the performance of the chosen certificated standards (SHA-256 and SHA-512) under the MATLAB environment. The time of hash computation from data that differed in size was considered as the performance measure.

The paper after introducing the idea of the approach and giving state-of-the-art, focuses on the presentation of the performance testing of the ANN-based hashing functions, comparing the obtained models with certificated ones.

2 Related Work

In this section hashing algorithms that are based on chaotic systems are described. In each article not only the chaotic model was considered, but also the core of the hashing algorithm.

In [11] authors proposed their own algorithm, which was based on the Lorenz Attractor. However, they incorporated some functions from the SHA-2 algorithm, for example rotations. In their research, similarly to the presented research, time of computation was considered as a performance measure. The algorithm core consisted of four iterations that were combining intermediate hash results and secret keys. The final results were compared with SHA-1. Even though the proposed algorithm was more efficient, SHA-1 is considered as an outdated function, and there were no comparisons with current standards, like SHA-2 or SHA-3.

The algorithm presented in [10] was not a classical hashing scheme but enabled checking the integrity of the data. The proposed procedure was based on huge numbers and their powers under the finite field, which makes the whole idea similar to the RSA ciphering scheme. The authors compared the efficiency of their solution with the Advanced Encryption Standard (AES) and concluded that the performance of their algorithm was slightly worse.

In [9] authors incorporated similar operations as described in the [10], a sponge function that was absorbing input data, and a hyper-chaotic Lorenz system. Because of the complexity of the algorithm, the authors noticed perturbations over time in the function performance. The solution was tested for 256 bit, and 512 bit hash lengths, but enabled returning 1024 bits of hash and more. Authors compared their proposition with SHA-2 and SHA-3 standards, however, did not test it for smaller hashes values.

The innovative idea was presented in [1], where the authors proposed their own equation for input data absorption. Their solution was based on the three-dimensional chaotic map and was excessively tested. The proposed function could return 128, 160, 256, or 512 bits of the hash. Results of the research were compared with SHA-1 and MD5. Both are considered as outdated.

Hashing algorithms may be created in many different ways. For example, in [8] some interesting hashing concepts that are based on evolutionary algorithms and genetic algorithms are presented. More information about hashing strategies is presented in the [17].

Chaotic attractors may be also used in different cryptographic areas. For example, in [6] authors proposed an image encryption scheme based on the Lorenz model. The core of the algorithm was utilizing crossover operations and sequences generated by the attractor. Even though the presented scheme was not a hashing function, research conducted by the authors proved the usefulness of chaotic systems in cryptographic solutions.

In contrast to described solutions, an idea presented in the paper enables the utilization of Artificial Neural Networks (ANNs) trained with the usage of the Lorenz Attractor as hashing models. The main advantage of the proposed scheme is a highly scalable ANNs output. The length of the hash returned by those networks can be adjusted with a precision of one bit (before the training process). Furthermore, the performance of ANNs was tested and results were compared with one of the most popular – and certified by the National Institute of Standards and Technology – Secure Hash Standards, that is SHA-256 and SHA-512. Efficiency comparison performed under MATLAB environment tends to be in favor of the presented networks. Further research will cover the security tests of networks and will also include a comparison with world standards.

3 The Chaotic Model Used

Chaotic equations are non-linear and dynamic systems (models), that are significantly vulnerable to the changes in their initial conditions [4]. The output of such models becomes non-deterministic over time. This phenomenon is also called deterministic chaos. One of the most iconic scientist that was investigating this topic was Edward Lorenz. He once described chaos as a situation [4]: *when the present determines the future, but the approximate present does not approximately determine the future.*

In our work the Lorenz Attractor was used for ANNs training data preparation. Two sets of data were prepared: input data containing binary strings representing messages to be hashed, and output data that contained hashes of those messages obtained with the usage of Lorenz Attractor. The idea was to code a message into attractors' initial conditions and then solve the model. The result was considered as a message hash.

The Lorenz Attractor is defined as presented in Eq. (1):

$$\begin{cases} \frac{dx_1}{dt} = a(x_2 - x_1) \\ \frac{dx_2}{dt} = cx_1 - x_2 - x_1x_3 \\ \frac{dx_3}{dt} = x_1x_2 - bx_3 \end{cases} \tag{1}$$

This model becomes chaotic when: $a = 10$, $b = 8/3$ and $c = 28$ [16]. Each input binary string $M = [m_1, m_2, m_3, ..., m_n]$ (where n denotes the desired hash length), was divided appropriately, converted into two real numbers that were used as first two initial conditions ($x_{1,0}$ and $x_{2,0}$). The algorithm of coding messages into initial parameters is presented below:

1. If n = 256 do Steps 2–4.
2. L11 = ctf($[m_1, ..., m_{64}]$), L21 = ctf($[m_{65}, ..., m_{128}]$).
3. L31 = ctf($[m_{129}, ..., m_{192}]$), L41 = ctf($[m_{193}, ..., m_{256}]$).
4. L1 = ctntf(L11, L21), L2 = ctntf(L31, L41).
5. If n = 512 do Steps 6–11.
6. L11 = ctf($[m_1, ..., m_{64}]$), L21 = ctf($[m_{65}, ..., m_{128}]$).
7. L31 = ctf($[m_{129}, ..., m_{192}]$), L41 = ctf($[m_{193}, ..., m_{256}]$).
8. L51 = ctf($[m_{257}, ..., m_{320}]$), L21 = ctf($[m_{321}, ..., m_{384}]$).
9. L31 = ctf($[m_{385}, ..., m_{448}]$), L41 = ctf($[m_{449}, ..., m_{512}]$).
10. L1 = ctntf(ctntf(L11, L21), ctntf(L31, L41)).
11. L2 = ctntf(ctntf(L51, L61), ctntf(L71, L81)).

12. $x^l_{1,0} = \text{L1}$, $x^l_{2,0} = \text{L2}$.

Construction of *ctf* and *ctnf* functions is presented in Listing (1.1):

Listing 1.1. Functions used for mapping messages into Lorenz Attractor initial conditions.

```
function res = ctf(temp)
    sum = 0;
    tabSize = size(temp, 2);
    for i = tabSize:-1:1
        sum = sum + (2^(i - 1)) * temp(tabSize + 1 -i);
    end
    while sum > 1
        sum = sum / 10;
    end
    res = sum;
end

function res = ctntf(a, b)
    sum = a + b;
    while sum > 1
        sum = sum / 10;
    end
    res = sum;
end
```

As it can be seen, two hash lengths were considered, namely: $n = 256$ and $n = 512$. These lengths are the most popular ones in SHA-2 and SHA-3 certificated hashing functions families. Third initial condition $x^l_{3,1}$ was a random real number from range $[0, 1]$. The model described in Eq. (1) was solved with the usage of the Runge-Kutta 4th Order method that can be represented as [18]:

$$k_1 = \Delta t * f(t, x_i) \tag{2}$$

$$k_2 = \Delta t * f(t + \frac{\Delta t}{2}, x_i + \frac{k_1}{2}) \tag{3}$$

$$k_3 = \Delta t * f(t + \frac{\Delta t}{2}, x_i + \frac{k_2}{2}) \tag{4}$$

$$k_4 = \Delta t * f(t + \Delta t, x_i + k_3) \tag{5}$$

$$x_{i+1} = x_i + \left(\frac{k_1 + 2k_2 + 2k_3 + k_4}{6}\right) \tag{6}$$

where f is representing equations described in (1), x_i is a vector containing solutions in all three dimensions (that is $x_i = [x_{1,i}, x_{2,i}, x_{3,i}]$) in $i - th$ algorithm iteration, t denotes a time in which calculation is done:

$$t = t_0 + i * \Delta t \tag{7}$$

t_0 is a moment when computation starts and is equal 0, and Δt is denoting a step (assumed time intervals in which computations are done), and was equal to 0.1. The parameter i was an iterator in interval $[0, 39999]$, thus always 40000 elements of solution in all three dimensions were generated. An example solution of a Lorenz Attractor for the following initial parameters: $\Delta t = 0.1$, $x_{1,0} = 0.4$, $x_{2,0} = 0.3$, and $x_{3,0} = 0.5$ is presented in Fig. 1.

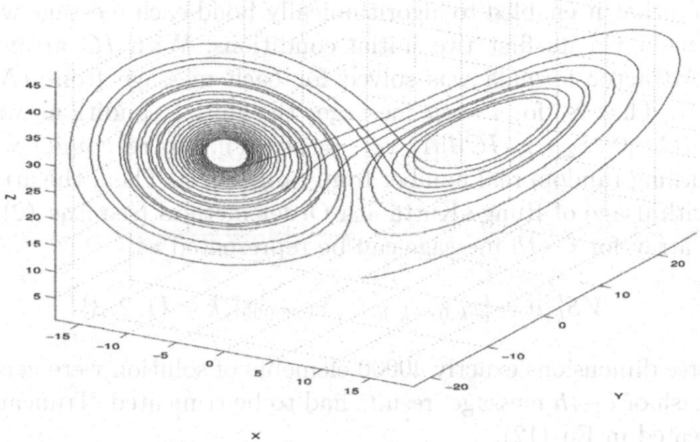

Fig. 1. Example solution of a Lorenz Attractor.

4 ANNs Training and Testing with Usage of Lorenz Attractor

In this section, the process of training and testing of Feed-Forward ANNs is described. Tested ANNs could be divided into two groups: returning 256 bits of hash and returning 512 bits of hash ($n \in \{256, 512\}$). The detailed algorithm of the whole process is presented below (with the assumption, that the value of parameter n was already chosen).

1. Input data preparation. Two sets of data were generated:

$$INPUT^{train}[i] = [b_1, b_2, ..., b_n], i = 1, ..., 10000. \quad (8)$$

$$INPUT^{test}[i] = [b_1, b_2, ..., b_n], i = 1, ..., 5000. \quad (9)$$

   Both $INPUT$ arrays represented bits of messages (denoted as $b \in \{0, 1\}$). In both cases, those bits were generated randomly and all created messages were unique (within the particular matrix as well as between matrices).

2. Target data preparation. To use the Lorenz Attractor, the formula presented in Eq. (1), messages from the training set ($INPUT^{train}$) had to be encoded as its initial conditions. Details related to the process of messages compression are described in Sect. 3. As a result, each message could have been represented as two real numbers from interval $[0,1]$:

$$IC[i] = [XL_i, YL_i], i = 1, ..., 10000; XL_i, YL_i \in \mathbb{R} \wedge XL_i, YL_i \in [0,1]. \quad (10)$$

IC is denoting an initial condition array. The main advantage of such solution is the fact, that it enabled to algorithmically bond each message with attractor's formula via its first two initial conditions. With IC array prepared, Lorenz Attractor formula was solved for each message from $INPUT^{train}$ separately. That is, for $i-th$ message, the initial conditions were set to: $[x_{1,0} = IC[i][0], x_{2,0} = IC[i][1], x_{3,0} = rand()]$, where $rand()$ was a function returning random real number from range $[0,1]$. Then, the attractor was solved with usage of Runge-Kutta 4th Order method (see Eqs. (2)–(6)). The solution array for $i-th$ message can be represented as:

$$VS'_k[i] = [x_{k,0}, x_{k,1}, ..., x_{k,39999}], k \in \{1,2,3\}. \quad (11)$$

In all three dimensions exactly 40000 elements of solution were generated. To form a hash of $i-th$ message, results had to be truncated. Truncated vectors are presented in Eq. (12).

$$VS_k[i] = [x_{k,1*step+1000}, x_{k,2*step+1000}, ..., x_{k,n*step+1000}], k \in \{1,2,3\}, \quad (12)$$

where:

$$step = \lfloor \frac{40000 - 1000}{n} \rfloor. \quad (13)$$

The first 1000 elements were skipped in all cases to avoid small distances (in the Euclidean sense) between solutions in 3D space. The step parameter was calculated to cover the whole solution space, and to avoid situations when two neighboring samples in a particular dimension are too close to each other. Neighboring samples might also form an ascending or descending slope, which was undesirable. After the truncation process, all vectors had to be binarized to form hashes. The general binarization formula is presented in Eq. (14).

$$BS[i][j]_k = \begin{cases} 1 \text{ if } VS[i][j]_k \geq AVG_k[j] \\ 0 \text{ otherwise} \end{cases} \quad (14)$$

Where $AVG_k[j]$ is the average value calculated from VS_k array ($k \in \{1,2,3\}$), for each column $j \in [1,2,...,n]$. At this stage of computation, every message from $INPUT^{train}$ array had three hashes candidates (each of length equal to n) stored in arrays BS_1, BS_2 and BS_3. To complete the target data preparation only one array of hashes had to be chosen. To determine which exactly should it be, statistical tests described in [19] were performed on arrays $BS_{1,2,3}$, and after analysis of results, only one was chosen.

3. ANNs training process. For every $n \in \{256, 512\}$ value, 7 networks were created. The structure of each network was the same and could be represented as: I-HL-OL-O. I was an input layer of size n where input messages were given. HL was a hidden layer containing sigmoid neurons. The number of neurons in this layer varied in networks within one group. OL was an output layer with n linear output neurons. O was a network output that form a hash. All ANNs were trained with the usage of the Scaled Conjugate Gradient method (SCG) [12]. The training set consisted of two arrays: $INPUT^{train}$ and $TARGET^{train}$. The first one, $INPUT^{train}$, was described in the step 1. The target set was an array containing results obtained from Lorenz Attractor that were truncated but not binarized (see Eq. (12)). Target array was selected in the process of statistical analysis of the three binarized and truncated versions of arrays (see Eq. (14), that is B.P.T., S.T., and C.T. tests described in [19] were performed. Selection of BS_i array in the statistical analysis process determined that VS_i was considered as a target set ($TARGET^{train} = VS_i$). Example results of such tests performed on a different set of networks is presented in [21].
4. Output data generation. Each trained network was used to generate output test data from $INPUT^{test}$ data. $OUTPUT^{test}$ data were used to prepare binarization vectors used in the performance analysis process (see Sect. 5).
5. ANNs evaluation process. The performance of hashing network was tested (in comparison with certificated standards). This was an independent stage in which different sets of data were used. All details and results are described in Sect. 5.

5 Analysis of Performance of the Hashing ANNs

In this section, analysis of the performance of hashing ANNs is presented as well as a comparison of the performance of hashing ANNs with MATLAB implementation of SHA-256 and SHA-512 functions. The performance measure was a time of hash computation. Hashes were calculated from data that differed in size. Accordingly to [3], the computational cost of one feed-forward network pass is $O(W)$, where W is the total number of weights. The training cost is equal to $O(W^2)$. Hashing algorithms have cost $O(1)$ for small messages, and $O(m)$ for longer messages where m is the message length. This implicates directly from their construction. Experiments' details are presented below:

- MATLAB implementation of SHA-512 and SHA-256 presented in [7] was used. This is the only implementation of these certificated hashing functions officially published on the MathWorks website (which is an official file exchange platform dedicated to MATLAB users).
- MATLAB functions were tested for the following data sizes: 512 b, 100 B, 1 kB, 10 kB, 50 kB, 100 kB, and 500 kB. As a *data* here strings of appropriate length were considered. Data representing 100 kB and 500 kB were not created directly as strings, but multiple hashing operations were performed on 50 kB data strings (2 times and 4 times, respectively).

- Hashing networks were tested for the following data sizes: 512 b, 128 B, 1 kB, 10 kB, 50 kB, 100 kB, 500 kB, 1000 kB, 5000 kB, and 50000 kB. As a *data* in this scenario, arrays of bits representing messages were considered. Every array had exactly n random bits in one row, and the appropriate number of rows. Hashing 1000 kB of data (and more) was tested as a multiple hashing of 500kB array. Performance measurements included the binarization process, but in this scenario, a binarization vector was used, which can be represented as:
$$BV = [AVG_1, AVG_2, ..., AVG_n], \quad (15)$$
where AVG_i is the average value calculated from the i-th column of the ANNs $OUTPUT^{test}$ array (for each network vector BV was generated separately).
- Experiments were conducted in the MATLAB2020 environment on the Personal Computer with 16 GB RAM, and AMD Ryzen 5 2600 Six-Core Processor (3.4 GHz).
- Notation $L\{n\}N\{HL\}$ denotes hashing network train with usage of the Lorenz Attractor, returning n bits of hash, and having HL neurons in the hidden layer.

Results of the experiments are presented in Tables 1, 2, and 3.

Table 1. Performance of SHA-256 and SHA-512 implemented in MATLAB [7].

	Data size (kB)						
	0.0625	0.098	1	10	50	100	500
	Time of computation (s)						
SHA-256	0.659	0.527	4.303	41.692	245.496	498.848	2436.608
SHA-512	0.926	0.832	6.999	64.525	358.101	716.745	3570.202

Table 2. Performance of Lorenz hashing networks returning 256 bits of hash.

	Data size (kB)									
	0.0625	0.125	1	10	50	100	500	1000	5000	50000
	Time of computation (s)									
L256N128	0.012	0.009	0.008	0.012	0.024	0.043	0.179	0.352	1.825	18.667
L256N192	0.013	0.008	0.009	0.012	0.026	0.044	0.202	0.433	2.099	20.883
L256N256	0.013	0.010	0.010	0.014	0.031	0.058	0.270	0.505	2.436	24.665
L256N320	0.014	0.009	0.009	0.013	0.033	0.051	0.247	0.510	2.719	26.020
L256N384	0.015	0.011	0.011	0.014	0.036	0.054	0.280	0.580	2.864	28.548
L256N448	0.014	0.011	0.011	0.015	0.034	0.071	0.326	0.598	3,149	30.699
L256N512	0.019	0.013	0.013	0.018	0.040	0.078	0.342	0.705	3.795	37.023

Table 3. Performance of Lorenz hashing networks returning 512 bits of hash.

	Data size (kB)									
	0.0625	0.125	1	10	50	100	500	1000	5000	50000
	Time of computation (s)									
L512N256	0.016	0.015	0.013	0.018	0.036	0.054	0.239	0.456	2.297	23.585
L512N384	0.020	0.016	0.016	0.019	0.037	0.060	0.249	0.493	2.688	26.531
L512N512	0.022	0.019	0.020	0.026	0.045	0.068	0.323	0.649	3.564	34.054
L512N640	0.024	0.032	0.026	0.032	0.059	0.099	0.392	0.787	3.919	35.527
L512N768	0.024	0.023	0.025	0.029	0.063	0.106	0.464	0.854	4.259	42.69
L512N896	0.028	0.026	0.026	0.032	0.062	0.115	0.491	0.872	4.259	44.082
L512N1024	0.031	0.029	0.029	0.036	0.073	0.134	0.578	1.132	5.735	56.269

Results from the Table 2 are visualized in Fig. 2, and from Table 3 in Fig. 3. Measurements were also approximated with the usage of a Bezier curve to make them more readable. In both figures and for each network, it can be observed that until about 50 kB data size threshold, the time of computation is growing slightly, and after this threshold, the growth takes the form of a linear function. In all cases, the time of computation for networks with a bigger number of neurons in the hidden layer is growing faster than for networks with a smaller number of neurons (note that the logarithmic scale is used for both: OX and OY axes). Networks returning 512 bits of the hash are also slower than networks returning 256 bits of the hash.

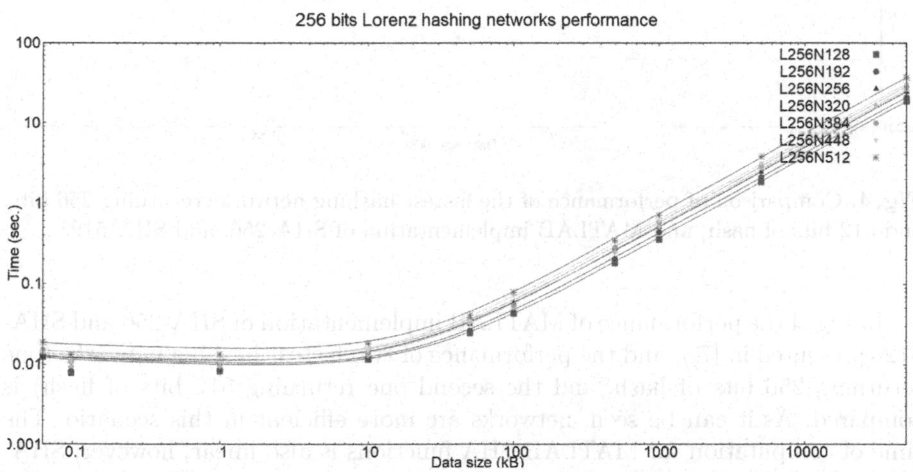

Fig. 2. Performance of hashing networks returning 256 bits of hash.

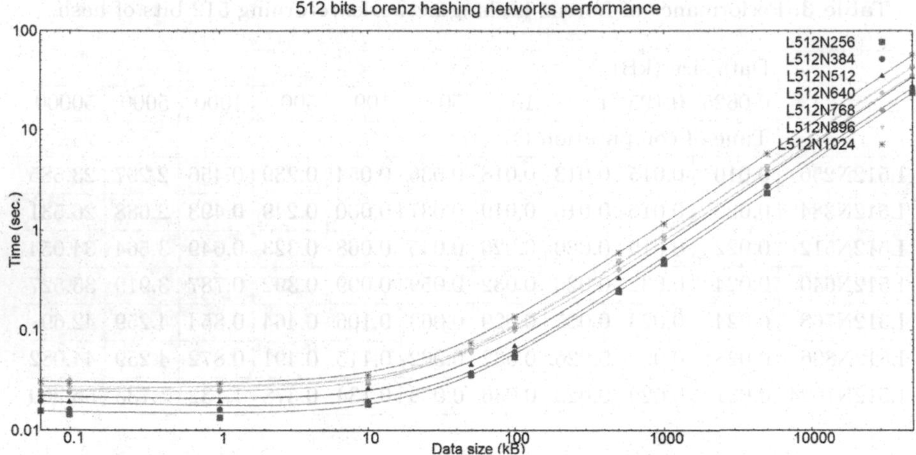

Fig. 3. Performance of hashing networks returning 512 bits of hash.

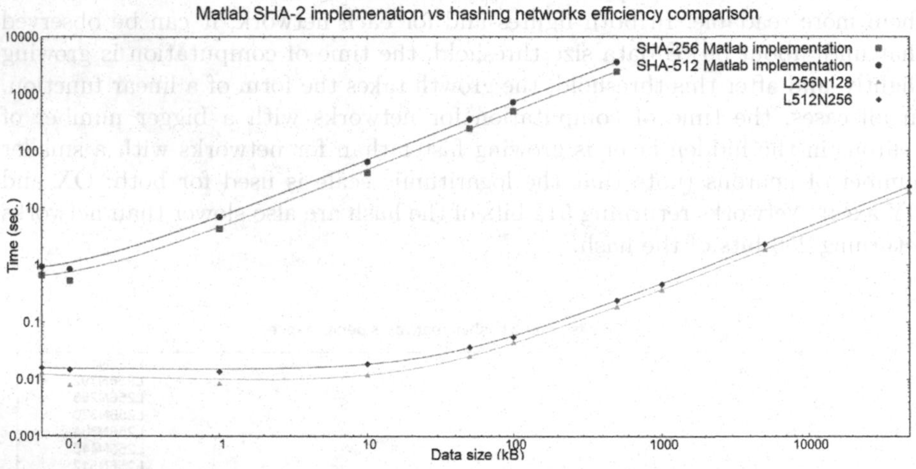

Fig. 4. Comparison of performance of the fastest hashing networks returning 256 bits, and 512 bits of hash, and MATLAB implementation of SHA-256, and SHA-512.

In Fig. 4 the performance of MATLAB implementation of SHA-256 and SHA-512 (presented in [7]), and the performance of two fastest hashing networks (one returning 256 bits of hash, and the second one returning 512 bits of hash) is compared. As it can be seen, networks are more efficient in this scenario. The time of computation for MATLAB SHA functions is also linear, however, SHA-512 algorithm is less efficient than SHA-256.

Table 4 shows a comparison of hashing efficiency of all networks and both MATLAB functions for popular data types. Assumed data sizes were [2]: 10 kB for a JPG image (JPG row), 19 kB for a PDF file (PDF row), 3.5 MB for an MP3

Table 4. Efficiency of popular data files hashing.

	Time (s)						
	L256N128	L256N192	L256N256	L256N320	L256N384	L256N448	L256N512
JPG	0.002	0.011	0.012	0.020	0.012	0.021	0.017
PDF	0.005	0.015	0.017	0.024	0.018	0.027	0.023
MP3	1.336	1.503	1.774	1.879	2.053	2.215	2.663
DVD	1565.721	1751.245	2068.131	2182.298	2394.243	2574.439	3105.545
BRM	9002.903	10069.623	11891.720	12548.144	13766.865	14802.953	17856.842
	L512N256	L512N384	L512N512	L512N640	L512N768	L512N896	L512N1024
JPG	0.006	0.012	0.028	0.075	0.027	0.015	0.042
PDF	0.010	0.017	0.034	0.081	0.035	0.023	0.052
MP3	1.691	1.908	2.461	2.612	3.077	3.164	4.063
DVD	1977.860	2225.143	2855.949	2977.090	3579.241	3696.065	4718.253
BRM	11372.694	12794.540	16421.609	17117.942	20580.548	21252.342	27129.811
	Time (different measure units)						
	MATLAB SHA-256			MATLAB SHA-512			
JPG	49.362 s			70.817 s			
PDF	93.248 s			135.104 s			
MP3	4.855 h			7.111 h			
DVD	236.715 days			346.754 days			
BRM	1361.110 days			1993.838 days			

file or an Ebook (MP3 row), 4 GB for a DVD movie (DVD row), and 23 GB for a Blue-Ray Movie (BRM row). Times presented in this table were not directly calculated, but interpolated with the usage of the linear interpolation performed on data presented in previous tables in this section. The obtained results show clearly that the cost of hashing using the ANN-based algorithms seems feasible even for very large commonly used files while comparing the obtained results with the classic implementation makes the latter much slower or actually unacceptable (even counted in days).

6 Conclusions

In this article, the concept of hashing artificial neural networks trained with the usage of the Lorenz Attractor was presented. The research was focused on the assessment of the performance of the proposed hash generators.

All the discussed networks had one hidden layer with sigmoid neurons and an output layer with n linear neurons (where n was denoting a hash length). Two values of the n parameter were considered, that is 256 and 512. For both values of n, seven networks were created that differed in the number of neurons in the hidden layer. All networks were trained with the usage of the Scaled Conjugate

Gradient method. The training set consisted of random messages (input data) and an appropriately prepared target set (hashes created from input messages with usage of the Lorenz Attractor).

ANNs presented significantly better time efficiency than SHA-256, and SHA-512 implemented in MATLAB. One of the biggest advantages of the proposed solution is a potentially scalable output of networks. The length of the returned hash can be established during the training process with a one-bit precision. Networks are also easy to replace, thus compromising one network can be easily fixed via training a new one.

Comparing the efficiency of our algorithms vs. the efficiency of MATLAB implementations of classic hashing algorithms showed, that for large, commonly used files, hashing times are still feasible for the ANN-based approach (reaching hours), while classic SHA implementations runtimes became unacceptable (reaching thousands of days).

Future work is aimed at including:

- testing different networks structures,
- testing different chaotic series,
- testing more values of n parameter,
- performing statistical tests as presented in [19]. Results of these tests will be also compared with results of the same tests performed on the certificated standards. A suite of tests presented in [19] may be also extended, for example by the Floyd-Marshall algorithm.

The example use case is presented in [21]. In the scenario described in [21], hashing networks are used to perform additional data integrity checking operations in the cloud environment. Because of variable output hash length and virtual machines' idle time, hash generation in such system has a very small computational overhead.

References

1. Akhavan Masoumi, A., Samsudin, A., Akhshani, A.: A novel parallel hash function based on a 3D chaotic map. EURASIP J. Adv. Signal Process. **2013**, 126 (2013). https://doi.org/10.1186/1687-6180-2013-126
2. Angela. Average file sizes. https://blog.online-convert.com/average-file-sizes/. Accessed Feb 2022
3. Bishop, C.M., Nasrabadi, N.M.: Pattern recognition and machine learning. Springer **4**(4), 246–249 (2006)
4. Boeing, G.: Visual analysis of nonlinear dynamical systems: chaos, fractals, self-similarity and the limits of prediction. Systems **4**(4) (2016). https://doi.org/10.3390/systems4040037
5. Gong, Z.: Survey on lightweight hash functions. J. Cryptol. Res. **3**(1), 1–11 (2016)
6. Guesmi, R., Ben Farah, M.A., Kachouri, A., Samet, M.: Hash key-based image encryption using crossover operator and chaos. Multim. Tools Appl. **75**(8), 4753–4769 (2015). https://doi.org/10.1007/s11042-015-2501-0

7. Khitish. Sha algorithms 160,224,256,384,512. https://nl.mathworks.com/matlabcentral/fileexchange/31795-sha-algorithms-160-224-256-384-512. Accessed Feb 2022
8. Kidoň, M., Dobai, R.: Evolutionary design of hash functions for IP address hashing using genetic programming. In: 2017 IEEE Congress on Evolutionary Computation (CEC), pp. 1720–1727 (2017). https://doi.org/10.1109/CEC.2017.7969509
9. Liu, H., Kadir, A., Liu, J.: Keyed hash function using hyper chaotic system with time-varying parameters perturbation. IEEE Access **7**, 37211–37219 (2019). https://doi.org/10.1109/ACCESS.2019.2896661
10. Marco, A., Martinez, A., Bruno, O.: Fast, parallel and secure cryptography algorithm using lorenz's attractor. Int. J. Mod. Phys. C **21** (2012). https://doi.org/10.1142/S0129183110015166
11. Medini, H., Sheikh, M., Murthy, D., Sathyanarayana, S., Patra, G.: Identical chaotic synchronization for hash generation. ACCENTS Trans. Inf. Secur. **2**, 16–21 (2016). https://doi.org/10.19101/TIS.2017.25002
12. Møller, M.F.: A scaled conjugate gradient algorithm for fast supervised learning. Neural Netw. **6**(4), 525–533 (1993). https://doi.org/10.1016/S0893-6080(05)80056-5
13. NIST. The secure hash algorithm validation system (SHAVS). Tech. rep. (2012). https://csrc.nist.gov/CSRC/media/Projects/Cryptographic-Algorithm-Validation-Program/documents/shs/SHAVS.pdf
14. NIST. Fips 180-4: Secure hash standard (SHS). Tech. rep. (2015). https://nvlpubs.nist.gov/nistpubs/FIPS/NIST.FIPS.180-4.pdf
15. NIST: Fips pub 202: Sha-3 standard: permutation-based hash and extendable-output functions. Tech. rep. (2015). https://nvlpubs.nist.gov/nistpubs/FIPS/NIST.FIPS.202.pdf
16. Peng, J., Jin, S.Z., Liu, H.I., Zhang, W.: A novel hash function based on hyperchaotic lorenz system. In: Cao, B., Li, T.F., Zhang, C.Y. (eds.) Fuzzy Information and Engineering, vol. 2, pp. 1529–1536. Springer, Heidelberg (2009). https://doi.org/10.1007/978-3-642-03664-4_162
17. Singh, M., Garg, D.: Choosing best hashing strategies and hash functions. In: 2009 IEEE International Advance Computing Conference, pp. 50–55 (2009). https://doi.org/10.1109/IADCC.2009.4808979
18. Süli, E., Mayers, D.F.: An Introduction to Numerical Analysis, pp. 328–329. Cambridge University Press, Cambridge (2003)
19. Tchórzewski, J., Jakóbik, A.: Theoretical and experimental analysis of cryptographic hash functions. J. Telecommun. Inf. Technol. (2019)
20. Tchórzewski, J., Jakóbik, A., Grzonka, D.: Towards ANN-based scalable hashing algorithm for secure task processing in computational clouds. In: 33rd European Conference on Modelling and Simulation, pp. 421–427 (2019)
21. Tchórzewski, J., Jakóbik, A., Iacono, M.: An ANN-based scalable hashing algorithm for computational clouds with schedulers. Int. J. Appl. Math. Comput. Sci. **31**(4), 697–712 (2021)
22. Wang, J., Zhang, T., Song, J., Sebe, N., Shen, H.T.: A survey on learning to hash. IEEE Trans. Pattern Anal. Mach. Intell. **40**(4), 769–790 (2018). https://doi.org/10.1109/TPAMI.2017.2699960

Application of the Hierarchic Memetic Strategy HMS in Neuroevolution

Mateusz Sokół and Maciej Smołka(✉)

Institute of Computer Science, AGH University of Science and Technology, Kraków, Poland

mateusz.sokol@outlook.com,
smolka@agh.edu.pl

https://www.informatyka.agh.edu.pl/en/

Abstract. Quite recently some noteworthy papers appeared showing classes of deep neural network (DNN) training tasks where rather simple one-population evolutionary algorithms (EA) found better solutions than gradient-based optimization methods. However, it is well known that simple single-population evolutionary algorithms generally suffer from the problem of getting stuck in local optima. A multi-population adaptive evolutionary strategy called Hierarchic Memetic Strategy (HMS) is designed especially to mitigate this problem. HMS was already shown to outperform single-population EAs in general multi-modal optimization and in inverse problem solving. In this paper we describe an application of HMS to the DNN training tasks where the above-mentioned single-population EA won over gradient methods. Obtained results show that HMS finds better solutions than the EA when using the same time resources, therefore proving the advantage of HMS over not only the EA but in consequence also over gradient methods.

Keywords: Neuroevolution · Deep neural networks · Evolutionary algorithm

1 Introduction

Deep Neural Networks (DNN) in their versatility have become one of the most popular techniques to be used in many areas, including advanced applications such as image recognition [6] or reinforcement learning [8]. In the DNN training domain gradient methods, such as Adam [4], are state-of-the-art techniques for problems such as the supervised learning. As the development advanced new methods emerged to solve restrictions emerging from existing approaches, including architectural solutions, e.g. deep residual learning [3]. One of the Deep Learning (DL) fields that has been in the spotlight for many years is reinforcement learning (RL). The specificity of optimization problems stated in this field,

This research was supported in part by PLGrid Infrastructure and by the funds of Polish Ministry of Education and Science assigned to AGH University of Science and Technology.

D. Groen et al. (Eds.): ICCS 2022, LNCS 13351, pp. 422–429, 2022.
https://doi.org/10.1007/978-3-031-08754-7_49

i.e. the multimodality and deceptiveness of the loss (objective) function, forced to search for new model training approaches in both gradient based methods [5] and those coming from other optimization techniques.

Neuroevolution, as an application of evolutionary algorithms in the neural networks domain, offers a multitude of approaches to be used: from architecture search to plain model training. Evolution as a substitution for gradient based training algorithms was recently successfully applied in RL settings [1,7,9]. In [9], which was the main inspiration for our work, a simple Genetic Algorithm (GA) was able to outperform state-of-the-art gradient methods, such as Deep Q-Learning (DQN) and Actor-Critic, for selected Atari games. In the same paper authors also considered the problem of getting stuck in local minima and environment deceptiveness in RL, proposing "Novelty Search" method for GA that rewards an agent for picking new actions, therefore enhancing exploration. Similar research was done for Evolution Strategies (ES) in [1], as ESs are even more prone to local minima traps, and a set of new "Novelty Search" methods for ESs was proposed. Another proposition described in [7] made one step further and showed that abandoning fitness maximization completely and focusing solely on rewarding novelty in actions also offers a viable approach for tackling environment deceptiveness.

Some advances in evolving both network topology and weights were also made. The work in [13,14] showed that as the human brain exhibits regularities and modularity in its structure a similar approach in DNN encoding can allow to encode trained neural networks previously unavailable due to their scale. In their first work [13] authors propose a graph based indirect encoding technique and apply this method in [14] to evolve both architecture and weights showing that DNN with 8 million connections can be effectively trained with an evolutionary algorithm.

One of the evolution-based methods that has not been applied in the DNN training is Hierarchic Memetic Strategy (HMS). HMS, introduced in [12] is a multi-population strategy with populations organized in a hierarchy according to the accuracy of performed search. It has been specifically designed to address ill-conditioned problems with high risk of getting stuck in local minima, with a special attention for the inverse problems solution. In this work we explore the application of HMS to training DNNs in the RL setting.

2 Hierarchic Memetic Strategy (HMS)

HMS is a complex multi-stage stochastic global optimizer and inverse problem solver. It has been developed to address the ill-conditioning of the considered problem, i.e., various kinds of multimodality of the objective function. It started with solving problems with multiple isolated local minima where it outperformed other global optimization methods (cf. [12]). Currently, the strategy is able to approximate the shape of sets of local minima in problems where those local minima sets have positive Lebesgue measure (cf. [11] and the references therein). The core of HMS is a multi-population evolutionary strategy. It is itself a global optimization method able to detect multiple isolated local minima. In this work we use only this evolutionary core, so we will concentrate on it in the description below. For the description of the full HMS as well as the analysis of its asymptotic properties we refer the reader to [10] and the references therein.

The HMS core is a multi-population strategy where single-population evolutionary algorithm instances (i.e., *demes*) form a fixed-height parent-child hierarchy with a single root and a different parametrization on each level. A single step of the strategy, called *metaepoch*, consists of a series of standard evolutionary epoch executions in each deme. HMS can utilize various GAs as deme engines. In the past it was used mostly with the simple evolutionary algorithm (SEA), but also with such strategies as the Covariance Matrix Adaptation Evolution Strategy (CMA-ES) or Non-dominated Sorting GA II (NSGA-II). In this work we utilize a no-crossover evolutionary algorithm used in [9]. In the sequel we shall call the algorithm *single-population GA* or simply *GA*.

After each metaepoch we decide whether to spawn new demes based on current state of the strategy. Spawning, in HMS framework called *sprouting*, creates a new child deme around the current local minimum found in the parent deme if a given sprouting condition is satisfied: see Fig. 1. A typical sprouting condition disallows spawning new demes in the proximity of already explored regions. The lifecycle of a deme is driven by Sprouting Condition and Local Stop Condition (LSC) to control search process and avoid unnecessary evaluations in low-quality regions. With Global Stop Condition (GSC) that determines when we should stop our search we end up with a full-fledged framework that allows us to perform broad search on high levels (i.e., closer to the root) to look for promising regions and high-precision search on the lower levels. In case a leaf deme gets stuck around a low-quality minimum a properly set LSC will purge it to avoid unnecessary evaluations. In this work the parametrization of each level consists of mutation power, population count, LSC and sprouting condition.

3 Application in Neuroevolution

In this section we will explore and explain the proposed method for incorporating HMS in the neuroevolution setting for RL environments. As shown in [9] the fact that a simple GA can outperform gradient methods suggests specific characteristics of the loss function, such as multimodality or deceptiveness. These

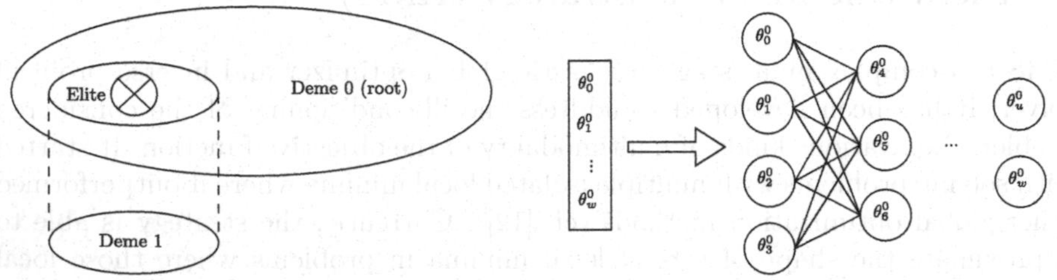

Fig. 1. Idea of HMS sprouting operation **Fig. 2.** Genotype direct encoding

are target issues for HMS, i.e., a method aimed at avoiding getting stuck in local minima, that are expected to show the advantage of HMS over single-population GA and consequently over gradient methods.

The first decision to be made when applying evolutionary algorithm in the DNN domain is the representation. As a neural network is not directly suitable for genotype operations we need an encoding technique to easily and unambiguously translate genotype into a network. In [2] a distinction was proposed into three types of encoding: direct, indirect and parametric. In this work we use a special kind of direct encoding with compression introduced in [9] for Atari problems. In the direct encoding a genotype determines only weights and biases. As shown in Fig. 2 genotype, which is an array of genes being floating point values, is unambiguously translated into a neural network. A gene θ_m^n at index m for some individual in epoch n represents a specific weight. This one-to-one direct encoding allows to interpret a gene as a concrete value for some weight or bias in the target DNN. From now on we will call this translation process a DNN materialization. It is worth noticing that in our experiments m exceeds 1.5 million, so the search domain is rather big.

In our experiments the evaluation of an individual is the process of running an individual in RL environment for a given number of iterations, determined by episode duration. For each step the model makes a decision which action to take by performing a forward pass for an input, which is a current observation. A reward r_i is given in i-th step and the final fitness of the model is the sum of all rewards G acquired through the episode: $G = r_1 + r_2 + \cdots + r_T$. The episode duration depends on the environment and selected actions: if in a game we lose all available lives the episode ends. Also the exact definition of r_i varies among games: from the collected item number to the time of staying alive.

4 Experiments

The methodology described in the previous section was applied to different RL environments, following the work in [9]. Main experiments were conducted in the Atari 2600 environment[1].

In all experiments we used the same type of LSC and sprout condition. The LSC stopped a deme after a given number of metaepochs without a significant improvement in the best objective value. The sprout condition prohibited sprouting when the nearest child-deme centroid was closer than a given minimal distance.

For Atari games GA parametrization and DNN architecture was the same as in [9] and only HMS parameters were adapted to conform with desired number of evaluations in total. As [9] compares GA to gradient methods, our work only covered HMS to GA comparison. It's also important to stress that GA algorithm to which we are comparing exactly follows the one proposed in [9]. Selected DNN architecture was originally proposed in [8] and is comprised of

[1] https://gym.openai.com/envs/#atari.

Table 1. Atari 2600: global parameters

Parameter	2-level HMS	Single-pop GA
Environment	Atari 2600	Atari 2600
GSC (nr of epochs)	40	40
Metaepoch length	3	–
Number of levels	2	1

Table 2. Eval. number

Game	GA	HMS
Frostbite	39000	35250
Kangaroo	39000	31950
Zaxxon	39000	33750
Venture	39000	31950
Asteroids	39000	34200

convolutional layers and fully connected ones with ReLU activation function. Also each experiment was performed for 40 epochs. The selection of parameters for the experiments are shown in Table 1 and 3.

In this work we report the results of experiments with selected five Atari games, which were also used in [9]. These are: Frostbite, Kangaroo, Zaxxon, Venture and Asteroids. Here, similarly to [9], we also got best results for the Frostbite game where the best individual was able to play the game and finish with a score on par with a human player. What is the most important, HMS was able to reach those scores much faster and using less iterations. The same applies to the Zaxxon game and also to some degree to the Kangaroo game, for which score was not that much better but resulting individual learned basic behaviour of dodging falling items. Results of these experiments are presented in Fig. 3. Each plot shows the medians and the 95% bootstrapped confidence intervals obtained from evaluating the best individual in a given epoch multiple times in the target environment, each time with a different initial seed. The maximum number of evaluations for both competing algorithms is shown in Table 2. For single-population GA the number was the same for each game, whereas in HMS this number varies but for each game it is less than in GA. The results show that even in the worst case (Asteroids) HMS performs at least as well as than GA. In remaining four cases HMS performs better and its advantage over GA is the most prominent in Frostbite and Zaxxon cases.

Table 3. Atari 2600: local parameters

Parameter	HMS level 0	HMS level 1	Single-pop GA
Mutation probability	1.0	1.0	1.0
Mutation power	0.05	0.002	0.002
Population count	600	150	1000
Number of promoted	45	20	20
LSC (no improvement)	–	3	–
Sprouting condition	0.5	–	–

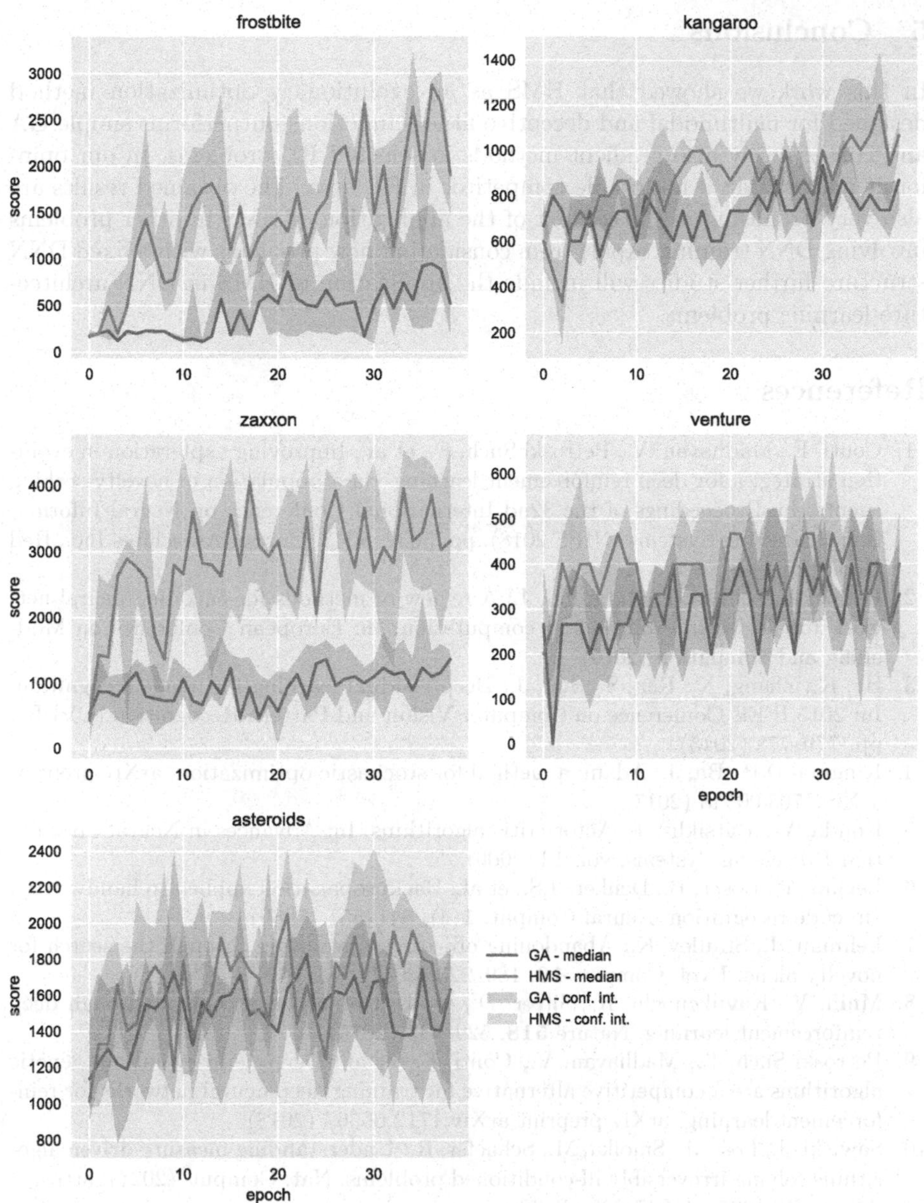

Fig. 3. Results for Atari games experiments.

The implemented framework for conducting experiments is available in a public GitHub repository[2]. A showcase playlist with recorded episodes performed by the best individuals for selected environments is also available[3].

[2] https://github.com/mtsokol/hms-neuroevolution.

[3] https://bit.ly/hms-neuroevolution-playlist.

5 Conclusions

In this work we showed that HMS as an evolutionary optimization method designed for multimodal and deceptive fitness functions outperforms simple GA and consequently also gradient methods in selected RL problems. In our opinion it makes HMS a noticeable competitor in RL area. The obtained results are also very promising in the context of the application of HMS in other problems involving DNN training. Apart from considering new problems with a fixed DNN structure further studies will include the application of HMS in DNN architecture learning problems.

References

1. Conti, E., Madhavan, V., Petroski Such, F., et al.: Improving exploration in evolution strategies for deep reinforcement learning via a population of novelty-seeking agents. In: Proceedings of the 32nd International Conference on Neural Information Processing Systems (NIPS 2018), pp. 5032–5043. Curran Associates Inc., Red Hook (2018)
2. Fekiač, J., Zelinka, I., Burguillo, J.: A review of methods for encoding neural network topologies in evolutionary computation. In: European Conference on Modelling and Simulation (2016)
3. He, K., Zhang, X., Ren, S., Sun, J.: Deep residual learning for image recognition. In: 2015 IEEE Conference on Computer Vision and Pattern Recognition (CVPR), pp. 770–778 (2015)
4. Kingma, D.P., Ba, J.: Adam: a method for stochastic optimization. arXiv preprint arXiv:1703.00548 (2017)
5. Konda, V., Tsitsiklis, J.: Actor-critic algorithms. In: Advances in Neural Information Processing Systems, vol. 12 (2000)
6. LeCun, Y., Boser, B., Denker, J.S., et al.: Backpropagation applied to handwritten zip code recognition. Neural Comput. **1**(4), 541–551 (1989)
7. Lehman, J., Stanley, K.: Abandoning objectives: evolution through the search for novelty alone. Evol. Comput. **19**, 189–223 (2011)
8. Mnih, V., Kavukcuoglu, K., Silver, D., et al.: Human-level control through deep reinforcement learning. Nature **518**, 529–533 (2015)
9. Petroski Such, F., Madhavan, V., Conti, E., et al.: Deep neuroevolution: genetic algorithms are a competitive alternative for training deep neural networks for reinforcement learning. arXiv preprint arXiv:1712.06567 (2018)
10. Sawicki, J., Łoś, M., Smołka, M., Schaefer, R.: Understanding measure-driven algorithms solving irreversibly ill-conditioned problems. Nat. Comput. (2021). https://doi.org/10.1007/s11047-020-09836-w
11. Sawicki, J., Łoś, M., Smołka, M., Schaefer, R., Álvarez-Aramberri, J.: Approximating landscape insensitivity regions in solving ill-conditioned inverse problems. Memetic Comput. **10**(3), 279–289 (2018). https://doi.org/10.1007/s12293-018-0258-5
12. Smołka, M., Schaefer, R., Paszyński, M., Pardo, D., Álvarez-Aramberri, J.: An agent-oriented hierarchic strategy for solving inverse problems. Int. J. Appl. Math. Comput. Sci. **25**(3), 483–498 (2015)

13. Stanley, K.: Compositional pattern producing networks: a novel abstraction of development. Genet. Program Evolvable Mach. **8**, 131–162 (2007)
14. Stanley, K.O., D'Ambrosio, D.B., Gauci, J.: A hypercube-based encoding for evolving large-scale neural networks. Artif. Life **15**(2), 185–212 (2009)

Biomedical and Bioinformatics Challenges for Computer Science

CXR-FL: Deep Learning-Based Chest X-ray Image Analysis Using Federated Learning

Filip Ślazyk[1,2], Przemysław Jabłecki[1,2], Aneta Lisowska[1], Maciej Malawski[1,2], and Szymon Płotka[1,3](✉)

[1] Sano Centre for Computational Medicine, Krakow, Poland
s.plotka@sanoscience.org
[2] AGH University of Science and Technology, Krakow, Poland
[3] Warsaw University of Technology, Warsaw, Poland

Abstract. Federated learning enables building a shared model from multicentre data while storing the training data locally for privacy. In this paper, we present an evaluation (called CXR-FL) of deep learning-based models for chest X-ray image analysis using the federated learning method. We examine the impact of federated learning parameters on the performance of central models. Additionally, we show that classification models perform worse if trained on a region of interest reduced to segmentation of the lung compared to the full image. However, focusing training of the classification model on the lung area may result in improved pathology interpretability during inference. We also find that federated learning helps maintain model generalizability. The pretrained weights and code are publicly available at (https://github.com/SanoScience/CXR-FL).

Keywords: Deep learning · Federated learning · Medical imaging

1 Introduction

Federated Learning (FL) is an effective privacy-preserving machine learning technique used to train models across multiple decentralized devices. It enables using a large amount of labeled data in a secure and privacy-preserving process [12] to improve the generalizability of the model [2]. Recent work on the application of federated learning in medical imaging shows promising results in dermoscopic diagnosis [3], volumetric segmentation [4] and chest X-ray image analysis [5]. In this paper, we evaluate the application of deep learning-based models to medical image analysis using the FL method. To gain insight into the impact of FL-related parameters on the global model, we conduct experiments with a variable number of clients and local training epochs. We explore utilisation of cascading approach, where medical image segmentation is performed prior to classification, for increased pathology classification interpretability. We compare our results with [1] in terms of explainable AI (XAI) and classification performance. We find faster convergence of the learning process for a greater fraction

D. Groen et al. (Eds.): ICCS 2022, LNCS 13351, pp. 433–440, 2022.
https://doi.org/10.1007/978-3-031-08754-7_50

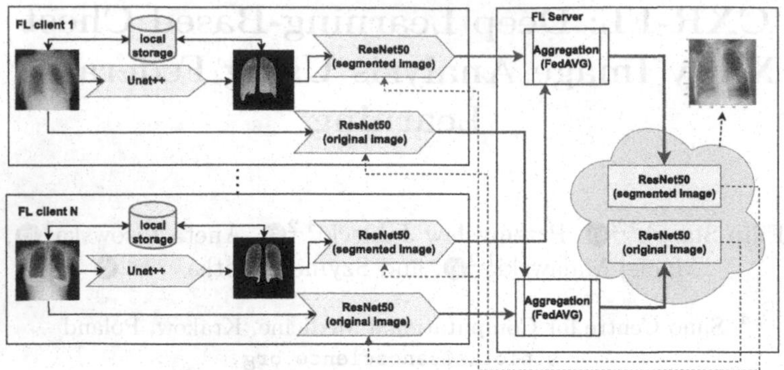

Fig. 1. Methodology: combining segmentation and classification in FL setting

of selected clients and a greater number of local epochs in the segmentation task. We show that federated learning improves the generalizability of the model and helps avoid overfitting in the classification task. We show that Grad-CAM explanations for classification models trained on segmented images may be more focused on the lung area than those trained on full images.

2 Method: FL for Segmentation and Classification

Our method consists of federated training of segmentation and classification models. First, we train segmentation models in a federated manner. For this purpose, we utilize the UNet++ model (with an EfficientNet-B4 backbone) that is later used to prepare the input for classification models. At the classification stage, we use the best segmentation model in terms of the chosen performance metric, and preprocess CXR images (from the training and testing set) to extract lung regions and reduce the impact of the background noise on the prediction. We subsequently train one model on full images and the second on segmented ones independently, all in a federated fashion. During each round of federated training, clients download the global model and fine-tune it with the use of locally stored data. Once all models are fine-tuned in the given round, the server aggregates weights and the next round begins. After the training phase, both types of models pass through the visual explanation step using GradCAM, as in [1]. We test two architectures: ResNet50 and DenseNet121, both commonly used in medical image data classification [10]. An overview of the proposed method for classification stage is depicted in Fig 1.

3 Experiments and Results

3.1 Datasets

Chest X-Ray Dataset: To train the UNet++ model in a federated manner, we use this data set, which is a union of two other data sets known as Chest X-Ray Images (Pneumonia) [8,9]. The dataset consists of 6380 CXR images.

RSNA 2018 Dataset : To evaluate our method, we use an open-source RSNA Pneumonia Detection Challenge 2018 chest X-ray data [7]. In total, the dataset consists of 26684 CXR images in the DICOM format. There are 3 classes in the dataset: "Normal" (8525 - train/326 - test), "No Lung Opacity/Not Normal" (11500 - train/321 - test) and "Lung Opacity" (5659 - train/353 - test).

3.2 Implementation Details

We implement our models in Python 3.8 with the PyTorch v.1.10.1 and Flower v.0.17.0 frameworks, based on our previous experience [6]. We train our models on 4 nodes of a cluster with 1 × NVIDIA v100 GPU each.

For the **segmentation task**, we use UNet++ with EfficientNet-B4 backbone pretrained on ImageNet. Adagrad is utilised as an optimizer for clients. We use a batch size of 2 and set learning rate and weight decay to $lr = 1 \times 10^{-3}$, $wd = 0$ respectively. We assess Jaccard score and BCE-Dice loss on a test set on the central server. The data set used to train the segmentation model was split into a training set and a test set with a 9:1 ratio, maintaining IID distribution of samples. Images are rescaled to 1024×1024 px and augmented with random flip and random affine transformations. The central model is evaluated on a server-side test set after each training round. For the **classification task**, we use Adam optimizer with learning rate $lr = 1 \times 10^{-4}$ and weight decay $wd = 1 \times 10^{-5}$, and set batch size to 8. Images are rescaled to 224×224 px and augmented with random flip and random affine transformations. We evaluate accuracy and CE loss on the test sets (segmented/non-segmented) on the central server. In both tasks, the models are pretrained on the ImageNet dataset. Such pretrained models are downloaded by clients during the first round of the process. We use the FedAvg [11] aggregation strategy and split data in the IID manner among FL clients both in segmentation and classification.

3.3 Segmentation Results

In order to find the optimal central segmentation model, we evaluate several configurations of parameters typical for FL such as the number of local epochs performed by each client during every training round and the fraction of clients selected by the server during each round. The process of training each model consists of 15 rounds. The Jaccard score and loss obtained by each model are presented in Fig 2. For each configuration, we check the number of rounds required to achieve a Jaccard score of 0.92 twice. Results are presented in Table 1. We identify that for a fixed number of local epochs, a greater fraction of selected clients results in a smaller number of rounds needed to exceed the score of 0.92, similarly to the trend observed in [11]. The highest score (0.924) is achieved by the model trained with 3 local epochs and 3 selected clients in the 15th round of training. This model is later used to generate masks for classification.

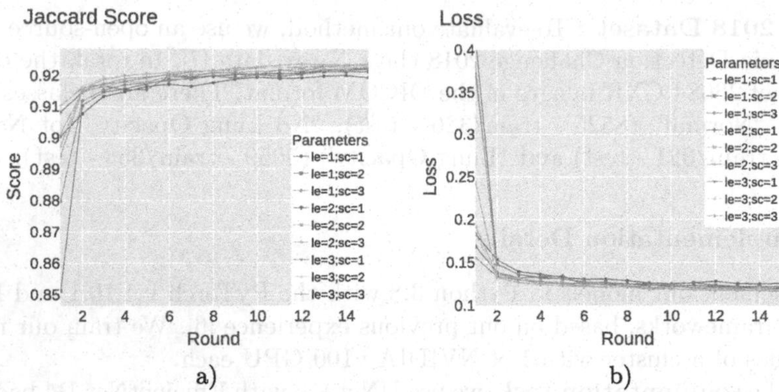

Fig. 2. (a) Jaccard score for the test dataset, achieved by segmentation models, and (b) loss of segmentation models for the test dataset, in successive rounds of training. "sc" - the number of clients selected by the server in each round, "le" - the number of local epochs performed by each client per round.

Table 1. Number of rounds needed by the segmentation model exceeded a Jaccard Score of 0.92 for the serverside test dataset. "sc" - the number of the clients selected by the server in each round, "le" - the number of local epochs performed by each client per round.

Configuration	Experiment 1	Experiment 2
le = 1 & sc = 1	13	14
le = 1 & sc = 2	11	10
le = 1 & sc = 3	**9**	**9**
le = 2 & sc = 1	9	9
le = 2 & sc = 2	7	7
le = 2 & sc = 3	**6**	**6**
le = 3 & sc = 1	6	6
le = 3 & sc = 2	5	5
le = 3 & sc = 3	**5**	**5**

3.4 Classification Results

In the case of the classification task, we evaluate how splitting the same amount of training data between 1, 2 and 3 clients impacts global model quality. Additionally, we assess differences between results obtained with ResNet50 and DenseNet121 architectures on full and segmented images. The accuracy score and loss for 10 rounds of training are presented in Fig. 4. It can be noted that the training process overfits in the case of 1 client and DenseNet121 model, both for segmented and full images, which is represented by a high loss value in the two last rounds for those configurations. The degradation of the global model quality can be also observed for DenseNet121 trained with full images on 2 and

3 clients. The lowest and most stable loss values are obtained for the ResNet50 model trained with 2 and 3 clients for full images and 1 to 3 clients for segmented images. Table 2 presents maximum accuracy and minimum loss values for each configuration of model architecture and dataset type.

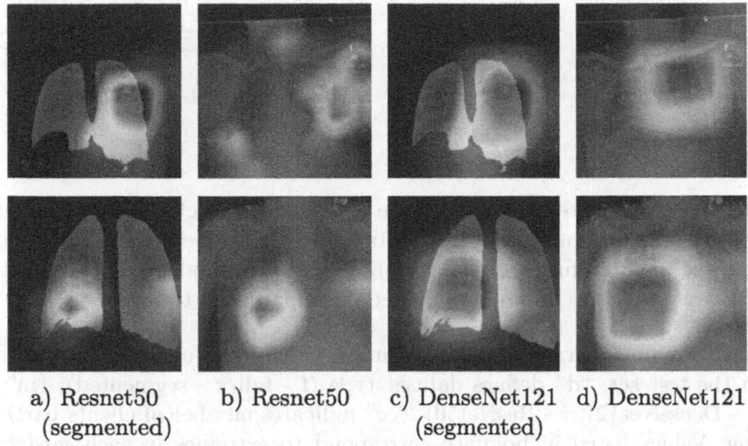

Fig. 3. Grad-CAM visualisations of Lung Opacity samples. In some instances, segmentation resulted in activations focused more on the lung area (upper sample). However, for a majority of cases, visualisation was comparable for segmented and full images (lower sample).

The best accuracy, 0.757, is achieved for ResNet50 model trained on two clients. The worst-performing model is DenseNet121 trained on full images on a single client. In general, the evaluation shows that training on a single client results in overall worse accuracy compared to training with 2 and 3 clients, which is reflected in Fig. 2. This leads to the conclusion that in this case, splitting the data among distinct clients and training the model in the FL manner helps maintain generalizability and avoid overfitting. We observe that models trained on segmented images perform consistently worse than models trained on full images, as is the case for [1]. There is one exception: for the DenseNet121 model the best accuracy is achieved for segmented images (0.742).

To understand qualitative differences in the classification of segmented and full images, we perform Grad-CAM visualisation for ResNet50 and DenseNet121 models. We identify samples that show that the use of segmented images leads to activations more focused on the lung area (as presented in the upper sample in Fig. 3), which is beneficial for model interpretability. However, it can be observed that samples in which the activations are already focused on regions with pathological lung changes, for both full and segmented images, are prevalent. We believe that the small difference in the quality of the models trained on full and segmented images can be explained by the common presence of that similarity.

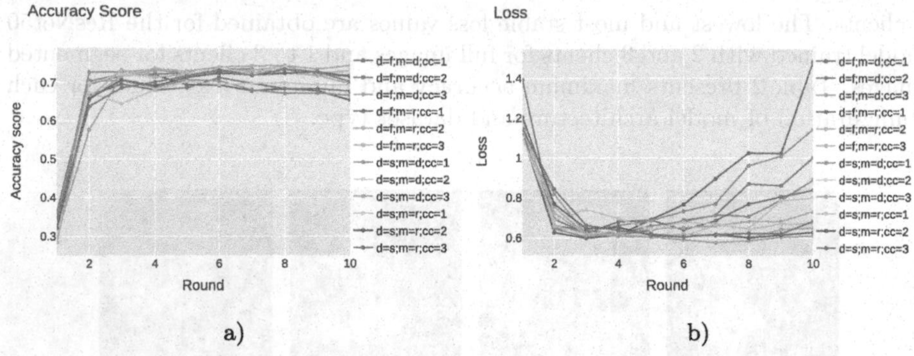

Fig. 4. (a) Accuracy score achieved for the test dataset by classification models, and (b) loss of classification models on test dataset, in successive rounds of training. "d" defines dataset type (f - full/s - segmented), "m" defines model (d - DenseNet121, r - ResNet50), "cc" indicates number of clients participating in training.

Table 2. Maximum accuracy and minimum loss values obtained for each classification model on the test set. "d" defines dataset type (f - full/s - segmented), "m" defines model (d - DenseNet121, r - ResNet50), "cc" indicates number of clients participating in training. Values listed in boldface correspond to extremes in each *model/dataset kind* subset.

Configuration	Max. accuracy	Min. loss
d = f & m = d & cc = 1	0.721	**0.599**
d = f & m = d & cc = 2	**0.737**	0.620
d = f & m = d & cc = 3	**0.737**	0.606
d = f & m = r & cc = 1	0.714	0.623
d = f & m = r & cc = 2	**0.757**	0.601
d = f & m = r & cc = 3	0.747	**0.579**
d = s & m = d & cc = 1	0.708	0.631
d = s & m = d & cc = 2	**0.742**	**0.612**
d = s & m = d & cc = 3	0.734	0.618
d = s & m = r & cc = 1	0.704	0.643
d = s & m = r & cc = 2	0.730	**0.602**
d = s & m = r & cc = 3	**0.736**	0.607

4 Conclusions

In this paper, we evaluated deep learning-based models in the context of CXR image analysis. We conducted experiments in a FL environment to understand the impact of FL-related parameters on the global model performance in segmentation and classification tasks. We also prepared Grad-CAM visualisations for classification models. We found that in the segmentation task, when the number of local epochs is fixed, the model reaches the desired quality faster with a greater fraction of selected clients. In addition, setting a greater number of local epochs for each client also leads to the same behaviour, which may contribute to lower network traffic in FL processes. Moreover, we conclude that splitting the same dataset among distinct FL clients may lead to improvements in classification for the tested models. We observed a higher accuracy score for full images compared to segmented images in the classification task. However, models trained on segmented images may be characterized by improved interpretability.

Acknowledgements. This publication is partly supported by the EU H2020 grant Sano (No. 857533) and the IRAP Plus programme of the Foundation for Polish Science. This research was supported in part by the PL-Grid Infrastructure. We would like to thank Piotr Nowakowski for his assistance with proofreading the manuscript.

References

1. Teixeira, L.O., et al.: Impact of lung segmentation on the diagnosis and explanation of COVID-19 in chest X-ray images. Sensors **21**, 7116 (2021). https://doi.org/10.3390/s21217116
2. Sheller, M.J., et al.: Federated learning in medicine: facilitating multi-institutional collaborations without sharing patient data. Sci. Rep. **10**, 12598 (2020). https://doi.org/10.1038/s41598-020-69250-1
3. Chen, Z., Zhu, M., Yang, C., Yuan, Y.: Personalized retrogress-resilient framework for real-world medical federated learning. In: de Bruijne, M., et al. (eds.) MICCAI 2021. LNCS, vol. 12903, pp. 347–356. Springer, Cham (2021). https://doi.org/10.1007/978-3-030-87199-4_33
4. Wu, Y., Zeng, D., Wang, Z., Shi, Y., Hu, J.: Federated contrastive learning for volumetric medical image segmentation. In: de Bruijne, M., et al. (eds.) MICCAI 2021. LNCS, vol. 12903, pp. 367–377. Springer, Cham (2021). https://doi.org/10.1007/978-3-030-87199-4_35
5. Dong, N., Voiculescu, I.: Federated contrastive learning for decentralized unlabeled medical images. In: de Bruijne, M., et al. (eds.) MICCAI 2021. LNCS, vol. 12903, pp. 378–387. Springer, Cham (2021). https://doi.org/10.1007/978-3-030-87199-4_36
6. Jabłecki, P., Ślazyk, F., Malawski, M.: Federated learning in the cloud for analysis of medical images - experience with open source frameworks. In: Oyarzun Laura, C., et al. (eds.) DCL/PPML/LL-COVID19/CLIP -2021. LNCS, vol. 12969, pp. 111–119. Springer, Cham (2021). https://doi.org/10.1007/978-3-030-90874-4_11
7. Shih, G., et al.: Augmenting the national institutes of health chest radiograph dataset with expert annotations of possible pneumonia. Radiol. Artif. Intell. **1**, e180041 (2019). https://doi.org/10.1148/ryai.2019180041
8. Kermany, D., et al.: Labeled optical coherence tomography (OCT) and chest x-ray images for classification. Mendeley Data **2** (2018). https://doi.org/10.17632/rscbjbr9sj.2
9. Cohen, J.P., et al.: COVID-19 image data collection: prospective predictions are the future. arXiv:2006.11988 [cs, eess, q-bio] (2020)
10. Tang, Y.-X., et al.: Automated abnormality classification of chest radiographs using deep convolutional neural networks. npj Digit. Med. **3**, 70 (2020). https://doi.org/10.1038/s41746-020-0273-z

11. McMahan, H.B., et el.: Communication-efficient learning of deep networks from decentralized data (2016). https://doi.org/10.48550/ARXIV.1602.05629
12. Kaissis, G.A., et al.: Secure, privacy-preserving and federated machine learning in medical imaging. Nat. Mach. Intell. **2**(6), 305–311 (2020)

POTHER: Patch-Voted Deep Learning-Based Chest X-ray Bias Analysis for COVID-19 Detection

Tomasz Szczepański[1], Arkadiusz Sitek[2], Tomasz Trzciński[1,3,4], and Szymon Płotka[1,2(✉)]

[1] Warsaw University of Technology, Warsaw, Poland
[2] Sano Centre for Computational Medicine, Cracow, Poland
s.plotka@sanoscience.org
[3] Jagiellonian University, Cracow, Poland
[4] Tooploox, Wroclaw, Poland

Abstract. A critical step in the fight against COVID-19, which continues to have a catastrophic impact on peoples lives, is the effective screening of patients presented in the clinics with severe COVID-19 symptoms. Chest radiography is one of the promising screening approaches. Many studies reported detecting COVID-19 in chest X-rays accurately using deep learning. A serious limitation of many published approaches is insufficient attention paid to explaining decisions made by deep learning models. Using explainable artificial intelligence methods, we demonstrate that model decisions may rely on confounding factors rather than medical pathology. After an analysis of potential confounding factors found on chest X-ray images, we propose a novel method to minimise their negative impact. We show that our proposed method is more robust than previous attempts to counter confounding factors such as ECG leads in chest X-rays that often influence model classification decisions. In addition to being robust, our method achieves results comparable to the state-of-the-art. The source code and pre-trained weights are publicly available at (https://github.com/tomek1911/POTHER).

Keywords: COVID-19 · Deep learning · Explainable AI

1 Introduction

The SARS-CoV-2 outbreak has claimed the lives of millions of people, and despite its onset in 2019, it remains a serious concern and a threat to public health. The gold standard for diagnosing COVID-19 disease is an RT-PCR test; however, it is expensive, necessitates specialised laboratories and requires the patient to wait relatively long for the outcome. For this reason, computer scientists and radiologists become interested in the computer-aided diagnosis (CAD) capabilities of chest X-rays (CXR). The automatic diagnosis of COVID-19 using chest X-ray images is challenging due to the high intra-class variations, superimposition of anatomical

D. Groen et al. (Eds.): ICCS 2022, LNCS 13351, pp. 441–454, 2022.
https://doi.org/10.1007/978-3-031-08754-7_51

structures, or implanted electronic devices [1]. Moreover, a significant limitation in developing reliable models for detecting pneumonia and COVID-19 is the lack of precisely annotated and rigorously collected data. These limitations may result in models learning various confounding factors in the CXRs. We investigate this phenomenon in our work.

Most of the approaches developed for the classification of the CXR images relay on global features [2,11,12,14]. However, these features may not accurately represent the complex nature of CXR images [22]. It should be verified when working with medical images, especially when the model's accuracy is very high [13,21], the reasons why algorithms perform so well to prevent developing algorithms which base their decision on confounding factors rather then medical pathology. Wang *et al.* [7] propose COVID-Net, the first lightweight capacity neural network dedicated for COVID-19 detection and introduce a novel COVIDx dataset. Authors create a feature extraction method tailored for COVID-19 and provide results of the model's decisions analysis with the use of GSInquire [24]. They create a map of lung areas important for COVID-19 detection and claim a production-ready solution. Considering the need for a critical approach to chest classification and analysis of biases [20,22,23], we carry out further research into the analysis of the deep learning model decisions using the COVIDx dataset.

Recently, a new patch-based learning technique [3,10,25] emerged as a successful method for robust model learning and generalization. Li *et al.* [3] propose a multi-resolution patch-based CNN approach for lung nodule detection on CXR image. The method achieves high accuracy by leveraging the local context of CXR images. Roy *et al.* [25] use the model's local training method for classification on the ICIAR-2018 breast cancer histology image dataset, which allows them to achieve state-of-the-art results on this dataset. Oh *et al.* [10], inspired by their statistical analysis of the potential imaging biomarkers of the CXRs, explore patch-based learning for COVID-19 detection. They propose random patch cropping, from which the final classification result is obtained by majority voting from inference results at multiple patch locations. Authors analyse model decisions using proposed probability gradient-weighted class activation map (GradCAM) [18] as a part of explainable AI (XAI) method [4], upon which they conclude their results are correlated with radiological findings. Similar to Oh *et al.*, our proposed method POTHER explores a patch-based learning approach in CXR images for reliably detecting COVID-19. In contrast, we do not use segmented lungs as input; instead, we propose a multi-task model that leverages segmentation task to extract valuable features. In addition, we limit the area from which we draw patches and reduce their size.

We use XAI methods to demonstrate that model decisions may rely on *confounding factors* rather than medical pathology. Degrave *et al.* [20] and Cabrera *et al.* [22] call them shortcuts, while Maguolo *et al.* [23] biases. An analysis of shortcuts in CXRs, which we call *confounding biases* (CBs), based on open-source datasets and global learning methods, is presented by [20]. It demonstrates the detrimental influence on models decisions caused by laterality tokens, i.e. L or R letter, meaning the left or right side of the image, the position of the clavicles and

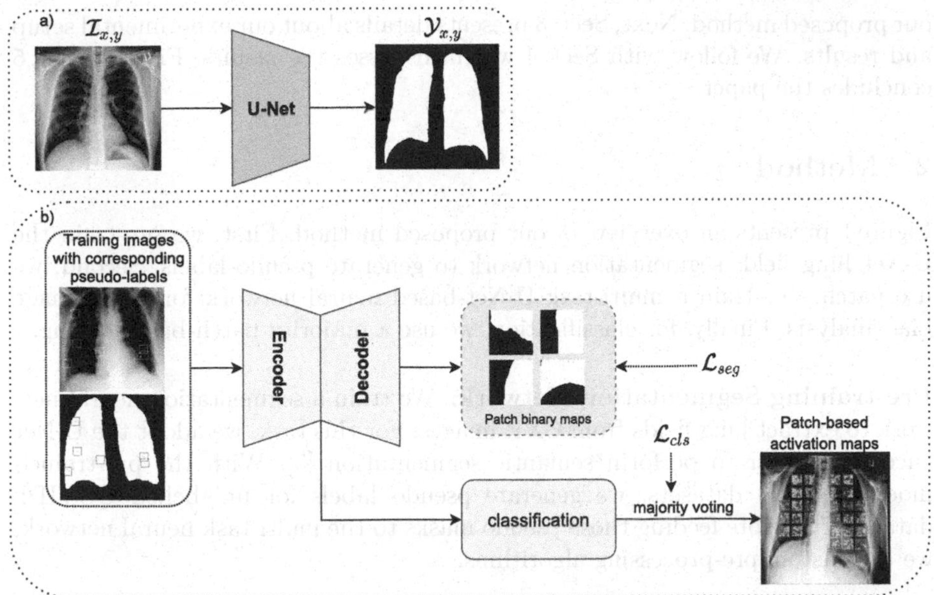

Fig. 1. The overview of the proposed POTHER framework for chest X-ray bias analysis. (a) Pre-trained segmentation network to generate pseudo-labels, (b) an encoder-decoder patch-based multi-task learning network to classify CXR images. We adopt majority patch-based voting for classification and employ patch-based activation maps to explain the results.

the presence of arms in the upper parts of the image. Authors present that machine learning (ML) models trained on CXR images may generalise poorly and owe the majority of their performance to the learning of shortcuts that may be consistently detected in both internal and external domains - making external validation alone insufficient to detect poorly behaved models. We analyse CBs in the frequently cited COVIDx dataset [7] and propose a multi-task learning COVID-19 detection method called POTHER that improves robustness to CBs.

The contributions of this paper are threefold. Firstly, we propose a novel multi-task patch-voted learning-based method called POTHER for chest X-ray image analysis for COVID-19 detection. Secondly, we analyse activation maps of currently available methods and sources of confounding biases in CXRs from the COVIDx dataset. Our analysis reveals a significant number of CBs. These biases should be considered when using the COVIDx dataset in future research. The main CBs are ECG leads, laterality tokens and hospital markings. Finally, we demonstrate that deep learning models learn how to classify pneumonia manifestations and lung morphology, focusing on the confounding mentioned above factors rather than the actual manifestation of the disease or lack thereof on lung parenchyma. In our methods, to counter CBs, we used segmentation as a helper task which allowed for an efficient feature extraction method that performs comparably to the state of the art. The paper is organized as follows. Section 2 details our proposed method. Next, Sect. 3 presents details about our experimental setup and results. We follow with Sect. 4 which discusses the results. Finally, Sect. 5 concludes the paper.

2 Method

Figure 1 presents an overview of our proposed method. First, we pre-train the U-Net lung fields segmentation network to generate pseudo-labels. Second, we use patches to train a multi-task U-Net-based neural network for CXR image bias analysis. Finally, for classification, we use a majority patch-based voting.

Pre-training Segmentation Network. We train a segmentation neural network to extract lung fields from CXR images. For this task, we adopt the U-Net encoder-decoder to perform semantic segmentation [5]. With the pre-trained model on [8,9] datasets, we generate pseudo labels for unlabelled COVIDx dataset [7]. Before feeding those pseudo masks to the multi-task neural network, we use custom pre-processing algorithms.

Multi-task Neural Network. Inspired by [15–17], we use an encoder-decoder based convolutional neural network (CNN) for simultaneous classification and segmentation of the lungs in the CXRs. Our network is U-Net-based with ImageNet pre-trained ResNet-50 as backbone encoder [6]. We use an encoder part to extract high-level X-ray image features. Every encoder block forwards feature maps and concatenates them with the corresponding decoder part. We employ an attention mechanism to the skip connections to learn salient maps suppressing irrelevant lung vicinity that may be a source of CBs. We extend an attention gate mechanism by aggregating multi-scale feature maps from the decoder to learn the local context of the lung feature maps representation. Each feature map is fed through an inception module that leverages convolutional filters of multiple kernels (i.e. 1×1, 3×3, 5×5, and 7×7) and stride sizes ($S = 1, S = 2$), which does not increase the number of parameters significantly. The amplitudes of features from the deeper layers are smaller than the shallow ones. To prevent shallow layers from dominating deeper ones, we normalise the weights of the features from multiple scales with the L2 norm before concatenation.

Patch-Based Learning. Motivated by [10], we adopt a patch-based learning method to train our multi-task neural network. Unlike Oh, we do not cut out the lungs with masks to avoid inductive bias. We use the lung masks instead as pseudo-labels for segmentation training. This helper task requires the model to learn new features necessary to recognise lung tissue boundaries. A pre-processed CXR image and corresponding mask are resized to 1024×1024. Then using a draw area based on the scaled-down, with a ratio of 0.9, whole lung mask, we randomly choose a patch centre. Its coordinates are drawn with a uniform distribution of non-zero pixels from the draw area mask. Figure 1(b) shows training images with corresponding pseudo-labels in the context of the entire CXR image, i.e. exemplary drawn image patches are marked with white squares in the CXR image and pseudo-labels with red squares on the corresponding mask. In our method, we use patches of size 80×80, in contrast to Oh uses patches of size 224×224. A region of the image and the corresponding mask are cut off using determined patch coordinates, resulting in an image patch and pseudo-label pair. Then the image patch and its mask are interpolated to 224×224. This pair is used to train a segmentation head, while the image patch and its corresponding class represent a training pair for the classification head. Thanks to the reduction of the draw area and the small size of the patch, the model's input never contains laterality tokens and hospital markings. At the same time, if chosen close to the lung edge, a patch covers a small lung boundary fragment, allowing the model to recognise lung tissue based on the segmentation task.

Majority Patch-Based Voting. Only one patch per image, per batch, is used in the training phase. However, we repeat the random draw multiple times for the inference to cover the whole lung field as the area of interest. Each time, based on the patch, the model makes a single classification called a single vote. The majority vote result, i.e. the class chosen based on the majority of the patches, is the final classification for that image.

3 Experiments and Results

In this section, we evaluate our methods on the COVIDx dataset. Next, we present gradient-weighted class activation maps that compare the decision bases of our proposed method with existing methods. We employ GradCAM to provide visual explanations of our method decision bases that focuses its attention on lung morphology and is less sensitive to CBs. Finally, we show quantitative results on the COVIDx test set.

3.1 Datasets

To pre-train a neural network for the segmentation of lungs, we use data from [8,9]. The datasets consist of 6380 2D CXR images. The annotations for each image provide a manual segmentation mask rendered as polygons, including the retrocardiac region. To evaluate our methods, we use an open-source **COVIDx dataset** [7]. In total, the dataset consists of 13970 2D CXR images (8806 normal, 5551 pneumonia and 353 COVID-19 cases). The authors constantly expand this dataset, and in order to compare with their published results, we use the same version of the dataset as they used at the time of publication. We create a validation set by splitting the training set with a 70:30 ratio, and the test dataset is built using the script provided by the COVIDx authors.

3.2 Data Pre-processing

Lung masks. We remove unnecessary objects from lungs masks like electronic devices, which are labelled with a shade of grey. We perform mask filtering to improve the masks or remove them from the dataset when the correction is not possible. We present a detailed algorithm in the supplement[1]. Finally, we resize the original image and the mask to a size of 1024 × 1024 pixels using linear interpolation. All images are resized to square regardless of their original aspect ratio.

COVIDx Dataset. The images for the training of our model are pre-processed with histogram equalisation. Soft augmentation methods consist of offset, scaling without preserved aspect ratio, rotation, horizontal flip, Contrast Limited Adaptive Histogram Equalisation (CLAHE) with random clip range and random grid range, brightness and contrast adjustment, sharpening and embossing.

3.3 Implementation Details

We implement our model in PyTorch deep learning framework and train on a workstation with a single GPU NVIDIA Titan RTX 24 GB until convergence over 100 epochs, with a mini-batch size of 16, an initial learning rate of 1×10^{-4} and a weight decay factor of 1×10^{-4}. We set Rectified Adam (RAdam) as the optimiser to minimise the loss function. To prevent overfitting, we apply various soft data augmentation techniques. During training, we perform the following transformations: horizontal flip, sharpen, emboss and CLAHE with $p = 0.5$. We also apply a random-weighted sampler. The weights are computed as an inverse class frequency. As loss function, we use Dice loss for segmentation task:

$$\mathcal{L}_{dice} = 1 - \frac{2\sum_i^N p_i g_i + \epsilon}{\sum_i^N p_i^2 + \sum_i^N g_i^2 + \epsilon}, \tag{1}$$

where p_i is the prediction pixel value, g_i is the ground truth, and ϵ is a numerical stability to avoid divide by zero errors, and Weighted Cross-Entropy (WCE) for classification task:

$$\mathcal{L}_{WCE} = -\frac{1}{N}\sum_{n=1}^{N} w r_n \log(p_n) + (1 - r_n)\log(1 - p_n), \tag{2}$$

where w is the class weight, p_n is the Softmax probability for the ith class, and r_n is the ground truth value of $\{0, 1\}$. Finally, the overall multi-task loss function can be formulated as the sum of both loss functions:

$$\mathcal{L} = \mathcal{L}_{Dice} + \mathcal{L}_{WCE}. \tag{3}$$

3.4 Gradient-Weighted Class Activation Mapping Results

We analyse probabilistic class activation maps for gradient weighted models that classify CXR images into three classes: normal, pneumonia and COVID-19 in the COVIDx dataset. It turns out that the globally [7] and the locally

[1] https://cutt.ly/rIB1JFQ.

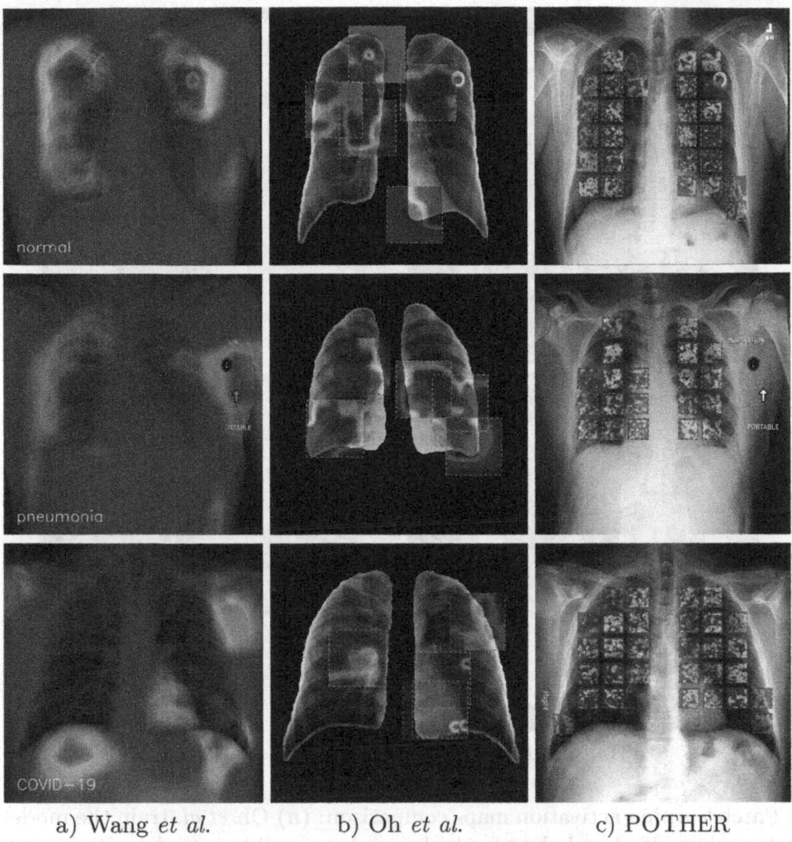

a) Wang *et al.* b) Oh *et al.* c) POTHER

Fig. 2. The activation maps of: **(a)** cropped image global training, **(b)** segmented lung patch-based local training, **(c)** our multi-task patch-based local training method - **POTHER**. The globally trained model focuses on artificial electronics, textual information, and other non-disease-related features such as shoulder position. The model trained on the segmented lung is spared textual information but uses lung contour for classification. Our proposed model does not use these spurious features for classification but lung morphology instead.

[10] trained model focus on non-disease related elements. In contrast, POTHER focuses on the morphological structure of lung tissue. The globally trained model concentrates on laterality tokens and ECG-leads artifacts even though input images are cropped. The locally trained model lacks textual information, but instead, it uses the lung mask's contour, supported by an edge's strong gradient because of the black background, to classify the image. The globally trained ResNet-50 model was fed with images pre-processed and augmented, as described in Wang's work. The pre-processing consists of cropping the top 8% of the image, and according to the authors, it is to mitigate commonly-found embedded textual information, which, as presented in Fig. 2, is not enough because there are texts localised too close to the lung to be cropped.

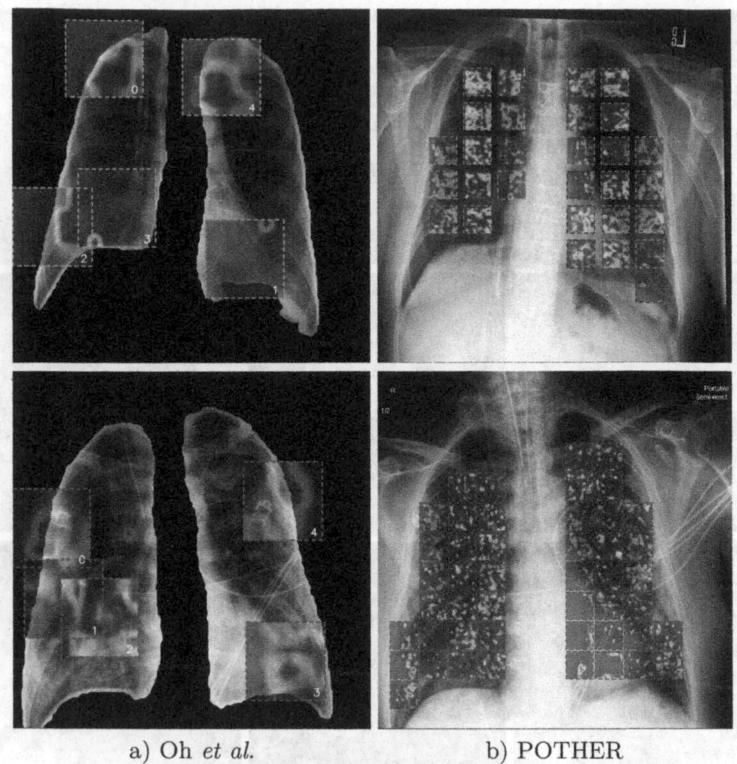

a) Oh *et al.* b) POTHER

Fig. 3. Patch learning activation maps comparison: **(a)** Oh *et al.* train the model using big patches which, if placed close to the lung edge, provide a model with awareness of lung contour, allowing it to learn that, (b) **POTHER** model focuses on fine-grained lung markings like, e.g. interstitial opacities.

Figure 3 compares activation maps of two models trained with a patch-based learning approach. We can see that Oh's patches cover a considerable percentage of the image, allowing the model to focus on the lung's corner or ECG leads. On the contrary, we train POTHER on the whole unsegmented images. It does not use pixels outside the lung because of small patches and reduced mask areas inside which patch centres can be located. Though some of the POTHER's patches are confused by cable on the healthy patient's image, thanks to their relatively small area, the vast majority of patches vote for a proper class during majority voting.

We analyse activation maps of the POTHER model based on the visible manifestations of pneumonia and COVID-19 on CXRs. Visible with an untrained eye, pneumonia and COVID-19 pneumonia signs are related to increased lung density, which may be seen as whiteness in the lungs, which obscures the lung markings that are typically seen depending on the severity of pneumonia [19]. During COVID-19, markings are partially obscured by the increased whiteness. These can be subtle and require a confirmation from a radiologist. COVID-19 pneumonia changes, like

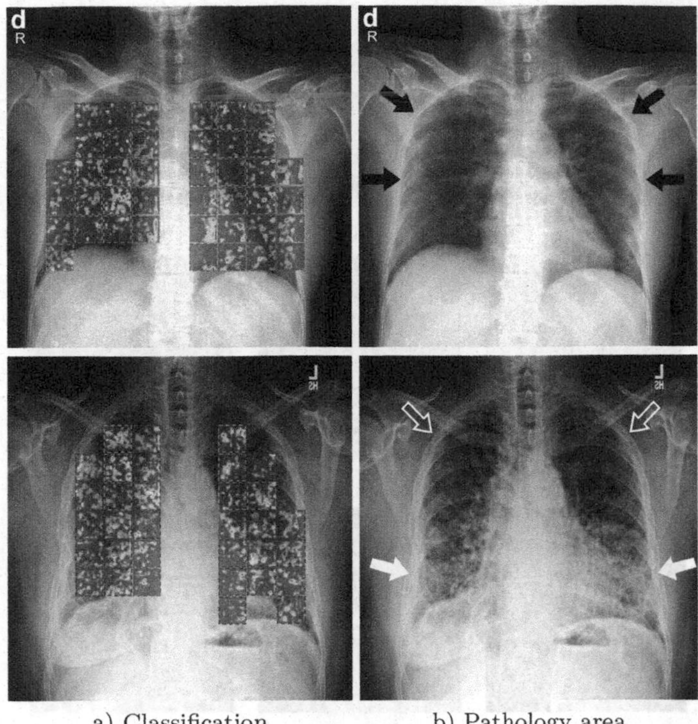

a) Classification b) Pathology area

Fig. 4. POTHER activation maps and per-patch classification results - (**a**) the model output class is coded with patch bounding box colour: *green - normal lung, blue - pneumonia, red - COVID-19.* (**b**) black arrows point to the area with white opacity characteristic of COVID-19, white arrows point to lower lobes opacity characteristic to different types of pneumonia, outlined arrows point to the area free of opacity. The areas indicated by the arrows and the POTHER classifications overlap, indicating that the model may consider manifestations specific to the classified diseases when making decisions. (Color figure online)

lung involvement, are mostly bilateral on chest radiograph. Areas of lungs - Fig. 4, where increased whiteness can be found are pointed with blue (pneumonia) and red (COVID-19) arrows, and the region free of it is marked with green arrows. It can be seen that POTHER's decisions about patches are correlated with them, which may suggest a classification based on the disease symptoms seen on the CXR.

The locations and neighbourhoods of ECG lead elements and other non-anatomical artifacts are analysed in more detail. Figure 5 presents generated activation maps of the analysed models, strongly indicating that electronic devices' presence in the image focuses heavily the model's attention and can be an important factor in decision-making. Using POTHER, we classify the exact locations,

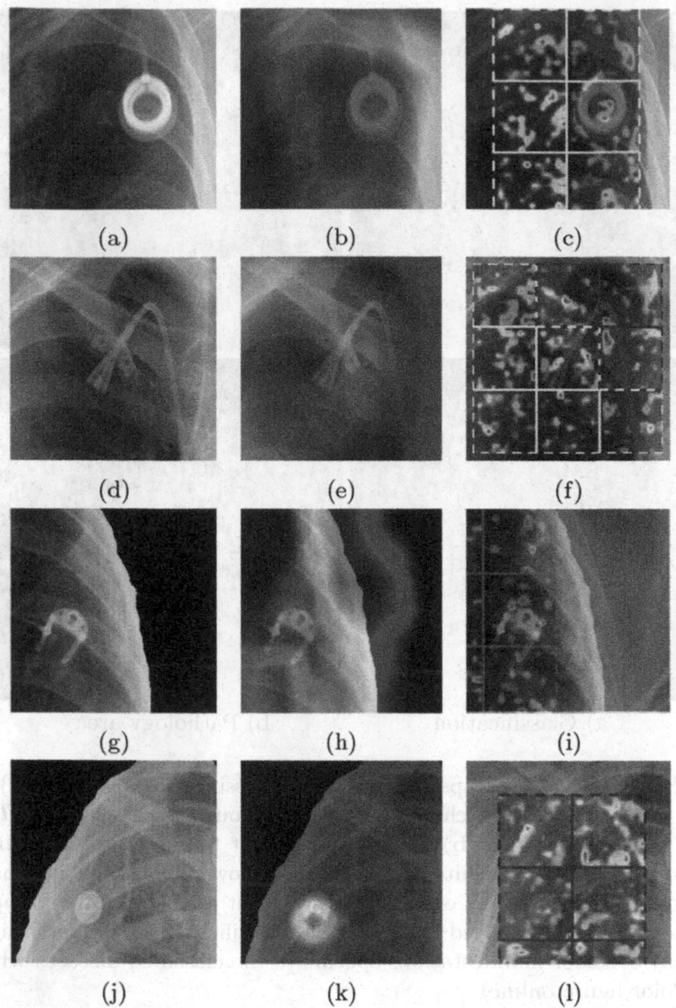

Fig. 5. Analysis of activation maps in the vicinity of elements influencing the model decision. Comparison of the models: **(b, e)** trained globally, **(h, k)** segmented-lung patch-based and **(c, f, i, l)** with our **POTHER** method. The left column of figures **(a, d, g, j)** shows fragments of the original CXR image, figures **(b, e, h, k)** a fragment with superimposed activation map and right column **(c, f, i, l)** activation maps of individual **POTHER** patches and their color-coded votes.

and its activation maps show no significant focus on artificial elements. Figure 5f shows that 3 out of 9 POTHER's patches change their output near the cable; however, thanks to the majority voting it does not influence the final decision.

3.5 Quantitative Analysis

To evaluate results, we use precision, recall, F1 score and accuracy. As a test set, we use a COVIDx introduced by Wang *et al.* [7]. It consists of 300 images equally distributed into three classes. A globally trained model's inference results in a single classification per image, while inference of a locally trained one requires majority voting before making the final decision. When we train Oh's patch-based learning model, we follow the same pre-processing and augmentations as the author and the only differences are the model used for lungs segmentation and our mask filtration algorithm. We compare the performance of four models that are trained on the COVIDx dataset. The results presented in Table 1 are comparable in terms of accuracy, which oscillates around 90%. However, COVID-Net's accuracy, as well as its precision of COVID-19 classification, stands out. Our method achieves the highest F1 score of 0.974 for the COVID-19 class. Oh's model performs similarly to the results reported in their work, where it scored an accuracy of 0.889 on their dataset.

Table 1. The performance comparison on COVIDx test set

Method	Class	Precision	Recall	F1	Accuracy
ResNet-50 [7]	Normal	0.882	0.970	0.924	0.906
	Pneumonia	0.868	0.920	0.893	
	COVID-19	0.988	0.830	0.902	
COVID-Net [7]	Normal	**0.905**	0.950	**0.927**	**0.933**
	Pneumonia	0.913	**0.940**	**0.926**	
	COVID-19	0.989	0.910	0.948	
Patch learning [10]	Normal	0.815	0.970	0.886	0.886
	Pneumonia	0.914	0.813	0.860	
	COVID-19	0.963	0.867	0.912	
Ours (POTHER)	Normal	0.790	**0.980**	0.875	0.903
	Pneumonia	**0.963**	0.780	0.862	
	COVID-19	**1.000**	**0.950**	**0.974**	

4 Discussion

In this work, we propose a novel learning method specifically designed to address difficulties inherent in CXR images. In addition, we analyse the model's biases when it is making classification decisions. The improvement of diagnostic capabilities and the reduction of the negative influence of the numerous CBs in the COVIDx dataset guide us during model development. Many researchers have achieved very high scores on CXR images when trying to improve COVID-19 classification, but unfortunately, very few have analysed what allows the model to achieve such high scores.

When trying to improve COVID-19 classification based on CXRs, great attention should be paid to understanding what features allow the model to achieve good classification. Otherwise, we may obtain a model with little diagnostic value, though it achieves seemingly high scores utilising CBs to classify on dataset that it was trained and tested. We reveal potential pitfalls in the CXR medical imaging domain, i.e. some types of clues irrelevant to disease recognition, like hospital markings and ECG leads, causing the model to miss the point of the disease symptoms. The list of CBs is not exhaustive as it cannot be guaranteed that the model will not shift its attention to something similarly undesired if known CBs are addressed. Therefore, a dataset specifically designed to eliminate as many detected CBs as possible is desirable. Annotation or segmentation of the pathological lung area by a radiologist visible in the CXR would also be valuable. According to [19], a chest radiograph can look normal in up to 63% of people infected with COVID-19, especially in the early stages. Using COVID-19 labelled radiograph with no noticeable disease-related changes in the lung image is very likely confusing to the model and encourages learning of undesirable confounding features.

5 Conclusions

In this paper, we proposed a novel multi-task patch-voted learning-based method called POTHER for CXR-image-based COVID-19 detection. Our model learns pneumonia manifestations and lung morphology features. We performed the activation map analysis and showed that the deep learning methods are susceptible to learning features unrelated to the pathology. The COVIDx dataset contains many images that can confuse the learned model, such as ECG leads, laterality tokens or other hospital-specific markings. To this end, POTHER's patches include the area of lung fields and only its closes vicinity to eliminate some of the sources of undesirable features. Using segmentation task with an attention mechanism provides an efficient feature extraction, allowing satisfactory results despite training with limited lung fragments. Thanks to training with small patches, our method is less sensitive to the CBs. In future work, we will apply our method with Vision Transformer (ViT) [26] on a large-scale CXR dataset to improve the model generalisation with more complex lung disease classes.

Acknowledgements. This work is supported in part by the European Union's Horizon 2020 research and innovation programme under grant agreement Sano No. 857533 and the International Research Agendas programme of the Foundation for Polish Science, co-financed by the EU under the European Regional Development Fund. This research was funded by Foundation for Polish Science (grant no POIR.04.04.00-00-14DE/18-00 carried out within the Team-Net program co-financed by the European Union under the European Regional Development Fund), National Science Centre, Poland (grant no 2020/39/B/ST6/01511).

References

1. Çallı, E., Sogancioglu, E., van Ginneken, B., van Leeuwen, K.G., Murphy, K.: Deep learning for chest X-ray analysis: a survey. Med. Image Anal. **72**, 102125 (2021). https://doi.org/10.1016/j.media.2021.102125
2. Minaee, S., Kafieh, R., Sonka, M., Yazdani, S., Jamalipour Soufi, G.: Deep-COVID: predicting COVID-19 from chest X-ray images using deep transfer learning. Med. Image Anal. **65**, 101794 (2020). https://doi.org/10.1016/j.media.2020.101794
3. Li, X., et al.: Multi-resolution convolutional networks for chest X-ray radiograph based lung nodule detection. Artif. Intell. Med. **103**, 101744 (2020). https://doi.org/10.1016/j.artmed.2019.101744
4. Tjoa, E., Guan, C.: A survey on explainable artificial intelligence (XAI): toward medical XAI. IEEE Trans. Neural Netw. Learning Syst. **32**, 4793–4813 (2021). https://doi.org/10.1109/TNNLS.2020.3027314
5. Ronneberger, O., Fischer, P., Brox, T.: U-Net: convolutional networks for biomedical image segmentation. In: Navab, N., Hornegger, J., Wells, W.M., Frangi, A.F. (eds.) MICCAI 2015. LNCS, vol. 9351, pp. 234–241. Springer, Cham (2015). https://doi.org/10.1007/978-3-319-24574-4_28
6. He, K., Zhang, X., Ren, S., Sun, J.: Deep residual learning for image recognition. In: 2016 IEEE Conference on Computer Vision and Pattern Recognition (CVPR), pp. 770–778. IEEE, Las Vegas, NV, USA (2016). https://doi.org/10.1109/CVPR.2016.90
7. Wang, L., Lin, Z.Q., Wong, A.: COVID-Net: a tailored deep convolutional neural network design for detection of COVID-19 cases from chest X-ray images. Sci. Rep. **10**, 19549 (2020). https://doi.org/10.1038/s41598-020-76550-z
8. Kermany, D., Zhang, K., Goldbaum, M.: Labeled optical coherence tomography (OCT) and chest X-ray images for classification, vol. 2 (2018). https://doi.org/10.17632/rscbjbr9sj.2
9. Cohen, J.P., Morrison, P., Dao, L., Roth, K., Duong, T.Q., Ghassemi, M.: COVID-19 image data collection: prospective predictions are the future. arXiv:2006.11988 [cs, eess, q-bio] (2020)
10. Oh, Y., Park, S., Ye, J.C.: Deep learning COVID-19 features on CXR using limited training data sets. IEEE Trans. Med. Imaging **39**, 2688–2700 (2020). https://doi.org/10.1109/TMI.2020.2993291
11. Al-Waisy, A.S., et al.: COVID-CheXNet: hybrid deep learning framework for identifying COVID-19 virus in chest X-rays images. Soft Comput., 1–16 (2020). https://doi.org/10.1007/s00500-020-05424-3
12. Rajaraman, S., Siegelman, J., Alderson, P.O., Folio, L.S., Folio, L.R., Antani, S.K.: Iteratively pruned deep learning ensembles for COVID-19 detection in chest X-Rays. IEEE Access **8**, 115041–115050 (2020). https://doi.org/10.1109/ACCESS.2020.3003810
13. Pham, T.D.: Classification of COVID-19 chest X-rays with deep learning: new models or fine tuning? Health Inf. Sci. Syst. **9**(1), 1–11 (2020). https://doi.org/10.1007/s13755-020-00135-3
14. Ucar, F., Korkmaz, D.: COVIDiagnosis-Net: deep Bayes-SqueezeNet based diagnosis of the coronavirus disease 2019 (COVID-19) from X-ray images. Med. Hypotheses **140**, 109761 (2020). https://doi.org/10.1016/j.mehy.2020.109761

15. Płotka, S., Włodarczyk, T., Klasa, A., Lipa, M., Sitek, A., Trzciński, T.: FetalNet: multi-task deep learning framework for fetal ultrasound biometric measurements. In: Mantoro, T., Lee, M., Ayu, M.A., Wong, K.W., Hidayanto, A.N. (eds.) ICONIP 2021. CCIS, vol. 1517, pp. 257–265. Springer, Cham (2021). https://doi.org/10.1007/978-3-030-92310-5_30
16. Płotka, S., et al.: Deep learning fetal ultrasound video model match human observers in biometric measurements. Phys. Med. Biol. **67**, 045013 (2022). https://doi.org/10.1088/1361-6560/ac4d85
17. Amyar, A., Modzelewski, R., Li, H., Ruan, S.: Multi-task deep learning based CT imaging analysis for COVID-19 pneumonia: classification and segmentation. Comput. Biol. Med. **126**, 104037 (2020). https://doi.org/10.1016/j.compbiomed.2020.104037
18. Chattopadhay, A., Sarkar, A., Howlader, P., Balasubramanian, V.N.: Grad-CAM++: generalized gradient-based visual explanations for deep convolutional networks. In: 2018 IEEE Winter Conference on Applications of Computer Vision (WACV), pp. 839–847. IEEE, Lake Tahoe, NV (2018). https://doi.org/10.1109/WACV.2018.00097
19. Cleverley, J., Piper, J., Jones, M.M.: The role of chest radiography in confirming Covid-19 pneumonia. BMJ **370**, m2426 (2020). https://doi.org/10.1136/bmj.m2426
20. DeGrave, A.J., Janizek, J.D., Lee, S.-I.: AI for radiographic COVID-19 detection selects shortcuts over signal. Nat. Mach. Intell. **3**, 610–619 (2021). https://doi.org/10.1038/s42256-021-00338-7
21. Nayak, S.R., Nayak, D.R., Sinha, U., Arora, V., Pachori, R.B.: Application of deep learning techniques for detection of COVID-19 cases using chest X-ray images: a comprehensive study. Biomed. Signal Process. Control **64**, 102365 (2021). https://doi.org/10.1016/j.bspc.2020.102365
22. López-Cabrera, J.D., Orozco-Morales, R., Portal-Díaz, J.A., Lovelle-Enríquez, O., Pérez-Díaz, M.: Current limitations to identify Covid-19 using artificial intelligence with chest X-ray imaging (part ii). The shortcut learning problem. Health Technol. **11**(6), 1331–1345 (2021). https://doi.org/10.1007/s12553-021-00609-8
23. Maguolo, G., Nanni, L.: A critic evaluation of methods for COVID-19 automatic detection from X-ray images. Inf. Fusion **76**, 1–7 (2021). https://doi.org/10.1016/j.inffus.2021.04.008
24. Lin, Z.Q., Shafiee, M.J., Bochkarev, S., Jules, M.S., Wang, X.Y., Wong, A.: Do explanations reflect decisions? A machine-centric strategy to quantify the performance of explainability algorithms. arXiv:1910.07387 [cs] (2019)
25. Roy, K., Banik, D., Bhattacharjee, D., Nasipuri, M.: Patch-based system for classification of breast histology images using deep learning. Comput. Med. Imaging Graph. **71**, 90–103 (2019). https://doi.org/10.1016/j.compmedimag.2018.11.003
26. Dosovitskiy, A., et al.: An image is worth 16 × 16 words: transformers for image recognition at scale. arXiv:2010.11929 [cs] (2021)

Modeling Contrast Perfusion and Adsorption Phenomena in the Human Left Ventricle

Evandro Dias Gaio, Bernardo Martins Rocha, and Rodrigo Weber dos Santos(✉)

Graduate Program in Computational Modeling, Universidade Federal de Juiz de Fora, Juiz de Fora, Minas Gerais, Brazil
rodrigo.weber@ufjf.edu.br

Abstract. This work presents a mathematical model to describe perfusion dynamics in cardiac tissue. The new model extends a previous one and can reproduce clinical exams of contrast-enhanced cardiac magnetic resonance imaging (MRI) of the left ventricle obtained from patients with cardiovascular diseases, such as myocardial infarct. The model treats the extra- and intravascular domains as different porous media where Darcy's law is adopted. Reaction-diffusion-advection equations are used to capture the dynamics of contrast agents that are typically used in MRI perfusion exams. The identification of the myocardial infarct region is modeled via adsorption of the contrast agent on the extracellular matrix. Different scenarios were simulated and compared with clinical images: normal perfusion, endocardial ischemia due to stenosis, and myocardial infarct. Altogether, the results obtained suggest that the models can support the process of non-invasive cardiac perfusion quantification.

Keywords: Myocardial perfusion · Adsorption · Ischemia · Left ventricle dynamics

1 Introduction

Cardiovascular diseases are one of the major causes of death worldwide [13]. These conditions include coronary atherosclerosis, aortic valve regurgitation, and left ventricle hypertrophy, which affect the myocardium perfusion (MP), reduce oxygen delivery (ischemia), cause tissue damage, and lead to infarct. Contrast-enhanced Magnetic Resonance Imaging (MRI) is an exam that seeks to characterize myocardial perfusion and detect scars or infarct regions by conducting a contrasting agent (CA) to the patient. On the images generated by a specific protocol, the CA assumes a specific contrast on poorly perfused regions, which allows its identification. The most used protocol is the Late Gadolinium Enhancement (LGE), a technique used in heart MRI for cardiac tissue characterization. Particularly, LGE allows the assessment of myocardial scar formation and regional myocardial fibrosis when the gadolinium, the CA, is perfused for about 600 s and is adsorbed in areas of excess of extracellular matrix.

D. Groen et al. (Eds.): ICCS 2022, LNCS 13351, pp. 455–468, 2022.
https://doi.org/10.1007/978-3-031-08754-7_52

Blood-tissue exchange investigation is a topic of long-standing interest to physiologists and was first modeled mathematically as a set of reaction-diffusion-advection equations by [4]. More recently, [15] has proposed a framework for the simulation of cardiac perfusion using Darcy's law within the idea of multi-compartment to represent the different blood vessel's spatial scale. In the field of medical image analysis, [7] has proposed to quantify the behavior of contrast agents in MR perfusion imaging. This work has used a simplified model of contrast agent transport and provided interesting insights on the design and selection of the appropriate CA for specific imaging protocol and post-processing method. Finally, [1,2] proposed similar tools that also uses mathematical models based on PDEs (Partial Differential Equations) and images, in this case, from contrast-enhanced MRI exams. In particular, the previous work presented in [1] evaluated the CA dynamics for three different scenarios: healthy, ischemic, and infarct in a 2D mesh slice from an MRI exam and performed a comparison with experimental data for the LGE protocol.

In this work, we present an extended model for describing the perfusion of the contrasting agent in cardiac tissue based on porous media flow, which is suitable for 3D geometries and patient-specific models generated through image segmentation from MRI [1,17]. The mathematical model uses Darcy's law for the extra- and intravascular regions and is coupled to a reaction-diffusion-advection equation for the CA dynamics. Through a series of numerical experiments, we show that the model can correctly reproduce clinical exams via computer simulations during normal perfusion and in the presence of ischemia or myocardial infarct. In addition, we also present the pipeline used for generating patient-specific finite element meshes appropriate for the simulations of the perfusion model. This study has a potential high impact since it combines information from two different exams: Fractional Flow Reserve (FFR) Measurement of the heart coronaries from CT and heart perfusion and topology from MRI scan [3,18].

2 Methods

2.1 Mathematical Modeling

There are many different ways to represent and deal with circulatory models in the literature [4,7,15]. In this work, the perfusion is modeled through a reaction-diffusion-advection equation in porous media. Even knowing that blood vessels have non-trivial topological features, a simplified representation of the domain can provide useful insights in many aspects and can be used to reproduce clinical exams of contrast-enhanced cardiac MRI of the whole heart.

A porous medium is a solid filled and connected by voids, where the ratio between the volume of void space and the total volume is called porosity, which is given by:

$$\phi = \frac{V_p}{V_t}, \tag{1}$$

where V_p is the volume of the void (or porous) space and V_t is the total volume.

Porous Media Flow in the Intravascular Domain. Darcy's law is used to describe flow in porous media, and can be expressed through the following equations for an incompressible fluid low:

$$\mathbf{v} = -\mathbf{K}\nabla p, \quad \text{in}\, \Omega, \tag{2}$$
$$\nabla \cdot \mathbf{v} = \alpha, \quad \text{in}\, \Omega, \tag{3}$$

where $\mathbf{v}$ is the velocity, p is the pressure, $\mathbf{K}$ is the permeability tensor, α is a source term, and Ω is the left ventricular domain (LV).

The tensor $\mathbf{K}$ must represent the anisotropy and heterogeneity of the intravascular domain due to the fibers presents in cardiac microstructure, as described in [1,7]. The heterogeneity is first represented by a transmural gradient of permeability (w) by solving the Laplace equation ($\nabla^2 w = 0$), using Dirichlet boundary condition at the epicardial ($w = 1$) and endocardial ($w = 0$) boundaries, respectively. The permeability tensor including anisotropy and heterogeneity can be described as:

$$\mathbf{K} = K_t \mathbf{I} + (K_l - K_t)\mathbf{f} \otimes \mathbf{f} \tag{4}$$

where $\mathbf{f}$ is the unit vector that represents the preferential direction of permeability which follows the myocardial fiber orientation, K_l is the permeability value along the fiber direction $\mathbf{f}$, and K_t is the permeability in the transversal direction. The values of these permeabilities are given by:

$$K_l = K_1(1 - w) + 2K_1 w, \quad K_t = K_2(1 - w) + 2K_2 w \tag{5}$$

where K_1 and K_2 are the permeabilities in outer boundary, with K_1 approximately 10.8 times higher than K_2 [7].

Contrast Agent Dynamics in the Intra- and Extravascular Domains. Contrast agent dynamics can be described by a system of diffusion-advection equations in a bidomain composed of the intra- and extravascular regions. The intravascular represents the combination of arteries and capillaries, whereas the extravascular represents the interstitial space, and possibly a region with fibrosis, when it is considered. The equations governing the dynamics of the concentrations of CA in the intra- and extravascular regions, denoted by C_i and C_e, respectively, are given by:

$$\frac{\partial(\phi C_i)}{\partial t} + \nabla \cdot (\mathbf{v} C_i) - \phi \nabla \cdot (\mathbf{D}_i \nabla C_i) + f = 0, \quad \text{in}\, \Omega_i, \tag{6}$$
$$\frac{\partial((1-\phi)\lambda C_e)}{\partial t} - (1-\phi)\lambda \nabla \cdot (\mathbf{D}_e \nabla C_e) - f + (1-\phi)\lambda k_e C_e + g = 0, \text{in}\, \Omega_e, \tag{7}$$

where $\mathbf{D}_i$ and $\mathbf{D}_e$ are diffusion tensors for the intra- and extravascular regions, respectively, f represents the communication between the domains which is given by:

$$f = \begin{cases} P(C_i - C_e), & \text{if} \;\; C_i > C_e, \\ 0, & \text{otherwise}, \end{cases} \tag{8}$$

where P is the endothelial permeability. The term $(1-\phi)\lambda k_e C_e$ models the flow from the interstitial space to the venous system, and λ represents the fraction of the extravascular domain that is occupied by the interstitial space.

Contrast Agent Adsorption. In addition, another equation and variable is needed to capture how the CA is trapped in the excess of extracellular matrix, which is the case of fibrosis or scar. This phenomenon, fluid (CA) attaching to a solid phase (extracellular matrix), is called adsorption, and is modeled by:

$$\frac{\partial((1-\phi)\lambda\lambda_f C_f)}{\partial t} + (1-\phi)\lambda\lambda_f k_f C_f - g = 0, \quad \text{em} \quad \Omega_f, \tag{9}$$

where C_f is the concentration of CA in the fibrosis network domain Ω_f, g is a exchange term between the fibrotic and extravascular domains, and the $k_f C_f$ term describes the flow from the fibrotic network to the venous system. The exchange term g is given by:

$$g = (1-\phi)\lambda\lambda_f k_{ef} C_e, \tag{10}$$

where k_{ef} is the rate at which the contrast moves from the interstitium to the fibrosis, and λ_f is the fraction of the interstitium occupied by the region with fibrosis.

Recirculation of the Contrast Agent. During MRI or CT scan, the cyclic behavior of the CA is captured. Part of the CA is retained by the kidneys for elimination, but a certain amount, after a while, returns by the blood flow itself and is infused into the cardiac tissue again by the coronary arteries, i.e., the intravascular domain.

To represent this behavior of the CA, a reaction-diffusion-advection equation in a one dimensional domain (of size L) is used. The total amount of CA in the intravascular domain of the myocardium is imposed as an inflow into the 1D domain. The parameters of the equation were defined so that the time and the amount of flux at the output of the 1D domain represent the physiological behavior. Therefore, the amount of flow at the exit of the 1D domain is imposed as a recirculation parameter $X(\mathbf{x}, t)$ for the boundary condition of the CA flow in the intravascular domain.

The recirculation is therefore described by the following equation:

$$\frac{\partial C_{out}}{\partial t} + \nabla \cdot v_{out} C_{out} - \nabla \cdot (D_{out} \nabla C_{out}) + k C_{out} = 0, \quad \text{in} \quad [0, L], \tag{11}$$

where v_{out} and D_{out} represents the velocity or convection term and the diffusion respectively of this re-circulatory system, discussed numerically in [1]. This equation is subject to the following conditions:

$$C(0,t) = \frac{\int_{\Omega_i} C_i \, d\Omega_i}{|\Omega_i|}, \quad X(\mathbf{x}, t) = C(L, t). \tag{12}$$

Initial and Boundary Conditions. The boundary conditions for the Darcy Eq. (2) are of the Dirichlet type, and are given by:

$$p = p_o, \quad \text{on} \quad \Gamma_{epi}, \tag{13}$$
$$p = p_i, \quad \text{on} \quad \Gamma_{endo}. \tag{14}$$

For CA dynamics subsystem, a convective and diffusive flow of CA inflow into the intravascular domain through the epicardium is imposed, controlled by a transient Gaussian function, which is given by:

$$\mathbf{v}C_i - \mathbf{D}_i \nabla C_i = \mathbf{v}Q(t), \quad \text{on} \quad \Gamma_{epi}, \tag{15}$$

with $Q(t)$ given by

$$Q(t) = \frac{1}{\sigma\sqrt{2\pi}} e^{-\frac{1}{2}\left(\frac{t - t_{peak}}{\sigma}\right)} + X(t, \vec{x}), \tag{16}$$

where σ^2 reflects the variance of the CA infusion, t_{peak} is the Gaussian mean which is the peak value of the function, and $X(t, \vec{x})$ is the additional term generated by CA recirculation due to the cyclic behavior of the blood system.

No-flux boundary conditions are used for the other boundaries, as follows:

$$\mathbf{D}_i \nabla C_i \cdot \mathbf{n} = 0 \quad \text{on} \quad \Gamma_{endo}, \tag{17}$$
$$\mathbf{D}_e \nabla C_e \cdot \mathbf{n} = 0 \quad \text{on} \quad \Gamma_{endo}, \tag{18}$$
$$\mathbf{D}_e \nabla C_e \cdot \mathbf{n} = 0 \quad \text{on} \quad \Gamma_{epi}. \tag{19}$$

2.2 Left Ventricular Geometry Models

In this work, two geometries for representing the left ventricle were used. The first one is the simplified representation of the LV as a family of truncated ellipsoids. Parametric equations were used to allow better geometric control, ideal for defining contours (endocardium and epicardium) and well-defined positioning of subdomains (fibrosis), as shown in Fig. 1A.

There are also heterogeneity and anisotropy in the properties of the models. Heterogeneity occurs since the subendocardial permeability is, on average, twice as high as the subepicardial permeability. Moreover, anisotropy is due to the fiber direction being the preferred direction of the microvascular system, that is, the preferential direction for perfusion. These characteristics are translated to the permeability tensor **K**.

To generate the fibers, the *Laplace-Dirichlet Rule-Based (LDRB)* algorithm [5] was used. The algorithm requires as inputs the helical angles for the fibers on endocardium and epicardium surfaces, marked subdomains (left ventricle (LV) and right ventricle (RV)), and contours (base, epicardium, and endocardium). It generates vectors describing the fiber direction, sheet direction, and normal direction as outputs. An illustrative example of the result of the LDRB algorithm for generating the fiber orientation field is shown in Fig. 1B.

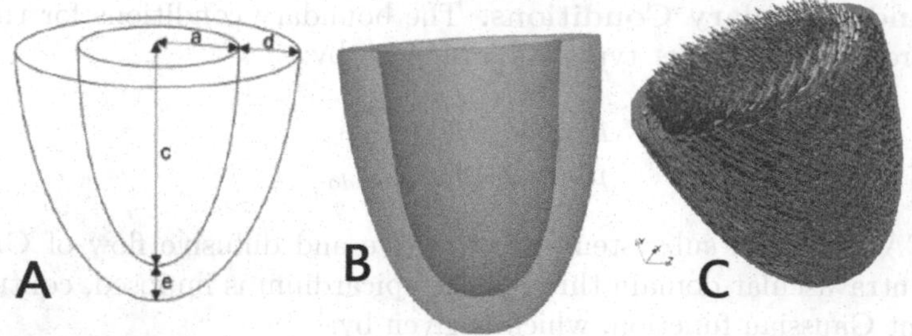

Fig. 1. **(A)** Parametric representation of the LV geometry: d,e = 1.0 cm, c = 6.0 cm, a = 2.0 cm. **(B)** Finite element mesh. **(C)** Fiber orientation field generated by the LDRB algorithm on a simplified LV mesh.

The study considers a patient-specific geometric model. For this study case, the following pipeline was considered. First, image segmentation of the LV from MRI is performed as described in [8]. This method produces a finite element mesh with a high spatial resolution. Second, scar regions are marked, which is required for the numerical studies carried out in this work (see Fig. 2B). The next step consists in generating a coarser mesh since the initial one is highly refined and would result in a huge computational effort for numerical simulation. Mesh processing techniques were used to get a coarser mesh keeping the same topology and reducing by about 90% the number of elements (see Fig. 2C). This step was carried out using the `meshtool` software [16]. Finally, the geometrical model is complemented with fiber and sheet orientation fields (see Fig. 2A and C).

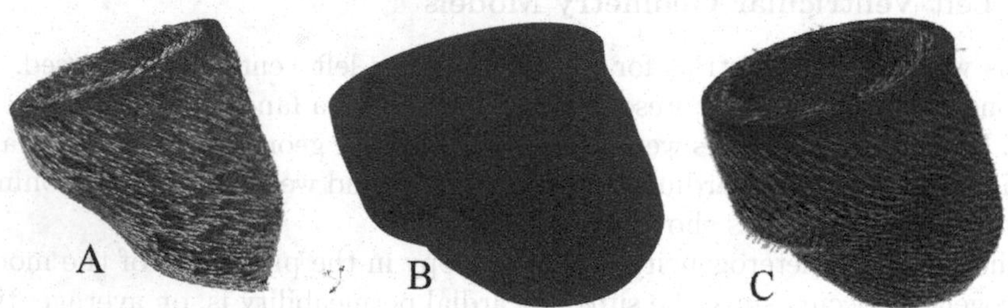

Fig. 2. Patient-specific meshes. **(A)** Fibers orientation, used in perfusion permeability tensor. **(B)** Mesh with marked scars and huge number of elements. **(C)** Used mesh for numerical simulations.

2.3 Numerical Methods

The differential equations presented in the mathematical modeling part were solved using the finite element method. In comparison with [1], where the reduced Darcy approximation was used, in this work, Darcy's problem is solved by using a mixed formulation [10] in terms of both pressure and velocity fields. For this goal, two discrete function spaces are needed to form a mixed function space and a stable choice of finite element spaces is the $\mathbf{H}(div)$ Brezzi-Douglas-Marini elements of polynomial order 1 and the discontinuous Lagrange elements $\mathbf{L}^2$ of order 0. In addition to ensuring the consistency of mass conservation at the macro scale level, this method is more suitable to impose flux as a boundary condition besides only pressure. More details about the numerical formulation may be found in [12].

The transient CA transport Eqs. (6), (7), and (11) were discretized with the Crank-Nicolson scheme, which is unconditionally stable and second-order accurate. The transport problem was discretized in space with the finite element method using first-order continuous Lagrange elements. Due to the presence of convective terms in the intravascular Eq. (6) and the recirculation Eq. (11), stabilization terms from the *Streamline upwind Petrov-Galerkin (SUPG)* method [6] were added for these equations.

3 Results and Discussion

In this section, we first present some numerical results with a simplified 3D model based on an idealized ellipsoid for comparisons and validation. Then, after a proper calibration of the mathematical model [1], we perform simulations on a patient-specific model constructed after image segmentation and processing.

All computer simulations were performed in a personal computer equipped with an 8th generation i7 3200 GHz processor and 16 GB of memory. The execution time to simulate a mesh with 47,662 tetrahedral elements and 10,232 nodes was around 3 h.

3.1 Simplified LV Model Study

The parameters used for the simulations are presented in Table 1 and were based on those reported in [1]. The ellipsoid simulation involves three different scenarios. The first is the Normal scenario which considers a healthy ventricle, represented by a uniform gradient pressure from epicardium (2 kPa) to endocardium (0 kPa). The second scenario represents an Ischemic case by assuming a restriction in blood supply from the coronaries, which was modeled here by a 30% of pressure drop (1.4 kPa) in a small region of the epicardium. Both Normal and Ischemic cases share the same physical tissue parameters. The third scenario represents an Infarct where a more significant pressure drop of 50% (1 kPa) is considered in the same region of the ischemic case. This case has an additional small infarcted region next to the endocardium at the same side where the pressure drop is applied. It is represented by a subdomain of dead tissue, which is modeled using parameters representing the infarct (see Table 1). The main differences between Healthy, Ischemic, and Infarct cases are the addition of this new subdomain and reducing the k_e parameter to represent the difficulty of flow from the interstitial space back to the venous system in a fibrotic region.

Table 1. Parameters used for the numerical simulations.

Parameter (unity)	Healthy/Ischemic	Infarct
$k_e\ (s^{-1})$	0.007	0.002
$P(s^{-1})$	1.0	1.0
$k_f(s^{-1})$	0	0.001
$k_{ef}(s^{-1})$	0	0.01
ϕ	0.10	0.10
λ	0.25	1.0
λ_f	0	0.5
$D(mm^2 s^{-1})$	0.01	0.01
σ	6.0	6.0
$t_{peak}(s)$	25	25
$v_{out}(mm.s^{-1})$	0.06	0.06
$D_{out}(mm^2 s^{-1})$	0.05	0.05
$k(s^{-1})$	0.01	0.01

Figure 3 shows the results of the LGE simulation obtained using the simplified LV geometry for the three cases: normal, ischemic, and infarct. This exam is used to reveal dead tissue by the absorption phenomena. Therefore, as expected, the results show that only the infarct case presents CA adsorption. These scenarios reproduced characteristics usually observed by clinicians. It is important to remark here that this preliminary study using 3D left ventricular meshes conforms with the one presented in [1], where simpler 2D cases were studied.

Figures 4 and 5 shows a quantitative comparison between the 3D idealized geometry and experimental clinical data with respect to the signal intensity (SI) of the contrast agent (sum of the CA concentration) for specific regions of interest (ROI) of the first pass (50 s) and LGE, respectively. In order to compare the CA dynamics, two ROIs were marked: one represents the injured region and the other represents a remote and healthy region. In addition, a linear relationship between the concentration of CA and its SI was considered [9]. The comparative curves between the simulated scenarios and the experimental data show that these results are in agreement with those found in the literature [9,19] for the first pass (Fig. 4), as well as for the LGE [3,11] (Fig. 5). The results are also in consonance to the results previously found by [1].

Figure 6(a) shows pressure field results of Darcy's problem for the idealized LV problem. The pressure gradient is imposed as boundary values and a smaller pressure (see Fig. 6(a), right of epicardium) is used to simulate pathologies (coronary artery diseases (CAD) or ischemic heart diseases (IHD)). Panel (b) from Fig. 6 presents the adsorption of CA in the ROI, which is characterized by a region with high contrast values as a result of the trapped gadolinium, revealing dead tissue.

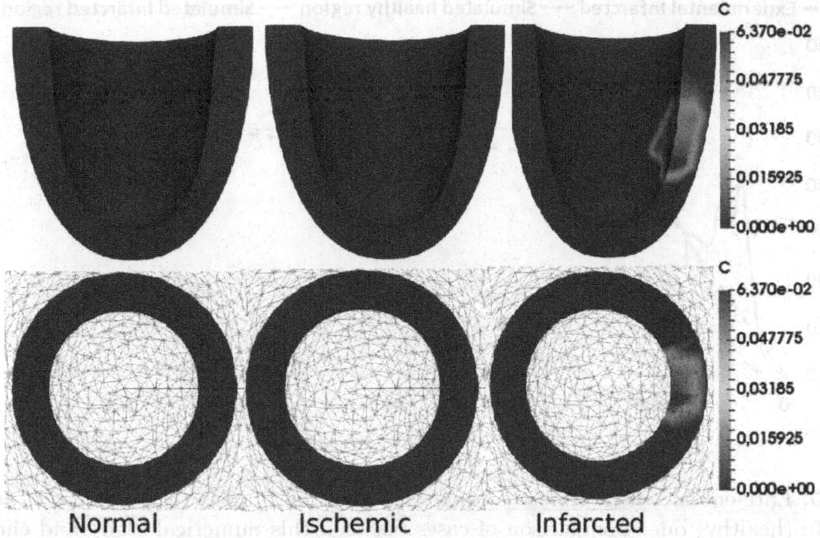

Fig. 3. Computer simulations of the LGE (600 s) for normal, ischemic, and infarcted left ventricles. After some time, the CA reveals adsorption in the infarcted region, as observed in the region with high contrast values on the right panel.

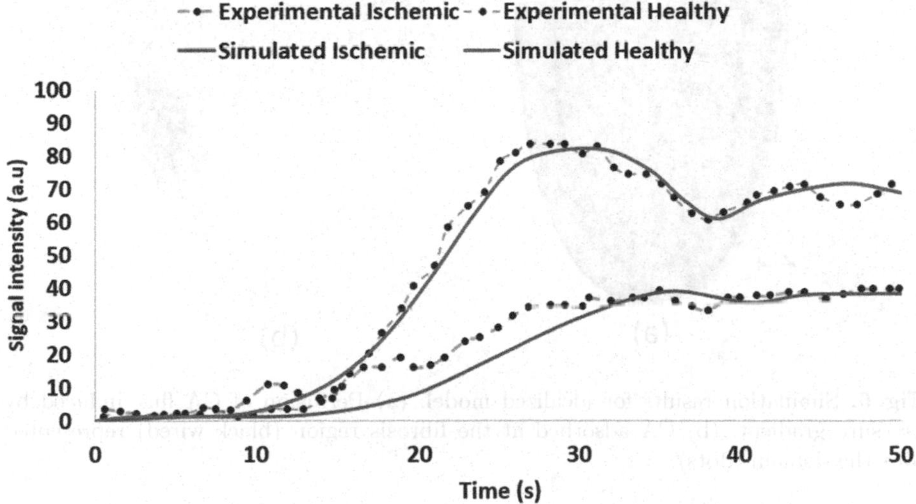

Fig. 4. First pass comparison of healthy and ischemic cases between this numerical study and experimental data reported in [9].

3.2 Patient Specific LV Study

The following study case explored a simulation using a patient-specific LV mesh. The same procedure was carried out here, but now with the difficulties of

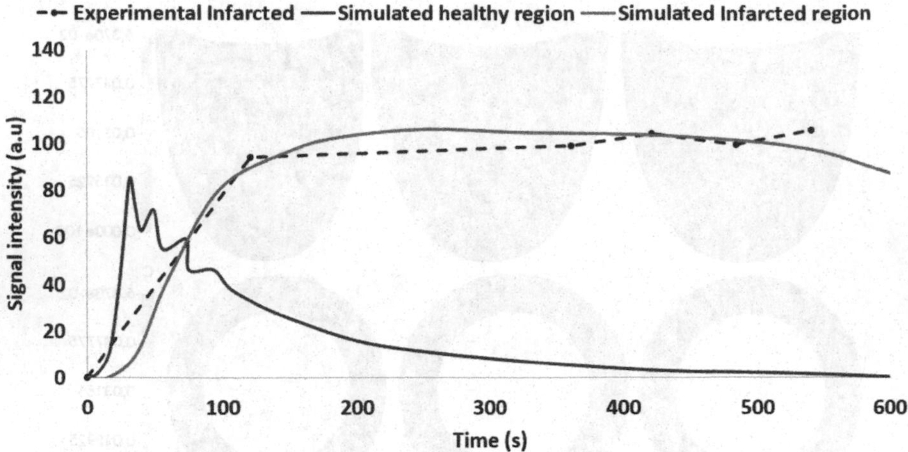

Fig. 5. Late-enhancement time evolution of CA in the fibrosis region (infarct) and a remote (healthy) one. Comparison of cases between this numerical study and clinical data from [11].

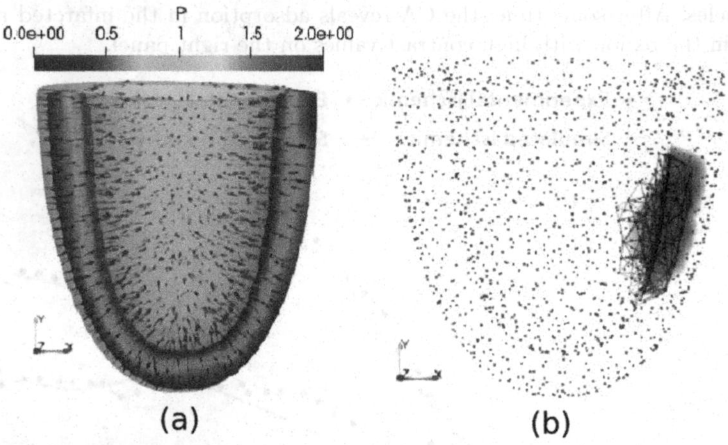

Fig. 6. Simulation results for idealized model. (a) Perfusion of CA flux induced by pressure gradient. (b) CA adsorbed at the fibrosis region (black wired) represented over the domain (dots).

handling a realistic and complex geometry. The LV mesh used for this case was processed from the geometries available at a database from Kings College London (more information is available in [14]). The parameters used were those presented in Table 1, except for $D = 0.05$, $P = 0.1$, $k_e = 0.0009$ and $\lambda = 0.75$.

Figure 7A shows the result of Darcy's problem for the patient-specific problem. The pressure gradient is imposed as boundary values and a pressure drop was used to simulate coronary artery disease. The arrows in the figure represent Darcy's flux, which is used in the convection term of the CA. Figure 7B presents the adsorption of CA in the fibrosis region (black wired). As expected, these results are similar to the ones presented in Fig. 6 for the simplified LV geometry.

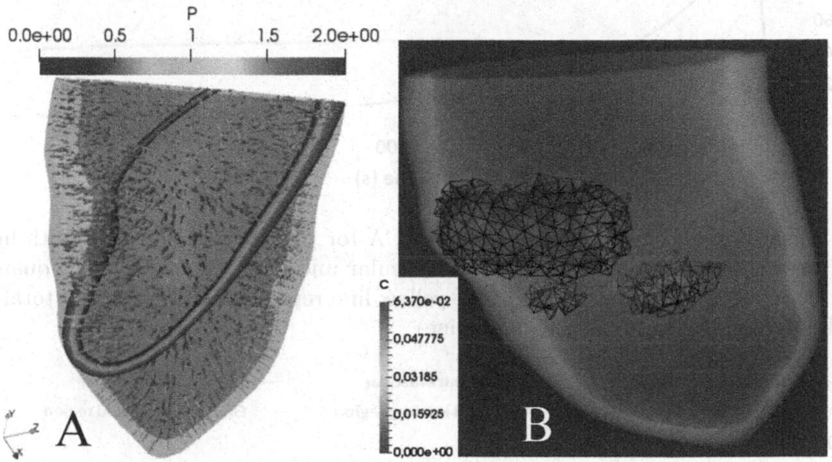

Fig. 7. Simulation results for the patient specific LV model. (a) Perfusion of CA flux induced by the pressure gradient. (b) CA adsorbed at the fibrosis region (black wired) represented over the domain.

Results for the LGE protocol are shown in Fig. 8, where we observe an accumulation of trapped CA in the fibrotic region. The results agree with the literature, demonstrating the validity of this numerical study for a patient-specific case.

Figure 9 presents a comparison between the results obtained with the simplified ellipsoidal model, the patient-specific model, and clinical data. One can observe qualitative similar dynamics. In addition, we observe that the amount of trapped CA in the infarct region is higher in the patient-specific model than in the other experiments. This is expected since the infarct region of the patient was observed to be larger than the one used in the ellipsoid-based model.

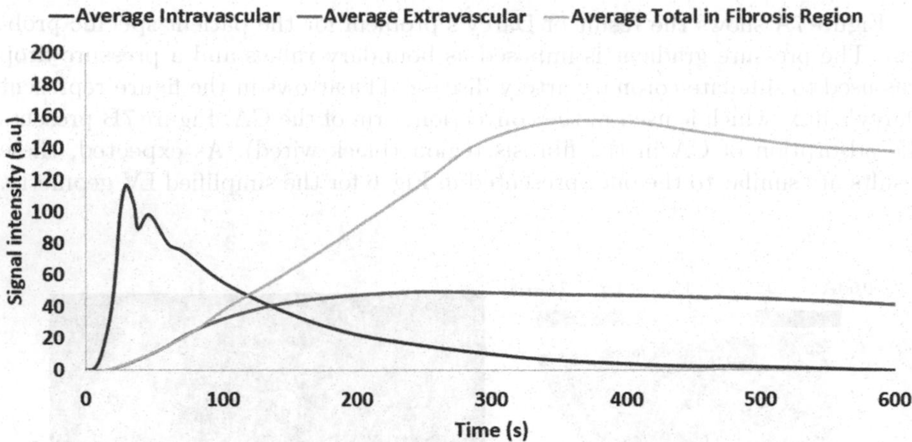

Fig. 8. Late-enhancement time evolution of CA for patient specific model with fibrosis. Blue and red line represents the intravascular and extravascular average quantity respectively in the healthy medium. The yellow line represents the average of total CA at the fibrosis domain. (Color figure online)

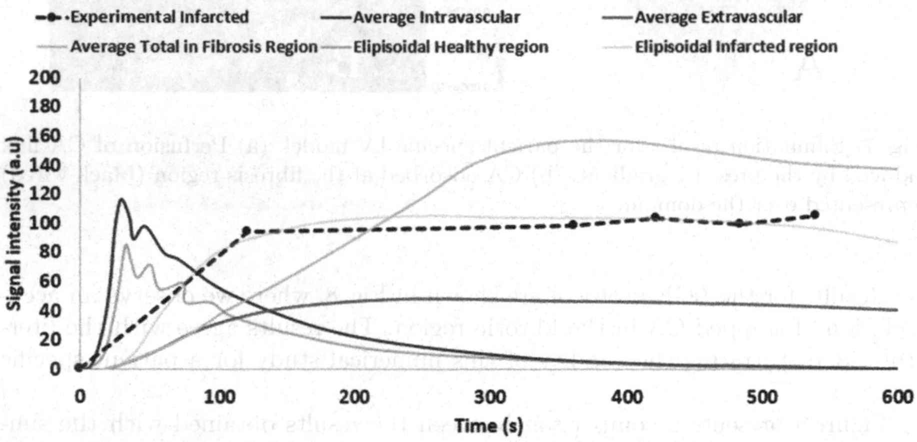

Fig. 9. Comparison of the late enhancement time evolution of CA between the simplified ellipsoidal model, the patient-specific LV model, and clinical data.

4 Conclusions

The main contribution of this study was to advance the computational models of cardiac perfusion and, more specifically, to apply them in realistic scenarios, such as clinical studies involving patient-specific models. For this goal, the mathematical model presented in [1] was used since it was able to reproduce clinical data.

Numerical experiments using a simplified 3D LV model were first performed to validate the mathematical model and the numerical approach since they were only tested in 2D cases before. The obtained results reproduced with great accuracy the clinical data. Therefore, our mathematical model and numerical methods are suitable for 3D realistic simulations of cardiac perfusion under different pathological conditions, such as ischemia and infarct.

Finally, we developed a patient-specific LV model for the description of cardiac perfusion. The precise shapes of the left ventricle and of the scar of the patient were obtained from MRI data. Distance and scars were evaluated and marked, and mesh operations were performed for the numerical optimization of the presented studies. The simulations of the 3D patient-specific model were performed and the results were found to be in agreement with the literature and with the available experimental data.

In the near future, we expect to extend the presented framework for the modeling of cardiac perfusion to other patients and to other cardiac diseases.

Acknowledgements. This work was supported by UFJF, CAPES, CNPq (Grants 310722/2021-7, 315267/2020-8), and FAPEMIG (Grants APQ-01340-18, APQ 02489/21).

References

1. Alves, J.R., de Queiroz, R.A., Bär, M., Dos Santos, R.W.: Simulation of the perfusion of contrast agent used in cardiac magnetic resonance: a step toward noninvasive cardiac perfusion quantification. Front. Physiol. **10**, 177 (2019)
2. Alves, J., de Queiroz, R., dos Santos, R.: Simulation of cardiac perfusion by contrast in the myocardium using a formulation of flow in porous media. J. Comput. Appl. Math. **295**, 13–24 (2016)
3. Arai, A.E.: The cardiac magnetic resonance (CMR) approach to assessing myocardial viability. J. Nucl. Cardiol. **18**(6), 1095–1102 (2011)
4. Bassingthwaighte, J., Wang, C., Chan, I.: Blood-tissue exchange via transport and transformation by capillary endothelial cells. Circ. Res. **65**(4), 997–1020 (1989)
5. Bayer, J.D., Blake, R.C., Plank, G., Trayanova, N.A.: A novel rule-based algorithm for assigning myocardial fiber orientation to computational heart models. Ann. Biomed. Eng. **40**(10), 2243–2254 (2012)
6. Brooks, A.N., Hughes, T.J.: Streamline Upwind/Petrov-galerkin formulations for convection dominated flows with particular emphasis on the incompressible Navier-Stokes equations. Comput. Methods Appl. Mech. Eng. **32**(1–3), 199–259 (1982)
7. Cookson, A.N., et al.: A spatially-distributed computational model to quantify behaviour of contrast agents in MR perfusion imaging. Med. Image Anal. **18**(7), 1200–1216 (2014)
8. Costa, C.M., et al.: Pacing in proximity to scar during cardiac resynchronization therapy increases local dispersion of repolarization and susceptibility to ventricular arrhythmogenesis. Heart Rhythm **16**(10), 1475–1483 (2019)
9. Daly, C., Kwong, R.Y.: Cardiac MRI for myocardial ischemia. Methodist Debakey Cardiovasc. J. **9**(3), 123 (2013)
10. Duran, O., Devloo, P.R., Gomes, S.M., Valentin, F.: A multiscale hybrid method for Darcy's problems using mixed finite element local solvers. Comput. Methods Appl. Mech. Eng. **354**, 213–244 (2019)

11. Knowles, B.R., et al.: Pharmacokinetic modeling of delayed gadolinium enhancement in the myocardium. Magn. Reson. Med. Official J. Int. Soc. Magn. Reson. Med. **60**(6), 1524–1530 (2008)
12. Logg, A., Mardal, K.A., Wells, G.: Automated Solution of Differential Equations by the Finite Element Method: The FEniCS Book, vol. 84. Springer, New York (2012). https://doi.org/10.1007/978-3-642-23099-8
13. Mehta, P.K., Wei, J., Wenger, N.K.: Ischemic heart disease in women: a focus on risk factors. Trends Cardiovasc. Med. **25**(2), 140–151 (2015)
14. Mendonca, C.C., Neic, A., et al.: A virtual cohort of twenty-four left-ventricular models of ischemic cardiomyopathy patients. King's College London Dataset (2020)
15. Michler, C., et al.: A computationally efficient framework for the simulation of cardiac perfusion using a multi-compartment Darcy porous-media flow model. Int. J. Numer. Methods Biomed. Eng. **29**(2), 217–232 (2013)
16. Neic, A., Gsell, M.A., Karabelas, E., Prassl, A.J., Plank, G.: Automating image-based mesh generation and manipulation tasks in cardiac modeling workflows using Meshtool. SoftwareX **11**, 100454 (2020)
17. Niederer, S.A., et al.: Creation and application of virtual patient cohorts of heart models. Philos. Trans. R. Soc. A Math. Phys. Eng. Sci. **378**(2173), 20190558 (2020)
18. Taylor, C.A., Fonte, T.A., Min, J.K.: Computational fluid dynamics applied to cardiac computed tomography for noninvasive quantification of fractional flow reserve: scientific basis. J. Am. Coll. Cardiol. **61**(22), 2233–2241 (2013)
19. Wang, J., et al.: Collateral circulation formation determines the characteristic profiles of contrast-enhanced MRI in the infarcted myocardium of pigs. Acta Pharmacologica Sinica **36**(4), 463–472 (2015)

Dense Temporal Subgraphs in Protein-Protein Interaction Networks

Riccardo Dondi[1], Mohammad Mehdi Hosseinzadeh[1(✉)], and Italo Zoppis[2]

[1] Università degli Studi di Bergamo, Bergamo, Italy
{riccardo.dondi,m.hosseinzadeh}@unibg.it

[2] Università degli Studi di Milano-Bicocca, Milan, Italy
zoppis@disco.unimib.it

Abstract. Temporal networks have been successfully applied to represent the dynamics of protein-protein interactions. In this paper we focus on the identification of dense subgraphs in temporal protein-protein interaction networks, a relevant problem to find group of proteins related to a given functionality. We consider a drawback of an existing approach for this problem that produce large time intervals over which temporal subgraphs are defined. We propose a problem to deal with this issue and we design (1) an exact algorithm based on dynamic programming which solves the problem in polynomial time and (2) a heuristic, based on a segmentation of the time domain and the computation of a refinement. The experimental results we present on seven protein-protein interaction networks show that in many cases our heuristic is able to reduce the time intervals with respect to those computed by the existing methods.

Keywords: Network mining · Temporal graphs · Protein protein interaction networks · Densest subgraphs · Algorithms

1 Introduction

The interactions between biological elements (genes or proteins, for example) are usually represented and analysed with a network/graph. Recently, the research focus has shifted to the evolution over time of the interactions. A model introduced to represent this evolution is that of temporal network (or temporal graph) [11,14]. This model enriches the classical graph one by defining when edges are active over a discrete sequence of timestamps. The analysis of temporal networks has provided valuable insights about their structure and properties, notable examples being community detection [4,8,16] and frequent subgraph discovery [15,19].

Protein-Protein Interaction (PPI) networks have been deeply studied in bioinformatics. In a PPI network, vertices represent gene coding proteins, edges interaction among proteins (for example co-expressed protein pairs). Since interactions among proteins change over time, their dynamic evolution has been analyzed considering temporal networks [9].

D. Groen et al. (Eds.): ICCS 2022, LNCS 13351, pp. 469–480, 2022.
https://doi.org/10.1007/978-3-031-08754-7_53

One of the most relevant problem in (static or temporal) networks is the identification of cohesive subgraphs. In PPI networks, cohesive subgraphs represent protein complexes involved in a biological process. Several models of cohesive subgraphs have been considered in literature, like cliques, s-plexes or s-clubs. A central model of cohesive subgraph applied in literature is that of densest subgraph, that defines a cohesive subgraph as one that maximizes the density (ratio between the number of edges and the number of vertices). The densest subgraph problem can be solved in polynomial-time with Goldberg's algorithm [10] and it can also be approximated within factor $\frac{1}{2}$ in linear-time [1,3].

In many network mining applications, finding a single dense subgraph doesn't provide enough information on the network structure. Therefore, other approaches have been proposed, a notable example being Top-k-Overlapping Densest Subgraphs, introduced in [7,17]. This problem asks for a set of k densest subgraphs, with $k \geq 1$, that may share vertices. The problem aims at maximizing an objective function that contains the sum of densities of the k subgraphs and a distance between them. The Top-k-Overlapping Densest Subgraphs problem has been recently applied to biological networks [12] and to dual networks [6].

The identification of densest subgraphs in temporal networks has been considered to analyze social networks in [17,18] has a key to identify interesting episodes, like groups of users of a platform highly interacting in a temporal interval. In [17,18] it is introduced the k-Densest-Temporal-Subgraphs problem whose goal is the identification of $k \geq 2$ densest temporal subgraphs in disjoint time intervals. In [17,18], it is shown that the problem can be solved in polynomial time via dynamic programming, and Goldberg's algorithm. However, due of its time complexity, the algorithm is not applicable even for medium-size temporal networks [17], hence heuristics and approximation algorithms have been considered [2,5,17].

One of the drawbacks of the k-Densest-Temporal-Subgraphs approach propsoed in [17] is that optimal solutions for the problem usually induce a segmentation, that is each timestamp of the time domain belongs to an interval of the solution. It follows that a solution of k-Densest-Temporal-Subgraphs: (1) May contain temporal subgraphs that are defined over large intervals; (2) Some vertices or edges may incorrectly be included in a subgraph of the solution.

Here we consider an approach (called k-Densest-Temporal-Refinement) that aims at correcting this drawback of the k-Densest-Temporal-Subgraphs model and that aims at identifying whether there exist temporal graphs of high density and defined over smaller intervals with respect to the ones of a solution of k-Densest-Temporal-Subgraphs. The idea is that if a dense temporal subgraph contains a temporal subgraph active in a smaller interval and with a close density, then this latter better represents the real cohesive group of elements.

We first present a dynamic programming algorithm that is able to solve the problem in polynomial time. However, the time complexity of this algorithm makes it not applicable even on small size graphs. Thus we design a heuristic for the problem and we present an experimental evaluation on a set of seven PPI networks. The experimental results show that in many cases our approach is able to reduce the length of the intervals, while maintaining the density significantly high.

The paper is organized as follows. First in Sect. 2 we give the definitions and we introduce the problems we are interested into. Then, in Sect. 3 we present a dynamic programming algorithm for the k-Densest-Temporal-Refinement problem. In Sect. 4 we present an efficient heuristic for the problem, while in Sect. 5 we present an experimental evaluation on seven PPI temporal networks.

2 Definitions

A time domain $\mathcal{T} = [t_1, t_2, \ldots, t_z]$, where $t_1 < t_2 < \cdots < t_z$, is a sequence of positive integer, each one called timestamp. An interval $[t_i, t_j]$ in $\mathcal{T} = [t_1, t_2, \ldots, t_z]$ is the sequence of timestamps between t_i, t_j, with $t_i \leq t_j$. $T' = [t_a, t_b]$ is a subinterval of $T = [t_c, t_d]$, denoted by $T' \subseteq T$, if $t_c \leq t_a \leq t_b \leq t_d$ and $t_c < t_a$ or $t_b < t_d$. Now, we can introduce the definition of temporal graphs.

Definition 1. *$G = (V, \mathcal{T}, E)$ is a temporal graph, where V is a set of vertices, $\mathcal{T}$ is a time domain and E is a set of triple $\{u, v, t\}$, where $u, v \in V$ and $t \in \mathcal{T}$.*

In the following we denote by z the number of timestamps in $\mathcal{T}$ (that is $z = |\mathcal{T}|$), by n the number of vertices in V (that is $n = |V|$) and by $|m|$ the number of edges in E (that is $m = |E|$). We give now the definition of induced temporal subgraph.

Definition 2. *Given a temporal graph $G = (V, \mathcal{T}, E)$, an induced temporal subgraph of G is defined as a pair $G[T, V']$, where*

- *T is an interval in $\mathcal{T}$ over which the induced temporal subgraph is defined*
- *$V' \subseteq V$ is the set of vertices of the induced temporal graph.*

Notice that the edge set of $G[T, V']$, denoted by $E(G[T, V'])$, is defined as follows:

$$E(G[T, V']) = \{\{u, v, t\} : u, v \in V', t \in T\}.$$

Given an induced temporal subgraph $G[T, V']$ and a timestamp t that belongs to T, we say that $G[T, V']$ covers t. The density of an induced subgraph $G[T, V']$, denoted by $dens(G[T, V'])$, is defined as follows:

$$dens(G[T, V']) = \frac{|E(G[T, V'])|}{|V'|}.$$

Given an interval I in $\mathcal{T}$, we denote by $MaxD(I)$ the density of a densest subgraph of G defined in interval I.

The first problem we consider, called k-Densest-Temporal-Subgraphs, introduced in [17], is defined as follows.

Problem 1. k-Densest-Temporal-Subgraphs
Input: A temporal graph $G = (V, \mathcal{T}, E)$, a positive integer $k \geq 2$.
Output: A set S of k temporal subgraphs, $S = \{G[T_1, V_1], \ldots, G[T_k, V_k]\}$, such that $T_1, \ldots, T_k$ are disjoint intervals and

$$\sum_{i=1}^{k} dens(G[T_i, V_i])$$

is maximized.

As discussed in [17], the density of an interval is a non decreasing function with respect to the length of an interval. Thus there exists an optimal solution of k-Densest-Temporal-Subgraphs that induces a segmentation of the time domain, that is each timestamp of the time domain $\mathcal{T}$ is covered by exactly one temporal subgraph of the solution. The methods proposed in literature [5,17] compute indeed solutions that induce a segmentation, and they may contain subintervals that gives a small contribution to the density.

Here we propose an approach to possibly shrink interval length. We start by introducing the definition of ε-refinement.

Definition 3. *Given an induced temporal subgraph $G[T, U]$ of a temporal graph $G = (V, \mathcal{T}, E)$, $G[T', U']$ is an ε-refinement of $G[T, U]$ if*

1. *T' is a subinterval of T*
2. *$dens(G[T', U']) \geq (1 - \varepsilon)\ dens(G[T, U])$*
3. *There is no ε-refinement of $G[T, U]$ in subintervals of T'*

Notice that by Point 3 of Definition 3, if $G[T', U']$ is an ε-refinement of $G[T, U]$ it is not possible to further reduce the interval T' in order to obtain an ε-refinement of $G[T, U]$.

The definition of ε-refinement can be extended to ε-complete refinement of a solution of k-Densest-Temporal-Subgraphs.

Definition 4. *Given a temporal graph $G = (V, \mathcal{T}, E)$, an ε-complete refinement $RF = \{R_1, \ldots, R_q\}$, where each R_i, $1 \leq i \leq q$, is a quadruple (V_i, T_i, V_i', T_i') such that $T_1, \ldots, T_q$ is a segmentation of $\mathcal{T}$, $G[V_i, T_i]$ is a densest subgraph in interval T_i and V_i', T_i' are defined as follows:*

1. *If there does not exist a ε-refinement of $G[V_i, T_i]$, then $V_i' = V_i$ and $T_i' = T_i$ and the profit of R_i, denoted by $p(R_i)$, is defined as $p(R_i) = (1 - \varepsilon)\ dens(G[T_i, V_i])$*
2. *If there exists an ε-refinement of $G[T_i, V_i]$, then $G[T_i', V_i']$ is an ε-refinement of $G[T_i, V_i]$ of maximum density; the profit of R_i, denoted by $p(R_i)$, is defined as $p(R_i) = dens(G[T_i', V_i'])$*

The profit of a ε-complete refinement of $R_1, \ldots, R_q$ of $G_1, \ldots, G_q$ is:

$$\sum_{i}^{k} p(R_i).$$

Problem 2. k-Densest-Temporal-Refinement
Input: A temporal graph $G = (V, \mathcal{T}, E)$, a positive integer $k \geq 2$.
Output: An ε-complete refinement RF having maximum density.

Algorithms for Computing a Densest Subgraph

A densest subgraph in a given static graph can be computed in $O(mn \log n)$ time by reducing the problem to min-cut with Goldberg's algorithm [10]. When $mn \leq n^3$, the time complexity of Goldberg's algorithm for unweighted graphs has been improved to $O(n^3)$ [13]. Finding a densest subgraph can be approximated within factor $\frac{1}{2}$ in linear time with a greedy algorithm (here called Charikar's algorithm) [3].

3 An Exact Algorithm for k-Densest-Temporal-Refinement

In this section, we present a dynamic programming algorithm for the k-Densest-Temporal-Refinement problem. First, define $P(j, h)$, $1 \leq j \leq z$ and $1 \leq h \leq k$, as a function that is equal to the maximum profit of an ε-complete refinement consisting of h subgraphs until timestamp $t_j \leq t_z$.

Given two timestamps t_i and t_j, with $1 \leq t_i \leq t_j \leq t_z$, we denote by $Pr([i, j])$ the maximum profit of a temporal subgraph of G defined in interval $[t_i, t_j]$. Now, $P(j, h)$ can be computed as follows:

$$P(j,h) = \begin{cases} \max_{1 \leq i < j} P(i, h-1) + Pr([i+1, j]) & \text{if } 2 \leq h \leq z \\ Pr([1, j]) & \text{if } h = 1 \end{cases} \tag{1}$$

The computation of $P(j, h)$ with Recurrence 1 requires the computation of $Pr([i, j])$, for each i and j, with $1 \leq i \leq j \leq z$. $Pr([i, j])$ can be computed as follows:

$$Pr([i,j]) = \max \begin{cases} (1-\varepsilon) MaxD([i,j]) \\ \max_{[a,b] \subseteq [i,j]} MaxD([a,b]) \end{cases} \tag{2}$$

Lemma 1. *$P(j, h) = q$ if and only if there exists an ε-complete refinement of G on interval $[1, j]$ consisting of h subgraphs of profit q.*

Proof. We prove the lemma by induction on h. In the base case, when $h = 1$, then the profit of a single temporal subgraph in the interval $[t_1, t_j]$ is equal to $Pr[1, j]$.

Assume that the lemma holds for each $1 \leq i < j$, we prove that it holds for j. Assume that $P(j, h) = q$, then by Recurrence 1 there must exists a value $i < j$ such that (1) $P(i, h-1) = q_1$ (2) $Pr([i+1, j]) = q_2$, where $q_1 + q_2$. By induction hypothesis there exists an ε-complete refinement consisting of $h-1$ subgraphs in interval $[1, i]$ of profit q_1 and by definition of Pr, there exists a subgraph in $[i+1, j]$ of profit q_2.

Assume that there exists an ε-complete refinement S consisting of h subgraphs in interval $[1, j]$ of profit q. Assume that among the subgraphs in S, the subgraph defined in the leftmost interval $[i+1, j]$ has profit $Pr([i+1, j]) = q_2$. Then there exists an ε-complete refinement consisting of $h-1$ subgraphs in interval $[1, i]$ of profit q_1, with $q_1 + q_2 = q$. By induction hypothesis, it holds that $P(i, h-1) = q_1$ and, by the first case of Recurrence 1, it holds that $P(j, h) = q$. □

Theorem 1. k-Densest-Temporal-Refinement *can be solved in* $O(z^4 mn \log(n))$ *time.*

Proof. By Lemma 1 it follows that an optimal solution of k-Densest-Temporal-Refinement on instance $G = (V, E, T)$ is equal to $P(z, k)$. Now, $P(j, h)$, with $1 \le j \le z$ and $1 \le h \le k$, consists of zk entries. Each entry $P(j, h)$ can be computed in time $O(z)$, assuming the values $Pr([i+1, j])$ are known. Thus, since $k \le z$, the entries $P(j, h)$, with $1 \le j \le z$ and $1 \le h \le k$, can be computed in $O(z^3)$ time, assuming the values $Pr([i+1, j])$ are known.

The values $Pr([i, j])$ are $O(z^2)$ and each value can be computed by applying the Goldberg's algorithm on the graph active in interval in $[a, b]$ with $[a, b]$ a subinterval of $[i, j]$. This requires, for each i, j, the computation of $O(z^2)$ densest subgraph via Goldberg's algorithm, so requiring time $O(z^2 mn \log n)$. The overall time complexity to compute an ε-complete refinement is then $O(z^4 mn \log n)$. □

4 A Heuristic of k-Densest-Temporal-Refinement

While k-Densest-Temporal-Refinement is solvable in polynomial time with the dynamic programming algorithm given in Sect. 3, the computational complexity of this algorithm makes it not practical even for small size temporal networks. Indeed, we applied it to the temporal PPI networks described in Sect. 5 and it was not able to produce any result within 10 h.

Thus, we have designed an efficient heuristic, called Reduce, for the k-Densest-Temporal-Refinement problem. Reduce first computes a solution $S = \{G[T_1, V_1], G[T_2, V_2], \ldots, G[T_k, V_k]\}$ of the k-Densest-Temporal-Subgraphs problem, by applying the method described in [5]. Notice that the temporal graphs in S induce a segmentation, of the time domain and that the solution S may not be an optimal solution of k-Densest-Temporal-Subgraphs.

For each temporal subgraph $G[T_i, V_i]$, $1 \le i \le k$, of S, with $dens(G[T_i, V_i]) = d$, Reduce looks for an ε-refinement by computing the following three temporal subgraphs:

1. $G[T_{i,1}, V_{i,1}]$ is computed by applying Charikar's algorithm on the temporal subgraph of $G[T_i, V_i]$ obtained by removing the first and the last timestamp of T_i
2. $G[T_{i,2}, V_{i,2}]$ is computed by applying Charikar's algorithm on the temporal subgraph of $G[T_i, V_i]$ obtained by removing the first timestamp of T_i

3. $G[T_{i,3}, V_{i,3}]$ is computed by applying Charikar's algorithm on the temporal subgraph of $G[T_i, V_i]$ obtained by removing the last timestamp of T_i

Now, if $G[T_{i,1}, V_{i,1}]$ has density at least $(1 - \varepsilon)d$ (hence is an ε-refinement of $G[T_i, V_i]$), the three previous steps are applied on $G[T_{i,1}, V_{i,1}]$. If $G[T_{i,1}, V_{i,1}]$ has density smaller then $(1-\varepsilon)d$, and one of $G[T_{i,2}, V_{i,2}]$ $G[T_{i,3}, V_{i,3}]$ has density at least $(1 - \varepsilon)d$ (thus there exists an ε-refinement of $G[T_i, V_i]$), the subgraphs having largest density between $G[T_{i,2}, V_{i,2}]$ $G[T_{i,3}, V_{i,3}]$ is selected and the three steps are applied on it.

If none of $G[T_{i,1}, V_{i,1}]$, $G[T_{i,2}, V_{i,2}]$, $G[T_{i,3}, V_{i,3}]$ have density at least $(1-\varepsilon)d$, the algorithm defines V'_i, T'_i in $R_i = (V_i, T_i, V'_i, T'_i)$ as follows: $V'_i = V_i$ and $T'_i = T_i$.

5 Experimental Analysis

In this section, we present the experimental results for our heuristics (Reduce) on real-world datasets. Since the real-world datasets we consider are over a limited number of timestamps (i.e. 36) we have defined small values of k, that is $k = 2$ and $k = 3$, in the k-Densest-Temporal-Refinement problem. We implemented Reduce in Python on MacBook-Pro (OS version 12.0.1) with processor 2.9 GHz Intel Core i5 and 8 GB 2133 MHz LPDDR3 of RAM, Intel Iris Graphics 550 1536 MB.

Table 1. PPI temporal networks informations

PPINs temporal networks	# Vertices	# Edges	# Timestamps
DPPIN-Babu	5,003	111,466	36
DPPIN-Gavin	2,541	140,040	36
DPPIN-Hazbun	143	1,959	36
DPPIN-Ho	1,548	42,220	36
DPPIN-Ito	2,856	8,638	36
DPPIN-Krogan(LCMS)	2,211	85,133	36
DPPIN-Lambert	697	6,654	36

For testing the performances of Reduce, we have considered seven PPI temporal networks, taken from [9]. The characteristics of the networks are summarized in Table 1. Notice that all the temporal networks we considered are defined over a time domain consisting of 36 timestamps, while the number of vertices and the number of edges vary significantly (the number of vertices ranges from 143 to 5003 and the number of edges ranges from 1959 to 140040).

In Table 2 we report the densities and the intervals of the solutions of k-Densest-Temporal-Subgraphs computed by Reduce (called **first phase**) and the refinement solutions, where $\varepsilon = 0.05$, computed by Reduce (called **second phase**). We report also the Jaccard's index similarity[1] of the solutions returned by the two phases of Reduce in order to measure the fraction of shared vertices.

Table 2 shows that for $k = 2$, when Reduce finds an ε-refinement, the subgraph computed by the second phase of Reduce has a Jaccard index of at least 0.94. Similar results are obtained for $k = 3$. Thus the ε-refinement computed in the second phase are very similar to the subgraphs computed in the first phase. On the other hand, there is a significant reduction in the length of the intervals over which the subgraphs returned by the second phase are defined, as discussed next.

In Table 3 we present, for each value of k and for each temporal network considered, the ratio between the densities and the interval lengths of the solutions returned by Phase 2 and Phase 1 of Reduce (if Phase 2 produces an ε-refinement). The results show that the ε-refinements significantly reduce the interval length. For instance for $k = 2$, the ε-refinements computed are defined over intervals of length at most 75% and at least 31% of the interval length of first phase solutions. For $k = 3$, the interval length of ε-refinement solutions are at most 81% and at least 9% of the interval length of first phase solutions.

In Table 4, we report the profit of the solutions of the k-Densest-Temporal-Refinement problem. We report in bold the densities for the cases where Phase 2 is able to computed an ε-refinement, while for the other cases the profit is $(1-\varepsilon)$ of the density of the subgraph returned by Phase 1.

Finally, we evaluate, for each network considered, the minimum value of ε so that Phase 2 of Reduce is able to compute an ε-refinement. In Table 5 we report the ratios between the densities of first and second phases of Reduce by varying ε from 5% to 37% (in bold the ε-refinement when $\varepsilon = 0.05$). Notice that, while the ε-refinements have a ratio close to 1 (for 12 of the 17 0.05-refinements computed the ratio is at least 0.99), for the cases where Reduce is not able to computed an ε-refinement the ratio is always smaller than 0.9. Furthermore, for three networks (DPPIN-Hazbun, DPPIN-Ito, DPPIN-Lambert), the second phase of Reduce was never able to compute an ε-refinement, while for the other networks this happens in at most one case. The three networks DPPIN-Hazbun, DPPIN-Ito, DPPIN-Lambert are also (considered as static graphs) those having lower density (they have a density of at most 13.7, the other graphs a density of at least 22). This suggest that the existence of an ε-refinement may be related to the density of the input graph.

[1] Jaccard's index measures the similarity between two sets by taking the ratio of intersection over union of the two sets.

Table 2. Refinement solution for $\varepsilon = 0.05$, $k = 2$ and $k = 3$ of the first phase.

Datasets		k = 2		k = 3		
DPPIN-Babu	**first phase**					
	density	78.58	75.99	78.58	60.87	72.15
	interval	(0,15)	(16,35)	(0,8)	(9,27)	(28,35)
	second phase					
	density	78.58	72.16	78.58	–	72.13
	interval	(0,8)	(23,31)	(8,8)	–	(28,31)
	Jaccard index	1	0.98	1	–	0.99
DPPIN-Gavin	**first phase**					
	density	123.41	119.74	123.38	81.07	114.33
	interval	(0,11)	(12,35)	(0,10)	(11,24)	(25,35)
	second phase					
	density	123.38	114.45	123.38	–	114.29
	interval	(0,8)	(23,31)	(0,8)	–	(25,31)
	Jaccard index	1	0.94	0.99	–	1
DPPIN-Hazbun	**first phase**					
	density	19.52	19.36	18.75	15.05	16.33
	interval	(0,13)	(14,35)	(0,10)	(11,20)	(21,35)
	second phase					
	density	–	–	–	–	–
	interval	–	–	–	–	–
	Jaccard index	–	–	–	–	–
DPPIN-Ho	**first phase**					
	density	39.94	39.25	38.91	33.97	38.83
	interval	(0,19)	(20,35)	(0,9)	(10,24)	(25,35)
	second phase					
	density	38.86	38.80	38.91	33.77	38.74
	interval	(0,8)	(31,35)	(5,8)	(18,19)	(6,6)
	Jaccard index	0.94	0.98	1	0.97	1
DPPIN-Ito	**first phase**					
	density	5.01	4.48	4.27	4.48	4.13
	interval	(0,20)	(21,35)	(0,12)	(13,24)	(25,35)
	second phase					
	density	–	–	–	–	–
	interval	–	–	–	–	–
	Jaccard index	–	–	–	–	–
DPPIN-Krogan (LCMS)	**first phase**					
	density	96.82	97.48	96.78	64.98	92.24
	interval	(0,15)	(16,35)	(0,10)	(11,30)	(31,35)
	second phase					
	density	96.71	92.47	96.71	–	92.14
	interval	(0,8)	(22,31)	(0,8)	–	(31,31)
	Jaccard index	1	0.95	1	–	0.99
DPPIN-Lambert	**first phase**					
	density	19.23	20.0	17.89	16.61	15.47
	interval	(0, 15)	(16, 35)	(0, 11)	(12, 24)	(25, 35)
	second phase					
	density	–	–	–	–	–
	interval	–	–	–	–	–
	Jaccard index	–	–	–	–	–

Table 3. Ratio between the densities and the interval lengths of the solutions returned by Phase 1 and Phase 2 of Reduce for $\varepsilon = 0.05$

Datasets		k = 2		k = 3		
DPPIN-Babu	density	1	0.95	1	–	0.99
	interval	0.56	0.45	0.11	–	0.5
DPPIN-Gavin	density	0.99	0.96	1	–	0.99
	interval	0.75	0.33	0.81	–	0.64
DPPIN-Hazbun	density	–	–	–	–	–
	interval	–	–	–	–	–
DPPIN-Ho	density	0.97	0.99	1	0.99	0.97
	interval	0.45	0.31	0.4	0.13	0.09
DPPIN-Ito	density	–	–	–	–	–
	interval	–	–	–	–	–
DPPIN-Krogan (LCMS)	density	0.99	0.95	0.99	–	0.99
	interval	0.56	0.5	0.81	–	0.2
DPPIN-Lambert	density	–	–	–	–	–
	interval	–	–	–	–	–

Table 4. Profit of the solutions for $\varepsilon = 0.05$ of the k-Densest-Temporal-Refinement problem

Datasets	K = 2		K = 3		
DPPIN-Babu	**78.58**	**72.16**	**78.58**	57.83	**72.13**
DPPIN-Gavin	**123.38**	**114.45**	**1**	77.02	**114.29**
DPPIN-Hazbun	18.54	18.39	17.81	14.30	15.51
DPPIN-Ho	**38.86**	**38.80**	**38.91**	**33.77**	**38.74**
DPPIN-Ito	4.76	4.26	4.06	4.26	3.92
DPPIN-Krogan (LCMS)	**96.71**	**92.47**	**96.71**	61.73	**92.14**
DPPIN-Lambert	18.27	19	16.99	15.78	14.70

Table 5. Ratio between the solutions (densities) of the first and second phases - varying ε from from 5% to 37% for refinement

Datasets	K = 2		K = 3		
DPPIN-Babu	**1**	**0.95**	**1**	0.88	**0.99**
DPPIN-Gavin	**0.99**	**0.96**	**1**	0.88	**0.99**
DPPIN-Hazbun	0.83	0.75	0.86	0.87	0.89
DPPIN-Ho	**0.97**	**0.99**	**1**	**0.99**	**0.97**
DPPIN-Ito	0.78	0.83	0.81	0.81	0.81
DPPIN-Krogan (LCMS)	**0.99**	**0.95**	**0.99**	0.86	**0.99**
DPPIN-Lambert	0.73	0.65	0.77	0.63	0.72

6 Conclusion

We have presented a problem for finding dense subgraphs in temporal networks, with an application to protein-protein interactions. We have designed an exact algorithm based on dynamic programming, that solves the problem in polynomial time, but it is not practical even for small size datasets. We have designed a heuristic, based on a segmentation of the time domain and the computation of a refinement, that allows us to produce dense subgraphs and to reduce the length of the intervals.

Future works include an extension of the experimental part to other PPI networks in order to verify if properties of the networks are related to the existence of an ε-refinement. It would be interesting to understand if the dense subgraphs identified are related to some functionality and, more generally, a biological analysis of the inferred temporal subgraphs. Finally, from an algorithmic point it would be interesting to design other heuristics for the problem.

References

1. Asahiro, Y., Iwama, K., Tamaki, H., Tokuyama, T.: Greedily finding a dense subgraph. J. Algorithms **34**(2), 203–221 (2000). https://doi.org/10.1006/jagm.1999.1062
2. Castelli, M., Dondi, R., Hosseinzadeh, M.M.: Genetic algorithms for finding episodes in temporal networks. Proc. Comput. Sci. **176**, 215–224 (2020)
3. Charikar, M.: Greedy approximation algorithms for finding dense components in a graph. In: Jansen, K., Khuller, S. (eds.) APPROX 2000. LNCS, vol. 1913, pp. 84–95. Springer, Heidelberg (2000). https://doi.org/10.1007/3-540-44436-X_10
4. Coscia, M., Giannotti, F., Pedreschi, D.: A classification for community discovery methods in complex networks. Stat. Anal. Data Min. ASA Data Sci. J. **4**(5), 512–546 (2011)
5. Dondi, R., Hosseinzadeh, M.M.: Dense sub-networks discovery in temporal networks. SN Comput. Sci. **2**(3), 1–11 (2021). https://doi.org/10.1007/s42979-021-00593-w
6. Dondi, R., Hosseinzadeh, M.M., Guzzi, P.H.: A novel algorithm for finding top-k weighted overlapping densest connected subgraphs in dual networks. Appli. Netw. Sci. **6**(1), 1–17 (2021). https://doi.org/10.1007/s41109-021-00381-8
7. Dondi, R., Hosseinzadeh, M.M., Mauri, G., Zoppis, I.: Top-k overlapping densest subgraphs: approximation algorithms and computational complexity. J. Comb. Optim. **41**(1), 80–104 (2021)
8. Fortunato, S.: Community detection in graphs. Phys. Rep. **486**(3–5), 75–174 (2010)
9. Fu, D., He, J.: Dppin: a biological repository of dynamic protein-protein interaction network data (2021). arXiv preprint, arXiv:2107.02168
10. Goldberg, A.V.: Finding a maximum density subgraph. Technical report, Berkeley, CA, USA (1984)
11. Holme, P.: Modern temporal network theory: a colloquium. The Eur. Phy. J. B **88**(9), 1–30 (2015). https://doi.org/10.1140/epjb/e2015-60657-4
12. Hosseinzadeh, M.M.: Dense Subgraphs in Biological Networks. In: Chatzigeorgiou, A., et al. (eds.) SOFSEM 2020. LNCS, vol. 12011, pp. 711–719. Springer, Cham (2020). https://doi.org/10.1007/978-3-030-38919-2_60

13. Kawase, Y., Miyauchi, A.: The densest subgraph problem with a convex/concave size function. Algorithmica **80**(12), 3461–3480 (2017). https://doi.org/10.1007/s00453-017-0400-7
14. Kempe, D., Kleinberg, J., Kumar, A.: Connectivity and inference problems for temporal networks. J. Comput. Syst. Sci. **64**(4), 820–842 (2002)
15. Kovanen, L., Karsai, M., Kaski, K., Kertész, J., Saramäki, J.: Temporal motifs in time-dependent networks. J. Stat. Mech: Theory Exp. **2011**(11), P11005 (2011)
16. Rossetti, G., Cazabet, R.: Community discovery in dynamic networks: a survey. ACM Comput. Surv. (CSUR) **51**(2), 35 (2018)
17. Rozenshtein, P., Bonchi, F., Gionis, A., Sozio, M., Tatti, N.: Finding events in temporal networks: segmentation meets densest subgraph discovery. Knowl. Inf. Syst. **62**, 1611–1639 (2019)
18. Rozenshtein, P., Gionis, A.: Mining temporal networks. In: Proceedings of the 25th ACM SIGKDD International Conference on Knowledge Discovery & Data Mining, pp. 3225–3226. ACM (2019)
19. Wackersreuther, B., Wackersreuther, P., Oswald, A., Böhm, C., Borgwardt, K.M.: Frequent subgraph discovery in dynamic networks. In: Proceedings of the Eighth Workshop on Mining and Learning with Graphs, pp. 155–162. ACM (2010)

Accelerating Edge Metagenomic Analysis with Serverless-Based Cloud Offloading

Piotr Grzesik(✉) and Dariusz Mrozek

Department of Applied Informatics, Silesian University of Technology, ul. Akademicka 16, 44-100 Gliwice, Poland
pj.grzesik@gmail.com, dariusz.mrozek@polsl.pl

Abstract. Third-generation nanopore sequencing technologies, along with portable devices such as MinION Nanopore and Jetson Xavier NX, allow performing cost-effective metagenomic analysis in a portable manner. At the same time, we observe the growth of the serverless computing paradigm that offers high scalability with limited maintenance overhead for the underlying infrastructure. Recent advancements in serverless offerings make it a viable choice for performing operations such as basecalling. This paper aims to evaluate if a combination of edge and serverless computing paradigms can be successfully used to perform the basecalling process, with the focus on acceleration of offline edge-based processing with serverless-based infrastructure. For the purposes of the experiments, we proposed a workflow in which DNA sequence reads are processed simultaneously at the edge with Jetson Xavier NX and in the cloud with AWS Lambda in different network conditions. The results of our experiments show that with such a hybrid approach, we can reduce the processing time and energy consumption of the basecalling process compared to fully offline or fully online processing. We also believe that while so far, the adoption of serverless computing for bioinformatic applications is not high, the recent improvements to platforms such as AWS Lambda make it a compelling choice for an increasing number of bioinformatics workflows.

Keywords: Nanopore sequencing · Edge computing · Edge analytics · Bioinformatics · Jetson Xavier NX · Cloud computing · Metagenomics · Serverless computing · AWS Lambda

1 Introduction

In recent years, we have seen the fast growth of the popularity of third-generation sequencing technologies. These technologies allow performing metagenomics analysis in a cost-effective manner, thanks to devices such as MinION Nanopore. MinION Nanopore is a sequencing device released by Oxford Nanopore Technologies (ONT), which, due to its small dimensions, weight, and costs, enables portable analysis in mobile laboratories, helping out with monitoring Ebola virus outbreak in Kenya [19], Lassa virus outbreak in Nigeria [23], performing early detection of

D. Groen et al. (Eds.): ICCS 2022, LNCS 13351, pp. 481–492, 2022.
https://doi.org/10.1007/978-3-031-08754-7_54

plant viruses in Africa [9], or monitoring sewage [8]. It even has been used on International Space Station [11] or during an ice cap traverse expeditions [15].

The use cases as listed above are taking advantage of edge computing paradigm[30], which enables processing data closer to its source, reduces the amount of data that needs to be sent to the cloud, and provides resilience in situations where Internet connection is unreliable or even unavailable at times. However, the metagenomics analysis at the edge is problematic due to the limited access to computational power at edge devices. A lot of the popular bioinformatics tools are written with multi-node clusters in mind, and they require significant computing power to run processing successfully. Based on the official documentation [4], the MinION device can produce up to 15 GB of data per day, which makes it challenging to take advantage of cloud computing power in cases where network connectivity is unstable, and network throughput is limited. Another challenge in such field applications is access to a reliable power supply. Due to that, it is essential to preserve energy and take advantage of techniques that allow maintaining a sustainable ratio of computational power to energy consumption. In our previous paper [17], we have determined that devices such as Jetson Xavier NX can be successfully used for portable metagenomics. However, performing such analysis in real-time is challenging if the experiments require increased accuracy.

At the same time, we observe the growing popularity of the serverless computing paradigm, which commercially started with the release of AWS Lambda [22] in 2014. The serverless paradigm allows reducing infrastructure maintenance in comparison to cloud servers or virtual machines while at the same time providing a highly scalable execution environment that supports parallel processing very well. While mainly being adopted for use cases such as Web APIs, based on our previous research [16], we can see that serverless computing is also gaining popularity for bioinformatic workflows. In one of our recent works [18], we have validated that a serverless-based solution can be successfully used for performing basecalling of nanopore sequencing data.

This paper aims to evaluate if and how a combination of edge and serverless computing paradigms can be successfully used for performing metagenomics analysis and what benefits such a hybrid approach can offer compared to performing the analysis in a fully offline edge or fully online serverless manner. In particular, we focus on a use case where the fully offline edge basecalling process is accelerated with serverless-based basecalling processing if available network conditions allow for it. Throughout experiments, we aim to highlight the potential reductions in processing time and energy consumption. We believe that the low maintenance overhead and high scalability offered by the serverless paradigm can make it a good fit for this particular use case. The main motivation behind the study is to evaluate how edge-based basecalling process can be potentially speed up without sacrificing energy-efficiency and keeping in mind the constraints of edge-based deployments.

The rest of the paper is organized as follows. In Sect. 2, we review the related works in the area. Section 3 describes the testing workflow and the environment used for experiments. Section 4 focuses on the testing methodology and the results of the performance experiments. Finally, Sect. 5 provides a summary and concludes the results of the paper.

2 Related Works

Portable metagenomics analysis has been gaining more interest in world research and scientific literature in recent years. Oliva et al. [27] presented an overview and benchmarks of bioinformatic tools that can be ported and used on an Android smartphone in order to evaluate if regular smartphones are powerful enough to support portable analytics. The authors considered 23 tools, but only 11 of them were successfully ported to work on an Android device. The only basecalling software that the authors managed to port was Nanocall [13], but the paper does not include a benchmark of basecalling with Nanocall. This research suggests that a new set of tools optimized for ARM architectures will be necessary to support portable analytics on regular smartphones reliably. Another example was presented by Grzesik et al. in [17], where the authors evaluated the feasibility of using a device such as Jetson Xavier NX for performing basecalling and classification operations in an edge computing manner. The authors developed a workflow based on Guppy basecaller and Kraken2 classification software, and throughout their experiments, they determined that Jetson Xavier NX can serve as an energy-effective and performant device that can be used for running metagenomic analysis in a portable manner. Yet another case of portable metagenomics was presented by D'Agostino et al. [14]. The authors proposed hybrid edge-cloud architecture for performing cost-effective metagenomic analysis. During the experiments, they evaluated a workflow that includes basecalling and classification steps using Deepnano and Kraken software. As the edge platform, they used Intel System-on-Chip boards. Based on the performed experiments, the authors suggested that while it is possible to run metagenomic analysis directly on selected devices, none supported the data processing in real-time. Similar research has been performed by Merelli et al. [26], where the authors described a fog computing architecture, based on low-powered portable devices, aimed at performing metagenomic analysis. However, Merelli et al. focused on energy consumption aspects and concluded that the system that would have to support real-time analysis could not be powered by batteries and would require multiple computing boards to process output from a single MinION device.

On the other hand, we also observe a growing interest in research related to the use of serverless computing for bioinformatics applications. Grzesik et al. [16] presented an overview of serverless techniques used for omics data analysis. The authors referred to multiple examples that consider using serverless computing for bioinformatic workflows in their work. One of the cases they mention is an API used for simulating the DNA sequencing data, proposed by Aboukhalil [7]. That solution is based on the "wgsim" tool for simulating sequence reads based on a provided reference genome. By taking advantage of "biowasm," the author managed to compile "wgsim" to WebAssembly to successfully run it on Cloudflare Workers Unbound, which is a serverless platform. Hung et al. [20] demonstrated how serverless computing could help with reducing computation time of RNA sequencing data analysis. In their application, the authors proposed a three-step architecture with split, merge, and align steps. They identified the merge step as the best one to be potentially parallelized and accelerated with the serverless approach. During the merge step, the human transcriptome reads are aligned by using the Burrows-Wheeler Aligner [25]. In the tested case, they sharded data into 60 MB files which resulted in a workflow that employed over 1,700 serverless functions in parallel. Taking advantage of such architecture reduced the total execution time of the workflow from 2.5 h for a cloud server to 6 min when using serverless functions. Another case where serverless computing allowed for reduced computing time is sBeacon [21], a serverless implementation of the Beacon protocol, proposed and implemented by The Commonwealth Scientific and Industrial Research Organization (CSIRO). By taking advantage of AWS Lambda and AWS S3, CSIRO managed to reduce the time required to upload new genomes into the database from 33 h when using a cloud server to only 22 s. The authors also mentioned that the selected architecture improves data privacy and allows reducing the costs of the infrastructure. CSIRO is one of the leading organizations involved in adopting serverless for bioinformatic workflows. In addition to sBeacon, it also proposed Serverless Variant Effect Predictor (sVEP) used for genomic variants prediction. Thanks to parallelization of the workflow being enabled by the use of AWS Lambda, the authors estimated that sVEP is 99% faster than traditional VEP implementations [29]. Yet another use case developed by CSIRO is GT-Scan [28], a web application that supports finding targets with minimal similar sequences in the genome. By taking advantage of the AWS Lambda and AWS DynamoDB, the authors managed to reduce the application costs from around \$700 to \$2.50 compared to a cloud server-based solution. The use of serverless for basecalling was validated by Grzesik et al. [17]. The authors implemented and evaluated the possibility of running a basecalling process of nanopore sequencing data. In their solution, they used AWS Lambda with Docker container support. During experiments, they determined that four Lambda functions running in parallel have enough computing power to support near real-time processing of data produced by a single MinION device. The authors also noted that in their experiments, they could scale up to 100 of such functions running simultaneously in less than a minute. In another case, Crespo-Cepeda et al., in their paper [12], analyse opportunities and challenges for using AWS Lambda for bioinformatic workflows. In their work, the authors propose an architecture for running CloudDmetMiner, based on AWS Lambda and AWS S3 services. Authors manage to run successful experiments and conclude the paper with suggestion that using serverless approach can reduce the time dedicated to managing and provisioning cloud infrastructure manually. Another successful instance of taking advantage of serverless computing for performing biomedical research has been presented by Kumanov et al., in their work [24].

The authors describe a proof of concept example of performing all-against-all pairwise comparison among 20,000 human protein sequences, implemented with Striped Smith-Waterman algorithm. According to the authors, use of serverless cloud computing allowed for increasing speed of execution time at a low cost. In the cited case, the experiment can be accomplished in about 2 min for a cost of less than one dollar, which is a speed up of about 250 times in comparison to running the experiment on a laptop computer. The authors also suggest that the similar approach could be effective for tasks such as protein-folding, deep-learning or sequence alignment.

Based on the above findings, we can conclude that there is a growing interest in both taking advantage of serverless infrastructures for bioinformatic workflows and performing metagenomic analytics in a portable manner. This paper aims to expand knowledge in both areas by evaluating how portable analytic workflows can benefit, in terms of processing time and energy consumption reduction, from integration with serverless-based infrastructure. This makes our solution a unique one, since, to our best knowledge, there is no paper yet that proposes and evaluates such a hybrid approach to basecalling.

3 Testing Workflow and Environment

For the purposes of the evaluation, we propose the workflow in which the offline edge-based basecalling process is enhanced with optional serverless-based cloud acceleration. In this workflow, the FAST5 files containing MinION Nanopore sequencing reads are split into two batches - one to be processed directly on the edge device and the second to be processed in a serverless manner in the cloud environment. Splitting files is based on the estimated processing time for both approaches, taking into account the available network upload speed. After splitting the files into batches, the basecalling process for the first batch is started locally, where the files from the second batch are sent to the AWS S3 bucket. As soon as files appear in the S3 bucket, for each of them, an AWS Lambda function is created to perform the data processing in a parallel manner. The AWS Lambda functions are responsible for downloading the file from the S3 bucket, running the basecalling step, and uploading the results of the process to a separate S3 bucket. An additional process is running locally on the edge device that is responsible for monitoring the S3 bucket with results and downloading them to the edge device. After all processing is done and all files with results are sent back to the edge device, the data is ready for further processing, e.g., for the classification step. The described workflow is presented in Fig. 1.

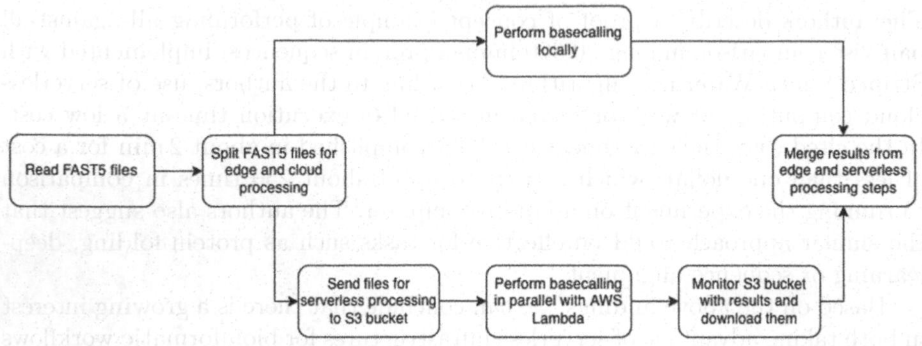

Fig. 1. Hybrid edge processing with serverless cloud offloading diagram.

As an edge device used during experiments, we selected Jetson Xavier NX. It has already been proven to be an effective and sufficient board for running basecalling experiments in our previous research, mainly thanks to its support for GPU acceleration. Another essential feature of Jetson Xavier NX is its energy efficiency and the capability to control power consumption with five distinct power consumption modes. The full technical specification of Jetson Xavier NX is presented below [3]:

- CPU - 6-core NVIDIA Carmel ARM®v8.2 64-bit CPU 6 MB L2 + 4 MB L3
- GPU - NVIDIA Volta™ architecture with 384 NVIDIA® CUDA® cores and 48 Tensor cores
- Memory - 8 GB 128-bit LPDDR4 × 51.2 GB/s
- OS Storage - SDHC card (32 GB, class 10)
- DB Storage - Solid State Drive, PNY 500 GB M.2 PCIe NVMe XLR8 CS3030
- OS - Ubuntu 18.04.5 LTS with kernel version 4.19.140-tegra

As the serverless platform of choice, we selected AWS Lambda, which we determined in our previous research to be an effective and feasible solution for running basecalling in a serverless manner, thanks to its support for Docker [1] containers and the ability to use up to 10,240 MB of RAM and up to 6 vCPU cores. In our experiments, each Lambda function used a Docker container based on Ubuntu 16.04 operating system with Node.JS script that downloaded the file from AWS S3, invoked the basecalling process on it, and uploaded the results to the output bucket. The Lambda function was configured and deployed with the use of Serverless Framework. [6] As the basecalling software, we used Guppy, which is a closed-source basecaller developed by Oxford Nanopore Technologies. It supports GPU acceleration that can take advantage of GPU on Jetson Xavier NX and support multiple basecalling models (fast and high accuracy). We also considered alternative basecallers such as Deepnano-blitz[10], Bonito [2], or Causalcall [32], but after preliminary testing, they all proven to either not work on AWS Lambda or on Jetson Xavier NX, or offer much lower performance in comparison to the Guppy basecaller. The whole processing was coordinated by a custom Python program running directly on Jetson Xavier NX, that was responsible for splitting the read files and passing them for further processing. Upload to AWS S3 was handled by s3cmd [5], which also offers possibility to throttle upload speeds, that we used to achieve different upload speeds in our experiments.

4 Performance Experiments

During experiments, we decided to evaluate the potential reduction in processing time and energy consumed during a basecalling process that is accelerated with serverless cloud offloading. Firstly, we ran a fully offline test where all processing was happening directly on the edge device. In the next step, we ran the fully online test for different upload speeds to measure the estimated processing time for processing all data only in the cloud. Throughout experiments, we ensured throttled upload speeds of 128 kB/s, 256 kB/s, and 512 kB/s, which can be achieved by using, e.g., 3G/4G Internet connection from a smartphone or dedicated board module. During experiments, we used a subset of benchmarking dataset of *Klebsiella pneumoniae* reads [31]. The used dataset had the size of 178 MB and consisted of 2,240 separate sequence reads files. The Guppy basecaller was configured with a "high accuracy" mode to ensure improved basecalling accuracy. The Jetson Xavier NX used the lowest power mode (2 Cores, 10 W) as well as the highest power mode (6 cores, 15W) to evaluate how the capabilities of edge device impact the results of the experiments.

In the preliminary testing, we determined that the fully offline edge processing of the prepared dataset took 320 s with an average power consumption of 11.96 W for lowest power mode and 240 s with an average power consumption of 15.3 W for highest power mode. The second workflow in which all files were uploaded to the cloud for online serverless processing took 700 s for 512kB/s upload speed, 980 s for 256 kB/s upload speed, and 1,520 s for 128 kB/s upload speed. For each of these scenarios, the average power consumption of edge device was measured to be around 4.25 W. Based on the outcomes of fully serverless processing for each of the average upload speeds, we determined the split ratio between files that should be processed at the edge and files that should be processed in the cloud that would result in processing all the data in the fastest manner, optimizing for execution speed. The ratio was obtained by comparing average processing speed of processed megabytes per second for edge and cloud processing. Based on that split ratio, we decided that for the upload speed of 128 kB/s, around 17.5% of the data was assigned for cloud processing for lowest power mode and 13.8% of the data for the highest power mode, for the upload speed of 256 kB/s, 24.9% of data was sent for cloud processing for lowest and 19% for highest power mode, and for the upload speed of 512 kB/s, 31% of data was assigned to be processed on the cloud side for lowest power mode and around 25% for highest power mode.

During experiments in which we evaluated the hybrid edge-serverless workflow with the mentioned split ratio of the files to be processed, we observed a reduction in processing time to 264 s and 206.9 s for the upload speed of 128 kB/s, to 240 s and 193.5 s for 256 kB/s, and to 220 s and 179.1 s and for upload speed of 512 kB/s for lowest and highest power modes, respectively. The results and comparison of all three experimental scenarios for lowest power mode are presented in Fig 2 while results for highest power mode are presented in Fig 3. During the hybrid edge processing with cloud offloading, the average power consumption was equal to 12.4 W for lowest power mode in all tested cases and 15.6 W for highest power mode, also for all tested cases. Figure 4 presents the energy consumed by the edge device during each considered processing scenario in lowest power mode, while Fig. 5 presents results for highest power mode.

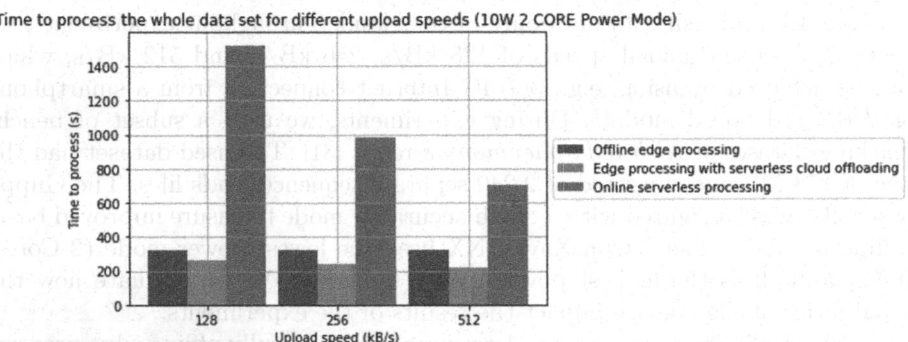

Fig. 2. Dependency between the basecalling execution time and upload speed for various strategies of data processing for lowest power mode of Jetson Xavier NX (offline edge processing vs. edge with cloud offloading vs. online serverless processing).

Based on the obtained results, we can observe that the proposed hybrid approach (edge processing with serverless cloud offloading) allows reducing consumed energy by 14% (lowest power mode) and 12% (highest power mode) as well as processing time by 17.3% (lowest power mode) and by 14% (highest power mode) for the upload speed of 128 kB/s. For 256 kB/s, we can achieve reductions by 22% (lowest power mode) and 20% (highest power mode) as well as 25% (lowest power mode) and 18% (highest power mode) for energy consumed and processing time, respectively, in comparison to the offline edge processing scheme. As we initially expected, we can observe the best results for upload speed of 512 kB/s, where we achieved a 28.7% (lowest power mode) and 24% (highest power mode) reduction of energy consumption as well as 31.2% (lowest power mode) and 25.5% (highest power mode) reduction of processing time compared to offline edge processing. In all considered cases,we see that fully

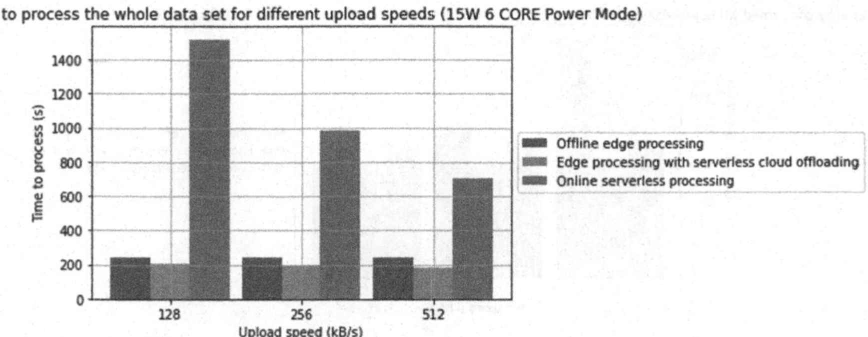

Fig. 3. Dependency between the basecalling execution time and upload speed for various strategies of data processing for highest power mode of Jetson Xavier NX (offline edge processing vs. edge with cloud offloading vs. online serverless processing).

online serverless processing was the slowest of all compared approaches. However, what is interesting, we observed that for the upload speed of 512 kB/s, the fully online serverless basecalling process is more energy-effective than the fully offline edge processing approach, offering a reduction of energy consumption by 22.2% (lowest power mode) and 19.1% (highest power mode), at the cost of processing time being increased to 700 s. This suggests that when energy preserving is more critical, a fully online basecalling process might potentially be more effective. We also observe that using highest power mode of Jetson Xavier NX does not necessarily means reduced energy-effectiveness, as in tested cases the energy consumption was similar in comparison to lowest power mode, while the processing time was significantly lower for highest power mode.

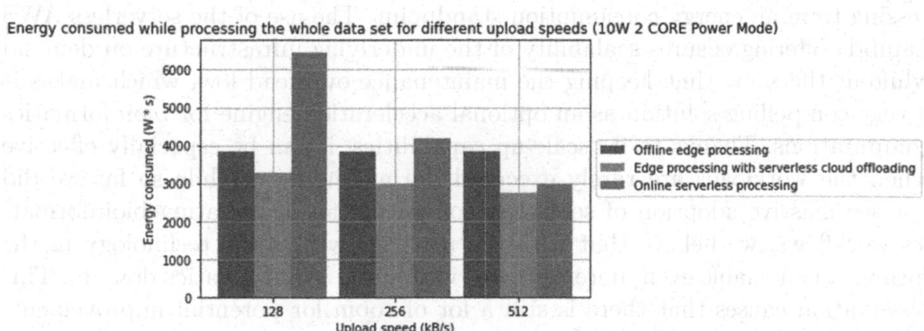

Fig. 4. Dependency between the energy consumed during basecalling for various strategies of data processing for lowest power mode of Jetson Xavier NX (offline edge processing vs. edge with cloud offloading vs. online serverless processing).

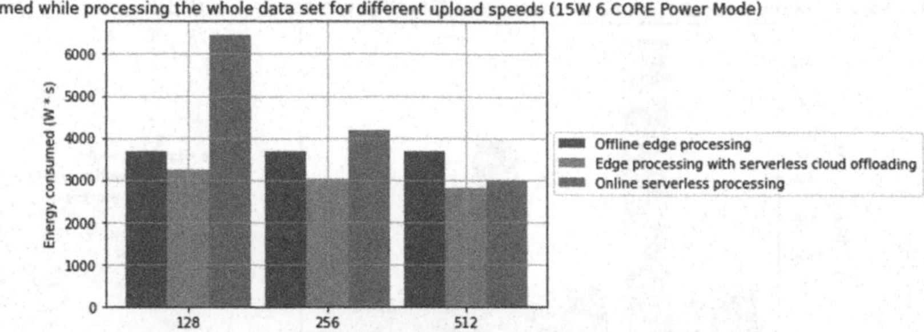

Fig. 5. Dependency between the energy consumed during basecalling for various strategies of data processing for highest power mode of Jetson Xavier NX (offline edge processing vs. edge with cloud offloading vs. online serverless processing).

5 Results Summary and Concluding Remarks

The works presented in the paper expand our previous research where we consider edge and serverless workflows totally separately. Considering the results of the experiments presented in the previous section, we can notice that the combination of offline edge processing with online serverless processing, given access to sufficient network connection, can be an effective strategy for reducing energy consumption and processing time when performing basecalling operations. In one of the cases, we managed to achieve a 31.2% reduction of processing time and 28.7% reduction of energy consumption in comparison to a fully offline process. Our approach may thus optimize the utilization of available computing resources. Throughout the experiments, we have also observed that in the tested scenarios, fully online serverless processing can be more effective than offline processing from an energy consumption standpoint. The use of the serverless AWS Lambda offering ensures scalability of the underlying infrastructure on demand while at the same time keeping the maintenance overhead low, which makes it a very compelling solution as an optional acceleration engine for bioinformatics computations. Thanks to its scale-up capabilities, it can be especially effective when the workloads are rarely executed during the day. While so far we did not see massive adoption of serverless computing for accelerating bioinformatics workflows, we believe that with the rapid growth of the technology in the future, it can enable even more use cases within the bioinformatics domain. This observation causes that there is still a lot of room for potential improvements and developments in this area.

Acknowledgments. The research was supported by the Polish Ministry of Science and Higher Education as a part of the CyPhiS program at the Silesian University of Technology, Gliwice, Poland (Contract No. POWR.03.02.00-00-I007/17-00) and by Statutory Research funds of Department of Applied Informatics, Silesian University of Technology, Gliwice, Poland (grant No BK/RAu7/2022).

References

1. AWS Lambda container image support. https://aws.amazon.com/blogs/aws/new-for-aws-lambda-container-image-support/, (accessed 5 February 2022),
2. Bonito basecaller repository on. https://github.com/nanoporetech/bonito, github (accessed 5 February 2022)
3. Jetson Xavier NX specification. https://developer.nvidia.com/embedded/jetson-xavier-nx-devkit, (accessed 5 February 2022)
4. Nanopore product comparison. https://nanoporetech.com/products/comparison, (accessed 5 February 2022)
5. s3cmd. https://s3tools.org/s3cmd, (accessed 5 April 2022)
6. Serverless framework. https://github.com/serverless/serverless, (accessed 5 April 2022)
7. Aboukhalil, R.: Serverless genomics - using WebAssembly and Cloudflare Workers to power genomics analysis. https://robaboukhalil.medium.com/serverless-genomics-c412f4bed726, (accessed 5 February 2022)
8. Acharya, K., Blackburn, A., Mohammed, J., Haile, A.T., Hiruy, A.M., Werner, D.: Metagenomic water quality monitoring with a portable laboratory. Water Res. **184**, 116112 (2020). https://www.sciencedirect.com/science/article/pii/S0043135420306497
9. Boykin, L.M., et al.: Tree lab: portable genomics for early detection of plant viruses and pests in sub-saharan africa. Genes **10**(9) 63 (2019). https://www.mdpi.com/2073-4425/10/9/632
10. Boža, V., Perešíni, P., Brejová, B., Vinař, T.: Deepnano-blitz: a AST base caller for minion nanopore sequencers. Bioinformatics (Oxford, England) **36**, 4191–4192 (2020)
11. Castro-Wallace, S.L., et al.: Nanopore DNA sequencing and genome assembly on the international space station. Sci. Rep. 7(1), 18022 (2017). https://doi.org/10.1038/s41598-017-18364-0
12. Crespo-Cepeda, R., Agapito, G., Vazquez-Poletti, J.L., Cannataro, M.: Challenges and opportunities of amazon serverless lambda services in bioinformatics. In: Proceedings of the 10th ACM International Conference on Bioinformatics, Computational Biology and Health Informatics, BCB 2019, pp. 663–668. Association for Computing Machinery, New York (2019). https://doi.org/10.1145/3307339.3343462
13. David, M., Dursi, L.J., Yao, D., Boutros, P.C., Simpson, J.T.: Nanocall: an open source basecaller for Oxford Nanopore sequencing data. Bioinformatics **33**(1), 49–55 (2016). https://doi.org/10.1093/bioinformatics/btw569
14. D'Agostino, D., Morganti, L., Corni, E., Cesini, D., Merelli, I.: Combining edge and cloud computing for low-power, cost-effective metagenomics analysis. Future Gener. Comput. Syst. **90**, 79–85 (2019). https://www.sciencedirect.com/science/article/pii/S0167739X18300293
15. Gowers, G.O.F., Vince, O., Charles, J.H., Klarenberg, I., Ellis, T., Edwards, A.: Entirely off-grid and solar-powered DNA sequencing of microbial communities during an ice cap traverse expedition. Genes 10(11), 902 (2019). https://www.mdpi.com/2073-4425/10/11/902
16. Grzesik, P., Augustyn, D.R., Wyciślik, L., Mrozek, D.: Serverless computing in omics data analysis and integration. Briefings Bioinform. 23(1) (2021). https://doi.org/10.1093/bib/bbab349, bbab349

17. Grzesik, P., Mrozek, D.: Metagenomic analysis at the edge with jetson xavier NX. In: Paszynski, M., Kranzlmüller, D., Krzhizhanovskaya, V.V., Dongarra, J.J., Sloot, P.M.A. (eds.) ICCS 2021. LNCS, vol. 12745, pp. 500–511. Springer, Cham (2021). https://doi.org/10.1007/978-3-030-77970-2_38
18. Grzesik, P., Mrozek, D.: Serverless nanopore basecalling with AWS Lambda. In: Paszynski, M., Kranzlmüller, D., Krzhizhanovskaya, V.V., Dongarra, J.J., Sloot, P.M.A. (eds.) ICCS 2021. LNCS, vol. 12743, pp. 578–586. Springer, Cham (2021). https://doi.org/10.1007/978-3-030-77964-1_44
19. Hoenen, T., et al.: Nanopore sequencing as a rapidly deployable EBOLA outbreak tool. Emerg. Inf. Dis. 22(2), 331–334 (2016). https://pubmed.ncbi.nlm.nih.gov/26812583, 26812583[pmid]
20. Hung, L.H., Niu, X., Lloyd, W., Yeung, K.Y.: Accessible and interactive RNA sequencing analysis using serverless computing. BioRxiv (2020). https://www.biorxiv.org/content/early/2020/10/03/576199
21. Jain, Y., et al.: sBeacon: cloud-native genomic data exchange. In: ABACBS-2020, vol. 2020, p. 1 (2020)
22. Jonas, E., et al.: Cloud programming simplified: a berkeley view on serverless computing. CoRR abs/1902.03383 (2019). http://arxiv.org/abs/1902.03383
23. Kafetzopoulou, L.E., et al.: Metagenomic sequencing at the epicenter of the Nigeria 2018 lassa fever outbreak. Science **363**(6422), 74–77 (2019). https://science.sciencemag.org/content/363/6422/74
24. Kumanov, D., Hung, L.H., Lloyd, W., Yeung, K.Y.: Serverless computing provides on-demand high performance computing for biomedical research (2018). https://arxiv.org/abs/1807.11659
25. Li, H., Durbin, R.: Fast and accurate short read alignment with Burrows-Wheeler transform. Bioinformatics **25**(14), 1754–1760 (2009). https://doi.org/10.1093/bioinformatics/btp324
26. Merelli, I., et al.: Low-power portable devices for metagenomics analysis: fog computing makes bioinformatics ready for the internet of things. Future Generat. Comput. Syst. **88**, 467–478 (2018). https://www.sciencedirect.com/science/article/pii/S0167739X17324123
27. Oliva, M., Milicchio, F., King, K., Benson, G., Boucher, C., Prosperi, M.: Portable nanopore analytics: are we there yet? Bioinformatics **36**(16), 4399–4405 (2020). https://doi.org/10.1093/bioinformatics/btaa237
28. What is "serverless" and "cloud-native" and when to use it?. https://bioinformatics.csiro.au/blog/converting-traditional-architecture-to-cloud-native-applications/, (accessed 5 February 2022)
29. Serverless VEP. https://bioinformatics.csiro.au/serverless-vep/, (accessed 5 February 2022)
30. Singh, S.: Optimize cloud computations using edge computing. In: 2017 International Conference on Big Data, IoT and Data Science, BID, pp. 49–53, December 2017
31. Wick, R.R., Judd, L.M., Holt, K.E.: Performance of neural network basecalling tools for oxford nanopore sequencing. Genome Biol. **20**(1), 129 (2019). https://doi.org/10.1186/s13059-019-1727-y
32. Zeng, J., Cai, H., Peng, H., Wang, H., Zhang, Y., Akutsu, T.: Causalcall: Nanopore basecalling using a temporal convolutional network. Front Genet. **10**, 1332 (2020). https://www.frontiersin.org/article/10.3389/fgene.2019.01332

Continuous-to-Continuous Data Model vs. Discrete-to-Discrete Data Model for the Statistical Iterative Reconstruction Method

Robert Cierniak(✉)

Department of Intelligent Computer Systems, Czestochowa University of Technology, Armii Krajowej 36, 42-200 Czestochowa, Poland
robert.cierniak@pcz.pl

Abstract. The article presents a comparison of two statistical approaches to the problem of image reconstruction from projections: the worldwide known concept based on a discrete-to-discrete data model and our original idea based on a continuous-to-continuous data model. Both reconstruction approaches are formulated taking into account the statistical properties of signals obtained by CT scanners. The main goal of this strategy is significantly improving the quality of the reconstructed images, so allowing a reduction in the x-ray dose absorbed by a patient during CT examinations. In the concept proposed by us, the reconstruction problem is formulated as a shift-invariant system. In consequence, that significantly improves the quality of the subsequently reconstructed images, and it allows to reduce the computational complexity compared to the reference method. The performed by us experiments have shown that our original reconstruction method outperforms the referential approach regarding the image quality obtained and the time of necessary calculations.

Keywords: Iterative reconstruction algorithms · Computed tomography · Statistical methods

1 Introduction

1.1 Motivation

Although computed tomography was invented many years ago, it continues to be a very appealing field of research. Every new generation of CT devices stimulates the development of reconstruction algorithms adapted for the new design. Put simply, we can say that since the first design made by Godfrey Newbold Hounsfield in 1971 all the most significant reconstruction algorithms have

The project financed under the program of the Polish Minister of Science and Higher Education under the name "Regional Initiative of Excellence" in the years 2019–2022 project number 020/RID/2018/19 the amount of financing 12,000,000 PLN.

D. Groen et al. (Eds.): ICCS 2022, LNCS 13351, pp. 493–506, 2022.
https://doi.org/10.1007/978-3-031-08754-7_55

used one of two basic approaches, depending on the signal processing methodology used in them: these are analytical methods (continuous-to-continuous data model), especially those based on convolution and back-projection [1], and methods based on the strategy called the algebraic strategy (discrete-to-discrete data model) [2]. It is worth underlining that apart from in the first scanner designed by Hounsfield, the EMI Mark I, and the prospective use of the newest concepts in reconstruction algorithms in high definition computed tomography (HDCT) devices, all other CT designs have been equipped with analytical reconstruction algorithms. The use of the algebraic method (ART - Algebraic Reconstruction Technique) in the first historical CT apparatus was presumably because there was no alternative at the time. After this "early mistake", the next generation of CT systems used only reconstruction algorithms based on analytical image processing methods. The main reason for this was the huge size of the matrices which appear in the algebraic reconstruction problem and the calculation complexity of the reconstruction method based on this methodology that this caused. The analytical (or transformation) methodology drastically simplifies the number of calculations needed and so is more appealing. It has been proven (e.g. [3]) that the frequency of cancerous diseases for patients who had a CT scan (at least one year after the scan) is about 24% higher than in the case of patients who had not had the scan. Due to the enormous prevalence of CT scans, any action aimed at reducing this impact are of fundamental importance, assuming of course, the further existence of this popular, cheap and effective diagnostics technique. For these reasons, but also for both social and commercial ones, manufacturers began a kind of competition to develop methods of reducing the X-ray dose absorbed by patients. The seemingly obvious solution, to simply reduce the radiation dose given during a scan, cannot be applied. This is because the required radiation dose is determined by the SNR, which defines the image quality. Thus, if the image quality is to remain high, the radiation intensity should stay at a defined level (which has a direct impact on the absorbed radiation dose during the scan).

1.2 Contribution

It is possible to improve the resistance of tomographic images to the measurement noise which occurs during image reconstruction by using appropriate statistical signal processing. This means that it is possible to decrease the radiation intensity applied, and so decrease the dose absorbed by patients. Recently, some commercial solutions of such systems have been developed, which perform reconstruction processing iteratively to decrease the noise in the images. These systems take into consideration the probabilistic conditions present in the measurement systems of CT scanners in order to limit the influence of noise on the images obtained from the measurements. The most interesting approach, called MBIR (Model-Based Iterative Reconstruction), is presented in such papers as [4,5], where a statistical model of the measurements is derived analytically, and, based on this, a statistical iterative reconstruction algorithm is formulated. The reconstruction problem formulated algebraically plays a crucial part in this approach. Indeed, the algebraic approach to the image reconstruction from projections problem is being intensively explored once more. This is because of one obvious reason - the measurement noises in it can be modelled relatively easily, because each measurement is considered separately. The reconstruction idea presented in the above publications is based on the maximum likelihood (ML) approach and a development of this concept - the maximum *a posteriori* probability (MAP) estimation approach (the iterative coordinate descent (ICD) algorithm described comprehensively in [6]) implements the MAP approach). Consequently, in 2013, this development had its debut under its commercial name Veo - CT Model-Based Iterative Reconstruction. This application of the algebraic reconstruction method, however, presents some significant technical difficulties in its practical realization, namely: the difficulty in establishing the coefficients of the forward model for 3D spiral cone-beam scanners [6,7]. The huge number of these coefficients in this model means that it is impossible to keep all of them in memory at the same time and the requirement for the simultaneous calculation of all voxels in the range of the reconstructed 3D image make the reconstruction problem extremely complex. Although, there have been attempts to decrease the calculation complexity of this approach, as presented for example in the paper [8], they have, as yet, only met with limited results. Moreover, this system uses a reconstruction problem model that has been shown to be extremely ill-conditioned. One can say that there are many solutions on the market in this area, but they are still insufficient when it comes to the limiting the radiation dose. Therefore, there is still room for improvement of such systems. It would be interesting to formulate a statistical reconstruction method which would take into consideration the statistical conditions of the measurement physics, as in the ICD algorithm, thereby eliminating most of the disadvantages of the algebraic scheme of signal processing methodology. We could avoid the above mentioned difficulties connected with using an algebraic methodology by using an analytical strategy for the reconstructed image processing. In previous papers, we have shown how to formulate the analytical reconstruction problem consistent with the ML methodology for scanners with parallel geometry [9,10], for fan-beams [11], and finally we have proposed a scheme of reconstruction method for the spiral cone-beam scanner [12]. Our approach has some significant advantages compared with algebraic methodology. Firstly, in our method, we establish certain coefficients, but this is performed in a much easier way than in comparable methods. Secondly, we perform the reconstruction process in only one plane in 2D space, greatly simplifying the problem. In this way, the reconstruction process can be performed for every cross-section image separately. After this, it is possible to reconstruct the whole 3D volume image from the set of previously reconstructed 2D images. And finally, because of the analytical methodology of the reconstruction process, we can perform most of the computationally expensive operations in the frequency domain (2D convolutions). Because it is a very much less computationally demanding approach, by using FFT, we make our reconstruction method independent of the dimensions of the reconstructed image, to an acceptable degree. This approach also outperforms the algebraic method regarding the better condition number at the level of problem formulation. This makes our method really competitive in terms of its resistance to the influence of measurement noise and errors in the forward model. The main motivation for this paper is to present a comparison between these two model-based approaches to the statistical reconstruction problem. In particular, we will present considerations and computer simulations that correspond with the optimization of the computational complexity in the case of the continuous-to-continuous method designed by us. We will also show how this optimization achieved by problem reformulation can impact the time of calculations obtained under actual conditions.

2 Statistical Reconstruction Approaches

We below present two approaches that can be directly applied to parallel beam tomography, but it is possible in an easy way to adapt them for a majority of all existing geometries of the CT scanners.

Let function $\mu(x, y)$ denote the unknown image representing a cross-section of an examined object (in medicine, a human body). Image $\mu(x, y)$ will be calculated using projections obtained by using the Radon transform. A diagram of a single projection measurement is depicted in Fig. 1.

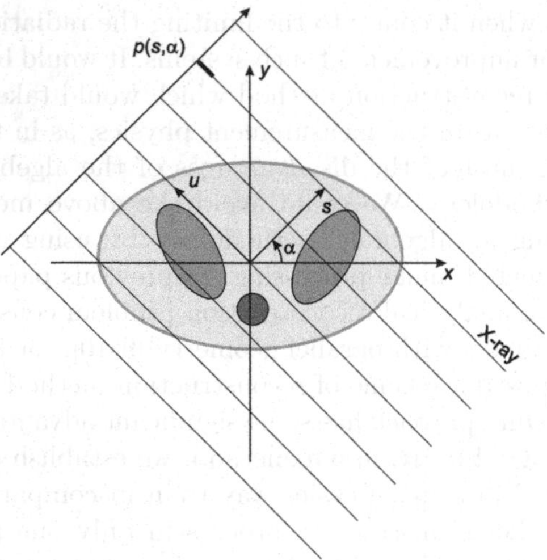

Fig. 1. The geometry of the projection system

The function $p(s, \alpha)$ is the result of a measurement carried out at a distance s from the origin when a projection is made at a specific angle α. This is called the Radon transform and is written mathematically as

$$p(s, \alpha) = \int_{-\infty}^{+\infty} \int_{-\infty}^{+\infty} \mu(x, y) \cdot \delta(x cos\alpha + y sin\alpha - s)\, dx dy. \tag{1}$$

Both reconstruction approaches can only make use of projections obtained at certain angles and measured only at particular points on the screen. Therefore, beams of x-rays reaches the individual detectors at points $l = -L/2, \ldots, L/2$, where L is a number of detectors placed on a screen. Values $s_l = l \cdot \Delta_s$ denote the distances on the screen between each ray and the origin, and Δ_s denotes the interval between detectors. In turn, parameters α_ψ denote discrete values of the projection angles indexed by the variable ψ, where $\psi = 0, \ldots, \Psi - 1$, where Ψ is the number of projections. Subsequent projections are carried out after a rotation by Δ_α. Following a discrete nature of available measurements, we will consider the discrete form of the image $\mu(i, j)$ as well, where $i = 1, \ldots, I$, $j = 1, \ldots, I$.

2.1 A Statistical Approach Based on the Discrete-to-Discrete Data Model

First, we consider a referential approach to the image reconstruction problem, in which a model-based iterative reconstruction method is based on a discrete-to-discrete data model. A forward model (a system of linear equations) of this approach can be presented as follows:

$$\mathbf{p} = \mathbf{A}\mu, \tag{2}$$

where: $\mathbf{p} = [p_m]$ is the projection vector with $m = 1, \ldots, L \cdot \Psi$; $\mathbf{A} = [a_{mn}]$ is a system matrix with dimensions $1, \ldots L \cdot \Psi \times 1, \ldots I^2$; $\mu = [\mu_n]$ is a vector representing a reconstructed image with dimension $n = 1, \ldots, I^2$. Practically, the elements a_{mn} can be interpreted as the contribution of a given image block (pixel) with parameters n to the formation of the projection value p_m, measured at the screen.

Defined above forward model was applied to formulate, according to statistical considerations (see e.g. [4] or [5]), the following iterative reconstruction method which is based on Maximum Likelihood estimation of the reconstructed image:

$$\mu_0 = \arg\min_{\mu} \left(\frac{1}{2} (\mathbf{p} - \mathbf{A}\mu)^{\mathrm{T}} \mathbf{D} (\mathbf{p} - \mathbf{A}\mu) \right), \tag{3}$$

where $\mathbf{D}$ is a diagonal matrix:

$$\mathbf{D} = \begin{bmatrix} d_1 & 0 & \cdots & 0 \\ 0 & d_2 & \cdots & 0 \\ \vdots & \vdots & \ddots & \vdots \\ 0 & 0 & \cdots & d_{L\cdot\Psi} \end{bmatrix} = \begin{bmatrix} \lambda_1 & 0 & \cdots & 0 \\ 0 & \lambda_2 & \cdots & 0 \\ \vdots & \vdots & \ddots & \vdots \\ 0 & 0 & \cdots & \lambda_{L\cdot\Psi} \end{bmatrix}, \tag{4}$$

wherein (see [13]):

$$d_m \cong \frac{1}{\sigma_{p_m}}, \tag{5}$$

where σ_{p_m} are the variances of the projection measurements p_m.

It is worth noting that formula (3) represents a Weighted Least Squares (WLS) problem and it can be solved using any gradient descent method. However, the huge number of a_{mn} coefficients in this model means that it is impossible to keep all of them in memory at the same time and the requirement for the simultaneous calculation of all voxels in the range of the reconstructed 3D image make the reconstruction problem extremely complex. Moreover, the computational complexity of the problem is approximately proportional to J^2, where J is the number of voxels in the reconstructed 3D image, and the iterative reconstruction procedure based on this conception necessitates simultaneous calculations for all the voxels in the range of this image. For a 3D geometry of the scanner (e.g. spiral cone-beam geometry), it means that the reconstruction for all $J = I^2 \times Z$ voxels is performed simultaneously, where Z is a number of examined cross-sections of a body. Therefore, the computational complexity of this approach is evaluated as $O\left(Z^2 I^4\right)$. The diagram of a basic form of this reconstruction algorithm is depicted in Fig. 2.

It is well-known that the form of the ML methodology expressed by (3) is ill-conditioned and, as described in the literature [5], is unstable in practice. That is why regularizing *a priori* terms are standardly introduced into the loss function. On the other hand, these additional terms cause an increase in the calculation demands during the optimization process, and lead to smoothing of the reconstructed image. It would be very appealing to use a reconstruction methodology based on a pure ML scheme, without any regularizing *a priori* term and so avoid these instabilities of the reconstruction process and the smoothing effect. It was proposed a modification of the loss function (3), in the following manner:

$$\mu_0 = \arg\min_{\mu} \left(\frac{1}{2} \left(\mathbf{p} - \mathbf{A}\mu\right)^{\mathrm{T}} \mathbf{D} \left(\mathbf{p} - \mathbf{A}\mu\right) \right) + \beta f\left(\mu\right), \tag{6}$$

where $f(\mu)$ is some scalar regularization term, whose introduction has the aim of penalizing local differences between elements of the reconstructed image; β is a constant coefficient. This regularization term may take different forms, however, an interesting approach to this method is the total variation (TV) regularization [14].

It is possible to implement this approach in practice, mainly thanks to the attempts to decrease the calculation complexity of this approach (for details of the ICD algorithm see [8]). Consequently, in 2013, this development had its debut under its commercial name Veo - CT Model-Based Iterative Reconstruction (GE Medical Systems).

2.2 A Statistical Approach Based on the Continuous-to-Continuous Data Model

Our reconstruction method also is based on the well-known maximum-likelihood (ML) estimation, where an optimization formula is consistent with the C-C data model, as follows:

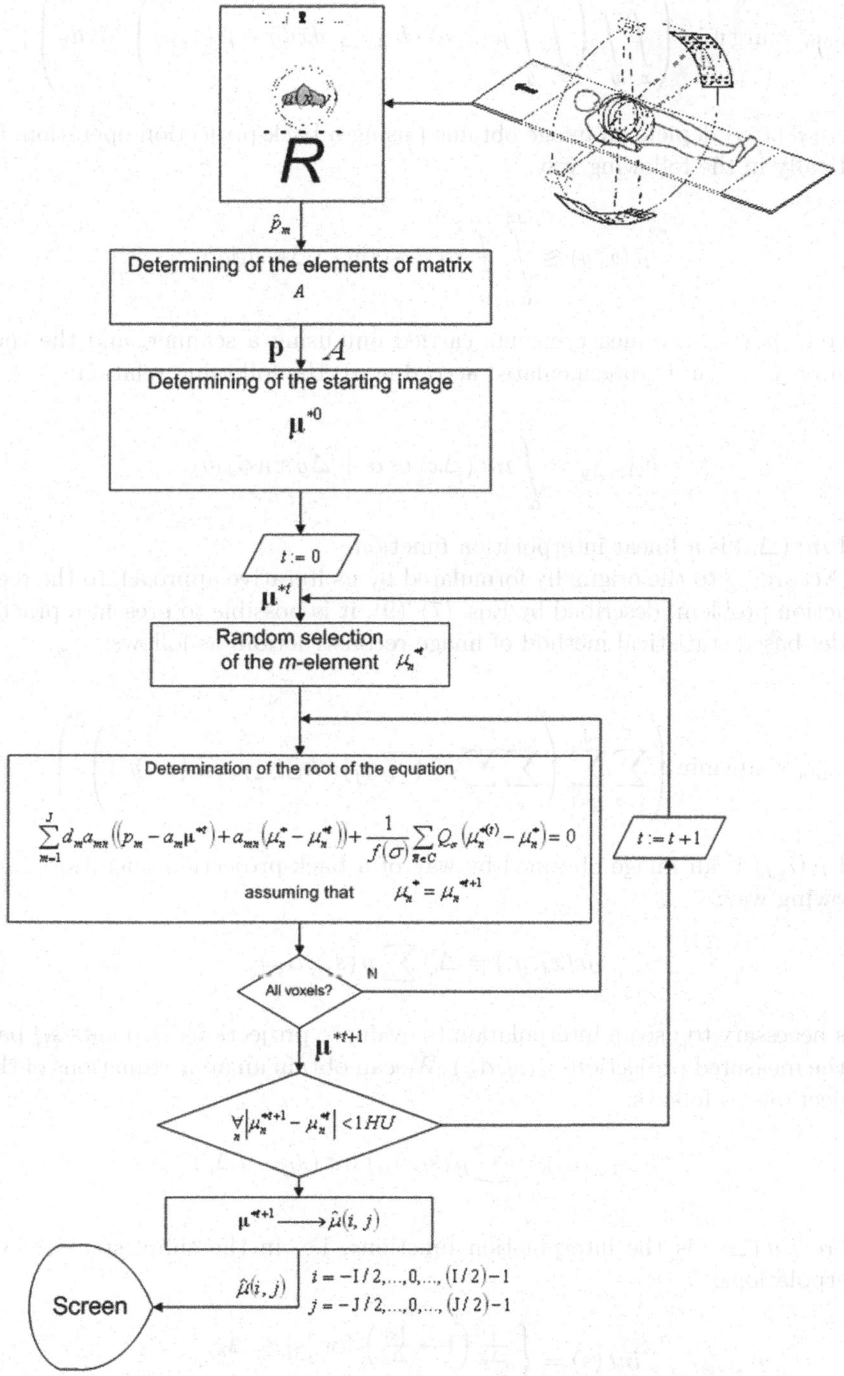

Fig. 2. The diagram of the approach based on the D-D data model

$$\mu_{\min} = \arg\min_{\mu} \left(\int_x \int_y \left(\int_{\bar{x}} \int_{\bar{y}} \mu(\bar{x}, \bar{y}) \cdot h_{\Delta x, \Delta y} d\bar{x} d\bar{y} - \tilde{\mu}(x, y) \right)^2 dx dy \right), \quad (7)$$

where $\tilde{\mu}(x, y)$ depicts an image obtained using a back-projection operation, theoretically in the following way:

$$\tilde{\mu}(x, y) \approxeq \int_0^{2\pi} \int_{-\infty}^{\infty} p(s, \alpha)\, int_L(\Delta s)\, d\beta d\alpha, \quad (8)$$

wherein $p(s, \alpha)$ are measurements carried out using a scanner, and the coefficients $h_{\Delta i, \Delta j}$ can be precalculated according to the following relation:

$$h_{\Delta x, \Delta y} = \int_0^{2\pi} int\,(\Delta x \cos \alpha + \Delta y \sin \alpha)\, d\alpha, \quad (9)$$

and $int\,(\Delta s)$ is a linear interpolation function.

According to the originally formulated by us iterative approach to the reconstruction problem, described by Eqs. (7)–(9), it is possible to present a practical model-based statistical method of image reconstruction, as follows:

$$\mu_{\min} = \arg\min_{\mu} \left(\sum_{i=1}^{I} \sum_{j=1}^{J} \left(\sum_{\bar{i}} \sum_{\bar{j}} \mu^*(x_{\bar{i}}, y_{\bar{j}}) \cdot h_{\Delta i, \Delta j} - \tilde{\mu}(x_i, y_j) \right)^2 \right), \quad (10)$$

and $\tilde{\mu}(i, j)$ is an image obtained by way of a back-projection operation, in the following way:

$$\tilde{\mu}(x_i, y_j) = \Delta_\alpha \sum_{\theta} \dot{p}(s_{ij}, \alpha_\psi). \quad (11)$$

It is necessary to use an interpolation to evaluate projections at points s_{ij} based on the measured projections $p(s_l, \alpha_\psi)$. We can obtain an approximations of these projections as follows:

$$\dot{p}(s_{ij}, \alpha_\psi) = \sum_{l} p(s_l, \alpha_\psi)\, int\,(s_{ij} - l\Delta_s), \quad (12)$$

where $int\,(\Delta s)$ is the interpolation functions, i.e. in the simplest case, linear interpolations:

$$int\,(s) = \begin{cases} \frac{1}{\Delta_s}\left(1 - \frac{|s|}{\Delta_s}\right) & \text{for } |s| \le \Delta_s \\ 0 & \text{for } |s| \ge \Delta_s \end{cases}. \quad (13)$$

In turn, the coefficients $h_{\Delta i,\Delta j}$ are determined according to the following formula:

$$h_{\Delta i,\Delta j} = \Delta_\alpha \sum_{\psi=0}^{\Psi-1} int\left(\Delta i \cos\psi\Delta_\alpha + \Delta j \sin\psi\Delta_\alpha\right), \tag{14}$$

wherein $int\,(\Delta s)$ is the same interpolation function as was used in the back-projection operation.

Same as before, the optimization problem (10) can be solved using any gradient descent method. Basically, in this case, the computational complexity of this problem is approximately proportional to J^2, where $J = I^2$ is the number of pixels in the reconstructed 2D image, where I is the image resolution, and the iterative reconstruction procedure based on this conception necessitates simultaneous calculations for all the pixels in the reconstructed 2D image, despite the geometry of a scanner. However, a shift-invariant system in the optimization problem (10) means that it is possible to transpose the most demanding computations into a frequency domain.

Therefore, it is necessary to transform two times a processed vector into a frequency domain, decreasing the computational complexity of the convolution from $O\left(I^4\right)$ to $O\left(I^2\right)$. Of course, each FFT costs $O\left(2log_2 I^2\right)$ operations, and we have to invert this transform every time. In total, that gives $O\left(8log_2 4I^2\right)$ operations per one iteration of the iterative reconstruction procedure (dimension of the image has to be doubled for the FFT processing). Figure 3 depicts this algorithm after discretization and implementation of the FFT that significantly accelerates the calculations.

Actually, this statistical reconstruction method consists of two steps, namely: a back-projection operation described by relation (8) and an iterative reconstruction procedure according to formula (7). In this case, the back-projection operation is not computationally demanding because there is no filtration during this operation, and it has a marginal influence on the real reconstruction time.

Although we have to establish certain coefficients in our method, this can be performed more easily than in a referential approach and the matrix containing these coefficients has relatively small dimensions, thus allowing it to be precalculated. Moreover, these coefficients can be transformed into the frequency domain and saved in memory in this form for further processing. In should be noted that this system is much better conditioned than the WLS problem present in the referential approach.

3 Experimental Results

We divided our experiments into two phases: first, we will try to show that the C-C data model gives better quality of the reconstructed images, and then, we will perform original tomographic data to evaluate both approaches regarding the time of calculations.

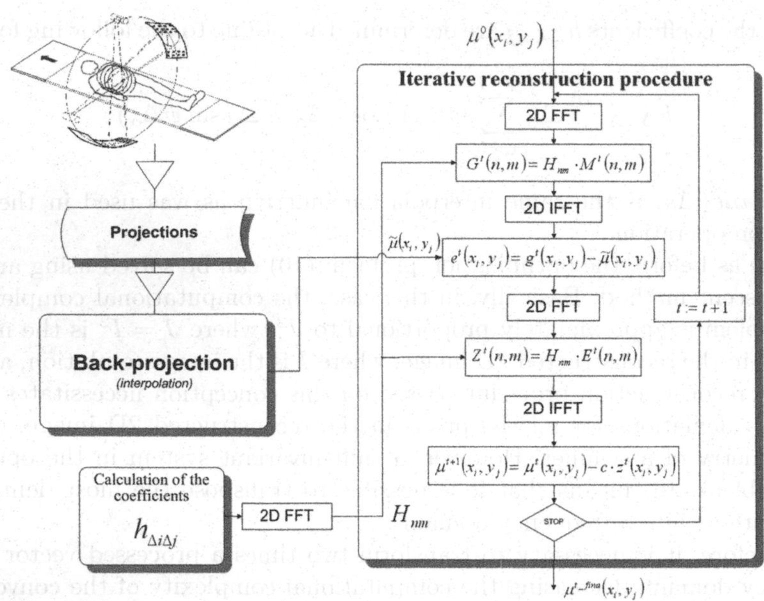

Fig. 3. The diagram of the approach based on the C-C data model

In this phase of our experiments, we have adapted the FORBILD[1], a mathematical phantom of the head. All the values of the attenuation coefficients placed in the original model were divided by a factor 10^{-3} in order to facilitate the calculations. This model was used to generate projections with noise with a Poisson probability distribution. During the simulations, we fixed L = 1024 measurement points (detectors) on the screen. The number of projections was chosen as $\Psi = 3220$ rotation angles per full-rotation and the size of the processed image was fixed at $\mathrm{I} \times \mathrm{I} = 1024 \times 1024$ pixels.

It was convenient to establish coefficients $h_{\Delta i,\Delta j}$ using relation (9) before we started the reconstruction process and these coefficients were fixed for the subsequent processing.

Having obtained the coefficients $h_{\Delta i,\Delta j}$, we can start the actual reconstruction procedure and perform the back-projection operation using relationships (8) to get a blurred image of the x-ray attenuation distribution in a given cross-section of the investigated object. We must use the linear interpolation function.

Evaluating a reconstruction procedure based only on a view of the reconstructed image is very subjective. For that reason the quality of the reconstructed image has been evaluated by an error measure defined as follows

$$MSE = \frac{1}{I^2} \sum_{i=1}^{\mathrm{I}} \sum_{j=1}^{\mathrm{J}} \left(\mu^{(t)}(i,j) - \mu(i,j) \right)^2, \tag{15}$$

[1] http://www.imp.uni-erlangen.de/forbild/deutsch/results/head/head.html.

where: $\mu^{*(t)}(i,j)$ is the reconstructed image after t iterations and $\mu(i,j)$ is the original image of the FORBILD phantom.

At this time, we have taken into account one form of regularization: the total variation (TV) prior [14]. The result obtained are shown in Table 1 (for all reconstructions, the starting image is a flat image $\mu^{*0} = 0.005$). For comparison, the original phantom image (Table 1.B), the image reconstructed by a standard FBP reconstruction method (Table 1.A) and by the referential ICD algorithm (Table 1.C) are also presented.

Table 1. Views of the images: reconstructed image using the standard FBP method with Shepp-Logan kernel (A); original image (B); reconstructed image using the D-D method described in the paper [8] (after 15 iterations) (C); reconstructed image using the C-C method described in this paper (after $t = 10^3$ iterations) (D).

	Image	Resolution phantom		Image	Resolution phantom
A	$MSE = 4.31 \cdot 10^{-9}$		E		
B	$MSE = 3.37 \cdot 10^{-9}$		F	$MSE = 3.86 \cdot 10^{-9}$	

Experiments in the next phase were carried out using projections obtained from a Somatom Definition AS+ scanner with parameters, as follows: reference tube potential at the level 120 kVp, quality reference effective at the level 200 mAs.

The size of the reconstructed image was fixed at 512 × 512 pixels. A discrete representation of the matrix $h_{\Delta x,\Delta y}$ was established in a computational way before the reconstruction process was started. These coefficients were fixed (transformed into the frequency domain) and used for the whole iterative reconstruction procedure. A result of an FBP reconstruction method was chosen as the starting point of the iterative reconstruction process.

A crucial parameter for the practical implementation of a reconstruction method is the actual computation time of the reconstruction procedure. We have implemented our iterative reconstruction procedure using some hardware configurations, namely: a computer with 10 cores, (Intel i9-7900X BOX/3800MHz processor), using different GPUs (see Table 3). It is worth noting that our iterative procedure was implemented at assembler level. In Table 2, we show time result for application which is working only on CPU which is develop in Assembler (special vector registers AVX 512 used). In turn, in Table 3, we present time result for application which is working only on GPU accelerators. There are compared those accelerators. It is worth noting that it is very stable time, because deviation is extremely small and that application is very susceptible to parallelisation, because time for one iteration it is getting smaller with on more CUDA Cores assembled in GPU Accelerator.

Table 2. Results of reconstruction on multi threading, i.e. CPU Intel i9-7900X (10-cores, 20-treads).

Threads:	4	8	10	16	20
Avg. time 30000 [ms]	63 724	33 571	29 836	30 532	27 905
Avg. time 20000 [ms]	42 482	22 380	19 890	20 354	18 603
Avg. time 10000 [ms]:	21 241	11 190	9 945	10 177	9 301
Time 1 iteration [ms]:	2,1241	1,1190	0,9945	1,0177	0,9301
HT effectiveness:	–	–	–	0,9094	0,9352
Median for 30000:	63 694	33 542	29 800	30 566	27 854
Deviation std.:	135,69	117,32	217,58	193,88	391,76

Table 3. Results of reconstruction on different models of GPUs accelerator.

GPU:	MSI GTX 1050	ASUS GTX 1080 Ti	nVidia Titan V
Avg. time 30000 [ms]	2 562 175,10	49 699,71	28 858,40
Avg. time 20000 [ms]	170 845,28	33 132,52	19 224,48
Avg. time 10000 [ms]:	85 467,24	16 593,00	9 616,75
Time 1 iteration [ms]:	8,540583	1,656657	0,961947
Median for 30000:	256 229,55	49 703,68	28 861,24
Deviation std.:	0,160806	0,310476	0,010239

According to an assessment of the quality of the obtained images by a radiologist, 7000 iterations are enough to provide an acceptable image for medical purposes. One can compare the results obtained by assessing the view of the reconstructed image in Fig. 4, where the quarter-dose projections were used.

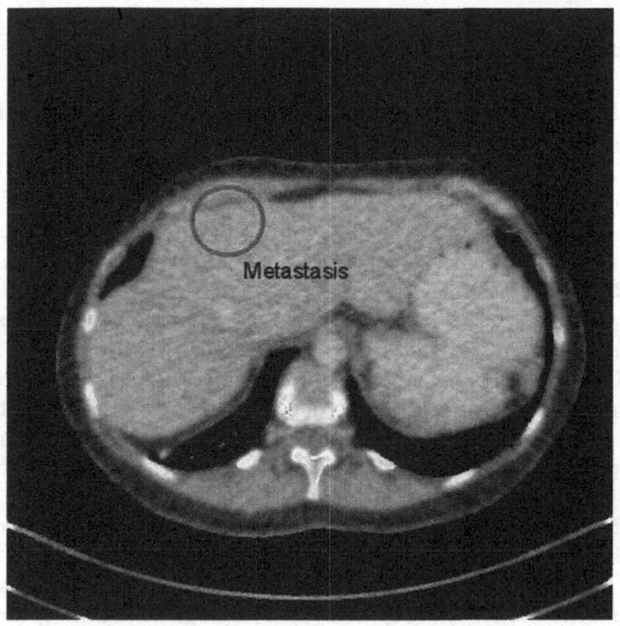

Fig. 4. Obtained image (a case with relative small pathological change in the liver) using quarter-dose projections with application of the statistical method presented in this paper.

4 Conclusion

Comprehensive experiments have been performed, which prove that our reconstruction method is relatively fast (thanks to the use of FFT algorithms) and gives satisfactory results with suppressed noise. It should be noted that approximately the same results were achieved for both hardware implementations: the iterative reconstruction procedure takes less than 7s, mainly thanks to the use of an FFT algorithm in the iterative reconstruction procedure and to the use of the efficient programming techniques. These are rewarding results regarding possibilities of the commercial Veo system (referential MBIR technique), where reconstruction times range between 10 to 90 min depending on the number of reconstructed slices [15]. It means an unacceptable delay between data acquisition and availability for interpretation for emergent indications. Additionally, the hardware used by us is relatively cheap (about 5000 USD) compared to the price of the equipment necessary for the referential solution. It should be emphasized that the designed by us statistical approach (formerly formulated for CT scanner with parallel beam geometry) can be adapted for helical scanners with various geometries, e.g. with cone-beams or with x-tube with flying focal spot.

References

1. Lewitt, R.M.: Reconstruction algorithms: transform methods. Proc. IEEE **71**, 390–408 (1983)
2. Censor, Y.: Finite SER-expansion reconstruction methods. Proc. IEEE **71**, 409–419 (1983)
3. Mathews, J.D., et al.: Cancer risk in 680 peolpe exposed to computed tomography scans in childhood or adolescence: data linkage study of 11 million Australians. Br. Med. J. **f2360**, 346–360 (2013)
4. Sauer, K., Bouman, C.: A local update strategy for iterative reconstruction from projections. IEEE Tran. Signal Proc. **41**, 534–548 (1993)
5. Bouman, C.A., Sauer, K.: A unified approach to statistical tomography using coordinate descent optimization. IEEE Tran. Image Proc. **5**, 480–492 (1996)
6. Thibault, J.-B., Sauer, K.D., Bouman, C.A., Hsieh, J.: A three-dimensional statistical approach to improved image quality for multislice helical CT. Med. Phys. **34**, 4526–4544 (2007)
7. DeMan, B., Basu, S.: Distance-driven projection and backprojection in three dimensions. Phys. Med. Biol. **49**, 2463–2475 (2004)
8. Zhou, Y., Thibault, J.-B., Bouman, C.A., Hsieh, J., Sauer, K.D.: Fast model-based X-ray CT reconstruction using spatially non-homogeneous ICD optimization. IEEE Tran. Image Proc. **20**, 161–175 (2011)
9. Cierniak, R.: A new approach to tomographic image reconstruction using a Hopfield-type neural network. Int. J. Artif. Intell. Med. **43**, 113–125 (2008)
10. Cierniak, R.: A new approach to image reconstruction from projections problem using a recurrent neural network. Int. J. Appl. Math. Comput. Sci. **183**, 147–157 (2008)
11. Cierniak, R.: New neural network algorithm for image reconstruction from fan-beam projections. Neurocomputing **72**, 3238–3244 (2009)
12. Cierniak, R., Pluta, P., Kaźmierczak, A.: A practical statistical approach to the reconstruction problem using a single slice rebinning method. J. Artif. Intell. Soft Comput. Res. **10**, 137–149 (2020)
13. Thibault, J.-B., Bouman, C.A., Sauer, K.D., Hsieh, J.: A recursive filter noise reduction in statistical iterative tomographic imaging. In: Proceedings of SPIE-IS&T Symposium on Electronic Imaging Science and Technology-Computational Imaging, vol. 6065, pp. 15–19 (2006)
14. Rudin, L.I., Osher, S., Fatemi, E.: Nonlinear total variation based noise removal algorithms. Physica D **60**, 259–268 (1992)
15. Geyer, L.L., et al.: State of the art: iterative CT reconstruction techniques. Radiology **276**, 339–357 (2017)

Musculoskeletal Model of Human Lower Limbs in Gait Simulation

Adrianna Bielak(✉), Radosław Bednarski, and Adam Wojciechowski

Institute of Information Technology, Lodz University of Technology,
Wólczańska 215, 90-924 Lodz, Poland
adrianna.bielak@dokt.p.lodz.pl
http://it.p.lodz.pl

Abstract. One of the most important issues in the context of contemporary research on a human body biomechanics is musculoskeletal models. It turns out that the need for biomechanically correct movement model is enormous not only in medicine and rehabilitation, but also in animations synthesis with computer game engines, where unsightly effects often occur between animation keys. Proposing a musculoskeletal model that could be efficiently adapted to a game development environment would speed up work and solve the difficulties in controlling movement sequence synthesis. The proposed model, using dynamic inversion simulation, includes Usik's equations which have been added to the 6DoF skeletal module. Usik uses the thermomechanics of a continuous medium, which takes into account the cross-effects of mechanical, electrical, chemical and thermodynamic phenomena in muscle tissue which makes it possible to better (closer to the nature) describe the reaction of muscles during movement. The validation value of the model correctness is the GRF (ground reaction force) where the simulated values are compared to the measured one. This work concerns human gait and is the basis for further development in the context of the identified problem in the field of computer game engines. Due to the more accurate GRF results against existing solutions, the proposed one is optimistic for further work on it.

Keywords: Musculoskeletal model · Biomechanic · Gait simulation

1 Introduction

Simulation of human gait is a complex phenomenon. It is investigated on many levels e.g.: character animations, medical diagnosis or rehabilitation. The most basic form of human ,locomotion is walking and this movement is analyzed in this article. Healthy gait occurs when the right and left parts of the human body perform similar movements in relation to the anatomical planes of the body. The gait can be analyzed by deriving flat dynamic models describing the movements occurring in the sagittal and frontal plane of the body [1,4,6].

The authors consider that simulation due to its use in blending computer animation. Modern game engines offer a number of functionalities for that operation. They hold the key animation frames of skeletal poses. There can still be

D. Groen et al. (Eds.): ICCS 2022, LNCS 13351, pp. 507–513, 2022.
https://doi.org/10.1007/978-3-031-08754-7_56

unsightly transitions between those keys. These types of problems are so far visible e.g. in sim racing (animation of the driver's hands) or in the movement of humanoid bots. It is necessary to resort to musculoskeletal modeling to solve it. The authors decided on the already existing 6DOF model because more degrees of freedom are not needed for the previously described situations. The method of placing the muscles is shown on the left side in the Fig. 1. The calculations for controlling movement are contained in the muscle module that the authors have implemented by using the Usik's mathematical model. It is a novelty because Usik's model has not been used before for this type solution.

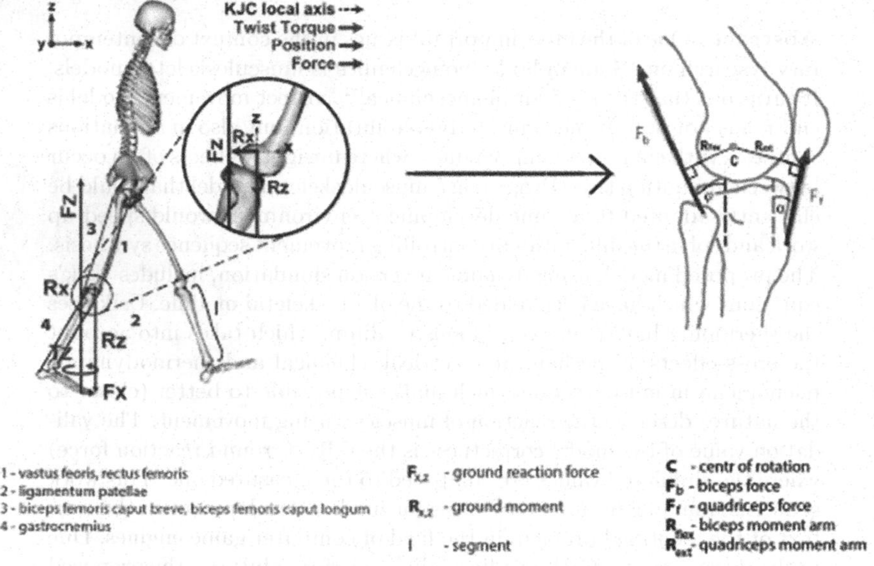

Fig. 1. (From the left) Forces in a lower limb and forces in a joint.

Dynamic inversion has been chosen for that solution. The input data were downloaded from the available database [2]. The output data was compared with the measurement one (Rigid Body Dynamic Library [2]) by comparing the ground reaction forces (vertical and horizontal one) (Fig. 1). This method of validation was chosen to compare the authors' results to other solutions from recent years such as Belaise et al. [5], Sartori M. et al. [8], Yamasaki et al. [7], Wojnicz W. [6], Moissenet's et al. [2]. The authors concluded that adding muscles using the Usik's model can improve simulation results, especially for the initial and final phase of gait. Thanks to that it gives an optimistic, potential solution to the problem of biomechanically correct blending for human movement animation.

2 Methodology

As part of this work the following were carried out:
- implementation of the 6DoF model in C++ with Vulkan API,
- creation a parser for the .c3d format, in which the input data was stored,
- creation of two gastrocnemius muscles based on the author's model,
- set the relationship between the 6DoF model and the muscle model,
- saving parameters on charts during a gait simulation.

2.1 Skeletal Model

The selected biomechanical model is defined by two 6DoF models: open for one-legged support and closed for two-legged support. The complete systems of the equation are contained in publication [6]. The first of them was presented in this work in Eq. 1 and 2 to indicate the occurrence of the influence of mass (as a parameter) on a given segment.

$$\begin{aligned} A_{11} \cdot \ddot{\alpha_1} + A_{12}(\alpha_1, \alpha_2) \cdot \ddot{\alpha_2} + A_{13}(\alpha_1, \alpha_3) \cdot \ddot{\alpha_3} \\ + A_{14}(\alpha_1, \alpha_4) \cdot \ddot{\alpha_4} + A_{15}(\alpha_1, \alpha_5) \cdot \ddot{\alpha_5} \\ + A_{16}(\alpha_1, \alpha_6) \cdot \ddot{\alpha_6} + A_{17}(\alpha_1, \alpha_7) \cdot \ddot{\alpha_7} \\ = M_1 - L_1 \cdot sin(\alpha_1) \cdot F_y + L_1 \cdot cos(\alpha_1) \cdot F_z \end{aligned} \tag{1}$$

where:

$$\begin{aligned} M_1 = M_{ex1} - M_{12} - cos(\alpha_1) \cdot (S_1 \cdot m_1 \cdot g + L_1 \cdot g \cdot \sum_{i=2}^{7} m_i) \\ + L_1 \cdot sin(\alpha_j - \alpha 1) \cdot (\dot{\alpha_j})^2 \cdot (S_j \cdot m_j + L_j \cdot \sum_{k=j+1}^{7} m_k) \\ + L_1 \cdot sin(\alpha_6 - \alpha 1) \cdot (\dot{\alpha_6})^2 \cdot S_6 \cdot m_6 \\ + L_1 \cdot sin(\alpha_7 - \alpha 1) \cdot (\dot{\alpha_7})^2 \cdot S_7 \cdot m_7 \\ for j = 2, 3, 4, 5 \\ \dots \end{aligned} \tag{2}$$

A_i - i-th segment, L_i - length of the segment, S_i - segment radius, M_i - moment derived from the ground reaction components, α - rotational displacement of the segment, $\dot{\alpha_i}$ - segment angular velocity, $\ddot{\alpha_j}$ - segment angular acceleration, m - mass of the segment.

2.2 Muscle Modeling

The process of the muscle modeling had a three steps: generate a muscle fiber representation using a B-spline solid, create a model with virtual muscle fibers, apply Usik's model as a behavior model [3].

2.3 3D Fiber Representation

The muscle fiber was chosen as the basic element and was implemented virtually as the volumetric B-spline solid (Fig. 2). A detailed description of the fiber formation method using by the authors was described in publication [3].

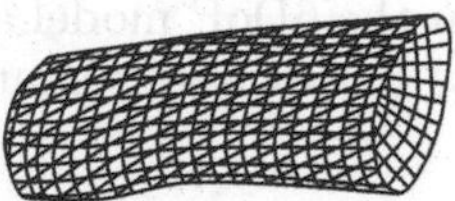

Fig. 2. A tesselated B-spline solid for a fiber muscle representation [3].

2.4 Muscle Model

For this solution, the gastrocnemius muscle was selected. It was created using the finite volume method with B-spline curves and Usik's equations. This volume was a fiber representation with two attachment points. Control point was represented by the Eq. 3:

$$V(u, v, w) = \sum_{i=0}^{l}\sum_{j=0}^{m}\sum_{k=0}^{n} B_i^u(u)B_j^v(v)B_k^w(w)C_{ijk} \tag{3}$$

where each $C_{ijk} \in R^3$.

$C = C_{ijk}$ - the set of points form a control point lattice which will influence the shape of the B-spline solid, V - a parametric solid given by the tritensor product of these B-spline basis functions (in this case, the polynomials $B_i^u(u)B_j^v(v)B_k^w(w)$) with the control points in C (Fig. 3).

2.5 Combination of Muscle Module with Skeletal Biomechanical Module

A module with two gastrocnemius muscles was added to the biomechanical skeletal model. During gait simulation, the forces were applying on each joint. They were as the base parameter of calf control points located at the muscle joints.

The use of the molecular weight of the k-th component is important in the proposed solution. This parameter is included in one of the equations from the Usik's model, specifically in the mass balance equation of the k-th component (Eqs. 4).

$$\begin{aligned} \rho\delta y_k^1 \div \delta t &= Q_k^{out} - Q_k^{in} \\ \rho\delta y_k^2 \div \delta t &= Q_k^{in} + \sum_{k=1}^{n} M_k V_{kj} I_j \end{aligned} \tag{4}$$

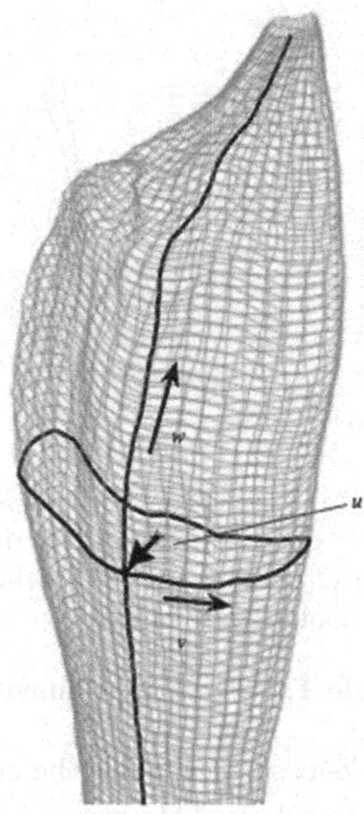

Fig. 3. A fragment of a gastrocnemius muscle with the u, v, w coordinates for any control point. Those values are the Cartesian coordinates that compose the boundary and volume of the solid

where k- k-th element, ρ - mass density, y- , t- time , Q^{out}- source density in exchange with the environment , Q^{in}- source density in inter-phase exchange, M- molecular weight, v- speed , I- chemical reaction rate.

K-th components are control points in the finite volumes of muscle fibers (Fig. 2). The sum of the molecules masses M_k from Eqs. 4 was treated as the mass m_i of the each knee joint segment in Eqs. 2 and 3 [9].

2.6 Results

Below is a summary of the author's charts with the calculated data obtained after adding the author's muscle. Table 1 and 2 contain values of the area under the curve of a given force during one walking cycle. The cycle has been divided into 100 compartments (Fig. 4).

Interpolation of 4 and 5 degree polynomial was used. The charts provide an illustrative course of validation parameters.

Table 1 and 2 compare the values of the fields below the chart. This method was used due to the fact that the graphs from the simulations intersected with the graph from the measurements only in 3 places, which gave ease in applying the method of counting the areas under the graph as the final validation method.

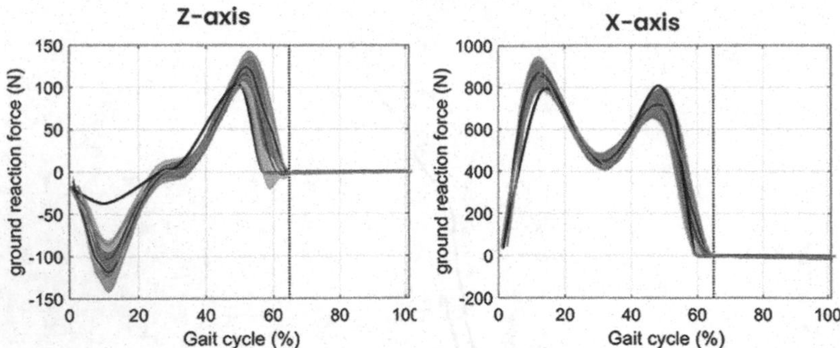

Fig. 4. The upper graph shows the ground force response values for the Z-axis and the lower graph for the X-axis. Black curve - measurement data from [6], blue curve - data from [6], red curve - data from [2], orange curve - authors' data. Red and blue areas are the possible limit of error. (Color figure online)

Table 1. Validation parameters

Reference	Z-axis - area under the graph		
model	Simulation	Measuring	Difference
[6]	48.56	42.24	6.32
[2]	46.15	42.24	3.91
Authors	40.12	42.24	2.12

Table 2. Validation parameters

Reference	X-axis - area under the graph		
model	Simulation	Measuring	Difference
[6]	56.50	51.43	5.07
[2]	54.70	51.43	3.27
Authors	49.88	51.43	1.55

3 Conclusions

It can be seen in Table 1 and 2 that the proprietary solution increased the correctness of the final results for the values of ground reaction forces. To further improve the results (bring the values from the simulation closer to the measured values), it is worth breaking the muscles into smaller sections, which would allow for a more accurate representation of the behavior of soft tissues.

Due to the better results, the next planned step is to create plug-in based on that solution for UE5 game engine as an open-source tool which would solve the problem described in the Introduction. If it turned out to be helpful, it would be worth extending this solution with other parts of the human body.

References

1. Ezati, M., Ghannadi, B., McPhee, J.: A review of simulation methods for human movement dynamics with emphasis on gait. Multibody Syst. Dyn. **47**(3), 265–292 (2019). https://doi.org/10.1007/s11044-019-09685-1
2. Moissenet, F., Bélaise, C., Piche, E., Michaud, B., Begon, M.: Centre National de Rééducation Fonctionnelle et de Réadaptation-Rehazenter. Luxembourg, Luxembourg (2019)
3. Pietka, E., Badura, P., Kawa, J., Wieclawek, W. (eds.): ITIB 2018. AISC, vol. 762. Springer, Cham (2019). https://doi.org/10.1007/978-3-319-91211-0
4. Bednarski, R., Bielak, A.: Use of motion capture in assisted of knee ligament injury diagnosis. J. Appl. Comput. Sci. **26**(1), 7–32 (2018)
5. Bélaise, C., Dal Maso, F., Michaud, B., Mombaur, K., Begon, M.: An EMG-marker tracking optimisation method for estimating muscle forces. Multibody Syst. Dyn. **42**(2), 119–143 (2017). https://doi.org/10.1007/s11044-017-9587-2
6. Wojnicz, W.: Biomechanical models of the human musculoskeletal system, Gdansk University of Technology (2018)
7. Yamasaki, T., Idehara, K., Xin, X.: Estimation of muscle activity using higher-order derivatives, static optimization, and forward-inverse dynamics. J. Biomech. **49**, 2015–2022 (2016). https://doi.org/10.1016/j.jbiomech.2016.04.024
8. Sartori, M., Farina, D., Lloyd, D.G.: Hybrid neuromusculoskeletal modeling to best track joint moments using a balance between muscle excitations derived from electromyograms and optimization. J. Biomech. **47**, 3613–3621 (2014). https://doi.org/10.1016/j.jbiomech.2014.10.009
9. Usik, P.I.: Continual mechanochemical model of muscular tissue. PMM **37**(3), 448–458 (1973)

Tissue Damage Control Algorithm for Hyperthermia Based Cancer Treatments

Gustavo Resende Fatigate[2], Rafael Felipe Coelho Neves[3], Marcelo Lobosco[1,2(✉)], and Ruy Freitas Reis[1]

[1] Departamento de Ciência da Computação, Universidade Federal de Juiz de Fora, Juiz de Fora, Brazil
marcelo.lobosco@ice.ufjf.br

[2] Pós-Graduação em Modelagem Computacional, Universidade Federal de Juiz de Fora, Juiz de Fora, Brazil

[3] Instituto Federal de Educação, Ciência e Tecnologia do Sul de Minas Gerais, Poços de Caldas, Brazil

Abstract. Cancer is a worldwide health problem. The fatality rate of some types of cancer motivates the scientific community to improve the standard techniques used to fight against this disease, as well as to investigate new forms of treatments. One of these emerging treatments is hyperthermia using the injection of magnetic nanoparticles into the tumour area. Its basic idea is to heat the target tumour tissue leading to its necrosis. This study simulates the bioheat processes using Pennes' model to evaluate the tissue damage *in silico*. Furthermore, the differential evolution optimisation technique is applied to suggest the optimal location of injection considering the minimisation of damage to the healthy tissue and the maximisation of the tumour necrosis. The results suggest that the proposed algorithm is a promising tool for aiding hyperthermia-based treatment planning.

Keywords: Hyperthermia · Cancer · Bioheat · Optimization · Differential evolution

1 Introduction

According to the World Health Organisation (WHO), cancer is the second biggest cause of death worldwide, and it is estimated that around 9.6 million people died in 2018 due to this disease [28]. Nearly 70% of the cancer cases occurs in low or middle-income countries. Furthermore, 30% of the deaths could be avoid with the early diagnosis and proper treatment [16].

Due to its high mortality rates, cancer is a public health concern worldwide, which motivates the scientific community to study and develop new strategies to fight against this disease. Among the methods to treat cancer patients, in this study we highlight hyperthermia which works as an adjuvant technique to

D. Groen et al. (Eds.): ICCS 2022, LNCS 13351, pp. 514–525, 2022.
https://doi.org/10.1007/978-3-031-08754-7_57

help existing treatments such as radiotherapy and chemotherapy. Recent studies show that hyperthermia is also presenting relevant results when combined with innovative clinical options such as hadron therapy for treating pancreatic cancer [4].

Hyperthermia procedure consists in overheating the tumour tissue over 43 °C degrees leading the tissue to necrosis [21]. The necrosis is obtained through the injection of nanoparticles into the tumour tissue. The nanoparticles produce heat through Brownian and Neelian relaxation when submitted to a magnetic field using low frequency [14]. This treatment has the advantage of being semi-invasive once it can reach the tumour intravenously or through direct injection.

The Pennes model [17] is commonly used to describe the heat transfer in living tissue, presenting similarities between the theoretical and experimental results [15,21,23,27]. A large number of works use this model in order to study the bioheat [6–8,23]. Moreover, the original Pennes model can be modified to include the hyperthermia cancer treatment process [2,13,18,25,26].

This study uses differential evolution (DE) to search for the best position for the injection of the magnetic nanoparticles to maximise the death of the tumour tissue and to minimise the healthy tissue affected by the hyperthermia process. DE is a heuristic-based algorithm used to find a set of parameters that minimise a known function [3,12,19,20]. Furthermore, DE is already applied in correlated studies of hyperthermia to optimise the value of radio frequency power, amplitude and/or phase [5,9,29]. The heat distribution is evaluated using the Pennes bioheat model through partial differential equations (PDE) via the finite difference method (FDM). To evaluate the objective function in DE we use the bioheat model to obtain the temperature distribution of the modelled tissue after 50 min of hyperthermia treatment. This information is used to measure the amount of healthy and tumour tissues influenced by the process.

We organise this paper as follows. Section 2 describes the bioheat model, numerical approximation and the optimisation strategy. The results are presented in Sect. 3 and discussed in Sect. 4. Finally, Sect. 5 presents the conclusions and plans for future work.

2 Methods

2.1 Mathematical Model

The choice of the Pennes model is due to its simplicity, reducing the computational cost of simulations. So, including the hyperthermia heat source the model is expressed as follows:

$$\begin{cases} \rho c \dfrac{\partial T}{\partial t} = \nabla \cdot k \nabla T + \omega_b \rho_b c_b (T_a + T) + Q_m + Q_r & \text{in } \Omega \times I \\ k \nabla T \cdot \vec{n} = 0 & \text{in } \partial\Omega \times I \\ T(\cdot, 0) = T_0 & \text{in } \Omega, \end{cases} \tag{1}$$

where $\Omega \subset \mathbb{R}^2$ is the equation spatial domain, $I \subset \mathbb{R}^+$ is the time domain, $T : \Omega \times I \to \mathbb{R}^+$ is the tissue temperature field; ρ, c and k are density, specific heat and thermal conductivity of the tissue, respectively; ρ_b, c_b and ω_b are density, specific heat of the blood and blood perfusion, respectively; T_a and T_0 are the blood temperature and the initial temperature, respectively; Q_m and Q_r are the metabolic heat source and the heat generated by the hyperthermia treatment, respectively.

It is important to emphasise the simplifications made by Pennes in the proposition of his model. The heat exchanges between blood and the tissue occur between capillaries and arterioles. The blood flow is assumed to be isotropic, *i.e.* there is no directional preference. The vascular geometry was disregarded, and the blood reaches the arterioles at a body temperature, in our case 37 °C [10].

The heat generated by the hyperthermia treatment (Q_r) is calculated by the specific absorption rate (SAR) through the Eq. (2) [22]. This equation models the heat generated considering a set of injections used in the hyperthermia process. In this study it was approximated by the following equation:

$$Q_r = \sum_{i=1}^{N_p} A e^{-r(\vec{x})_i^2 / r_{0,i}^2}, \tag{2}$$

where N_p is the number of injections points in the tissue; A is the energy maximum strength of the volumetric heat generation rate, $r(\vec{x})_i^2$ is the Euclidean distance to the injection point, *i.e.* $r = ||\vec{x} - \vec{x_0}||_2$; x_0 is the injection position; r_0 is the radius of coverage of hyperthermia.

2.2 Numerical Scheme

The numerical method used to solve Eq. (1) is the Finite Difference Method (FDM), considering a heterogeneous medium. We consider the closed domain Ω discretised into a set of regular points defined by $S_s = \{(x_i, y_j); i = 0, 1, \cdots, N_x; j = 0, 1, \cdots, N_y\}$, where N_x and N_y are the number of intervals of length $h_x = h_y = h$. Moreover, the time discretisation of the time domain I is partitioned into N_t equal time intervals of length h_t, *i.e.* $S_t = \{(t_n); n = 0, 1, \cdots, N_t\}$. To obtain the discrete form of the model, we employ an FTCS (Forward-Time Central-Space), resulting in an explicit numerical method. This scheme has convergence order $O(h^2, h_t)$ [11].

$$\begin{aligned} T_{i,j}^{n+1} &= \frac{h_t}{\rho c}\left[\frac{k_{i+1/2,j}(T_{i+1,j}^n - T_{i,j}^n) - k_{i-1/2,j}(T_{i,j}^n - T_{i-1,j}^n)}{h^2}\right. \\ &\quad + \frac{k_{i,j+1/2}(T_{i,j+1}^n - T_{i,j}^n) - k_{i,j-1/2}(T_{i,j}^n - T_{i,j-1}^n)}{h^2} \\ &\quad \left. + \rho_b c_b \omega_b \left(T_a - T_{i,j}^n\right) + Q_m + Q_r\right] + T_{i,j}^n. \end{aligned} \tag{3}$$

2.3 Differential Evolution

Differential evolution (DE) is a stochastic-heuristic algorithm for global optimisation [24]. DE is based on natural evolution, presenting generations, selections, mutations and the capacity of an individual to survive the environment [1].

As an evolutionary algorithm, DE works in generations and presents a population with a fixed number of individuals. Given an initial population, the next generation (or offspring) is formed considering mutation and the crossing between the individuals of the same generation. This process continues until it achieves the convergence of results or the maximum number of generations.

To maximise the damage to the tumour tissue and to minimise the damage to the healthy tissue, we employed the following minimisation problem:

$$\min O(p) = 300 - N_t - (100 - N_h) - 100\beta, \tag{4}$$

where p is the set of points to be estimated. $N_t \in [0, 100]$ is the percentage of tumour tissue necrosis and $N_h \in [0, 100]$ is the percentage of healthy tissue necrosis. $\beta \in \{0, 1\}$ is a variable that assumes 1 when entire tumour reaches 43°C or more and 0 otherwise.

We consider the mutation strategy *best/1/bin*, *i.e.* the best individual (X_{best}) and two more random individuals are chosen (X_a and X_b). The random individuals are subtracted and multiplied by the mutation factor F. The result is added to X_{best}, originating the mutation vector X_p as shown in Eq (5):

$$X_p^{i+1} = X_{best}^i + F * (X_a^i - X_b^i). \tag{5}$$

The next step is the crossover operation applied to individuals of the same generation. This operation considers the mutation vector and a target vector, randomly chosen from the population. A new vector of individuals is created and its content depends on a random number generated for each position of the vector. If the number is smaller than the crossing constant C, the value is taken from the mutation vector, otherwise, the value is taken from the target vector, *i.e.*:

$$U^{i+1} = \begin{cases} X_p^{i+1}, & \text{if } r_i \leq C \\ X_r^i, & \text{otherwise.} \end{cases} \tag{6}$$

Finally, the fitness of the trial vector is calculated considering the objective function and compared with the target vector. The individuals with smaller fitness are considered to pass to the next generation.

3 Results

All results were obtained using of Google Colab platform and C/C++ programming language.

We consider a two-dimensional squared domain of lengths equal to 0.1 m representing the simulated tissue and circular tumours with radii equal to 0.01 m. We perform three scenarios: 1) one tumour and one injection point (see Fig. 1A), 2) two tumours and two injections points (see Fig. 2A), and 3) three tumours and three injections points (see Fig. 3A). All tumours were randomly positioned in the mesh. The simulated domain is represented in Figs. 1–3 A. Besides, Eq. (1) is solved using the parameters described in Tables 1 and 2, and the initial condition $T_0 = 37.0$ is used for all performed simulations.

Table 1. Parameters used to solve Eq. (1) for both tumour and healthy tissues.

Parameters	Unit	Healthy tissue	Tumour tissue
k	W/m^oC	0.51	0.64
ω_b	s^{-1}	5.0×10^{-4}	1.25×10^{-3}
ρ	Kg/m^3	1000.0	1000.0
ρ_b	Kg/m^3	1000.0	1000.0
Q_m	W/m^3	420.0	4200.0
c	J/Kg^oC	4200.0	4200.0
c_b	J/Kg^oC	4200.0	4200.0

Table 2. Parameters used in the hypertermia treatment (see Eq. (2)).

Parameters	Unit	Value
N_p	–	3
A	W	0.05×10^6
r_0	m	1.9×10^{-2}

For each scenario, the optimisation process was performed 10 times, considering different seeds to ensure that the algorithm does not converge to a local minimum. Figures 4, 5 and 6 present a boxplot of all performed simulation results, considering one, two and three tumours, respectively. It is important to note that all points presented a standard deviation of the order of 10^{-3}.

Figures 1C, 2C and 3C show the simulation of the hyperthermia process considering the best result obtained by the optimisation process. Moreover, in all three scenarios, we perform a naive tentative considering the injection point in the centre of the tumour, which is illustrated by Figs. 1B, 2B and 3B. These results represents the temperature distribution of the tissue after 50 min of treatment. The black solid line delimits the entire part of the tissue that reached 43° or more, *i.e.* the region of the tissue that suffered necrosis, and the black dots represents the injection points.

In the first scenario, we consider a tumour centred at $(0.050, 0.050)$ (Fig. 1A). Figure 1C presents the result of the optimisation, and the best position found is in the centre of the tumour, *i.e.* the same as the naive approach (Fig. 1B).

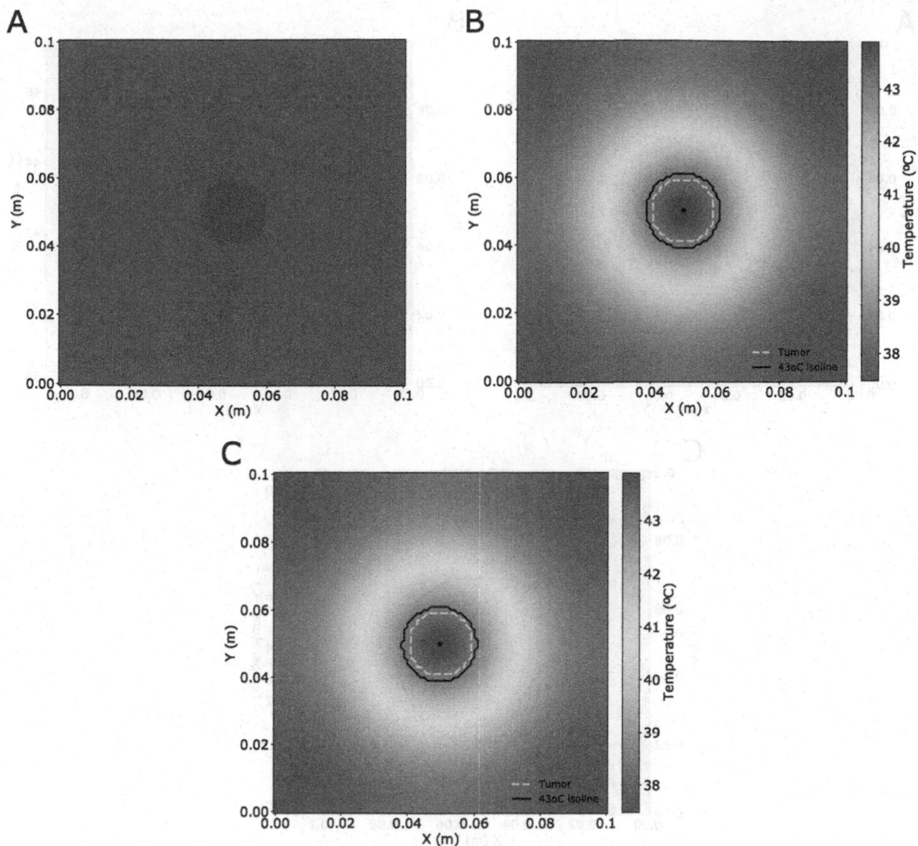

Fig. 1. The simulated tissue and the results for the first scenario. In Panel A the red area represents the tumour tissue, and the blue area the healthy tissue. The tumour has a radius of 0.01 m, and its centre is positioned at (0.050, 0.050). Panels B and C present the results of the numerical solution of Eq. (1) at $t = 50$ min. In Panel B, the simulation considers $P_1 = (0.050, 0.050)$, *i.e.* the centre of the tumour. In Panel C, the optimisation processes found $P_1 = (0.0498984, 0.0497171)$. The black dots represent the position of P_1, the solid black contour highlights the portion of the domain that reaches $T \geq 43°$, and the dashed grey contour indicates the tumour location.

Figure 2 shows the results of the second scenario, *i.e.* the tissue with two tumours centred at (0.050, 0.030) and (0.050, 0.060), respectively. In this case, the optimisation result is different from the naive tentative: the amount of healthy tissue affected by the treatment is reduced from 16.51% (Fig. 2 B) to 15.18% (Fig. 2 C). In other words, the optimisation reduced the damage to the healthy tissue.

Finally, the third scenario was performed in the tissue with three tumours (Fig. 3A). In this case, the first tumour has its centre located at (0.035, 0.025), the second tumour at (0.065, 0.050), and the third tumour at (0.035, 0.075). In

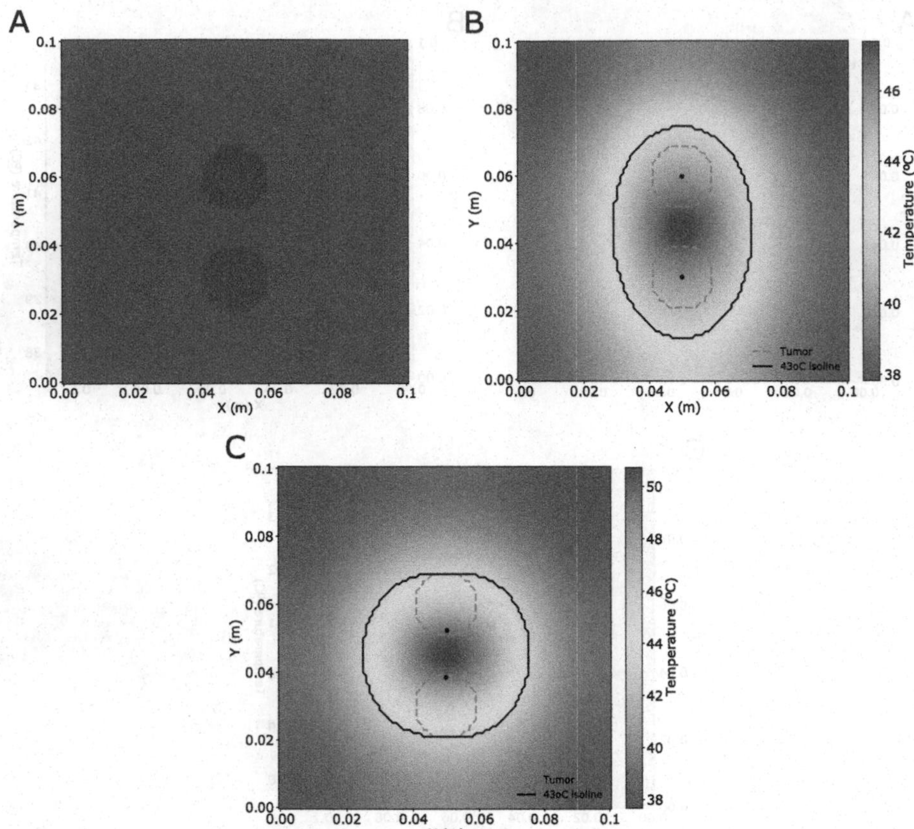

Fig. 2. The simulated tissue and the results for the second scenario. In Panel A, the red area represents the tumour tissue, and the blue area the healthy tissue. The tumours have a radius of 0.01 m each and their centres are positioned at $(0.050, 0.030)$, and $(0.050, 0.060)$, respectively. Panels B and C present the results of the numerical solution of Eq. (1) at $t = 50$ min. In Panel B, the simulation considers $P_1 = (0.050, 0.030)$, and $P_2 = (0.050, 0.060)$, *i.e.* the centre of each tumour. In Panel C, the optimisation processes found $P_1 = (0.0500118, 0.0384764)$, and $P_2 = (0.0503467, 0.0523711)$. The black dots represent the position of P_1 and P_2, the solid black contour highlights the portion of the domain that reaches $T \geq 43°$, and the dashed grey contour indicates the tumour location.

the third scenario, the optimised solution is different from the naive tentative, similarly to the observed in the second scenario. In this case, the naive tentative result in 40% of healthy tissue necrosis while the optimised one affects only about 30% of the healthy tissue.

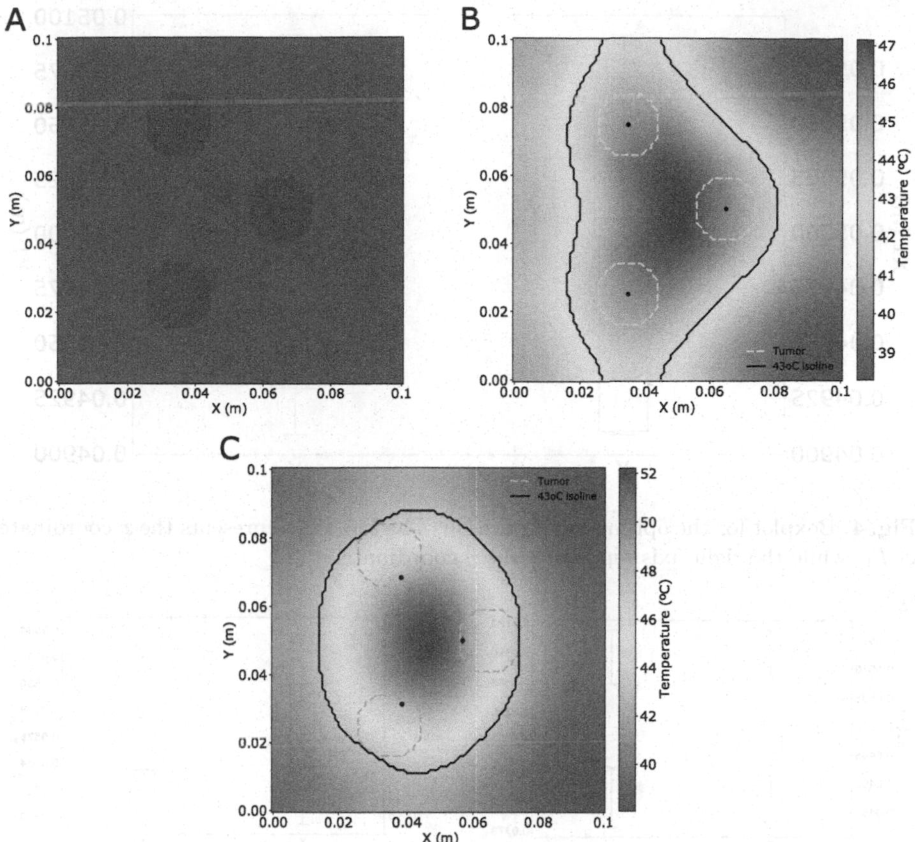

Fig. 3. The simulated tissue and the results for the third scenario. In Panel A, the red area denotes the tumour tissue, and the blue area the healthy tissue. The tumours have a radius of 0.01 m each and their centres are positioned at (0.035, 0.025), (0.065, 0.050) and (0.035, 0.075). Panels B and C present the results for numerical solution of Eq. (1) at $t = 50$ min. In Panel B, the simulation considers $P_1 = (0.035, 0.025)$, $P_2 = (0.065, 0.050)$, and $P_3 = (0.035, 0.075)$, *i.e.* the centre of each tumour. In Panel C, the optimisation processes found $P_1 = (0.0313107, 0.0614578)$, $P_2 = (0.0500641, 0.0432409)$, and $P_3 = (0.0684779, 0.0617488)$. The black dots represent the position of P_1, P_2 and P_3, the solid black contour highlights the portion of the domain that reaches $T \geq 43°$, and the dashed grey contour indicates the tumour location.

4 Discussion

Figure 4, 5 and 6 show a boxplot for 10 executions of the optimisation process. All executions succeed in obtaining the necrosis of the entire tumour tissue, suggesting that the proposed scheme offers a robust algorithm for hyperthermia-based cancer treatments. Moreover, Fig. 4, 5 and 6 show that even considering

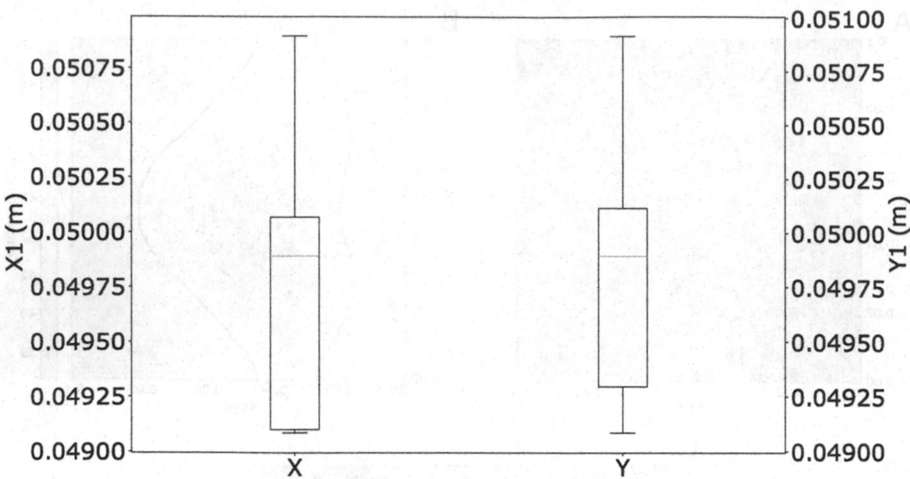

Fig. 4. Boxplot for the optimised P_1 position. The left axis represents the x coordinate of P_1, while the right axis represents the y coordinate.

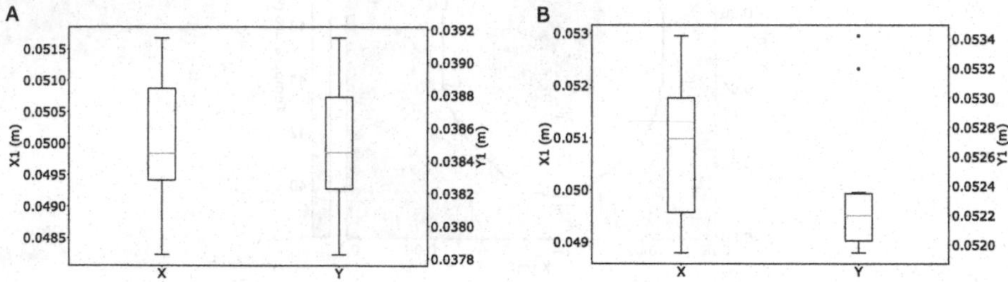

Fig. 5. Panels A and B show the boxplots for the optimised P_1 and P_2 positions, respectively, considering all the ten tests. The left axis represents the x coordinate of the point, while the right axis represents the y coordinate.

multiple executions of the proposed scheme, the solution converges for a similar result. Moreover, the results present a standard deviation smaller than 2%.

Furthermore, the proposed strategy proved to be useful, once the optimisation process results in a non-trivial solution that minimises tissue damage. For example, Fig. 3 B demonstrate that the trivial guess does succeed in destroying the tumour, but produces higher damage to the healthy tissue. The proposed strategy succeeds in causing less damage to the healthy tissue by formulating a minimisation problem that penalises damage to healthy tissue and rewards total tumour destruction. It is worthwhile to notice that none of the injection points has direct contact with the tumours. A possible explanation is that cancerous tissues have higher values of thermal conductivity, blood perfusion and metabolic heat.

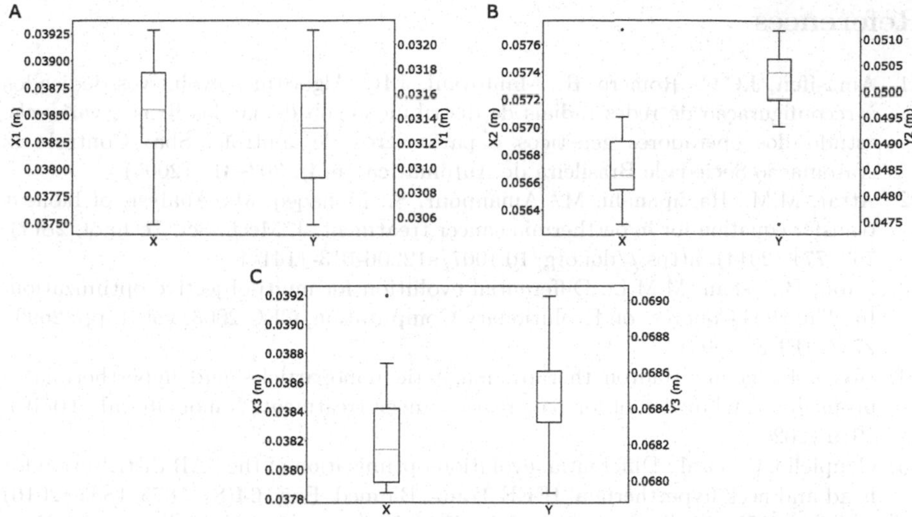

Fig. 6. Panels A, B and C show the boxplots for the optimised P_1, P_2 and P_3 positions, respectively, considering all the ten tests. The left axis represents the x coordinate of the point, while the right axis represents the y coordinate.

5 Conclusions and Future Works

This work presents a tool for aiding hyperthermia-based treatment planning. The proposed algorithm suggests that it is possible to find optimal locations for injecting the nanoparticles and for inducing the heat on hyperthermia treatment considering both damages of the entire tumour and minimal damage to healthy tissue. Moreover, this study also indicates that the optimal position for injecting the nanoparticles might be in a non-trivial location such as in the healthy tissue near the target area instead of injecting inside the tumour site.

For future works, we are planning to expand the simulation to a more realistic tissue, considering a three-dimensional domain as well as patient-based tumour formats. Moreover, a parallel strategy is necessary, once the average time for obtaining each optimised site is 2 h 31 m. So, the reduction of computational time is going to be even more relevant when considering the three-dimensional model.

Acknowledgments. The authors would like to express their thanks to CAPES (Finance Code 001), CNPq, FAPEMIG (APQ-01226-21) and UFJF for funding this work.

References

1. Amasifen, J.C.C., Romero, R., Mantovani, J.R.: Algoritmos evolutivos dedicados à reconfiguração de redes radiais de distribuição sob demandas fixas e variáveis: estudo dos operadores genéticos e parâmetros de controle. Sba: Controle & Automação Sociedade Brasileira de Automatica **16**(3), 303–317 (2005)
2. Attar, M.M., Haghpanahi, M., Amanpour, S., Mohaqeq, M.: Analysis of bioheat transfer equation for hyperthermia cancer treatment. J. Mech. Sci. Technol. **28**(2), 763–771 (2014). https://doi.org/10.1007/s12206-013-1141-4
3. Babu, B., Jehan, M.M.L.: Differential evolution for multi-objective optimization. In: The 2003 Congress on Evolutionary Computation, CEC 2003, vol. 4, pp. 2696–2703. IEEE (2003)
4. Brero, F., et al.: Hadron therapy, magnetic nanoparticles and hyperthermia: a promising combined tool for pancreatic cancer treatment. Nanomaterials **10**(10), 1919 (2020)
5. Cappiello, G., et al.: Differential evolution optimization of the SAR distribution for head and neck hyperthermia. IEEE Trans. Biomed. Eng. **64**(8), 1875–1885 (2016)
6. Charny, C.K.: Mathematical models of bioheat transfer. In: Advances in Heat Transfer, vol. 22, pp. 19–155. Elsevier (1992)
7. Ezzat, M.A., AlSowayan, N.S., Al-Muhiameed, Z.I., Ezzat, S.M.: Fractional modelling of Pennes' bioheat transfer equation. Heat Mass Transf. **50**(7), 907–914 (2014)
8. Ferrás, L.L., Ford, N.J., Morgado, M.L., Nóbrega, J.M., Rebelo, M.S.: Fractional Pennes' bioheat equation: theoretical and numerical studies. Fractional Calculus Appl. Anal. **18**(4), 1080–1106 (2015)
9. Gas, P., Miaskowski, A.: Sar optimization for multi-dipole antenna array with regard to local hyperthermia. Przeglad Elektrotechniczny **1**(95), 17–20 (2019)
10. Jiji, L.M.: Heat Conduction. Springer, Heidelberg (2009)
11. LeVeque, R.J.: Finite Difference Methods for Ordinary and Partial Differential Equations: Steady-State and Time-Dependent Problems. SIAM (2007)
12. Liu, R., Fan, J., Jiao, L.: Integration of improved predictive model and adaptive differential evolution based dynamic multi-objective evolutionary optimization algorithm. Appl. Intell. **43**(1), 192–207 (2015). https://doi.org/10.1007/s10489-014-0625-y
13. Miaskowski, A., Sawicki, B.: Magnetic fluid hyperthermia modeling based on phantom measurements and realistic breast model. IEEE Trans. Biomed. Eng. **60**(7), 1806–1813 (2013)
14. Moros, E.: Physics of Thermal Therapy: Fundamentals and Clinical Applications. CRC Press, Boca Raton (2012)
15. Ng, E.Y.K., Kumar, S.D., et al.: Physical mechanism and modeling of heat generation and transfer in magnetic fluid hyperthermia through néElian and Brownian relaxation: a review. Biomed. Eng. Online **16**(1), 1–22 (2017)
16. OPAS/OMS: Organização mundial da saúde (2022). https://www.paho.org/pt/topicos/cancer. Accessed 02 Feb 2022
17. Pennes, H.H.: Analysis of tissue and arterial blood temperature in the resting human forearm. J. Appl. Phisiol. **1**, 93–122 (1948)
18. Reis, R.F., dos Santos Loureiro, F., Lobosco, M.: A parallel 2D numerical simulation of tumor cells necrosis by local hyperthermia. J. Phys. Conf. Ser. **490**(012138) (2014)

19. Rogalsky, T., Kocabiyik, S., Derksen, R.: Differential evolution in aerodynamic optimization. Can. Aeronaut. Space J. **46**(4), 183–190 (2000)
20. Ronkkonen, J., Kukkonen, S., Price, K.V.: Real-parameter optimization with differential evolution. In: 2005 IEEE Congress on Evolutionary Computation, vol. 1, pp. 506–513. IEEE (2005)
21. Salloum, M., Ma, R., Zhu, L.: Enhancement in treatment planning for magnetic nanoparticle hyperthermia: optimization of the heat absorption pattern. Int. J. Hyperth. **25**(4), 309–321 (2009)
22. Salloum, M., Ma, R., Zhu, L.: An in-vivo experimental study of temperature elevations in animal tissue during magnetic nanoparticle hyperthermia. Int. J. Hyperth. **24**(7), 589–601 (2008)
23. Shih, T.C., Yuan, P., Lin, W.L., Kou, H.S.: Analytical analysis of the Pennes bioheat transfer equation with sinusoidal heat flux condition on skin surface. Med. Eng. Phys. **29**(9), 946–953 (2007)
24. Storn, R., Price, K.: Differential evolution-a simple and efficient heuristic for global optimization over continuous spaces. J. Global Optim. **11**(4), 341–359 (1997)
25. Suleman, M., Riaz, S.: 3D in silico study of magnetic fluid hyperthermia of breast tumor using fe3o4 magnetic nanoparticles. J. Therm. Biol. **91** (2020)
26. Tucci, C., Trujillo, M., Berjano, E., Iasiello, M., Andreozzi, A., Vanoli, G.P.: Pennes' bioheat equation vs. porous media approach in computer modeling of radiofrequency tumor ablation. Sci. Rep. **11**(1), 1–13 (2021)
27. Valente, A., Peters, F.C., de Souza, R.V.M., Mansur, W.J.: 3D numerical simulation of real-time temperature field in a hyperthermia cancer treatment using octree meshes. J. Braz. Soc. Mech. Sci. Eng. **43**(1), 1–11 (2021)
28. WHO: World health organization (2021). http://www.who.int/. Accessed 20 Dec 2021
29. Xu, L., Wang, X.: Comparison of two optimization algorithms for focused microwave breast cancer hyperthermia. In: 2018 International Applied Computational Electromagnetics Society Symposium-China (ACES), pp. 1–2. IEEE (2018)

Resting-State EEG Classification for PNES Diagnosis

Chiara Zucco[1(✉)], Barbara Calabrese[1], Rossana Mancuso[1], Miriam Sturniolo[2], Antonio Gambardella[2], and Mario Cannataro[1]

[1] Data Analytics Research Center, Department of Medical and Surgical Sciences, University "Magna Græcia" of Catanzaro, Catanzaro, Italy
{chiara.zucco,calabreseb,cannataro}@unicz.it

[2] Clinic of Neurology, Policlinico Mater Domini, University "Magna Græcia" of Catanzaro, Catanzaro, Italy

Abstract. Psychogenic Non-Epileptic Seizure (PNES) represents a neurological disorder often diagnosed and pharmacologically treated as epilepsy. PNES subjects show the same symptoms as epileptic patients but do not have an EEG characterized by ictal patterns during psychogenic seizures. Diagnosis requires an EEG video, but this methodology is very time-consuming and dispensable in both time and cost. Our paper aims to define a novel methodology to support the clinical diagnosis of PNES by analyzing electroencephalographic (EEG) signals obtained in resting conditions. In this case, it is unnecessary to induce seizures in the subjects. A software pipeline was implemented based on robust feature extraction methods used in quantitative EEG analysis in the clinical setting, integrating them with machine learning classifiers. Unlike other similar works, the methodology was tested on a large dataset consisting of 225 EEGs (75 healthy, 75 PNES and 75 subjects with epilepsy), showing that it has a classification accuracy greater than 85%.

Keywords: Epilepsy · PNES · EEG · Data mining · Classification

1 Introduction

There is a growing body of literature on the COVID-19 pandemic's impact on patients with chronic neurological conditions. These studies include the direct effect of infection with the novel SARS-CoV-2 virus and the wide-reaching societal implications of the pandemic. This global crisis has had a profound psychological impact, perhaps more severe in people with seizures. Patients with more frequent attacks at baseline were more susceptible to worsening and increased stress, and barriers to care appeared to play significant roles in their deterioration. In [13], the authors reported an aggravation of the seizure frequency in PNES (Psychogenic Non-Epileptic Seizures) patients, a vulnerable group of people during this pandemic COVID-19. Among a cohort of 18 subjects with PNES, 22.2% reported an improvement in seizure control during the peak of the COVID-19 pandemic in New York City [11].

D. Groen et al. (Eds.): ICCS 2022, LNCS 13351, pp. 526–538, 2022.
https://doi.org/10.1007/978-3-031-08754-7_58

Psychogenic non-epileptic seizures are sudden behavioural changes simulating epileptic seizures but without EEG ictal patterns, caused by psychic alterations [1,3,7,12].

The gold standard for PNES diagnosis is video-electroencephalography (video-EEG), during which seizures are recorded spontaneously or provoked by stimulation techniques.

EEG recordings alone are insufficient to diagnose PNES because an ictal scalp EEG may reveal no epileptic characteristics during simple partial seizures or mesial frontal lobe seizures. In addition, the discrimination between non-epileptic seizures and healthy subjects can be challenging. Moreover, differential diagnosis cannot rely only on clinical features of PNES because most of the signs are associated with epileptic seizures. PNES patients simulate the different types of epileptic seizures but no epileptic seizures EEG patterns.

A wrong diagnosis with epilepsy may direct to treatments through anti-epileptic drugs. However, it has been assessed that the correct diagnosis of PNES is usually postponed for an average of seven years [5], with a profound consequence on patients' and caregivers' quality of life [10].

Additionally, EEG video-monitoring is highly time-consuming and labour intensive and, therefore, relatively expensive and limited in availability; thus, alternative diagnostic procedures are necessary to support neurologist diagnosis and proper pharmacological treatment.

Despite many efforts made, no bio-marker of PNES has yet been identified. However, in [15], the authors sustain patients with PNES have a stable frequency of rhythmic movements, about (5 Hz).

Continuous wavelet transform is used in [6] to process controls (CNT) and PNES EEG signals. In [14], a novel machine learning (ML) pipeline for classifying EEG epochs of PNES and healthy controls is described. The authors propose a semi-automatic signal processing technique and a supervised ML classifier to support the discriminative clinical diagnosis of PNES. In addition, they extracted statistical features like the mean, standard deviation, kurtosis, and skewness from a power spectral density (PSD) map split up into the five EEG bands. Finally, they compared three different supervised ML algorithms, namely, the Support Vector Machine (SVM), Linear Discriminant Analysis (LDA), and Bayesian network (BN), to classify control vs PNES subjects. The authors tested the proposed methodology on a small dataset of 20 EEG signals (10 PNES and 10 control), reaching an average accuracy above 90%.

To the best of our knowledge, only a few studies have investigated semi-automatic or automatic machine learning-based approaches for discrimination between healthy, epileptic and PNES subjects by only considering EEG recordings. Furthermore, the current study is one of the few in which an EEG dataset without any correlated video-EEG PNES marker has been analyzed to discriminate PNES via EEG.

Specifically, this paper proposes a novel and semi-automatic pipeline to discriminate between healthy, PNES and epileptics subjects based on the extraction of spectral features from EEG signals and classification through Machine

Learning-based approaches. To design a software pipeline that allows discrimination, we have implemented different classifiers, i.e. Light Gradient Boosting Machine, Random forest, Decision tree and Linear Discriminant Analysis. We analyzed 75 EEGs for each class (healthy, PNES and epileptic) in our work. The results achieved in terms of classification accuracy are higher than 80%.

The paper is organized as follows: Sect. 2 introduces the implemented EEG methodologies; Sect. 3 presents and discusses the results obtained from the proposed software pipeline. Finally, Sect. 4 concludes the paper.

2 Methods

In this section we describe the methodologies implemented for EEG analysis. Figure 1 illustrates the main steps that are:

- **EEG acquistion**: This module performs standard EEG acquisition according to the 10/20 international standard;
- **Pre-processing**: This module performs digital filtering and EEG segmentation into epochs;
- **Features extraction**: The Power Spectral Density (PSD) function is estimated for each channel using the classical Welch method. From PSD functions, cumulative power coefficients in clinical EEG bands are calculated to create a features vector to feed in input to the classifier module;
- **EEG classification**: the framework implements a pool of classifiers, such as Linear Discriminant Analysis, Decision Tree, Random Forest and Light Gradient Boosting Machine to discriminate healthy, PNES and epileptic EEGs acquired in resting conditions.

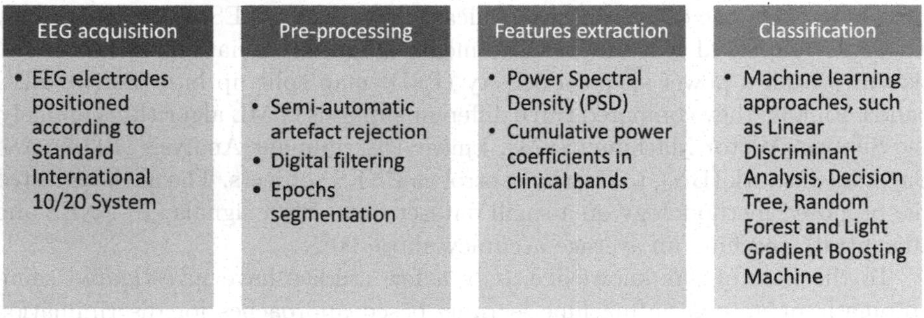

Fig. 1. EEG data analysis pipeline.

2.1 EEG Pre-processing

In general, EEG signals acquisition is very difficult because the signals are very weak and are contaminated by environmental noise or distorted by physiological artefacts (i.e. ocular and muscle artefacts). Therefore, removing noise is a fundamental step in EEG signals processing and classification. A proper data cleaning may improve the signal to noise ratio and allow for the discrimination of the most meaningful features from the EEG signals. In clinical practice, the artefacts' detection is performed visually by trained neurologists by discarding contaminated EEG epochs. Therefore, the pre-processing stage is operator-dependent, monotonous and time-consuming.

In this study, each EEG recording was inspected by a qualified neurologist to mark noise and artefact corrupted epoch (see Fig. 2). Afterwards, all EEG data were pre-processed using digital filtering techniques. Specifically, we have employed a Butterworth band-pass filter (0.1–70 Hz) and a notch filter (cut-off 50 Hz) to reduce high-frequency artefacts and power-line interference (see Fig. 3).

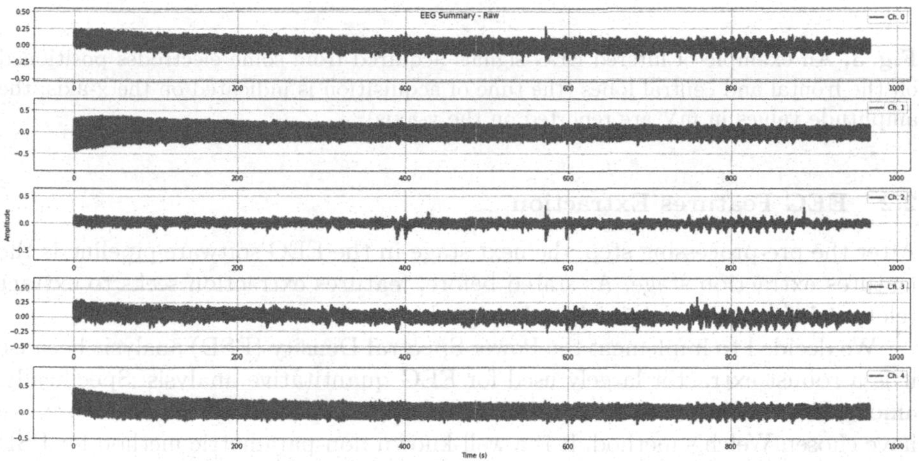

Fig. 2. An example of EEG signals acquired from some electrodes positioned on the frontal and central lobes (the time of acquisition is indicated on the x-axis; the amplitude values in mV are reported on the y-axis).

After noise removal, EEG signals were segmented in EEG epochs of 10 s in order to apply the subsequent features extraction methods on each epoch. Moreover, the segmentation of the signal in epochs is helpful as it allows a more accurate analysis of the variations of the EEG signal at the local level. However, the EEG signal is strongly stationary. Therefore, to apply the subsequent spectral analysis, it is necessary to segment it into epochs rather than analyze it for its entire duration. This operation also extends the dataset without resorting to artificial methods, improving the classification process.

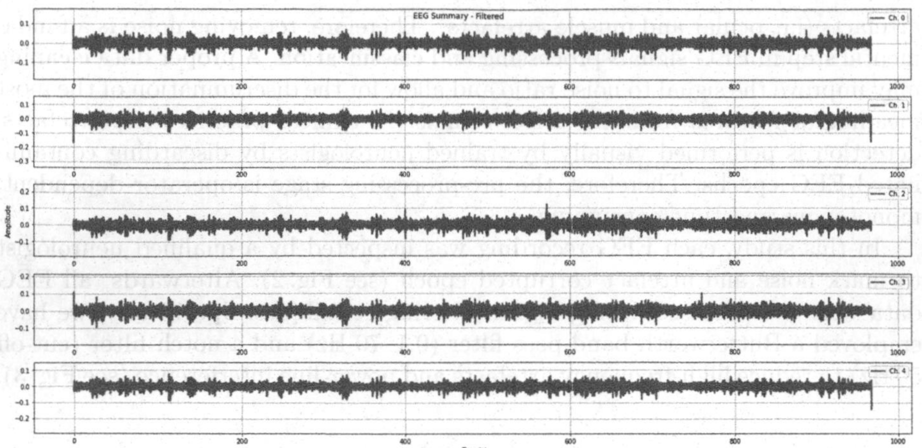

Fig. 3. An example of filtered EEG signals acquired from some electrodes positioned on the frontal and central lobes (the time of acquisition is indicated on the x-axis; the amplitude values in mV are reported on the y-axis).

2.2 EEG Features Extraction

After the pre-processing step, the next stage in the EEG software pipeline is the features extraction stage. As stated before, features extraction seeks to extract relevant information retained in the signals.

We decided to implement the Power Spectral Density (PSD) analysis because it is a robust extractor largely used for EEG quantitative analysis. Specifically, among the several methods for PSD estimation reported in the literature, we have chosen Welch's method. It is a well-known non-parametric method for PSD computation. Let $x[n],\ n = 0, \Delta\Delta\Delta, N-1$ be the samples from an EEG epoch. The evaluation of PSD by using Welch's method consists of the following steps:

- the original EEG epoch is divided into N sections (possibly overlapped O) of equal lengths M;

$$x\lfloor n\rfloor = x\lfloor n + iO\rfloor \qquad i = 0, \dots K-1, \text{ and } n = 0, \dots N-1 \tag{1}$$

- a window is applied to each section, and then the periodogram on the windowed sections is calculated. The periodogram is defined as:

$$\tilde{S}_{xx}(k) = \frac{1}{N}\left|\sum_{n=0}^{N-1} x(n)e^{-\frac{2\Pi jkn}{N}}\right|^2. \tag{2}$$

- the periodograms are averaged from the K sections in order to obtain an estimator of the spectral density

$$P_{yx}(f) = \frac{1}{K}\sum_{i=0}^{K-1} P_i(f) \tag{3}$$

where P_{yx} estimates the cross power spectral density of two discrete-time signals, x and y, the Welch method eliminates the tradeoff between spectral resolution and variance by allowing the segments to overlap. If a high-frequency resolution is needed, the record could split into a small number N of segments of length L. Our analysis uses a segment with a 50% overlap for the first step and the Hamming window in the second step.

In this paper, we extracted power spectral density (PSD) of classical frequency bands from 1 Hz 70 Hz. Specifically, delta (1–4 Hz), theta (4–8 Hz), alpha (8–13 Hz), beta (13–30 Hz), gamma (30–70 Hz) bands have been considered.

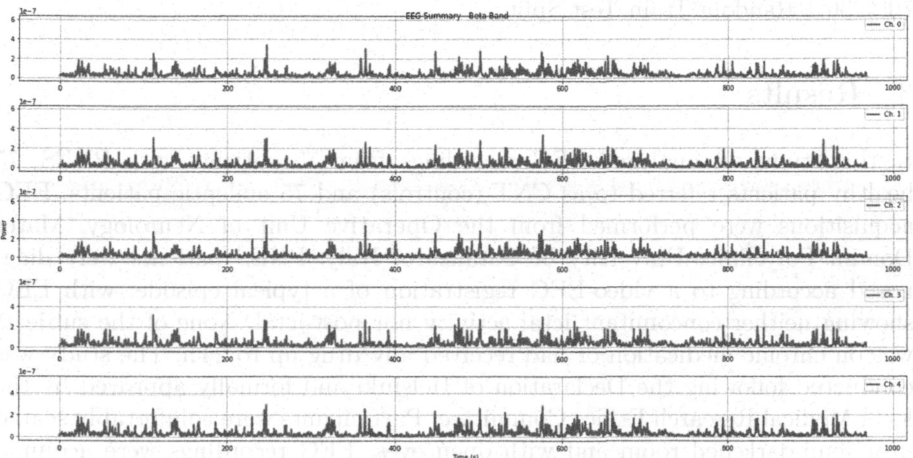

Fig. 4. EEG signals power in the beta band evaluated for all epochs.

The main processing steps of our feature extraction approach can be summarized as:

- Power Spectral Density (PSD) was estimated through Welch method;
- from PSD matrix output, we evaluated cumulative power for all EEG frequencies;
- from PSD matrix output, we selected five frequency sub-bands;
- for each band (delta, teta, alpha, beta, gamma) we computed cumulative power.

Figure 4 shows an example of cumulative power in beta band for all epochs extracted from an EEG signal.

All features extracted (six power cumulative coefficients for all epochs for all EEGs) are arranged into the features vector.

2.3 EEG Classification

Several supervised Machine Learning algorithms and ensemble techniques have been tested to discriminate CNT (control) EEG (healthy) from PNES EEG and epileptic EEG: LDA, Decision Trees, Random Forest and Gradient Boosting.

LDA classifier finds an optimal linear transformation that maximizes the class separability. LDA generates a linear combination of data sets that permits the largest mean differences between the desired classes. It works well when the feature vector is multivariate normally distributed in each class group, and different groups have a common covariance.

In decision trees algorithm [9] knowledge is gained through a set of rules, structured in a tree form. Random forest [4] is a bagging ensemble of decision trees while in Light Gradient Boosted Machine (LGBM) [8] a boosting ensemble of decision trees is built by minimizing a differentiable loss function through gradient descent optimization algorithm. Evaluation was performed through a 70%/30% Random Train Test Split.

3 Results

In this study, we analyzed EEG recordings from 75 patients with PNES, 75 healthy patients referred to as CNT (controls) and 75 epileptic patients. EEG acquisitions were performed from the Operative Unit of Neurology, Mater Domini Polyclinic, University of Catanzaro, Italy. PNES patients were diagnosed according to a video-EEG registration of a typical episode, with EEG showing neither concomitant ictal activity nor post-ictal. None of the subjects was on chronic medication or had received any drug up to 24 h. The study was conducted following the Declaration of Helsinki and formally approved by the local Medical Research Ethics Committee. Participants were comfortably seated in a semi-darkened room and with open eyes. EEG recordings were acquired using 19 Ag/AgCl surface electrodes positioned according to the International 10/20 System (see Fig. 5). Recordings were performed with an Xltek Brain Monitor EEG Amplifier with a sampling rate 256 Hz, a high-pass filter at 0.5 Hz, a low-pass filter 70 Hz, and 50 Hz notch filter.

All the electrode-skin impedance has been kept below 5 KΩ. The EEG data were recorded in a resting condition. The average EEG acquisition duration is between 10 and 20 min. The signals have been segmented in epochs lasting 10 s. So an average number of 105 epochs has been obtained for each channel for each subject.

Then for each EEG epoch, the six cumulative power coefficients (introduced in Sect. 2) were evaluated. The features vector has the following dimension: the number of cumulative power coefficients (6) times the number of epochs (an average of 105 epochs for each EEG) times the number of channels (19) times the number of subjects (225).

The training and testing of the different machine learning models described in Sect. 2 was conducted through the python PyCaret library (available online at https://pycaret.org.)

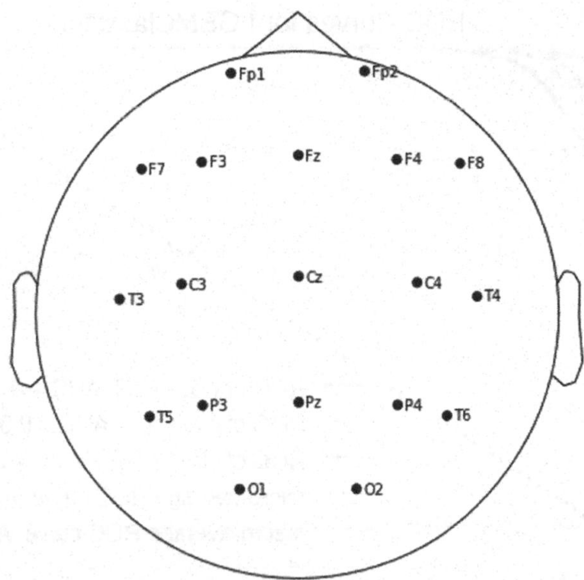

Fig. 5. EEG electrodes montage according to the 10/20 International Standard.

For the evaluation of the multiclass predictive models, several comparative metrics were chosen, including accuracy and AUC (The Area Under the Curve) with a One vs Rest (OvR) strategy [2].

Accuracy represents the proportion of instances correctly predicted by the algorithm. For example, in the case of multiclass classification with OvR strategy, we indicated with $TCNT$, $TEPI$ and $TPNES$, the number of CNT, EPI and PNES subjects that the model correctly predicts, and $FCNT$, $FEPI$ and $FPNES$, the number of CNT, EPI and PNES subjects misclassified by the model. Therefore, we can define the accuracy as:

$$Accuracy = \frac{TCNT + TEPI + TPNES}{TCNT + TEPI + TPNES + FCNT + FEPI + FPNES}$$

AUC is one of the evaluation criteria of the ability of a classifier to distinguish between classes. It is used as a summary of the Receiver Operating Characteristic (ROC) curve. The ROC curve reveals the discrimination capability by plotting the True Positive Rate against the False Positive Rate in threshold values. The area under the ROC curve (AUC) provides an aggregate measure of ROC performance. In an OvR strategy, the ROC curve was considered separately for each class (see Fig. 6, where we plotted ROC curves for the LGBM classifier and each class). Micro and macro-average ROC curves were also considered as global measures.

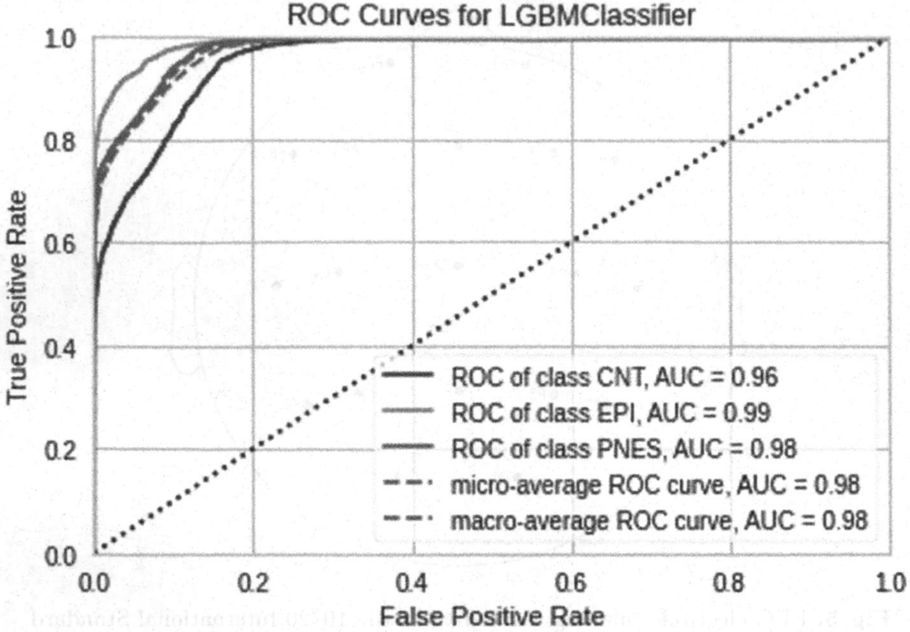

Fig. 6. ROC curves for LGBM classifier.

In Table 1 Accuracy and micro-average Area Under the Curve (AUC) of the classifier models are compared. Best values are highlighted in bold. The results show that LGBM reached the best performance.

Table 1. Accuracy and AUC of the considered models.

Model	Accuracy	AUC
Light gradient boosting machine	**0.86**	**0.98**
Random forest	0.85	0.95
Decision tree	0.74	0.81
Linear discriminant analysis	0.45	0.65

The good performances of the LGBM model are also confirmed by the precision and recall curve, reported in Fig. 7. It shows a good tradeoff between false positive and false negative rates.

Further insights regarding the most discriminative features and the separability of each of the classes are shown for the most performing model, i.e. LGBM. In particular, by combining the results reported in the confusion matrix w.r.t. the test set (Table 2), the OvR ROC curves and the micro and macro average ROC curves in Fig. 6, and the percentage of errors committed by class, as shown

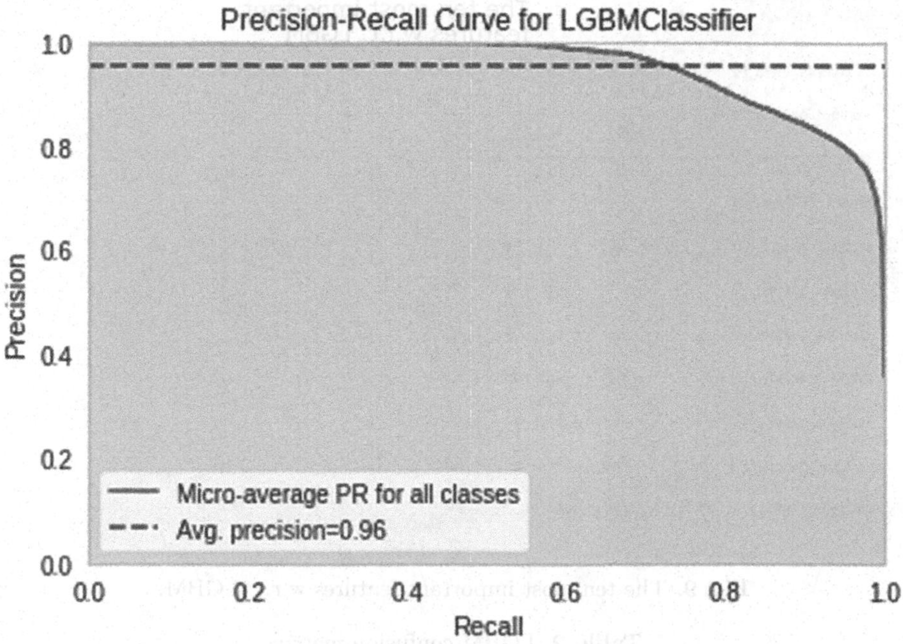

Fig. 7. Precision-Recall curve.

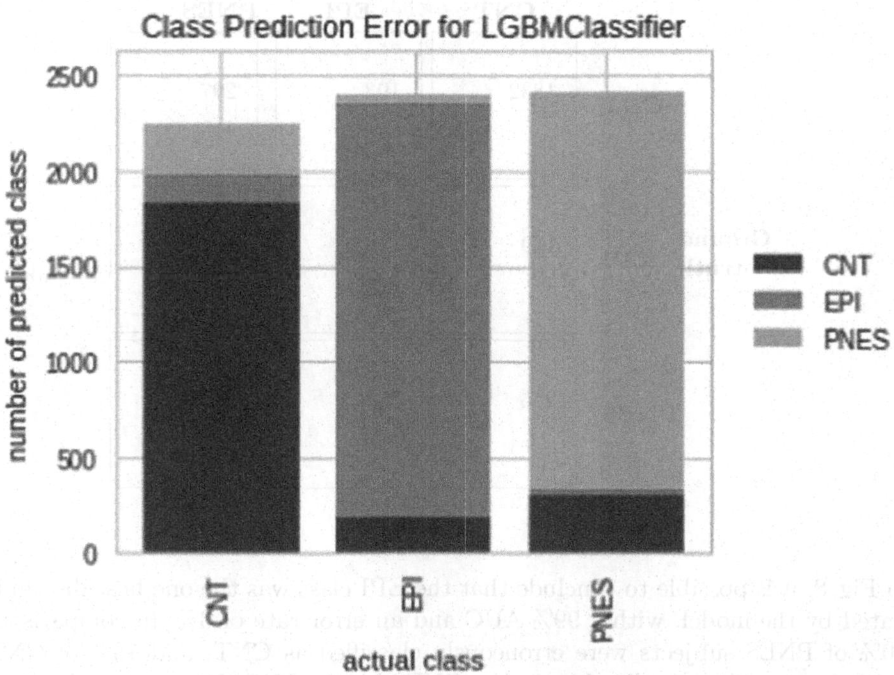

Fig. 8. Class predition errors for LGBM classifier.

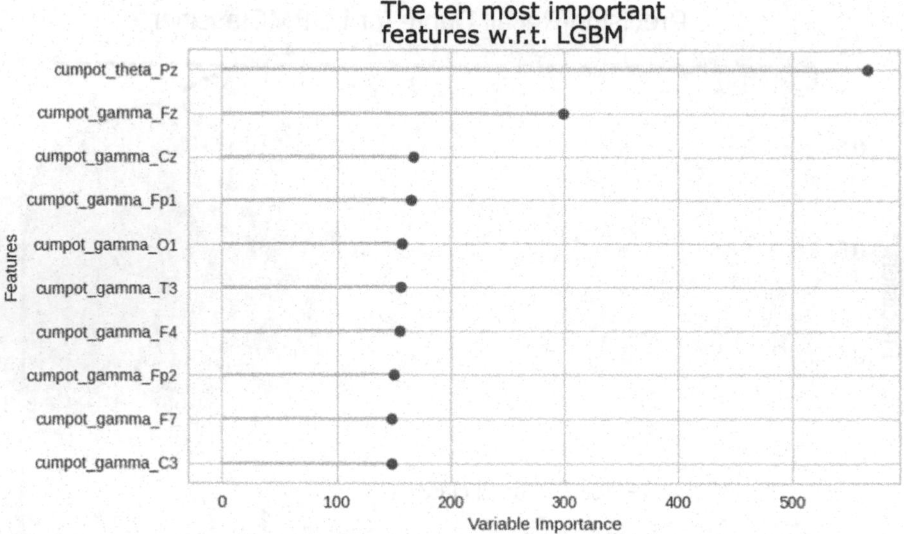

Fig. 9. The ten most important features w.r.t. LGBM.

Table 2. LGBM confusion matrix.

		Predicted Class		
		CNT	**EPI**	**PNES**
Ground truth	**CNT**	1832	193	297
	EPI	149	2168	19
	PNES	256	35	2099

in Fig. 8, it is possible to conclude that the EPI class was the one best discriminated by the model, with a 99% AUC and an error rate of 3%. In comparison, 10% of PNES subjects were erroneously classified as CNT, and 13% of CNT subjects were erroneously classified as PNES by the LGBM model1.

According to the results shown in Fig. 9, the most significant features of the LGBM classifier are the cumulative power in the theta band in the Pz electrode and the gamma band.

4 Conclusions

In the neurological field, the diagnosis of PNES seizures requires considerable effort and time, as PNES patients show the same symptoms as people with epilepsy. Generally, PNES patients are diagnosed as epileptic and pharmacologically treated for several years before the correct diagnosis is made, with significant consequences for the patient. Therefore, we are trying to define new methodologies to support clinical diagnosis to reach an accurate diagnosis of PNES quickly. This work proposes a method for analyzing the EEG acquired in resting conditions with combined signal processing and machine learning techniques. Our analysis pipeline was tested on a massive dataset of approximately 225 subjects and produced classification accuracy results between control subjects, PNES and epileptics greater than 85%.

The results also show that the best model well discriminates EPI subjects among CNT and PNES, while 10% of PNES subjects were erroneously classified as CNT and 13% of CNT subjects as PNES.

The results show that the most significant features of the best performing model are the cumulative power in the theta band in the Pz electrode and the gamma band.

Further efforts will be made in future work to assess the impact of feature engineering on predictive performance by also considering Deep Learning algorithms.

References

1. Baslet, G., Roiko, A., Prensky, E.: Heterogeneity in psychogenic nonepileptic seizures: understanding the role of psychiatric and neurological factors. Epilepsy Behav. **17**(2), 236–241 (2010)
2. Bishop, C.M., Nasrabadi, N.M.: Pattern Recognition and Machine Learning, vol. 4. Springer, Heidelberg (2006). https://doi.org/10.1007/978-1-4615-7566-5
3. Bodde, N.M.G., et al.: Psychogenic non-epileptic seizures-definition, etiology, treatment and prognostic issues: a critical review. Seizure **18**(8), 543–553 (2009)
4. Breiman, L.: Random forests. Mach. Learn. **45**(1), 5–32 (2001)
5. Cianci, V., et al.: Rating scale for psychogenic nonepileptic seizures: scale development and clinimetric testing. Epilepsy Behav. **21**(2), 128–131 (2011)
6. Gasparini, S., et al.: Management of psychogenic non-epileptic seizures: a multidisciplinary approach. Eur. J. Neurol. **26**(2), 205-e15 (2019)
7. Gasparini, S., et al.: Information theoretic-based interpretation of a deep neural network approach in diagnosing psychogenic non-epileptic seizures. Entropy **20**(2), 43 (2018)
8. Ke, G., et al.: LightGBM: a highly efficient gradient boosting decision tree. Adv. Neural Inf. Process. Syst. **30** (2017)

9. Quinlan, J.R.: Induction of decision trees. Mach. Learn. **1**(1), 81–106 (1986)
10. Reuber, M., Kral, T., Kurthen, M., Elger, C.E.: New-onset psychogenic seizures after intracranial neurosurgery. Acta Neurochir. **144**(9), 901–7 (2002). discussion 907
11. Rosengard, J., Ferastraoaru, V., Donato, J., Haut, S.: Psychogenic nonepileptic seizures during the COVID-19 pandemic in New York city - a distinct response from the epilepsy experience. Epilepsy Behav. **123**, 37–37 (2021). https://doi.org/10.1016/j.yebeh.2021.108255
12. Uliaszek, A.A., Prensky, E., Baslet, G.: Emotion regulation profiles in psychogenic non-epileptic seizures. Epilepsy Behav. **23**(3), 364–369 (2012)
13. Valente, K., Alessi, R., Baroni, G., Marin, R., dos Santos, B., P.A.: The COVID-19 outbreak and PNES: the impact of a ubiquitously felt stressor. Epilepsy Behav. **117** (2021)
14. Varone, G., et al.: A comprehensive machine-learning-based software pipeline to classify EEG signals: a case study on PNES vs. control subjects. Sensors **20**(4) (2020)
15. Vinton, A., et al.: Convulsive nonepileptic seizures have a characteristic pattern of rhythmic artifact distinguishing them from convulsive epileptic seizures. Epilepsia **45**(11), 1344–1350 (2004)

Machine Learning Approaches in Inflammatory Bowel Disease

Ileana Scarpino[1], Rosarina Vallelunga[2(✉)], Francesco Luzza[2], and Mario Cannataro[1]

[1] Data Analytics Research Center, Department of Medical and Surgical Sciences, University of Catanzaro, Catanzaro, Italy
{ileana.scarpino,cannataro}@unicz.it

[2] Department of Health Science, University of Catanzaro, Catanzaro, Italy
{rosarina.vallelunga,luzza}@unicz.it

Abstract. The great flow of clinical data can be managed with efficiency and effectiveness, improving the speed of interpretation of information, through Machine Learning (ML) methodologies, aimed at overcoming the barriers present in the diagnosis and treatment processes of patients, such as those affected by Inflammatory Bowel Disease (IBD). In this paper we survey relevant ML applications used for managing the large flow of clinical data and for overcoming the barriers present in the diagnosis and treatment processes of patients, with special focus on IBD. In IBD settings, main data sources include cohort study data, administrative databases, e-Health applications, Electronic Health Records (EHR), medical image data, Omics data, Clinical trial data and social media data. Potential applications for overcoming barriers in the field of IBD are also discussed.

Keywords: Machine Learning · Inflammatory Bowel Disease · Natural Language Processing

1 Introduction

Medicine needs technological revolution, which allows the identification of new interesting markers that can't be identified with statistical methods. Inflammatory Bowel Disease (IBD), which includes Ulcerative Colitis (UC) e Crohn's Disease (CD), is a complex multifactorial inflammatory disease with common symptoms such as abdominal pain, diarrhea, rectal bleeding, fatigue, and extraintestinal manifestations of the disease [1]. Machine Learning (ML) application in IBD represents a path of research to improve patient health outcomes since it offers patients greater opportunities to access treatment, to understand their state of health, to evaluate prevention, and to receive early diagnoses. IBD gave birth to new challenges that traditional scientific methods have failed to address [2,3]. The present paper discusses the possibility of ML applications for the characterization of the IBD disease through the extraction of topics from public and private sources, such as clinical reports and clinical notes. The rest of the paper

D. Groen et al. (Eds.): ICCS 2022, LNCS 13351, pp. 539–545, 2022.
https://doi.org/10.1007/978-3-031-08754-7_59

is organized as follows. Section 2 presents an overview of Data Sources and available public databases in IBD. In Sect. 3 the background on ML approaches applied on IBD data is introduced. Finally Sects. 4 and 5 present discussion and conclusion of the paper, opening new challenges of future works.

2 Data Sources in IBD

In the field of IBD, the use of big data has allowed medical researchers to understand the disease and the models related to it and to obtain more information that allow to progress in the clinical practice. The most important data sources in IBD include study cohorts, clinical studies, administrations, medical and electronic health record (EHR) databases, reported results databases, medical imaging. For example, imaging modalities such as colonoscopy, gastroscopy, abdominal ultrasound allow the evaluation of any structural changes in the affected districts. Large data volumes collect both administrative databases and clinical notes, representing structured and unstructured data respectively. Medical data sources include biomarker data, medical images, clinical trials registries, electronic medical records, epidemiological studies, patient-reported health data, omics data, biometric data, data from social media and the Internet [4,5].

Variety of data sources will continue to grow and the challenges will increase. In the field of IBD research data can be acquired from Administrative databases that are the most straightforward sources. For storing data collected during clinic, hospital, laboratory or pharmacy visits, many countries have developed large databases [6]. Typically, EHRs include both structured and unstructured data [7]. Data analytics in IBD is enhanced by extracting raw data that are processed to be stored, analysed and manipulated, while structured data are in the form of patient demographics, diagnosis codes, laboratory data, vital signs and similar material [8].

2.1 IBD Databases

With the purpose of homogenizing data, many databases collecting a growing number of information in the field of IBD were established. Medical and socio-demographic information from all hospital care and outpatient drug reimbursements can be extracted by Système National d'Information InterRégimes de l'Assurance Maladée (SNIIRAM) and Programme de Médicalisation des Systèmes d'Information (PMSI) [9–12]. Numerous successful databases have been implemented at European and world level.

General Practice Research Database (GPRD) includes information about incident diagnoses, hospitalizations and surgeries, owing to incomplete records [13].

National Patient Register (NPR) contains data on specialized hospital-based outpatient care as well as data on diagnoses of IBD [14].

Swedish Quality Register (SWIBreg) contains clinical data missing in NPR [15].

Both SWIBreg and NPR have been validated in clinical studies related diagnoses of IBD [16]. Table 1 shows some publicly available clinical databases.

Table 1. Open access public available clinical database and ontologies in IBD

Description	Resource
Database of public and private clinical trials	ClinicalTrials.gov(1)
GWAS Catalog	Genome-Wide Association Study(2)
Genome-Phenome dataset	Database of Genotypes and Phenotypes(3)
PheWAS Catalog	GPhenome-Wide Association Studies(4)
Ontology related to inflammatory bowel disease	Mondo Disease Ontology(5)
Inflammatory Disease Ontology Browser	Disease Ontology(6)

(1)https://clinicaltrials.gov/ct2/results?term=gastroenterology
(2)https://www.ebi.ac.uk/gwas
(3)https://www.ncbi.nlm.nih.gov/gap/?term=gastroenterology
(4)https://phewascatalog.org/phewas
(5)http://purl.obolibrary.org/obo/MONDO_000525
(6)http://www.informatics.jax.org/disease/612241

The availability of these data could be hampered by several factors, such as intellectual property, fears of different conclusions, confidentiality concerns and lack of resources [17]. When data are analysed, personal information is de-identified but the possibility of recognizing individuals still exists [18].

3 Need for Machine Learning in IBD

Computational techniques can be used to solve problems related to storage, analysis and interpretation caused by enormous amounts of omics data [19,20].

The arrival of ML into IBD clinical research has allowed researchers to capture complex associations and to increase understanding of disease mechanisms; therefore, ML could play an important role for improving diagnosis. ML algorithms require input data useful for training phase. In IBD input data are those patient biological as gene expression, biomarkers of inflammation in the tissue and blood, gut microbiota composition, endoscopic imaging and histologic imaging [21–25]. ML uses the ability of algorithms to detect predictive patterns, simplifying the interpretation of models at the base of complex medical conditions as IBD.

Table 2 shows some specific types of clinical technologies on IBD in which ML approaches have been applied.

Table 2. Machine learning application on IBD clinical setting

ML method	Application	PubMed ID (PMID)
NLP of EHR data	Identification of surveillance colonoscopy in IBD	23086115 [26]
NLP of EHR data	Improving case definition of IBD in EHR using NLP approach	23567779 [27]
Gaussian Bayesian Network	A probabilistic methods for classification of IBD	29048458 [28]
Bayesian ML model using clinical data	Bayesian Machine Learning Techniques for revealing complex interactions in IBD patients	28269885 [29]
ML using data from clinical trial	Predictive modeling of endoscopic remission in IBD	29359519 [30]
ML model using data set from IBD Genetics Consortium	Advanced machine-learning technique for risk prediction in IBD	23731541 [21]
ML model using EHR data	Prediction of outpatient corticosteroid use and hospitalization	29272474 [30]
ML model using EHR data	Validation of a Thiopurine Monitoring Algorithm on the SONIC Clinical Trial Dataset	28838785 [31]

4 Discussion

Data Mining (DM) and ML algorithms are computational approaches with the aim of extracting knowledge in the medical field. They are used to predict remission in patients with IBD and to analyze if remission predicted by algorithm leads to fewer clinical events [32]. For example Natural Language Processing (NLP) is used to identify arthralgia in electronic health records and to compare the risk of arthralgia between patients with IBD taking vedolizumab and those receiving anti-TNF agents [33]. There are many strengths and limitations of potential data sources from which big data analytics could draw from, in the field of IBD. One of the main challenges is the heterogeneity of data represented by social media posts and unstructured electronic health record notes. Since clinical information is spread across thousands of electronic documents, NLP approaches can reduce the time to organize this information, by overcoming the limitations in understanding documents. Some data sources raise questions of patient privacy and of corrupted, duplicate, missing or inaccurate data that require security solutions. Finally, a discussion about bioinformatics and computational sciences that are essential to adequately manage and integrate data from these components and other sources is reported here [34–36].

5 Conclusions and Future Work

Available data sources in the clinical setting represent the input to apply different analytical methods ranging from traditional statistical methods and advanced methods such as data mining, machine learning, clustering, text analysis and image analytics, allowing to improve IBD knowledge and to fill the gaps present in this area. IBD can benefit from ML methodologies useful to understand behavioral drivers and undertake predictive therapeutic approaches. Various ML methods could discover the hidden nature of big data in the gastroenterology field, helping to subtype chronic and complex diseases within the bowel diseases. Performance improvement of NLP will be essential to organize, interpret and recognize patterns from textual data [7], such as unstructured clinical reports. These insights can lead to new discoveries through the extraction of information from medical records and the application of NLP techniques, in particular Text Mining (TM) approaches that can improve the characterization of bowel diseases. For instance, pharmacovigilance can be improved by using text mining, to obtain data on adverse drug events from medical notes [37]. The advantage is that in addition to the PDF format, clinical reports, including endoscopic ones, are often available in plain text and can be processed for NLP analysis. Among the possible future developments and challenges, we are working on the application of NLP and TM techniques to extract useful information from medical records as well as from medical questionnaires, with the aim of a better diagnosis in clinical practice.

References

1. Weersma, R.K., et al.: Multiomics analyses to deliver the most effective treatment to every patient with inflammatory bowel disease. Gastroenterology **155**(5), e1–e4 (2018)
2. Bernstein, C.N.: Treatment of IBD: where we are and where we are going. Official J. Am. College Gastroenterol. ACG **110**(1), 114–126 (2015)
3. Actis, G.C., Pellicano, R., Rosina, F.: Inflammatory bowel diseases: Current problems and future tasks. World J. Gastrointestinal Pharmacol. Ther. **5**(3), 169 (2014)
4. Rumsfeld, J.S., Joynt, K.E., Maddox, T.M.: Big data analytics to improve cardiovascular care: promise and challenges. Nat. Rev. Cardiol. **13**(6), 350–359 (2016)
5. Lee, C.H., Yoon, H.J.: Medical big data: promise and challenges. Kidney Res. Clin. Pract. **36**(1), 3 (2017)
6. Hashimoto, R.E., Brodt, E.D., Skelly, A.C., Dettori, J.R.: Administrative database studies: goldmine or goose chase? Evid. Based Spine Care J. **5**(02), 074–076 (2014)
7. Ross, M., Wei, W., Ohno-Machado, L.: "Big data" and the electronic health record. Yearbook Med. Inform. **23**(01), 97–104 (2014)
8. Raghupathi, W., Raghupathi, V.: Big data analytics in healthcare: promise and potential. Health Inf. Sci. Syst. **2**(1), 1–10 (2014)
9. Bezin, J., et al.: The national healthcare system claims databases in France, Sniiram and EGB: powerful tools for pharmacoepidemiology. Pharmacoepidemiol. Drug Saf. **26**(8), 954–962 (2017)

10. Moulis, G., Lapeyre-Mestre, M., Palmaro, A., Pugnet, G., Montastruc, J.L., Sailler, L.: French health insurance databases: what interest for medical research? Rev. Med. Interne **36**(6), 411–417 (2015)
11. Tuppin, P., De Roquefeuil, L., Weill, A., Ricordeau, P., Merlière, Y.: French national health insurance information system and the permanent beneficiaries sample. Rev. Epidemiol. Sante Publique **58**(4), 286–290 (2010)
12. Tuppin, P., et al.: Value of a national administrative database to guide public decisions: from the système national d'information interrégimes de l'assurance maladie (sniiram) to the système national des données de santé (snds) in france. Rev. Epidemiol. Sante Publique **65**, S149–S167 (2017)
13. Lewis, J.D., Brensinger, C., Bilker, W.B., Strom, B.L.: Validity and completeness of the general practice research database for studies of inflammatory bowel disease. Pharmacoepidemiol. Drug Saf. **11**(3), 211–218 (2002)
14. Ludvigsson, J.F., et al.: External review and validation of the Swedish national inpatient register. BMC Public Health **11**(1), 1–16 (2011)
15. Jin, L., et al.: Pathway-based analysis tools for complex diseases: a review. Genom. Proteom. Bioinform. **12**(5), 210–220 (2014)
16. Jakobsson, G.L., et al.: Validating inflammatory bowel disease (IBD) in the Swedish national patient register and the Swedish quality register for IBD (SWIBREG). Scand. J. Gastroenterol. **52**(2), 216–221 (2017)
17. Bertagnolli, M.M., et al.: Advantages of a truly open-access data-sharing model. N. Engl. J. Med. **376**(12), 1178–1181 (2017)
18. Genta, R.M., Sonnenberg, A.: Big data in gastroenterology research. Nature Rev. Gastroenterol. Hepatol. **11**(6), 386–390 (2014)
19. Schultze, J.L., Rosenstiel, P., et al.: Systems medicine in chronic inflammatory diseases. Immunity **48**(4), 608–613 (2018)
20. Gedela, S.: Integration, warehousing, and analysis strategies of omics data. In: Mayer B. (ed.) Bioinformatics for Omics Data, pp. 399–414. Springer (2011). https://doi.org/10.1007/978-1-61779-027-0_18
21. Wei, Z., et al.: Large sample size, wide variant spectrum, and advanced machine-learning technique boost risk prediction for inflammatory bowel disease. Am. J. Hum. Genet. **92**(6), 1008–1012 (2013)
22. Isakov, O., Dotan, I., Ben-Shachar, S.: Machine learning-based gene prioritization identifies novel candidate risk genes for inflammatory bowel disease. Inflamm. Bowel Dis. **23**(9), 1516–1523 (2017)
23. Iadanza, E., Fabbri, R., Bašić-ČiČak, D., Amedei, A., Telalovic, J.H.: Gut microbiota and artificial intelligence approaches: a scoping review. Heal. Technol. **10**(6), 1343–1358 (2020)
24. Mossotto, E., Ashton, J., Coelho, T., Beattie, R., MacArthur, B., Ennis, S.: Classification of paediatric inflammatory bowel disease using machine learning. Sci. Rep. **7**(1), 1–10 (2017)
25. Chen, P., et al.: Serum biomarkers for inflammatory bowel disease. Front. Med. **7**, 123 (2020)
26. Hou, J.K., et al.: Automated identification of surveillance colonoscopy in inflammatory bowel disease using natural language processing. Dig. Dis. Sci. **58**(4), 936–941 (2013)
27. Improving case definition of Crohn's disease and ulcerative colitis in electronic medical records using natural language processing: a novel informatics approach
28. Han, L., et al.: A probabilistic pathway score (props) for classification with applications to inflammatory bowel disease. Bioinformatics **34**(6), 985–993 (2018)

29. Menti, E., et al.: Bayesian machine learning techniques for revealing complex interactions among genetic and clinical factors in association with extra-intestinal manifestations in IBD patients. In: AMIA Annual Symposium Proceedings, vol. 2016, p. 884. American Medical Informatics Association (2016)
30. Waljee, A.K., et al.: Predicting corticosteroid-free endoscopic remission with vedolizumab in ulcerative colitis. Aliment. Pharmacol. Ther. **47**(6), 763–772 (2018)
31. Waljee, A.K., Sauder, K., Zhang, Y., Zhu, J., Higgins, P.D.: External validation of a thiopurine monitoring algorithm on the sonic clinical trial dataset. Clin. Gastroenterol. Hepatol. **16**(3), 449–451 (2018)
32. Waljee, A.K., et al.: Machine learning algorithms for objective remission and clinical outcomes with thiopurines. J. Crohns Colitis **11**(7), 801–810 (2017)
33. Cai, T., et al.: The association between arthralgia and vedolizumab using natural language processing. Inflamm. Bowel Dis. **24**(10), 2242–2246 (2018)
34. De Souza, H.S., Fiocchi, C., Iliopoulos, D.: The IBD interactome: an integrated view of aetiology, pathogenesis and therapy. Nat. Rev. Gastroenterol. Hepatol. **14**(12), 739–749 (2017)
35. Fiocchi, C.: Integrating omics: the future of IBD? Dig. Dis. **32**(Suppl. 1), 96–102 (2014)
36. Chuong, K.H., Mack, D.R., Stintzi, A., O'Doherty, K.C.: Human microbiome and learning healthcare systems: integrating research and precision medicine for inflammatory bowel disease. OMICS J. Integr. Biol. **22**(2), 119–126 (2018)
37. Harpaz, R., et al.: Text mining for adverse drug events: the promise, challenges, and state of the art. Drug Saf. **37**(10), 777–790 (2014)

A Machine Learning Framework for Fetal Arrhythmia Detection via Single ECG Electrode

Dawlat Al-Saadany, Omneya Attallah(✉), Khaled Elzaafarany, and A. A. A. Nasser

College of Engineering and Technology, Department of Electronics and Communications Engineering, Arab Academy for Science, Technology, and Maritime Transport, Alexandria, Egypt

o.attallah@aast.edu

Abstract. Fetal Arrhythmia is an abnormal heart rhythm caused by a problem in the fetus's heart's electrical system. Monitoring fetal ECG is vital to delivering useful information regarding the fetus's condition. Acute fetal arrhythmia may result in cardiac failure or death. Thus the early detection of fetal arrhythmia is important. Current approaches use several electrodes to acquire abdomen ECG from the mother, which causes discomfort. Moreover, ECG signals acquired are extremely noisy and have artifacts from breathing and muscle contraction, which hardens ECG extraction. In this study, a machine learning framework for fetal arrhythmia detection. The proposed framework uses only a single abdomen ECG. It employs multiple filtering techniques to remove noise and artifacts. It also extracts 16 significant features from multiple domains, including (time, frequency, and time-frequency features. Finally, it utilizes four machine learning classifiers to detect arrhythmia. The highest accuracy of 93.12% is achieved using Boosted decision tree classifier. The performance of the proposed method shows its competing ability compared to other methods.

Keywords: Discrete wavelet transform (DWT) · Electrocardiography (ECG) · Peak energy envelop (PEE) · Shannon energy envelope (SEE) · Machine learning

1 Introduction

Arrhythmia is an aortic condition characterized by alterations in the normal heartbeat, including rhythm [1]. Fetal arrhythmia refers to irregular fetal heartbeats which may be too fast or too slow [2]. In terms of hazard, this aortic abnormality may impair cardiac pumping by not being coordinated by the heart muscle. Infant arrhythmias occur in only 1–2% of all pregnancies and are classified according to frequency and regularity. Almost all arrhythmias fall into one of three categories: irregular, tachycardia, or bradycardia. Normal fetal heart rates range from 120–160 bpm at 30 weeks gestation and 110–150 bpm at term [3]. Bradycardia is a heart rate of fewer than 100 beats per minute, and tachycardia is a heart rate greater than 180 beats per minute. The most often used diagnostic method to detect an arrhythmia is via cardiac electrocardiogram (ECG). Several studies have used ECG signals to detect and classify arrhythmias [4–6]. On the

D. Groen et al. (Eds.): ICCS 2022, LNCS 13351, pp. 546–553, 2022.
https://doi.org/10.1007/978-3-031-08754-7_60

MITDB dataset, Rooijakkers et al. [5] suggested a method for R- peak identification of fetal ECG (f- ECG) signals utilizing a discrete-time- continuous wavelet transform with an error detection rate of 0.22 percent [6]. other hand, Apsana et al. [7] utilized an Independent Component Analysis (ICA)-based algorithm to detect fetal arrhythmia with a 93.71 percent accuracy, whereas Devika et al. [8] used ICA to diagnose myocardial infarction with a 96.77 percent accuracy. [9], global and temporal adaptive techniques for detecting fetal beats from abdominal ECG were used. Besides, Lo et al. [10] used a temporal frequency technique, such as the short-time Fourier transform, for feature extraction, followed by classification using a convolutional neural network (CNN) [11] [12], to identify arrhythmia with a 92.65% accuracy. Acharya et al. [13] suggested a CNN-based deep learning system for automated diagnosis of congestive heart failure using ECG data. Singh et al. [14] suggested a Short-Term Long Memory (LSTM) network with an accuracy of 88.1 percent for detecting and classifying arrhythmia. The above-mentioned deep learning architectures [10, 14] need a large amount of data and computing time to train the suggested models. As a result, a novel real-time strategy has been presented to decrease processing time. Using a single abdomen ECG electrode, this study proposed a machine learning framework for fetal arrhythmia detection. It uses multiple filtering techniques and several features extraction methods from multiple domains. It employs 4 classifiers to detect fetal arrhythmia. The following is the layout of this paper. Section 2 describes the proposed technique in detail, including the MIT-BIH dataset, data description, DWT, SEE, feature extraction, selection, and learning and classification algorithms. Section 3 presents the findings. Section 4 provides the conclusion.

2 Materials and Methods

2.1 Discrete Wavelet Transform (DWT)

DWT analyzes data in the time-frequency domain by using orthogonal basis functions. commonly used in medical applications. For 1-D signals, the DWT convolves the input signal with low and high pass filters. Afterward, a downsampling procedure is done. Two clusters of coefficients are generated after this process involving details and approximation coefficients. The coefficients are further analyzed using low and high pass filters for multi-level DWT analysis.

2.2 Data Description

The NIFEA DB (non-invasive fetal ECG Arrhythmic Database) comprises recordings of 12 fetal arrhythmias ECGs and 14 regular rhythmical ECGs. This data was acquired from pregnant women during routine medical visits. The fetal median gestational age for ECG signals with arrhythmia is 36 weeks (22–41 weeks). For normal ECGs, the fetal average gestational age is 21 weeks (20–36 weeks).

2.3 Proposed Model

The proposed method consists of four steps, including preprocessing of raw data, segmentation, feature extraction, and finally, classification. In the first step, several filters are used to remove noise and artifacts. Next, ECG signals are segmented using a fixed window size. Afterward, 11 features were extracted from each segment in time, frequency, and time-frequency domains then added 5 different features to increase the accuracy. Finally, four machine learning classifiers are used to detect fetal arrhythmia. Figure 1 represents the block diagram of the proposed method of arrhythmia identification.

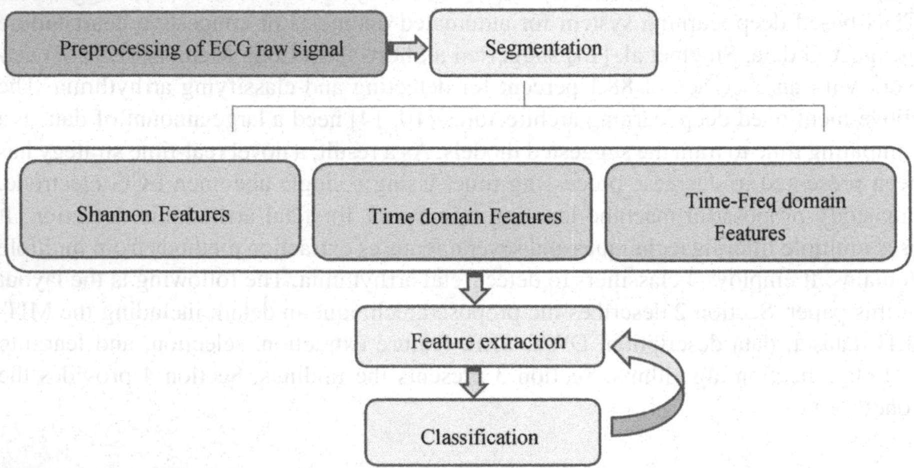

Fig. 1. Flowchart of the proposed method for classifying normal and arrhythmic ECG signals.

2.3.1 Preprocessing of the Raw Signal

Initially, a notch filter is used to eliminate the 50 Hz powerline interference. Afterward, a 2nd order low pass Butterworth filter at 0.5 Hz cut-off frequency is used to remove artifacts. Finally, a 4th order Butterworth bandpass filter is utilized to filter R-peaks.

2.3.2 Segmentation

Segmentation is the process of dividing a signal into many parts that have similar statistical properties, such as amplitude and frequency. The statistical characteristics of ECG signals change over time, which makes them non-stationary. Segmenting an ECG signal necessitates the identification of the numerous waves contained in it, such as the P wave, ORS complex, and S waves. These are called fiducial points. Many QRS detection techniques are based on the differentiation procedure. Therefore, the first and second derivatives are used in this study. The R-peak locations were obtained from the dataset to obtain these heartbeat segments. The average heartbeat consists of around 355 samples, almost including the R-peak and samples around it. An asymmetrical fixed window equivalent to 400 ms is used in this study as such window size contains the fiducial points per heartbeat. The authors identify the R peak on a scale of one and the P and T waves on scales of three and four, respectively, using a Db1 wavelet. Finally, the db4 waveform is used to detect QRS complexes [15].

2.3.3 Feature Extraction

This study extracts features from time and time-frequency domains. This section will explain the feature extraction methods used and how features are calculated. In the time domain, five features are calculated, including heart rate variability (HRV) [16], mean R-R interval, root means square of R-R interval, average heart rate, and the standard deviation of R-R interval. A 4 dB 1-D DWT is applied to each segment for the time-frequency domain, and the four detail coefficients are collected in one vector. Afterward, five features are calculated from this vector, including root mean square (RMS), mean, standard deviation, skewness, and kurtosis. Furthermore, some features based on Shannon theory are computed involving Shannon energy envelope (SEE), peak energy envelope (PEE), and final R-peak. Figure 2 shows the SEE. Figure 3 shows the PEE, and 4 shows the final R peak.

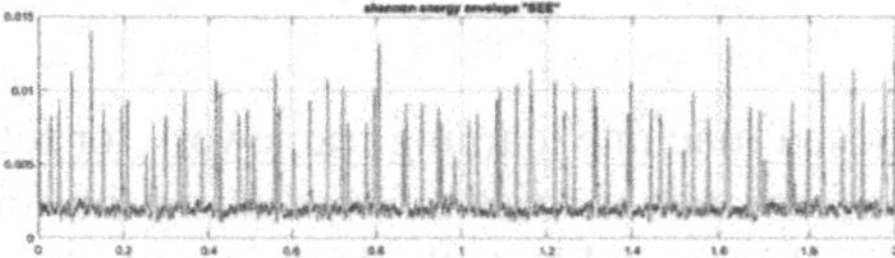

Fig. 2. The shannon energy envelop (SEE).

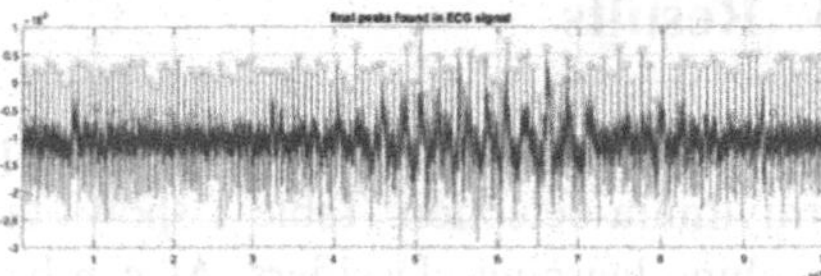

Fig. 3. The final R peaks

2.3.4 Classification

In the classification step, four machine learning classifiers are used involving a decision tree with 100 n-estimators, two minimum samples split, one minimum sample leaf, and seven levels of the decision tree that acts to boost the algorithm because it can be implemented with if conditions not like SVM kernel function. Boosted DT, k- nearest neighbor (k NN), and ensemble subspace K-NN. For the k-NN, the number k is 1, and the Euclidean distance metric is used. For DT, the Gini-index splitting matrix is utilized. 5-fold cross-validation is employed in this study to decrease the estimated variance for classifiers. The data set was divided into five subsets. Each time, the test set is drawn from one of the five subsets, while the training set is drawn from the other four. Each fold's test data was utilized four times as train data and once as test data.

2.4 Performance Metrics

Several tests have been conducted to validate the suggested strategy. These metrics include sensitivity (SEN), specificity (SPF), accuracy (ACC), positive predictive value (PPV), negative predictive value (NPV), Matthews Correlation Coefficients (MCC), and the F1 score. Where TP is the true positive, TN is the true negative, and FP and FN are the false positive and negative, respectively.

$$SEN = \frac{TP}{TP + FN} \times 100(\%) \tag{1}$$

$$\mathrm{SPF} = \frac{\mathrm{TN}}{\mathrm{TN} + \mathrm{FP}} \times 100(\%) \tag{2}$$

$$\mathrm{ACC} = \frac{\mathrm{TP} + \mathrm{TN}}{\mathrm{TP} + \mathrm{FN} + \mathrm{TN} + \mathrm{FP}} \times 100(\%) \tag{3}$$

$$\mathrm{PPV} = \frac{\mathrm{TP}}{\mathrm{TP} + \mathrm{FP}} \times 100(\%) \tag{4}$$

$$\mathrm{NPV} = \frac{\mathrm{TN}}{\mathrm{TN} + \mathrm{FN}} \times 100(\%) \tag{5}$$

$$\mathrm{F1} - \mathrm{score} = 2\frac{PPV \times SEN}{PPV + SEN} = \frac{2 \times TP}{2 \times TP + FN + FP} \times 100(\%) \tag{6}$$

$$\mathrm{MCC} = \frac{\mathrm{TP} \times \mathrm{TN} - \mathrm{FN} \times \mathrm{FP}}{\sqrt{(\mathrm{TP} + \mathrm{FN})(\mathrm{TP} + \mathrm{FP})(\mathrm{TN} + \mathrm{FN})(\mathrm{TN} + \mathrm{FP})}} \times 100(\%). \tag{7}$$

3 Results

The section discusses the results of the proposed machine learning framework for fetal arrhythmia detection. Table 1 shows the performance metrics attained using the four machine learning classifiers. As can be noticed from Table 1, the highest accuracy is achieved using the boosted DT classifier. The boosted DT classifier achieved an accuracy of 93.12%, a sensitivity of 99.14%, specificity of 82.72%, PPV of 90.77%, NPV of 98.25%, and MCC of 81.35%, and F1-score of 94.77%. The confusion matrix of the boosted DT is shown in Fig. 5. The DT classifier obtained a slightly lower accuracy of 92.13% than the boosted DT. A sensitivity of 94.5%, specificity of 88.07%, PPV of 93.14% NPV of 90.33%, MCC of 93.02%, and F1-score of 93.82% are obtained using DT classifier. The receiving operating characteristic (ROC) curve and the area under ROC (AUC) are displayed in Figs. 3 and 4.

Table 1. Classifiers Evaluation (%) of the proposed method

Classifier	SEN	SPF	ACC	PPV	NPV	MCC	F1
Boosted DT	99.14	82.72	93.12	90.77	98.25	81.35	94.77
Ensemble-subspace KNN	90.54	85.82	88.80	91.63	84.11	76.04	91.08

The proposed method's results are compared with other related studies and shown in Table 2 to further validate the proposed framework's performance. The proposed method provides an average classification accuracy of 93.12% based on five-fold cross-validation. This accuracy is obtained with only one ECG channel, which is not the case in the other studies. The outstanding performance of the proposed framework verifies its competing ability compared to other related studies.

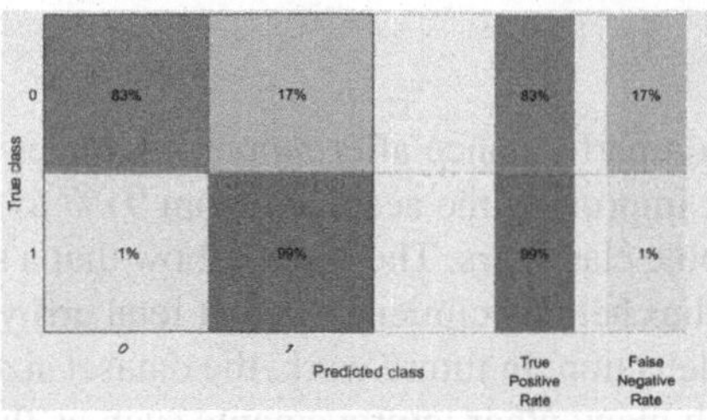

Fig. 4. Confusion matrix (Boosted DT).

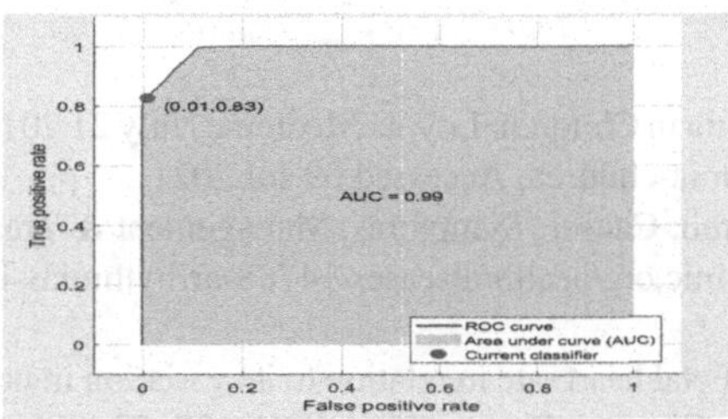

Fig. 5. ROC curve (Ensemble boosted trees).

Table 2. (Compared different related studies with the same dataset and others)

Study ref.	Method	Channel	Accuracy
[17]	2-stages filtering with SVM & gaussian kernel method	Single ECG channel	83.3%
[18]	Filtering, ECG suppression, and R-peak detection	Abdomen channels	94.1%
[19]	ICA-RLS-EMD method for extracting the neonatal ECG from the abdomen	Six abdomen channels	92.7%
[20]	Two-stage improved non-linear adaptive filter for ECG extraction	Six abdomen channels	94.5%
[21]	LMS and PHS indexes compared	Single ECG channel	87.7%
[22]	Deep learning layers with rate 0.001 input learning layer	ECG and one abdomen channel	95.76%
Our work	Proposed method	Single ECG channel	93.12%

Abbreviations: (ICA) independent component analysis, (RLS) recursive least squares, (EMD) empirical mood decomposition, (SVM) support vector machine, (LMS) local maxima similarity, (PHS) pulse harmonic strength.

4 Conclusion

In this study, the detection performance after numerous stages of the proposed framework is quite promising, improving the accuracy from 91% to 93.12% using real-time decision trees and ensemble classifiers. The results show that a single ECG channel can detect fetal arrhythmia, thus helping clinicians detect fetal arrhythmia rapidly and more accurately than manual detection. In future work, the dataset size may be increased, and deep learning, along with many other improvements such as the feature selection approach, may be used to provide more promising results. A variety of other optimization techniques can also be used to improve the evaluation results.

References

1. Thomas, K.: Arrhythmia in Children. Loyola Medicine, July 21 2015. https://loyolamedicine.org/pediatrics/arrhythmia-children. Accessed 09 Jul 2021
2. Arrhythmias in Children; Causes, Symptoms, Management & Treatment. Cleveland Clinic. https://my.clevelandclinic.org/health/diseases/14788-arrhythmias-in-children. Accessed 09 Jul 2021
3. Nijhuis, I.J.M., et al.: Fetal heart rate in relation to its variation in normal and growth retarded fetuses. Eur. J. Obstet. Gynecol. Reprod. Biol. **89**(1), 27–33 (2000). https://doi.org/10.1016/S0301-2115(99)00162-1
4. Lamesgin, G., Kassaw, Y., Assefa, D.: Extraction of fetal ECG from abdominal ECG and heart rate variability analysis. In: Abraham, A., Krömer, P., Snasel, V. (eds.) Afro-European Conference for Industrial Advancement. AISC, vol. 334, pp. 65–76. Springer, Cham (2015). https://doi.org/10.1007/978-3-319-13572-4_5
5. Rooijakkers, M., Rabotti, C., Bennebroek, M., van Meerbergen, J., Mischi, M.: Low-complexity R-peak detection in ECG signals: A preliminary step towards ambulatory fetal monitoring. In: Proceedings of the 2011 Annual International Conference of the IEEE Engineering in Medicine and Biology Society, pp. 1761–1764 (2011). https://doi.org/10.1109/IEMBS.2011.6090503
6. Moody, G.B., Mark, R.G.: The impact of the MIT-BIH arrhythmia database. IEEE Eng. Med. Biol. Mag. **20**(3), 45–50 (2001). https://doi.org/10.1109/51.932724
7. Apsana, S., Suresh, M.G., Aneesh, R.P.: A novel algorithm for early detection of fetal arrhythmia using ICA. In: Proceedings of the 2017 International Conference on Intelligent Computing, Instrumentation and Control Technologies (ICICICT), July 2017, pp. 1277–1283 (2017). https://doi.org/10.1109/ICICICT1.2017.8342753
8. Devika, M.G., Gopakumar, C., Aneesh, R.P., Nayar, G.R.: Myocardial infarction detection using hybrid BSS method. In: Proceedings of the 2016 International Conference on Communication Systems and Networks (ComNet), July 2016, pp. 167–172 (2016). https://doi.org/10.1109/CSN.2016.7824008
9. Rodrigues, R.: Fetal beat detection in abdominal ECG recordings: Global and time adaptive approaches. Physiol. Meas. **35**(8), 1699–1711 (2014). https://doi.org/10.1088/0967-3334/35/8/1699
10. La, F.W., Tsai, P.Y.: Deep learning for detection of fetal ECG from Multi-channel abdominal leads. In: Proceedings of the 2018 Asia-Pacific Signal and Information Processing Association Annual Summit and Conference, APSIPA ASC 2018, pp. 1397–1401 (March 2019). https://doi.org/10.23919/APSIPA.2018.8659503

11. Ganguly, B., Biswas, S., Ghosh, S., Maiti, S., Bodhak, S.: A deep learning framework for eye melanoma detection employing convolutional neural network. In: Proceedings of the 2019 International Conference on Computer, Electrical Communication Engineering (ICCECE), pp. 1–4 (2019). https://doi.org/10.1109/ICCECE44727.2019.9001858
12. Ganguly, B., et al.: Wavelet kernel-based convolutional neural network for localization of partial discharge sources within a power apparatus. IEEE Trans. Industr. Inf. **17**(3), 1831–1841 (2021). https://doi.org/10.1109/TII.2020.2991686
13. Acharya, U.R., et al.: Deep convolutional neural network for the automated diagnosis of congestive heart failure using ECG signals. Appl. Intell. **49**(1), 16–27 (2018). https://doi.org/10.1007/s10489-018-1179-1
14. Singh, S., Pandey, S., Pawar, U., Janghel, R.: Classification of ECG arrhythmia using recurrent neural networks. Procedia Comput. Sci. **132**, 1290–1297 (2018). https://doi.org/10.1016/j.procs.2018.05.045
15. Ktata, S., Ouni, K., Ellouze, N.: ECG signal maxima detection using wavelet transform. In: Proceedings of the 2006 IEEE International Symposium on Industrial Electronics, vol. 1, pp. 700–703 (2006). https://doi.org/10.1109/ISIE.2006.295547
16. Finley, J.P., Nugent, S.T.: Heart rate variability in infants, children and young adults. J. Auton. Nerv. Syst. **51**(2), 103–108 (1995). https://doi.org/10.1016/0165-1838(94)00117-3
17. Pavel, M.S.R., Islam, M.R., Siddiqee, A.M.: Fetal arrhythmia detection using fetal ECG signal. In: Proceedings of the 2019 IEEE International Conference on Telecommunications and Photonics (ICTP) December 2019, pp. 1–4 (2019). https://doi.org/10.1109/ICTP48844.2019.9041789
18. Surya, K., Abdul Majeed, K.K.: Multichannel probabilistic framework for prenatal diagnosis of fetal arrhythmia using ECG. In: Palesi, M., Trajkovic, L., Jayakumari, J., Jose, J. (eds.) Second International Conference on Networks and Advances in Computational Technologies. TCSCI, pp. 141–150. Springer, Cham (2021). https://doi.org/10.1007/978-3-030-49500-8_13
19. Barnova, K., et al.: A novel algorithm based on ensemble empirical mode decomposition for non-invasive fetal ECG extraction. PLoS ONE **16**(8), e0256154 (2021). https://doi.org/10.1371/journal.pone.0256154
20. Krupa, A.J.D., Dhanalakshmi, S., Kumar, R.: An improved parallel sub-filter adaptive noise canceler for the extraction of fetal ECG. Biomedical Engineering / Biomedizinische Technik **66**(5), 503–514 (2021). https://doi.org/10.1515/bmt-2020-0313
21. Corino, V.D.A., Iozzia, L., Scarpini, G., Mainardi, L.T., Lombardi, F.: A simple model to detect atrial fibrillation via visual imaging. Biomed. Eng. / Biomed. Tech. **65**(6), 721–728 (2020). https://doi.org/10.1515/bmt-2019-0153
22. Gowtham, A., Anirudh, L., Sreeja, B., Aakash, B., Adittya, S.: Detection of arrhythmia using ECG waves with deep convolutional neural networks. In: Proceedings of the 2020 4th International Conference on Electronics, Communication and Aerospace Technology (ICECA), November 2020, pp. 1390–1396 (2020). https://doi.org/10.1109/ICECA49313.2020.9297467

Computational Collective Intelligence

Consensus Algorithm for Bi-clustering Analysis

Paweł Foszner(✉), Wojciech Labaj, Andrzej Polanski, and Michal Staniszewski

Department of Computer Graphics, Vision and Digital Systems, Faculty of Automatic Control, Electronics and Computer Science, Silesian University of Technology, Akademicka 2A, 44 -100 Gliwice, Poland
pawel.foszner@polsl.pl

Abstract. Bi-clustering is an unsupervised data mining technique, which involves concurrent clustering of rows and columns of a two-dimensional data matrix. It has been demonstrated that bi-clustering may allow accurate and comprehensive mining of information, important for many practical applications. Numerous algorithms for data bi-clustering were proposed in the literature, based on different approaches and leading, in general, to different outputs. In this paper we propose a consensus method for combining outputs of many bi-clustering algorithms for improved quality of predictions. The proposed algorithm includes two steps. The first step, "assignment", leads to detecting groups of bi-clusters of high similarity and the second step, "trimming", results in transforming a group of similar bi-clusters into one bi-cluster of high quality. We demonstrate, on the basis of both simulated and real datasets, that using our algorithm highly improves quality of bi-clustering. We also provide an easy to use software tool, which includes implementations of several bi-clustering algorithms and our consensus method.

Keywords: Bi-clustering · Machine learning · Consensus methods · Ensemble methods

1 Introduction

Bi-clustering (or co-clustering) is a data analysis and data mining approach, which involves simultaneous clustering of rows and columns of a data matrix [13,21,22]. In recent years interest in algorithms for bi-clustering of data has substantially grown due to many new areas of applications, e.g., text analysis [9], pattern recognition [15], signal analysis [23] bioinformatics [35]. There is a large literature in the area, which can be roughly divided into parts concerning formulations of bi-clustering problems, developing bi-clustering algorithms and verifying and comparing their outcomes.

Supplementary Information The online version contains supplementary material available at https://doi.org/10.1007/978-3-031-08754-7_61.

D. Groen et al. (Eds.): ICCS 2022, LNCS 13351, pp. 557–570, 2022.
https://doi.org/10.1007/978-3-031-08754-7_61

There are several versions (formulations) of bi-clustering problems. Bi-clusters can be of different types, constant values, constant rows or columns, coherent values (scaled, shifted, plaid etc.) [19]. Bi-clustering can involve continuous, discrete or binary data. Discrete or binary bi-clusters can either originate from analyses of continuous data, where thresholding was applied as a preprocessing step [4], or can involve data of discrete origin [30]. Formulation of the bi-clustering problem also includes specifying the number of bi-clusters in the data matrix, extent of overlappings between rows and columns of bi-clusters and the level of noise in the data [1].

Several studies in the literature present surveys of bi-clustering algorithms and evaluate and compare their efficiencies [20,26]. Methods to evaluate results of bi-clustering algorithms, in principle similar to those used when evaluating classification or clustering algorithms, can be divided into three groups, ground truth (applicable for artificially generated data, where the true structure of bi-clusters is known), internal (efficiency of the bi-clustering algorithm is measured by some quality indexes of the obtained bi-clusters, like correlations or mean square errors) and external (efficiency of the bi-clustering algorithms is measured by the significance and quality of the conclusions in the application area, e.g., biology or text processing). Compared to classification or clustering, there is an additional difficulty in evaluating quality of bi-clustering, related to more complicated output. Prelić, et al. (2006) [28] introduced a methodology for evaluating results of bi-clustering algorithms, suitable for the case where the ground truth is available, based on using a distance measure between bi-clusters given by Jaccard index and on solving the optimal assignment problem.

Each of large number of bi-clustering algorithms applied to a dataset leads, in general, to a different outcome. Estimates of bi-clusters returned by bi-clustering algorithms are sensitive to the choice of initial conditions for recursive procedures and to the choice of parameters. Therefore, an important issue becomes possibility of integrating results of different algorithms and the question whether integrating results of different algorithms can improve quality of bi-clustering. By analogy, in clustering and classification, consensus algorithms for aggregating results of clustering algorithms [32,34] and ensemble algorithms for aggregating classification algorithms [2,33], were proven to lead to improvements in accuracy and robustness of clustering or classification.

Surprisingly, methods for integrating/combining results of bi-clustering algorithms are very rare in the literature. We have found only one such method called ensemble method for bi-clustering, published by Hanczar and Nadif [12]. Hanczar and Nadif [12] propose the ensemble bi-clustering algorithm based on tri-clustering of auxiliary, $N \times M \times RK$ binary matrix L_{ijk} generated on the basis of outputs of R bi-clustering algorithms, each returning K bi-clusters. They demonstrate (see their Fig. 3) improvement of quality of bi-clustering obtained by using their ensemble method.

More detailed analysis reveals possibilities of further improvement related to some losses of information in the Hanczar and Nadif algorithm. The first information loss is related to using only binary data. Using continuous rather than binary values of the data matrix entries may potentially improve quality. The second information loss is related to neglecting the relation between bi-clusters and algorithms, which generated them. After the binary matrix L_{ijk} has been generated there is no possibility, for any bi-cluster, to restore information on which algorithm produced it. Summing up, it seems that it is possible to construct more efficient method for integrating/combining results of bi-clustering algorithms by including information, which was neglected in the Hanczar and Nadif algorithm.

In this paper we propose a new approach for combining outputs of bi-clustering algorithms. The proposed algorithm, which we call consensus bi-clustering algorithm includes two steps. In the first step, called assignment step, the generalized assignment algorithm [27] is used to obtain groups of bi-clusters of high similarity. In the second step, called trimming step, each group of bi-clusters is turned into one bi-cluster of high quality. The first step of our algorithm utilizes information on the relations between bi-clusters and algorithms. The second step uses continuous rather than binary data, which leads to improved predictions. We demonstrate, on the basis of both simulated and real datasets, that application of our algorithm improves quality of bi-clustering compared to the method of Hanczar and Nadif. The obtained improvement increases with the increase of complexity of bi-clustering problem and with the increase of the level of noise in the bi-clustering problem.

The effectiveness of the proposed method we evaluated on specially prepared for this purpose synthetic data and adequately well-described real data from literature. Synthetic data were prepared in such a way as to cover the widest spectrum of possible scenarios occurring in bi-clustering. These cases relate to various aspects of the data matrix such as noise, number of bi-clusters, internal data structure and the degree of overlapping of clusters within both dimensions. The real data has been chosen so as to be able to clearly interpret the quality of the results obtained. Data from the work of Monica Chagoyen [5] were selected, where bi-clusters should represent groups of genes (associated with the specific gene ontology) and words. The quality of founded bi-clusters were assessed by measuring quality of assigment of bi-clusters with ontologies established at the data matrix design level.

2 Description of the Proposed Consensus Algorithm

We assume that the desired number of bi-clusters is given, denoted by K, and that we already have outputs of L bi-clustering algorithms, each containing K (hypothetical) bi-clusters. Each of the L bi-clustering algorithms is called a component bi-clustering algorithm and bi-clusters returned by component bi-clustering algorithms are called component bi-clusters. The bi-clusters returned by our consensus algorithm are called resultant or consensus bi-clusters.

2.1 Notation

We use similar notation to that introduced by Madeira and Oliveira [19]. We denote a $n \times m$ data matrix by A. Following Madeira and Oliveira [19] we write $A = (X, Y)$, where $X = x_1, x_2, \cdots, x_n$ denotes the set of indexes of rows of A and $Y = y_1, y_2, \cdots, y_n$ stands for the set of indexes of columns of A. One assumes that there is an underlying function $a(i, j)$, which returns entry of the matrix given indexes of its row and column, $a_{ij} = a(i, j)$. Then the precise notation would be $A = a(X \times Y)$, where $\times$ stands for Cartesian product. However, with slight abuse of notation, we shall write $A = (X, Y)$ instead of $A = a(X \times Y)$. Such notation, introduced in [19], does not lead to ambiguities and is convenient for describing bi-clusters and bi-clustering algorithms.

Bi-cluster B is defined as a pair

$$B = B_{I,J} = (I, J), I \subseteq X, J \subseteq Y. \tag{1}$$

Subsets I and J, in the above, are called sets of attributes, or attributes of the bi-cluster $B_{I,J}$. If necessary we add superscripts to index different bi-clusters, e.g., $B^{l,k} = B^{l,k}_{I,J} = (I^{l,k}, J^{l,k})$ stands for $k - th$ bi-cluster, returned by $l - th$ component bi-clustering algorithm.

The output of the l-th component bi-clustering algorithm is denoted as

$$R^l = \{B^{l,1}, B^{l,2}, \cdots, B^{l,K}\} \tag{2}$$

where

$$B^{l,i} = (I^{l,i}, J^{l,i}), \tag{3}$$

$i \in 1, \cdots, K$.

We combine outputs $R^1, R^2, ..., R^L$ of the component algorithms for data bi-clustering in two steps. In the first step, called assignment step, we use the generalized assignment algorithm [27] to obtain K groups, each containing L bi-clusters, of highest possible intra-group similarities. Jaccard index between pairs of bi-clusters is used to define the intra-group similarity measure. In the second step, called trimming step, each group of component bi-clusters obtained in the first step, is transformed into one resultant bi-cluster. Applying the second step to each of the K groups of bi-clusters results in obtaining K consensus bi-clusters.

2.2 Assignment Step

Each of the component bi-clustering algorithms returns K component bi-clusters, but no relations between bi-clusters returned by different component algorithms are known a priori. However, since component bi-clustering algorithms are applied to the same data set it is likely that the component bi-clusters will exhibit similarities stemming from the true structure of the dataset. We therefore search for K groups of component bi-clusters with possibly high intra-group similarity in each group. This problem can be formulated as a generalized assignment problem involving minimization of the index

$$I = \sum cost(G_k), \tag{4}$$

where $cost(G_k)$ is the cost of grouping component bi-clusters $B^{1,k^{(1)}}, B^{2,k^{(2)}}, \cdots, B^{L,k^{(L)}}$ into one group

$$G_k = \{B^{1,k^{(1)}}, B^{2,k^{(2)}}, \cdots, B^{L,k^{(L)}}\}, \tag{5}$$

such that no pair of the bi-clusters within the group G_k comes from the same component bi-clustering algorithm. In (5) $B^{l,k^{(l)}}$ denotes $k^{(l)}$ bi-cluster returned by the l-th component bi-clustering algorithm.

The cost function $cost(G_k)$ is computed on the basis of the Jaccard indexes between pairs of component bi-clusters. More specifically, for each pair of bi-clustering algorithms (l_1, l_2) we define a similarity matrix $S(l_1, l_2)$ between their returned sets of bi-clusters

$$S(l_1, l_2) = \begin{bmatrix} S_{Jacc}(B^{l_1,1}, B^{l_2,1}) & S_{Jacc}(B^{l_1,1}, B^{l_2,2}) & \cdots & S_{Jacc}(B^{l_1,1}, B^{l_2,K}) \\ S_{Jacc}(B^{l_1,2}, B^{l_2,1}) & S_{Jacc}(B^{l_1,2}, B^{l_2,2}) & \cdots & S_{Jacc}(B^{l_1,2}, B^{l_2,K}) \\ \cdots & \cdots & \cdots & \cdots \\ S_{Jacc}(B^{l_1,K}, B^{l_2,1}) & S_{Jacc}(B^{l_1,K}, B^{l_2,2}) & \cdots & S_{Jacc}(B^{l_1,K}, B^{l_2,K}). \end{bmatrix} \tag{6}$$

In the above $S_{Jacc}(B^{l_1,q}, B^{l_2,r})$ - denotes the value of the Jaccard similarity index between the two bi-clusters - $q'th$ bi-cluster from result l_1 and $r'th$ bi-cluster from result l_2 $(q, r = 1, \cdots, K)$. Entries of similarity matrices (6) are then used to compute the cost of grouping component bi-clusters $B^{1,k^{(1)}}, B^{2,k^{(2)}}$ $\cdots$, $B^{L,k^{(L)}}$, $cost(G_k)$

$$cost(G_k) = \sum_{l_1=1}^{L} \sum_{l_2=l_1+1}^{L} S_{Jacc}(B^{l_1,k^{l_1}}, B^{l_2,k^{l_2}}). \tag{7}$$

The summation goes over all possible pairs (l_1, l_2) of component bi-clustering algorithms.

Algorithms for solving the generalized assignment problem have been developed in several papers in the literature, e.g., [24,27]. In [27] a branch and bound algorithm for minimizing 5 was proposed. However, computational complexity of these algorithm scale exponentially with KL and therefore their application is difficult for larger datasets. There are several time efficient algorithms, which provide sub-optimal solutions to generalized assignment problems, [7,11]. In our implementations of the consensus bi-clustering algorithm we have options of either using the branch and bound algorithm for minimizing 5 or replacing it by a suboptimal heuristic, greedy search, described in detail in the Supplementary Materials, which is much faster. In computational experiments we have verified that replacing branch and bound algorithm for minimizing formula 5 by a heuristic, greedy search does not decrease performance of the whole consensus bi-clustering algorithm.

2.3 Trimming Step

In the trimming step each group of the component bi-clusters (5) is transformed into one resultant bi-cluster B^k_{trimed}. Trimming is designed in such a way that unreliable outputs of component bi-clustering algorithms, e.g., resulting from poorly chosen initial conditions, are corrected. One can notice that, for each group of component bi-clusters, the problem of trimming can be understood as searching for one bi-cluster in the subset of the data matrix A given by the union of component bi-clusters.

There are many possible approaches to the problem of searching for one bi-cluster in the data, e.g., [29]. In our implementation of the trimming step we use a heuristic algorithm, similar to that described in [29], designed for maximal robustness against noise in the data. The iterative procedure starts from the initial condition given by the union of all bi-clusters within G_k and is designed in such a way that the value of the ACV index (Average Corelation Value [31]) of the resulting bi-cluster must increase in each step. If there is no possibility to increase the values of thee ACV index the procedure stops.

3 Comparisons of Performances of Bi-clustering Algorithms

We compare performance of our consensus bi-clustering algorithms to performances of component bi-clustering algorithms and to performance of Hanczar and Nadif's ensemble bi-clustering algorithm [12] for both synthetically created data and for several real datasets. For evaluation of performance of different algorithms we use ground truth method (for synthetic data) internal quality index ACV (for synthetic data and for real data) and suitably defined external indexes (for real data).

In the case of analysis of synthetic data we use a ground truth methodology described by Prelić, et al. (2006) [28] based on solving the optimal assignment problem between the computed and the known bi-clusters in the first step and computing difference measure between bi-clusters given by the Jaccard index in the second step.

In the case of analysis of real data we use an internal quality index for evaluating quality of bi-clustering. There are several indexes suitable for evaluating quality of bi-clustering proposed in the literature e.g., mean square residue (MSR [6]), average Spearman's rho (ASR [3]), average correlation value (ACV [31]). Here we use the ACV index of the set of bi-clusters, recommended in many papers, defined as follows:

$$ACV(B) = max\left\{\frac{\sum_{i=1}^{n}\sum_{j=1}^{n}|r\_row_{ij}| - n}{n^2 - n}, \frac{\sum_{k=1}^{m}\sum_{l=1}^{m}|r\_col_{kl}| - m}{m^2 - m}\right\} \quad (8)$$

where

$$ACV(B) \in [0, 1] \quad (9)$$

where:

- r_row_{ij} – is the value of the Pearson correlation between the $i'th$ and $j'th$ rows
- r_col_{kl} – is the value of the Pearson correlation between $k'th$ and $l'th$ columns.

The higher the value of ACV(B) the better quality of the bi-cluster B. One should also notice flexibility of the ACV index. ACV index is suitable for constant, additive and multiplicative bi-clusters.

3.1 Synthetic Data

Our aim when creating synthetic data for comparing bi-clustering algorithms was covering diversity of possible bi-clustering data. Our data sets contain every important combinations of bi-cluster structures, the degree to which overlap the rows and columns and noise level that was introduced into the bi-cluster. Data consist of matrices with one of four major structure each. Additionally every matrix represents single structure appears in one of eight variants regarding to bi-clusters overlapping over rows and columns. Finally, the last parameter used to create the data matrix was the noise level. To produce the above described matrices, BiBench software by Kemal Eren [10] was used.

To summarize and point out the details of each category we have:

1. Regarding to level of overlapping test set consist of various number for matrices with different variants of bi-clusters positions in data matrix. We distinguish data matrices with:

(a) Bi-clusters with separate sets of rows and columns,
(b) Bi-clusters with separate sets of rows and overlapping columns with different degree of overlaps (25%, 50%, 75%),
(c) Bi-clusters with separate sets of columns and overlapping rows with different degree of overlaps (25%, 50%, 75%)

2. Each structure described above appears in four different variants of bi-clusters values. Regarding to bi-clusters structure we distinguish data with (every single matrix contains only one of the following):

- Constant data
- Constant data up-regulated
- Plaid data
- Shift and scale data

3. The last, third dimension of parameters is the noise level that has been introduced into the bi-cluster. This value comes from the set (0, 0.001, 0.25, 0.5, 1)

4. Number of bi-clusters in the range $< 2 - 10 >$

All matrices are of size 500×500. The background noise is generated as a uniformly distributed random variable in the range $< 0 - 100 >$.

To sum this up we have sixteen different data sets regarding to bi-cluster position and four regarding to bi-cluster structure. Also, each matrix is present in five versions depending on the level of noise introduced into the bi-clustering. Noise is introduced to bi-cluster accordingly to BiBench algorithm. Finally, all these variants come in 9 different versions due to the number of bi-clusters. The final set consisting of 2880 matrices, each having a different structure, different distribution and number of the bi-clusters inside data matrix and different noise level.

3.2 Text Mining Data

We have analyzed the text mining dataset, which we have created using the same scenario as described in the article by Monica Chagoyen, et al. [5]. Authors of [5] analyzed 7080 scientific papers from the PubMed database devoted to genetics of the Saccharomyces cerevisiae species. To each of the papers they assigned a set of genes, selected from the lists of 575 genes from Saccharomyces cerevisiae genome, using a semi-automatic procedure supported by human experts.

Functions of genes analyzed in [5] were also summarized by eight broad biological processes described by the following GO Ontology terms: cell cycle (GO:0007049), cell wall organization and biogenesis (GO:0007047), DNA metabolism (GO:0006259), lipid metabolism (GO:0006629), protein biosynthesis (GO:0042158), response to stress (GO:0006950), signal transduction (GO: 0007165), transport (GO:0006810).

We implemented the algorithm described in [5] including the following steps: 1. Downloading 7080 texts of scientific papers using lists of their PubMed ids published by [5]; 2. Extracting words in these papers; 3. Applying filters removing colloquial (most frequent, frequency > 80%) and vary rare words (frequency < 4%). As a result we got a word - occurrences matrix, whose rows corresponded to genes and columns to word. Entries of this matrix are balanced term frequencies, D_{ij}, of term j in document i, defined as:

$$D_{ij} = tf_{ij} * IDF_j \tag{10}$$

In the above formula IDF_j stands for the inverse document frequency of term j [48],

$$IDF_j = log\left(\frac{T}{t_j}\right) \tag{11}$$

where:

- T – total number of documents in set,
- t_j – number of documents that contains document j

When analyzing the word - occurrences matrix we search for 8 bi-clusters, each corresponding to on of 8 GO terms of eight broad biological processes, listed above.

4 Results and Discussion

For a single data matrix all single results are given to the input of the consensus algorithm described in Sect. 2, as well as the input of the tri-clustering algorithm [12]. Synthetic data were assessed both from the perspective of the quality of the resulting bi-clusters and their mapping in a set of expected bi-clusters. Real data were evaluated using only quality index [31]. The similarity measure is obtained as follows:

- Expected and Founded set are presented as cost matrix where each row is represented by different bi-cluster form expected set and each column by bi-cluster from founded set. The values of this matrix are Jaccard indices computed on both clusters:

$$c_{i,j} = \frac{\frac{I^i \cap I^j}{I^i \cup I^j} + \frac{J^i \cap J^j}{J^i \cup J^j}}{2} \tag{12}$$

- Hungarian algorithm is performed on this cost matrix in order to find perfect assignment
- Sum of Jaccard indexes pointed out in previews point is divided by the number of founded bi-clusters

Both similarity measure and quality index range from $< 0 - 1 >$

4.1 Synthetic Data

The advantage of having synthetic data is the fact that we have ground truth. This allows for calculating accurate quality measures based on it. To compare the results of synthetic data, we decided to use independent measures of quality usually used for synthetic data: Recovery and Relevance. Measures were introduced by Kemal Eren [10] and are quite useful in terms of comparing data that provides ground truth.

$$Re(R^1, R^2) = \frac{1}{|R^1|} \sum_{b_1 \in R^1} \max_{b_2 \in R^2} S_{Jacc}(b_1, b_2) \tag{13}$$

Equation 13 apply for both Recovery and Relevance. It takes founded set of bi-clusters and expected set of bi-clusters at the input. Recovery: Re(Expected, Founded) can be interpreted as checking if the algorithm found all of the expected bi-clusters. Relevance: Re(Founded, Expected) can be interpreted as checking if all the found bi-clusters were expected. Both measures take values from the range $< 0, 1 >$. Figure 1 presents a graphical summary of the entire synthetic data set. A single point on the chart means the average value for Recovery and Relevance from the point of view of a single algorithm for all synthetic matrices.

For comparison have been selected algorithms specializing in different types of input data, as well as algorithms with different assumptions as to data representation. The final set of methods was as follows: BiMax [28], Floc [36], ITL

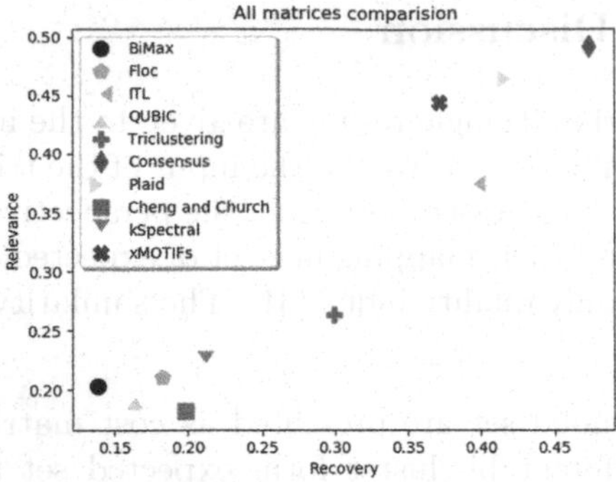

Fig. 1. Averaged results from the entire synthetic data set.

[8], QUBIC [18], Triclustering [12], Consensus, Plaid [17], Cheng and Church [6], kSpectral [16], xMOTIFs [25]. The above list presents a wide spectrum of approaches in the field of bi-clustering. The results for classic methods were obtained using the implementation from the matlab mtba package [14]. The results for the ensemble methods (Triclustering and Consensus) were obtained by providing the results of the other methods on their input. It is clear here that the method proposed in this work improves significantly the outcome based on the results of specialized methods.

In the Fig. 1 we can distinguish the following groups of algorithms: (1) in the lower left corner we have algorithms performed for a specific data type (BiMax, Floc, QUBIC, Cheng&Church, kSpectral). Therefore, on average for all cases, they do not work well. Then just above them we see firts ensemble method - Triclustering, which can improve on average for most of the results. Next we see the methods (xMOTIFs, ITL, Plaid) that do quite well regardless of the given data structure, noise, etc. On top of this we have our own Consensus algorithm that improves the outcome regardless of the quality of the results given at its input.

Figure 2 shows a detailed comparison between the algorithm proposed by Hanchar [12] (red) and the algorithm proposed in this paper (green). In the this figure we can read a few things that were to be expected. Such as increasing noise or increasing bi-clusters numbers, which reduces quality. The expected effect was also a very high result for "const_upreguleted" data. The bi-clusters there are easily found because they can be seen even on heat maps of the data matrix. It clearly shows that regardless of the aspect selected, the algorithm proposed in this work achieves better results. All results with input data matrices have been published on the dedicated website https://aspectanalyzer.foszner.pl/

4.2 Real Data

Obtained bi-clusters of genes for each bi-clustering method were compared biologically based on the Gene Ontology database [1]. Along with the data there

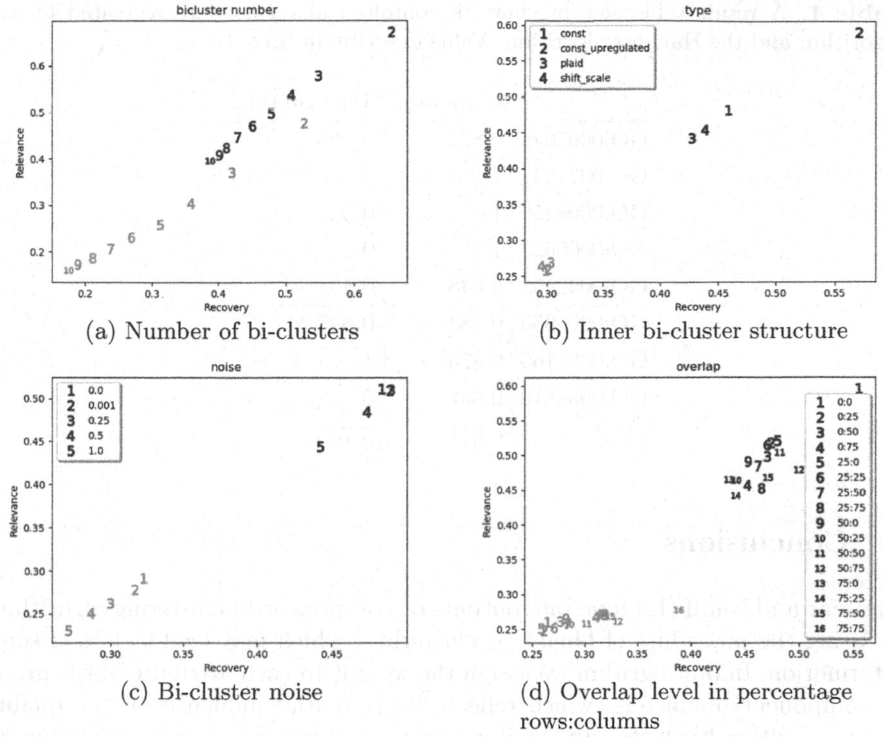

(a) Number of bi-clusters

(b) Inner bi-cluster structure

(c) Bi-cluster noise

(d) Overlap level in percentage rows:columns

Fig. 2. Detailed comparison of Consensus and Triclustering due to various aspects of the data matrix. (Color figure online)

were 8 GO terms from the biological process ontology provided, which served us as a reference for 8 bi-clusters (one GO term per one bi-cluster). In order to compare our results with the reference GO terms, first a functional analysis was performed for each genes bi-cluster using the elimination algorithm [2] and Fisher's exact test. The elimination algorithm was used to reduce the number of redundant GO terms, by taking into account only GO terms with the largest information content (IC). The level of statistical significance was set at 0.05. Received list of statistically significant GO terms for each bi-cluster was compared with each of the reference GO term. For this purpose a semantic similarity measure, Wang method [3] which takes into account hierarchy of GO terms in the ontology tree, was used. Next, combining of semantic similarity measures was performed, checking for each reference GO term and each bi-cluster of GO terms what is the maximum value of similarity measure. As a result, a similarity matrix was obtained where our result bi-clusters were in the rows and reference bi-clusters in the columns. Finally, the Hungarian algorithm was used to assign bi-clusters to each other, maximizing the measure of similarity. By summing up the received measures for all bi-clusters, we obtain a measure of quality, which takes into account the biological information flowing from the data.

Table 1. A result table showing how the ontological terms were recreated by our algorithm and the Hanczar algorithm. Value describe in Sect. 4.2

	Consensus	Triclustering
GO:0007049	0.872	0.698
GO:0071555	1	0
GO:0006259	1	0.944
GO:0006629	1	0
GO:0042158	0.648	0.45
GO:0006950	0.584	0.643
GO:0007165	0.875	1
GO:0006810	0.632	0
SUM	6.611	3.735

5 Conclusions

Hanczar and Nadif [12] integrate outputs of component bi-clustering algorithms by using the procedure of binary tri-clustering, which may lead to loss of some information. In our algrithm we assign the weight to each attribute of the group of component bi-clusters, which reflects its potential influence on the quality of the resulting bi-cluster. In the iterative procedure of consensus trimming we remove attributes characterized by least weights.

As shown by numerical results of Sect. 4 of the consensus algorithm is much better than direct competition in the form of Tri-clustering algorithm. First of all, it was proved that the proposed method meets the basic assumptions for ensemble methods. As Fig. 1 shows, the result was better than any of the methods given on its input. Next results were checked in detail for various aspects of the data matrix. For noise, the number of bi-clusters, internal structure of bi-cluster and for the cluster overlapping rate - the proposed method achieved very good results (Fig. 2). Finally, the method was evaluated for real data. We selected well described in the literature text mining data, where bi-clusters consist of words and genes. The quality of the method was assessed based on a comparison of the results with the original intentions of the creators of the data matrix. As Table 1 shows, the results were twice as good as the other ensemble method.

As for the development of the method described here, it has two aspects. The first is to make this method available as a service. It will be possible to perform a bi-clustering experiment with algorithms from the literature, as well as conduct a consensus (ours or Hanczar). On the scientific level, we would like to focus on the automatic recognition of the number of bi-clusters present in the input data. We believe that it can be done by analyzing the quality of bi-clusters from various executions.

All and detailed results can be found at https://aspectanalyzer.foszner.pl.

Acknowledgements. This work was funded by Polish National Science Centre, OPUS grant 2016/21/B /ST6/02153 (AP,WL); research project (RAU-6, 2020) and projects for young scientists of the Silesian University of Technology (Gliwice, Poland) (PF,MS).

References

1. Aguilar-Ruiz, J.S.: Shifting and scaling patterns from gene expression data. Bioinformatics **21**(20), 3840–3845 (2005)
2. Avidan, S.: Ensemble tracking. IEEE Trans. Patt. Anal. Mach. Intell. **29**(2), 261–271 (2007)
3. Ayadi, W., Elloumi, M., Hao, J.K.: A biclustering algorithm based on a bicluster enumeration tree: application to DNA microarray data. BioData Mining **2**(1), 9 (2009)
4. Benabdeslem, K., Allab, K.: Bi-clustering continuous data with self-organizing map. Neural Comput. Appl. **22**(7–8), 1551–1562 (2013)
5. Chagoyen, M., Carmona-Saez, P., Shatkay, H., Carazo, J.M., Pascual-Montano, A.: Discovering semantic features in the literature: a foundation for building functional associations. BMC Bioinf. **7**(1), 1 (2006)
6. Cheng, Y., Church, G.M.: Biclustering of expression data. In: ISMB, vol. 8, pp. 93–103 (2000)
7. Cohen, R., Katzir, L., Raz, D.: An efficient approximation for the generalized assignment problem. Inf. Process. Lett. **100**(4), 162–166 (2006)
8. Dhillon, I.S., Mallela, S., Modha, D.S.: Information-theoretic co-clustering. In: Proceedings of the Ninth ACM SIGKDD International Conference on Knowledge Discovery and Data Mining, pp. 89–98. ACM (2003)
9. Diaz, A.K.R., Peres, S.M.: Biclustering and coclustering: concepts, algorithms and viability for text mining. Revista de Informática Teórica e Aplicada **26**(2), 81–117 (2019)
10. Eren, K., Deveci, M., Küçüktunç, O., Çatalyürek, Ü.V.: A comparative analysis of biclustering algorithms for gene expression data. Brief. Bioinf. **14**(3), 279–292 (2013)
11. Fleischer, L., Goemans, M.X., Mirrokni, V.S., Sviridenko, M.: Tight approximation algorithms for maximum general assignment problems. In: Proceedings of the Seventeenth Annual ACM-SIAM Symposium on Discrete Algorithm, pp. 611–620. Society for Industrial and Applied Mathematics (2006)
12. Hanczar, B., Nadif, M.: Ensemble methods for biclustering tasks. Pattern Recogn. **45**(11), 3938–3949 (2012)
13. Hartigan, J.A.: Direct clustering of a data matrix. J. Am. Stat. Assoc. **67**(337), 123–129 (1972)
14. Gupta, J.K., Singh, S., Verma, N.K.: Mtba: matlab toolbox for biclustering analysis, pp. 94–97. IEEE (2013)
15. Kerr, G., Ruskin, H.J., Crane, M., Doolan, P.: Techniques for clustering gene expression data. Comput. Biol. Med. **38**(3), 283–293 (2008)
16. Kluger, Y., Basri, R., Chang, J.T., Gerstein, M.: Spectral biclustering of microarray data: coclustering genes and conditions. Genome Res. **13**(4), 703–716 (2003)
17. Lazzeroni, L., Owen, A.: Plaid models for gene expression data. Statistica Sinica, 61–86 (2002)
18. Li, G., Ma, Q., Tang, H., Paterson, A.H., Xu, Y.: Qubic: a qualitative biclustering algorithm for analyses of gene expression data. Nucleic Acids Res., gkp491 (2009)

19. Madeira, S.C., Oliveira, A.L.: Biclustering algorithms for biological data analysis: a survey. IEEE/ACM Trans. Comput. Biol. Bioinf. (TCBB) **1**(1), 24–45 (2004)
20. Maind, A., Raut, S.: Comparative analysis and evaluation of biclustering algorithms for microarray data. In: Perez, G.M., Mishra, K.K., Tiwari, S., Trivedi, M.C. (eds.) Networking Communication and Data Knowledge Engineering. LNDECT, vol. 4, pp. 159–171. Springer, Singapore (2018). https://doi.org/10.1007/978-981-10-4600-1_15
21. Mirkin, B.: Mathematical classification and clustering, volume 11 of nonconvex optimization and its applications (1996)
22. Morgan, J.N., Sonquist, J.A.: Problems in the analysis of survey data, and a proposal. J. Am. Stat. Assoc. **58**(302), 415–434 (1963)
23. Moussaoui, S., et al.: On the decomposition of mars hyperspectral data by ICA and bayesian positive source separation. Neurocomputing **71**(10), 2194–2208 (2008)
24. Munkres, J.: Algorithms for the assignment and transportation problems. J. Soc. Ind. Appl. Math. **5**(1), 32–38 (1957)
25. Murali, T., Kasif, S.: Extracting conserved gene expression motifs from gene expression data. In: Biocomputing 2003, pp. 77–88. World Scientific (2002)
26. Padilha, V.A., Campello, R.J.: A systematic comparative evaluation of biclustering techniques. BMC Bioinf. **18**(1), 55 (2017)
27. Pierskalla, W.P.: Letter to the editor-the multidimensional assignment problem. Oper. Res. **16**(2), 422–431 (1968)
28. Prelić, A.: A systematic comparison and evaluation of biclustering methods for gene expression data. Bioinformatics **22**(9), 1122–1129 (2006)
29. Rangan, A.V.: A simple filter for detecting low-rank submatrices. J. Comput. Phys. **231**(7), 2682–2690 (2012)
30. Rodriguez-Baena, D.S., Perez-Pulido, A.J., Aguilar, J.S., et al.: A biclustering algorithm for extracting bit-patterns from binary datasets. Bioinformatics **27**(19), 2738–2745 (2011)
31. Teng, L., Chan, L.: Discovering biclusters by iteratively sorting with weighted correlation coefficient in gene expression data. J. Signal Process. Syst. **50**, 1520–1527 (2010)
32. Topchy, A., Minaei-Bidgoli, B., Jain, A.K., Punch, W.F.: Adaptive clustering ensembles. In: Proceedings of the 17th International Conference on Pattern Recognition, ICPR 2004, vol. 1, pp. 272–275. IEEE (2004)
33. Tsoumakas, G., Vlahavas, I.: Random k-Labelsets: an ensemble method for multilabel classification. In: Kok, J.N., Koronacki, J., Mantaras, R.L., Matwin, S., Mladenič, D., Skowron, A. (eds.) ECML 2007. LNCS (LNAI), vol. 4701, pp. 406–417. Springer, Heidelberg (2007). https://doi.org/10.1007/978-3-540-74958-5_38
34. Vega-Pons, S., Ruiz-Shulcloper, J.: A survey of clustering ensemble algorithms. Int. J. Pattern Recogn. Artif. Intell. **25**(03), 337–372 (2011)
35. Xie, J.: Qubic2: a novel and robust biclustering algorithm for analyses and interpretation of large-scale rna-seq data. Bioinformatics **36**(4), 1143–1149 (2020)
36. Yang, J., Wang, H., Wang, W., Yu, P.S.: An improved biclustering method for analyzing gene expression profiles. Int. J. Artif. Intell. Tools **14**(05), 771–789 (2005)

Collective of Base Classifiers for Mining Imbalanced Data

Joanna Jedrzejowicz[1] and Piotr Jedrzejowicz[2](✉)

[1] Institute of Informatics, Faculty of Mathematics, Physics and Informatics, University of Gdansk, 80-308 Gdansk, Poland
joanna.jedrzejowicz@ug.edu.pl

[2] Department of Information Systems, Gdynia Maritime University, 81-225 Gdynia, Poland
p.jedrzejowicz@umg.edu.pl

Abstract. Mining imbalanced datasets is a challenging and difficult problem. In this paper we adress it by proposing GEP-NB classifier based on the oversampling technique. It combines two learning methods - Gene Expression Programming and Naïve Bayes, which cooperate to produce a final prediction. At the pre-processing stage a simple mechanism for generating synthetic minority class examples and balancing the training set is used. Next, two genes g1 and g2 are evolved using Gene Expression Programming. They differ by applying in each case a different procedure for selecting synthetic minority class examples. If the class prediction by g1 agrees with the class prediction made by g2, their decision is final. Otherwise the final predictive decision is taken by the Naïve Bayes classifier. The approach is validated in an extensive computational experiment. Results produced by GEP-NB are compared with performance of several state-of-the-art classifiers. Comparisons show that GEP-NB offers a competitive performance.

Keywords: Imbalanced data · Oversampling · Gene expression programming

1 Introduction

Datasets with an unequal distribution of classes are commonly referred to as imbalanced ones. Unequal distribution of classes is encountered in numerous real-life situations such as, for example, fault diagnosis, medical diagnosis, fraud detection, credit rating, and many other critical applications. During the last two decades numerous approaches, techniques and algorithms have been proposed to deal with mining imbalanced datasets. Our research goal is to extend the range of available approaches for mining imbalanced datasets by proposing and validating an effective new classifier based on a novel oversampling procedure and GEP and NB learners integrated using a semi-ensemble architecture. In such an architecture at least two out of three base learners have to agree when taking the predictive decision.

D. Groen et al. (Eds.): ICCS 2022, LNCS 13351, pp. 571–585, 2022.
https://doi.org/10.1007/978-3-031-08754-7_62

Algorithms and methods proposed for mining imbalanced datasets can be broadly categorized as data-level, algorithm-level, and hybrid approaches. Data-level approaches can be further divided into oversampling and undersampling methods. Their goal is to transform the dataset used for learning prior to applying some learners. Such transformation usually leads to achieving the balanced or, at least, a better balanced distribution of classes.

Our motivation for using GEP and NB for inducing base classifiers has been based on the earlier performance of both techniques in data mining applications. A review of numerous successful GEP applications in machine learning can be found in [13]. Naïve Bayes is a probabilistic classifier that can achieve a high accuracy level [10]. Besides, NB learners are scalable and require some parameters linear in the number of variables in a learning problem. Both types of learners, that is GEP and NB are based on different philosophies, which makes their prediction fairly independent. The above feature plays an important role in the proposed approach where NB has a decisive role in the case when GEP induced base learners produce different predictions.

The rest of the paper is organized as follows. Section 2 contains a concise overview of the related work. Section 3 presents the proposed learner named GEP-NB. Section 4 discusses results of an extensive computational experiment carried out to validate the approach. Final Sect. 5 contains conclusions and suggestions for future research.

2 Related Work

In this Section we will briefly review several learners, including those used for mining imbalanced datasets, that are used for comparison and validation purposes in Sect. 4.

The simplest approach to balancing imbalanced datasets are random undersampling (RUS) and random oversampling (ROS). RUS works through random elimination of instances from majority class, and ROS through random replication of minority class instances. Both approaches have disadvantages - RUS may eliminate potentially informative examples and ROS may cause an overfitting.

One of the most often used approaches for mining imbalanced datatests is SMOTE - an oversampling technique proposed in [3]. In SMOTE the minority class is oversampled by introducing synthetic instances selected randomly and iteratively along the line segments joining some of the k minority class nearest neighbors until the balance between classes is achieved. Well known extension of SMOTE is the ADASYN method [11]. In [24] an approach named LLE for enhancing the SMOTE by incorporating the locally linear embedding algorithm was proposed. Another improvement of SMOTE obtained by introducing the PCA framework was proposed in [20]. Further, numerous, extensions and modification of SMOTE are reviewed in [6].

An approach to oversampling strategy using a rough-granular computing approach (RGA) was proposed in [2]. Another approach based on the rough set theory was proposed in [5]. The authors proposed a method for feature selection for imbalanced datasets using the neighborhood rough set theory. The approach assumes that imbalanced distribution of classes reflects the definition of the feature significance. A discernibility-matrix-based feature selection method is next defined and used in the feature selection algorithm (RSFSAID). Finally, a particle swarm optimization algorithm is suggested to optimize parameters of the algorithm.

Recently, an approach for enhancing the performance of oversampling methods for class imbalance classification was proposed in [16]. The authors propose a novel hybrid technique named ant colony optimization resampling (ACOR) to overcome class imbalance.

In [27] the authors claim that oversampling methods are often disrupted by noise when data are not well separated. As a remedy they propose the framework using the Laplacian eigenmaps (EIGEN FRAMEWORK) to find an optimal dimensional space, where the data are well separated and the generation of noise by SMOTE based oversampling methods can be avoided or minimized.

Tomek Link (TL) is an undersampling technique originating from [21]. One of the oldest approaches to undersampling is the Edited Nearest Neighbors (ENN) algorithm based on Wilsons rules [26]. The default behavior of ENN is to remove examples from the majority class that are misclassified by their k nearest neighbors. The Repeated ENN (RENN) runs the ENN algorithm repeatedly until all instances remaining have a majority of their neighbors with the same class [25]. One Side Selection (OSS) algorithm proposed in [14] is another undersampling technique. The algorithm starts with constructing a 1-NN classifier from dataset containing all minority class instances and a single, randomly drawn, majority class instance. Next, it appends misclassified instances from the set of remaining ones and removes borderline and noisy instances using Tomek links. An improvement of the OSS was proposed in [15]. The proposed Neighbouring Cleaning Rule (NCR) algorithm is similar to OSS, except that to identify uninformative and noisy data the edited nearest-neighbor rule is used instead of the TL.

Undersampling approach for learning Naïve Bayes classifiers for mining imbalanced datasets (NBU) was presented in [1].

In the recent years several undersampling algorithms using clustering have been proposed. One of the first was the algorithm Fast-CBUS proposed by [19]. The idea was to group majority instances from the training set into clusters. A separate classifier is then trained for each cluster. An unlabeled instance is classified as the majority class if it does not fit into any of the clusters. Otherwise, separate classifiers induced earlier on are used to return the classification results, and the results are weighted by the inverse-distance from the clusters.

Well performing Clustering-based undersampling (CBU) was proposed in [17]. The authors introduce two undersampling strategies aiming at reducing the number of instances in the majority class to balance the training dataset. The idea is to partition majority class instances into clusters. The number of clusters is set to the number of instances in the minority class. The first strategy is to use cluster centroids as the majority class representation, while the second strategy uses their nearest neighbors instead. During the learning phase the AdaBoost with C4.5 ensemble classifier was induced.

Combining a clustering-based undersampling based with instance selection was the idea of [22]. The cluster based instance selection (CBIS) uses two components. The first, groups instances from the majority class into clusters, and the second filters out unrepresentative ones from each cluster. For clustering the affinity propagation (AP) algorithm proposed in [9] is used and for instance selection, either a genetic algorithm, or IB3, or DROP3 can be used (for instance selection algorithms see [25]).

An effective approach to mining imbalanced datasets is using the ensemble learning techniques. The idea is to combine several base-learners into ensembles of classifiers. One of the first was the ensemble oversampling algorithm named SMOTEBoost, proposed in [4].

Combining undersampling and oversampling techniques in an ensemble learner was proposed in [23]. The learner known as UnderBagging and OverBagging (UOBag) works as follows: In UnderBagging, several subsets of instances are created by undersampling majority class randomly to construct classifiers. In a similar way, OverBagging creates subsets of instances by oversampling minority classes randomly. When a new instance arrives the majority vote decides on class prediction.

Ensemble classifier for imbalanced data based on feature space partitioning and hybrid metaheuristics (AdaSSGACE) was proposed in [18].

3 GEP-NB Classifier

3.1 General Idea of the GEP-NB

The proposed GEP-NB classifier is based on the oversampling technique. It combines two learning methods - Gene Expression Programming and Naïve Bayes classifier which are used to produce a collective learner responsible for the final prediction. GEP-NB can be used for solving binary classification problems. At the pre-processing stage a simple mechanism for generating a synthetic minority class examples and balancing the training set is used. For balancing purposes, available minority class examples from the training set are supplemented by some synthetic minority class examples to produce an expanded minority class (EMC) training dataset consisting of original plus synthetic examples. During the first phase of constructing the EMC dataset, original minority examples are randomly replicated and attached to the current EMC. The number of minority class instances drawn in such a way, denoted as Ms, is set by the user. Each instance from the minority class can be drawn many times. The number of replicated minority instances plus the original set of minority instances in the training set should exceed the number of majority instances in the training set. The size of the EMC is controlled by the parameter Ms. All replicated minority instances are subject to mutation.

For each replicated instance, the mutation procedure starts with randomly selecting (based on the uniform distribution) subset of features (without class labels), which will undergo a mutation. Selected features are modified according to the following heuristic rules: Boolean values are reversed, integer values are changed by randomly modifying (adding or subtracting) x percent of their value, and taking the integer part of the result, real values are changed by randomly adding or subtracting x percent of their value, where x is a parameter, set by the user. The idea is to produce some synthetic minority class instances constructed from the original minority class samples using the proposed random mutation procedure. The mutation scale is controlled by the parameter x, and the quality of thus produced synthetic samples is controlled by the subsequent selection procedures used for achieving a balance between minority and majority classes. In numerous applications including evolutionary computations, genetic programming, and population-based meta-heuristics, mutation procedures are used as means for improving diversification of solutions, assuring better convergence, and helping to escape from local optima. In our case, the role of mutation is to diversify synthetic samples and still keeping them somehow similar to original minority instances. To avoid the negative influence of the outliers and to keep samples fairly uniformly distributed we propose two specialized selection procedures. Our approach has been inspired by population-based techniques that have proven effective for solving a variety of difficult problems.

Next, two genes $g1$ and $g2$ are evolved using Gene Expression Programming. Both learners have the form of expression trees induced under the criterion of geometric mean (G) value which should be maximized. The choice of geometric mean as the main criterion is motivated by the fact that G is one of the most often used metrics for evaluating the performance of learners designed for mining imbalanced datasets. Besides, the value of G is closely correlated with values of other metrics commonly used in the case of imbalanced datasets mining. The learners $g1$ and $g2$ differ by applying in each case a dedicated selection procedure for reducing the EMC to balance minority and majority datasets:

- In the case of $g1$, the centroid of the EMC is identified, and the Euclidean distance between the centroid and each of the instances in the EMC is calculated. At this point we apply an instance reduction procedure to obtain the reduced EMC with fairly uniform distribution of instances in the solution space as shown in Fig. 2. Reduction aims at balancing majority and minority classes.
- In the case of $g2$, the centroid of the majority class is identified, and the distance between this centroid and each of the instances in the EMC is calculated. At this point we apply an instance reduction procedure to obtain the reduced EMC by discarding instances that are close to the centroid of the majority class as shown in Fig. 3, until majority and minority classes become balanced.

At the learning stage, classifiers $g1$ and $g2$, and a Naïve Bayes classifier play the role of base learners. Naïve Bayes learner is induced using a subset of the training set involving instances for which predictions produced by $g1$ and $g2$ have differed. The above learners are expected to maximize the value of the respective geometric mean. Classifiers produced by GEP have the form of expression trees.

When classifying instances belonging to the test set, if for an instance the class prediction by $g1$ agrees with that of $g2$, their decision is final. Otherwise, the final predictive decision is made by the Naïve Bayes classifier. An example of an expression tree, a formal description of the approach and its computational complexity analysis is given in the next subsections.

3.2 Formal Description of the Approach

Gene Expression Programming (GEP), introduced by Ferreira [8] is a metaheuristic which can be used in several areas, classification included. It combines the idea of genetic algorithms and genetic programming and makes use of a population of genes. Each gene is a linear structure divided in two parts. The first part, head, contains functions and terminals while the second part, tail, contains only terminals. For this study terminals are of type $(oper, attr, const)$, where the value of $const$ is in the range of attribute $attr$ and $oper$ is a relational operator from $\{<, \leq, >, \geq, =, \neq\}$. Functions are from the set $\{AND, OR, NOT, XOR, NOR\}$. For a fixed instance x from the dataset, the value $g(x)$ of a gene g is boolean and thus a gene can be treated as a binary classifier.

Learning the best gene classifier is an iterative process which starts with a random population of genes. In each iteration the population is subjected to operations such as: mutation, root transposition, transposition of insertion sequence, 1-point and 2-point recombination. Each operation is performed with a probability which is a parameter of the process. More details on applying GEP can be found in [12].

Considering our hybrid classifier, the first step is to oversample the minority class by mutating random rows, as described in Fig. 1. The mutation of an attribute is defined as:

Require: data from minority class MinC, parameter Ms
Ensure: expanded minority class EMC of size Ms.
EMC=$\emptyset$
for $i = 0$ to Ms **do**
  draw random data row rw from MinC
  draw random subset AT of attributes
  for all $at \in AT$ **do**
    mutate $rw(at)$ to $\bar{rw}(at)$ applying (1)
  end for
  add $\bar{rw}$ to EMC
end for
return EMC

Fig. 1. Oversampling to generate expanded minority class

$$r\bar{w}(at) = \begin{cases} 1 - at & \text{if } at \text{ boolean} \\ (int)(rand(at * (+/-(1+x))) & \text{if } at \text{ integer} \\ rand(at * (+/-(1+x))) & \text{otherwise} \end{cases} \quad (1)$$

In the next step one of two different selection procedures is applied to EMC to balance majority and minority sets. The procedures are given in Fig. 2 and Fig. 3, respectively. Finally, the algorithm shown in Fig. 4 is applied to learn the classifier, test it and calculate entries in confusion matrix and the respective performance measures.

```
Require: expanded minority class EMC , data from majority class MajC
Ensure: balanced dataset
calculate centroid CN of MajC
for all x ∈EMC do
   calculate dist(x, CN)
end for
discard from EMC instances closest to CN to balance with MajC
return EMC ∪ MajC
```

Fig. 2. Instance reduction with equal distribution

```
Require: expanded minority class EMC, data from majority class MajC
Ensure: balanced data set.
calculate centroid CN of EMC
define quartiles for {dist(x, CN) x ∈ EMC}
for i = 1, 2, 3, 4 do
   let n_i=number of elements in quartile i
   Q_i = n_i/(n_1 + n_2 + n_3 + n4)
end for
repeat
   t=random
   let (t > Q_{i−1} ∧ (t ≤ Q_i))
   discard random instance from quartile i
until EMC and MajC are balanced
return EMC ∪ MajC
```

Fig. 3. Instance reduction with centroids

As far as computational complexity is concerned, for Fig. 1 it is $O(|EMC|)$, where EMC is the expanded minority class. For Fig. 2 and Fig. 3 it is $O(|EMC|^2)$, and finally, for Fig. 4 it is bounded by the complexity of Fig. 2 and the complexity of learning the best gene which is $O(nIt \times popSize \times |dataset|)$, where nIt is the number of iterations in GEP, $popSize$ is the population size and $|dataset|$ is the size of the dataset.

Require: minority class MinC, majority class MajC, testing data Test
Ensure: measures of classification quality
use algorithm from Fig.1 to expand minority class MinC to extended minority class EMC
use algorithm from Fig. 2 to generate balanced training set $Train1$
generate best possible gene $g1$ with $Train1$
use algorithm from Fig. 3 to generate balanced training set $Train2$
generate best possible gene $g2$ with $Train2$
for all $(x, c) \in$ Test **do**
  calculate $g1(x) = v1$ and $g2(x) = v2$
  if $v1 = v2$ **then**
    $class = v1$
  else
    apply Naïve Bayes classifier to x to define $class$
  end if
  compare $class$ with c and modify TP, FN, FP, TN respectively
end for
calculate quality measures from TP, TN, FP, FN
return quality measures

Fig. 4. Learning best genes and testing

4 Computational Experiment

To validate the proposed approach we have carried out an extensive computational experiment. It has covered 100 of the imbalanced datasets available in KEEL Dataset Repository (https://sci2s.ugr.es/keel/imbalanced.php). Datasets originate from [6] and [7]. Full information about the above datasets including dataset names, number of instances, number of features and value of the imbalanced ratio can be found in the KEEL Dataset Repository.

In the experiment we have applied 5-folds cross validation procedure which has been repeated 10 times. All the reported values are averages from the above scheme. Geometric Mean (G) and Area Under the ROC Curve (AUC) were selected as performance metrics.

GEP-NB has been run with the following settings, identical for all considered datasets:

- GEP population size: 100
- Number of iterations in GEP: 100
- Selection rule: tournament from the pair
- Mutation probability: pm = 0.5
- Root Insertion Sequence Transposition probability: pris = 0.2
- Insertion Sequence probability: is = 0.2
- 1-point recombination probability: pr1 = 0.2
- 2-points recombination probability: pr2 = 0.2
- Size of the EMC: twice the number of majority instances in the training set
- The value of $x = 5\%$.

5 Comparative Analysis

To evaluate the proposed approach several comparisons with the state-of-the-art algorithms for mining imbalanced datasets have been carried out. Table 1 shows average geometric mean (G) obtained using the following learners: Naïve Bayes Undersampling (NBU), Edited Nearest Neighbors (ENN), Neighbouring Cleaning Rule (NCR), One Side Selection (OSS), Repeated Edited Nearest Neighbors (RENN), Random Undersampling (RUS), Synthetic Minority Oversampling Technique (SMOTE), Tomek Links (TL), the proposed oversampling scheme denoted Population-based Oversampling (PBO), and GEP-NB proposed in this paper. In all cases values of the respective metric have been calculated as an average from several runs of the 5-cross-validation scheme. Results for NBU, ENN, NCR, OSS, RENN, RUS, SMOTE and TL are taken from [1]. In all of the above cases, as well as in the case of PBO, final results were obtained using Naïve Bayes classifier. To determine whether there are any significant differences among results from Table 1, produced by different classifiers we used the Friedman ANOVA by ranks test. The null hypothesis state that there are no such differences. With Friedman statistics equal to 75, 39 and p-value equal to 0.00000 the null hypothesis should be rejected at the significance level of 0.05. However, the Kendall concordance coefficient expressing the simultaneous association (relatedness) between the considered samples, with the value of 0.1948 tells that there is a limited degree of relatedness between the considered samples.

Another comparison involved GEP-NB and the following ensemble learners: Random Undersampling with Boosting (RUSBoost), ensemble classifier based on feature space partitioning with hybrid metaheuristics (AdaSSGACE) using KNN and SVM base classifiers, and Underbagging and Overbagging Ensemble Learner (UOBag). The performance metric used in the comparison is the area under the ROC curve (AUC). In all cases values of the respective metric have been calculated as an average from several runs of the 5-cross-validation scheme. Results for RUSBoost, AdaSSGACE and UOBag are taken from [18]. The respective results are shown in Table 2. Table 3 shows results obtained by GEP-NB, clustering based undersampling (CBU) proposed in [17] and two versions of the ensemble classifier proposed in [22], and known as undersampling by combining clustering analysis and instance selection (CBIS). Both version of this algorithm use clustering by passing messages between data points technique [9] and IB3 instance selection algorithm [25]. First version uses boosting for constructing ensembles, and the second one uses bagging for the same purpose. All results have been obtained using the 5-CV scheme. Results for CBU and CBIS are taken from the original sources. In all cases the performance metric is the area under the ROC curve (AUC). To determine whether there are any significant differences among results from Table 3, we again used the Friedman ANOVA by ranks test. The null hypothesis state that there are no such differences. With Friedman statistics equal to 7.301 and p-value equal to 0.0629 the null hypothesis should not be rejected at the significance level of 0.05. The Kendall concordance coefficient expressing the simultaneous association (relatedness) between the considered samples, with the value of 0.0811 tells that there

Table 1. Comparison of the average geometric mean (G) values obtained by the analyzed classifiers.

Dataset	NBU	ENN	NCR	0SS	RENN	RUS	SMOTE	TL	PBO	GEP-NB
abalone19	0,659	0,695	0,695	0,695	0,695	0,708	0,679	0,694	0,713	**0,746**
dermatology6	0,988	0,966	0,966	0,966	0,879	0,966	0,975	0,966	0,997	**0,999**
ecoli-0-1-4-6-VS-5	0,880	0,858	0,858	0,853	0,858	0,861	0,862	0,858	0,934	**0,949**
ecoli-0-1-4-7-VS-2-3-5-6	**0,960**	0,922	0,924	0,926	0,915	0,908	0,927	0,925	0,860	0,875
ecoli-0-1-4-7-VS-5-6	**0,958**	0,948	0,948	0,945	0,948	0,826	0,943	0,949	0,953	0,928
ecoli-0-2-3-4-vs-5	0,900	0,856	0,858	0,869	0,856	0,816	0,874	0,863	0,912	**0,957**
ecoli-0-3-4-6-vs-5	0,922	0,849	0,850	0,853	0,849	0,766	0,854	0,849	0,879	**0,951**
ecoli-0-3-4-7-vs-5-6	**0,941**	0,925	0,925	0,919	0,925	0,928	0,906	0,925	0,911	0,920
ecoli-0-3-4-vs-5	0,894	0,835	0,844	0,860	0,835	0,750	0,860	0,844	0,908	**0,927**
ecoli-0-4-6-vs-5	0,872	0,846	0,846	0,853	0,846	0,848	0,857	0,846	0,917	**0,949**
ecoli-0-6-7-vs-5	0,915	0,874	0,879	0,871	0,874	0,838	0,883	0,883	0,914	**0,950**
ecoli1	0,885	0,850	0,852	0,839	0,853	0,872	0,842	0,848	0,892	**0,931**
ecoli2	**0,941**	0,928	0,929	0,928	0,927	0,921	0,942	0,928	0,845	0,939
ecoli3	0,894	0,890	0,892	0,920	0,873	0,923	0,908	0,907	0,857	**0,924**
ecoli4	0,950	0,916	0,916	0,920	0,916	0,916	0,934	0,917	0,879	**0,982**
glass-0-1-4-6-vs-2	0,720	0,685	0,690	0,680	0,699	0,618	0,703	0,696	0,702	**0,737**
glass0	0,826	0,771	0,800	0,810	0,755	0,800	0,793	0,804	0,799	**0,875**
glass1	0,660	0,663	0,662	0,669	0,702	0,702	0,637	0,656	0,678	**0,761**
glass4	0,830	0,723	0,726	0,714	0,723	0,760	0,752	0,728	0,766	**0,969**
glass6	0,864	0,854	0,854	0,825	0,888	0,855	0,827	0,856	0,869	**0,971**
haberman	0,656	0,670	0,676	0,658	0,671	0,611	0,645	0,657	0,712	**0,731**
iris0	**1,000**	**1,000**	**1,000**	**1,000**	**1,000**	**1,000**	**1,000**	**1,000**	**1,000**	1,000
new-throid1	**1,000**	**1,000**	**1,000**	**1,000**	**1,000**	**1,000**	**1,000**	**1,000**	0,945	0,974
new-thyroid2	**1,000**	**1,000**	**1,000**	**1,000**	**1,000**	**1,000**	**1,000**	**1,000**	0,966	0,978
page-blocks-1-3-vs-4	0,992	0,902	0,908	0,908	0,902	0,902	0,909	0,909	0,913	**0,999**
page-blocks0	**0,954**	0,936	0,935	0,928	0,937	0,935	0,932	0,932	0,905	0,893
pima	**0,815**	0,799	0,809	**0,815**	0,789	0,814	0,815	0,813	0,814	0,748
poker-8-vs-6	0,531	0,437	0,439	0,427	0,437	0,456	0,500	0,577	0,588	**0,801**
segment0	**0,987**	0,982	0,982	0,982	0,982	0,982	0,980	0,982	0,981	0,984
vehicle0	**0,904**	0,806	0,805	0,823	0,801	0,809	0,820	0,811	0,834	0,901
vehicle1	0,740	0,709	0,713	0,717	0,708	0,719	0,717	0,713	0,726	**0,754**
vehicle2	0,920	0,861	0,859	0,850	0,857	0,834	0,849	0,857	0,789	**0,926**
vehicle3	0,762	0,701	0,700	0,698	0,700	0,697	0,699	0,698	0,798	**0,926**
winequality-red-4	**0,694**	0,659	0,653	0,651	0,660	0,626	0,653	0,650	0,616	0,679
winequality-red-8-vs-6-7	0,695	0,713	0,717	0,669	0,721	0,674	0,651	0,713	0,674	**0,815**
wisconsin	0,978	0,975	0,977	**0,993**	0,974	0,983	0,983	0,982	0,979	0,973
yeast-0-2-5-6-vs-3-7-8-9	0,816	0,761	0,762	0,762	0,764	0,742	0,755	0,760	0,776	**0,848**
yeast-0-2-5-7-9-vs-3-6-8	0,932	0,916	0,916	0,888	0,916	0,907	0,816	0,915	0,923	**0,939**
yeast-0-3-5-9-vs-7-8	0,721	0,695	0,690	0,712	0,680	0,713	0,731	0,698	0,698	**0,727**
yeast-1-vs-7	0,801	0,802	0,800	0,800	**0,805**	0,792	0,776	0,800	0,703	0,737
yeast-2-vs-4	0,864	0,833	0,835	0,833	0,831	0,822	0,834	0,839	0,837	**0,956**
yeast-2-vs-8	0,838	0,835	0,836	0,828	0,835	**0,850**	0,793	0,836	0,818	0,833
yeast5	**0,989**	0,987	0,986	0,986	0,987	0,976	0,982	0,986	0,935	0,978
Average	0,862	0,833	0,835	0,834	0,832	0,824	0,832	0,835	0,840	**0,892**

Table 2. Comparison of the average geometric mean (G) values obtained by the analyzed classifiers.

Dataset	GEP-NB	RUSBoost	AdaSSGACEKNN.40	AdaSSGACESVM.40	UOBag
ecoli1	**0,933**	0,884	0,890	0,872	0,876
ecoli3	**0,927**	0,840	0,864	0,785	0,886
iris0	**1,000**	0,990	0,999	0,841	0,970
page-blocks0	0,917	**0,956**	0,931	0,751	0,953
pima	**0,751**	0,725	0,738	0,589	0,730
vehicle1	0,757	**0,786**	0,778	0,651	0,745
yeast1	**0,722**	0,701	0,711	0,688	0,720
yeast3	**0,945**	0,919	0,897	0,891	0,919
glass1	**0,772**	0,780	0,750	0,639	0,739
glass6	**0,971**	0,921	0,903	0,641	0,901
glass-0-1-6_vs_2	**0,773**	0,700	0,708	0,611	0,629
ecoli4	**0,982**	0,896	0,940	0,907	0,867
glass-0-1-6_vs_5	0,960	0,954	0,867	0,733	**0,963**
glass5	0,966	0,949	0,804	0,675	**0,988**
dermatology6	**0,999**	0,966	0,966	0,749	0,938
shuttle-6_vs_2-3	**1,000**	0,902	0,965	0,843	0,948
poker-9_vs_7	**0,886**	0,590	0,740	0,636	0,556
yeast-2_vs_8	**0,840**	0,747	0,801	0,737	0,778
yeast4	**0,867**	0,827	0,799	0,509	0,763
led7digit-0-2-4-5-6-7-8-9_vs_1	**0,917**	0,894	0,856	0,785	0,881
ecoli-0-1-3-7_vs_2-6	**0,959**	0,896	0,848	0,588	0,867
winequality-red-8_vs_6	**0,827**	0,815	0,589	0,528	0,700
winequality-white-9_vs_4	**0,914**	0,893	0,645	0,576	0,714
yeast6	**0,914**	0,851	0,875	0,515	0,814
poker-8-9_vs_6	**0,835**	0,915	0,623	0,557	0,534
winequality-white-3-9_vs_5	**0,683**	0,674	0,561	0,531	0,576
shuttle-2_vs_5	**1,000**	**1,000**	0,986	0,672	**1,000**
winequality-red-3_vs_5	**0,848**	0,644	0,608	0,580	0,615
poker-8-9_vs_5	0,616	0,547	**0,631**	0,557	0,618
poker-8_vs_6	0,817	**0,915**	0,623	0,469	0,534
Average	**0,877**	0,836	0,797	0,670	0,791

is a limited degree of relatedness between the considered samples. It is worth noting that the above findings are not contradictory to the fact that the average performance of GEP-NB using the AUC metric and calculated over the sample of 29 datasets as shown in Table 3, is better than the performance of the remaining learners.

Table 3. Comparison of the area under the ROC curve obtained by CBU, CBIS and GEP-NB classifiers

Dataset	CBU	CBIS		GEP-NB
		AP+IB3boost	AP+IB3boost	
Abalone9-18	0,831	0,849	**0,894**	0,808
Abalone19	0,728	0,624	0,617	**0,771**
Ecoli-0-vs-1	0,982	0,975	0,982	**0,993**
Ecoli-0-1-3-7-vs-2-6	0,804	0,877	0,879	**0,959**
Ecoli1	0,927	**0,958**	0,957	0,933
Glass0	0,873	**0,888**	0,885	0,876
Glass-0-1-2-3-vs-4-5-6	0,970	**0,980**	0,966	0,961
Glass-0-1-6-vs-2	**0,790**	0,775	0,713	0,773
Glass-0-1-6-vs-5	0,964	0,894	**0,987**	0,960
Glass1	0,824	0,812	**0,847**	0,772
Glass2	0,760	0,741	0,766	**0,822**
Glass4	0,853	0,944	**0,971**	0,970
Glass5	0,949	**0,994**	**0,994**	0,966
Glass6	0,905	0,951	0,934	**0,971**
Haberman	0,603	0,646	0,648	**0,737**
Iris0	0,990	0,990	0,990	**1,000**
New-thyroid1	0,973	0,979	**0,997**	0,975
New-thyroid2	0,924	0,976	**0,994**	0,978
Page-blocks0	0,986	**0,987**	**0,987**	0,917
Page-blocks-1-3-vs-2	0,992	0,998	0,997	**0,999**
Pima	0,758	0,771	**0,805**	0,751
Segment0	0,996	**0,999**	0,993	0,984
Shuttle-0-vs-4	**1,000**	**1,000**	**1,000**	**1,000**
Shuttle-2-vs-4	0,988	**1,000**	**1,000**	**1,000**
Yeast-1-2-8-9-vs7	0,692	**0,818**	0,775	0,661
Yeast-1-4-5-8-vs7	0,627	**0,777**	0,605	0,709
Yeast4	0,874	0,857	**0,914**	0,867
Yeast5	**0,987**	0,967	0,970	0,978
Yeast6	0,909	0,881	0,884	**0,914**
Average	0,878	0,893	0,895	**0,897**

6 Conclusion

The paper contributes by proposing a new classifier for mining imbalanced datasets. The classifier named GEP-NB, has the following main features:

- At the preprocessing stage it uses an original oversampling method to balance minority and majority sets of instances. The approach is based on producing synthetic minority class instances by replication, mutation, and selection of instances from the original set of minority instances.

- At the learning stage a semi-ensemble strategy is used. It consists of developing two complex expression trees using the Gene Expression Programming paradigm. Both are supported by the Naïve Bayes learner used in the case when the class prediction from both genes differs.

An extensive computational experiment has shown that the performance of the proposed learner is competitive. Comparison with the state of the art single learners for mining imbalanced datasets proves that GEP-NB outperforms all of them. Comparison with the state of the art ensemble learners shows that GEP-NB either outperforms them or is at least as good as the best of the considered ensemble classifiers. The above findings allow to state that GEP-NB is a worthy addition to the family of classifiers dedicated to mining imbalanced dataset.

Competitive performance of GEP-NB can be attributed to a synergetic effect produced by three following factors - learners, oversampling procedure, and semi-ensemble architecture. Ensemble components including GEP and Naïve Bayes learners are themselves good performers, which have been confirmed by numerous studies. The proposed oversampling procedure has been constructed using the population-based paradigm where mutation of the population members and selection of better-fitted individuals play a vital role in the search for high-quality solutions. Finally, using the concept of semi ensemble learning with three base classifiers could mask some prediction errors. The main novelty and, at the same time, the solution of the research problem tackled by the paper is the integration of the above factors into a specialized learner producing good quality predictions when mining imbalanced datasets.

Future research will focus on extending the approach, possibly by integrating oversampling and undersampling approaches for balancing minority and majority class instances with a view to increasing effectiveness of the approach. Another direction of studies could bring some improvements in the process of generating and selecting the synthetic minority instances.

References

1. Aridas, C.K., Karlos, S., Kanas, V.G., Fazakis, N., Kotsiantis, S.B.: Uncertainty based under-sampling for learning Naive Bayes classifiers under imbalanced data sets. IEEE Access **8**, 2122–2133 (2020)
2. Borowska, K., Stepaniuk, J.: A rough-granular approach to the imbalanced data classification problem. Appl. Soft Comput. **83** (2019)
3. Chawla, N.V., Bowyer, K.W., Hall, L.O., Kegelmeyer, W.P.: SMOTE: synthetic minority over-sampling technique. J. Artif. Intell. Res. **16**, 321–357 (2002)
4. Chawla, N.V., Lazarevic, A., Hall, L.O., Bowyer, K.W.: SMOTEBoost: improving prediction of the minority class in boosting. In: Lavrač, N., Gamberger, D., Todorovski, L., Blockeel, H. (eds.) PKDD 2003. LNCS (LNAI), vol. 2838, pp. 107–119. Springer, Heidelberg (2003). https://doi.org/10.1007/978-3-540-39804-2_12
5. Chen, H., Li, T., Fan, X., Luo, C.: Feature selection for imbalanced data based on neighborhood rough sets. Inf. Sci. **483**, 1–20 (2019)
6. Fernández, A., García, S., Galar, M., Prati, R.C., Krawczyk, B., Herrera, F.: Learning from Imbalanced Data Sets. Springer, Heidelberg (2018). https://doi.org/10.1007/978-3-319-98074-4

7. Fernández, A., del Jesus, M.J., Herrera, F.: Hierarchical fuzzy rule based classification systems with genetic rule selection for imbalanced data-sets. Int. J. Approx. Reason. **50**(3), 561–577 (2009)
8. Ferreira, C.: Gene expression programming: a new adaptive algorithm for solving problems. Complex Syst. **13**(2) (2001)
9. Frey, B.J., Dueck, D.: Clustering by passing messages between data points. Science **315**(8), 972–976 (2007)
10. Hand, D.J., Yu, K.: Idiot's Bayes: not so stupid after all? Int. Stat. Rev. Rev. Internationale de Statistique **69**(3), 385–398 (2001). http://www.jstor.org/stable/1403452
11. He, H., Bai, Y., Garcia, E.A., Li, S.: ADASYN: adaptive synthetic sampling approach for imbalanced learning. In: IEEE International Joint Conference on Neural Networks (IEEE World Congress on Computational Intelligence), IJCNN 2008, pp. 1322–1328 (2008)
12. Jedrzejowicz, J., Jedrzejowicz, P.: Experimental evaluation of two new GEP-based ensemble classifiers. Expert Syst. Appl. **38**(9), 10932–10939 (2011)
13. Jedrzejowicz, J., Jedrzejowicz, P.: Gene expression programming as a data classification tool. A review. J. Intell. Fuzzy Syst. **36**(1), 91–100 (2019)
14. Kubat, M., Matwin, S.: Addressing the curse of imbalanced training sets: one-sided selection. In: Proceedings of the Fourteenth International Conference on Machine Learning (ICML 1997), Nashville, Tennessee, USA, 8–12 July 1997, pp. 179–186 (1997)
15. Laurikkala, J.: Improving identification of difficult small classes by balancing class distribution. In: Quaglini, S., Barahona, P., Andreassen, S. (eds.) AIME 2001. LNCS (LNAI), vol. 2101, pp. 63–66. Springer, Heidelberg (2001). https://doi.org/10.1007/3-540-48229-6_9
16. Li, M., Xiong, A., Wang, L., Deng, S., Ye, J.: ACO resampling: enhancing the performance of oversampling methods for class imbalance classification. Knowl. Based Syst. **196**, 105818 (2020)
17. Lin, W.C., Tsai, C.F., Hu, Y.H., Jhang, J.S.: Clustering-based undersampling in class-imbalanced data. Inf. Sci. **409**, 17–26 (2016)
18. Lopez-Garcia, P., Masegosa, A.D., Osaba, E., Onieva, E., Perallos, A.: Ensemble classification for imbalanced data based on feature space partitioning and hybrid metaheuristics. Appl. Intell. **49**(8), 2807–2822 (2019). https://doi.org/10.1007/s10489-019-01423-6
19. Ofek, N., Rokach, L., Stern, R., Shabtai, A.: FAST-CBUS: a fast clustering-based undersampling method for addressing the class imbalance problem. Neurocomputing **243**, 88–102 (2017)
20. Tang, S., ping Chen, S.: The generation mechanism of synthetic minority class examples. In: Proceedings of International Conference on Information Technology and Applications in Biomedicine, pp. 444–447 (2008)
21. Tomek, I.: Two modifications of CNN. IEEE Trans. Syst. Man Cybern. **SMC-6**(11), 769–772 (1976)
22. Tsai, C.F., Lin, W.C., Hu, Y.H., Yao, G.T.: Under-sampling class imbalanced datasets by combining clustering analysis and instance selection. Inf. Sci. **477**, 47–54 (2019)
23. Wang, S., Yao, X.: Diversity analysis on imbalanced data sets by using ensemble models. In: 2009 IEEE Symposium on Computational Intelligence and Data Mining, CIDM 2009-Proceedings, pp. 324–331 (2009)

24. Wang, Z., Li, Y., Li, D., Zhu, Z., Du, W.: Entropy and gravitation based dynamic radius nearest neighbor classification for imbalanced problem. Knowl. Based Syst. **193**, 105474 (2020)
25. Wilson, D.R., Martinez, T.R.: Reduction techniques for instance-based learning algorithms. Mach. Learn. **38**(3), 257–286 (2000)
26. Wilson, D.L.: Asymptotic properties of nearest neighbor rules using edited data. IEEE Trans. Syst. Man Cybern. **2**, 408–421 (1972)
27. Ye, X., Li, H., Imakura, A., Sakurai, T.: An oversampling framework for imbalanced classification based on Laplacian eigenmaps. Neurocomputing **399**, 107–116 (2020)

Impact of Clustering on a Synthetic Instance Generation in Imbalanced Data Streams Classification

Ireneusz Czarnowski[1(✉)] and Denis Mayr Lima Martins[2]

[1] Department of Information Systems, Gdynia Maritime University, Morska 83, 81-225 Gdynia, Poland
i.czarnowski@umg.edu.pl

[2] Department of Information Systems, University of Münster, ERCIS, Leonardo-Campus 3, 48149 Münster, Germany
denis.martins@wwu.de

Abstract. The goal of the paper is to propose a new version of the Weighted Ensemble with one-class Classification and Over-sampling and Instance selection (WECOI) algorithm. This paper describes WECOI and presents the alternative approach for over-sampling, which is based on a selection of reference instances from produced clusters. This approach is flexible on applied clustering methods; however, the similarity-based clustering algorithm has been proposed as a core. For clustering, different methods may also be applied. The proposed approach has been validated experimentally using different clustering methods and shows how the clustering technique may influence synthetic instance generation and the performance of WECOI. The WECOI approach has also been compared with other algorithms dedicated to learning from imbalanced data streams. The computational experiment was carried out using several selected benchmark datasets. The computational experiment results are presented and discussed.

Keywords: Classification · Learning from data streams · Imbalanced data · Over-sampling · Clustering

1 Introduction

Data analysis is a key component of the effective decision-making processes as well as modern, innovative, and autonomous decision-making systems. Prediction is one of the main tasks of data analysis and is very often solved using machine learning tools. Prediction can also be considered through the prism of classification and regression. Considering in this paper a classification task, supervised machine learning tools train models using labelled data instances, finally obtaining a classifier that is able to work with unlabelled data and unseen data. After that, the task of the trained classifier is to assign appropriate decision classes to new instances flowing into the system [4].

A current trend in machine learning research (as well as in the field of data-driven decision-making) focuses on learning models from data streams. However, decision-making on streaming data is not an easy task, since high data volumes are continuously

D. Groen et al. (Eds.): ICCS 2022, LNCS 13351, pp. 586–597, 2022.
https://doi.org/10.1007/978-3-031-08754-7_63

flowing into the system [3], especially in cases where the machine learning model must deal with data that are imbalanced in nature.

The problem of learning from imbalanced data streams is seen to be more complicated and complex than learning from static data where all the decision classes are known and balanced in the training dataset [1]. This problem arises in many real scenarios such as social media analytics, text classification tasks, fraud detection, technical and manufacturing systems etc. However, despite the increasing number of studies addressing imbalanced data using different methods, more attention needs to be given to the problem of dealing with streaming data when the data exhibit a changing and imbalanced class ratio [1, 2].

When the imbalanced data problem is considered, the popular approach proposed to eliminate an unfavorable imbalance between instances from different decision classes is to apply the SMOTE (Synthetic Minority Over-sampling Technique) algorithm [5]. The SMOTE algorithm produces new, synthetic instances located between random instances of the minority class. Alternatively, for generating synthetic instances that are closer to the model's decision boundary between the minority instances and other classes (i.e., the majority class), the borderline-SMOTE algorithm has been proposed [12]. These two algorithms are well-known examples among a myriad of the existing algorithms addressing imbalanced data and belong to the family of over-sampling algorithms. Opposite to the SMOTE class of algorithms, which create instances from the least frequent class only, under-sampling techniques for removing instances from the most frequent class have also been proposed. In both cases, the aim is to reach a good balance between instances belonging to all considered classes.

In general, the approaches for solving (i.e., eliminating) the class imbalance problem are divided into the following categories: data-level approaches, algorithm-level approaches, and hybrid approaches. A comprehensive discussion on this matter is included in [6].

The problem of imbalanced data needs more attention when the data streams are considered. A comparative study of selected algorithms dedicated to solving a class imbalance problem has been presented in [7]. More updated reviews of different techniques for imbalanced data streams are included in [8] and [9]. In [8] the authors underlined that the approaches that are able to learn from imbalanced data streams can be divided into two groups: passive and active approaches. This taxonomy is based on the possibility of the algorithms to detect drift. Among the discussed algorithms are: RLSACP, ONN, ESOS-ELM, an ensemble of neural networks, OnlineUnderOverBagging, OnlineSMOTE-Bagging, OnlineAdaC2, OnlineCSB2, OnlineRUSBoost and OnlineSMOTEBoost, ARFRE, RebalanceStream, OOB, UOB, WEOB1 and WEOB2. As an alternative, the meta-strategy Continuous-SMOTE (C-SMOTE) is proposed and compared with four other strategies which are able to deal with class imbalance in data streams, including ARFRE, RebalanceStream, OOB and UOB (see [8]).

In addition to an overview of different techniques dealing with imbalanced data streams, in [9] the Online-MC-Queue (OMCQ) algorithm is proposed. This algorithm is based on a framework for online learning in multi-class imbalanced data streams. The algorithm utilises a queue as a resampling method. The queue is dynamically created for instances belonging to each considered class. They are updated independently for each considered class, continuously assuring a balance between the instances belonging to the considered classes.

The core of this paper is also based on the framework for online learning presented previously by the author in [10]. The framework has been designed to work with imbalanced data streams. The WECOI (Weighted Ensemble with one-class Classification and Over-sampling and Instance selection) algorithm has been based on the implementation of over- and under-sampling techniques to eliminate imbalance between the instances belonging to minority and majority classes.

This paper is an extension of the work presented in [10] and considers the problem of clustering and its impact on the quality of the over-sampling process within WECOI. In this paper WECOI has been extended in a new approach to synthetic instances generation, where they are generated with respect to the reference instances representing clusters produced on the instances belonging to the majority class. In this algorithm, the way the clustering is done may influence the synthetic instance generation. Thus, the main aim of the paper is to formulate the answers to the following questions: (1) whether the clustering technique can have an impact on the quality of the over-sampling implemented within WECOI, and (2) whether the number of clusters influences the process of synthetic instances generation.

This paper is organised as follows. In the next section the WECOI-based framework is presented. Section 3 presents in detail the process of synthetic instances generation and presents the new approach. The computational experiment and results are included in Sect. 4. The last section points out to a few conclusions and directions for future research.

2 A Framework for Learning from Imbalanced Data Streams

The approach discussed in this paper is dedicated to learning from data streams. This approach, originally called Weighted Ensemble with one-class Classification and Over-sampling and Instance selection (WECOI), is also based on the decomposition of a multi-class classification problem into a set of sub-problems involving one-class classification. WECOI also uses mechanisms to eliminate the negative effect of the imbalance between minority and majority instances in data streams. The framework is also based on a drift detection concept within data streams and the final decision output is produced using a weighted ensemble classification model. The basic components of the framework are shown in Fig. 1.

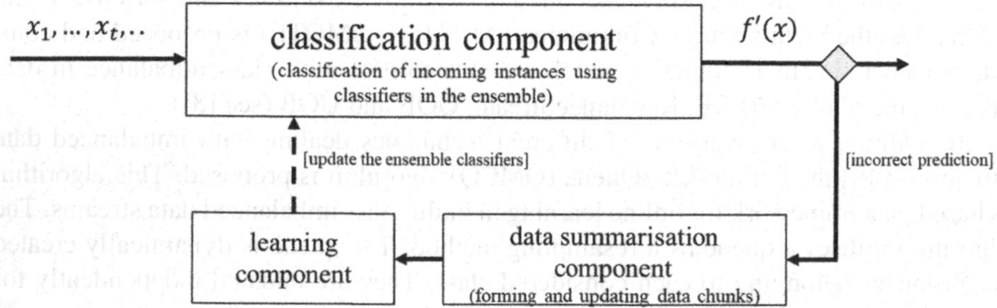

Fig. 1. A framework for learning from imbalanced data streams [10].

The main task of the classification component is to classify the incoming instances using ensemble classifiers. The approach discussed is also based on an assumption that the correctness of the output class will be known in the future. Instances for which the decision classes have not been established in the correct way are redirected to the data summarisation component.

The data summarisation component forms data chunks from instances incoming from the classification component. The component constantly updates the data chunks, so the data chunks consist of possible current instances. Based on the data chunks, base classifiers of the ensemble are formed. Each modification (updating) of the data chunk entails an update of the base classifier, i.e. also the ensemble.

A basic assumption is that the data are considered with respect to each class independently. This means that the framework is based on decomposition of the multi-class classification problem into a set of one-class classification problems. What it also means is that each detected decision class is considered independently. From the algorithmic point of view, this means that data chunks are formed independently for each decision class. Thus, a given data chunk consists of positive instances for the considered decision class, while other data chunks are in opposition consisting of negative instances.

All the current data chunks formed by the data summarisation component are used to induce base classifiers, a process that is carried out within the learning component. The learning process is done independently for each considered decision class using the current positive and negative instances.

Each new induced base classifier replaces an older base classifier in the ensemble. This also means that the framework is based on remembering earlier-induced classifiers and the number of remembered base classifiers is a parameter of the algorithm. Finally, new incoming instances are classified using ensemble classifiers and the prediction result is determined through, for example, the weighted majority vote.

In the proposed approach, much attention is paid to updating and forming data chunks. When a new instance arrives, the current data chunks are updated in the following way:

- When the size of the data chunk is smaller than the defined threshold, then a new instance is added to the data chunk;
- When the data chunk is completed, i.e. it consists of a number of instances equal to the defined threshold, the data chunk is updated.

In the second of the above cases, to decide whether instances should be added to the current set different techniques can be used. WECOI, adopting the one-class classification problem, uses two different nearest neighbour-based methods, i.e. CNN-d and ENN-d. Both of these algorithms are dedicated to instance selections between instances belonging to the same decision class and independently from other considered decision classes. The pseudocode for CNN-d and ENN-d is given in [11]. In other words, the process of updating the data chunks is carried out using under-sampling techniques.

WECOI is not free from additional challenges. When the learning component starts the induction of the classifier, the existing problem of imbalanced data in the data chunks must be eliminated. This is especially the case when, while the system is working, more of the incoming instances belong to one decision class rather than the others. This means that the number of instances included in the available data chunks is not equal. To sum up, in the considered system the problem of imbalanced data must be eliminated and for this an over-sampling procedure is applied.

In the next section of the paper the procedure for over-sampling is discussed in more detail, and a new approach for reducing the negative effect of the imbalance between minority and majority instances in data streams is also proposed. The further discussion is limited to the case of a binary classification, i.e., where only two decision classes in streams are possible.

3 Over-Sampling Procedure for Imbalanced Data Streams

3.1 Basic Over-Sampling Procedure

The over-sampling procedure described in [10] and implemented in the WECOI starts by clustering of the instances belonging to the data chunk consisting of the majority class. The centres of the produced clusters are next used as reference instances and, for each of them, nearest neighbours belonging to the minority are selected. The number of the neighbours in this case is a parameter of the algorithm. Next, a synthetic instance is generated between all the selected nearest neighbours. This procedure is repeated until a balance between the minority and majority classes has been reached.

Originally, the algorithm was implemented using k-means, and for a small number of clusters only. Based on the experiment results (see [11]), we observe that such an approach can be promising and ensure acceptable results for the over-sampling, but not for all possible cases. An example of such a case is when only one of the produced clusters exists close to the decision borderline and the others are distant from this decision boundary or even behind the first cluster. In this case, the reference instances belonging to the remote clusters have no real impact on the process of synthetic instance generation, which means that they have no significant impact on the target process. The basic procedure can have also an negative impact on the computational cost, when the clustering is repeated without influence on the quality of the synthetic instance generation.

The above observations have been included in the modified version described in the next subsection.

3.2 A Modified Procedure for Over-Sampling

In this paper we propose an alternative approach to the generation of synthetic instances within a minority set of instances and to eliminate imbalanced data within the data chunks used by WECOI. In comparison to the basic procedure described in Sect. 3.1, this procedure assures the generation of synthetic instances with respect to each reference instance from the majority set of instances independently. The extended modified procedure gives more competence to reference instances located closer to the decision borderline than to reference instances located farther away. In other words, the reference instances lying closer to the decision borderline have more real impact on the process of the elimination of the imbalance between instances from the minority and majority class.

The pseudocode of the discussed over-sampling procedure is given in Algorithm 1.

Algorithm 1 Over-sampling procedure for the minority set of instances

Input: T, Y – data chunks; k_1, k_2 – parameters of the procedure;
Begin
Let T be a set of minority instances;
Let Y be a set of majority instances;
Set $q = false$;
Map instances from Y into clusters using a selected clustering algorithm;
Let Y_i $(i = 1, \dots, n)$ denote the obtained clusters and n denote the number of clusters;
For each $Y_{i:i=1,\dots,n}$ find its centres, called reference instances;
Let $y_{i:i=1,\dots,n}$ denote the reference instances for $Y_{i:i=1,\dots,n}$;
Let Y^* denote a set of the reference instances;
Let t denote the reference instance for T;
Repeat
For i:=1 to n do
Select from $(Y^* \cup \{t\}) \backslash \{y_i\}$ k_1-nearest reference instances of y_i;
Let Y_i^* denote a set of the nearest reference instances calculated for y_i;
Let P_i be a subset of Y_i^* such that $P_i \subset T$;
If $|P_i| \neq \emptyset$ **then**
Select k_2-nearest neighbours of y_i from T;
Let P_i^* denote a set of the nearest neighbours of y_i from T;
Generate randomly a synthetic instance x_a located between instances from P_i^*;
$T = T \cup \{x_a\}$;
$q = true$;
End if
End for
If $q = false$ **then** k_1++;
Until $|T| \cong |Y|$;
End

The procedure of synthetic instance generation, shown as Algorithm 1, has implemented an adaptation mechanism, which extends the possibilities of the procedure in the event of an unfavourable data distribution and excludes the complete lack of the possibility of selecting so-called nearest reference instances from a class of minority instances. Nevertheless, the number of nearest reference vectors k_1 and the number of nearest neighbours k_2 remain the input parameters of the procedure.

The modified procedure of over-sampling is independent of the clustering algorithm and gives more freedom to decide which clustering algorithm to use. However, it is best to use a clustering algorithm that is free from any parameterization and that can produce clusters in an automatic way.

Representative here may be the similarity-based clustering algorithm (SCA) (see for example [13]). The SCA starts the clustering from the calculation of a value of the similarity coefficient independently for each instance from the data set. The number of clusters is determined by the value of the similarity coefficient. It is done based on the assumption that the similarity coefficient of all instances within a cluster is equal. In such an approach, the clusters and their number are set automatically.

Algorithm 2 SCA clustering procedure

Input: X – data set;

Begin

Calculate for each instance from X the similarity coefficient $s_{i:i=1,\ldots,N}$, where N is a number of instances;

Map instances from X into k clusters denoted as $X_i, i = 1, \ldots, k$ in such way that each cluster contains instances with an identical value of the similarity coefficient s_i, where k is the number of different values of s_i;

End

4 Computational Experiment

The aim of the computational experiment was to evaluate the new approach for the balancing of instances between minority and majority classes in a data stream.

The research question was: whether the choice of the clustering algorithm and selection of other clustering parameters influence the quality of the over-sampling and performance of WECOI, and whether WECOI can be considered as a competitive approach compared to other tools dedicated to learning from imbalanced data streams.

WECOI based on the new over-sampling procedure has been evaluated using: SCA (as a basic clustering approach), k-means, XMeans and Kernel-based fuzzy C-means (KFCM). SMOTE and borderline-SMOTE algorithms have also been applied for balancing instances between minority and majority data chunks within the WECOI framework. WECOI based on the new over-sampling procedure has been denoted as WECOI', WECOI based on SCA as WECOI'$_{\text{SCA}}$, WECOI'$_{\text{k-means}}$ – for k-means clustering etc. When WECOI uses a SMOTE version for generating synthetic instances, it is denoted as WECOI'$_{\text{SMOTE}}$, when the borderline-SMOTE algorithm it is noted as WECOI'$_{\text{BR-SMOTE}}$. Results provided based on using of the primary over-sampling procedure are denoted as WECOI.

For all versions of WECOI to generate the base classifiers a POSC4.5 algorithm has been applied. In the computational experiment discussed, as a base approach to undersampling within WECOI, the ENN method has been used. All WECOI versions have been run with the number of nearest reference vectors k_1 set as two.

The results obtained have also been compared with other algorithms proposed for data streams: Oversampling and Undersampling Online Bagging (OUOB) [19], the ensemble learning algorithm based on Online Bagging (OB) [20], and the Learn++.NIE algorithm [21]. The algorithms were implemented as extensions of the Massive Online Analysis (MOA) package [18] within the WEKA environment [22]. For OUOB, OB and Learn ++.NIE, the Hoeffding Tree [23] has been selected for the base classifiers and the base classifier pool equals 10. The results have been also compared with results obtained by the Accuracy Weighted Ensemble (AWE) [24], using only the Hoeffding Option Tree (HOT), iOVFDT (Incrementally Optimised Very Fast Decision Tree) [18, 23].

The performance was measured based on the classification accuracy, defined as the percentage of all instances correctly identified and using the test-then-train paradigm. All experiments were repeated 30 times on each data benchmark and the results are shown as the mean from these repetitions.

The computational experiment was performed using synthetic and real data sets. Among the data benchmarks of real data streams were: electricity, airlines, and an ozone. A summary of the characteristics of the real data sets is presented in Table 1. The synthetic data streams were generated using the MOA framework [18] using: SEA [16], HYPERPLANE [17] and AGRAWAL [16]. In general, the SEA generator was used based on 10% of noise, a sudden concept drift and without class balancing. The HYPERPLANE (Hyp) generator was used with standard MOA parameters; however, a rotation of the decision boundary was set for each concept, assuring also incremental concept drift (i.e. -*t* 0.01). The AGRAWAL (Agr) generator was set to obtain a sudden concept drift and without class balancing. The number of instances was set to 10,000,000 with two decision classes and the number of attributes was set to ten with all attributes with drift. Table 1 also shows the size of the threshold which defined the size of the data chunk for the following datasets. In the case of the synthetic data, the size equals 1000 instances.

The values of the classification accuracy obtained by WECOI (and all its considered versions) as well as the other algorithms compared are shown in Table 2[1]. For WECOI the results are presented as a best obtained by the algorithm independently on parameter settings.

Table 1 Real data streams and their characteristics

Dataset	Source	#instances	#attributes	#classes	Threshold (as # of instances)
Electricity	[14]	45312	8	2	1000
Airlines	[14]	539383	7	2	1000
Ozone	[15]	2534	72	2	250

[1] The best solution obtained by the compared algorithms is indicated in bold. The underline indicates the best solution obtained by the WECOI' and its considered versions.

Table 2 Average accuracy for all data set (in %)

Algorithm	Electricity	Airlines	Ozone	Hyp	Agr	SEA
WECOI	76,5	64,82	81,56	85,78	91,68	88,45
WECOI'$_{SCA}$	76,7	63,48	82,21	**85,91**	92,71	88,79
WECOI'$_{k\text{-means}}$	76,31	64,15	81,04	84,17	92,24	86,5
WECOI'$_{Xmeans}$	73,51	62,48	78,21	79,48	87,45	82,14
WECOI'$_{KFCM}$	77,2	64,86	81,74	84,62	92,37	87,68
WECOI'$_{SMOTE}$	73,45	62,48	79,41	79,3	83,12	81,64
WECOI'$_{BR\text{-}SMOTE}$	75,8	63,57	80,92	80,62	87,84	86,4
OUOB	75,42	65,42	**82,31**	72,27	92,48	86,69
OB	**77,66**	63,21	76,56	84,7	81,67	83,74
Learn++.NIE	70,7	**66,8**	82,1	84,51	92,42	**89,3**
AWE	71,06	63,4	66,54	70,15	80,2	77,81
HOT	74,02	62,34	81,03	80,07	**94,99**	88,03
iOVFDT	72,52	63,52	81,25	81,56	93,67	83,02

The results in Table 2 demonstrate that none of the evaluated approaches is best for all the datasets. However, true is also conclusion that WECOI' is a competitive algorithm to other compared, including also algorithms proposed for learning from data streams. When the WECOI' is compared with WECOI, i.e. with algorithm, where the primary over-sampling procedure has been used, it can be notice, that more advanced approach for over-sampling assure better results. It is also important to notice that WECOI' based on SCA (WECOI'$_{SCA}$) is more accurate than others its versions based on another clustering algorithms. WECOI'$_{SCA}$ assured the best results in four cases. Alternative clustering approach is KFCM, assuring also competitive classification results. Both algorithms, that is SCA and KFCM, outperform - in terms of the resulting accuracy, two remaining clustering algorithms (k-means and XMeans).

For proper working of proposed algorithm two parameters also play an important role. The average number of nearest reference vectors k_1 during the experiments equalled 4. On the start of the algorithm the parameter was set on 2. It means that the proposed algorithm increased it to find adequate number of nearest reference instances and analysed the structure of the nearest neighbour clusters in the data space to select the best. The second parameter, i.e. the number of nearest neighbours k_2 equalled from 6 to 8. This parameter decided on number of nearest neighbours where between them a synthetic instance was generated.

The clustering approach for over-sampling is also more promising than SMOTE approach. In general, the SMOTE approach guaranteed results on the level k-means.

In one case using the borderline-SMOTE was most attractive. Although in all other cases the results produced by WECOI'BR-SMOTE have been comparable to the

WECOI' based on clustering, thus it can be concluded that it is also a competitive approach for the reduction of imbalanced classes in the data stream.

One additional conclusion may be also formulated, that the proposed WECOI is flexible on the implementation of different algorithms for over-sampling.

5 Conclusions

Learning from imbalanced data streams is one of the key challenges in supervised learning, especially when data streams flow into the system and when their distribution changes over time. To address the imbalanced data in such data streams, this paper proposes a novel cluster-based approach called WECOI as an alternative approach for over-sampling. The paper discusses the proposed approach and compares it with others.

The computational experiments showed that the novel approach is competitive to other existing methods dedicated to address imbalanced data elimination in streams. The effectiveness of oversampling based on clustering needs an approach that automatically sets the appropriate number of clusters. Such properties were demonstrated by the SCA and KFCM algorithms. The paper shows also that WECOI belongs to the category of open algorithms and can be implemented using different methods for over-sampling and under-sampling.

Future research includes a more comprehensive statistical analysis of the results, particularly targeting additional datasets and benchmarks. Additionally, further research focusing on studying the influence of the size of the data chunks is required. In on-going research, the problem of diversification of selection of nearest neighbours from minority data is solved. Based on the computational experiments, it has been observed that this can have an impact on the WECOI performance. A deeper investigation of such an impact is outlined for future work.

References

1. Charte, F., Rivera, A.J., del Jesus, M.J., Herrera, F.: MLSMOTE: Approaching imbalanced multilabel learning through synthetic instance generation. Knowl.-Based Syst. **89**, 385–397 (2015)
2. Zyblewski, P., Sabourin, R., Woźniak, M.: Preprocessed dynamic classifier ensemble selection for highly imbalanced drifted data streams. Inform. Fusion **66**, 138–154 (2021)
3. Sahel, Z., Bouchachia, A., Gabrys, B., Rogers, P.: Adaptive mechanisms for classification problems with drifting data. In: Apolloni, B., Howlett, R.J., Jain, L. (eds.) KES 2007. LNCS (LNAI), vol. 4693, pp. 419–426. Springer, Heidelberg (2007). https://doi.org/10.1007/978-3-540-74827-4_53
4. Duda, R., Hart, P., Stork, D.: Pattern Classification, 2nd edn. John Wiley & Sons (2000)
5. Chawla, N.V., Bowyer, K.W., Hall, L.O., Kegelmeyer, W.P.: SMOTE: Synthetic minority over-sampling technique. J. Artif. Intell. Res. **16**, 321–357 (2002)
6. Sharma, S., Gosain, A., Jain, S.: A review of the oversampling techniques in class imbalance problem. In: Khanna, A., Gupta, D., Bhattacharyya, S., Hassanien, A.E., Anand, S., Jaiswal, A. (eds.) International Conference on Innovative Computing and Communications. AISC, vol. 1387, pp. 459–472. Springer, Singapore (2022). https://doi.org/10.1007/978-981-16-2594-7_38

7. Nguyen, H.M., Cooper, E.W., Kamei, K.: A comparative study on sampling techniques for handling class imbalance in streaming data. In: Proceedings of the 6th International Conference on Soft Computing and Intelligent Systems, and the 13th International Symposium on Advanced Intelligence Systems, pp. 1762–1767 (2012). doi:https://doi.org/10.1109/SCIS-ISIS.2012.6505291
8. Bernardo, A., Gomes, H.M., Montiel, J., Pfahringer, B., Bifet, A., Valle, E.D.: C-SMOTE: Continuous synthetic minority oversampling for evolving data streams. In: Proceedings of 2020 IEEE International Conference on Big Data (Big Data), pp. 483–492 (2020). doi: https://doi.org/10.1109/BigData50022.2020.9377768
9. Sadeghi, F., Viktor, H.L.: Online-MC-queue: Learning from imbalanced multi-class streams. In: Proceedings of the Third International Workshop on Learning with Imbalanced Domains: Theory and Applications, PMLR, vol. 154, pp. 21–34, ECML-PKDD, Bilbao (2021)
10. Czarnowski, I.: Learning from imbalanced data streams based on over-sampling and instance selection. In: Paszynski, M., Kranzlmüller, D., Krzhizhanovskaya, V.V., Dongarra, J.J., Sloot, P.M.A. (eds.) ICCS 2021. LNCS, vol. 12744, pp. 378–391. Springer, Cham (2021). https://doi.org/10.1007/978-3-030-77967-2_32
11. Czarnowski, I., Jędrzejowicz, P.: Ensemble online classifier based on the one-class base classifiers for mining data streams. Cybern. Syst. **46**(1–2), 51–68 (2015). https://doi.org/10.1080/01969722.2015.1007736
12. Han, H., Wang, W.-Y., Mao, B.-H.: Borderline-SMOTE: A new over-sampling method in imbalanced data sets learning. In: Huang, D.-S., Zhang, X.-P., Huang, G.-B. (eds.) ICIC 2005. LNCS, vol. 3644, pp. 878–887. Springer, Heidelberg (2005). https://doi.org/10.1007/11538059_91
13. Czarnowski, I., Jędrzejowicz, J., Jędrzejowicz, P.: Designing RBFNs structure using similarity-based and kernel-based fuzzy C-means clustering algorithms. IEEE Access **9**, 4411–4422 (2021). https://doi.org/10.1109/ACCESS.2020.3048104
14. Harries, M.: Splice-2 comparative evaluation: Electricity pricing. Technical Report 1, University of New South Wales, Sydney (1999)
15. Asuncion, A., Newman, D.J.: UCI Machine Learning Repository. University of California, School of Information and Computer Science, Irvine (2007) http://www.ics.uci.edu/~mlearn/MLRepository.html
16. Agrawal, R., Imilielinski, T., Swani, A.: Database mining: A performance perspective. IEEE Trans. Knowl. Data Eng. **5**(6), 914–925 (1993)
17. Hulten, G., Spencer, L., Domingos, P.: Mining time-changing data streams. In: Proceedings of the 7th ACM SIGKDD International Conference on Knowledge Discovery and Data Mining, pp. 97–106. ACM (2001)
18. Bifet, A., Holmes, G., Kirkby, R., Pfahringer, B.: MOA: Massive online analysis. J. Mach. Learn. Res. **11**, 1601–1604 (2010)
19. Wang, S., Minku, L.L., Yao, X.: Dealing with multiple classes in online class imbalance learning. In: Proceedings of the 25th International Joint Conference on Artificial Intelligence (IJCAI 2016), July 2016
20. Oza, N.C.: Online bagging and boosting. In: Proceedings of the 2005 IEEE International Conference on Systems, Man and Cybernetics, Waikoloa, 10–12 October 2005, vol. 2343, pp. 2340–2345 (2005)
21. Ditzler, G., Polikar, R.: Incremental learning of concept drift from streaming imbalanced data. IEEE Trans. Knowl. Data Eng. **25**(10), 2283–2301 (2013). https://doi.org/10.1109/TKDE.2012.136
22. Frank, E., Hall, M.A., Witten, I.H.: The WEKA workbench. Online Appendix for Data Mining: Practical Machine Learning Tools and Techniques, 4th edn. Morgan Kaufmann (2016)

23. Bifet, A.: Adaptive learning and mining for data streams and frequent patterns. PhD thesis, Universitat Politecnica de Catalunya (2009)
24. Wang, H., Fan, W., Yu, P.S., Han, J.: Mining concept-drifting data streams using ensemble classifiers. In: Proceedings of 9th ACM SIGKDD International Conference on Knowledge Discovery and Data Mining, pp. 226–235 (2003). https://doi.org/10.1145/956750.956778

Enhancing Decision Combination in Classifier Committee via Positional Voting

Jacek Trelinski and Bogdan Kwolek(✉)

AGH University of Science and Technology, 30 Mickiewicza, 30-059 Krakow, Poland
{tjacek,bkw}@agh.edu.pl

Abstract. In this work, we propose an approach for aggregating classifiers using positional voting techniques. We extend the positional voting by optimizing weights of the preferences to better aggregate the committee classifiers. Staring from initial weights determined by a voting algorithm the aggregating weights are optimized by a differential evolution algorithm. The algorithm has been evaluated on a human action dataset. We demonstrate experimentally that on SYSU 3DHOI dataset the proposed algorithm achieves superior results against recent algorithms including skeleton-based ones.

Keywords: Classifier committee · Collective intelligence · Voting rules

1 Introduction

Ensemble techniques combine a number of diverse models to build a composite model that improves generalizability/robustness over each of them alone, either by using many different modeling algorithms or using different training datasets. They involve aggregating multiple models with the aim of decreasing both bias and variance. A classifier committee (or ensemble) is a classifier constructed by combining the predictions of multiple classifiers [1].

Classifier committees tend to yield better results if the individual classifiers work as independently as possible, i.e. when there is a significant diversity among the models [2]. The conventional ensemble methods include bagging, boosting, and stacking-based methods. These methods have been well studied in recent years and applied widely in different applications [3]. More recent approaches for ensemble learning such as XGBoost and LightGBM [4] permit achieving very competitive results on commonly used benchmark datasets. In the last decade, due to availability computational power that permits training large ensembles in a reasonable time, the number of ensemble-based applications has grown increasingly.

In ensemble learning we can distinguish three phases [5]. The first one consists in generating a set of models, and aims at obtaining a pool of models. In the second step a single classifier or a subset of best classifiers is selected. In the last phase a strategy to combine individual models is defined. The combination

D. Groen et al. (Eds.): ICCS 2022, LNCS 13351, pp. 598–609, 2022.
https://doi.org/10.1007/978-3-031-08754-7_64

of classifiers can be achieved using class labels, such as in the majority voting scheme, which calculates the prediction on the basis of the most frequent class assignment, or by utilizing scores of individual classifiers. In [6], a weighted voting ensemble was employed to improve the classification model's performance by combining classification results and selecting a group with the highest vote based on the weights given to the single classifiers. The impact of ensemble size with majority voting and optimal weighted voting aggregation rules has recently been discussed in [7]. Ranked voting approaches are recommended for combining classifiers if and when the classifiers can rank the order of the classes [8]. Borda count is a rank-based combination technique in which each classifier ranks the classes according to their potentiality to be the correct class [9]. Weights are linearly proportional to position in the ordering. It is considered to be one of the simplest scoring rules. Recently, in [10] a feature selection using election methods and ranker clustering for ensembles has been proposed.

Human action recognition is an important component in many applications including but not limited to ambient intelligence [11], human-computer interaction systems, and video surveillance. Little research has been done in the area of human action recognition on raw depth maps [12]. Recently, in [13] an algorithm for human action classification on depth motion images and Laplacian pyramids as structured multi-scale feature maps has been proposed. In a recent work [14], voting rules as an aggregation technique for classifier combination have been utilized to improve human action recognition on raw depth maps. In this work, an approach to aggregating classifiers through positional voting techniques is proposed. The proposed optimized positional voting achieves better results in comparison to results achieved by Borda count, Coombs, Bucklin, and Copeland.

2 Algorithm

The architecture of the classifier committee is based on architecture discussed in [15]. With the feature extraction in mind, the main difference is that in this work a Dynamic Time Warping (DTW) is utilized instead of the shapelets. In Sect. 3 we compare results achieved by both versions of the algorithm. In Subsect. 2.1 we outline main ingredients of classifier committee. Afterwards, in Subsect. 2.2 we present a DTW-based extraction of action features. Finally, in Subsect. 2.3 we introduce the optimized positional voting.

2.1 Main Ingredients of Classifier Committee

Having on regard that most frequently used benchmark datasets for human action recognition on depth maps contain limited number of depth map sequences, we utilize a multiple classifier system. The recognition of human actions is performed on raw depth maps only. We learn various features in different domains, like single depth map, time-series of embedded features, time-series represented by DTW-based features. A Siamese neural network as well as a convolutional autoencoder (CAE) are learned on single frames to extract features.

The multivariate time-series of Siamese features are fed to the DTW in order to extract the features representing actions. The multivariate time-series of CAE features are fed to a 1D CNN in order to extract another representation of action features. The discussed features are common for all classes. We extract also class-specific action features using TimeDistributed and LSTM layers (TD-LSTM). Multi-class logistic classifiers are trained on concatenated class-specific features and features common for all classes, see Fig. 1. The most discriminative classifiers are selected using a differential evolution (DE) [15]. The final decision is made through aggregating classifiers on the basis of voting.

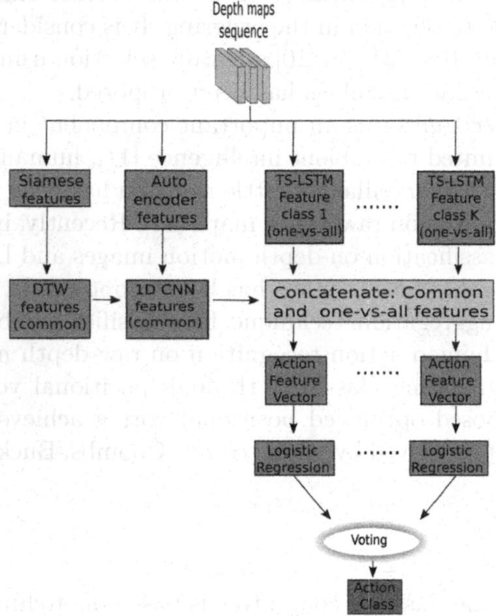

Fig. 1. Flowchart of the classifier committee operating on learned features, and aggregating decisions of selected classifiers via optimized positional voting.

2.2 Dynamic Time Warping-Based Action Features

In order to compactly represent actions we learn features on representative depth maps. A Siamese neural network is trained on pairs of single depth maps as a representation of the whole depth map sequences. The network trained on such a compact representation of depth map sequences is used to extract frame-features on depth map sequences. In contrast to the CAE it has been trained only on frontal depth maps. In the current implementation, for each sequence from the training subset a middle depth map as a representation the whole depth map sequence has been selected and then included in a training subset for the Siamese neural network. The Siamese neural network operates on depth maps of size $1 \times 64 \times 128$. It consists of 64 Conv2D filters of size 5×5 followed by max-pooling, 32 Conv2D filters of size 5×5 followed by max-pooling, which in turn are followed by the flattening layer and then a dense layer consisting of 128 neurons. The neural network has been trained using the contrastive loss [16]. The neural network trained in such a way was used to extract features on every depth map from a given input sequence. A human action represented by a number of depth maps is described by a multivariate time-series of length equal to number of frames in the sequence and dimension equal to 128.

In time-series analysis, the DTW can be employed for measuring similarity between two temporal sequences, which may vary in speed. Let us assume that our aim is to measure the distance between two time-series: $\mathbf{a} = \{a_1, a_2, \ldots, a_n\}$ and $\mathbf{b} = \{b_1, b_2, \ldots, b_n\}$. Let us denote by $M(\mathbf{a}, \mathbf{b})$ the $n \times n$ pointwise distance matrix between $\mathbf{a}$ and $\mathbf{b}$, where $M_{i,j} = (a_i - b_i)^2$. A warping path $P = (u_1, v_1), (u_2, v_2), \ldots, (u_s, v_s)$ is a set of pairs of indexes that define a traversal of M. The valid warping path must satisfy: $(u_1, v_1) = (1, 1)$ and $(u_s, v_s) = (n, n)$ as well as $0 \leq u_{i+1} - u_i \leq 1$ and $0 \leq v_{i+1} - v_i \leq 1$ for all $i < n$. Let p_i stand for the distance between elements with indexes u_i and v_i for the i-th pair of points. The distance for a path P is: $D_P(\mathbf{a}, \mathbf{b}) = \sum_{i=1}^{s} p_i$. The DTW path P^* has the minimum distance

$$P^* = \min_{P \in \mathcal{P}}(D_P(\mathbf{a}, \mathbf{b})) \tag{1}$$

over all possible paths $\mathcal{P}$, which can be found by dynamic programming (DP). For a given depth map sequence we calculated the DTW distances with respect to all remaining depth maps sequences in the training subset. For each depth map sequence the DTW distances between multivariate time-series have been determined for the discussed above Siamese features. The distances between a given sequence and all remaining sequences from the training set have then been utilized as features. This means that the resulting feature vector is of size equal to $n_t \times 2$, where n_t denotes the number of training depth map sequences.

2.3 Optimized Positional Voting

In a few papers the voting-based aggregation techniques have been utilized to improve the performance of ensembles [10]. Recently, in [14] Borda counts, Bucklin and Coombs have been investigated in terms of improving the performance of action classification using a classifier committee. In this work, we examine positional voting [17], in which points are allocated to the candidates under consideration based on the order they were ranked. In such voting systems, the voters order the candidates from best to worst, and a pool of winners is chosen on the basis of positions of the candidates in the preference order. The rank position of each voter preference has allocated a specific fixed weighting [17]. A candidate with the most points overall wins. Borda counting is an important example of positional voting techniques. Figure 2 illustrates the process of positional voting in an example classifier committee.

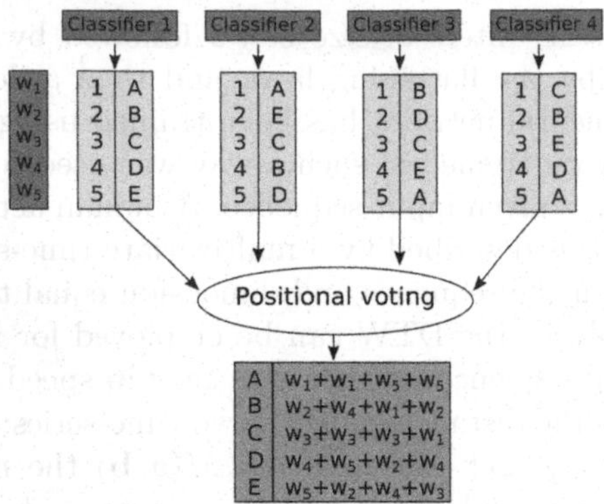

Fig. 2. Positional voting.

In this work, we extend the positional voting by optimizing weights of the preferences. Area Under the ROC (Receiver Operating Characteristics Curve), commonly called AUC has been utilized in the objective function. Direct optimization of AUC can lead to solving an NP-hard problem since it can be cast into a combinatorial optimization problem. Recently, in [18] a fast stochastic AUC optimization with $O(1/n)$ convergence rate has been proposed. In this work, the objective function mentioned above has been optimized using the differential evolution. The differential evolution initiated its search from a population of weights determined by the Borda count algorithm. The convergence is reached when the standard deviation of the fitness function for each individual in the population, normed by the average, is smaller than the given tolerance value. The weights determined in such a way have been utilized in making the final decision by the classifier committee.

Figure 3 presents a schematic diagram of algorithm steps. In the considered toy example, we assume that a dataset consists of five samples, whereas the classifier committee comprises four classifiers. The dataset is split into a training subset on which the classifiers are trained and a validation subset on which the weights are optimized. This means that for training of the classifiers as well as optimization of the weights of the preferences the training data was further divided into the training subset and the validation subset. In the considered example, all four classifiers are trained on the training subset. The validation subsets are fed into the trained classifiers, whose outputs are stored for the subsequent optimization. The optimization of the weights is conducted using the fitness function with AUC components for all validation samples.

The classifiers have been trained on the training subset, the weights of the preferences have been optimized on the validation data, whereas the performance of the optimized classifier committee has been judged on the test data. Such an approach is often called holdout validation. Figure 4 depicts data split on a toy example. As illustrated on discussed figure, a 5-fold data split has been employed for the training and optimization of the classifier committee.

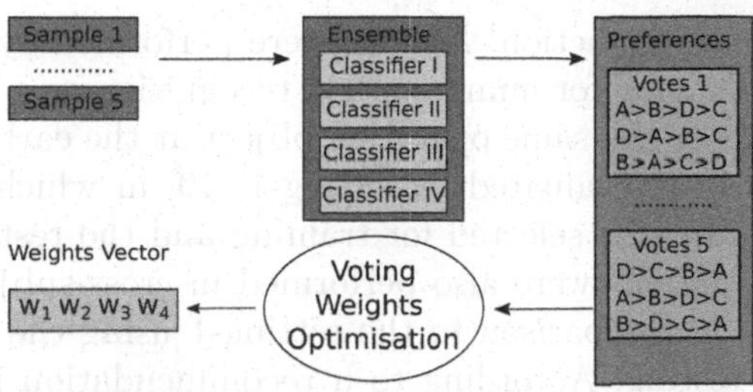

Fig. 3. Optimized positional voting.

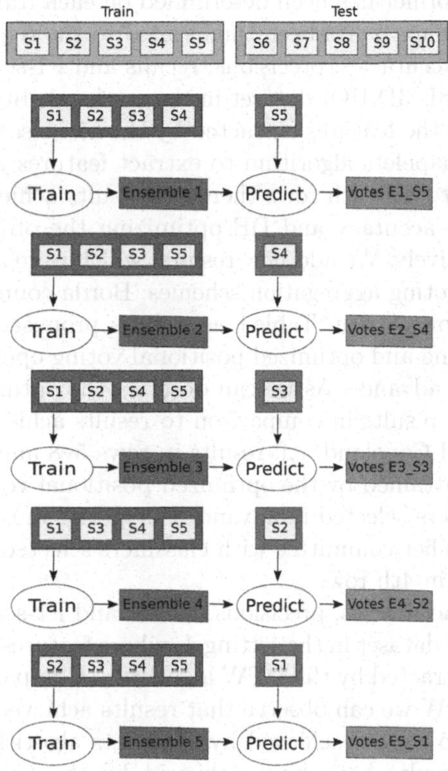

Fig. 4. Data split for training classifiers and optimizing weights of the preferences.

3 Experimental Results

The performance of proposed algorithm has been determined on publicly available SYSU 3D Human-Object Interaction Set (SYSU 3DHOI) [19]. The dataset consists of 480 RGB-D image sequences with 12 action classes that include calling with a cell phone, playing with a cell phone, drinking, pouring, moving a chair, sitting on a chair, packing a backpack, wearing a backpack, sweeping, mopping, taking something out from the wallet and taking out a wallet. Each activity is a kind of human-object interaction. Actions were performed by forty performers. This dataset is challenging for human action recognition as a number of actions have similar motions or the same operating object at the early temporal stages. The algorithm has been evaluated in setting-1 [19] in which for each activity class, half of the samples is selected for training and the rest samples are used in testing. The evaluations were also performed in cross-subject setting, which is more challenging in comparison to the setting-1 using the same subjects for both training and testing. According to a recommendation in [19], the evaluations were done on thirty training/testing splits. Because the performers are not extracted from the background, they have been extracted by us. A window surrounding the performer has been determined on each frame and then used to crop the raw depth map. It has been then scaled to the required input shape.

Table 1 presents accuracies, precisions, recalls and F1-scores achieved by our algorithm on the SYSU 3D HOI dataset in the setting-1. In discussed configuration of the algorithm the features extracted by the Siamese neural network have been processed by shapelets algorithm to extract features representing actions, same as in [15]. In 3rd and 4th rows there are results achieved by DE optimizing the classification accuracy and DE optimizing the objective function proposed in [20], respectively. We add new results, which were achieved by classifier committee built on voting aggregation schemes: Borda count, Coombs, Bucklin, and Copeland, see rows 5–8 in Table 1 as well as proposed in this work: optimized positional voting and optimized positional voting operating on a subset of classifiers selected in advance. As we can observe, the optimized positional voting achieves superior results in comparison to results achieved by Borda count, Coombs, Bucklin and Copeland, c.f. results in rows 5–8 and row 9. The classification performance attained by the optimized positional voting, which operates on outputs of classifiers selected in advance is superior in comparison to results obtained by the classifier committee with classifiers selected in advance, see also experimental results in 4th row.

Table 2 presents accuracies, precisions, recalls and F1-scores achieved by our algorithm on 3D HOI dataset in the setting-1, where features representing human actions have been extracted by the DTW algorithm. Comparing results achieved by shapelets and DTW we can observe that results achieved by DTW are superior in comparison to results achieved by shapelets algorithm. Among election methods the best results have been achieved by the Copeland. The results achieved by classifier committee based on the Copeland are slightly better in comparison to results achieved by the hard voting and the soft voting, which are frequently used as aggregation techniques in the ensembles. The `DE-sel. ens.` algorithm selected seven classifiers and the results achieved by classifier committee built on such a pool of the classifiers are worse in comparison to results obtained by discussed methods. The results achieved by the optimized positional voting are better in comparison to results achieved by classifier committee with `DE-acc`-based classifier selection. As we can observe, the best results have been achieved by the optimized positional voting, which operates on outputs of classifiers selected in advance.

Table 1. Recognition performance on SYSU 3DHOI dataset (setting-1) using shapelets.

Voting	num. class.	Accuracy	Precision	Recall	F1-score
Hard voting	12	0.9167	0.9217	0.9167	0.9171
Soft voting	12	0.9079	0.9102	0.9079	0.9071
DE-acc.	12	0.9079	0.9110	0.9079	0.9073
DE-sel. ens.	7	0.9254	0.9271	0.9254	0.9246
Borda count	12	0.9079	0.9141	0.9079	0.9079
Coombs	12	0.9035	0.9113	0.9035	0.9029
Bucklin	12	0.9035	0.9097	0.9035	0.9030
Copeland	12	0.9035	0.9097	0.9035	0.9030
opt. pos. voting	12	0.9167	0.9223	0.9167	0.9166
DE-sel. ens., opt. pos. voting	7	**0.9254**	**0.9300**	**0.9254**	**0.9259**

Table 2. Recognition performance on SYSU 3DHOI dataset (setting-1) using DTW.

Voting	num. class.	Accuracy	Precision	Recall	F1-score
Hard voting	12	0.9341	0.9298	0.9341	0.9298
Soft voting	12	0.9341	0.9298	0.9341	0.9298
DE-acc.	11	0.9266	0.9211	0.9266	0.9211
DE-sel. ens.	7	0.9340	0.9298	0.9340	0.9298
Borda count	12	0.9304	0.9254	0.9304	0.9254
Coombs	12	0.9230	0.9167	0.9230	0.9167
Bucklin	12	0.9336	0.9298	0.9336	0.9298
Copeland	12	0.9342	0.9374	0.9342	0.9341
opt. pos. voting	12	0.9386	0.9410	0.9386	0.9385
DE-sel. ens., opt. pos. voting	7	**0.9386**	**0.9414**	**0.9386**	**0.9384**

Table 3 presents accuracies, precisions, recalls and F1-scores achieved by our algorithm on SYSU 3D HOI dataset in the cross-subject setting (setting-2). As in the setting-1, the features extracted by the Siamese neural network have been further processed by shapelets algorithm in order to extract features representing actions. As previously, among election methods the best results have been achieved by the Copeland. The optimized positional voting permits achieving better classification performance in comparison to performance obtained by the Copeland. Once again, the best results have been achieved through selecting the most discriminative classifiers by the DE and then executing optimized positional voting on outputs determined by such a pool of best classifiers.

Table 3. Recognition performance on SYSU 3DHOI dataset (setting-2, cross-subject) using shapelets.

Voting	num. class.	Accuracy	Precision	Recall	F1-score
Hard voting	12	0.8991	0.9079	0.8991	0.8990
Soft voting	12	0.9035	0.9098	0.9035	0.9036
DE-acc.	2	0.9123	0.9175	0.9123	0.9119
DE-sel. ens.	2	0.9211	0.9259	0.9211	0.9209
Borda count	12	0.8904	0.8933	0.8904	0.8896
Coombs	12	0.8904	0.8925	0.8904	0.8895
Bucklin	12	0.8947	0.8967	0.8947	0.8941
Copeland	12	0.8991	0.9079	0.8991	0.8990
opt. pos. voting	12	0.9079	0.9107	0.9079	0.9077
DE-sel. ens., opt. pos. voting	2	**0.9211**	**0.9219**	**0.9211**	**0.9201**

Table 4 presents results obtained in the cross-subject setting (setting-2), where features representing actions have been extracted by the DTW algorithm. Amongst the election-based methods the best results have been achieved by the Coombs. A classifier committee built on only three best classifiers selected by the DE algorithm, achieved superior results in comparison to results discussed above. The number of the most discriminative classifiers selected by the DE optimizing the accuracy is far larger and the resulting classification performance is smaller. The classification performance achieved by the optimized positional voting is better than performance achieved by common election methods. It is smaller in comparison to performance achieved by the `DE-sel. ens.` algorithm. As we can observe, the best results have been achieved by the optimized positional voting, operating on outputs of classifiers selected in advance.

In summary, in both settings, both with shapelets and DTW-based algorithms, the best classification performances have been achieved by the optimized positional voting operating on subset of classifiers selected in advance. In all considered cases the optimized positional voting achieved better results than Borda count, Coombs, Bucklin and Copeland. The results achieved by DTW-based algorithm are better in comparison to shapelets-based algorithm [15].

Table 4. Recognition performance on SYSU 3DHOI dataset (setting-2, cross-subject) using DTW.

Voting	num. class.	Accuracy	Precision	Recall	F1-score
Hard voting	12	0.9230	0.9211	0.9230	0.9211
Soft voting	12	0.9230	0.9211	0.9230	0.9211
DE-acc.	11	0.9204	0.9167	0.9204	0.9167
DE-sel. ens.	3	0.9291	0.9254	0.9291	0.9254
Borda count	12	0.9159	0.9133	0.9159	0.9123
Coombs	12	0.9202	0.9167	0.9202	0.9167
Bucklin	12	0.9152	0.9123	0.9152	0.9123
Copeland	12	0.9079	0.9098	0.9079	0.9074
opt. pos. voting	12	0.9211	0.9242	0.9211	0.9207
DE-sel. ens., opt. pos. voting	3	**0.9291**	**0.9343**	**0.9291**	**0.9295**

Table 5 presents action recognition accuracies that are achieved by recent algorithms. As we can observe, the proposed algorithm achieves superior results against all recent algorithms on challenging 3D HOI dataset. It outperforms all recent algorithms on both settings. As far as we know, on SYSU 3DHOI dataset the best classification accuracies among skeleton-based algorithms achieves a recently published SGN algorithm [21]. The proposed algorithm achieves far better classification accuracy on the discussed dataset.

Table 5. Comparative recognition performance of the proposed method with recent algorithms on 3D HOI dataset.

Method	Modality	Setting	Acc. [%]
LGN [22]	skel.	II	83.33
SGN [21]	skel.	II	86.90
MSRNN [23]	depth+RGB+skel	II	79.58
LAFF [24]	depth+RGB	II	80.00
PTS [25]	depth+skeleton	II	87.92
bidirect. rank p. [26]	Depth	I	76.25
bidirect. rank p. [26]	Depth	II	75.83
D3C [14]	Depth	I	88.75
D3C [14]	Depth	II	92.98
HAR [15]	Depth	I	93.54
HAR [15]	Depth	II	92.11
Proposed method	Depth	I	**93.86**
Proposed method	Depth	II	**92.91**

4 Conclusions

In this paper, we presented an approach to aggregating classifiers through positional voting techniques. The proposed optimized positional voting achieved better results in comparison to results achieved by Borda count, Coombs, Bucklin and Copeland, which have been previously used in classifier committees to combine decisions. We demonstrated experimentally that significant gains in classification performance can be obtained by executing the proposed optimized positional voting on decisions of classifiers selected in advance by the DE algorithm.

Acknowledgment. This work was supported by Polish National Science Center (NCN) under a research grant 2017/27/B/ST6/01743.

References

1. Kuncheva, L.I.: Combining Pattern Classifiers: Methods and Algorithms. Wiley, New York (2004)
2. Kuncheva, L.I., Whitaker, C.J.: Measures of diversity in classifier ensembles and their relationship with the ensemble accuracy. Mach. Learn. **51**(2), 181–207 (2003)
3. Wozniak, M., Grana, M., Corchado, E.: A survey of multiple classifier systems as hybrid systems. Inf. Fusion **16**, 3–17 (2014)
4. Ke, G., et al.: LightGBM: a highly efficient gradient boosting decision tree. In: Proceedings of the 31st International Conference on Neural Information Processing Systems, NIPS 2017, pp. 3149–3157 (2017)
5. Sagi, O., Rokach, L.: Ensemble Learning: A Survey. Wiley Interdisciplinary Reviews: Data Mining and Knowledge Discovery, vol. 8 (2018)
6. Osamor, V.C., Okezie, A.F.: Enhancing the weighted voting ensemble algorithm for tuberculosis predictive diagnosis. Sci. Rep. **11**(1), 14806 (2021)
7. Bonab, H., Can, F.: Less is more: a comprehensive framework for the number of components of ensemble classifiers. IEEE Trans. Neural Netw. Learn. Syst. **30**(9), 2735–2745 (2019)
8. Polikar, R.: Ensemble based systems in decision making. IEEE Circuits Syst. Mag. **6**(3), 21–45 (2006)
9. van Erp, M., Vuurpijl, L., Schomaker, L.: An overview and comparison of voting methods for pattern recognition. In: Proceedings of Eighth International Workshop on Frontiers in Handwriting Recognition, pp. 195–200 (2002)
10. Drotár, P., Gazda, M., Vokorokos, L.: Ensemble feature selection using election methods and ranker clustering. Inf. Sci. **480**, 365–380 (2019)
11. Haque, A., Milstein, A., Fei-Fei, L.: Illuminating the dark spaces of healthcare with ambient intelligence. Nature **585**(7824), 193–202 (2020)
12. Wang, L., Huynh, D.Q., Koniusz, P.: A comparative review of recent Kinect-based action recognition algorithms. IEEE Trans. Image Process. **29**, 15–28 (2020)
13. Li, C., Huang, Q., Li, X., Wu, Q.: A multi-scale human action recognition method based on Laplacian pyramid depth motion images. In: Proceedings the 2nd ACM International Conference on Multimedia in Asia. ACM (2021)
14. Treliński, J., Kwolek, B.: Decision combination in classifier committee built on deep embedding features. In: Nguyen, N.T., Iliadis, L., Maglogiannis, I., Trawiński, B. (eds.) ICCCI 2021. LNCS (LNAI), vol. 12876, pp. 480–493. Springer, Cham (2021). https://doi.org/10.1007/978-3-030-88081-1_36

15. Treliński, J., Kwolek, B.: Human action recognition on raw depth maps. In: VCIP. IEEE (2021)
16. Wang, F., Liu, H.: Understanding the behaviour of contrastive loss. In: IEEE/CVF Conference on Computer Vision and Pattern Recognition (CVPR), pp. 2495–2504 (2021)
17. Saari, D.G.: Basic Geometry of Voting. Springer, Cham (2015). https://doi.org/10.1007/978-3-642-57748-2
18. Liu, M., Zhang, X., Chen, Z., Wang, X., Yang, T.: Fast stochastic AUC maximization with $o(1/n)$-convergence rate. In: Proceedings of the 35th International Conference on Machine Learning, PMLR, pp. 3189–3197 (2018)
19. Hu, J., Zheng, W., Lai, J., Zhang, J.: Jointly learning heterogeneous features for RGB-D activity recognition. In: CVPR, pp. 5344–5352 (2015)
20. Zhou, Z.H., Wu, J.X., Jiang, Y., Chen, S.F.: Genetic algorithm based selective neural network ensemble. In: Proceedings of the 17th International Joint Conference on Artificial Intelligence, vol. 2, pp. 797–802 (2001)
21. Zhang, P., Lan, C., Zeng, W., Xing, J., Xue, J., Zheng, N.: Semantics-guided neural networks for efficient skeleton-based human action recognition. In: IEEE/CVF Conference on Computer Vision and Pattern Recognition, pp. 1109–1118. IEEE (2020)
22. Ke, Q., Bennamoun, M., Rahmani, H., An, S., Sohel, F., Boussaid, F.: Learning latent global network for skeleton-based action prediction. IEEE Trans. Img. Proc. **29**, 959–970 (2020)
23. Hu, J., Zheng, W., Ma, L., Wang, G., Lai, J., Zhang, J.: Early action prediction by soft regression. IEEE Trans. PAMI **41**(11), 2568–2583 (2019)
24. Hu, J.-F., Zheng, W.-S., Ma, L., Wang, G., Lai, J.: Real-time RGB-D activity prediction by soft regression. In: Leibe, B., Matas, J., Sebe, N., Welling, M. (eds.) ECCV 2016. LNCS, vol. 9905, pp. 280–296. Springer, Cham (2016). https://doi.org/10.1007/978-3-319-46448-0_17
25. Wang, X., Hu, J.F., Lai, J.H., Zhang, J., Zheng, W.S.: Progressive teacher-student learning for early action prediction. In: CVPR, pp. 3551–3560 (2019)
26. Ren, Z., Zhang, Q., Gao, X., Hao, P., Cheng, J.: Multi-modality learning for human action recognition. Multimedia Tools Appl. **80**(11), 16185–16203 (2020). https://doi.org/10.1007/s11042-019-08576-z

Competition and Cooperation Mechanisms for Collective Behavior in Large Multi-agent Systems

Franciszek Seredyński[1], Tomasz Kulpa[1(✉)], and Rolf Hoffmann[2]

[1] Cardinal Stefan Wyszyński University, Warsaw, Poland
{f.seredynski,tomasz.kulpa}@uksw.edu.pl
[2] Technische Universität Darmstadt, Darmstadt, Germany
hoffmann@ra.informatik.tu-darmstadt.de

Abstract. We consider a 2-dimensional discrete space modeled by Cellular Automata consisting of $m \times n$ cells that can be occupied by agents. There exist several types of agents which differ in their way of behavior related to their own strategy when they interact with neighbors. We assume that interaction between agents is governed by a spatial Prisoner's Dilemma game. Each agent participates in several games with his neighbors and his goal is to maximize his payoff using own strategy. Agents can change their strategies in time by replacing their own strategy with a more profitable one from its neighborhood. While agents act in such a way to maximize their incomes we study conditions of emerging collective behavior in such systems measured by the average total payoff of agents in the game or by an equivalent measure–the total number of cooperating players. These measures are the external criteria of the game, and players acting selfishly are not aware of them. We show experimentally that collective behavior in such systems can emerge if some conditions related to the game are fulfilled. We propose to introduce an income-sharing mechanism to the game, giving a possibility to share incomes locally by agents. We present the results of an experimental study showing that the sharing mechanism is a distributed optimization algorithm that significantly improves the capabilities of emerging collective behavior measured by the external criterion of the game.

Keywords: Collective behavior · Competition · Distributed optimization · Income sharing · Multi-agent systems · Spatial Prisoner's Dilemma game

1 Introduction

Competition and cooperation are the two most observed phenomena in the world of the living organisms (people, animals, insects, viruses). Organisms compete for living space, food, rights for reproduction, and other goods to maximize their own income defined in some way. While the personal income maximization

D. Groen et al. (Eds.): ICCS 2022, LNCS 13351, pp. 610–623, 2022.
https://doi.org/10.1007/978-3-031-08754-7_65

of a member of a society is a driving force for the existence of a society, the productivity (welfare) of a society measured by the total income is an important measure of the efficiency of a society. The selfish behavior of individuals usually leads to improvements both in personal income and in welfare, but if the level of competition is not adequate for a real situation it may lead to a significant decrease in personal and societal measures of welfare. We can observe that in such situations, the phenomenon of cooperation often emerges and results in an improvement of personal and societal welfare measures due to lower risk and a higher level of stability.

The recent advances in modern computer-communication technologies (e.g. IoT systems, fog/edge/cloud computing systems [12]) oriented toward collecting and processing information by a vast number of simple units unexpectedly show similarities with the natural social systems and that many problems related to providing high performance of these distributed systems can be solved by applying mechanism observed in social systems. Therefore, we propose to consider and study a framework based on a large-scale multi-agent system approach, where agents are capable of solving problems in a distributed way. We believe that the results of the study can be used directly for solving management problems in distributed systems with a massive number of components. The principle requested from such multi-agent systems is the ability of collective behavior which emerges as a result of local interactions of a considerable number of simple components (see, e.g. [4]).

In this paper, we will consider a multi-agent Cellular Automata (CA) – based system working in the framework of CA [11], where an interaction between players is described in terms of non-cooperative game theory [6] using the Spatial Prisoner's Dilemma (SPD) game. We will expect a global collective behavior in such a system measured by the total number of cooperating players, i.e., maximizing the average total payoff of agents of the system. The phenomenon of emerging cooperation in systems described by the SPD game has been a subject of current studies [1,7]. They show that it depends on many factors, such as payoff function parameters, the type of learning agent, or the way of interaction between agents.

Recently we have shown [10] that the second-order CA based on the concept "adapt to the best neighbor" [5] can be successfully applied as simple learning machines in solving problems of collective behavior. In this paper, we extend a set of agents with different types of behavior and apply a new mechanism of interaction between players, based on the possibility of a local income sharing by agents participating in the game. Our objective is to study the conditions of emerging collective behavior in such a system with local income sharing, i.e., its ability to maximize the average total income in a distributed way.

The structure of the paper is the following. In the next section, SPD game is presented and discussed in the context of collective behavior. Section 3 contains a description of the CA-based multi-agent system acting in the SPD game environment. Section 4 presents a basic mechanism of the game, including income sharing. Section 5 presents some results of the experimental study, and the last section concludes the paper.

2 Iterated Spatial Prisoner's Dilemma Game and Collective Behavior

We consider a 2D spatial array of size $m \times n$. We assume that a cell (i, j) will be considered as an agent–player participating in the SPD game [3,5]. We assume that a given player's neighborhood is defined in some way (see, the next Section). Players from this neighborhood will be considered opponents in the game. At a given discrete moment, each cell can be in one of two states: C or D. The state of a given cell will be considered as an action C (cooperate) or D (defect) of the corresponding player against an opponent from its neighborhood. The payoff function of the game is given in Table 1.

Table 1. Payoff function of a row player participating in SPD game.

Player's action	Opponent's action	
	Cooperate (C)	Defect (D)
Cooperate (C)	$R = 1$	$S = c$
Defect (D)	$T = b$	$P = a$

Each player playing a game with an opponent in a single round (iteration) receives a payoff equal to R, T, S or P, where $T > R > P > S$. We assume that $R = 1$, $T = b$, $S = c$, and $P = a$, and values of a, b and c can vary depending on the purpose of an experiment.

Let us consider the game with the following values of the payoff table parameters: $R = 1$, $b = 1.4$, $c = 0$, and $a = 0.3$. If a player (i, j) takes action $s_{ij} = C$ and the opponent (i_k, j_k) from a neighborhood also takes action $s_{i_k j_k} = C$, then the player receives payoff $u_{ij}(s_{i,j}, s_{i_k j_k}) = 1$. If the player takes the action D and the opponent player still keeps the action C, the defecting player receives payoff equal to $b = 1.4$. If the player takes the action C while the opponent takes action D, the cooperating player receives payoff equal to $c = 0$. When both players use the action D, then both receive payoff equal to $a = 0.3$.

It is worth noticing that choosing by all players the action D corresponds to the Nash equilibrium point (NE) [6], and it is considered a solution to the one–shot game. Indeed, if all players select the action D, each of them receives a payoff equal to a, and there is no reason for any of them to change the action to C while the others keep their actions unchanged because it would result in decreasing its payoff to value 0.

We assume that players are rational and act in such a way to maximize their payoff defined by the payoff function. To evaluate the level of collective behavior achieved by the system, we will use an external criterion (not known for players) and ask whether it is possible to expect from players selecting such actions s_{ij} which will maximize the average total payoff (ATP) $\bar{u}()$ of the whole set of players:

$$\bar{u}(s_{11}, s_{12}, ..., s_{mn}) = \frac{1}{mn}\sum_{j=1}^{m}\sum_{i=1}^{n}\sum_{k=1}^{n_{ij}} u_{ij}(s_{ij}, s_{i_k j_k})/n_{ij}, \tag{1}$$

where n_{ij} is the number of opponents in the neighborhood.

Game theory predicts the behavior of players oriented towards achieving NE, i.e., choosing the action D by all players. ATP at NE is equal to a, and we will call it the price of NE. In our game example, this value of ATP is low and equal to $a = 0.3$. The maximal value of ATP is equal to $R = 1$ and corresponds to selecting by all players the action C. We will call this value of ATP the *maximal price point*. The set of players' actions corresponding to the maximal price point is not NE; therefore, it is challenging to expect to reach global cooperation. The purpose of this study is to discover conditions for the emergence of a high level of cooperation between players in iterated spatial games.

3 CA–Based Players

Cells (i, j) of a 2D array are considered as CA–based players. At a given discrete moment t, each cell is either in state D or C, which is used by the player as an action with an opponent player. Initial states of CA cells are assigned randomly according to a predefined probability of initial cooperation. For each cell, a local neighborhood is defined. We apply a cyclic boundary condition to avoid irregular behavior at the borders. We will assume the Moore neighborhood with eight immediate neighbors. It means that each player has eight ($N_k = 8$) opponents in the game.

At discrete moments, CA–based players will use their current states as actions to play games with opponents, they will receive payoffs, and next they will change their states applying *rules* (also called *strategies*) assigned to them. We will be using some set of socially motivated rules among which one of them will be initially randomly assigned to a given CA cell, so we deal with a non-uniform CA.

We will consider the following set of six rules: *all–C*: always cooperate; *all–D*: always defect; *k–D*: cooperate until not more than k neighbors defect, otherwise defect; *k–C*: cooperate until not more than k neighbors cooperate, otherwise defect; *k–DC*: defect until not more than k neighbors defect, otherwise cooperate; *pC*: cooperate with probability p_C.

The parameter k plays the role of a threshold of tolerance, and each player using a strategy with tolerance k has its personal value that is assigned initially according to some predefined scheme, and it has a value within the range $0 \leq k \leq 7$. The probability p_C for the strategy *pC* is defined as a random number from the range $[z - \triangle, z + \triangle]$, where z and $\triangle$ are predefined parameters and $0 < p_C < 1$ and $0 \leq \triangle \leq 0.5$.

4 Competition and Income Sharing Mechanisms

To study the possibility of the emergence of CA-based players' global collective behavior, we will modify the concept of CA and introduce some local mechanisms of interaction between players.

The first mechanism is *competition* that is based on the idea proposed in [5]. It is based on the principle "adapt to the best neighbor" and differs from the classical concept of CA. It assumes that each player associated with a given cell plays in every single round a game with each of its neighbors, and in this way collects some total score. If the competition mechanism is turned *on*, after q number of rounds (iterations), each agent compares its total payoff with the total payoffs of its neighbors. If a more successful player exists in the neighborhood, this player replaces its own rule with the most successful one. This mechanism converts a classical CA into a *second-order* CA, which can adapt in time.

The second mechanism called *income sharing mechanism* (ISM) that we propose provides a possibility of sharing payoffs by players in a neighborhood. Hard local sharing based on mandatory sharing was successfully used [8] in the context of 1D evolutionary and learning automata games. Here we propose soft sharing, where a player decides to use it or not on the base of competition for income. It is assumed that each player has a tag indicating whether it wishes (*on*) or not (*off*) to share its payoff with players from the neighborhood. The sharing works in such a way that if two or more players from a neighborhood wish to share, each of them: (a) calculates an amount of his payoff $u_{ij}^{to\text{-}send} = u_{ij}()/(1+n_{ij}^{k})$, where n_{ij}^{k} is a number of neighbors with tags *on*, to send to these neighbors, (b) sends this amount to corresponding neighbors, (c) receives from the corresponding parts of their payoffs. Before starting the iterated game, each player turns on its tag with a predefined probability p_{shar}. Due to the competition mechanism, the most successful rules, together with their tags containing information about willing (or not willing) to share incomes, are spreading over the system during its evolution.

5 Experimental Results

The purpose of the conducted experiments was to find out what are possibilities of emerging collective behavior in the large multi-agent systems when different subsets of strategies are used in the game, and in particular the influence of the ISM mechanism. A 2D array of size 100×100 with 10 000 alive cells was used, with an initial state C or D (player actions) of each cell set with a probability of 0.5. Strategies were assigned initially to CA cells with equal probabilities, except for an experiment with the full set of strategies, where the strategy pC was assigned with a probability equal to 0.001 and the remaining strategies with equal probabilities. To an agent with a "k rule, an integer value k was randomly selected from the range $[0 \ldots 7]$. When an agent was using the strategy pC the probability of cooperation p_C was defined as a real random number from the range $[z - \triangle, z + \triangle]$, where z and $\triangle$ are predefined parameters and the values $z = 0.5$ and $\triangle = 0.4$ were used.

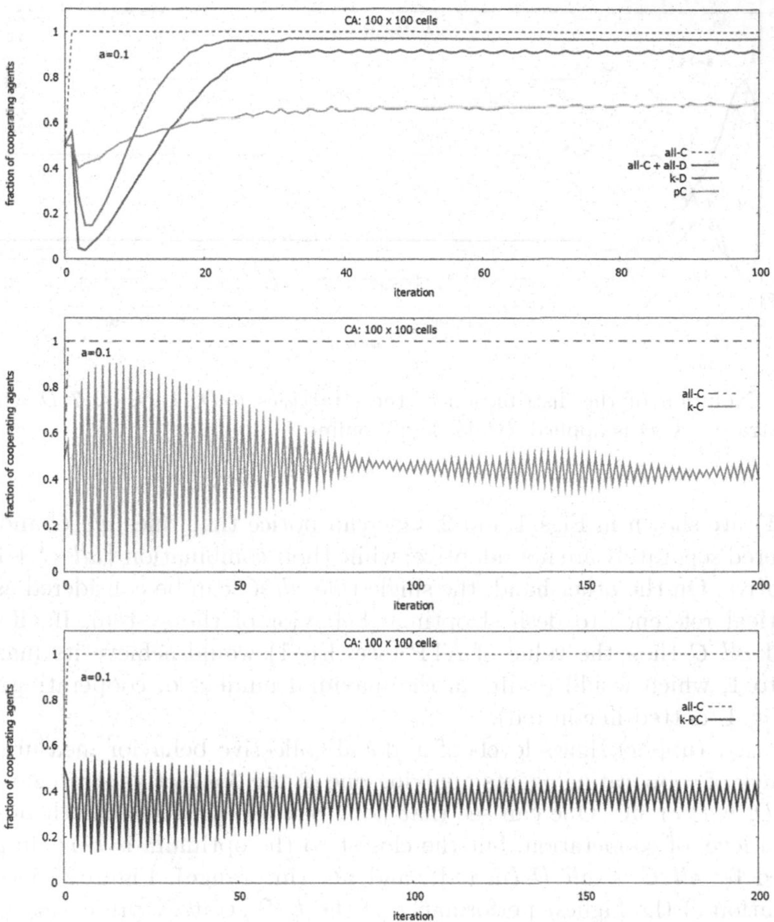

Fig. 1. Fraction of cooperating agents in games with a single adaptive strategy: strategies $\{all\text{–}C + all\text{–}D\}$, $k\text{–}D$ and p_C (upper), strategy $k\text{–}C$ (middle), and strategy $k\text{–}DC$. (lower). (Color figure online)

Experiments were conducted with the following parameters of the payoff function: $b = 1, 2$, $c = 0$ and the parameter a was a subject of change. Updating rules of agents by the competition mechanism was conducted after each iteration ($q = 1$). The number of iterations of each run lasted from 100 to 1500 depending on the current set of parameters. When it was necessary, the results presented below were averaged on the base of 20 runs.

5.1 Competition in Systems with a Single Adaptive Strategy

The purpose of the first set of experiments was to see what is the behavior of the multi-agent system when it is homogenous, i.e. all agents use the same adaptive strategy from the set of the 6 available strategies, and results for single runs (for

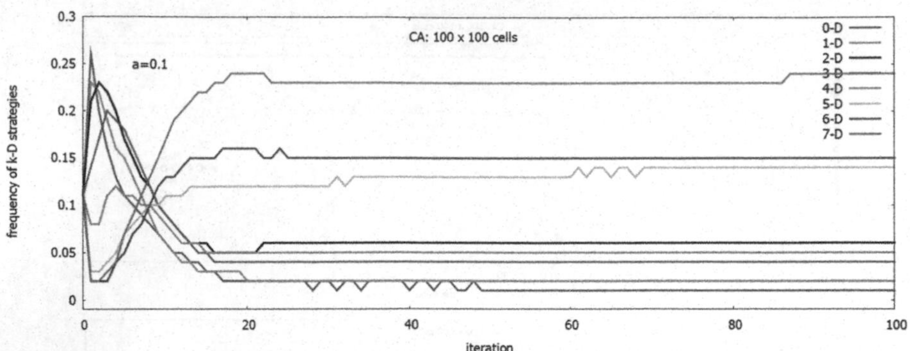

Fig. 2. Evolution of the distribution of the strategies *0–D, 1–D, ... 7–D* when the initial strategy *k–D* is applied. (Color figure online)

$a = 0.1$) are shown in Figs. 1 and 2. One can notice that rules *all–C* and *all–D* considered separately are not adaptive, while their combination {*all–C* + *all–D*} is adaptive. On the other hand, the single rule *all–C* can be considered as some theoretical reference to desired optimal behavior of the system. If all agents applied *all–C* then the value of ATP (see, Eq. 1) would achieve its maximum equal to 1, which would results in the maximal number of cooperating agents (see, Fig. 1, dotted line in red).

Figure 1 (upper) shows levels of a global collective behavior measured by a frequency of cooperating agents applying one of the 3 adaptive strategies: *all–C* + *all–D*, *k–D* or *pC*. One can see that none of them provides the theoretically optimal level of cooperation, but the closest to the optimum is *k–D* (in green), followed by *all–C* + *all–D* (in red), and *pC* (in orange). Figure 2 gives some explanation of the highest performance of the *k–D* strategy presenting the distribution of the final frequencies related to the value of *k*. Figure 2 gives some explanation of the highest performance of the *k–D* strategy presenting the evolution of the distribution of the frequencies related to the value of *k* (see the final frequencies values). The key factor of the high level of cooperation is a high level of tolerance for defecting neighbors. One can see that the population of agents is dominated by agents having strategies *7–D* (in violet), *6–D* (in blue), and *5–D* (in yellow) highly tolerating defection. The final frequencies of the remaining strategies are equal to or close to 0. Figure 1 (middle and lower) shows levels of a global collective behavior of two remaining adaptive strategies: *k–C* and *k-DC.* One can see that the behavior of multi-agent systems with any of these two strategies has two phases: the phase with high oscillation of the frequency of cooperating agents, and the phase of reaching some equilibrium with the value of frequency of cooperating agents far from the optimum. We can judge that these strategies can not be used directly as single ones for achieving a high level of collective behavior.

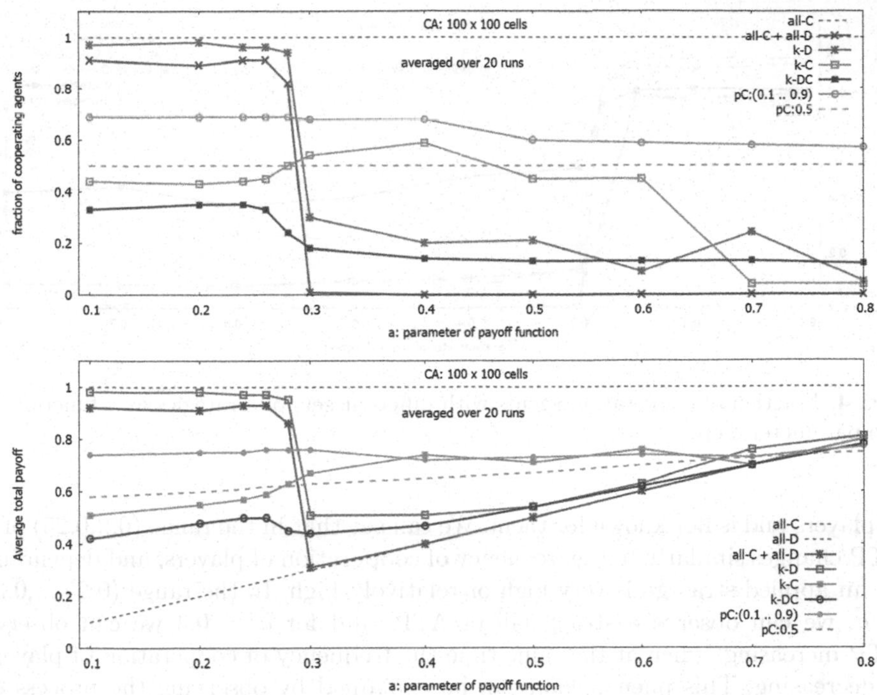

Fig. 3. Behavior of players in games with a single adaptive strategy for different values of the parameter a: fraction of cooperating players (upper), average total payoff of players (lower).

Figure 3 (upper) shows (results averaged on 20 runs) how the level of cooperation in the multi-agent system with the above considered adaptive rules depends on the value a of the payoff function, and Fig. 3 (lower) presents the corresponding values of ATP. One can see that the highest level of cooperation close to 97% is achieved when players use the strategy *k–D* for the values of a from the range (0,...,0.26). We can notice a similar behavior of the system when the strategy *all–C* + *all–D* is used, with a lower level of cooperation close to 91%. However, for $a \geq 0.3$ we can see a strong drop of the levels of cooperation, especially for *all–C* + *all–D* when cooperation disappears. For the remaining strategies, the performance of the system in the range (0,...,0.25) is significantly lower than for the first two strategies. For $a \geq 0.3$ we can also observe a drop in the level of cooperation for these strategies, but the level of cooperation remains only slightly lower, and in the case of *k–C* we can see some improvement in the level of the cooperation.

We can better understand the behavior of players by looking at Fig. 3 (lower) presenting values of ATP of the system as a function of the parameter a, which defines a payoff when two players participating in the game defeat. It is necessary to remember that players take their decisions only based on their personal payoffs, and the value of ATP characterizes the performance of the whole team

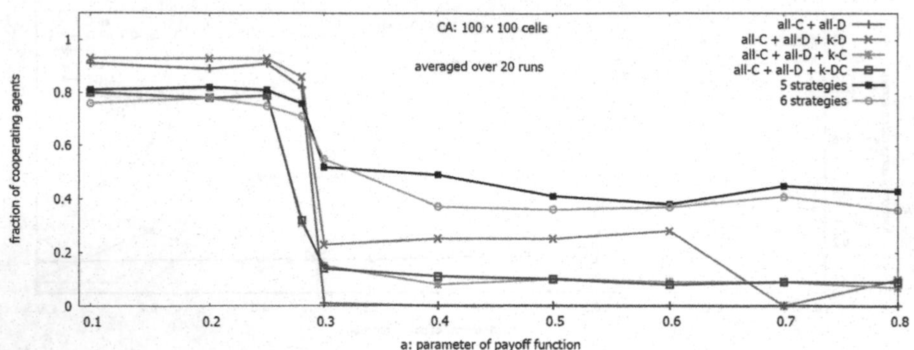

Fig. 4. Fraction of cooperating agents with different sets of strategies as a function of the parameter a.eps

of players and is not known for them. We can see that in the range (0,..,0.25) of a ATP changes similarly to the frequency of cooperation of players, and depending on an applied strategy is very high or relatively high. In the range (0.25,...,0.3) of a, we can observe a strong fall of ATP, and for $a > 0.3$ we can observe ATP increasing, when at the same time the frequency of cooperation of players is decreasing. This phenomenon can be explained by observing the process of escaping two players from the mutual cooperation when both receive the payoff equal to 1, and coming to a situation when one of them defeats and receives a payoff $b = 1.2$ (the other one still cooperates and receives the payoff equal to 0), which causes defeating also of the second player and results in receiving by both payoffs equal to a, which constitutes a Nash equilibrium point. Due to the nature of NE, returning from it to mutual cooperation is much more difficult than changing mutual cooperation. When the value a is small ($a < 0.25$) escaping from mutual cooperation which leads to NE is not attractive for a player. However, when the value of a increases and players are in NE with a relatively high payoff corresponding to a, returning to mutual cooperation will be more and more problematic. In summarizing, we can see (Fig. 3 (lower)) that ATP, under the given values of a and the remaining parameters of the payoff function, is limited by the maximal value of ATP equal to 1 when all players cooperate (dotted line in red) and the value a (dotted line in blue) corresponding to NE.

5.2 Competition in Heterogeneous Systems

The purpose of the next set of experiments was to study the behavior of the multi-agent system when it is heterogeneous, i.e. agents can use different rules available from a subset of the basic 6 rules, and averaged results (20 runs) are shown in Fig. 4. We considered 3 subsets of strategies consisting of 3 rules: {*all–C* + *all–D* + *k–D* } (plot in green), {*all–C* + *all–D* + *k–C* } (plot in light blue), and {*all–C* + *all–D* + *k–DC* } (plot in violet), a subset with 5 rules {*all–C* + *all–D* + *k–D* + *k–C* + *k–DC* } (plot in graphite), and a subset with all 6 rules (plot

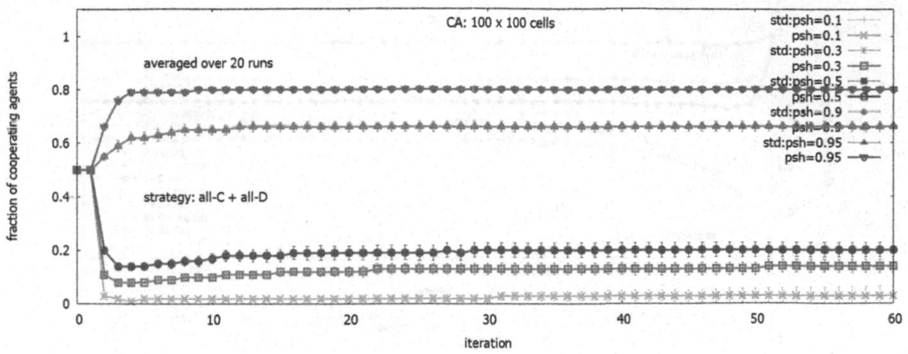

Fig. 5. Games with income sharing ($a = 0.4$): developing of cooperation for different values of an initial value of probability *psh* of *RTS*. (Color figure online)

in orange). Like in Fig. 3 we can see similar three regions of a for the behavior of the system, but we can also observe differences. The main differences for the region $a < 0.25$ are the following: (a) the 3 rules subset containing *k–D* rule (plot in green) is the best performing one, (b) remaining strategies composed of 3, 5 or 6 rules have a good performance, and the 5-rules strategy set (plot in graphite) is the best performing among them, (c) the performance of the 3-rules strategies with rules *k–C*, *k–DC* is much higher in comparison when they were used as single adaptive strategies. For $a > 0.3$ we can observe: (a) similar performance of the system composed of 3 rules like it was observed in experiments with single rules, (b) significant increasing of the performance of the system composed of 5 and 6 rules; the increase of the performance is the result of collective behavior caused by including rules *k–C* and *k–DC* to the pool, which were not efficient working separately.

5.3 Competition and Cooperation: A Case Study

Results of experiments conducted until now have shown that when the mechanism of competition was used, we could observe a global collective behavior close to the optimal in only some short windows of values of the parameter a under some subsets of agent rules. The main question which arises is the following: can we increase the performance of the system in observed short windows of the parameter a, and probably a much more important question is – whether we can both, significantly extend the window of values of the parameter a to yield a high level of agents' cooperation and also to increase the level of global collective behavior. For this purpose we introduced to the game an additional mechanism – the *income sharing mechanism* (ISM) presented in Subsect. 4. We will apply this mechanism to the worst-performing subset of rules *all–C* + *all–D* (see, Fig. 3).

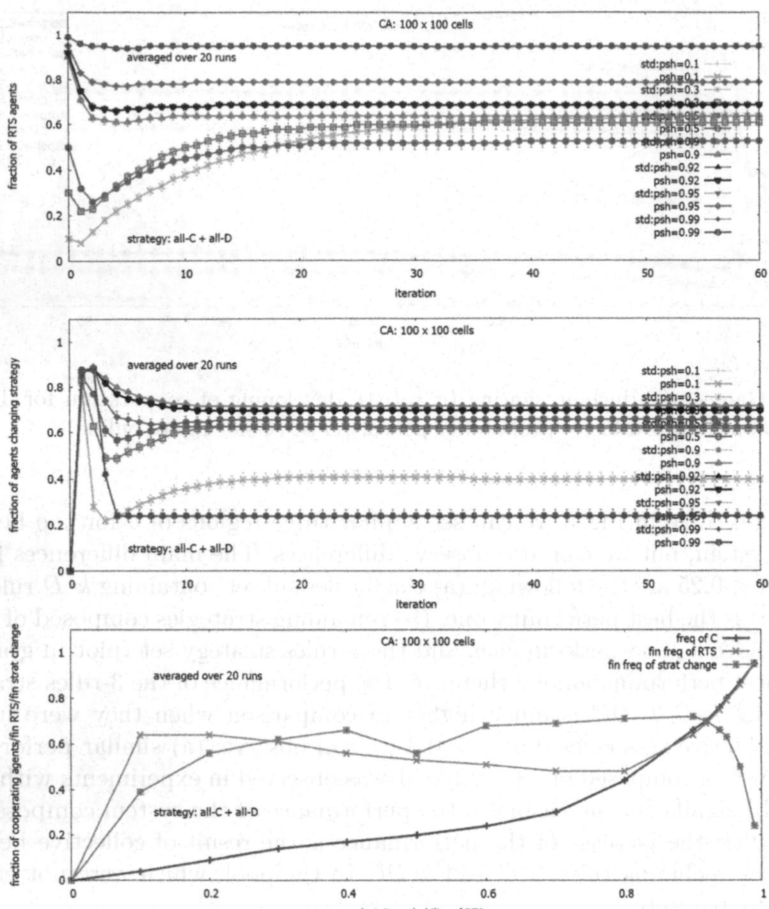

Fig. 6. Games with income sharing ($a = 0.4$): fraction of RTS agents for different values of psh (upper), fraction of agents changing their strategies for different values of psh (middle), final values of the basic income sharing model parameters as a function of psh (lower). (Color figure online)

Figure 5 shows how the fraction of cooperating agents develops in time when the value $a = 0.4$ and ISM are introduced to the game. We can see that the level of global cooperation can be significantly greater than 0 (like it was when ISM was not used) and it ranges from 0.04 to 0.80 and depends on the value of initial probability psh which defines readiness (RTS) of an agent to share income with neighbors. Plots presented in Fig. 6 give some insight into the process of emerging cooperation under ISM.

Figure 6 (upper) shows how the fraction of RTS agents changes in time and how it depends on *psh*, Fig. 6 (middle) shows the behavior of the other parameter - the frequency of agents changing their strategies due to the local competition between agents, and Fig. 6 (lower) summarizes the influence of these both parameters on developing of cooperation in the system.

We can notice (see, Fig. 6 (upper)) two ways of changing a number of RTS agents which depend on an initial value of probability *psh*. For $psh < 0.9$ the number of agents wishing to cooperate decreases initially, and after that increases to reach a value from the range (0.53,...,0.66) independently on an initial value of *psh*. For $a \geq 0.9$ the frequency of RTS agents reaches a high level, lower than the corresponding *psh*, and it is in the range (0.63,...,0.95). Figure 6 (lower) shows that RTS frequency (in blue) is slowly decreasing for nearly all ranges of values of *psh* and depends near linearly on it until reaching the value of *psh* from the range (0.8,...,0.9). After reaching this area it starts quickly increasing.

Observing Fig. 6 (middle) we can notice two phases of changing the frequency of agents changing their strategies. In the first phase, the value of this parameter grows very fast and independently on the value of *psh*, and reaches the level around 90% and after that decreases achieving a level depending on *psh*. Figure 6 (lower) shows that a value of this parameter (in orange) grows slowly near linearly till around $psh = 0.9$, and after this, it drops very fast, which corresponds to a fast increase of the frequency of cooperation (in red) of agents.

Figures 7 and 8 give some additional insight into the complex issue of emerging of cooperation of agents caused by ISM. They present final spatial distributions of RTS agents, and final spatial distribution of strategies/states respectively for different values of *psh*. We already know that a relatively high level of RTS final frequency can be observed for a wide range of values of *psh* and we can observe it in Fig. 7 (in red) but a strong increase of a level of global cooperation can be observed for $a \geq 0.9$. This figure shows that a reason for this phenomenon can be the spatial structure of agents that do not participate in income sharing. We can observe a significant increase in the level of global cooperation if a block structure of non-sharing agents reaches a high level of granularity when big-size blocks are replaced by a number of small-size blocks distributed over the spatial area of agents.

Figure 9 summarizes our results concerning the influence of ISM on a global collective behavior. It shows how the frequency of cooperating agents depends on *psh* for a whole range of the parameter *a* of the payoff function. We can see a significant extension of *a* range where a high level of cooperation can be reached, and a near-optimal value of global cooperation can be achieved by introducing ISM, and where this level depends on the value of *psh*.

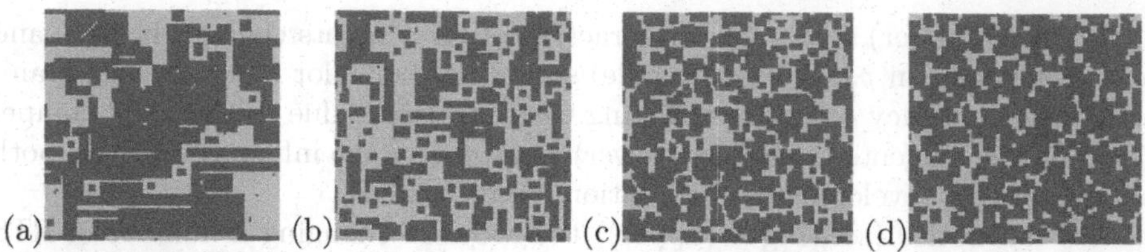

Fig. 7. Distribution of RTS agents (in red) and not RTS agents (in orange) ($a = 0.4$): (a) $psh = 0.1$, (b) $psh = 0.9$. (c) $psh = 0.9$, and (d) $psh = 0.95$. (Color figure online)

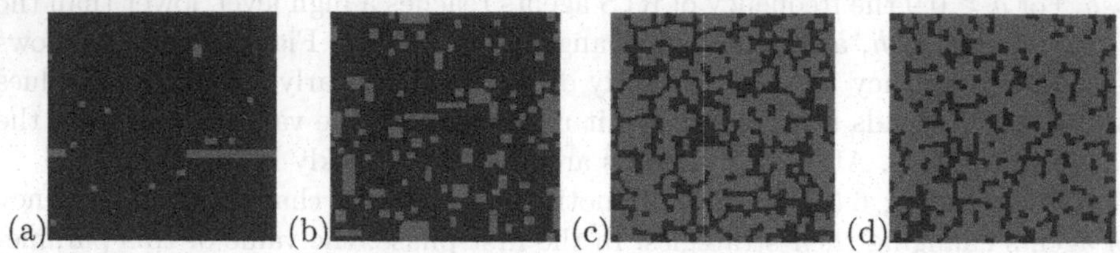

Fig. 8. Distribution of agents strategies/states (*all–C*, in red, *all–D*, in blue) ($a = 0.4$): (a) $psh = 0.1$, (b) $psh = 0.9$. (c) $psh = 0.9$, and (d) $psh = 0.95$. (Color figure online)

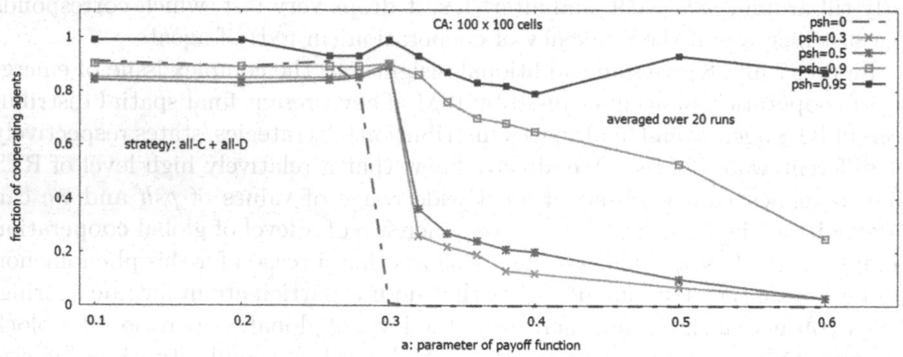

Fig. 9. Games with income sharing: fraction of cooperating agents for different values of *psh* as a function of the parameter *a*.

6 Conclusions

In this paper, we have studied the conditions of the emergence of collective behavior in large CA-based multi-agent systems with agents interacting according to the principles of SPD games. We have shown that two basic mechanisms - competition and cooperation have the essential importance for the collective behavior of such systems measured by the number of cooperating agents and the average total payoff. While competition provides a high level of collective behavior for systems with a different number of agents' strategies, the level of cooperation can be significantly improved by introducing a mechanism of voluntary income sharing. Our current studies are oriented on recognizing and more deeply understanding processes of emerging cooperation due to income sharing and also on the application of these results for solving optimization problems in emerging computer-communication technologies. The concepts of collective behavior of automata were already successfully used to solve problems of scheduling related to cloud computing [2] and lifetime optimization in wireless sensor networks [9].

References

1. Fernández Domingos, E., et al.: Emerging cooperation in N-person iterated prisoner's dilemma over dynamic complex networks. Comput. Inform. **36**, 493–516 (2017)
2. Gąsior, J., Seredyński, F.: Security-aware distributed job scheduling in cloud computing systems: a game-theoretic cellular automata-based approach. In: Rodrigues, J.M.F., et al. (eds.) ICCS 2019. LNCS, vol. 11537, pp. 449–462. Springer, Cham (2019). https://doi.org/10.1007/978-3-030-22741-8_32
3. Katsumata, Y., Ishida, Y.: On a membrane formation in a spatio-temporally generalized prisoner's dilemma. In: Umeo, H., Morishita, S., Nishinari, K., Komatsuzaki, T., Bandini, S. (eds.) ACRI 2008. LNCS, vol. 5191, pp. 60–66. Springer, Heidelberg (2008). https://doi.org/10.1007/978-3-540-79992-4_8
4. Khaluf, Y., Ferrante, E., Simoens, P., Huepe, C.: Scale invariance in natural and artificial collective systems: a review. J. R. Soc. Interface **14**, 20170662 (2017)
5. Nowak, M.A., May, R.M.: Evolutionary games and spatial chaos. Nature **359**, 826 (1992)
6. Osborne, M.: An Introduction to Game Theory. Oxford Univ. Press, New York (2009)
7. Peleteiro, A., Burguillo, J.C., Bazzan, A.L.: Emerging cooperation in the spatial IPD with reinforcement learning and coalitions. In: Bouvry, P., González-Vélez, H., Kołodziej, J. (eds.) Intelligent Decision Systems in Large-Scale Distributed Environments. Studies in Computational Intelligence, vol. 362, pp. 187–206. Springer, Cham (2011). https://doi.org/10.1007/978-3-642-21271-0_9
8. Seredyński, F.: Competitive coevolutionary multi-agent systems: the application to mapping and scheduling problems. J. Parallel Distrib. Comput. **47**(1), 39–57 (1997)
9. Gąsior, J., Seredyński, F., Hoffmann, R.: Towards self-organizing sensor networks: game-theoretic ϵ-learning automata-based approach. In: Mauri, G., El Yacoubi, S., Dennunzio, A., Nishinari, K., Manzoni, L. (eds.) ACRI 2018. LNCS, vol. 11115, pp. 125–136. Springer, Cham (2018). https://doi.org/10.1007/978-3-319-99813-8_11
10. Seredyński, F., Gąsior, J., Hoffmann, R.: The second order CA-based multiagent systems with income sharing. In: Gwizdałła, T.M., Manzoni, L., Sirakoulis, G.C., Bandini, S., Podlaski, K. (eds.) Cellular Automata. ACRI 2020, vol. 12599, pp. 134–145. Springer, Cham (2020). https://doi.org/10.1007/978-3-030-69480-7_14
11. Wolfram, S.: A New Kind of Science. Wolfram Media (2002)
12. Östberg, P., Byrne, J., et al.: Reliable capacity provisioning for distributed cloud/edge/fog computing applications. In: 2017 European Conference on Networks and Communications (EuCNC), pp. 1–6, June 2017

Divergence of an Observed User Profile and a Simulated Real State of User Due to Social Communication

Marcin Maleszka(✉)

Wroclaw University of Science and Technology, st. Wyspianskiego 27, 50-370 Wroclaw, Poland
marcin.maleszka@pwr.edu.pl

Abstract. In this paper, we extend our previous modular model of a social collective with hierarchical knowledge structure, and we propose an additional step in tuning information retrieval systems after dataset experiments and before testing them with real users. For this purpose, we add a simulation of the group of users as a social collective model and run the information retrieval system on it. Here, we take first steps in that direction, by focusing on a subsystem of a recommender system and simulating what type of social collective it is most effective with. We present details on the social collective model and the information retrieval subsystem model, as well as how to put them together in one simulation. We run several experiments and present some initial findings. In our opinion, this overall approach could be used to greatly enhance further tests with real users.

Keywords: Collective intelligence · Recommender system · Group modeling · Multi-agent simulation

1 Introduction

Information retrieval is one of the most known and most researched areas of computer science. One of the typical problems considered in the area is recommendation, i.e. providing a user with items that may be potentially relevant to him at a given moment. Interestingly, even without the modern focus on taking into account social networks, it was determined that no universal solutions are applicable [18]. With the social graph, the situation becomes more complex, as external influences come in real time from many directions, and the context is harder to determine. *Computational collective intelligence* is one of several approaches that formalize the very broad and multidisciplinary research area of collective intelligence, also known as the wisdom of the crowds. One of the research directions in this area is modelling groups of intelligent agents working towards a common purpose or simply exchanging knowledge related to some specific unsolved problem. A complete model of such a group could be used to show the process of knowledge diffusion or opinion formation in a social

D. Groen et al. (Eds.): ICCS 2022, LNCS 13351, pp. 624–637, 2022.
https://doi.org/10.1007/978-3-031-08754-7_66

network, connecting these two research areas. There is no significant overlap between information retrieval and computational collective intelligence.

In this paper, we extend the previous modular model of the social collective with a new variant of the knowledge structure – a hierarchical one based on a predefined thesaurus. This structure is identical to the one we have previously used in research in the information retrieval area, which was motivated by the idea to use them together. Information retrieval systems are usually tested with historical data, but sometimes require also live user input. In the second case, such tests are very time consuming and require a long preparation time. A series of simulations that pretune some parameters of the system would help with the quality of this preparation. This is the place where having a good model of the user group, instead of just a single user, is helpful. While the idea may appear similar to adversarial learning, it is very distinct, as only one model is tuned against the other.

With such a motivation, the final aim is to have a good model of a social group that also participates in the information retrieval activity (e.g., uses a recommender system) and then adapt the IR system to offer the best recommendations to the users in the modeled group. After that, less work would be needed when working with the real world group. As an initial step of this research, we start with adapting only parts of the information retrieval system. Additionally, we set up the group model to be similar to the one used in IR in terms of the knowledge structure used and then set up the parameters of the model to better fit the information retrieval one. While this would not be done in the final system, it serves as a demonstration if such an approach is even viable.

This paper is organized as follows: in Sect. 2 we describe the research works that are relevant to both the social collective model and the collaborative filtering model that are used in this paper; Sect. 3 contains details on both models, with the first being in the Facebook-based variant, and the second in the modified median variant; in Sect. 4 we describe the setup of both models in a simulation environment and the results of those simulations; finally in Sect. 5 we give some concluding remarks and present our intended further work in combining both research areas.

2 Related Works

One part of our research in this paper is based on a combination of work done in the areas of social influence (in sociology) and collective intelligence (in computer science). It has been observed that in real-world groups there are various levels of influence in a collective depending on individual and group factors, e.g., subjective norms, group norms, or social identity [4]. As a result, individuals may have different resistance to changes in their internal knowledge (internalization or induction), from no resistance at all to full resistance to any inputs [7]. Different aspects may further modify this, for example known competence of the speaker or the previous knowledge level of the person listening [17]. As an interesting example, the less a person knows, the more they can learn – but also the less they are willing to learn. The collective intelligence was best introduced in Surowiecki's popular book in 2004 [20] and, among other elements, describes the necessary postulates for wise crowds. Some of the works following this are based on consensus theory [15], where the informal descriptions of Surowiecki are formalized into mathematical postulates. Still, the entire area of research is not applicable in many cases and has come under criticism recently due to being unable to represent real world groups [19]. This problem is addressed by our research combining the mathematical aspect of collective intelligence with real-world based social influence theory, making it more closely resemble real world groups.

Representations of groups in a similar manner are also often done from the point of view of social network research. What we developed in our research and use in this paper may be understood as an influence maximization model in these terms. These are usually understood as a type of predictive models, i.e., they are created with the aim of predicting the network behavior when a new node is added to the network. This may be based on the probability of a person sending a message to other group members [16] or observing how the content of the messages in the group changes over time [1]. Alternatively, in threshold-based approaches, the authors may consider the number of messages concerning a given topic before the receiver is influenced by them [3]. The final group of social network models to consider are explanatory influence models where classifying a person to a specific subgroup (e.g., active or passive members) helps to determine which of them have the most influence on the entire group [2].

The other part of our research in this paper combines more classical information retrieval systems with collective intelligence. The general idea we use as a basis are recommender systems, which have been described as being able to record user actions and use them to refine further results [9]. While the initial approach to those is nowadays limited to very formulaic systems, the modern approaches are based on analysis of the communities they are applied to [10]. Even in those, the user query remains the main tool of the user and the main source of information for the algorithms. The consistent problem is that it may still not reflect real user needs, or even the same query may have a different meaning for different users. One of the classic approaches [14] is using different knowledge structures to represent user profiles (models) and then personalizing the results, but these days an effect similar to overfitting leads to the problem of echochambers [8]. In case of a new user, the profile is still to be built. Instead of making no recommendations at that time, a type of default profile may be instead given to the user temporarily [6]. This profile is then adapted using different models towards real user knowledge [21].

3 Group Models

The overall idea we study in this paper is that there are two models of a group working concurrently. The first model is a collective with members communicating and sharing opinions, and the second is as an additional system that models each user and builds a centroid for the whole group. The centroid is then used for other purposes, e.g., content recommendation, cold-start problem avoidance, anomaly detection, etc. We created a model of the first part and applied it as input to the modeling system operating on the basis of a different representation of the group. Both parts were implemented and run simultaneously in a simulation environment with their end results compared (see Sect. 4).

3.1 Model of the Social Collective

The model of the communicating group is based on our previous research on modeling a social collective as an agent system [12]. We constructed those models to intentionally resemble online social networks. The general model of the social collective allows for flexibility in certain parts, which in the case of this paper is used to model the Facebook social network. The three main components are the knowledge structure used, the types of communication used, and the methods of knowledge internalization.

In this paper, we use the hierarchical structure of agent knowledge as a novel variant. This approach was originally used in the second model and we use it in this one to remove any unnecessary difference between them. The hierarchical structure of knowledge assumes that there exists exactly one graph representing the relations between all individual knowledge statements. The internal representation of this knowledge for each agent is always a subgraph of this representation. In implementation terms, all agents always have the full representation of the knowledge graph, but if the associated weight is 0, then the agent does not *know* this statement. As the graph structure is common between agents, the hierarchical structure of individual knowledge statements is stored in a vector format, with a weight associated with each. This weight is interpreted as a numerical representation of either the sentiment towards the issue (a higher value represents a more positive sentiment) or the agents certainty that the knowledge statement is true. The knowledge and sentiment of each agent a about each issue k_i is represented by a set of pairs $k_i = \{< k_i^a, w_i^a >\}$ (in implementation: only a set of weights in range $[0, W]$). Any message an agent sends is a single pair $\{< k_i^a, w_i^a >\}$, which is sufficient for the receiver due to the common structure.

The model of communication based on Facebook social network may be described in general terms as four main modes of communication: a private chat between users, a public post, public comment under a post or another comment, or a public reinforcement of a post or comment. Other methods of communication are derivative of those. In the presented model, all of those are allowed. An agent may use any of them with a probability of P_i^c. As even public messages are initially only displayed to that persons *friends*, in the model, each agent has a limited list of agents that he can communicate with (a list of vertices that connect it to other agents – nodes in the social graph). Specifically, the communication modes are:

- Each time communication may occur, an agent may randomly select to send a message containing one statement from the hierarchical knowledge structure to one or more agents it is connected to, selected at random (with the

possibility of any one agent being selected multiple times). This represents posts on Facebook *wall*, where people may skip a message or read it several times.
- Each time communication may occur, an agent may randomly select to send a message, that is a copy of any previous message it received, to one or more agents it is connected with, selected at random (with the possibility of any one agent being selected multiple times). This represents people using the *Like* function of some Facebook wall posts. Again, people may see the *Like* many times, or not see it at all.
- Each time communication may occur, an agent may randomly select to send a message, that is a copy of any previous message it received, but with the weight modified to the agents own opinion on the statement, to one or more agents it is connected with, selected at random (with the possibility of any one agent being selected multiple times). This represents people commenting on posts on Facebook *wall* (including commenting on other comments). Again, people may see the comment many times, or not see it at all.
- Each time communication may occur, an agent may also randomly select to send a message containing one statement from the hierarchical knowledge structure to exactly one agent it is connected to, selected at random, but concurrently the other agent will send back the message containing the same knowledge statement, but with its own associated weight. Processing the messages is done after the bidirectional communication occurs, so agents are only informed of the others opinion prior to communication and not already influenced by it. This type of communication represents discussions via the chat option between two different people.

In reaction to incoming communication, agents in this model internalize the incoming knowledge by changing their own. This is done by applying specific knowledge integration algorithms that we have based on various research in different areas of literature, from psychology to purely mathematical models. A message may be internalized immediately, or it may be stored in the agent's memory until several messages on the same topic are accumulated. Algorithmically speaking, the input is one or more pair $k_i = \{< k_i^a, w_i^a >\}$ and the agent's initial knowledge state, while the output is the agent's final knowledge state. In this paper, we do not use all integration algorithms (which we call integration strategies in our research), but only those most similar to the other model used:

- Substitute – the idea for this integration algorithm is derived from the sociological concept of *no resistance to induction*. An agent (person) internalizes all incoming opinions or knowledge. The strategy works by immediately changing the weight of the knowledge statement given in any incoming message from its own weight to the weight provided in the incoming message.
- Extend – is a modification of the *Substitute* strategy, in which the changes are only applied to previously neutral (unknown to the agent) knowledge statements. If the agent already has any knowledge (opinion) on some topic, then it remains as is.

- Immediate consensus (Merge) – the idea for this integration algorithm is derived from the mathematical principles of consensus theory. It is severely limited, as it only applies to one incoming message (instead of several, like the following four strategies). Following the concepts of consensus theory, the approach to determining the new numerical value from two different weights assigned to knowledge statements, is the average of those values. For neutral knowledge (unknown to the agent), that means that the weight is halved from the incoming message.
- Delayed voting (Majority) – the four algorithms named *delayed* are also based on consensus theory. The basis of all the algorithms is accumulating several messages concerning one knowledge statement, before integration is conducted and the internal value of the weight associated with the knowledge statement is changed. In case of *delayed voting*, determining a new knowledge state is done by selecting the most common one among the gathered messages.
- Delayed weighted average consensus – as previously, a message buffer is used. To calculate the new state of knowledge that the agent will have on the given topic, in this strategy, the *average* of received (and own) knowledge or opinions is calculated. This means that the algorithm looks for a centroid that has the smallest sum of squared distances to all other elements.
- Delayed nonweighted median consensus – as previously, a message buffer is used. In this integration strategy, all gathered knowledge states are sorted according to a predefined ordering and the middle state is selected. For an even number of opinions, one of the middle states is selected at random.
- Delayed weighted median consensus – as previously, a message buffer is used. The algorithm is similar to the previous one, but messages with higher weights are copied multiple times in the sorted list, corresponding to their weights (normalized for a maximum of ten repetitions for the maximum allowed weight).

In some of our research [13] we have also considered agents forgetting knowledge (here: the weight is changed to 0: neutral/unknown), but this aspect is not considered in the current paper.

3.2 Model of the Collaborative Filtering Group

The second model was based on the research in the area of collaborative recommendation that was presented in [11]. It was created as a means to limit the *cold start problem*, i.e. new users in the system being presented with bad or no recommendations. The proposed solution was to find other users similar to the new one (based on short initial searches or other information, e.g. demographic), then determine a centroid representative of this group. Such centroid would be used as the initial representation of the new user, until sufficient proper information about him has been gathered. In effect, the centroid of the group represents its average opinion as gathered in the user profiles, and should be very close to the average of real opinions as represented in real user preferences. The paper

[11] specifically distinguishes what is the real user preference and how the system represents him as a user profile, but uses the same knowledge structure and agent representation for both. Here, we use the same knowledge structure, but for the user preferences agents are represented as in Sect. 3.1, and for the user profile, as in [11].

The idea of the user profile as a hierarchical structure is based on a common thesaurus and weights reflecting the frequencies of particular terms in user queries. This is based on the well-known assumption that when the user is interested in a term, he includes this term in his queries and the more interested he is, the more often the term is contained in a query. The hierarchical thesaurus allows determining relationships between terms and better grouping users due to the possible generalization of their interests. Based on the assumption that only one term from each path between a leaf and root (excluding the root itself) may occur in a query, such profile is further limited by the following constrains:

- Total Frequency Minimum (K1) – as each user query contains at least one term, the sum of the frequencies of queried terms in the tree (i.e. weights) should be no smaller than 1. This comes from normalization when queries consist only of one term.
- Path Frequency Maximum (K2) – as each element on a path in the thesaurus may occur only once in each query, the sum of weights on a single path should be no larger than 1.
- Total Frequency Maximum (K3) – as user queries consist of a limited number of terms, the total sum of weights in the tree should not be larger than some constant b (e.g. [5] states that typical user query has at most 3 terms, therefore both in this paper and in [11] we assumed $b = 3$).

We build a centroid by gathering observations of user activity, then calculating the frequencies and storing them as weights of appropriate nodes in the hierarchical structure.

Based on the given assumptions on the user profile, there are several postulates formulated which a satisfactory centroid should satisfy:

- Reliability (Re) – this postulate requires the centroid profile to satisfy K1, K2, and K3 constraints on the profile. Otherwise, the profile is not valid.
- 1-Optimality (O1) – this postulate requires that the result of the integration should minimize the sum of distances to all profiles in the group (according to some metric). In practical terms, it is the median of the weights of each node in the hierarchical structure.
- 2-Optimality (O2) – this postulate requires that the result of the integration should minimize the sum of squared distances to all profiles in the group (according to some metric). In practical terms, it is the average of the weights of each node in the hierarchical structure.
- Conflict Solving (CS) – a conflict occurs in a constructed centroid when the difference between the weights of terms that have the same parent is greater than some assumed threshold. This postulate requires that all conflicts are

solved. This is done by increasing the preference of the parent and reducing the preference of the elements in conflict, in the following way:

$$\Big(w_{t_i}(v_x) < (\psi - \tfrac{\gamma}{2})\Big) \wedge \Big(w_{t_j}(v_x) > (\psi + \tfrac{\gamma}{2})\Big)$$
$$\wedge\Big(w_{t_i}(v_y) < (\psi - \tfrac{\gamma}{2})\Big) \wedge \Big(w_{t_j}(v_y) > (\psi + \tfrac{\gamma}{2})\Big)$$
$$\rightarrow \Big(w_{t_{cs}}(v_x) = 0\Big) \wedge \Big(w_{t_{cs}}(v_y) = 0\Big) \wedge \Big(w_{t_{cs}}(v_p) > max(w_{t_i}(v_p), w_{t_j}(v_p))\Big)$$

where v_x and v_y are nodes in conflict, with a common parent v_p; the initial weights come from trees t_i and t_j, and the integrated tree is t_{cs}.

Unfortunately, it is mathematically impossible for a centroid to satisfy all four postulates in any nontrivial situation. For all combinations, we always require at least Reliability and one other postulate to be satisfied.

In this paper, for the purpose of building the integrated centroid, we apply one of algorithms from [11]: the Closest Tree Algorithm (CTA). The algorithm has two main phases. In the first, we determine the median of weights for each node, which makes it satisfy some of the conditions laid down for the desired centroids. The second phase is focused on another condition and optimizes the weight values towards it, in effect transferring some value of the weights towards the root of the tree. The effect on the conditions is that at the end two are satisfied (Re and CS) and one is almost satisfied (O1). The complexity of this algorithm is between $O(|V|\cdot N)$ and $O(|V|\cdot|V|\cdot N)$ (no conflict and only conflicts, respectively, to solve in the second phase). The full details of the algorithm can be found in [11].

3.3 Combining the Models

In this paper, we create a system where both models work concurrently, with some flow of information between them, as shown in Fig. 1. The social collective simulates a group of real users, while the function of the collaborative filtering model is identical as in the full information retrieval system (i.e. building centroids that are a good representative of a group of users).

Initially, both models start with the same set of agents (with randomly generated knowledge, see Sect. 4). The social collective model treats them as members of the collective, while the CF model interprets them as initial profiles of users. Thus, at initialization, both user preference (real needs of users) and user profile (a representation of the user in CF model) are identical. In a real world system, this would only be the case for users with no initial interest, therefore of limited applicability (e.g., elearning platforms). The collaborative filtering model immediately builds an initial centroid out of those profiles.

Following the initialization, the social collective model works through its iterations, with agents communicating and exchanging knowledge. Meanwhile, the CF model observes nonprivate communication (here: all but *chat* messages) and updates user profiles. Periodically, it also updates the representative centroid.

At some chosen point, the concurrent run of both models ends. We calculate the mean knowledge of all agents in the collaborative model and then compare it with the last centroid calculated by the collaborative filtering model. Thus, we observe how both models diverge over time from the initial identical start. Following this, we could determine if the social collective approach is viable as a tool for pretuning information retrieval systems for further tests with real users.

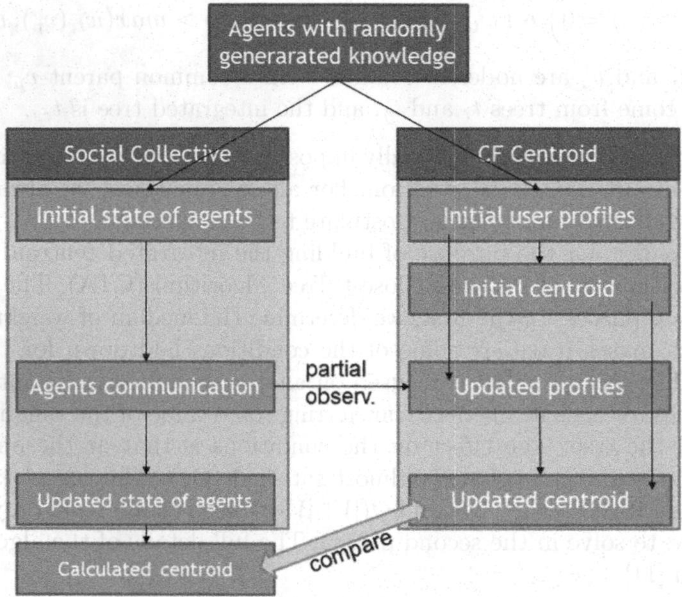

Fig. 1. Schema of experimental system with the flow of information between both models.

4 Evaluation of the Model

In our previous research, we were using different measures to evaluate each of the models. Collaborative filtering model was initially developed as a part of a larger information retrieval system, and as such, traditional measures could be used to measure its influence on the overall system, e.g., precision, recall, and f-measure. Additionally, in [11] this specific part of the overall system was evaluated using a criterion defined as BQC (Basic Quality Criterion), which states that if the profiles of each group member are in some neighbourhood of size ϵ, then the centroid solution would also be inside. Similarly, for the social collective model, we formulated a measure of drift based on the sociological literature, to calculate the average change of all opinions (weights associated with knowledge) in the group overtime. Following, the criterion for a good collective was that it had a small drift when there was no outside influence, and a larger one with such influence exerted. Unfortunately, neither of those can be applied directly for measuring the divergence of both models.

Instead, the divergence of the collective models was instead evaluated in a series of runs in a simulation environment. We have implemented both models and run them concurrently, as described in detail in the previous section. The collective model of the social collective was simulating a group of real social network (Facebook) users, and the collaborative recommendation model was used to build to create a centroid profile of the group. The initial knowledge structures were identical, and after the end of the simulation run, we compared the calculated median of agents in the social collective (real group centroid) with the centroid as calculated by the collaborative recommendation part.

The parameters of the social collective model have been set up as follows:

- Number of possible knowledge statements: 128,
- Initial number of statements for each agent (with nonzero wights): 16,
- Range of allowed weights: [0,1],
- Weight distribution: uniform in the entire range,
- Maximum number of agents in *friend* relation: 10,
- Probability of starting communication by an agent in each time moment: 0.2,
- Maximum number of receivers for a message: 5,
- Size of message buffer for delayed strategies T^{dv}: 11,
- Probability of different type of communication: depending on experimental series.

The parameters of the collaborative filtering model have been set up as follows:

- Parameter for K3 postulate $b = 3$;
- Parameter for Conflict Solving criterion $\epsilon = 5$
- Parameter for Conflict Solving Ciretion $\psi = 0.3$

The simulation was run for 1000 agents for 1000 iterations (time moments, during which interaction between agents is possible). We have run the simulation for every applicable integration strategy (as described in Sect. 3) and for varying probabilities of using different communication modes, repeating each permutation 100 times, and averaging the results. Some of the visually clearest of the aggregated results have been shown in Fig. 2 and Fig. 3, which present the difference between both models in the most generic term in the hierarchical structure and in the most specific one, respectively.

General results of the experiments show that there is a difference of ca. 0.02–0.05 between the social collective and collaborative filtering models after the simulation run (within the range $[0, 1]$). The CF model follows the social collective model quite closely, but it is never exact and the difference is statistically significant.

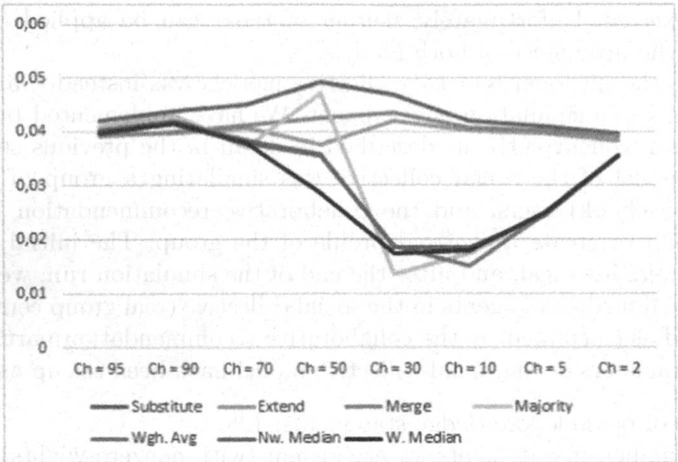

Fig. 2. Divergence between two models after the simulation. Difference measured by opinion weight for all simulated variants of the social collective model with different probabilities of observed and unobserved user behavior (*Ch* value represents probability of agent using unobserved communication). Figure presents divergence for the issue at the root of the tree.

When agents use integration strategies that are based on gathering more outside information, instead of changing the previous knowledge state immediately, the parameters of the communication in the model start having an influence on how good is the CF model. Generally, the more communication may be observed, the closer the CF model is to the social collective one. This is an expected result in any system of the type. However, after some threshold, the model is overfitted and does not react in time to changes in the agent. In our observations, it happened mostly with more than ca. 90% of all messages were observed (i.e., less than 10% were of the private chat type). Barring some strategy-specific cases, we have observed that the CF model is the closest when it can observe 70–90% of interactions between users (i.e., chat messages consist of no more than 10–30% of them). This can be seen to happen for every node in the tree, as seen in the root in Fig. 2, and in the leafs in Fig. 3.

We also performed statistical tests if the difference between the typical situation and the one with *optimized* number of observable interaction in terms of the aggregated value is statistically significant and we have done so for the data resulting in all presented averages. The p-values were smaller than 10^{-15}, therefore we could abandon the null hypothesis that the data is from the same distribution.

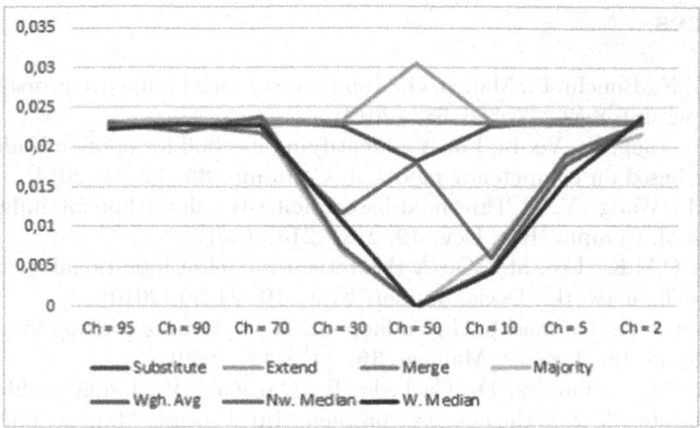

Fig. 3. Divergence between two models after the simulation. Difference measured by opinion weight for all simulated variants of the social collective model with different probabilities of observed and unobserved user behavior (*Ch* value represents probability of agent using unobserved communication). Figure presents divergence for the issue in a leaf of the tree.

5 Conclusions

As stated in the introduction, this paper represents only the initial stages of the overall idea to have a good model of a social group that also participates in the information retrieval activity (e.g., uses a recommender system) and then adapt the IR system to offer the best recommendations to the users in the modeled group. This would create a situation where less work is needed when working with the real world group. This paper shows that the models are not identical, but may be modified to be more similar, so the approach is viable.

The experiments performed in this paper were opposite to the final idea, i.e., we modified the parameters of the social collective to observe how the information retrieval model diverges from it. We have chosen such an approach, as it was possible to use two existing models that were already proved to be good in their own specific areas, and change only the social collective model by using a different knowledge structure in its modular construction, so that we could use both of them in the experiments. Any future research we plan will start from building a model of the group that best represents the one occurring in the application area and tune it, then create a fitting model for the recommender system, conduct finetuning of the second in the simulations, and finally, start the experiments with the group of real users. This will be a direct follow-up study after this paper, which will allow to estimate the possible profit from using this approach.

References

1. Barbieri, N., Bonchi, F., Manco, G.: Topic-aware social influence propagation models. Knowl. Inf. Syst. **37**, 555–584 (2012)
2. Chen, B., Tang, X., Yu, L., Liu, Y.: Identifying method for opinion leaders in social network based on competency model. J. Commun. **35**, 12–22 (2014)
3. Chen, H., Wang, Y.T.: Threshold-based heuristic algorithm for influence maximization. J. Comput. Res. Dev. **49**, 2181–2188 (2012)
4. Cheung, C.M.K., Lee, M.K.O.: A theoretical model of intentional social action in online social networks. Decis. Support Syst. **49**, 24–30 (2010)
5. Clarkea, C.L.A., Cormackb, G., Tudhope, E.A.: Relevance ranking for one to three term queries. Inf. Process. Manage. **36**, 291–311 (2000)
6. Formoso, V., Fernandez, D., Cacheda, F., Carneiro, V.: Using profile expansion techniques to alleviate the new user problem. Inf. Process. Manage. (2012). http://dx.doi.org/10.1016/j.ipm.2012.07.005
7. Kelman, H.C.: Interests, relationships, identities: three central issues for individuals and groups in negotiating their social environment. Annu. Rev. Psychol. **57**, 1–26 (2006)
8. Ge, Y., et al.: Understanding echo chambers in e-commerce recommender systems. In: Proceedings of the 43rd International ACM SIGIR Conference on Research and Development in Information Retrieval (2020)
9. Li, W., Ganguly, D., Jones, G.J.F.: Enhanced information retrieval using domain-specific recommender models. In: Amati, G., Crestani, F. (eds.) ICTIR 2011. LNCS, vol. 6931, pp. 201–212. Springer, Heidelberg (2011). https://doi.org/10.1007/978-3-642-23318-0_19
10. Li, D., et al.: Interest-based real-time content recommendation in online social communities. Knowl.-Based Syst. **28**, 1–12 (2012)
11. Maleszka, M., Mianowska, B., Nguyen, N.T.: A method for collaborative recommendation using knowledge integration tools and hierarchical structure of user profiles. Knowl. Based Syst. **47**, 1–13 (2013)
12. Maleszka, M.: An intelligent social collective with Facebook-based communication. In: Paszynski, M., Kranzlmüller, D., Krzhizhanovskaya, V.V., Dongarra, J.J., Sloot, P.M.A. (eds.) ICCS 2021. LNCS, vol. 12744, pp. 428–439. Springer, Cham (2021). https://doi.org/10.1007/978-3-030-77967-2_36
13. Maleszka, M.: A generic model of a social collective. In: 2021 IEEE International Conference on Fuzzy Systems (FUZZ-IEEE), pp. 1–7 (2021). https://doi.org/10.1109/FUZZ45933.2021.9494407
14. Montaner, M.: A taxonomy of personalized agents on the internet. Technical report, University of Girona (2001)
15. Nguyen, N.T.: Inconsistency of knowledge and collective intelligence. Cybernet. Syst. Int. J. **39**(6), 542–562 (2008)
16. Saito, K., Nakano, R., Kimura, M.: Prediction of information diffusion probabilities for independent cascade model. In: Lovrek, I., Howlett, R.J., Jain, L.C. (eds.) KES 2008. LNCS (LNAI), vol. 5179, pp. 67–75. Springer, Heidelberg (2008). https://doi.org/10.1007/978-3-540-85567-5_9
17. Pratkanis, A.R.: Social influence analysis: an index of tactics. In: Pratkanis A.R. (ed.) Frontiers of Social Psychology. The Science of Social Influence: Advances and Future Progress, pp. 17–82. Psychology Press (2007)
18. Schafer, J.B., Konstan, J.A., Riedl, J.: E-commerce recommendation applications. Data Min. Knowl. Disc. **5**, 115–153 (2001)

19. Søilen, K.S.: Making sense of the collective intelligence field: a review. J. Intell. Stud. Bus. **9**(2), 6–18 (2019)
20. Surowiecki, J.: The Wisdom of Crowds: Why the Many are Smarter Than the Few and How Collective Wisdom Shapes Business, Economies, Societies, and Nations. 1st Doubleday Books, New York (2004)
21. Treur, J., Umair, M.: An agent model integrating an adaptive model for environmental dynamics. Int. J. Intell. Inf. Database Syst. **5**(1), 201–228 (2012)

Temporal-Attribute Inference Using Dynamic Bayesian Networks

Lihi Idan(✉)

Yale University, New Haven, CT 06511, USA
lihi.idan@yale.edu

Abstract. As social networks continue to grow in popularity, it is essential to understand what can be learned about private attributes of social-network users by mining social-network data. Previous work focused on the inference of time-invariant attributes such as personality traits. By contrast, in this paper we focus on the inference of dynamic, time-varying attributes. We present a new approach to modeling social-network users and mining time-varying attributes using dynamic bayesian networks (DBNs). We then explore the extent to which such temporal models can improve the inference results of various dynamic attributes. This work is the first to take a DBN-based approach to the task of private-attribute inference in social networks.

1 Introduction

Knowledge of social-network users' intentions has immense potential to improve the design of recommendation systems, ad-targeting mechanisms, public-health campaigns, and other social and commercial endeavors. At the same time, such knowledge can have a detrimental effect on users' privacy. In this paper, we are interested in inferring intentions of social-network users using public data extracted from their social-network profiles.

Problem Description. Let u be a social-network user and S_u be the set of social networks on which u has accounts. We use $\xi_{(u,s)}$ to denote user u's account on network s. Each account has a private portion $\xi^{pr}_{(u,s)}$ and a public portion $\xi^{pu}_{(u,s)}$. The private portion contains data that only u's ties and the social-network provider can see, while the public portion contains data that can be seen by everyone. In addition to data that u publishes, $\xi^{pu}_{(u,s)}$ contains metadata information about $\xi_{(u,s)}$ such as the mere existence of $\xi_{(u,s)}$ and the visibility levels of different attributes in $\xi_{(u,s)}$. The goal of this work is to infer *offline* behavioral intentions of a social-network user u using only the public portions, $\{\xi^{pu}_{(u,s)}\}_{s\in S_u}$, of u's *online* social-network accounts. We focus on present or near-future behavioral intentions, *i.e.*, on decisions to perform certain actions within short periods of time after the decisions are made.

D. Groen et al. (Eds.): ICCS 2022, LNCS 13351, pp. 638–652, 2022.
https://doi.org/10.1007/978-3-031-08754-7_67

The paper makes the following contributions:

A New Approach to Modeling Social-Network Users Using Dynamic Bayesian Networks. We present a new approach to modeling social-network users and mining time-varying attributes using DBNs. We evaluate our models when used for the inference of different dynamic attributes given temporal, real-world social-network data. This work is the first to take a DBN-based approach to the task of attribute inference in social networks and the first to offer a DBN-based representation of social-network users.

A Unique Focus on Offline, Time-Varying Behavioral Intentions. Unlike other existing works that tackle the task of attribute inference, ours is the first work that aims at inferring *offline and dynamic*, non-politically-related behavioral intentions of social-network users (*i.e.*, a *user-centric* approach) solely based on *public social-network data.* Other works either focus on online intentions or time-invariant preferences; use private data or data that is not obtained from social networks; or take an "object-centric" approach by trying to infer the intention associated with a single, standalone and contextless social-network "object" such as a post or a tweet. Furthermore, some of the behavioral intentions that we consider in this paper have never been studied in any prior machine learning (ML) or social-network-related work.

A New Multidisciplinary Methodology for the Inference of Behavioral Attributes. We introduce a novel, multidisciplinary methodology for the inference of behavioral attributes such as decisions and intentions. We design modular bayesian-network (BN) models that are able to capture the evolving nature of the human decision-making process by combining data and priors from multiple domains. Our methodology handles common challenges in social-network mining such as incomplete datasets, unlabeled data and bidirectional influence between features and the target variable.

2 Related Work

Inference of personal attributes using social-network data has been extensively researched. Inferring users' personality type was investigated in [8] using regression models and Twitter/Facebook data, respectively. Youyou *et al.* [29] showed that automatic inference methods that rely on Facebook likes achieve better prediction accuracy than those achieved by asking the users' friends. Staiano *et al.* [27] used data gathered through smartphones such as calls and texts; their results significantly vary across different personality dimensions.

Demographic attributes' inference is another well-studied topic, with age and gender being the most researched attributes [14,15]. A related stream of research focuses on psychological and mental conditions. Depression is the most researched condition, followed by anxiety and stress [9,18].

The common denominator of all the above works is that they focus on attributes that are either static (their values rarely change), non-self controlled, or both.

Inference of self-controlled attributes has also been extensively studied. However, such works focus on the inference of opinions and attitudes rather than behavioral attributes [4,25]. While a substantial amount of work does study different types of behavioral attributes, their goals are different than ours. Such works study general correlations between network or linguistic features and a given behavior, identify the prevalence of a certain behavior among the general population, or classify social-network textual objects such as tweets or posts. For example, while there exists a considerable amount of work about the use of social networks for monitoring public health, none of those works aims at inferring vaccination intent of a given social-network user at a given point in time. Rather, existing works analyze collective sentiment towards vaccinations [19], track the spread of infectious diseases and monitor online discussions concerning widespread diseases [23], or perform classification of stand-alone social-network objects according to vaccination attitudes of the object's creator [1].

Inference of time-varying, behavioral attributes using public social-network data has therefore been hardly researched, with two exceptions: voting intentions and *online* purchase intentions. There are several key differences between this work and prior ML work on PI. First, the majority of existing works examine general buying preferences rather than time-varying PIs [30]. Other works try to infer PI of stand-alone social-network objects (content-centric) rather than PI of social-network users (user-centric), an approach which is inherently biased [2,10]. The remaining works that do try to infer a user-centric, time-varying PI use data derived solely from E-commerce platforms. Such data is both platform-specific and oftentimes considered private, unlike our use of public social-network data [21]. The closest work to ours is [17] which infers PI of Pinterest users using temporal features and a logistic regression model. However, they only consider online purchases and do not differentiate between different product categories.

3 Methodology

"Intentions are people's decisions to perform particular actions" [24]. In this paper, we aim at understanding to what extent we can infer behavioral intentions of social-network users. In order to do that, we build a BN model that leans on intentions' most influential factors as shown in behavioral psychology literature [7,11,24]. We split those factors into two groups: static factors, such as personality, demographic attributes and self-efficacy, and dynamic factors such as emotions, interest and opinion. A significant challenge, however, is the fact that the values of some of those determinants (*e.g.* personality) can not be directly obtained from the user's social-network profiles ("latent variables"). Therefore, we enrich the model with various observed network features which may assist in both inferring the target intention and inferring its latent determinants.

Though different intentions are influenced by the same high-level factors, their associated BNs still differ in their qualitative, quantitative and temporal specifications. To reflect those differences, we build on our general intention-inference BN and create, for each behavioral intention, an intention-specific DBN. This is achieved using a multistage process: first, we identify the best

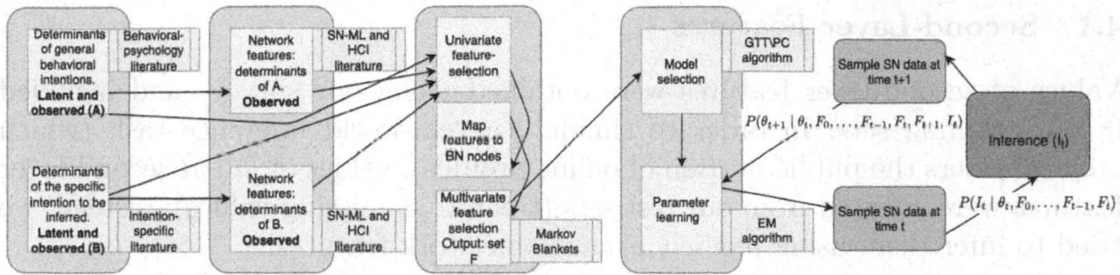

Fig. 1. An illustration of our behavioral-intention-inference methodology

set of determinants of general behavioral intentions using existing behavioral-psychology literature. Second, for each intention, we identify its unique determinants using existing literature which investigates that specific intention. Third, we identify the set of network features that are known to have a strong relation to the set of general and intention-specific determinants described above; the priors used for the third step, collected from existing literature, are not as strong as the priors used for stages 1 and 2 but are still informative—especially those collected from prior attribute-inference works and human-computer interaction (HCI) literature. Fourth, the final feature set of each intention is determined using priors, feature-selection methods, or both. Fifth, the set of selected features is mapped into network nodes; this includes aggregation, state-elicitation and discretization strategy. Sixth, the temporal structure of each intention-specific DBN is specified using priors, structure-learning methods, or both. Lastly, The DBN's temporal parameters are quantified using a combination of prior information and data. The diagram in Fig. 1 illustrates our approach as detailed above.

In Sect. 6 we use the above methodology to infer the values of five dynamic attributes—behavioral intentions, using real-world, social-network datasets. The behavioral intentions that we consider are weight-loss intentions (WI), vaccination intentions (VI), travel-purchase intentions (PI), borrowing intentions (BI) and job-searching intentions (JI).

4 Features

A high-level diagram of our intention-inference model is shown in Fig. 2. Note that Fig. 2 is brought for illustration purposes and thus only features edges between layers; edges between specific nodes must be determined separately for each intention according to its own unique priors, intention-specific features and results of feature selection methods applied to it. To avoid a large conditional probability table (cpt), we used a layering-divorcing technique and created a layered network model: The first layer contains the target intention nodes that we aim at inferring. The second layer contains either latent or partially-observed nodes which represent external and internal factors that are highly influential on the formation of behavioral intentions. The third layer contains observable network features. They serve two purposes: assisting in inferring the behavioral intentions, and serving as observed predictors for second layer's latent variables.

4.1 Second-Layer Features

Values of second-layer features were obtained using our surveys and included in our training sets. In order to simulate a real-world inference task (which only considers the public portion of online profiles), values of *latent* second-layer features were omitted from our test sets (treated as missing values); instead, we tried to infer them using *publicly available* network features.

Personality. This variable represents five broad dimensions of personality obtained from the "Big Five" model of personality dimensions. The big five model distills personality into five traits: neuroticism, extraversion, agreeableness, conscientiousness, and openness to experience. To measure the Big Five personality traits among survey participants we used a short version of the Big Five Inventory based on BFI-10 [22].

Demographic Attributes. We considered the following demographic attributes: age, gender, ethnicity (and country of origin), marital status, occupation group, income (latent variable). Only a subset of those attributes was used in each model.

Situational Variables. Events that might trigger a behavioral intention. Those events include personal-life transitions, professional-life transitions, external events (such as a holiday or an election), *etc.*. Priors were obtained for some intentions. For instance, life transitions were shown to have an important impact on weight-loss intentions [3].

Emotions. Different emotions may serve as either the cause of a behavioral intention or as its effect. Therefore, we went beyond the binary emotion-representation (positive-negative) and also considered fine-grained emotions. The most studied model of discrete emotions is the Ekman model [6] which posits the existence of six basic emotions: anger, disgust, fear, joy, sadness and surprise. Since momentary emotion ratings are not particularly indicative of the behavioral intentions explored in this work, survey participants were presented with eight emotion categories (six basic emotions and two positive-negative emotion categories) and were asked to rate their feelings over the past week/month/three months *in general.*

Interest, Opinion. Those variables represent the user's level of interest and opinion regarding topics related to a given intention.

4.2 Network Features

The value of a given network feature was included in our datasets only if it was part of the public portion of one of the user's social-network profiles.

Numeric Features (NUMERIC). We considered statistics about the user's activity (number of posts, status updates, number of uploaded photos *etc.*), reactions to the user's content (number of tagged photos, for instance) and the user's reactions to others' content. The latter measure was sparse, as both Facebook

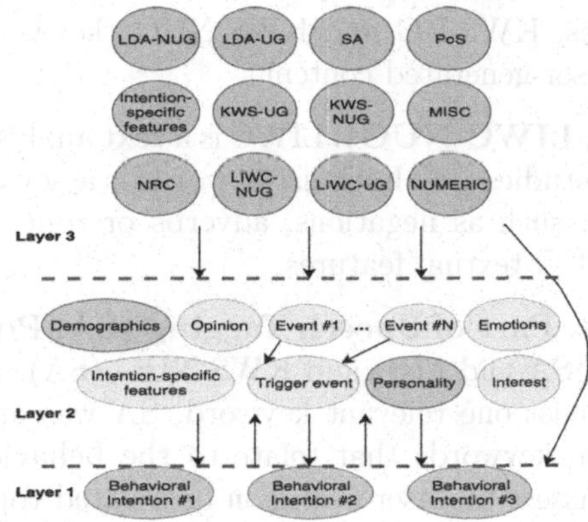

Fig. 2. A diagram of our static network model

and Instagram limit the visibility of such reactions. We also considered basic statistics about the users' network, but we limit ourselves to statistics that are both publicly visible and can be directly extracted from the user's own social-network profile/s (number of friends, followers-following ratio, public-figures-non-public-figures following ratio, *etc.*).

Raw Textual Features (TEXT). Textual features were classified as either user-generated (UG) features (including textual content that was written by the user), or non-user-generated (NUG) features (textual features that were not written by the user such as likes (Facebook) or hashtags (Instagram)). We limit ourselves to textual content that is both publicly visible and was either produced by the target user, or can be directly extracted from the user's own social-network profile/s. Raw textual features were not directly fed to our models. Instead, each textual feature was analyzed using various linguistic methods as described below; Textual-features-related-nodes in our network represent averaged score (frequency) of a given category of a given linguistic feature among the user's raw textual features. Such nodes represent the prevalence of a specific linguistic category among the entire set of raw textual features.

Miscellaneous Features (MISC). Miscellaneous features include features that are neither numeric nor textual, such as the mere existence of various social-network accounts, visibility level/s that the user has chosen to apply to her social-network accounts, profile attributes from which demographic attributes can be extracted, *etc.*. MISC features can be seen as social-network accounts' metadata rather than data itself (NUMERIC, TEXT).

Linguistic Features. We use a broad range of linguistic features, created based on our raw textual features.

Keyword-Search (KWS-UG, KWS-NUG). For a given intention or an event, $\mathcal{A}$, we manually identified the most prominent keywords related to $\mathcal{A}$. We then performed a keyword search on our textual features. This resulted in two groups of features, KWS-UG and KWS-NUG (keyword search applied to user-generated/non-user-generated content).

LIWC (LIWC-UG, LIWC-NUG). LIWC is a text analysis tool that is widely used in psychological studies [28]. Each list of words is associated with a semantic or syntactic category, such as negations, adverbs or tone. LIWC analysis was applied to UG and NUG textual features.

Sentiment Analysis, Part-of-Speech Tagging (SA, PoS). These were only applied to KWS-UG (SA and PoS) and KWS-NUG (SA), *i.e.*, items that were found to contain at least one relevant keyword. SA was applied to items that were found to contain keywords that relate to the behavioral intention to be inferred, in order to assess the user's opinion on related topics. The use of PoS tagging was more implicit; it was applied to items that were found to contain keywords that are related to events in order to assess whether an event is relevant to each inference task (use of first-person writing, tensed verbs *etc.*).

Topic Modeling (LDA-UG, LDA-NUG). Topics were extracted using Latent Dirichlet Allocation (LDA). Shorter features (such as likes) and longer features (such as posts) were considered separately using different parameters.

Emotions (NRC, LIWC). We automatically quantify emotions from our UG textual features using LIWC and NRC. NRC is a publicly available lexicon of words associated with different emotions, as well as general positive and negative sentiment [20]. We assign a predicted emotion to each UG textual feature and then average across all users' features.

5 Temporal Modeling Using Dynamic Bayesian Networks

A Dynamic Bayesian Network is a sequence of T static bayesian networks. Each BN represents a time slice ("slice") of the DBN, $i \in T$, corresponding to one instance of time. A DBN adds three components to a static BN: temporal variables, temporal edges and temporal evidence. For instance, if a static BN contains the variables $\{X_j\}_{j \in D}$, a DBN contains variables that can take different values in different time slices, *e.g.* $\{X_{j,i}\}_{j \in D, i \in T}$, as well as temporal edges between them. Formally, a DBN is defined as a pair (B_0, B_t) where B_0 defines the prior $P(X_1)$ and B_t is a two-slice temporal BN that defines $P(X_i|X_{i-1})$ by means of a directed acyclic graph ($PA(X_{j,i})$ represents $X_{j,i}$'s parents):

$$P(X_i|X_{i-1}) = \prod_{j \in D} P(X_{j,i} \mid PA(X_{j,i})) \tag{1}$$

A DBN-Based Approach to Modeling Social-Network Users: Each social-network user is modeled using a *set of Dynamic Bayesian Networks*. Specifically, let u be a social-network user, and let $|\mathcal{K}| = K$ be the set of u's dynamic attributes we aim at inferring. u is represented by the set $\{(D_u^{X_k}, T(X_k))|k \in [\mathcal{K}]\}$. $D_u^{X_j}$ corresponds to a DBN which aims at inferring an attribute X_j^u, the attribute X_j of a user u, and $D_u^{X_j}[i]$ corresponds to the i'th slice of the DBN. $T(X_j)$ refers to X_j's unique "sampling rate": the rate in which data for each attribute is sampled from $\{\xi_{(u,s)}^{pu}\}_{s \in S_u}$. The sampling rate $T(X_k)$ associated with a DBN $D_u^{X_k}$ should be determined according to the unique attribute to be inferred. For instance, if our target attributes are various behavioral intentions, the sampling rate of each intention's DBN should be determined according to the intention-behavior (IB) interval [24] of intention X_k; the shorter the IB interval of an intention X_k is, the higher $T(X_k)$ should be.

After determining its feature set and its structure, as we describe in the following subsection, each of the user's DBNs can be used to perform temporal inference of its associated attribute at any point in time. For an inference performed at time $i = 0$, before any training data has been collected, inference of $X_{j,0}^u$ will be done solely based on $D_u^{X_j}[0]$ as a "prior network"—where cpts are solely determined according to prior information. At time point $i = t$ s.t. $t > 0$ inference of $X_{j,t}^u$ is done by training a new slice of DBN_j, $D_u^{X_j}[t]$ using an up-to-date set of training records where each record is composed of $t - 1$ sets of historical features $\{F_i | i < t\}$, a set of current features $\{F_t\}$, and a set of historical labels, $\{I_i | i < t\}$.

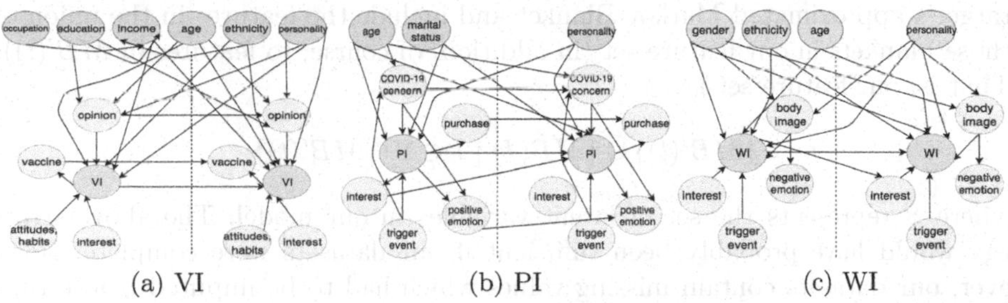

(a) VI (b) PI (c) WI

Fig. 3. A DBN representation of various intentions.

In both cases, inference of attribute $X_{j,i}^u$ will be done using $\{F_t^u\}$, the sampled feature sets of u at time t; $\{F_i^u | i < t\}$, sampled feature sets of the user from prior points in time; and historical labels (if exist), $\{I_i^u | i < t\}$. In addition, when the target attribute to be inferred is a behavioral intention, we can input the inference algorithm with a set of historical *behaviors*, $\{B_i^u | i < t\}$. A behavior at time i may suggest on an associated intention at time $i-1$ or $i-2$ thus allowing us to retroactively update the network's parameters to reflect the new insights.

5.1 Feature Selection and Model Selection

We designed a two-level, hybrid feature-selection method. Due to the high number of correlations between features, we opted for a multivariate feature-selection method based on bayesian networks. However, solely relying on a DBN-based feature selection method may lead to overfitting. Hence, we employed a hybrid feature-selection approach. First, a simple, univariate feature selection method was applied to a subset of the features on which we didn't have strong priors. For that purpose, we used a mutual information-based feature-selection method and removed all the features that received a score below a certain threshold. The resulting features, as well as the set of latent/high-prior features were the input for the second phase which uses two structure-learning algorithms: Greedy Thick-Thinning and the PC algorithm [26]. This phase aimed at identifying the best features using Markov Blankets.

A Markov Blanket of a variable t is a minimal variable subset conditioned on which all other variables are probabilistically independent of t. The Markov Blanket of a DBN node, $MB(t)$ is the set of its parents, $P(t)$; children, $C(t)$; and spouses, $U(t)$ as encoded by the structure of the DBN. As shown in [13], the Markov Blanket of a given target variable is the theoretically optimal set of variables to predict its value. However, simply considering all the features in the Markov Blanket of the behavioral intention node is unsatisfactory in our case, due to the existence of latent variables. Thus, a better strategy would be to first find an "approximated" Markov Blanket of the target node, $MB'(t)$ which includes the variables in the sets $P(t)$, $C(t)$ and $U(t)$ as discussed above. Then, identify the Markov Blanket of *each* latent variable that is also a member of the target's approximated Markov Blanket and include the features in the union of those blankets in our feature set (in addition, of course, to features in $MB'(t)$). That is, our feature set is:

$$\{MB'(t)\} \cup \{MB(I) \mid I \in S \cap MB'(t)\}$$

where S represents the set of latent variables in our model. The above strategy would have probably been sufficient if our datasets were complete. However, our datasets contain missing values which had to be imputed before running the structure-learning algorithm. Hence, for some variables we consider an "extended" notion of a Markov Blanket which also includes certain variables that belong to the variable's second-degree Markov Blanket. Specifically, if a given variable, v, represents an observed attribute with more than 50% missing values ($m()$) and for which we do not have a strong prior ($p()$), we consider a restricted notion of v's second degree Markov Blanket, and add both its direct parents, $P(v)$, and its direct children, $C(v)$, to our feature set. Let F be our variable-set before applying feature selection, and O the set $F \setminus S$. Our final feature set is:

$$\{MB'(t)\} \cup \{MB(I) \mid I \in S \cap MB'(t)\} \cup$$
$$\{P(I) \mid I \in O \cap MB'(t) \wedge m(I) > 50\% \wedge p(I) = false\} \cup$$
$$\{C(I) \mid I \in O \cap MB'(t) \wedge m(I) > 50\% \wedge p(I) = false\}$$

Model Selection and Parameter Learning: The approach described above not only yields a feature set but also a network structure, comprised of the nodes in the feature set and the edges connecting features in the feature set. Some edges were corrected in order to reflect strong prior information. The balance between automatic structure-learning algorithms and the use of priors for structure elicitation, as well as the initial parameters for the structure-learning algorithms (when applicable) were validated using cross-validation. Note that while information gathered from prior literature would have probably been sufficient to model most of the meaningful dependency relations between an intention and second-layer features, relations between third-layer features and other features, as well as between third-layer features and the target intentions can not be captured solely using priors, as those types of relations are not as extensively studied as behavioral intentions-second-layer features relations. Parameter learning was performed using the Expectation-Maximization (EM) algorithm [5]. Hence, we were able to combine both labeled and unlabeled data in our training sets as explained in Sect. 6 as well as use the *original* training datasets which contain missing values. We believed that since our test datasets include a large number of missing values, training the DBN on incomplete datasets will allow the DBN to learn relations between missing and observed values of different features. Prior information was combined in the model using a Dirichlet prior.

5.2 Intention-Specific Models

Figure 3 presents a high-level overview of three intention-specific DBNs (DBNs for JI and BI, as well as third-level features are omitted due to lack of space). As can be seen, a temporal link is created between variables that represent our target intentions in consecutive time slices. $P(intention_{i+1} \mid intention_i, U)$ represents the intention's evolution over time, given changes in other temporal variables in the network (U).

Interest-intention is an interesting relation. First, we see that interest may serve as either a cause or an effect of different intentions. Second, interest seems to be a cyclic process as can be concluded from $P(WI_i \mid interest_i, U)$ and $P(interest_{i+1} \mid WI_i)$, for example. Such a temporal relation might be attributed to the fact that interest in a certain topic assists in forming a behavioral intention related to that topic. After the intention has been formed, a new level of interest is formed, aimed at understanding how to fulfill that intention. In addition, $P(PI_{i+1} \mid interest_i, U)$ and $P(interest_{i+1} \mid PI_{i+1})$ show that both prior interest-level and current interest-level are important determinants of some intentions. Such historical data can assist in identifying a sudden increase in the user's interest level.

"Opinion" is another interesting variable; it is influenced by multiple factors such as personality traits and demographics as demonstrated by VI's $P(opinion_{i+1} \mid opinion_i, education, age, personality)$. Note that this cpt also contains $opinion_i$. This represents the fact that oftentimes, opinion is a self-propelling process: opinion at a given point in time, in addition to other factors, influences opinion at future points in time. A similar cpt is seen in "COVID-19 concern".

Fine-grained emotions were not used in any model. Furthermore, we weren't able to extract from the data meaningful inter-slice relations between different emotions and the target intentions. We attribute that difficulty to the fact that unlike other features, emotions change quickly. Thus understanding emotions' temporal evolvement mechanism for each intention requires the use of finer-grained sampling rates

In Sect. 7 we show our inference results (Lauritzen-Spiegelhalter algorithm [16]) when using a two-slice DBN and social-network data sampled twice.

Table 1. Datasets' statistics

Intention	VI	WI	BI	PI	JI
% Intending, first-wave dataset	.58	38	.17	.24	.19
% Intending, second-wave dataset	.66	.4	.13	.19	.23

6 Data Collection

We designed and distributed a comprehensive survey, created and hosted using Qualtrics survey platform. The first part of our survey contained questions about the participants' personal attributes, as discussed in Sect. 4. The second part contained the following statements, which users were asked to rank (as well as dummy statements about unrelated intentions): "I am planning to start a weight-loss regime within the next 1–4 weeks" and "I am currently trying to lose weight" (weight-loss intentions); "I am planning to look for a new job within the next 1–4 weeks" and "I am currently looking for a new job" (job-searching intentions); "I am planning to apply for a loan within the next 1–4 weeks" (borrowing intentions); "I received a flu vaccine this season"—depending on the answer to that statement the following statement was presented for either the upcoming (2020–2021) flu season or the next season (2021–2022): "I am planning to get vaccinated against influenza this upcoming fall-winter/next year" (vaccination intentions); "I am planning to make a travel-related purchase within the next 1–4 weeks" (travel-purchase intentions). All survey data was anonymized after collection. We informed participants that their responses would be used for academic research. We implemented several methods for identifying and excluding data from participants who answered unreliably, as extensively discussed in [12].

Datasets. Survey data was collected in two waves with a three-month lag. Training and test datasets include data obtained from Amazon Mechanical Turk (MTurk), Facebook, Instagram and Linkedin. Our datasets include both labeled and unlabeled data; unlabeled data is specifically important when using multi-wave data, as a considerable number of participants dropped out after the first wave: from 1300 respondents who participated in our first-wave survey, only 803 respondents participated in our second-wave survey (0.617 response rate). In order to both reduce non-response bias and create a bigger training dataset, we chose a subset of our partially-labeled data records which belong to participants who dropped out (missing attributes were treated as missing values) and added it to our training set. Our training datasets, $\mathcal{D}_j^1$ (first-wave data for intention j) and $\mathcal{D}_j^2$ (second-wave data for intention j) consist of 780 and 592 labeled and unlabeled data records, respectively. Our test datasets, $\mathcal{D}_j^3$ (first-wave data for intention j) and $\mathcal{D}_j^4$ (second-wave data for intention j) consist of 520 and 361 labeled data records, respectively.

Table 2. Results of the DBN models presented in this paper

Intention	Micro F1, (1)	Macro F1, (1)	Micro F1, (2)	Macro F1, (2)
Vaccination intentions	.732	.73	.75	.741
Weight-loss intentions	.832	.815	.831	.82
Borrowing intentions	.663	.54	.662	.526
Travel-purchase intentions	.763	.691	.812	.732
Job-searching intentions	.704	.623	.747	.699

7 Experimental Results

For a given intention, j, we tested its DBN, DBN_j, using our datasets as follows: in the first stage, (1), we trained DBN_j using $\mathcal{D}_j^1$ and tested it on $\mathcal{D}_j^3$. Only the first DBN's slice was affected in this stage. In the second stage, (2), we trained DBN_j using $\mathcal{D}_j^2$ (implicitly using $\mathcal{D}_j^1$ as well due to the use of priors) and tested it on $\mathcal{D}_j^4$, using evidence data from $\mathcal{D}_j^3$ as well. Hence, inference results in (2) were obtained based on data and parameters from two slices of the DBN. Note that "evidence data" contains only historical values of publicly available features, and *does not include historical labels* as in most cases exact information on historical labels for all prior test sets will not be available in real time. As seen in Table 1 (% intending), some of our datasets are *highly imbalanced*. Moreover, each dataset contains a large number of missing values—attributes that the user has not publicly revealed on her social-network accounts. Those facts make the inference task highly challenging.

Table 2 provides a detailed summary of our results. We report Micro F1 and Macro F1 scores for each intention-specific DBN. In addition, we compare our ROC AUC scores to those achieved by a Support Vector Machine (SVM) and a Decision-Tree Ensemble (boosted decision trees, BDT).

As can be seen from Table 2, different intentions achieved significantly different Micro F1 and Macro F1 scores. BI's score is the lowest, whereas WI's score is the highest. A possible explanation for BI's performance is that applying for a loan is an intention that is oftentimes not publicly shared on social networks. However, other non-publicly shared intentions such as JI scored significantly better than BI. This can be attributed to the fact that we were able to find other strong predictors for JI which don't depend on user-generated content, whereas for BI we failed to do so.

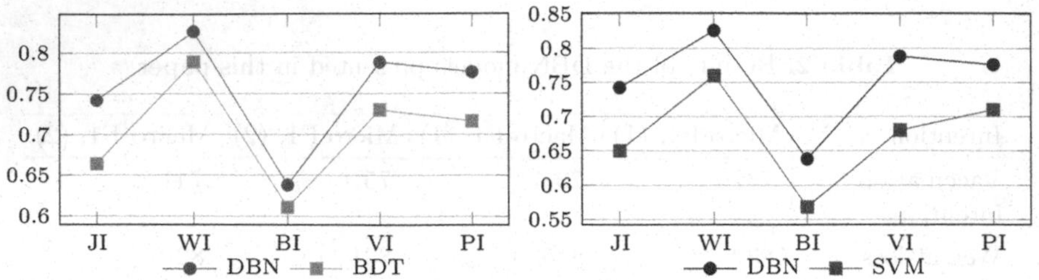

Fig. 4. Average ROC AUC scores

Figure 4 compares our ROC AUC scores to those achieved by two different types of classifiers: BDT and SVM (RBF kernel). Hyperparameters were tuned using a grid search over a large grid covering at least 7 options for each numeric hyperparameter. Imputation of missing values was done using Scikit-Learn's IterativeImputer (using a random-forest regressor), a multivariate imputation method. As seen in Fig. 4, our models outperform both SVM and BDT on all five intentions, though the differences in results vary between intentions. A possible explanation is the varying number of missing values within the unique set of features of each intention, or the varying number of latent variables in each DBN. Another possible explanation is the varying number of temporal dependencies *between* features of each target intention.

When comparing Micro F1 and Macro F1 scores achieved in different stages ((1) and (2)) using the same DBN, we can see that the differences are more pronounced for PI and JI. This can be attributed to the underlying differences between different intentions. As evidenced by our data, intentions such as WI and VI can be seen as "continuous intentions" in the sense that their intention-behavior interval is longer than for other intentions; the persistence rate of such intentions is significantly higher than rates reported for PI or JI. Another explanation for the varying differences is the different set of determinants of each intention. While the importance of some of those determinants stems from their intra-slice values (that is, their values at a given point in time), the importance of others is derived from a combination of intra-slice values *and* inter-slice change patterns between slices. For instance, various features related to *non* user-generated content serve as excellent predictors of PI in (2), but only as solid predictors in (1).

Acknowledgments. The author was supported by US National Science Foundation (CNS-1407454 and CNS-1409599) and William and Flora Hewlett Foundation (2016-3834).

References

1. Aramaki, E., Maskawa, S., Morita, M.: Twitter catches the flu: detecting influenza epidemics using twitter. In: EMNLP 2011 (2011)
2. Atouati, S., Lu, X., Sozio, M.: Negative purchase intent identification in twitter. In: WWW 2020 (2020)
3. Brink, P.J., Ferguson, K.: The decision to lose weight. West. J. Nurs. Res. **20**(1), 84–102 (1998)
4. Conover, M., et al.: Predicting the political alignment of twitter users. In: SocialCom 2011 (2011)
5. Dempster, A., Rubin, D.: Maximum likelihood from incomplete data via the EM algorithm. J. Roy. Stat. Soc. **39**(1), 1–22 (1977)
6. Ekman, P.: An argument for basic emotions. Cogn. Emot. **6**(3–4), 169–200 (1992)
7. Fishbein, M.: The role of theory in HIV prevention. AIDS Care **12**(3), 273–278 (2000)
8. Golbeck, J., et al.: Predicting personality with social media. In: CHI (2011)
9. Guntuku, S.C., et al.: Understanding and measuring psychological stress using social media. In: ICWSM 2019 (2019)
10. Gupta, V., et al.: Identifying purchase intent from social posts. In: ICWSM (2014)
11. Hampshire, S., Hart, H.: Decision, intention and certainty. Mind **67**, 1–12 (1958)
12. Idan, L., Feigenbaum, J.: Show me your friends, and I will tell you whom you vote for: predicting voting behavior in social networks. In: ASONAM 2019 (2019)
13. Koller, D., Sahami, M.: Toward optimal feature selection. In: ICML 1996 (1996)
14. Kulshrestha, J., et al.: Web routineness and limits of predictability: investigating demographic differences using web tracking data. In: ICWSM 2021 (2021)
15. Lampos, V., Aletras, N., Geyti, J.K., Zou, B.: Inferring the socioeconomic status of social media users based on behaviour and language. In: ECIR 2016 (2016)
16. Lauritzen, S., Spiegelhalter, D.: Local computations with probabilities on graphical structures and their application to expert systems. J. Roy. Stat. Soc. **50**, 157–194 (1988)
17. Lo, C., Frankowski, D., Leskovec, J.: Understanding behaviors that lead to purchasing: a case study of pinterest. In: KDD 2016 (2016)
18. Mann, P., et al.: See and read: detecting depression symptoms in higher education students using multimodal social media data. In: ICWSM 2020 (2020)
19. Mitra, T., Counts, S., Pennebaker, J.W.: Understanding anti-vaccination attitudes in social media. In: ICWSM 2016 (2016)
20. Mohammad, S.M., Turney, P.D.: Crowdsourcing a word-emotion association lexicon. Comput. Intell. **29**, 436–465 (2013)
21. Mokryn, O., et al.: Will this session end with a purchase? Inferring purchase intent of anonymous visitors. Electron. Commer. Res. Appl. **34**, 100836 (2019)
22. Rammstedt, B., John, O.: Measuring personality in one minute: a 10-item version of the big five inventory. J. Res. Pers. **41**, 203–212 (2007)
23. Šćepanović, S., et al.: The healthy states of America: creating a health taxonomy with social media. In: ICWSM 2021 (2021)
24. Sheeran, P.: Intention–behavior relations: a conceptual and empirical review. Eur. Rev. Soc. Psychol. **12**, 1–36 (2002)

25. Silva, A., et al.: On predicting personal values of social media users using community-specific features and personal value correlation. In: ICWSM (2021)
26. Spirtes, P., et al.: Causation, prediction, and search. Lecture Notes in Statistics (1993)
27. Staiano, J., et al.: Friends don't lie: inferring personality traits from social network structure. In: ACM Conference on Ubiquitous Computing (2012)
28. Tausczik, Y., Pennebaker, J.W.: The psychological meaning of words: LIWC and computerized text analysis methods. J. Lang. Soc. Psychol. **29**(1), 24–54 (2010)
29. Youyou, W., et al.: Computer-based personality judgments are more accurate than those made by humans. PNAS **112**(4), 1036–1040 (2015)
30. Zhang, Y., Pennacchiotti, M.: Predicting purchase behaviors from social media. In: WWW 2010 (2010)

Fuzzy Logic Framework for Ontology Instance Alignment

Bogumiła Hnatkowska, Adrianna Kozierkiewicz(✉), and Marcin Pietranik

Faculty of Information and Communication Technology, Wroclaw University of Science and Technology, Wybrzeze Wyspianskiego 27, 50-370 Wroclaw, Poland
{bogumila.hnatkowska,adrianna.kozierkiewicz,marcin.pietranik}@pwr.edu.pl

Abstract. The widely addressed topic of ontology alignment to this day contains several open research questions that remain either unanswered or only vaguely tackled. One of them is designating alignments of concept instances, which according to the literature are addressed in a handful of publications. Therefore, in this paper we propose a formal framework based on fuzzy logic that can be used to determine such mappings. We provide several similarity functions and a set of inference rules for combining them. The approach has been experimentally verified using widely accepted datasets provided by the Ontology Alignment Evaluation Initiative, yielding promising results.

Keywords: Ontology alignment · Instance mappings · Fuzzy logic · Knowledge management

1 Introduction

Ontology alignment, a frequently researched topic, is the task of asserting a sound communication between separate computer systems which utilize different ontologies as their knowledge bases. To deliver such means of communication, a specific ontology alignment tool needs to select which elements from two ontologies refer to the same (or sufficiently similar) elements of the real world. When such selection is done, the final ontology alignment is a set of pairs of such elements along with a confidence degree to which these elements can be aligned and a relationship that holds between them.

Many different ontology alignment tools are based on computing a variety of different similarity measures between ontology elements, which are eventually combined into a single value used to judge whether or not the considered pair of elements can form a valid mapping. However, a plethora of the aforementioned ontology alignment tools focus mainly on designating alignments of concepts (also referred to as TBox alignment), while two remaining levels of ontology abstractions, relations, and instances, are frequently treated neglectfully [4,16]. There is very little research available in the literature that addresses these issues.

D. Groen et al. (Eds.): ICCS 2022, LNCS 13351, pp. 653–666, 2022.
https://doi.org/10.1007/978-3-031-08754-7_68

In our previous publication [11] we addressed the level of relations. The main idea came from analyzing earlier approaches to the task, which (as aforementioned) we based on calculating different similarity measures. A naive approach to combining them would include calculating their average value. However, we wanted to include another layer of experts' knowledge concerning ontology alignment in the process. We achieved this goal by successfully applying the fuzzy logic to the task, in the form of fuzzy inference rules. This approach has been proved useful, rendering good results obtained from experimental verification we conducted using a state of the art datasets provided by the Ontology Alignment Evaluation Initiative [16]. The obtained results became a straight inspiration for the next research.

The following paper is devoted to the level of instances. This level can be treated as the actual reflection of the objects taken from the real world, expressed using concepts definitions. In other words, while the level concepts contain abstract descriptions of the world (e.g. *a Person* or *a Book*), the instance level contains materializations of the real objects (e.g. stating that "John" is *a Person*). Therefore, the level of instances expresses the real knowledge about the assumed universe of discourse, and not generic properties.

Thus, the main contribution of the following paper is twofold. Firstly, we provide a set of functions that can be used to calculate a similarity between two instances from independent ontologies. Secondly, we formulated a set of fuzzy inference rules that could be used to reason about how close two instances describe the same elements from the real world.

The article is structured as follows. In the next section, an overview of the related research done in the field is given. Section 3 contains basic notions used throughout the paper, while Sect. 4 describes our approach to ontology alignments on the instance level. Section 5 is split into two subsections - the first walk the reader through the experimental procedure we designed to verify our framework. The second contains the experimental results gathered during the process. The last section is a summary and a brief overview of our upcoming research plans.

2 Related Works

The increasing number of methods available for schema or ontology matching mandate consensus for evaluation of these methods. The Ontology Alignment Evaluation Initiative (OAEI) is a coordinated international initiative to forge this consensus OAEI [16]. Since 2004, OAEI organises evaluation campaigns aiming at evaluating ontology matching technologies. Organizers provide benchmark datasets consisting of a set of pairs of ontologies with their corresponding alignment, which is supposed to be treated as the correct one. Most of these datasets are devoted to schema matching, while instance matching is treated more neglectfully. Nevertheless, it is possible to find datasets such as: IIMB, Sabine, Doremus, SPIMBENCH, Sandbox [7]. In our work we especially focus on IIMB dataset because it contains 80 ontologies. Each ontology has been created by systematically applying a set of transformations to the reference ontology such as: data value transformation, data structure transformation and data semantics transformation. IIMB dataset has been used in OAEI campaigns held in 2009, 2010, 2011, 2012 and 2018.

In the last years the interest in participating in the OAEI instance matching competition has not been prominent. Among the systems which participated the one which stand out are out LogMap [14], AML [8], Codi [12], SBUEI [15] and semsim. Performance of those systems have been verified in 2011–2018 years based on IIMB datasets.

AML (AgreementMakerLight) [8] approach to instance matching is build on Data Property values of individuals and the relations between individuals. In the newer version, the AML added to its instance matching arsenal the same lexical-based strategy it was already using for class and property matching based on ontologies annotations. However, the efficiency of the system has not been satisfying because the authors were unable to properly configure this matching strategy and ensure its efficiency.

The better results were achieved by LogMap [14]. LogMap since 2011 has evolved from a logic-based tool to an advanced system that applies additional features like lexical indexation, propositional horn reasoning, axiom tracking, local repair, and semantic indexation. LogMap performance is very high, in particular for schema matching tasks. However, its performance for instance mapping leaves space for improvement, especially in the case of the Recall measure.

CODI (CombinatorialOptimization for Data Integration) [12] uses terminological structure for ontology matching. The current implementation produces mappings between concepts, properties, and individuals. The system combines lexical similarity measures with schema information. Authors assume that they have one common TBox and two different ABoxes, where both TBoxes have been integrated beforehand. The efficiency of CODI is not high, because in benchmark ontologies some individuals (instances) are not assigned to any concepts, hence no TBoxes are available.

SBUEI addresses two issues - instance matching and schema matching [15]. It utilizes schema matching results in instance matching tasks in order to track direct matching on the schema level. Similar to CODI, SBUEI is not efficient if the instances are not associated with any concepts.

To the best of our knowledge there is no formal information about semsim system. The system participated in OAEI 2018 competition, however no documentation has been provided.

The multitude of alignment systems forces the improvement of their effectiveness by applying different techniques. The most popular are string- and language-based methods (i.e. LogMap, RIMOM, FALCON, SAMBO, AML, etc.) [1,2]. Some systems incorporate external sources like WordNet (i.e. LogMap, YAM, SAMBO, etc.). Others, like i.e. SEMINT, ProbaMap, LSD, MoTo apply the newest achievements from the machine learning and artificial intelligence fields. It is possible to find an application of Naive Bayes classifiers, neural networks,

SVM, or clustering techniques for correspondence determination. To the best of our knowledge, the fuzzy logic-based approach has not been widely investigated.

The papers [6,9] present a fuzzy-based approach of concept alignment. However, both works are in the preliminary stage and focus only on the concept level of the ontology. Our previous works [11] partially fills this gap. We incorporated fuzzy rules for designating ontology alignments on the relation level. We claim that this is the very first research that shows the usefulness of the fuzzy logic-based framework for ontology instance alignment.

3 Basic Notions

Before presenting our fuzzy based approach to instance alignment we will introduce some basic notions important to understand our ideas. Our ontology model is defined as a quintuple. Let *(A, V)* be a pair, where A is a set of attributes describing objects and V is a set of valuations of such attributes (their domains) such that $V = \bigcup_{a \in A} V_a$, where V_a is a domain of a particular attribute. The (A, V)-based ontology is represented as follow:

$$O = (C, H, R^C, I, R^I) \tag{1}$$

where:

- C is a finite set of concepts,
- H is a concepts' hierarchy, that may be treated as a distinguished relation between concepts,
- R^C is a finite set of binary relations between concepts $R^C = \{r_1^C, r_2^C, ..., r_n^C\}$, $n \in N$, such that every $r_i^C \in R^C$ $(i \in [1, n])$ is a subset of a cartesian product, $r_i^C \subset C \times C$,
- I denotes a finite set of instances' identifiers,
- $R^I = \{r_1^I, r_2^I, ..., r_n^I\}$ is used to denote a finite set of binary relations between concepts' instances.

In our previous works like [10] or [11] we describe in details the ontology model and their components. In this work, we will focus only on concepts and instances level.

Instances are understood as a specific materialisation of concepts. Instances can not exists without belonging to concepts, Thus, firstly, we will introduce the concepts level of an ontology:

$$c = (id^c, A^c, I^c) \tag{2}$$

where:

- id^c is an identifier of the concept c,
- A^c is a set of its attributes,
- I^c is a set of concepts' c instances.

By $a \in c$ we denote the fact, that the attribute a belongs to the concept's c set of attributes A^c. The limitations of many ontology model is the lack of information about attributes semantic. For example, the same attribute *address* may carry different meanings while included in the *Home* concept and completely different when incorporated in the *Personal Website* concept. Our ontology model is based on a notion of attributes' semantics, which gives explicit meanings to attributes when they are included in different concepts. Thus, we need to define a sub-language of the sentence calculus denotes as L_S^A. The set L_S^A consists of an atomic description of attributes from the set D_A and logical operators of conjunction, disjunction, and negation. A partial function:

$$S_A : A \times C \to L_S^A \tag{3}$$

allows us to assign a logical sentence from L_S^A to attributes within a specific concept. The context of concept c is defined as a conjunction of semantics of each of its attributes: $ctx(c) = S_A(a_1, c) \wedge S_A(a_2, c) \wedge ... \wedge S_A(a_n, c)$.

The function S_A allows us to formally define relation: *equivalency* (denoted by $\equiv$), *generalization* (denoted by $\leftarrow$) and *contradiction* (denoted by $\sim$) between attributes:

- Two attributes $a \in A^{c_1}, b \in A^{c_2}$ are semantically equivalent $a \equiv b$ if the formula $S_A(a, c_1) \Leftrightarrow S_A(b, c_2)$ is a tautology for any two $c_1 \in C_1, c_2 \in C_2$.
- The attribute $a \in A^{c_1}$ in concept $c_1 \in C_1$ is more general than the attribute $b \in A^{c_2}$ in concept $c_2 \in C_2$ (denoted by $a \leftarrow b$) if the formula $S_A(b, c_2) \Rightarrow S_A(a, c_1)$ is a tautology for any two $c_1 \in C_1, c_2 \in C_2$.
- Two attributes $a \in A^{c_1}, b \in A^{c_2}$ are in semantical contradiction $a \sim b$ if the formula $\neq (S_A(a, c_1) \wedge S_A(b, c_2))$ is a tautology for any two $c_1 \in C_1, c_2 \in C_2$.

The paper is devoted to designating an alignment between two ontologies on the level of instances. Thus, we broadly describe this ontology level. For given a concept c, its instances from the set I^c are defined as a tuple:

$$i = (id^i, v_c^i) \tag{4}$$

where

- id^i is an instance identifier,
- v_c^i is a function with a signature:

$$v_c^i : A^c \to 2^V \tag{5}$$

According to the equation above, the valuation of a particular attribute within an instance can be multivalued. In other words, it can be represented as a set with repetitions of atomic values taken from the domain. Such approach is cohesive with variety of ontology representation formats (e.g. OWL).

For simplicity, we write $i \in c$ which can be understood that the instance i belongs to the concept c. By $I = \bigcup\limits_{c \in C} \{id^i | (id^i, v_c^i) \in I^c\}$ we denote the set of all instances' identifiers. In the further part of this paper we used two auxiliary functions. First of them, returns a set containing identifiers of instances assigned to a given concept c: $Ins(c) = \{id^i | (id^i, v_c^i) \in I^c\}$. The second one, gives a set of concepts to which an instance with some identifier belongs: $Ins^{-1}(i) = \{c | c \in C \wedge i \in c\}$.

To simplify a notation by $i \in r^I$ we will denote a situation in which an instance i participates in a relation $r^I \in R^I$ with some other unspecified instance. Formally, we can define this as $i \in r^I \implies \exists_{i' \in I} : (i, i') \in r^I \vee (i', i) \in r^I$. Additionally, we introduce a helper function rng used during processing of instances and their relations. It is used to designate sets of instances with which some instance is connected through certain relation:

$$rng(i, r^I) = \{i' \in I | (i, i') \in r^I\} \tag{6}$$

Assuming the existences of two ontologies, the integration of them is possible only in case of existing alignment. Such alignment, can allows us to "translate" content of one (source) ontology to the content of some other ontology (target). Formally speaking, between two independent (A, V)-based ontologies $O_1 = (C_1, H_1, R^{C_1}, I_1, R^{I_1})$ and $O_2 = (C_2, H_2, R^{C_2}, I_2, R^{I_2})$ there exist a set of correspondences, called alignment, defined in the following way:

$$Align(O_1, O_2) = \{Align_C(O_1, O_2), Align_I(O_1, O_2), Align_R(O_1, O_2)\} \tag{7}$$

The main aim of this work is determination of $Align_I(O_1, O_2)$. Due the fact, that instances are tightly connected with concepts, two instances can be mapped if belong to the mapped concepts. Thus, to determine an instance alignment we need as an input a concepts alignment $Align_C(O_1, O_2)$ defined in the following way:

$$Align_C(O_1, O_2) = \{(c_1, c_2, \lambda_C(c_1, c_2)) | c_1 \in C_1 \wedge c_2 \in C_2 \wedge \lambda_C(c_1, c_2) \geq T_C\} \tag{8}$$

where:

- c_1, c_2 are concepts from O_1 and O_2 respectively,
- $\lambda_C(c_1, c_2)$ is a value of a degree to which concept c_1 can be aligned into the concept c_2, a vast majority of alignments between two ontologies include only mappings of concepts that are equivalent with 100% certainty. The value of $\lambda_C(c_1, c_2)$ can be calculated in different way i.e. like in our previous works [11] or taken directly from other alignment systems.
- T_C represents an assumed threshold

The ontology alignment on the instance level is a set of sets of alignments of instances belonging to two already aligned concepts and can be formulated as:

$$Align_I(O_1, O_2) = \{(i_1, i_2) | i_1 \in I^{C_1} \wedge i_2 \in I^{C_2}\} \tag{9}$$

We consider only relation pairs that have been processed and eventually selected by fuzzy-based alignment algorithm described in the further sections.

4 Fuzzy Based Approach to Instance Alignment

The main aim of our work is to determine the mappings between two ontologies on the instance level. It is performed with the use of a fuzzy system, which has one output variable connection (*con*) with two possible values: *independent* and *equivalent*. We distinguish seven input variables presented on Table 2 and eight inference rules to decide if two instances represent the same (equivalent) or different (independent) phenomenon. The rules are presented in Table 1.

In our fuzzy framework, we use the Mamdani type rule inference, the centroid of gravity method to defuzzify the output variable, and the maximum operator to accumulate the activated terms.

Table 1. Fuzzy inference rules

ID	Rule
1	IF PRSIM IS very high AND CPR IS high AND CCO IS high THEN con IS equivalent
2	IF PRSIM IS very high AND CCO IS high AND (PARTSIM IS high OR MAXSIM IS high) THEN con IS equivalent
3	IF PRSIM IS high AND CCO IS high AND MAXSIM IS high THEN con IS equivalent
4	IF MAXSIM IS high AND ASIM IS high THEN con is equivalent
5	IF ASIM IS high AND RSIM IS high TIIEN con is equivalent
6	IF CCO IS high AND (MAXSIM IS high AND ASIM IS high) THEN con is equivalent
7	IF PRSIM IS medium AND (NOT MAXSIM IS high AND NOT CPR IS high) THEN con IS independent
8	IF PRSIM IS medium AND (NOT CCO IS high AND (NOT RSIM IS high OR NOT ASIM IS high)) THEN con IS independent

Property similarity ($PRSIM$) is the most important input variable. How it is calculated is shown below. The definition is long because it requires the provision of auxiliary elements. The computation of the other input variables either relies on $PRSIM$ or is easily explained in natural language. The details are shown in Table 2.

The function λ_{value} takes as an input two sets (x and y) of elementary values (e.g. strings) and uses a similarity function *sim_function* for comparing atomic values. It calculates an overall similarity between the given sets:

$$\lambda_{value}(x,y) = \frac{\sum_{v \in x} \max_{v' \in y} sim\_function(v,v') + \sum_{v \in y} \max_{v' \in x} sim\_function(v,v')}{|x| + |y|} \tag{10}$$

The *sim_function* can be any arbitrary given similarity function for particular datatypes. Since all of the values found in ontologies, during the experiment described in the next section, were cast on text type, in further parts of the article the widely known Longest Common Subsequence similarity is used as the *sim_function*.

By *to_value_representation* we denote a function which takes as an input an instance and converts it in a set of values of its attributes. If an instance does not contain any attributes, then the function returns a single-element set containing this instance identifier:

$$to\_value\_representation(i) = \begin{cases} \bigcup_{c \in Ins^{-1}(i)} \bigcup_{a \in c} \{v_c^i(a)\}, & \text{if } \exists c \in Ins^{-1}(i) : A^c \neq \phi \\ \{id^i\} & \text{, otherwise} \end{cases} \tag{11}$$

We introduce an auxiliary set A_{AR} containing four kinds of explicitly given alignments of two ontologies O_1 and O_2. Its elements represent different types of connections established between two ontologies on a level of concepts and relations, thus they include attribute-attribute mappings, attribute-relation mappings, relation-attribute mappings and relation-relation mappings. The set A_{AR} is formally defined below:

$$\begin{aligned} A_{AR}(O_1,O_2) = & \{(a_1, r^{C_2}) | a_1 \in A, r^{C_2} \in R^{C_2}\} \cup \{(r^{C_1}, a_2) | a_2 \in A, r^{C_1} \in R^{C_1}\} \cup \\ & \{(a_1,a_2) | a_1, a_2 \in A\} \cup \{(r^{C_1}, r^{C_2}) | r^{C_1} \in R^{C_1}, r^{C_2} \in R^{C_2}\} \end{aligned} \tag{12}$$

In order to process two instances in the context of the alignment of their mutual characterising properties (attributes and relations) we use a helper function which for given two instances i_1 and i_2 from two ontologies O_1 and O_2 returns a subset of $A_{AR}(O_1,O_2)$ with alignments somehow connected with the given instances $a \in \bigcup_{c \in Ins^{-1}(i_1)} A^c$.

$$\tilde{A}_{AR}(i_1, i_2, O_1, O_2) = \{(a_1, r^{C_2}) | (a_1, r^{C_2}) \in A_{AR}(O_1, O_2), a_1 \in \bigcup_{c_1 \in Ins^{-1}(i_1)} A^{c_1}, i_2 \in r^{C_2}\} \cup \{(r^{C_1}, a_2) | (r^{C_1}, a_2) \in A_{AR}(O_1, O_2), a_2 \in \bigcup_{c_2 \in Ins^{-1}(i_2)} A^{c_2}, i_1 \in r^{C_1}\} \cup \{(a_1, a_2) | (a_1, a_2) \in A_{AR}(O_1, O_2), a_1 \in \bigcup_{c_1 \in Ins^{-1}(i_1)} A^{c_1}, a_2 \in \bigcup_{c_2 \in Ins^{-1}(i_2)} A^{c_2}\} \cup \{(r_1, r_2) | (r_1, r_2) \in A_{AR}(O_1, O_2), i_1 \in r_1, i_2 \in r_2\}\} \quad (13)$$

Final version of the property relatedness ($PRSIM$) calculation is presented on Algorithm 1.

Algorithm 1: Calculate property relatedness

Input : $i_1 \in I_1$, $i_2 \in I_2, O_1, O_2$
Output: $PRSIM(i_1, i_2) \in [0, 1]$

$partial := 0$
foreach $(e_1, e_2) \in \tilde{A}_{AR}(i_1, i_2, O_1, O_2)$ **do**
    foreach $(c_1, c_2) \in Ins^{-1}(i_1) \times Ins^{-1}(i_2)$ **do**
        if $e_1 \in A \wedge e_2 \in A$ **then**
            $v = v^i_{c_1}(e_1)$
            $v' = v^i_{c_2}(e_2)$
        end
        else if $e_1 \in R^{C_1} \wedge e_2 \in R^{C_2}$ **then**
            $v = \bigcup_{i \in rng(i_1, e_1)} to\_value\_representation(i)$
            $v' = \bigcup_{i' \in rng(i_2, e_2)} to\_value\_representation(i')$
        end
        else if $e_1 \in A \wedge e_2 \in R^{C_2}$ **then**
            $v = v^i_{c_1}(e_1)$
            $v' = \bigcup_{i' \in rng(i_2, e_2)} to\_value\_representation(i')$
        end
        else if $e_1 \in R^{C_1} \wedge e_2 \in A$ **then**
            $v = \bigcup_{i \in rng(i_1, e_1)} to\_value\_representation(i)$
            $v' = v^i_{c_2}(e_2)$
        end
        $partial = partial + \lambda_{value}(v, v')$
    end
end
if $|\tilde{A}_{AR}(i_1, i_2, O_1, O_2)| > 0$ **then**
    return $\frac{partial}{|\tilde{A}_{AR}(i_1, i_2, O_1, O_2)|}$
end
else
    return 0
end

Table 2. Input variables with calculation description

Variable	Comment	Calculation description
PRSIM	Similarity of properties (attributes and relations)	PRSIM (Algorithm 1)
MAXSIM	Max similarity of properties	Ratio of mapped properties defined in two instances i_1 and i_2 with maximal (1.0) similarity to the number of all their properties
PARTSIM	Partial similarity of attributes	Ratio of mapped attributes defined in two instances i_1 and i_2 with maximal (1.0) similarity to the number of all their attributes
ASIM	Similarity of attributes	PRSIM limited to attributes
RSIM	Similarity of relations	PRSIM limitted to relations
CPR	Consistency of properties	$p * \lambda_c(c_1, c_2)$ where p – Ratio of mapped properties defined in two instances i_1 and i_2 to the number of all of their properties for such $c_1 \in Ins^{-1}(i_1)$ and $c_2 \in Ins^{-1}(i_2)$ with the highest $\lambda_c(c_1, c_2)$
CCO	Consistency of concepts	Ratio of mapped concepts two instances i_1 and i_2 belong to calculated as: $\frac{\lvert(Ins^{-1}(i_1) \times Ins^{-1}(i_2)) \cap Align_C(O_1, O_2)\rvert}{\lvert Ins^{-1}(i_1) \times Ins^{-1}(i_2)\rvert}$

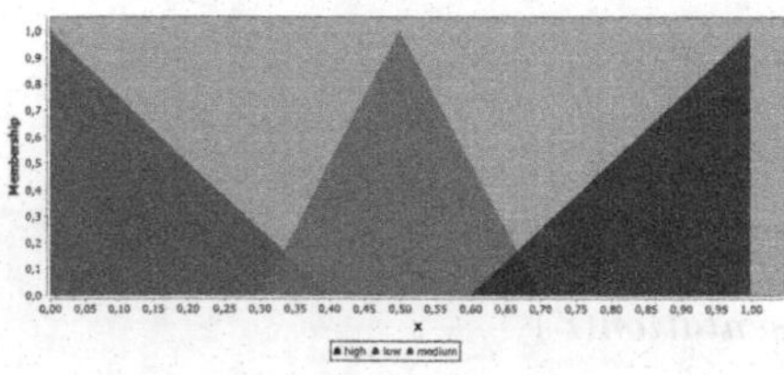

Fig. 1. ASIM fuzzy variable

Fig. 2. CCO fuzzy variable

5 Experimental Verification

Our fuzzy based approach to instance alignment, described in Sect. 4, has been implemented and verified against a widely accepted and incorporated benchmark dataset provided by the Ontology Alignment Evaluation Initiative (OAEI), aforementioned in Sect. 2. We have developed a special dedicated java tool that is able to parse the ontologies expressed in OWL and mappings between concepts expressed in RDF. Our tool (which is online available [18]) uses the jfuzzylogic library [5] to handle fuzzy logic computations.

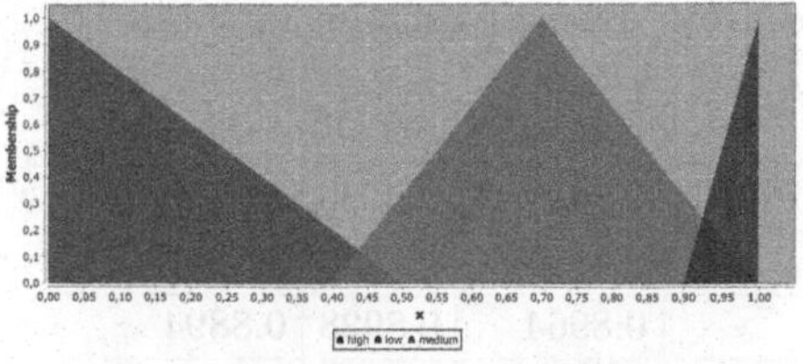

Fig. 3. CPR fuzzy variable

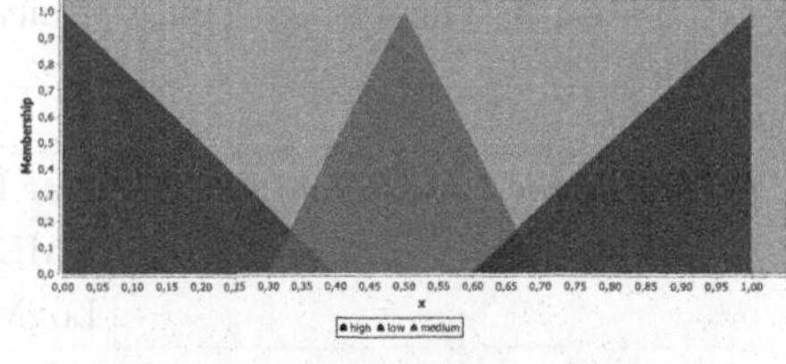

Fig. 4. MAXSIM fuzzy variable

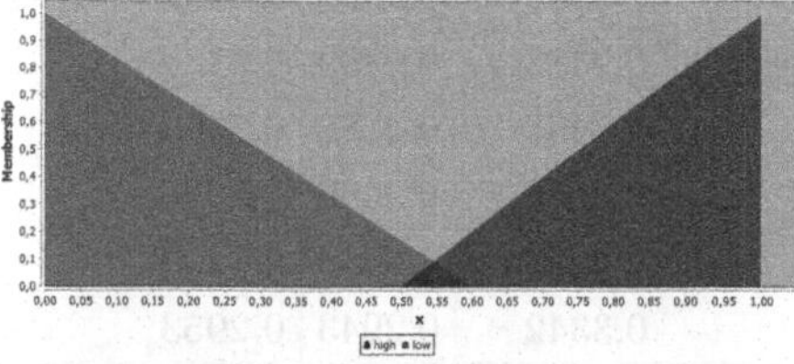

Fig. 5. PARTSIM fuzzy variable

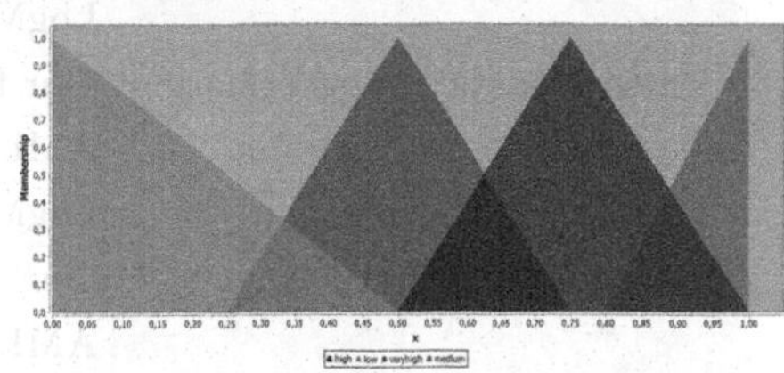

Fig. 6. PRSIM fuzzy variable

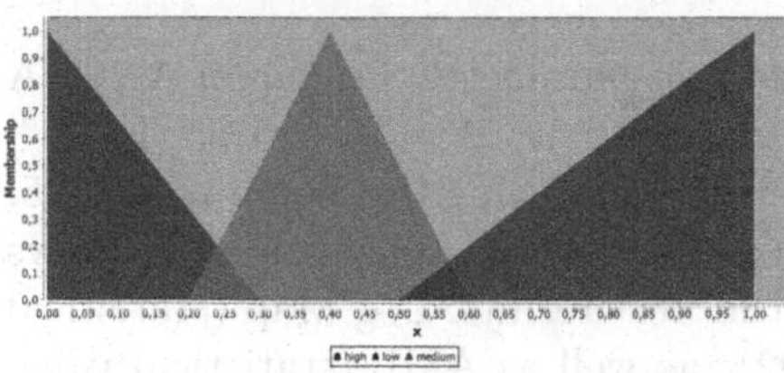

Fig. 7. RSIM fuzzy variable

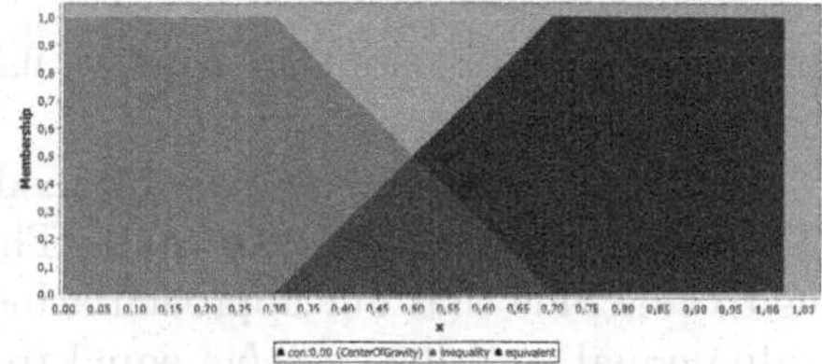

Fig. 8. CON fuzzy variable

The results of the instance alignment calculated by our framework were then compared with the expert alignment provided by OAEI along with ontology datasets. Basic measures like Precision, Recall and F-measure were used for this purpose. The IIMB benchmark dataset was divided into four separate test suits – one with data value transformations, one with data structure transformations, one with data semantics transformations and the final one with mixed transformations. According to this division, our experiment was also divided into four stages.

We wanted to verify a hypothesis that the performance of our approach is better or at least not worse than the existing alignment system. All of the collected data can be found in [18]. They have been compared with results presented in the website [17]. A summary of the results is presented in Table 3.

Thus we obtain nine samples – Precision, Recall, and F-measure for our framework, AML, and LogMap, respectively. All the analysis was made with a significance level $\alpha = 0.05$. Before selecting the appropriate test, we analyzed the distribution of all samples using the Shapiro-Wilk test. None of the samples

Table 3. The average values of measures from different experiment stages

		Precision	Recall	F-Measure
Data value transformation	Our framework	0.9630	0.5280	0.6319
	AML	0.8933	0.7890	0.8280
	LogMap	0.8964	0.8928	0.8894
Data structure transformation	Our framework	0.9241	0.4585	0.4811
	AML	0.4194	0.4329	0.4241
	LogMap	0.9343	0.9856	0.9592
Data semantics transformation	Our framework	0.9488	0.9469	0.9388
	AML	0.7466	0.8888	0.7964
	LogMap	0.8546	0.9466	0.8926
Mixed transformations	Our framework	0.9243	0.0680	0.1095
	AML	0.3342	0.2943	0.2953
	LogMap	0.9199	0.7581	0.8195

have a normal distribution, thus we used the non-parametric Friedman ANOVA test for further analysis.

For Precision samples, we obtained the Friedman value test equal to 43.075. The *p-value* is less than 0.000001. The Dunn Benferroni's post-hoc test allows us to conclude that our approach achieves the best results. Log Map (statistical value equal to 3.499, *p-value* equal to 0.002), as well as AML (statistical value equal to 6.56, *p-value* less than 0.000001), generate smaller number o correct correspondences than our framework.

The result of the Friedman test for Recall samples is equal to 59.299 and the *p-value* is less than 0.000001, which means that at least one system works differently from another. Dunn Benferroni's test chosen as post-hoc points out that our approach is not worse than the AML system (statistic value equal to 2.25, *p-value* equal to 0.073) and there is a statistical difference between Recall value obtained by our approach and the LogMap system (statistic value equal to 5.23, *p-value* less than 0.000001).

A similar conclusion is a result of statistical analysis for F-measure samples. The statistical value of the Friedman test is equal to 28.825 and *p-value* less than 0.000001. The Dunn Bonferroni confirm that there is no significant difference for the result obtained by our and AML system (statistic value equal to 0.533, *p-value* equal to 1). In terms of F-measure LogMap is better than both (our and AML approach) - statistic value equal to 4.348, *p-value* less than 0.000041.

The results of experiments are promising. We have noticed that our framework deals with data semantics transformation perfectly. In other test cases, the performance of our approach is also very good - statistically not worse than AML, and in some cases even better than AML or LogMap. The proposed framework copes with incoherence, which entails reducing the number of false-positive results.

In our experiments, for all test cases, we obtained a very high value of Precision. This means that almost all correspondences found were correct. The lower value of Recall is caused by the non-uniform quality of expected, automatically-generated correspondences. The expected correspondences were created by applying a sequence of transformations of various lengths (i.e., number of transformations) and complexity (i.e., strength of data manipulations applied) [2]. Correspondences in benchmark alignments are often not intuitive and more difficult to agree with than to detect.

6 Future Works and Summary

The following paper is devoted to finding ontology alignment on the instance level. The proposed solution is a fuzzy-logic-based framework built on a set of functions that calculate similarities between concept instances. The values of these functions are eventually treated as fuzzy variables, which are then incorporated into a set of inference rules used for determining the final instance mappings.

The entire fuzzy framework has been experimentally verified utilizing a state-of-the-art dataset provided by the Ontology Alignment Evaluation Initiative. The results obtained are promising and in many cases outperform competing ontology alignment solutions in terms of assumed quality measures. We claim that adjusting fuzzy variables and inference rules can further improve the quality of alignments collected by our framework.

In the upcoming future, we plan to conduct more extensive experiments using different datasets created by the Ontology Alignment Evaluation Initiative, focusing on the scalability of our framework. Furthermore, we will extend the framework to work on the level of concepts, a recent area of interest that has not beed addressed in our previous research.

References

1. Aguirre, J.L., et al.: Results of the ontology alignment evaluation initiative 2012. In: Proceedings of the 7th International Ontology Matching Workshop, Boston (MA, US), pp. 73–115 (2012)
2. Algergawy, A., et al.: Results of the ontology alignment evaluation initiative 2018. In: Proceedings of the 13th International Workshop on Ontology Matching Co-located with the 17th ISWC (OM 2018), vol. 2288, pp. 76–116 (2018)
3. Ardjani, F., Bouchiha, D., Malki, M.: Ontology-alignment techniques: survey and analysis. I.J. Mod. Educ. Comput. Sci. **11**, 67–78 (2015). https://doi.org/10.5815/ijmecs.2015.11.08
4. Cheatham, M., Pesquita, C., Oliveira, D., McCurdy, H.B.: The properties of property alignment on the semantic web. Int. J. Metadata Semant. Ontol. **13**(1), 42–56 (2018)
5. Cingolani, P., Alcalá-Fdez, J.: jFuzzyLogic: a java library to design fuzzy logic controllers according to the standard for fuzzy control programming. Int. J. Comput. Intell. Syst. **6**(Suppl.), 61–75 (2013)

6. de Lourdes Martínez-Villaseñor, M., González-Mendoza, M.: Fuzzy-based approach of concept alignment. In: Ochoa, S.F., Singh, P., Bravo, J. (eds.) UCAmI 2017. LNCS, vol. 10586, pp. 172–180. Springer, Cham (2017). https://doi.org/10.1007/978-3-319-67585-5_18
7. Daskalaki E., Flouris G., Fundulaki I., Saveta T.: Instance matching benchmarks in the era of linked data. J. Web Semant. **39**, 1–14 (2016)
8. Faria, D., et al.: Results of AML participation in OAEI 2018. In: Proceedings of the 13th International Workshop on Ontology Matching Co-located with the 17th International Semantic Web Conference, vol. 2288 (2018)
9. Fernández, S., Velasco, J.R., López-Carmona, M.A.: A fuzzy rule-based system for ontology mapping. In: Yang, J.-J., Yokoo, M., Ito, T., Jin, Z., Scerri, P. (eds.) PRIMA 2009. LNCS (LNAI), vol. 5925, pp. 500–507. Springer, Heidelberg (2009). https://doi.org/10.1007/978-3-642-11161-7_35
10. Hnatkowska, B., Kozierkiewicz, A., Pietranik, M.: Semi-automatic definition of attribute semantics for the purpose of ontology integration. IEEE Access **8**, 107272–107284 (2020). https://doi.org/10.1109/ACCESS.2020.3000035
11. Hnatkowska, B., Kozierkiewicz, A., Pietranik, M.: Fuzzy based approach to ontology relations alignment. In: 2021 IEEE International Conference on Fuzzy Systems (FUZZ-IEEE), pp. 1–7. IEEE (2021)
12. Huber, J., Sztyler, T., Noessner, J., Meilicke, C.: CODI: combinatorial optimization for data integration-results for OAEI 2011. In: Proceedings of the 6th International Conference on Ontology Matching, vol. 814, pp. 134–141 (2011)
13. Pietranik, M., Nguyen, N.T.: Semantic distance measure between ontology concept's attributes. In: König, A., Dengel, A., Hinkelmann, K., Kise, K., Howlett, R.J., Jain, L.C. (eds.) KES 2011. LNCS (LNAI), vol. 6881, pp. 210–219. Springer, Heidelberg (2011). https://doi.org/10.1007/978-3-642-23851-2_22
14. Ruiz, E.J., Grau, B.C., Zhou, Y., Horrocks, I.: Large-scale interactive ontology matching: algorithms and implementation. In: The 20th European Conference on Artificial Intelligence (ECAI 2012) (2012)
15. Taheri, A., Shamsfard, M.: SBUEI: results for OAEI 2012. In: Ontology Matching (2012)
16. http://oaei.ontologymatching.org/
17. http://islab.di.unimi.it/content/im_oaei/2018/
18. https://github.com/bhnatkowska/FuzzyLogicInstanceAlignment

Neuro-Symbolic Models for Sentiment Analysis

Jan Kocoń(✉), Joanna Baran, Marcin Gruza, Arkadiusz Janz, Michał Kajstura, Przemysław Kazienko, Wojciech Korczyński, Piotr Miłkowski, Maciej Piasecki, and Joanna Szołomicka

Department of Artificial Intelligence, Wrocław University of Science and Technology, Wrocław, Poland
jan.kocon@pwr.edu.pl

Abstract. We propose and test multiple neuro-symbolic methods for sentiment analysis. They combine deep neural networks – transformers and recurrent neural networks – with external knowledge bases. We show that for simple models, adding information from knowledge bases significantly improves the quality of sentiment prediction in most cases. For medium-sized sets, we obtain significant improvements over state-of-the-art transformer-based models using our proposed methods: Tailored KEPLER and Token Extension. We show that the cases with the improvement belong to the hard-to-learn set.

Keywords: Neuro-symbolic sentiment analysis · plWordNet · Knowledge base · Transformers · KEPLER · HerBERT · BiLSTM · PolEmo 2.0

1 Introduction

Sentiment analysis is an NLP task performed in industrial or marketing solutions. It aims to determine how customers (authors of textual opinions) react to given products or services. In the classical *symbolic approach*, a text is evaluated using external knowledge bases, e.g., sentiment dictionaries [3,4]. Then, words from the text are linked to positive, negative, or neutral polarization derived from such dictionaries. The final sentiment is an aggregation over all words. State-of-the-art sentiment analysis methods are mainly based on transformers. Such language models contain millions of parameters but also require large computational resources. Hence, their simplified methods, e.g., BiLSTM [15,19], are often used in practice. We refer to both of these approaches as our *baselines*.

This work was funded by the National Science Centre, Poland, project no. 2021/41/B/ST6/04471 (PK) and 2019/33/B/HS2/02814 (MP); the Polish Ministry of Education and Science, CLARIN-PL; the European Regional Development Fund as a part of the 2014–2020 Smart Growth Operational Programme, CLARIN – Common Language Resources and Technology Infrastructure, project number POIR.04.02.00-00C002/19; the statutory funds of the Department of Artificial Intelligence, Wrocław University of Science and Technology.

D. Groen et al. (Eds.): ICCS 2022, LNCS 13351, pp. 667–681, 2022.
https://doi.org/10.1007/978-3-031-08754-7_69

In this paper, we present and validate neuro-symbolic solutions to sentiment analysis that combine both approaches: deep neural networks and symbolic inference. These methods use vector representations of text from deep language models and external knowledge bases in the form of, e.g., lexicons (sentiment), knowledge graphs (WordNet), and lexico-syntactic patterns (sentiment modification rules). Our main contributions are: (1) design or adaptation of multiple neuro-symbolic methods, (2) comparing our approaches against methods without knowledge bases; (3) prove that for simpler models, the knowledge base significantly improves the prediction quality; (4) showing specific cases of medium-sized sets for which knowledge base information significantly improves the prediction quality for the current transformer-based SOTA models; (5) evidence that neuro-symbolic approaches improve reasoning mainly for hard-to-learn cases.

2 Related Work

Sentiment analysis (SA) is a standard classification task aiming to decide whether the presented text has a positive, negative or neutral polarity. Some works treat SA as a multi-class prediction problem when data are focused on ranking system (e.g. 5-star). In the past, standard machine learning methods were applied to SA such as decision tree, SVM, Naive Bayes or random forest. However, in recent years we are observing the growing popularity of deep-learning (DL) models which proved to be very succesfull.

Standard Deep-Learning Approach. Different types of DL architectures were exploited in sentiment classification. We can mention here CNN, LSTM, RNN, GRU, Bi-LSTM and their variations with attention mechanism [11]. Most of then were trained in a supervised setting. However, despite the promising results achieved by these models, vulnerabilities have been observed such as poor knowledge propagation of cross-domain sentiment analysis in online systems [2], mainly due to lack of enough manual annotated datasets for all domains.

Neuro-Symbolic Approach. Many lexicon resources for various languages have been developed. Princeton WordNet (PWN) is a major one for English but similar knowledge bases were created for other languages too. Some contain emotive annotations for specific word meanings assigned by people (e.g. SentiWordNet). In addition, NLP tools were created to analyse data in a manner similar to human understanding (POS – part-of-speech tagger, WSD – word sense disambiguation). Given the complexity of the SA task, which combines natural language processing, psychology, and cognitive science, using such external knowledge processed according to human logic could improve results of standard DL methods. Moreover, it can imply more explainable predictions. Some works have been done in that field. [8] incorporated graph-based ontology ConceptNet into sentiment analysis enriching the text semantics. Apart from knowledge graph, [25] added a WSD process into social media posts processing. A context-aware sentiment attention mechanism acquiring the sentiment polarity of each word with its POS tag from SentiWordNet was studied in [13]. The pre-training process very rarely respects sentiment-related knowledge. If so, the problem of proper representation of sentiment words and aspect-sentiment pairs needs to be solved. To address it Sentiment Knowledge Enhanced Pre-training (SKEP) [24] has been proposed. It uses sentiment masking and constructs three sentiment knowledge prediction objectives to embed this information at the word- and aspect-level into a pre-trained representation.

3 Datasets

3.1 plWordNet and plWordNet Emo

plWordNet is a very large lexico-semantic network for Polish constructed on the basis of the corpus-based wordnet development method, according to which *lexical units*[1] (henceforth LUs) are the basic building blocks of the wordnet [7]. LUs of very similar meaning are grouped into synsets (sets of synonyms) – each LU belongs to only one synset. The most recent version describes ≈295k LUs for ≈194k lemma of four PoS (part of speech) grouped into ≈228k synsets[2] [1].

Emotive annotation was performed on the level of LUs and LU use examples [27]. Context-independent emotive characterisation of an LU was obtained by comparing its authentic use in text corpora. The main distinction is between *neutrality* vs *polarity* of LUs. Polarised LUs are assigned the *intensity* of the sentiment polarisation, *basic emotions* and *fundamental human values*. The latter two help to determine the sentiment polarity and its intensity expressed in the 5 grade scale: *strong* or *weak* vs *negative* and *positive*, plus the ambiguous tag. Annotator decisions are supported by text examples that must be included in the annotations. Due to the compatibility with other wordnet-based annotations, eight basic emotions[3] recognised by Plutchik [20] were used. One LU can be assigned more than one emotion and, as a result, complex emotions are represented by using the same eight-element set. The 12 fundamental human values[4] postulated by Puzynina [21] link the emotive state of the speaker to the *evaluative attitude*. The annotations were done by two annotators each (a linguist and a psychologist) according to the 2+1 scheme.

3.2 PolEmo

PolEmo 2.0 dataset [12,15] is a sentiment analysis task benchmark dataset. It consists of more than 8,000 consumer reviews, containing more than 57,000 sentences. Texts come from four domains: hotels, medicine, products, and school. Each review was annotated with sentiment in a 2+1 scheme at the text level and the sentence level. In this work, only text level examples were used. There are the following sentiment classes: positive, negative, neutral, and ambivalent.

[1] Triples: lemma, Part of Speech (PoS) and sense identifier.

[2] http://plwordnet.pwr.edu.pl.

[3] *Joy*, *fear*, *surprise*, *sadness*, *disgust*, *anger*, *trust* and *anticipation*.

[4] *Utility*, *truth*, *knowledge*, *beauty*, *happiness*, *futility*, *harm*, *ignorance*, *error*, *ugliness*.

The obtained Positive Specific Agreement (PSA) [9] was 90% at the text level and 87% at the sentence level. PolEmo 2.0[5] is available under an MIT open license.

3.3 Preprocessing

All texts from PolEmo were tokenized, lemmatized, and tagged using CLARIN PoS tagger[6]. Word sense disambiguation [10] (WSD[7]) was performed to identify the appropriate LU belonging to that token. Next plWordNet Emo was used to annotate words with sentiment, basic emotions and fundamental human values (*valuations*). Additionally, we also propagated sentiment and emotion annotations from wordnet to words that originally did not have this annotation in the plWordNet Emo. It required training a regressor based on fastText model [6] using emotive dimensions from plWordNet Emo aggregated per lemma (emotions propagated). Data annotation statistics are presented in Table 1.

The example pipeline for combining text with a knowledge base is shown in Fig. 1. It tokenizes text and matches words with their correct meanings in Wordnet. Furthermore, information on sentiment and emotions from Wordnet annotations (WordnetEmo) is added to the text at the *word sense* level using the EMOCCL tool. The emotional Wordnet annotation is aggregated at the *word lemmas level* and added to the text (lemma lexicon).

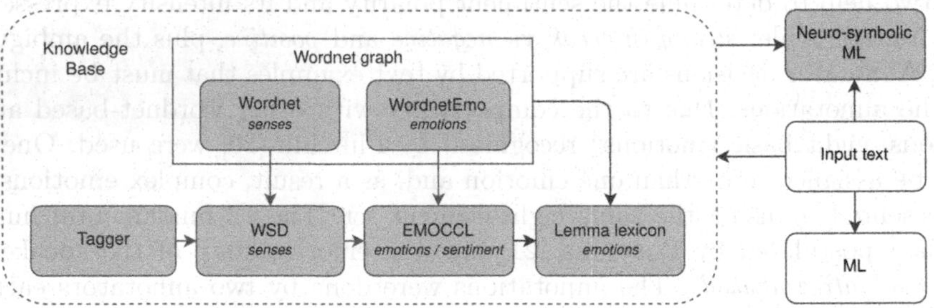

Fig. 1. Baseline approach (ML) vs. neuro-symbolic approach (neuro-symbolic ML). The blue colour on the diagram indicates neuro-symbolic part of the method. (Color figure online)

[5] https://clarin-pl.eu/dspace/handle/11321/710.
[6] http://ws.clarin-pl.eu/tager.
[7] http://ws.clarin-pl.eu/wsd.

Table 1. Token annotation coverage in preprocessed PolEmo2.0

Feature	Train	Dev	Test
Sentiment (all)	28.3%	28.7%	28.4%
Sentiment (pos/neg)	9.2%	9.4%	9.4%
Basic emotions	8.3%	8.5%	8.5%
Valuations	8.6%	8.8%	8.7%
Emotion propagated	99.9%	99.9%	99.9%

4 Neuro-Symbolic Models

4.1 HB: HurtBERT Model

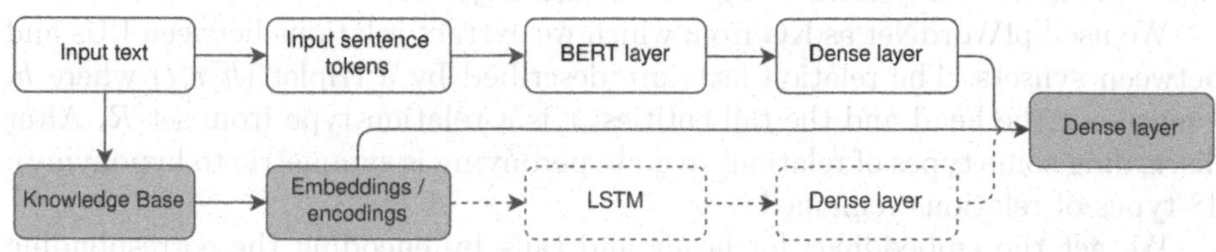

Fig. 2. HB: HurtBERT model.

HurtBERT [16] (Fig. 2) was proposed for the abusive language detection task. Apart from the standard transformer-based text representation, it incorporates knowledge from a lexicon [5]. Additional features are processed by a separate branch and then are concatenated with a text representation before the classification layer. Lexical information can be utilized in two ways: (1) HB-enc: HB-encoding using a simple frequency count for the lexicon categories; (2) HB-emb: HB-embedding obtained with a LSTM network. The second method is more expressive, as it takes token order into account. As the number of categories in plWordNet differs from the ones used in the original paper, we modified the dimensionality of sentiment embedding layer accordingly.

4.2 TK: Tailored KEPLER Model

Tailored KEPLER model (Fig. 3) is an adaptation of KEPLER [26] which incorporates information from a knowledge graph (KG) into a pretrained language model (PLM) like BERT during fine-tuning. It is different to the original KEPLER model where extra KE knowledge is used during pretraining stage (unsupervised masked language modeling). Our Kepler approach is tailored to single task, it utilizes extra knowledge during fine-tuning. To harness knowledge from KG, its entities representation is obtained by encoding their text

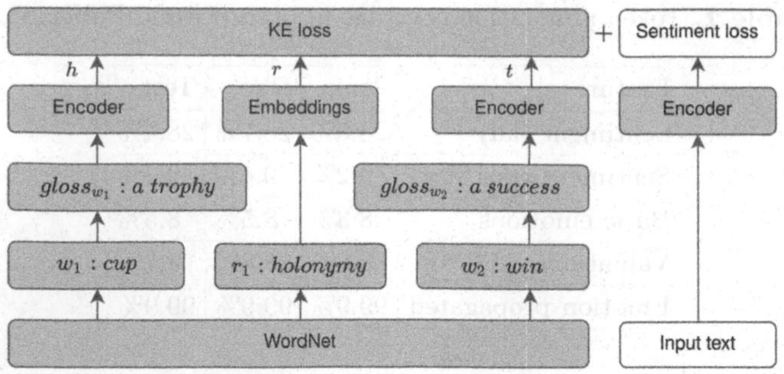

Fig. 3. Tailored KEPLER model. The same encoder is used to obtain embeddings for KE loss and for the downstream task.

descriptions with PLM. Thus, PLM can be additionally learned with Knowledge Embedding (KE) objective along with a task objective.

We used plWordNet as KG from which we extract relations between LUs and between synsets. The relation facts are described by a triplet (h, r, t) where h, t represent the head and the tail entities; r is a relation type from set $\mathcal{R}$. After discarding some types of relations (e.g., hyperonymy is symmetric to hyponymy), 48 types of relations remained.

We get the embeddings for heads and tails by encoding the corresponding LUs descriptions with PLM. The relation types are encoded by a randomly initialized, learnable embedding table. As KE loss, the loss from [22] is used. It adopts negative sampling [18] and tries to minimize TransE distance for the entities being in the relation and to maximize it for negative samples triplets.

To fine-tune the pretrained model, we applied multitask loss $\mathcal{L} = \mathcal{L}_{\mathrm{KE}} + \mathcal{L}_{\mathrm{NLP}}$ where $\mathcal{L}_{\mathrm{NLP}}$ is loss for a downstream NLP task. We used only those triplets which LUs are present in the downstream task training set and we clipped the number of steps in each epoch to the size of the downstream task training set.

4.3 TE: Token Extension Model

The benefits of additional knowledge bases are best seen in simple language models [17]. For this reason, fastText model for Polish language [14] and BiLSTM model [15] working on the basis of embeddings per token derived from it has been taken into consideration (Fig. 4). This approach allows to use the knowledge base at the level of each token. Thus, we propose 3 variants: (1) baseline - which uses token embedding only, (2) TE-original – where additional knowledge (as a vector) from the wordnet is concatenated to the token embedding, and (3) TE-propagated – using propagated data (Sect. 3.3) on all words in text.

4.4 ST(P): Special Tokens (with Positioning) Model

In transformer with **S**pecial **T**okens (ST) model (Fig. 5) we added special BERT tokens corresponding to emotions and sentiments. They are put after a word

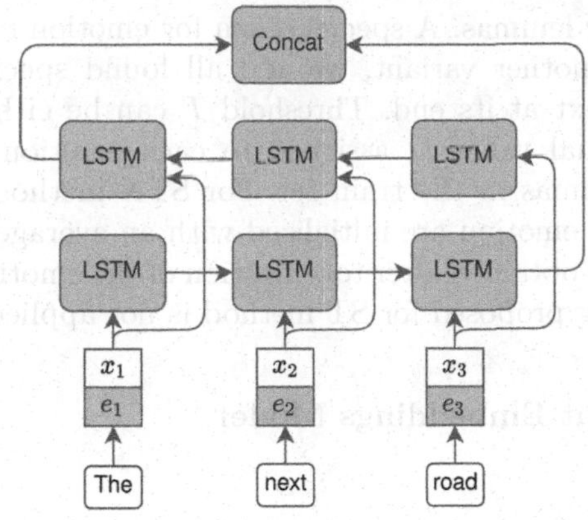

Fig. 4. TE: Token Extension model.

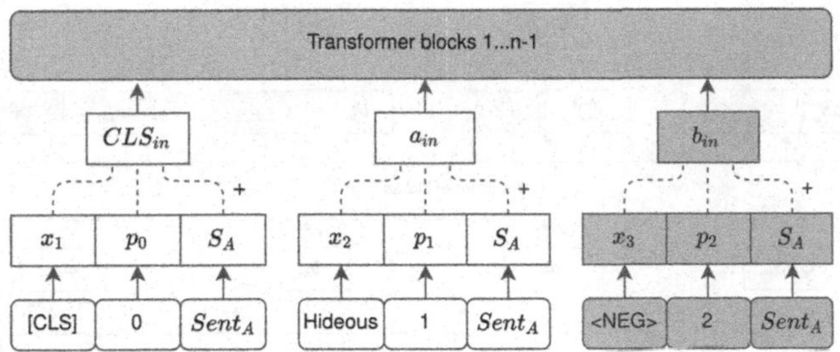

Fig. 5. ST: Special Tokens model.

which lemma is annotated with emotion or sentiment in plWordNet. It is a way to harness emotive knowledge from plWordNet to Transformer. Exemplary input can be in a form of: *She was still weeping [SAD], despite the happy [JOY] end of the movie.* Since emotion tokens are marked as special tokens, they will not be broken down into word pieces by tokenizer and their embedding vectors will be initialized randomly. Since adding new tokens to the text breaks its sequentiality, we test additional version of the model (STP: Special Tokens with Positioning) in which we adjust the emotion token position indexes so that they are equal to the lemma token position indexes they correspond to (e.g. $Happy_{idx=1}$ $[JOY]_{idx=1}$ $and_{idx=2}$ $amazed_{idx=3}$ $[SURPRISED]_{idx=3}$ $girl_{idx=4}$.). With this adjustment, the emotional tokens will have the same positional embeddings as their corresponding lemmas.

4.5 STN: Special Tokens from Numeric Data Model

STN method is an extension of ST method (same model as in Fig. 5) designed for the cases when a lemma is annotated by many annotators. Lemma intensity of emotion e can be expressed as fraction $\alpha_e \in (0, 1)$ of annotation with emotion e. Since not all LUs are annotated, a regression model is used to propagate these values to other lemmas. A special token for emotion e is put after a word if its $\alpha_e > T$. In another variant, we add all found special tokens (without replacement) in a text at its end. Threshold T can be either the same for all emotions or individual value T_e assigned to each emotion e as a quantile of all α_e values for lemmas in the train set. For STN method, the special token embeddings for each emotion are initialized with an average of the embeddings of all subword tokens obtained after tokenization of the emotion name. Adjusting positional embedding proposed for ST method is not applied for STN.

4.6 SE: Sentiment Embeddings Model

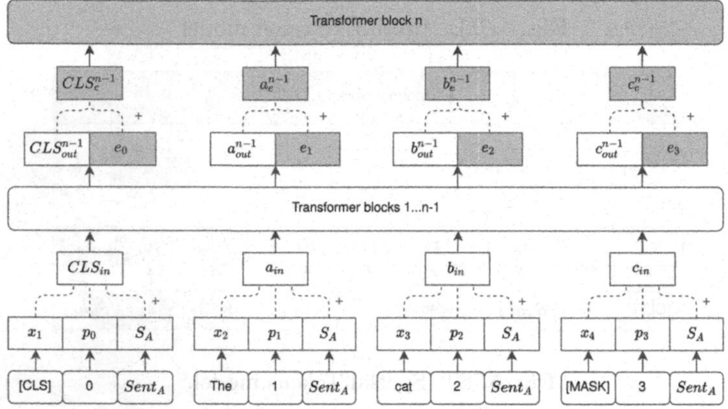

Fig. 6. SE: Sentiment Embeddings model.

Both HurtBERT-embedding and HurtBERT-encoding aggregate additional information at text level, which can limit the interaction between the text and features obtained from plWordNet. To incorporate token-level lexical annotations into a transformer, we add trainable sentiment embeddings as hidden representations before the last transformer layer (Fig. 6). If the word consists of multiple BPE parts, we add the embedding to all subword tokens. Augmented representations are then passed to a classifier to compute the probability of each sentiment class. The classifier consists of a dense layer followed by a softmax activation function. During the pretraining phase of HerBERT, there is no additional lexical information. Adding the sentiment token in the second to last layer of the transformer could corrupt the token representations. We consider two variants: (1) SE: the last transformer block's weights are left unchanged and (2) SE-reset: the last transformer block's weights are randomly initialized. Random reinitialization of the last BERT layer is a common practice [28] and can make it easier for the model to utilize additional features.

5 Experimental Setup and Results

For each experimental setup, we compare a baseline neural model with its neuro-symbolic extension. In each method (excluding TE), we used HerBERT as a SOTA baseline for sentiment analysis performed on PolEmo 2.0 dataset. We test the method on selected undersampled training datasets of different sizes. Both baseline and neuro-symbolic models are trained using the same hyperparameters. Some of the methods are adapted from other papers, so the baselines are not identical in different setups in terms of hyperparameters. For each configuration, the experiments are repeated 10 times.

5.1 TK: Tailored KEPLER Model

Fine-tuning is performed for 4 epochs with learning rate 5e-5, batch size 4 and weight decay 0.01. Maximum sequence length is 256 and 32 for *PolEmo texts* and for *entities text* representations, respectively. Results are presented in Fig. 8. The statistical gains are obtained for the smaller training sets what shows that the extra knowledge from KG helps when an amount of data is limited.

For the case where the difference between the baseline and TK was significant, both models were compared using the cartography method [23]. It uses

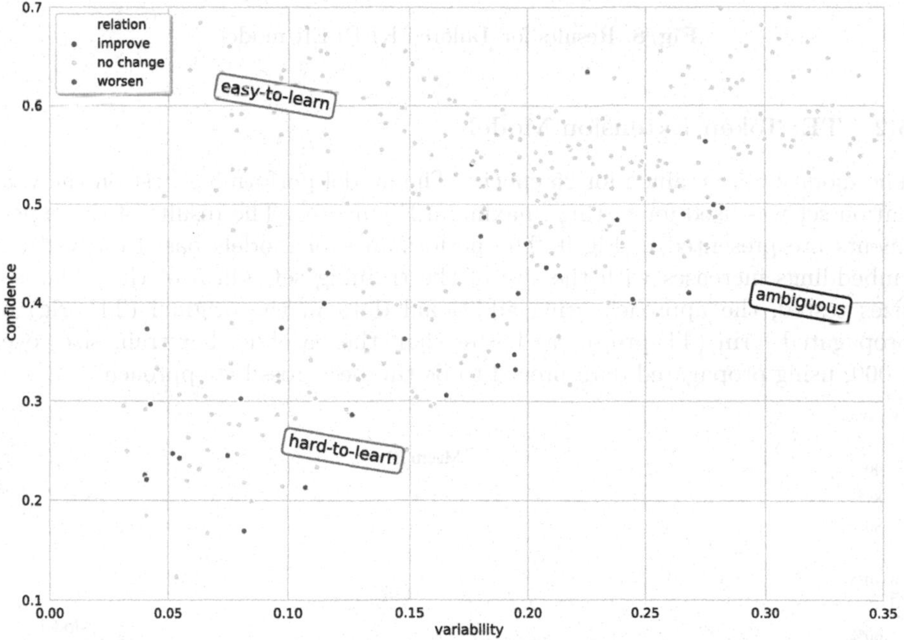

Fig. 7. Datamap [23] for the baseline model in Tailored Kepler (TK) setup. Green colour indicates cases for which correctness c_K of TK has a higher value than correctness c_H for the baseline. Gray examples have $|c_K - c_H| \leq 0.3$ (small or no change). Red instances mean that $c_K < c_H$. (Color figure online)

model confidence, variability, and correctness over epochs to find which texts are hard, easy or ambiguous to learn. The correctness specifies the fraction of times the true label is predicted. The confidence is the mean probability of the ground truth between epochs. The variability measures how indecisive the model is. Figure 7 shows datamap for HerBERT. Colours of the points on the diagram indicate if the instance is easier to learn for Tailored KEPLER than the baseline (HerBERT). The diagram shows that adding extra knowledge improves correctness for far more cases than it worsens. Moreover, the examples, which are affected belong to hard-to-learn and ambiguous classes only.

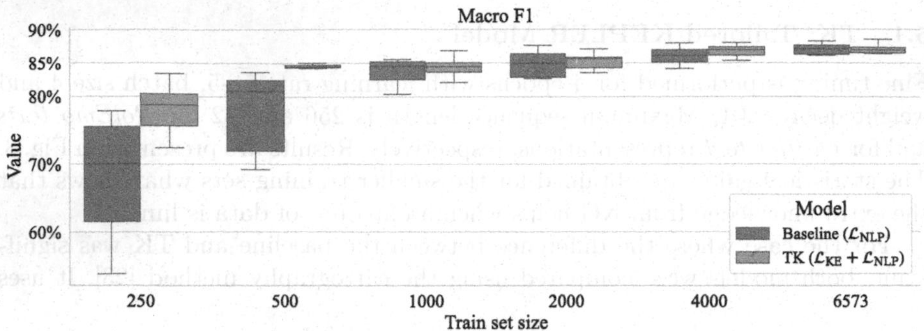

Fig. 8. Results for Tailored KEPLER model.

5.2 TE: Token Extension Model

The models were trained for 25 epochs. The model performing best on the validation set was used for testing (maximum F1-macro). The results of the experiments are presented in Fig. 9. The performance of models based on fastText embeddings increases with the size of the training set. On 5 of the 6 dataset sizes tested, the approach using additional data in the original (TE-orig) or propagated form (TE-prop) was better than the baseline. For train sizes over 1,000, using propagated data proved to be the best possible approach.

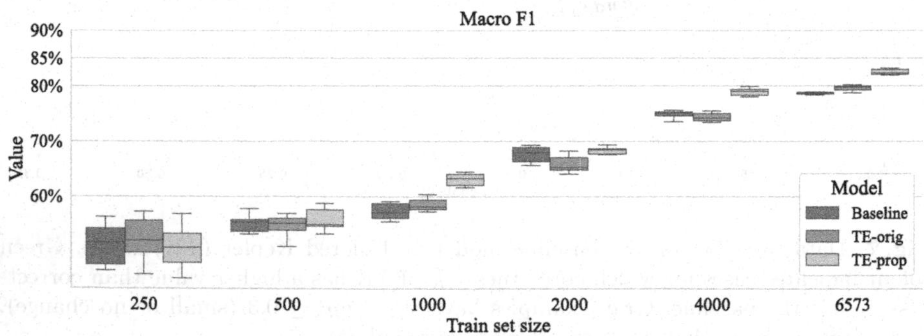

Fig. 9. Results for Token Extension model.

It is important to compare the running time of the TE model with that of the example transformer-based (SE) model in Fig. 10. The performance of the TE model (macro F1: 83%) is significantly worse by about 4 p.p. relative to the SE model (macro F1: 87%). However, the inference time of the TE model for the test set (3.6 s) is almost four times shorter than that of the SE model. https://www.overleaf.com/project/620b8ae3cda06ae4691ba512.

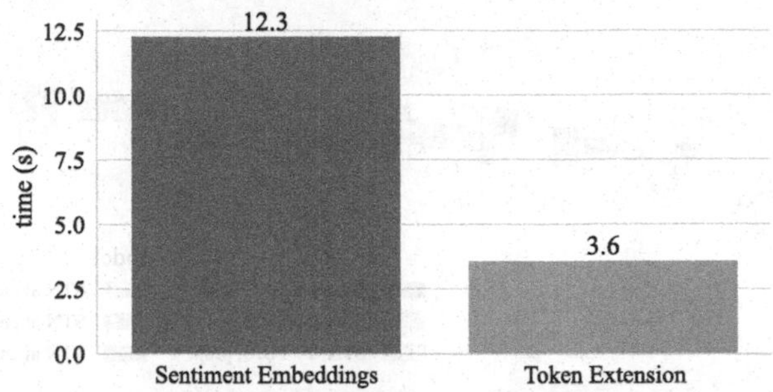

Fig. 10. Inference time of Sentiment Embeddings (transformer-based) and Token Extension (BiLSTM+fastText) neuro-symbolic models.

5.3 ST(P): Special Tokens (with Positioning) Model

The maximum tokenizer text length has been set to 512, so that adding new emotional tokens does not require to truncate the text. The batch size was set to 20. We used the Adam optimizer with the learning rate set to 2e-5 during training. The models were trained for 5 epochs and the model with the smallest validation loss was checkpointed and tested. The results are presented in Fig. 11.

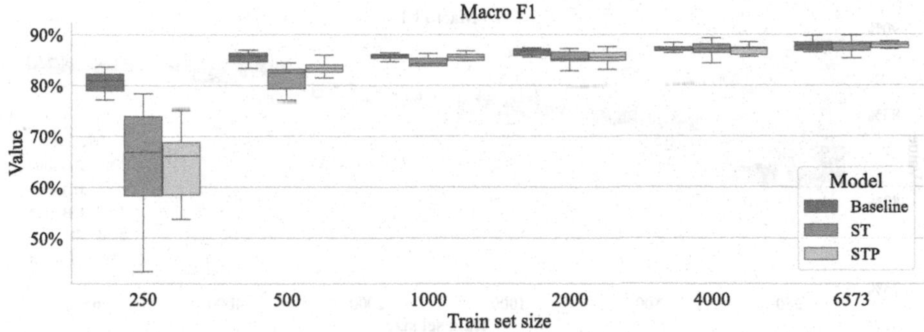

Fig. 11. Special Token Model (ST) and ST with positional embeddings (STP).

ST and STP models achieve worse results than the baseline for smaller train datasets (250 and 500 samples). For bigger train datasets, there are no significant differences between the models.

5.4 STN: Special Tokens from Numeric Data

We consider the following variant with *in-text* and *at-end* special tokens: (1) no propagated data, $T = 0.5$; (2) propagated data, $T = 0.6$; (3) propagated data, individual threshold T_e equal to 0.75 qunatile. In each case, fine-tuning is performed for 4 epochs with learning rate 5e-5, batch size 16, weight decay 0.01, and maximum sequence length 512. Results are presented in Fig. 12.

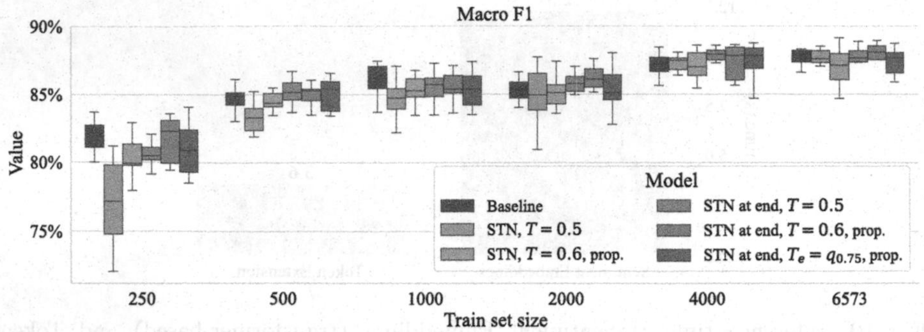

Fig. 12. Results for Special Tokens for Numeric Data model.

The results do not show significant improvements for each STN method. In the case of in-text special tokens, the results are usually worse. For at-end-of-text special tokens performance is very similar to the baseline.

5.5 HB: HurtBERT Model and SE: Sentiment Embeddings Model

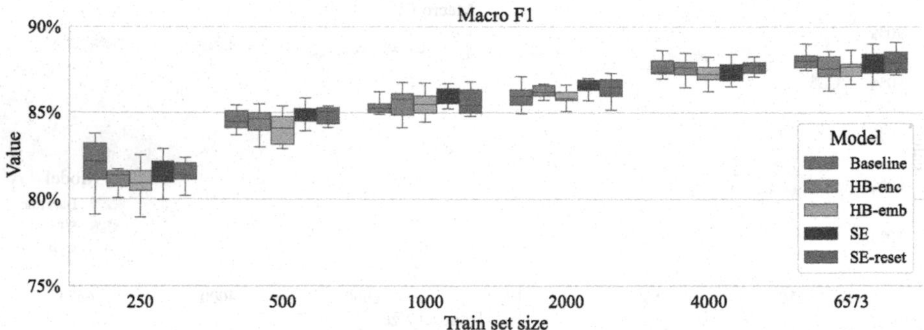

Fig. 13. Results for HurtBERT-encoding, HurtBERT-embeddings, Sentiment Embedding and Sentiment Embedding Reset models.

Models are fine-tuned using AdamW optimizer with learning rate 1e-5, linear warmup schedule, batch size 32, and maximum sequence length 256 for 30 epochs and the best model is chosen according to a validation F-score. Results are presented in Fig. 13. In lower data regimes (250 and 500 samples), there may not be enough data to learn embeddings of sentiment features, hence the similar performance. For larger datasets, the additional information from a knowledge base is outweighed by a textual information. Our experiments do not show a significant improvement over a baseline, both for HurtBERT and the proposed SE method. Texts in PolEmo dataset are complex and aggregating additional lexical features on the level of a whole text is not sufficient.

6 Conclusions

We designed and adapted multiple neuro-symbolic methods. The additional knowledge in most transformer-based neuro-symbolic models does not lead to improvement in most cases. For the smallest variants of datasets (training dataset: 250 texts), it can even make the training process more unstable and degrade the model quality (ST*, HB*, SE*). Adding special tokens inside the text is not beneficial for pretrained BERT models because it damages the natural structure of the text. It is not the case for tokens added at the end of the text, but still no performance gain is observed. It can be caused by the fact that the considered PolEmo dataset has high PSA, so the knowledge encompassed in the pretrained HerBERT model is sufficient to obtain very good results.

However, for small and medium-sized datasets, our Tailored KEPLER neuro-symbolic transformer-based model produced statistically significant gains. It also allowed to obtain better and more stable results. Analysis of these cases shows performance gains for examples belonging to *ambivalent* sentiment class. We examined in which cases additional knowledge improved the quality of inference, Fig. 7. The vast majority of these cases were identified by the baseline model as hard-to-learn.

A key finding of the study is that the knowledge base significantly improves the quality of simple models such as Token Extension, Fig. 10. Compared to transformer-based models, we obtain an almost fourfold reduction in inference time, at the cost of a significant but relatively small decrease in quality (4 pp.). For the TK model, the quality gain due to additional knowledge was significant for most cases. This shows that with very little computational cost, the inference quality can be significantly improved for such models.

References

1. plWordNet 4.5 (2021). http://hdl.handle.net/11321/834. CLARIN-PL
2. Al-Moslmi, T., Omar, N., Abdullah, S., Albared, M.A.: Approaches to cross-domain sentiment analysis: systematic lit. Review. IEEE Access **5**, 16173–16192 (2017)
3. Augustyniak, L., Kajdanowicz, T., Kazienko, P., Kulisiewicz, M., Tuliglowicz, W.: An approach to sentiment analysis of movie reviews: lexicon based vs. classification. In: Polycarpou, M., de Carvalho, A.C.P.L.F., Pan, J.-S., Woźniak, M., Quintian, H., Corchado, E. (eds.) HAIS 2014. LNCS (LNAI), vol. 8480, pp. 168–178. Springer, Cham (2014). https://doi.org/10.1007/978-3-319-07617-1_15

4. Augustyniak, Ł., et al.: Simpler is better? Lexicon-based ensemble sentiment classification beats supervised methods. In: ASONAM 2014, pp. 924–929 (2014)
5. Bassignana, E., Basile, V., Patti, V.: Hurtlex: a multilingual lexicon of words to hurt. In: CLiC-it 2018, vol. 2253, pp. 1–6. CEUR-WS (2018)
6. Bojanowski, P., Grave, E., Joulin, A., Mikolov, T.: Enriching word vectors with subword information (2017)
7. Dziob, A., Piasecki, M., Rudnicka, E.: plWordNet 4.1 - a linguistically motivated, corpus-based bilingual resource. In: The 10th Global Wordnet Conference, pp. 353–362. Global Wordnet Association, July 2019
8. Ghosal, D., Hazarika, D., Roy, A., Majumder, N., Mihalcea, R., Poria, S.: Kingdom: knowledge-guided domain adaptation for sentiment analysis. arXiv:2005.00791 (2020)
9. Hripcsak, G., Rothschild, A.: Agreement, the f-measure, and reliability in information retrieval. J. Am. ER. Med. Inform. Ass. (JAMIA) **12**(3), 296–298 (2005)
10. Janz, A., Piasecki, M.: A weakly supervised word sense disambiguation for polish using rich lexical resources. Poznan Stud. Cont. Ling. **55**(2), 339–365 (2019)
11. Joseph, J., Vineetha, S., Sobhana, N.: A survey on deep learning based sentiment analysis. Mater. Today Proc. **58**, 456–460 (2022)
12. Kanclerz, K., Miłkowski, P., Kocoń, J.: Cross-lingual deep neural transfer learning in sentiment analysis. Procedia Comput. Sci. **176**, 128–137 (2020)
13. Ke, P., Ji, H., Liu, S., Zhu, X., Huang, M.: SentiLARE: sentiment-aware language representation learning with linguistic knowledge. arXiv:1911.02493 (2020)
14. Kocoń, J., Gawor, M.: Evaluating KGR10 Polish word embeddings in the recognition of temporal expressions using BiLSTM-CRF. Schedae Informaticae **27** (2018)
15. Kocoń, J., Miłkowski, P., Zaśko-Zielińska, M.: Multi-level sentiment analysis of PolEmo 2.0: extended corpus of multi-domain consumer reviews. In: CoNLL2019, pp. 980–991. ACL, November 2019
16. Koufakou, A., Pamungkas, E.W., Basile, V., Patti, V.: HurtBERT: incorporating lexical features with BERT for the detection of abusive language. In: The 4th Workshop on Online Abuse and Harms, pp. 34–43. ACL, November 2020
17. Ma, Y., Peng, H., Cambria, E.: Targeted aspect-based sentiment analysis via embedding commonsense knowledge into an attentive LSTM. In: AAAI 2018, vol. 32 (2018)
18. Mikolov, T., Sutskever, I., Chen, K., Corrado, G., Dean, J.: Distributed representations of words and phrases and their compositionality. In: NIPS 2013, pp. 3111–3119 (2013)
19. Kocoń, J., Miłkowski, P., Kanclerz, K.: MultiEmo: multilingual, multilevel, multidomain sentiment analysis corpus of consumer reviews. In: Paszynski, M., Kranzlmüller, D., Krzhizhanovskaya, V.V., Dongarra, J.J., Sloot, P.M.A. (eds.) ICCS 2021. LNCS, vol. 12743, pp. 297–312. Springer, Cham (2021). https://doi.org/10.1007/978-3-030-77964-1_24
20. Plutchik, R.: EMOTION: A Psychoevolutionary Synthesis. Harper & Row (1980)
21. Puzynina, J.: Język wartości [The language of values]. Polish Scientific Publishers PWN (1992)
22. Sun, Z., Deng, Z.H., Nie, J.Y., Tang, J.: RotatE: knowledge graph embedding by relational rotation in complex space. In: The International Conference on Learning Representations (ICLR) (2019)
23. Swayamdipta, S., et al.: Dataset cartography: mapping and diagnosing datasets with training dynamics. In: EMNLP 2020, pp. 9275–9293. ACL, November 2020
24. Tian, H., et al.: SKEP: sentiment knowledge enhanced pre-training for sentiment analysis (2020)

25. Vizcarra, J., Kozaki, K., Torres Ruiz, M., Quintero, R.: Knowledge-based sentiment analysis and visualization on social networks. NGC **39**(1), 199–229 (2021)
26. Wang, X., Gao, T., Zhu, Z., Liu, Z., Li, J.Z., Tang, J.: KEPLER: a unified model for knowledge embedding and pre-trained language representation. Trans. Assoc. Comput. Linguist. **9**, 176–194 (2021)
27. Zaśko-Zielińska, M., Piasecki, M.: Towards emotive annotation in plWordNet 4.0. In: The 9th Global Wordnet Conference, pp. 153–162. Global WordNet Association (2018)
28. Zhang, T., Wu, F., Katiyar, A., Weinberger, K.Q., Artzi, Y.: Revisiting few-sample BERT fine-tuning. arXiv:2006.05987 (2020)

A Unified Sense Inventory for Word Sense Disambiguation in Polish

Arkadiusz Janz(✉), Agnieszka Dziob, Marcin Oleksy, and Joanna Baran

Department of Computational Intelligence, Wrocław University of Science and Technology, Wrocław, Poland
{arkadiusz.janz,joanna.baran}@pwr.edu.pl

Abstract. We introduce a comprehensive evaluation benchmark for Polish Word Sense Disambiguation task. The benchmark consists of 7 distinct datasets with sense annotations based on *plWordNet*–4.2. As far as we know, our work is a first attempt to standardise existing sense annotated data for Polish. We also follow the recent trends of neural WSD solutions and we test transfer learning models, as well as hybrid architectures combining lexico-semantic networks with neural text encoders. Finally, we investigate the impact of bilingual training on WSD performance. The bilingual model obtains new State of the Art performance in Polish WSD task.

Keywords: WSD · Knowledge bases · Neural models · Benchmarking

1 Introduction

For over 50 years, the interest in Word Sense Disambiguation (WSD) has not changed. A great progress has been made on improving general quality of WSD data and algorithms. Still, WSD is an open problem for low-resource languages due to the lack of sense annotated datasets. Modern WSD algorithms require great volumes of data to successfully disambiguate word meanings in a multi-domain setting. On the contrary, multilingual language models and transfer learning methods have been proved to be quite effective when annotated data is unavailable [15].

Polish WSD data seemed to be quite scarce, especially for lemmas in verb and adjective categories. This issue has always been seen as a limiting factor for Polish language, slowing the development of supervised WSD approaches. On the other hand, *plWordNet* with its lexico-semantic structure and sense descriptions has been rebuilt and significantly extended with more data [4,8,19,22]. One of the main issues is that, until now, Polish WSD data has resided in various sources and formats. In this paper, we decided to explore all necessary requirements to develop an effective WSD solution for Polish by following a data-centric approach. The main contributions of this paper are as follows:

D. Groen et al. (Eds.): ICCS 2022, LNCS 13351, pp. 682–689, 2022.
https://doi.org/10.1007/978-3-031-08754-7_70

- We cleaned, updated, and substantially extended existing sense annotated corpora, and we unified them under one modern knowledge-base – *plWordNet*–4.2. We introduced 2 manually annotated WSD datasets of significant size. Overall, 7 distinct datasets covering more than 60% of senses from Polish wordnet have been collected. Finally, we propose a first WSD evaluation framework for Polish in one package with all resources integrated with Princeton WordNet [12] and BabelNet [14] indices.
- We evaluated crosslingual architectures as well as monolingual neural models and hybrid solutions on Polish data at scale. Such evaluation has been so far impossible due to the lack of the necessary linguistic resources and annotated data of considerable size. We selected two modern neural architectures to train for WSD task: XL-WSD solution [15] and EWISER [2].
- We investigated bilingual training and its impact on Polish WSD for the first time. We show that bilingual training is not necessarily a harmful process [21]. The investigated models raised Polish WSD performance to a new level.

2 Related Work

Resources. Princeton WordNet (PWN) [12] is a lexical database used in various NLP applications. It was also a foundation for wordnets created in other languages, often as a translation from English. The eXtended WordNet project [6] aimed to enrich PWN with information found in sense definitions and usage examples, on the basis of which Princeton WordNet Gloss Corpus[1] was later developed. Words extracted from the definitions (also called "glosses") in synsets were manually linked to their context-appropriate sense in WordNet. This created a major training dataset for the WSD task for a long time. Another extension of PWN was proposed in [17] where the authors utilised external knowledge resources such as *Wikipedia* to increase the general density of semantic relations in wordnet, leading to higher WSD performance. With this assumption the BabelNet [14] was created, taking the role of main WSD resource.

Raganato et al. [18] introduced a unified WSD evaluation framework for English. It merged all Senseval and SemEval data into a single dataset containing 7,253 instances to disambiguate. Pasini et al. [15] proposed XL-WSD – a cross-lingual dataset for the WSD task. It stands as a key semantic benchmark not only in terms of size, but also in terms of coverage. Unfortunately, it does not contain any test data for Polish. Leveraging the resources of Open Multilingual WordNet [20] and BabelNet [14] allowed to create new multilingual evaluation benchmarks.

The work on the creation of Polish WSD corpora began about 10 years ago. `KPWr` [3] and `Składnica` [10] were the first evaluation resources for Polish WSD. However, they were developed at the time of the official announcement of *plWordNet*-2.1. In [7] the development of sense annotated data included partial update of Składnica and KPWr data to *plWordNet*-3.2. Still, the available WSD resources were of moderate size.

[1] https://wordnetcode.princeton.edu/glosstag.shtml.

Methods. We can distinguish three main approaches to WSD classification task: *knowledge-based*, *supervised* and *hybrid* combining the previous two. Despite their large coverage of senses, the knowledge-based approaches are easily outperformed by supervised solutions when sufficient training data is provided [18]. The key factor of their performance is based on assumption that sense descriptions in the knowledge base are reflecting the natural context of sense occurrences in the corpora. This applies to both textual descriptions as well as their lexico-semantic structure in the knowledge-graph. [9,10] represent the adaptations of knowledge-based methods such as [1,13] to Polish language. Recently, with the growing popularity of deep neural language models, new architectures such as EWISER [2] (*hybrid*) or XL-WSD [15] (*supervised*) have been proposed. They proved that the multilingual models can be successfully used to prepare an effective solution for languages other than English.

3 Polish WordNet as a Knowledge Base

plWordNet (pol. *"Słowosieć"*; *plWN*) is a large wordnet of Polish built from scratch and manually mapped onto PWN [19], in which the central semantic entity is the *lexical unit* (LU). It is integrated with SUMO ontology [16], Wikipedia [11], Polish valency lexicon Walenty [5], and enriched with an extensive emotive annotation [22]. Starting from version *4.2* (see Table 1) (2020) all efforts have focused on increasing the density of lexico-semantic structure and revising sense granularity. With all of its extensions, plWordNet could possibly improve WSD performance in other languages when integrated properly with WSD models.

Table 1. Descriptive statistics of wordnet-based sense inventories for Polish. The average polysemy ratio was computed for polysemous lemmas only.

Feature	*plWN 2.1*	*plWN 3.2*	*plWN 4.2*
LU	206 567	286 804	294,842
Multi-word LU	53 752	70 019	71 133
Synsets	151 252	221 101	227 369
LU with gloss or usage example	37 207	145 901	155 290
Average length of utterance	12.56	11.54	11.08
Average number of senses per lemma	2.79	2.96	3.05

4 Polish WSD Inventory

Each corpus was prepared according to the following data pre-processing pipeline. All texts were segmented, tokenized and underwent morpho-syntactic analysis. We added an additional annotation layer with automatically recognized multi-word expressions (MWE) existing in plWordNet[2]. Most of our datasets were annotated in 2 + 1 system - two linguists working independently supported by third super-annotator to resolve inconsistencies. Lastly, the tokens representing open-class words were selected for manual sense correction or annotation. All corpora described below were manually updated to plWN 4.2[3].

`Składnica (SK)` is a sense-annotated treebank [5] used in the past as an evaluation set for knowledge-based WSD approaches for Polish [10]. Re-introduced at *PolEval's WSD competition Task 3* [7], recently has been trending as a training set. The sentences in `Składnica` were carefully parsed and manually annotated. `KPWr-N` annotation was based on a lexical sampling approach for a small set of words [3,10] – here we present an updated version with manually extended semantic annotation. `Sherlock Holmes: The Adventure of The Speckled Band (SPEC)` by Sir Arthur Conan Doyle has been translated to Polish by a team of professionals as a part of The NTU Multilingual Corpus [20], and manually tagged both with morphological information and WSD. Unlike the first annotation of the `KPWr-N` corpus, in `KPWr-100` the process was aimed at full-text sense annotation. Documents from various sources and representing different functional styles and genres were manually tagged. `SPEC` and `KPWR-100` were introduced in [7] as a test framework for the competition – in this work we updated all of its sense annotations from *plWordNet*–3.2 version to 4.2 (Table 2).

Table 2. The overall number of tokens and lemmas in our corpora. We also provide a percentage of covered senses with respect to *plWordNet*–4.2.

Feature	Dataset						
	`SK`	`SPEC`	`EmoGLEX`	`WikiGLEX`	`KPWr-N`	`KPWr-100`	`GLEX`
All tokens	136 075	9 087	290 313	96 308	438 505	32 522	5 020 817
All lemmas	17 065	2 212	22 927	13 996	31 142	6 317	170 200
Sense coverage	5.58%	0.85%	4.53%	3.49%	1.31%	2.30%	60.53%
Avg. polysemy in corpus	2.6	2.3	2.6	2.5	2.9	2.5	–
Avg. polysemy in plWN	4.2	5.1	4.2	4.3	3.9	4.5	–
Sense annotations	43 776	3 947	N/A	N/A	14 429	14 004	333 254
After mapping	31 294	3 113	N/A	N/A	10 603	11 334	292 119

`GLEX` is a corpus of glosses and usage examples. It includes synsets which contain the lexical units having at least one natural language uterrance. Among the whole corpus, we distinguished two distinct subcorpora with full-text sense annotation. The first one, `WikiGLEX`, contains glosses and usage examples of senses that represent the intersection of `plWordNet` and Wikipedia. The second

[2] https://clarin-pl.eu/dspace/handle/11321/508.
[3] https://clarin-pl.eu/dspace/handle/11321/891.

one, EmoGLEX, was created from synsets with at least one lexical unit containing sentiment annotation [22] and emotive examples obtained in [8] project. The aforementioned corpora were cleaned, accurately annotated and compiled together within our framework.

5 Evaluation

We use the datasets described in Sect. 4 as a basis for our evaluation framework. We decided to designate default data splits for training and evaluation. Sense distribution and their coverage are usually seen as key WSD factors to consider when building representative training set. We investigated the following scenarios.

1. **Zero-shot** setting evaluated multilingual language models fine-tuned on English WSD data only. Training set consisted of SemCor mixed with PWN definitions and usage examples. We adapted XLM RoBERTa Large to the task as it was proposed in XL-WSD.
2. **Monolingual** approach was focused on assessing monolingual models as they are expected to perform better than transfer learning when sufficient amount of data is available. To prepare this model we trained EWISER architecture on GLEX corpus only due to its large vocabulary and sense coverage. The remaining available Polish corpora were used in evaluation step.
3. **Bilingual** setting examined the impact of bilingual training on Polish WSD performance. We trained the same model as in *monoligual* scenario on extended corpora consisting of SemCor data, PWN Gloss Corpus and our GLEX corpus.

Parameter Settings. The models and their parameters were tuned on validation data using early stopping strategy with validation loss as a core metric. The number of epochs was set to 30. In the zero-shot setting, SemEval's 2015 dataset was used as a validation set following XL-WSD research. In other settings, we used a sample of GLEX corpora as our validation data.

Data Preprocessing. To train and evaluate the models we mapped all of sense annotations in our corpora onto BabelNet as the EWISER architecture requires all of the data to be compatible with BabelNet indices. This mapping was done by taking existing interlingual links between plWordNet and PWN including *i-synonyms*, *i-hypernyms*, *i-hyponyms*, preferring *i-synonyms* at first place. We also prepared a joint sense inventory with Polish and English lemmas and their candidate meanings mapped onto BabelNet indices.

As a baseline solution, we considered WoSeDon – a knowledge-based model with PageRank algorithm at its core [1,10]. Table 3 presents a summary of our experimental part. We can notice that the baseline architecture was outperformed by all of tested neural architectures. The zero-shot model has proved to be quite effective on frequent-sense data such as SPEC, SK and KPWr-100. However, for non-trivial data e.g. KPWr-N with diverse sense distribution the results are less optimistic. As expected, the monolingual model performs slightly better than zero-shot solution. The bilingual solution has achieved the greatest performance among all models being evaluated. However, the results for KPWr-N dataset suggest that a monolingual model might be better when dealing with rare senses.

Table 3. F1-scores of tested architectures. Asterisk represents development data.

Architecture	Datasets				
	SK	SPEC	KPWr-N	KPWr-100	GLEX
WoSeDon	—	62.30	—	64.72	*
Zero-shot (XLMR-L)	70.07	73.07	47.81	70.24	68.31
Monolingual (EWISER)	70.42	72.05	51.52	70.51	*
Bilingual (EWISER)	72.52	75.55	50.41	72.48	*

6 Conclusions

We proposed a new evaluation benchmark for Polish WSD task. We summarized all of its resources including newly obtained corpora as well as the knowledge base used as its sense inventory. We evaluated modern language models on our benchmark data achieving a new State-of-the-Art performance in Polish WSD. The results suggest that neural models can be successfully utilised to prepare a good enough WSD solution for Polish. Still, the evaluation for rare senses uncovers the main issue of existing data sources – the most frequent sense bias. In further work we plan to improve language models in terms of their adaptability to new and rare senses. We release presented data and code for public use (available at https://github.com/CLARIN-PL/polish-wsd-datasets).

Acknowledgment. This work was co-financed by (1) the Polish Ministry of Education and Science, CLARIN-PL; (2) the European Regional Development Fund as a part of the 2014–2020 Smart Growth Operational Programme, project number POIR.04.02.00-00C002/19; and (3) by the National Science Centre, Poland, grant number 2018/29/B/HS2/02919.

References

1. Agirre, E., López de Lacalle, O., Soroa, A.: The risk of sub-optimal use of open source NLP software: UKB is inadvertently state-of-the-art in knowledge-based WSD. In: Proceedings of the Workshop for NLP Open Source Software (NLP-OSS), Melbourne, Australia (2018)
2. Bevilacqua, M., Navigli, R.: Breaking through the 80% glass ceiling: raising the state of the art in word sense disambiguation by incorporating knowledge graph information. In: Proceedings of the 58th Annual Meeting of the Association for Computational Linguistics, pp. 2854–2864 (2020)

3. Broda, B., Marcińczuk, M., Maziarz, M., Radziszewski, A., Wardyński, A.: KPWr: towards a free corpus of Polish. In: Proceedings of the 8th International Conference on Language Resources and Evaluation. Istanbul, Turkey, May 2012
4. Dziob, A., Piasecki, M., Rudnicka, E.K.: plWordNet 4.1 - a linguistically motivated, corpus-based bilingual resource. In: Proceedings of the 10th Global Wordnet Conference, pp. 353–362
5. Hajnicz, E.: Lexico-semantic annotation of składnica treebank by means of PLWN lexical units. In: Proceedings of the 7th Global Wordnet Conference, Tartu, Estonia, January 2014
6. Harabagiu, S., Moldovan, D.: Knowledge processing on an extended wordnet. WordNet Electron. Lexical Database **305**, 381–405 (1998)
7. Janz, A., Chlebus, J., Dziob, A., Piasecki, M.: Results of the PolEval 2020 shared task 3: word sense disambiguation. In: Proceedings of the PolEval 2020 Workshop, p. 65
8. Janz, A., Kocon, J., Piasecki, M., Zasko-Zielinska, M.: plWordNet as a basis for large emotive lexicons of Polish. In: Proceedings of Human Language Technologies as a Challenge for Computer Science and Linguistics Poznan, pp. 189–193 (2017)
9. Janz, A., Piasecki, M.: Word sense disambiguation based on constrained random walks in linked semantic networks. In: Proceedings of the International Conference on Recent Advances in Natural Language Processing, Varna, Bulgaria (2019)
10. Kędzia, P., Piasecki, M., Orlińska, M.: Word sense disambiguation based on large scale Polish CLARIN heterogeneous lexical resources. Cogn. Stud. (15) (2015)
11. Maziarz, M., Piasecki, M., Rudnicka, E., Szpakowicz, S., Kędzia, P.: plWordNet 3.0-a comprehensive lexical-semantic resource. In: Proceedings of COLING 2016, pp. 2259–2268 (2016)
12. Miller, G.A.: WordNet: An Electronic Lexical Database. MIT Press, Cambridge (1998)
13. Moro, A., Raganato, A., Navigli, R.: Entity linking meets word sense disambiguation: a unified approach. Trans. Assoc. Comput. Ling. **2**, 231–244 (2014)
14. Navigli, R., Ponzetto, S.P.: BabelNet: building a very large multilingual semantic network. In: Proceedings of the 48th Annual Meeting of the Association for Computational Linguistics, pp. 216–225 (2010)
15. Pasini, T., Raganato, A., Navigli, R.: XL-WSD: an extra-large and cross-lingual evaluation framework for word sense disambiguation. In: Proceedings of the AAAI Conference on Artificial Intelligence. AAAI Press (2021)
16. Pease, A.: Ontology - A Practical Guide. Articulate Software Press, Angwin (2011)
17. Ponzetto, S.P., Navigli, R.: Knowledge-rich word sense disambiguation rivaling supervised systems. In: Proceedings of the 48th Annual Meeting of the Association for Computational Linguistics, pp. 1522–1531 (2010)
18. Raganato, A., Camacho-Collados, J., Navigli, R.: Word sense disambiguation: a unified evaluation framework and empirical comparison. In: Proceedings of the 15th Conference of the European Chapter of the Association for Computational Linguistics, vol. 1. pp. 99–110 (2017)
19. Rudnicka, E., Maziarz, M., Piasecki, M., Szpakowicz, S.: A strategy of mapping polish wordnet onto Princeton wordnet. In: Proceedings of COLING 2012, pp. 1039–1048 (2012)
20. Tan, L., Bond, F.: Building and annotating the linguistically diverse NTU-MC (NTU-multilingual corpus). In: Proceedings of the 25th Pacific Asia Conference on Language, Information and Computation, Singapore (2011)

21. Wang, Z., Lipton, Z.C., Tsvetkov, Y.: On negative interference in multilingual models: findings and a meta-learning treatment. In: Proceedings of the 2020 Conference on Empirical Methods in Natural Language Processing (EMNLP), pp. 4438–4450 (2020)
22. Zaśko-Zielińska, M., Piasecki, M.: Towards emotive annotation in plWordNet 4.0. In: Proceedings of the 9th Global Wordnet Conference, pp. 153–162 (2018)

Sentence-level Sentiment Analysis Using GCN on Contextualized Word Representations

Huyen Trang Phan[1], Ngoc Thanh Nguyen[2(✉)], Zygmunt Mazur[2], and Dosam Hwang[1(✉)]

[1] Department of Computer Engineering, Yeungnam University, Gyeongsan, South Korea
huyentrangtin@ynu.ac.kr, dshwang@yu.ac.kr

[2] Department of Applied Informatics, Wroclaw University of Science and Technology, Wroclaw, Poland
{Ngoc-Thanh.Nguyen,zygmunt.mazur}@pwr.edu.pl

Abstract. Sentiments expressed in opinions on social networks have played an increasingly significant impact in solving various social problems. Improving the effectiveness of sentiment analysis methods on social networks is still of interest to several scientists. A notable and robust development direction is sentiment analysis methods based on graph convolutional networks (GCNs). This paper introduces a model called contextual-based GCN for sentence-level sentiment analysis by considering the following steps: (i) Sentences are converted into contextualized word representation vectors based on the combination of the bidirectional encoder representations from the transformer model and bidirectional long short-term memory. (ii) The contextualized word representations are used to construct a sentence graph as a feature of nodes. (iii) A GCN model with two convolutional layers was used to learn the structure-aware node representations on the sentence graph. (iv) The softmax classifier was used for the sentence-level sentiment analysis. Experimental results on benchmark datasets showed that, unlike other methods, the proposed method can extract more context information from the opinions to obtain a better representation of the graph structure and learn better structure-aware nodes represented on the graph. The proposed method has improved the performance in terms of accuracy of the conventional methods from 2.2 to 3.2% points.

Keywords: Sentence-level sentiment analysis · Graph convolutional network · BERT-BiLSTM · Contextual-based GCN

1 Introduction

Since 2004, social networks have grown exponentially, and they have yet to reach the peak of their popularity. The latest statistics on social networks show that

D. Groen et al. (Eds.): ICCS 2022, LNCS 13351, pp. 690–702, 2022.
https://doi.org/10.1007/978-3-031-08754-7_71

3.78 billion users worldwide have used social media in 2021, and this number will continue to grow over the next few years[1]. Social media platforms are now a significant source of news and information for researchers, governments, and businesses. By extracting and analyzing public moods and views on this data source, businesses, governments, and researchers can gain insight into trade, policies, and proposals, and make better decisions.

The sentiment analysis (SA) is a process of collecting, processing, analyzing, inferring, and synthesizing subjective sentiments contained in texts. SA of opinions is considered an exciting trend for artificial intelligence on social media. In the past, SA was considered a powerful tool for extracting and identifying the polarities of the emotions expressed in texts regarding opinions or evaluations about certain entities [13]. Today, as the economy is gradually transforming into a digital economy, SA is a significant step used in several systems automatically, such as stance detection, recommendation systems, decision-making, and fake news detection on social media. Three main levels of SA exist: document level, sentence level, and aspect level. Document-level SA aims to determine the sentiment polarity of the entire text without dividing sentences, and sentence-level SA seeks to determine the sentiment polarity of separate sentences. In comparison, aspect-level SA aims to determine the sentiment polarity regarding the aspects of entities appearing in the text. Sentence-level SA is the focus of this study. It is used to identify the sentiment in a sentence as positive, negative, or neutral. For instance, consider the sentence *"The phone color is not nice, but its style is so modern."* The sentiment in this sentence is positive.

Various approaches have been proposed for sentence-level SA, such as deep-learning-based, machine-learning-based, lexicon-based, and, most recently, graph convolutional network (GCN)-based approaches. GCN [11] is a graphically structured lattice and is a category of graph neural network. Consider a graph $G = \{V, E, A\}$, where V is a set of nodes; E is a set of edges, and A represent an adjacency matrix. GCNs use convolutional layers to learn node representations well by aggregating knowledge from the neighbours node on graph G [26]. They have achieved promising results in various tasks, particularly in natural language processing [1,14]. The use of GCNs for SA has recently attracted considerable attention and has begun to produce expected results. However, previous GCN-based methods for SA often use GCNs with the limitations as follows.

- Most GCN-based methods focus on aspect-level SA, such as [11,28]; rare approaches [2] consider sentence-level SA.
- Previous GCN-based sentence-level SA methods directly used GCNs without considering contextualized word representation.

The above two justifications motivated us to propose a novel sentence-level SA using GCN over contextualized word representations. The bidirectional encoder representations from the transformer (BERT) model is an embedding model that uses attention models as transformers to establish relationships between words via an encoder at the input and a decoder at the output. Unlike

[1] https://www.statista.com/.

other embedding models that take one word at a time as input, BERT can take the entire sentence as input at once based on transformers. Therefore, BERT can learn the real meaning hidden between words [7]. Unlike basic grammar-based methods, only statistical characteristics that ignore context information are considered. Bidirectional long short-term memory (BiLSTM) can learn contextual information, which is suitable for the logic of human language [3]. Meanwhile, GCNs have considerable expressive power for learning graph representations [29]. The proposed method includes the following steps: First, words in sentences are converted into word vectors, called BERT embeddings, using the pretrained BERT model [7]. Second, contextualized word vectors are created based on BiLSTM over BERT embeddings [8]. Third, the obtained vectors are used as word node features to build the sentence graph. Subsequently, the sentence graph is fed into the GCN model with two convolutional layers to create a sentence representation. Finally, the sentiment of the sentences is classified using the softmax classifier on the sentence representation. Experiments on two benchmark datasets illustrate that our proposed model showed an improvement in performance without using GCN over BERT-BiLSTM methods.

The remainder of this paper is organized as follows. The second section reviews related works on deep-learning-based and graph-based SA methods. The third section describes the research problem of this study. The fourth section presents the proposed method. The fifth section presents the experimental results and evaluations. The final section discusses the conclusions and future work.

2 Related Works

The previous approaches considered sentences in terms of types [20,21], such as conditional and comparison sentences, or linguistics, such as a string of words with a dot symbol at the end. In this study, we considered sentences in terms of linguistics. This refers to considering sentences as short documents for the analysis. Deep-learning-based approaches are correctly oriented in the SA area, and using deep learning algorithms over graph structures is an interesting approach. In this section, we discuss some studies related to SA using deep-learning-based and graph-based methods.

The convolutional neural network (CNN) model was first proposed by Collobert for a semantic role labeling task [5]. In another attempt, Collobert [4] used a CNN model by serving a syntactic parse. A significant development in the direction of CNN-based SA methods is the study by Kim [9] regarding a simple CNN model with one convolution layer to create feature maps and until now it is often used as a strong baseline for various SA methods. Several variants have been introduced based on the CNN-based model of Kim, such as a densely connected CNN [27] for text classification. In addition, Poria et al. [23] applied a CNN for extracting document features and then fed them into a multiple-kernel learning model for SA. In another study [22], the authors used an extended long short-term memory (LSTM) for context information extraction from the surrounding sentences.

Graph structures were first used for SA methods as a representation step in various studies. Minkov et al. [17]built a labeled directed graph to represent words as nodes and the syntactic relation between the words as edges. They then proposed a path-constrained graph walk SA method on the built graph. This algorithm performs better and is more scalable than other methods. Similarly, Violos et al. [26] presented an SA approach by extracting feature vectors from a word graph. Meanwhile, Bijari et al. [2] proposed a GCN-CNN model for sentence-level SA by considering semantic and term relations when text representation includes stopping words. This method marks a new direction for sentence-level SA using deep learning methods over graph structures.

The aforementioned studies show the performance of the deep learning and GCN methods for sentence-level SA. We consider whether combining GCNs and deep learning methods, such as BERT and BiLSTM, improves the performance of sentence-level SA. We aim to verify this hypothesis in this study.

3 Research Problem

3.1 Problem Definition

Consider a set of n opinions $O = \{o_1, o_2, ..., o_n\}$. For $o_i \in O$, let $S = \{s_1, s_2, \ldots, s_n\}$ be a set of sentences in opinions O. For $s_i \in S$, let $s = \{w_1, w_2, \ldots, w_m\}$ be a set of words of one sentence s_i and let c_i be contextualized word embeddings of sentence s_i. The objective of this study is to construct a GCN over the BERT-BiLSTM model for sentence-level SA. This objective can be formalized by finding a mapping function $F : (c_i) \rightarrow \{negative, neutral, positive\}$ such that:

$$F(c_i) = \begin{cases} positive, & if\ sentiment\ expressed\ in\ o_i\ is\ positive, \\ neutral, & if\ sentiment\ expressed\ in\ o_i\ is\ neutral, \\ negative, & if\ sentiment\ expressed\ in\ o_i\ is\ negative \end{cases} \quad (1)$$

3.2 Research Questions

The main objective of this study is to propose a GCN for contextualized word representations (contextual-based GCN) for sentence-level SA. Therefore, we attempt to answer the following research questions:

- How can sentences be converted into contextualized word representations by combining the BERT and BiLSTM models?
- How can we build a sentence graph based on contextualized word representations?
- How can we construct a contextual-based GCN model by using a GCN with two convolutional layers over the sentence graph?
- How can the contextual-based GCN model be used for sentence-level SA?

4 Proposed Method

In this section, we describe the concept and flow of the contextual-based GCN model. The proposed method is illustrated in Fig. 1. The contextual-based GCN model consists of the following main steps:

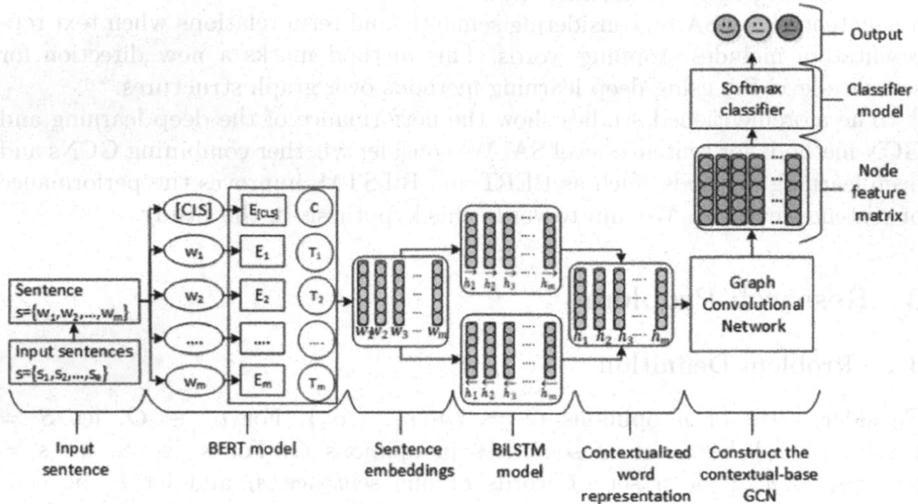

Fig. 1. Overall framework for proposed method

- Sentences are converted into contextualized word representation vectors based on the combination of the BERT and BiLSTM models.
- The contextualized word representations are used to construct a sentence graph as a feature of nodes.
- The GCN model with two convolutional layers is used to learn structure-aware node representations on the sentence graph.
- The softmax classifier is used for sentence-level SA.

The details of the approach are presented in the following sections.

4.1 Creating Sentence Vectors Based on BERT

The BERT model automatically inserts a [CLS] symbol at the beginning of the sentence to indicate the beginning of each sentence. In this study, we intend to obtain the sentence vector; therefore, the BERT model uses the corresponding output vector to the symbol as the representation vector of the entire sentence as follows:

$$C = BERT(s) \in R^{m \times d_w} \tag{2}$$

where $BERT(s)$ is the vector corresponding to sentence s extracted from the pretrained BERT[2] [25], called BERT embeddings, and d_w is the dimension of the word vector.

4.2 Creating the Contextualized Word Representations

In this study, we create the contextualized word representations using the BiLSTM model over the sentence vectors. Unlike basic grammar-based methods, only statistical features are considered, ignoring context information. BiLSTM can learn information about context, consistent with the logic of human language [3,19]. The BiLSTM model was constructed based on the phases as follows:

Input Layer: The BiLSTM model takes the sentence vector of the BERT model, where each word refers to a row of matrix C to which the weight matrix $W^a \in R^{d_a \times m}$ is added, as its input. This layer is formulated as follows:

$$a_i = \sigma(W^a.C_i + b^a) \in R^{d_a} \tag{3}$$

where σ is the sigmoid activation function. $i = [1, m]$, where m is the size of the sentence vector C_i, and b^a is the bias vector with dimension d_a.

BiLSTM Layer: This layer learns the contextual information from the input layer via both directions for words. This layer consists of a forward LSTM and a backward LSTM to encode the sentence from left to right and vice versa, respectively. Therefore, from the word vector a_i, the BiLSTM layer creates a pair of hidden vectors $\overrightarrow{h}_i$ and $\overleftarrow{h}_i$ as follows:

$$\overrightarrow{h_i} = \overrightarrow{lstm}(a_i) \in R^{d_h}, i = [1, m] \tag{4}$$

$$\overleftarrow{h_i} = \overleftarrow{lstm}(a_i) \in R^{d_h}, i = [m, 1] \tag{5}$$

$$h_i = [\overrightarrow{h_i}, \overleftarrow{h_i}] \tag{6}$$

where $\overrightarrow{h}_i$ and $\overleftarrow{h}_i$ are the hidden states of $\overrightarrow{lstm}$ and $\overleftarrow{lstm}$, respectively; h_i is the contextualized representation of the word w_i; for the i-th word in sentence s, $\overrightarrow{lstm}$ and $\overleftarrow{lstm}$ are the forward and backward LSTMs, respectively, and they are performed as follows:

For $\overrightarrow{lstm}(a_i)$:

$$\overrightarrow{G_i} = \begin{bmatrix} \overrightarrow{h}_{i-1} \\ a_i \end{bmatrix} \tag{7}$$

$$f_i = \sigma(W^f.\overrightarrow{G_i} + b^f) \tag{8}$$

$$in_i = \sigma(W^{in}.\overrightarrow{G_i} + b^{in}) \tag{9}$$

$$o_i = \sigma(W^o.\overrightarrow{G_i} + b^o) \tag{10}$$

2 https://github.com/google-research/bert.

$$c_i = f_i \odot c_{i-1} + in_i \odot tanh(W^c.\overrightarrow{G_i} + b^c) \quad (11)$$

$$\overrightarrow{h_i} = o_i \odot tanh(c_i) \quad (12)$$

where h_{i-1} is the previous hidden state of h_i and $h_0 = 0$.

For $\overleftarrow{lstm}(a_i)$:

$$\overleftarrow{G_i} = \begin{bmatrix} \overleftarrow{h}_{i+1} \\ a_i \end{bmatrix} \quad (13)$$

$$f_i = \sigma(W^f.\overleftarrow{G_i} + b^f) \quad (14)$$

$$in_i = \sigma(W^{in}.\overleftarrow{G_i} + b^{in}) \quad (15)$$

$$o_i = \sigma(W^o.\overleftarrow{G_i} + b^o) \quad (16)$$

$$c_i = f_i \odot c_{i-1} + in_i \odot tanh(W^c.\overleftarrow{G_i} + b^c) \quad (17)$$

$$\overleftarrow{h_i} = o_i \odot tanh(c_i) \quad (18)$$

where σ is the sigmoid activation function. h_{i+1} is the next hidden state of h_i, and $h_{m+1} = 0$. c, o, in, f are the operations used in LSTM model. $\odot$ is element-wise multiplication. $W^c, W^o, W^f, W^{in} \in R^{d_h \times (d_h + d_a)}$, $b^f, b^{in}, b^o, b^c \in R^{d_h}$ are LSTM parameters, and d_h is the dimension of the hidden vectors.

Output Layer: From the sentence vector matrix $C \in R^{m \times d_a}$, we obtain the contextualized word matrix $H = (h_1, h_2, ..., h_m) \in R^{m \times d_h}$, where d_h is the dimension of the contextualized vector.

4.3 Building the Contextual-Based GCN Model

We build the contextual-based GCN model, which consists of the following two steps:

Building the Sentence Graph: A sentence graph is denoted by $G = (V, E, A)$, where V is a set of nodes, E is a set of edges, and A is an adjacency matrix. The nodes correspond to words in the sentence. The edges represent the dependencies of adjacent node pairs in the syntactic dependency tree[3]. The matrix $A \in R^{m \times m}$ represents the weights between nodes and are determined as follows:

$$A_{ij} = \begin{cases} 1, & if\ v_i, v_j \in V,\ and\ e_{ij} \in E, \\ 1, & if\ v_i = v_j, \\ 0, & otherwise \end{cases} \quad (19)$$

Additionally, graph G has a node feature matrix $Q = [H] \in R^{m \times d_h}$, where each row Q_i represents the contextualized vector of word node $v_i \in V$.

[3] https://nlp.stanford.edu/software/stanford-dependencies.html.

Creating the Contextual-based GCN: Using the sentence graph, the contextual-based GCN is built by using two convolutional layers over the sentence graph as follows. Each node v_i on the sentence graph is represented by a vector $h_i \in R^{d_h}$. The entire graph yields a node feature matrix $Q \in R^{m \times d_h}$ and an adjacency matrix $A \in R^{m \times m}$. Subsequently, we feed matrices A and Q into a simple two-layer GCN proposed by Kipf et al. [11] as follows:

$$N^1 = \sigma(B \cdot N^0 \cdot D^1 + b^1) \tag{20}$$

Hence,

$$N^2 = \sigma(B \cdot N^1 \cdot D^2 + b^2) \tag{21}$$

That means:

$$X = N^2 = \sigma(\sigma(B \cdot N^0 \cdot D^1 + b^1) \cdot D^2 + b^2)) \tag{22}$$

where $X \in R^{m \times d_h}$; $N^0 = Q$. σ is an activation function, such as $ReLU$. $D^1 \in R^{d_h \times m}$ and $D^2 \in R^{m \times d_h}$ are the weight matrices of two convolutional layers. b^1 and b^2 are the biases of two layers, respectively.

$$B = P^{-0.5} A P^{-0.5} \tag{23}$$

B is the symmetrically normalized matrix of A; P is the degree matrix of A, where:

$$P_{ii} = \sum_j A_{ij} \tag{24}$$

4.4 Sentence-level Sentiment Classifier

This step determines the sentiment polarity of a sentence based on the node representations. The sentiment classifier is defined as follows:

$$\hat{y} = \sigma(M \cdot X + b) \tag{25}$$

where σ is an activation function of Softmax. $M \in R^{l \times m}$ and $b \in R^l$ are a weight matrix and a bias of the σ, respectively, where l is the number of sentiment labels.

The GCN on the contextualized word representation model is trained by minimizing the cross-entropy error of the predicted and true label distributions using the following equation:

$$L = -\sum_i^l y_i log(\hat{y}_i) + \lambda \|\theta\|^2 \tag{26}$$

where y_i is the i-th real label distribution, and $\hat{y}_i$ is the i-th predicted label probability. λ is the coefficient of L_2 regulation. θ is the parameter set from the previous layers. The steps to train the GCN on contextualized word representations model are illustrated as Fig. 2:

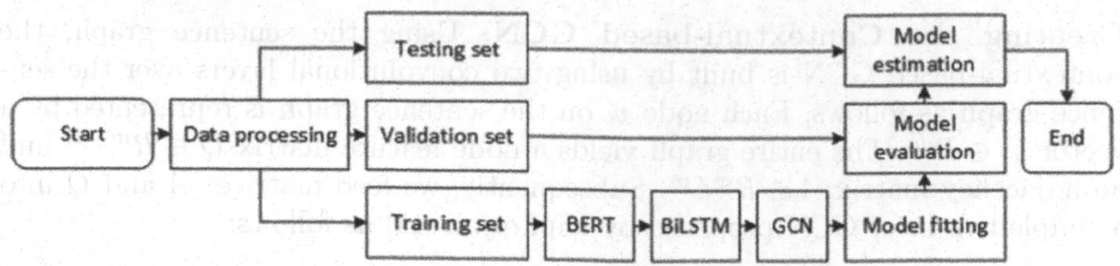

Fig. 2. The training process of the GCN over BERT-BiLSTM model

5 Experimental Evaluation

5.1 Dataset and Experimental Setup

In this study, to demonstrate the performance of our model and to ensure a fair comparison with other methods, we used benchmark datasets, such as IMDB[4] and Financial PhraseBank[5] (FPB) are respectively from Kaggle[6]. Detailed information on the databases is shown in Table 1.

Table 1. Datasets used in experiments

Class	IMDB-Train	IMDB-Validation	IMDB-Test	FPB-Train	FPB-Validation	FPB-Test
Positive	4140	1035	5176	2558	639	320
Neutral	–	–	–	4862	1215	608
Negative	4134	1033	517	773	193	97
Total	8274	2068	5693	8193	2047	1025

The following parameters were set for the proposed model. For the BERT model, we used a pretrained BERT[7] and set the dimensions to 768. All the model weights were initialized with a uniform distribution. The dimensions of the hidden state vectors were set to 300. The Adam optimizer [10] was used with a learning rate of 0.001. The value of λ was 10^{-5}, and the batch size was 64. Moreover, the number of GCN layers was set to 2. The value of parameters was set up via the implementation process.

The results of the experimental process were obtained by averaging five-run results with random initialization, where the accuracy and loss were measured as the evaluation metrics [24]. We also compared the accuracy and loss with those of the baseline methods to confirm the improved performance of our proposed method.

[4] https://www.kaggle.com/lakshmi25npathi/imdb-dataset-of-50k-movie-reviews/version/1.

[5] https://www.kaggle.com/ankurzing/sentiment-analysis-for-financial-news/version/5?select=all-data.csv.

[6] https://www.kaggle.com/.

[7] https://github.com/google-research/bert.

5.2 Baseline Methods

To demonstrate the improved performance of our model compared with that of other models, we performed three different methods, namely, the proposed method and two baseline methods, on two datasets.

- The ensemble of LSTM and CNN [16] is an SA model based on LSTM to capture the temporal information and on CNN to extract the local structure from the data.
- CNN + GloVe and LSTM + GloVe [16] are variants of CNN and LSTM, respectively, using GloVe embeddings as the input layer.
- A modified GCN is a GCN-based SA with two convolutional layers.

5.3 Results and Discussion

The performances of the SA methods on the given datasets are shown in Tables 2 and 3.

Table 2. Performance of our proposal on Train and Validation (Val) datasets

IMDB						
Epoch	Train accuracy	Train loss	Val accuracy	Val loss	Test accuracy	Test loss
1	0.8850	0.2801	0.8504	0.3240	0.8366	0.3492
2	0.9402	0.1751	0.8704	0.3027	0.8512	0.3254
3	0.9631	0.1092	0.8759	0.3471	0.8579	0.3523
4	0.9850	0.0563	0.8686	0.3801	0.8579	0.4018
5	0.9931	0.0307	0.8723	0.4676	0.8546	0.4778
FPB						
Epoch	Train accuracy	Train loss	Val accuracy	Val loss	Test accuracy	Test loss
1	0.9240	0.2128	0.8704	0.3218	0.8839	0.2792
2	0.9757	0.0764	0.8869	0.3279	0.9580	0.1186
3	0.9872	0.0400	0.8631	0.4790	0.9733	0.0913
4	0.9887	0.0288	0.8686	0.5028	0.9713	0.1153
5	0.9930	0.0186	0.8723	0.4459	0.9726	0.0849

From Table 2, we can observe that, although the number of samples of the two datasets is the same, the performance of the proposed method on the IMDB dataset is slightly better than that on the FPB dataset. This is mainly because samples in the IMDB dataset are classified into two classes, whereas samples in the FPB are divided into three classes. This indicates that the samples in the IMDB dataset are denser than those in the FPB dataset. The proposed method can significantly improve this result by constructing a dataset that ensures a better balance between sentiment classes.

Table 3 presents a performance comparison of the models. The proposed method obtained better results than the baseline methods for the IMDB dataset.

Table 3. Performance comparison of models on IMDB dataset (%)

Method	Accuracy
CNN + GloVe	89.3
LSTM + GloVe	89.0
Ensemble of LSTM and CNN	90.0
Modified GCN	90.0
Proposed method	92.2

The proposed method improved the accuracy of the LSTM + Glove model by 3.2% points, that of the CNN + Glove model by 2.9% points, and that of the LSTM + CNN model and the modified GCN model by 2.2% points. Let us consider why the proposed method can enhance the accuracy of the baseline methods. In this study, the BERT model can capture the semantics of the text well, and the BiLSTM model can accurately extract the context of the sentence. Moreover, the combination of BERT and BiLSTM has considerable effectiveness in capturing contextualized representations. The results once again confirm that the use of GCN for contextual information representation significantly impacts the accuracy of sentence-level SA methods.

6 Conclusion and Future Works

This paper proposed a method to improve the performance of sentence-level SA based on a GCN on a contextualized word representation model. Experimental results showed that the proposed method significantly improves the performance of sentence-level SA on two benchmark datasets. However, this method does not consider all contextual factors, semantic relations, and emotional knowledge simultaneously when building text representation graphs. In the future, we will focus on the following directions: (i) Building graphs simultaneously represent contextual factors, semantic relations [18], and sentimental knowledge to improve the performance of SA methods. (ii) Constructing a consensus-based [6], user profiles tuning-based [15], and symbolic knowledge representation-based [12] GCNs for ALSA are also an interesting approaches.

References

1. Bastings, J., Titov, I., Aziz, W., Marcheggiani, D., Sima'an, K.: Graph convolutional encoders for syntax-aware neural machine translation. arXiv preprint arXiv:1704.04675 (2017)
2. Bijari, K., Zare, H., Kebriaei, E., Veisi, H.: Leveraging deep graph-based text representation for sentiment polarity applications. Expert Syst. Appl. **144**, 113090 (2020)
3. Cai, R., et al.: Sentiment analysis about investors and consumers in energy market based on BERT-BiLSTM. IEEE Access **8**, 171408–171415 (2020)

4. Collobert, R.: Deep learning for efficient discriminative parsing. In: Proceedings of the Fourteenth International Conference on Artificial Intelligence and Statistics, pp. 224–232. JMLR Workshop and Conference Proceedings (2011)
5. Collobert, R., Weston, J., Bottou, L., Karlen, M., Kavukcuoglu, K., Kuksa, P.: Natural language processing (almost) from scratch. J. Mach. Learn. Res. **12**(ARTICLE), 2493–2537 (2011)
6. Daniłowicz, C., Nguyen, N.T.: Consensus-based methods for restoring consistency of replicated data. In: Intelligent Information Systems. Advances in Soft Computing, vol. 4, pp. 325–335. Physica, Springer, Heidelberg (2000). https://doi.org/10.1007/978-3-7908-1846-8_29
7. Devlin, J., Chang, M.W., Lee, K., Toutanova, K.: Bert: Pre-training of deep bidirectional transformers for language understanding. arXiv preprint arXiv:1810.04805 (2018)
8. Graves, A., Schmidhuber, J.: Framewise phoneme classification with bidirectional LSTM and other neural network architectures. Neural Netw. **18**(5–6), 602–610 (2005)
9. Kim, Y.: Convolutional neural networks for sentence classification. CoRR abs/1408.5882 (2014). http://arxiv.org/abs/1408.5882
10. Kingma, D.P., Ba, J.: Adam: a method for stochastic optimization. arXiv preprint arXiv:1412.6980 (2014)
11. Kipf, T.N., Welling, M.: Semi-supervised classification with graph convolutional networks. arXiv preprint arXiv:1609.02907 (2016)
12. Kolaczek, G., Pieczynska-Kuchtiak, A., Juszczyszyn, K., Grzech, A., Katarzyniak, R.P., Nguyen, N.T.: A mobile agent approach to intrusion detection in network systems. In: Khosla, R., Howlett, R.J., Jain, L.C. (eds.) KES 2005. LNCS (LNAI), vol. 3682, pp. 514–519. Springer, Heidelberg (2005). https://doi.org/10.1007/11552451_69
13. Liu, B.: Sentiment analysis and opinion mining. Synth. Lect. Hum. Lang. Technol. **5**(1), 1–167 (2012)
14. Marcheggiani, D., Titov, I.: Encoding sentences with graph convolutional networks for semantic role labeling. arXiv preprint arXiv:1703.04826 (2017)
15. Mianowska, B., Nguyen, N.T.: Tuning user profiles based on analyzing dynamic preference in document retrieval systems. Multimed. Tools Appl. **65**(1), 93–118 (2013)
16. Minaee, S., Azimi, E., Abdolrashidi, A.: Deep-sentiment: sentiment analysis using ensemble of CNN and Bi-LSTM models. arXiv preprint arXiv:1904.04206 (2019)
17. Minkov, E., Cohen, W.: Learning graph walk based similarity measures for parsed text. In: Proceedings of the 2008 Conference on Empirical Methods in Natural Language Processing, pp. 907–916 (2008)
18. Nguyen, N.T.: Processing inconsistency of knowledge on semantic level. J. Univers. Comput. Sci. **11**(2), 285–302 (2005)
19. Phan, H.T., Nguyen, N.T., Hwang, D.: Convolutional attention neural network over graph structures for improving the performance of aspect-level sentiment analysis. Inf. Sci. **589**, 416–439 (2022)
20. Phan, H.T., Nguyen, N.T., Tran, V.C., Hwang, D.: A method for detecting and analyzing the sentiment of tweets containing conditional sentences. In: Nguyen, N.T., Gaol, F.L., Hong, T.-P., Trawiński, B. (eds.) ACIIDS 2019. LNCS (LNAI), vol. 11431, pp. 177–188. Springer, Cham (2019). https://doi.org/10.1007/978-3-030-14799-0_15

21. Phan, H.T., Nguyen, N.T., Van Cuong, T., Hwang, D.: A method for detecting and analyzing the sentiment of tweets containing fuzzy sentiment phrases. In: 2019 IEEE International Symposium on Innovations in Intelligent Systems and Applications (INISTA), pp. 1–6. IEEE (2019)
22. Poria, S., Cambria, E., Hazarika, D., Majumder, N., Zadeh, A., Morency, L.P.: Context-dependent sentiment analysis in user-generated videos. In: Proceedings of the 55th Annual Meeting of the Association for Computational Linguistics (volume 1: Long papers), pp. 873–883 (2017)
23. Poria, S., Chaturvedi, I., Cambria, E., Hussain, A.: Convolutional MKL based multimodal emotion recognition and sentiment analysis. In: 2016 IEEE 16th International Conference on Data Mining (ICDM), pp. 439–448. IEEE (2016)
24. Schütze, H., Manning, C.D., Raghavan, P.: Introduction to Information Retrieval, vol. 39. Cambridge University Press, Cambridge (2008)
25. Turc, I., Chang, M.W., Lee, K., Toutanova, K.: Well-read students learn better: on the importance of pre-training compact models. arXiv preprint arXiv:1908.08962v2 (2019)
26. Violos, J., Tserpes, K., Psomakelis, E., Psychas, K., Varvarigou, T.: Sentiment analysis using word-graphs. In: Proceedings of the 6th International Conference on Web Intelligence, Mining and Semantics, pp. 1–9 (2016)
27. Wang, S., et al.: Densely connected CNN with multi-scale feature attention for text classification. In: IJCAI, pp. 4468–4474 (2018)
28. Yao, L., Mao, C., Luo, Y.: Graph convolutional networks for text classification. In: Proceedings of the AAAI Conference on Artificial Intelligence, vol. 33, pp. 7370–7377 (2019)
29. Zhang, S., Tong, H., Xu, J., Maciejewski, R.: Graph convolutional networks: a comprehensive review. Comput. Soc. Netw. **6**(1), 1–23 (2019). https://doi.org/10.1186/s40649-019-0069-y

Machine Learning for Bus Travel Prediction

Łukasz Pałys[1], Maria Ganzha[1(✉)], and Marcin Paprzycki[2]

[1] Warsaw University of Technology, Warsaw, Poland
maria.ganzha@pw.edu.pl

[2] Systems Research Institute Polish Academy of Sciences, Warsaw, Poland
marcin.paprzycki@ibspan.waw.pl

Abstract. Nowadays, precise data of movements of public transport can be collected. Specifically, for each bus, geoposition can be regularly obtained and stored. In this context, an attempt to build a model to represent behavior of busses, and predict their delays, is discussed.

1 Introduction and Related Work

Recently, Warsaw, Poland, opened its data resources, including regularly collected, location of city buses. Based on this information an attempt was made to predict delays in public transport. This problems has been discussed in the literature. Applied methods can be divided into groups. First, *statistical methods*, such as k-NN [2,7], regression [2], prediction based on sequence patterns [9] or time series [10], and application of Kalman filter [6]. Second, *methods based on observations*, which include the historical means approach [1,5]. Next, *methods based on machine learning*, which include (see, also, [8]): back propagation neural network [11], radial basis function network [12], multilayer perceptron [1,5]. The last group consists of *hybrid methods* that combine multiple algorithms into a single model [4,13]. Let us now present selected results, while [15] contains detailed discussion.

The Pattern Sequence-based Forecasting was used in [9], to model bus transport in Chennai (India). Overall, K means algorithm delivered best performance. Predictions have been tested on sections with low, medium and high travel time variability. For sections with large time variance, results were not "satisfatory".

Time series approach was used in [10] to forecast bus travel time in Lviv (Ukraine). The average error, for travel time towards city center was approximately 3 min, and 2 min in the opposite direction.

In [12], authors used Radial Basis Function Neural Network (RBFNN) to model bus transport in Dalian (China). The model was trained on historical data, with additional correction of results (using Kalman Filter). Overall, the best reported mean absolute percentage error (MAPE) was at 7.59%.

Work described in [13], is based on deep learning and data fusion. Models used multiple features, including: stop ID, day of the week, time, bus speed (based on GPS), stopping time at a stop, travel time between stops. Data came from Guangzhou and Shenzhen (China) for single line in each city. Predictions

D. Groen et al. (Eds.): ICCS 2022, LNCS 13351, pp. 703–710, 2022.
https://doi.org/10.1007/978-3-031-08754-7_72

were compared with historical averages. Proposed model outperformed other approaches with MAPE of 8.43% (Gangzhou). Moreover, MAPE for the peak hours was 4–7% lower than that reported for other solutions.

In [1], dynamic model was developed, to predict bus arrival times at subsequent stops. GPS data, from Macea (Brazil), for a bus line with 35 stops, was used. Here, *Historical Average (HA)*, *Kalman Filtering* and *Artificial Neural Network (ANN)* were tried. The 3-layer perceptron achieved best MAPE (18.3%).

In [2], review of methods modeling bus transport, found in [3], was extended. Authors used: *k-NN, ANN*, and *Super Vector Regression (SVR)* to predict bus travel times in Trondheim (Norway). To study influence of individual time intervals, 423 different data sets were created. Depending on data set, (best) MAE varied between 61 and 86 s. Separately, additional attributes, e.g.: weather, football matches, tickets, were tried, with no visible improvement.

In [4], an ANN was used, with historical GPS data and an automatic toll collection system data. Moreover, impact of intersections with traffic lights was taken into account. Data came from Jinan (China), for a single bus line. To deal with travel time variation, a hybrid ANN (HANN) was developed, with separate subnets, trained for specific time periods, e.g. working days, weekends, peak hours. Overall, ANN and HANN were more reliable than the Kalman Filter. Moreover, HANN was better suited for short-distance prediction.

Separately, comparison of methods modeling tram travel in Warsaw (Poland), on the basis of historical GPS data, is presented in [5].

Main findings from the literature can be summarized as follows. (a) **Reported results** concern a single city. Only in one case two cities have been studied. (b) In all cases a **single bus line** was used. (c) There is **no "benchmark" data** for the problem. (d) The **main methods** that have been tried were: (i) statistical methods – k-NN, regression model, Kalman filter, (ii) historical observation methods (HA), (iii) machine learning methods – BPNN, RBFN, multilayer perceptron (MLP), and (iv) hybrid methods combining the above algorithms into one model (HANN). Among them, best results have been reported for HANN, HA, RBFN and MLP. However, it was reported that HANN requires substantially larger datasets. (e) **Quality of predictions** has been measured using: (i) mean absolute error (MAE; in seconds); (ii) mean percentage absolute error (MAPE); (iii) standard deviation (STD; in seconds). (f) The **simplest approaches** used GPS data alone. Other popular data elements were: information about sold tickets and about bus speed. Additionally, effects of non-travel events (e.g. games or weather) have been (unsuccessfully) tried. (g) **Best accuracy** was reported in [2], where MAE was 40s. However, this result was obtained for 1 to 16 stops only. For a "longer bus line" best MAE was of order 60s-70s. Finally, for long Warsaw tram lines (more than 40 stops), best MAE was at 123s.

2 Data, Its Preprocessing, and Experimental Setup

As a part of the project *Open data in Warsaw*[1], exact location of public buses, reported in real time, is available. From there, 30 days of data reporting bus

[1] https://api.um.warszawa.pl/.

movements was harvested (total of about 10 GB of data). Data was filtered, retaining: (1) line number, (2) departure time from the last stop, (3) current percentage of distance traveled between adjacent stops, (4) time of the last GPS signal, (5) current time, (6) driving direction. Additionally, file containing timetable of buses, their routes, including list, and GPS coordinates, of stops, and departure times from each stop, was downloaded from the ZTM site[2]. After preprocessing, file with: line number, time, vehicle and brigade number, driving direction, number of the next stop to visit on the route, information whether the vehicle is at the stop, the percentage of distance traveled between consecutive stops, was created. All preprocessed data is available from: https://github.com/lukaspal97/predicting-delays-in-public-transport-in-Warsaw-data.

From available data, 29 bus lines have been selected. Based on "manual" analysis of their routes (in the context of Warsaw geography), the selected bus lines have been split into eight semi-homogeneous groups:

1. Long routes within periphery North-South: bypassing the City Center, running on the western side of the Wisła River; lines: 136, 154, 167, 187, 189.
2. Long routes within periphery West-East: bypassing the City Center, crossing the Wisła River; lines 112, 186, 523.
3. Centre-periphery: routes with one end in the City Center, running to the peripheries; not crossing the river; lines 131, 503, 504, 517, 518.
4. Long routes through Center (with ends on peripheries); lines 116, 180, 190.
5. Express: a fairly straight long lines with small number of stops (typically around 1/3 stops of "normal" lines); lines 158, 521, 182, 509.
6. Centre-Praga: short routes starting in the City Center; crossing the Wisła River; lines 111, 117, 102.
7. Short lines within peripheries: in Western Warsaw; lines 172, 191, 105.
8. Short lines within the Center: not crossing the river; lines 128, 107, 106.

Travel time distribution differs between days of the week (e.g. working days vs. weekends). Therefore, following [2], all models were trained on data from the same day of the week, from three weeks, and tested on data from the last (fourth) week. For each model, the best results are reported. In general, methods that use "extra features" were compared to these that use only GPS location. Accuracy was measured using MAE and STD.

3 Experimental Results and Their Analysis

Let us now summarize the experimental results[3]. We start with **Total Travel Time Prediction.** Here, two methods have been tried: "recursive" and "long distance". In the *recursive* method, the model is trained on data that includes travel time to the nearest stop. Hence, estimating travel time from stop n to $n+k$ consists of predicting k-steps using the trained model, i.e. result from the

[2] https://www.wtp.waw.pl.

[3] All reported models have been implemented in Python, using Keras library.

previous stop is included, as the input data, for the next prediction. The *long distance* method is based on prediction of travel time to a specific stop. Here, training dataset includes information about total travel time. The assumption was that this approach would be worse for short distances, but better for long(er) trips.

The comparison was made for bus line 523. Training data originated from: March 11, 18, 25. Testing data was from April 1. The number of records in the training set was over 1 million, and in the test set over 300,000. Both approaches used MLP with two hidden layers consisting of 6 and 24 neurons, with ReLU activation function. Results are presented in Table 1. As can be seen, for travel longer than 4 stops, the long distance method was more accurate than the recursive method. Moreover, for distances longer than 8 stops, most results were more than twice as accurate. For 1–3 stops, results of both methods were comparable.

Table 1. Prediction of total travel time (bus line 523)

Using the recursive method								
Distance	Time							
	7:00–10:00		10:00–14:00		14:00–19:00		19:00–23:00	
	MAE	STD	MAE	STD	MAE	STD	MAE	STD
1–3	58.24 s	62.76 s	60.01 s	79.94 s	59.51 s	75.74 s	47.02 s	54.01 s
4–8	125.52 s	129.44 s	130.89 s	141.44 s	131.74 s	124.96 s	92.10 s	88.65 s
9–15	211.36 s	165.19 s	214.10 s	179.41 s	240.66 s	170.14 s	156.74 s	121.10 s
16–20	294.83 s	188.44 s	310.42 s	219.57 s	344.49 s	201.74 s	222.66 s	137.57 s
21–27	360.19 s	200.56 s	398.48 s	266.61 s	407.73 s	217.13 s	277.52 s	154.40 s
Using long distance method								
1–3	56.72 s	66.06 s	59.97 s	69.67 s	59.07 s	64.44 s	54.82 s	51.11 s
4–8	81.07 s	97.31 s	85.59 s	106.85 s	78.52 s	87.18 s	66.44 s	68.93 s
9–15	107.01 s	121.46 s	101.93 s	129.62 s	105.89 s	113.19 s	85.65 s	87.78 s
16–20	130.41 s	144.66 s	140.38 s	170.57 s	130.76 s	138.80 s	113.58 s	106.45 s
21–27	153.70 s	161.04 s	170.72 s	205.03 s	155.09 s	152.63 s	172.30 s	135.29 s

Architecture Comparison. Here, effectiveness of four different *RBFN* and five *MLP* architectures was compared. Training dataset included the same working days of the week from three consecutive weeks (e.g. March, 10, 17 and 24). The test data was from the following week (March 29 to April 2). Only data from working days was used. The RBFN architectures were implemented using the RBF [14] code. The hidden layer used Gaussian radial basis function: $exp(-\beta r^2)$. The tested architectures had $M = 10, 15, 25, 35$ neurons in the hidden layer (denoted as RBFN M). As can be seen in [15], for short routes (*Center–Praga, short within Center, short within periphery*) and *Express* routes, the most accurate results were obtained by "smaller" RBFN architectures (RBFN 10 and

RBFN 15). For the remaining routes, RBFN 25 and RBFN 35 performed better. Overall, if one model is to be selected, RBFN 25 seems to be the "best architecture".

For MLP, five architectures have been tried: networks with 2, or 3, hidden layers and: [6, 12], [12, 32], [6, 8, 12], [8, 8] neurons. Here, *ReLU* and *tanh* were also compared, as activation functions. The results are presented in Table 2.

Overall, the most effective, and stable, architecture, for all groups, had two hidden layers (12 and 32 neurons), with ReLU as the activation function. Since, the [12, 32] ReLU was the "overall winner", its performance is reported in what follows. Additional results, comparing performance of RBFN 25 and MLP [12, 32] ReLU, can be found in [15].

Table 2. MLP-based prediction results

Grupa	Model									
	MLP 6, 12 ReLU		MLP 12, 32 ReLU		MLP 12, 32 ReLU tanh		MLP 6, 8, 12 ReLU		MLP 8, 8 tanh	
	MAE	STD	MAE	STD	MAE	STD	MAE	STD	MAE	STD
Center–Praga	112.33 s	143.9 s	96.43 s	119.94 s	**94.08 s**	**118.30 s**	114.5 s	147.78 s	115.93 s	146.75 s
Center–periphery	**101.54 s**	**133.53 s**	106.9 s	135.26 s	110.86 s	146.74 s	110.69 s	145.8 s	143.31 s	180.73 s
Long within periphery North-South	139.06 s	178.95 s	119.79 s	**159.51 s**	**114.39 s**	164.76 s	120.05 s	164.01 s	136.25 s	173.61 s
Long within periphery East–West	122.47 s	167.79 s	**113.67 s**	**151.78 s**	133.6 s	181.72 s	124.23 s	162.63 s	154.59 s	196.67 s
Long through the centre	109.35 s	146.14 s	**97.71 s**	**128.49 s**	118.17 s	153.16 s	105.89 s	139.59 s	109.16 s	143.21 s
Express	127.52 s	168.45 s	**92.46 s**	**123.74 s**	118.71 s	157.53 s	116.61 s	154.02 s	132.97 s	174.49 s
Short within center	97.02 s	130.73 s	**95.42 s**	**121.13 s**	108.07 s	139.06 s	106.31 s	132.19 s	101.55 s	135.38 s
Short within periphery	107.13 s	133.08 s	**90.24 s**	**122.65 s**	96.44 s	124.07 s	99.89 s	129.8 s	104.84 s	139.98 s
All groups	114.55 s	150.32 s	**101.57 s**	**132.81 s**	111.78 s	148.16 s	112.27 s	146.97 s	124.82 s	161.35 s

Next experiment used HA method [1,5], which applies average travel times from previous days to estimate the current travel time. For each bus line, data from the training sets (i.e. all working days from March, 8–26) was divided into 20-min groups (i.e. records from 10:00–10:20 belonged to one group). The average travel times between current locations and all stops, till the end of the route, were calculated. Average travel times, determined using this algorithm, have been stored as the training sets. Next, for each bus line, for all test sets (i.e. data for working days from March, 29 to April, 2), travel time predictions were calculated using the HA algorithm. In addition, analogous calculations were carried out, while the average travel times were calculated for the same day of

Table 3. Prediction MAE for timetable, HA for same days, HA for working days.

Group	Time table	HA(the same day)	HA(all days)
Long within periphery North-South	79.22 s	81.53 s	**71.94 s**
Long within periphery West-East	80.85 s	75.02 s	**62.38 s**
Center–periphery	64.76 s	62.02 s	**61.62 s**
Long through the centre	71.58 s	68.75 s	**67.64 s**
Express	60.38 s	64.65 s	**52.70 s**
Center–Praga	64.56 s	52.89 s	**50.22 s**
Short within periphery	55.87 s	50.92 s	**35.65 s**
Short within centre	60.41 s	53.02 s	**49.02 s**

the week only. For example, travel time predictions for March 31 (Wednesday) were based on data from March 10, 17, 24. Table 3 summarizes the MAE values of travel time predictions using the HA method. As can be seen, HA based on data from all working days provides better accuracy then when using data from the same day of the week only.

Next, a hybrid model was developed, consisting of: (1) RBFN 25 or MLP [12, 32], using predicted travel time based on schedule data and delay at the last stop; and (2) RBFN 25 or MLP [12, 32], using estimated travel time using HA. When making predictions, depending to which group given bus line belonged, and the distance for which the prediction was performed, the hybrid approach used model that was expected to be the best for that combination (bus+distance).

The created hybrid model was compared with: MLP and RBFN models that used the basic set of features, the HA algorithm, and predictions based on data from timetables only. Figure 1 represents comparison of MAE values (in seconds) of the predictions made by the hybrid model with the remaining methods, depending on the distance (number of stops). Due to space limitation, only two, very different, cases are reported. Bus lines belonging to the same group had different lengths of routes. Therefore, black triangles mark distances for which the prediction of travel time ends for specific bus lines, belonging to the group described by the graph. For this reason, the graphs report rapid changes in MAE for adjacent distances, as in the case of *Long within periphery* group, where lines have total of 30, 35, 36 and 38 stops. Results represented in both figures, and the remaining experimental results (see, [15]), show that for short-distance predictions, for all groups, the HA algorithm combined with distribution times, delivered better accuracy than the hybrid model.

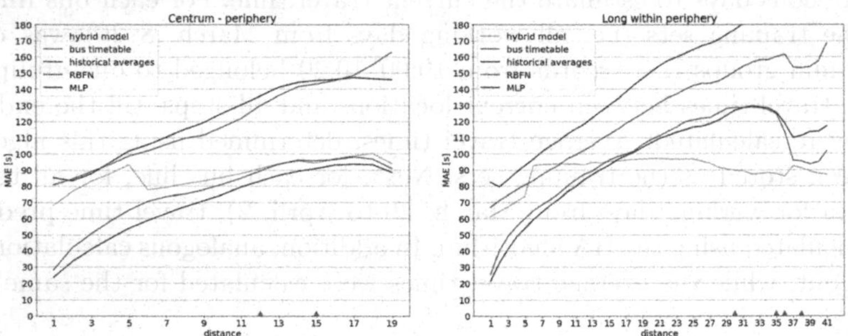

Fig. 1. Comparison of the MAE value of prediction methods at different distances

4 Concluding Remarks

Now, let us consider top 5 lowest MAE values found in the literature vis-a-vis results reported here: (1) ANN from [2]; MAE at 40 s; for distances of up to 16 stops. (2) Hybrid model reported here; MAE at 67.17s; for the *Short through the center* group, for time interval 19:00 to 23:00. (3) ANN from [1]; MAE at

70 s. (4) BP from [11]; MAE at 125 s; for morning rush hours. (5) MLP from [5]; MAE at 138.40 s; for travel time of trams for noon hours.

In this context recall that in each article, results were obtained for bus (or tram) lines from different cities, with very different road and traffic structures, and public transport characteristics. For example, in more populated cities, or those with less developed infrastructure, there may be more delays due to the heavy traffic, which may influence the accuracy. Moreover, Trondheim (city population is around 200 thousands, while metropolitan population is around 280,000) is much smaller than other cities. Further, Warsaw is split by a river, with limited number of bridges. Besides, analyzed bus/tram lines had different lengths. Finally, no other work used data for multiple bus lines "jointly" (combined into groups). Taking this into account, it can be argued that the results reported in this contribution (and in [15]) are very competitive and worthy further explorations.

Separately note that additional experiments confirmed that use of auxiliary features, e.g. number of busses moving between stops, or number of crossings with lights, did not visibly improve accuracy of prediction for any of the tried models. This fact seems to be somewhat counter intuitive, but it supports results reported in [2].

References

1. Fan, W., Gurmu, Z.: Dynamic travel time prediction models for buses using only GPS data. Int. J. Transp. Sci. Technol. **4**(4), 353–366 (2015). https://www.sciencedirect.com/science/article/pii/S204604301630168X
2. Dahl, E., Sjåfjell, A., Skogen, S.: On implementations of bus travel time prediction utilizing methods in artificial intelligence, NUST (2014)
3. Sjåfjell, A., Dahl, E., Skogen, S.: Intelligent transportation systems and artificial intelligence - a state of the art review, NUST (2013)
4. Lin, Y., Yang, X., Zou, N., Jia, L.: Real-time bus arrival time prediction: case study for Jinan, China. J. Transp. Eng. **139**(11), 1133–1140 (2013)
5. Zychowski, A., Junosza-Szaniawski, K., Kosicki, A.: Travel time prediction for trams in Warsaw. In: Kurzynski, M., Wozniak, M., Burduk, R. (eds.) CORES 2017. AISC, vol. 578, pp. 53–62. Springer, Cham (2018). https://doi.org/10.1007/978-3-319-59162-9_6
6. Yang, J.S.: Travel time prediction using the GPS test vehicle and Kalman filtering techniques. In: Proceedings of the 2005, American Control Conference, 2005, pp. 2128–2133. IEEE (2005)
7. Jiwon, M., Kim, D., Kho, S., Park, C.: Travel time prediction using k nearest neighbor method with combined data from vehicle detector system and automatic toll collection system. Transp. Res. Rec. J. Transp. Res. Board **2256**, 51–59 (2012)
8. Haykin, S.: Neural Networks: A Comprehensive Foundation. Prentice Hall PTR, Upper Saddle River, NJ, USA (1998)
9. Shaji, H.E., Tangirala, A.K., Vanajakshi, L.: Prediction of trends in bus travel time using spatial patterns. Transp. Res. Procedia **48**, 998–1007 (2020)
10. Comi, A., Zhuk, M., Kovalyshyn, V., Hilevych, V.: Investigating bus travel time and predictive models: a time series-based approach. Transp. Res. Procedia **45**, 692–699 (2020)

11. Fei, J., Lu, Y., Guo, Y., Zhang, H.: Predicting bus arrival time using BP neural network and dynamic transfer. Procedia Comput. Sci. **174**, 95–100 (2019)
12. Wang, L., Zuo, Z., Fu, J.: Bus arrival time prediction using RBF neural networks adjusted by online data. Procedia. Soc. Behav. Sci. **138**, 67–75 (2014)
13. Wang, L., Zuo, Z., Fu, J.: Bus dynamic travel time prediction: using a deep feature extraction framework based on RNN and DNN. Electronics **9**(11), 1876 (2020). https://doi.org/10.3390/electronics9111876. https://www.mdpi.com/2079-9292/9/11/1876
14. Vidnerová, P.: RBF-Keras: an RBF Layer for Keras Library (2019). https://github.com/PetraVidnerova/rbf_keras
15. Pałys, Ł., Ganzha, M., Paprzycki, M.: Applying machine learning to predict behavior of bus transport in Warsaw, Poland. https://arxiv.org/submit/4254888

Prediction of Ether Prices Using DeepAR and Probabilistic Forecasting

Andras Ferenczi and Costin Bădică(✉)

University of Craiova, Craiova, Romania
ferenczi.andras.h5f@student.ucv.ro, costin.badica@edu.ucv.ro

Abstract. Ethereum is a decentralized public blockchain powered by its native cryptocurrency, the Ether (Ξ), which is second to Bitcoin (BTC) in market capitalization. To ensure the integrity of the network, Ethereum requires a fee for every transaction. This fee is called gas (by analogy to the fuel used by cars) and can fluctuate based on supply and demand. This volatility stepped up a number of initiatives to predict future gas prices. The paper proposes a novel solution beyond current state-of-art using DeepAR. This is a model built on Recurrent Neural Networks with the ability leverage hundreds of related time series to produce more accurate predictions. Our implementation uses features extracted from the Ethereum MainNet as well as off-chain data to achieve accurate predictions.

Keywords: Ethereum · Blockchain · Gas price · Proof of work · Machine learning · Prediction · Probabilistic forecasting · DeepAR

1 Introduction

Ethereum is one of the most popular blockchains and its cryptocurrency is ever increasing in value. Just as other blockchains, such as Bitcoin, it uses a consensus algorithm called Proof of Work (PoW), a.k.a. mining, to maintain the integrity of the blockchain and to prevent double spend. The blockchain provides financial incentives to miners to perform PoW in the form of newly-minted coins and transaction fees paid by users wanting to perform transactions. In Ethereum these fees are called "gas".

The term comes from the analogy that a car needs fuel to run and gas is the fuel that helps recording of transactions on the distributed ledger. Gas is the unit of measurement of computational power required for a miner to process a transaction and is measured in $WEI = 10^{-18} ETH$. The price of the execution of a transaction, a contract, or a deployment for a smart contract is $GasCost \cdot GasPrice$ [10]. Just as fuel prices in real world, gas price may vary being subject of a negotiation process. The sender of a transaction specifies the maximum amount they are willing to pay, just as the miner has the options of accepting, partially refunding, or rejecting the offer.

D. Groen et al. (Eds.): ICCS 2022, LNCS 13351, pp. 711–718, 2022.
https://doi.org/10.1007/978-3-031-08754-7_73

Ethereum is also a hotbed for innovative Decentralized Applications fueled by Smart Contracts [21], the foundation for tokens representing digital assets or even real-world objects [14,18]. Token transfers, as much as cryptocurrency transactions have an impact on gas prices, and hence provide valuable information for our prediction model.

2 Background and Related Works

2.1 Background

The Ethereum Blockchain. A block in the blockchain contains a header and several Merkle Patricia Trie structures [19], including one that has the transactions in it. Our model uses a limited number of fields as presented in Table 1.

Table 1. Block header fields

Item	Description	Source
Timestamp	When block was assembled.	Block Header
Gas used	Total gas used for block	Block Header
Block number	Index of block	Block Header
Hash	Transaction hash	Transaction
Gas	Gas paid by sender	Transaction
Gas price	Price in Wei	Transaction

The Estimation Model. For this experiment we decided to use DeepAR [13]. Amazon SageMaker DeepAR is a tool that implements an unsupervised forecasting model based on autoregressive recurrent neural networks (RNN).

Unlike other forecasting methods, such as autoregressive integrated moving average (ARIMA) and exponential smoothing (ETS), DeepAR can learn a global model from multiple time series. The empirical experimental results produced by the Amazon team [13] show an improvement on standard metrics of up to 15% compared to state-of-the-art methods, such as Facebook's Prophet [16].

2.2 Related Works

Three methods to analyze and predict gas prices are highlighted in [9]. The first method assumes the analysis of pending transactions in large Mempools [6]. Mempool is a buffer area where pending transactions sent by Ethereum clients are stored before they are added to the Ethereum blockchain. This method proves to be resource-intensive and complex to implement as it requires access to multiple Mempools and also it assumes that the owners of these Mempools are honest.

A second method analyzes recently committed blocks using oracles. These are systems that connect the blockchain to the outside world. Specifically, gas price oracles provide guidance to users regarding the gas price to pay to ensure that miner will accept the fees and commit the submitted transactions into subsequent blocks [15]. Examples include Ethereum client, Geth [8], EthGasStation [3], Gas Station Express [4].

A forecasting model based on Gated Recurrent Unit (GRU) [7] and a Gas Recommendation engine that leverages the output of the forecasting model was proposed in [20]. The approach used a Neural Net model that also included an additional parameter that reflects the urgency of the transaction (the higher the gas price, the faster the transaction is committed). The model reduced fees by more than 50% while increasing the waiting time by 1.3 blocks, when compared to the GETH oracle.

Rawya Mars et al. [12] evaluated the LSTM, GRU and Prophet models [16] to anticipate gas prices. An empirical evaluation resulted in better outcomes from LSTM and GRU models than Prophet model and the GETH oracle.

A Gaussian process model to infer the minimum gas price is presented in ChihYun et al. [10]. Gaussian process is a non-parametric Bayesian approach to estimate a posterior over functions based on prior over functions using test data. This model performs better than GasStation-Express and Geth only when gas prices fluctuate widely. For this reason, they propose a hybrid solution combining GasStation-Express with their model.

3 Experiments and Results

3.1 Data Collection and Pre-processing

According to Salinas et al. [13], the covariates can be item and/or time dependent. For collecting the historical blockchain data, we used the Kaggle Ethereum Blockchain Complete live historical data (BigQuery) [2], as well as live minute-by-minute Ether (ETH or Ξ) and Polygon **MATIC** prices from cryptodatadownload.com [1], as seen in Table 2. Our Jupyter notebook [11] prepared the data for the training.

Table 2. Features used for training and inference (minute-by-minute intervals). The target time series are the Gas Prices and there are 6 additional dynamic features. During our experiments we found MATIC prices not to be helpful

Feature	Description	Source
Gas prices	Gas prices by transaction	BigQuery *transactions* [2]
Ξ prices	Ether to USD price minute-by-minute	Binance exchange data [1]
MATIC prices	MATIC to USD price minute-by-minute	Binance exchange data [1]
Transaction values	Value transfer by transaction	BigQuery *transactions* [2]
Committed transactions	Transactions in blocks	BigQuery *blocks* [2]
Token transfers	# ERC20 token transfers	BigQuery token_transfers [2]
Contract events	# ERC20 token events	BigQuery *logs* [2]
Gas used	# units of gas per transaction	BigQuery *traces* [2]

For the training and validation phase, we processed the mean of all the time series for every 20 min. After experimenting with various time series frequencies, we chose 20 min intervals as the best for this type of data. Ethereum gas prices fluctuate widely and hence the data is very noisy. In spite of not smoothing the data by eliminating outliers, our model performed very well. In the end, we had data processed at 20 min intervals for 291 days (January 1, 2021 to October 18, 2021). We used 80% of the data for training and 20% for validation.

3.2 Experimental Setup

We built a Python Jupyter notebook [11] and used Gluon Time Series (GluonTS) [5] for probabilistic time series modeling. The *DeepAREstimator* is an implementation of the model described by Salinas et al. [13]. We have configured the estimator as follows:

- Prediction length of 40, thus providing $40 \cdot 20 = 800\,\text{min} = 13\,\text{h}$ and 20 min.
- Architecture of 4 layers with 40 cells per each layer.
- Dropout rate = 0.1.
- Context length (number of steps) of 80 (double of prediction length). Context length is the number of points provided to the model to make the prediction.
- Cell type GRU. Note that we experimented with LSTM cells as well, although we did not notice any significant improvement, but rather a slight slow-down of the training.
- The learning rate callback had the following settings: patience = 10, base LR $= 10^{-3}$, decay factor = 0.5.
- Training was configured to run for 200 epochs.
- We selected the checkpoints from 2 models, based on the best metric values.

The experiments were performed on a desktop computer equipped with Intel Corporation Xeon E3-1200 v6/7th GenCore Processor with 32 GB RAM and 240 GB SSD, NVIDIA GeForce GTX 1080 TI GPU, running *Ubuntu 18.04*.

3.3 Experimental Results

The tests were run for various date/time targets, and we noticed empirically an overall improvement in metrics of predictions as we added more features.

To find the best combination of features, we performed a greedy approach. First, we selected the feature with the highest impact by running the algorithm 7 times. Since using the MATIC prices gave worse results than using 0 dynamic features, we dropped this data from subsequent tests. Once we found the feature with best results, we ran the training and inference with the remaining 5 feature data and selected again the one giving best results. We continued the process for the rest of the features, thus performing a total of $7 + 5 + 4 + 3 + 2 + 1 = 22$ trials.

Given y_i as the observed value, $\hat{y}_i$ the predicted value and n the number of samples, we computed the following metrics using *sklearn* package:

1. Mean Absolute Error (MAE): $MAE = \frac{\sum_{i=1}^{n}|y_i - \hat{y_i}|}{n}$
2. Quantile Loss (QL) for a given quantile q, defined as: $L(\hat{y_i}, y_i) = \max\{q(\hat{y_i} - y_i), (q-1)(\hat{y_i} - y_i)\}$. This value is averaged across all predictions. We compared the values obtained for the following quantiles: $q \in \{0.1, 0.5, 0.9\}$.
3. Mean Squared Error (MSE) and Root Mean Squared Error (RMSE) defined as: $MSE = \frac{\sum_{i=1}^{n}(y_i - \hat{y_i})^2}{n}$ and $RMSE = \sqrt{\frac{\sum_{i=1}^{n}(y_i - \hat{y_i})^2}{n}} = \sqrt{MSE}$. To calculate the RMSE metric, we used data normalization by re-scaling the test and predicted data to have a mean of 0 and variance of 1.

Our best performance result with 5 dynamic features is presented in Fig. 1. Table 3 shows the values obtained for each metrics, depending on number of features.

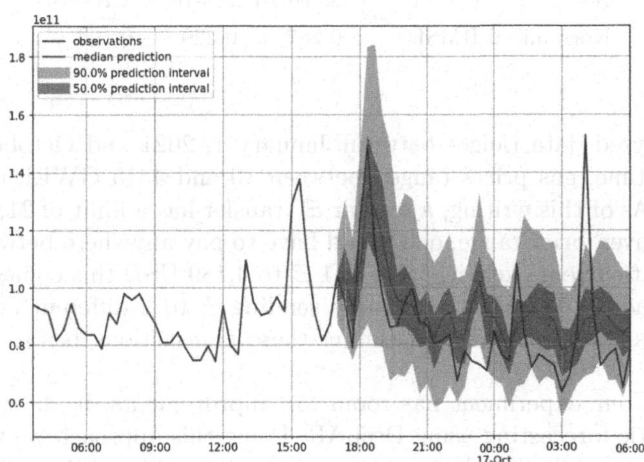

Fig. 1. Prediction with 5 dynamic feature inputs: Ξ prices, Transaction Values, Committed Transactions, Token Transfers, and Gas Used

3.4 Discussion

Time series forecasting is one of the most important tools used by businesses, and there are a number of frameworks data scientists can use for this purpose. As expected, there is no "silver bullet" for any problem. The empirical research conducted by Žunić et al. [17] concludes that DeepAR (AWS) models "show superiority over classical methods only when they have a large number of signals over which to create a model, and in the case of articles with a short history". Since cryptocurrency prices in general and Ξ in particular have numerous covariates that can be used, one can perform accurate predictions with shorter history, a benefit that low-power (e.g. IOT) edge devices can take advantage of when deciding on the timing for submitting transactions to the blockchain.

Table 3. Comparison of metrics by number of dynamic features involved. Values are represented in WEI ($1\Xi = 10^{18} WEI$).

Metric	0 feats.	3 feats.	5 feats.
MAE	7.6e+9	6.3e+9	4.9e+9
MSE	6.2e+20	3.9e+20	2.9e+20
Quantile Loss [0.1]	2.9e+11	1.9e+11	1.7e+11
Coverage Quantile [0.1]	0.0	0.225	0.250
Quantile Loss [0.5]	7.7e+11	6.4e+11	5e+11
Coverage Quantile [0.5]	0.1	0.82	0.77
Quantile Loss [0.9]	4.7e+11	3.7e+11	2.8e+11
Coverage Quantile [0.9]	0.625	0.975	0.975
RMSE	2.5e+10	2e+10	1.7e+10
Normalized RMSE	0.287	0.229	0.194

Our analyzed data ranges between January 1, 2021 and October 18, 2021. During this time, gas prices ranged between 10 and 4315 GWEI, or a 431.5% fluctuation. As of this writing, a regular Ξ transfer has a limit of 21,000 units of gas. At the given price range, one would have to pay anywhere between 0.00021 and 0.9 Ξ. At current exchange rate of 1 Ξ to 4,189 USD this comes to roughly between 1 and 3,770 USD for a simply sending Ξ to a different address. This clearly shows the importance of timing these transactions based on accurate predictions.

Although our experiment has room for improvements, it shows the power of probabilistic forecasting using DeepAR. Using this approach we were able to obtain accurate predictions in spite of a noisy dataset. DeepAR requires minimal feature engineering. We performed down sampling but did not remove outliers. We did, nevertheless, have to perform normalization, in spite of suggestions in the literature otherwise. Our model did not converge without normalizing the features first. By adding dynamic features, we achieved improvements in the prediction metrics (see Table 3). As the figures show, DeepAR's Monte Carlo sampling-based quantile estimates are accurate and can be very useful in practice.

4 Conclusions and Future Works

Empirical analysis of Ethereum gas price prediction with DeepAR proves that carefully chosen covariates can improve model performance. Gas prices are impacted by various factors, including seasonality, volume of transactions, transaction values, number of token transactions, Ξ price, amount of gas used per block. Our focus in the future will be to identify additional features that can improve the performance of our model, by researching factors that have an impact on the supply and demand for gas. Such examples may include off-chain data, such as twitter or other social media events that may influence the volume of transactions on the Ethereum blockchain and, hence indirectly, gas and/or Ξ prices.

As DeepAR performs better than Facebook's Prophet with a smaller amount of sales data [17], we will research potentials for deployment of our models on low-power connected devices.

References

1. Binance ETH to USD minute-by-minute. https://www.cryptodatadownload.com
2. Ethereum blockchain: Complete live historical ethereum blockchain data (BigQuery). https://www.kaggle.com/bigquery/ethereum-blockchain
3. Ethereum gas station. https://github.com/ethgasstation
4. Gas station express oracle. https://github.com/ethgasstation/gasstation-express-oracle
5. Alexandrov, A., et al.: GluonTS: Probabilistic time series models in Python. ArXiv abs/1906.05264 (2019)
6. Blocknative: Gas estimation for builders, by builders. https://www.blocknative.com/gas-platform
7. Cho, K., et al.: Learning phrase representations using RNN encoder-decoder for statistical machine translation. In: Moschitti, A., Pang, B., Daelemans, W. (eds.) Proceedings of the 2014 Conference on Empirical Methods in Natural Language Processing, EMNLP 2014, A meeting of SIGDAT, a Special Interest Group of the ACL, pp. 1724–1734. ACL (2014). https://doi.org/10.3115/v1/d14-1179
8. Chuang, C.: Official go implementation of the ethereum protocol. https://github.com/ethereum/go-ethereum/
9. Chuang, C.: A practical and economical of gas price prediction. https://medium.com/getamis/a-practical-and-economical-of-gas-price-prediction-d9abe955ac63
10. Chuang, C.Y., Lee, T.F.: A practical and economical Bayesian approach to gas price prediction. In: Awan, I., Benbernou, S., Younas, M., Aleksy, M. (eds.) Deep-BDB 2021. LNNS, vol. 309, pp. 160–174. Springer, Cham (2022). https://doi.org/10.1007/978-3-030-84337-3_13
11. Ferenczi, A.: Using probabilistic forecasting for gas price prediction in ethereum blockchain. https://github.com/andrasfe/ethgas
12. Mars, R., Abid, A., Cheikhrouhou, S., Kallel, S.: A machine learning approach for gas price prediction in ethereum blockchain. In: 2021 IEEE 45th Annual Computers, Software, and Applications Conference (COMPSAC), pp. 156–165 (2021). https://doi.org/10.1109/COMPSAC51774.2021.00033
13. Salinas, D., Flunkert, V., Gasthaus, J., Januschowski, T.: DeepAR: probabilistic forecasting with autoregressive recurrent networks. Int. J. Forecast. **36**(3), 1181–1191 (2020). https://doi.org/10.1016/j.ijforecast.2019.07.001
14. Simionescu, S.M., Bădică, A., Bădică, C., Ganzha, M., Paprzycki, M.: On the role of blockchain in evolving the online business landscape. In: 2021 Conference on Information Communications Technology and Society (ICTAS), pp. 85–90. IEEE (2021). https://doi.org/10.1109/ICTAS50802.2021.9395045
15. Smith, C.: Ethereum oracles. ethereum.org (2022). https://ethereum.org/en/developers/docs/oracles/
16. Taylor, S.J., Letham, B.: Forecasting at scale. PeerJ Preprints **5**, e3190v2 (2017). https://doi.org/10.7287/peerj.preprints.3190v2

17. Žunić, E., Korjenić, K., Delalić, S., Šubara, Z.: Comparison analysis of Facebook's Prophet, Amazon's DeepAR+ and CNN-QR algorithms for successful real-world sales forecasting. Int. J. Comput. Sci. Inf. Technol. **13**(2), 67–84 (2021). https://doi.org/10.5121/ijcsit.2021.13205
18. Wackerow, P.: ERC-20 token standard. ethereum.org (2021). https://ethereum.org/en/developers/docs/standards/tokens/erc-20/
19. Ward, C.: Merkle patricia tree (2020). https://eth.wiki/fundamentals/patricia-tree
20. Werner, S.M., Pritz, P.J., Perez, D.: Step on the Gas? A better approach for recommending the ethereum gas price. In: Pardalos, P., Kotsireas, I., Guo, Y., Knottenbelt, W. (eds.) Mathematical Research for Blockchain Economy. SPBE, pp. 161–177. Springer, Cham (2020). https://doi.org/10.1007/978-3-030-53356-4_10
21. Zheng, Z., et al.: An overview on smart contracts: challenges, advances and platforms. Future Gener. Comput. Syst. **105**, 475–491 (2020). https://doi.org/10.1016/j.future.2019.12.019

Purchasing Decisions on Alternative Fuel Vehicles Within an Agent-Based Model

Arkadiusz Jędrzejewski[1(✉)], Katarzyna Sznajd-Weron[2], Jakub Pawłowski[2], and Anna Kowalska-Pyzalska[1]

[1] Department of Operations Research and Business Intelligence, Faculty of Management, Wrocław University of Science and Technology, Wrocław, Poland
arkadiusz.jedrzejewski@pwr.edu.pl

[2] Department of Theoretical Physics, Faculty of Fundamental Problems of Technology, Wrocław University of Science and Technology, 50-370 Wrocław, Poland

Abstract. We develop an empirically grounded agent-based model to explore the purchasing decisions of mutually interacting agents (consumers) between three types of alternative fuel vehicles. We calibrate the model with recently published empirical data on consumer preferences towards such vehicles. Furthermore, running the Monte Carlo simulations, we show possible scenarios for the development of the alternative fuel vehicle market depending on the marketing strategies employed.

Keywords: Agent-based model · Alternative fuel vehicle · Diffusion

1 Introduction

According to experts, the achievement of the goals of sustainable transport requires an increase in the share of vehicles powered by alternative fuels (AFV) in road traffic [2,7]. Among these cars, battery electric vehicles (BEVs), plug-in electric vehicles (PHEVs), and hybrid electric vehicles (HEVs) are mainly included. Although the market share of AFV is constantly increasing worldwide, its smooth diffusion encounters a number of barriers including lack of sufficient charging infrastructure, high prices, and safety issues [13].

There are a number of studies on the technical, economic, social, and psychological factors that influence the choice of vehicle type when deciding to purchase it. The authors use various stated preferences methods to analyze what factors determine the decision to buy a certain type of a vehicle [1,6]. Apart from that, a large number of simulating and modeling studies, making usage of agent-based modeling (ABM), have been recently published [3,4,11]. ABM allows us, among others, to investigate how the individual decisions of the agents (i.e., households, consumers, etc.) and their social interactions lead to effects on the macroscopic level (e.g., market penetration of a given good). Based on [8–11], we propose an

Supported by the National Science Center (NCN, Poland) through Grants Nos. 2018/29/B/HS4/00069 and 2019/35/B/HS6/02530.

D. Groen et al. (Eds.): ICCS 2022, LNCS 13351, pp. 719–726, 2022.
https://doi.org/10.1007/978-3-031-08754-7_74

empirically grounded agent-based model with explicitly introduced interactions with local and global neighborhoods. Our proposed approach allows the model to be easily developed in the future, for example, taking into account space heterogeneity, individual consumer preferences, or interactions in online social networks. However, here, we focus on the zero-level approach, in which the environment is represented by a regular grid, and the agents are homogeneous in terms of preferences. This allows us to determine what is the importance of external factors such as global marketing or different government policies. What is even more important, it allows other researchers to replicate the results, which in our opinion is necessary for reliable model verification.

2 The Model

We consider a square $L \times L$ lattice with periodic boundary conditions. The linear size of the system $L = 100$ is taken in most of our simulations. Each node is occupied by exactly one agent, and thus the total number of agents in the system equals $N = L^2$. Each agent has exactly four neighbors due to the network structure, and it can own exactly one car.

Initially, the agents do not have cars. We use a random sequential updating scheme, which mimics continuous time. This means that in an elementary update, only one agent is selected randomly from all N agents, and a single Monte Carlo step consists of N elementary updates (note that not necessarily all agents are updated in a single time step). The selected agent buys one car picked from the alternative fuel vehicle choice set. Motivated by the empirical studies [5], in our simulations this set includes in total 75 cars, 25 of each type (HEV, PHEV, and BEV). All cars are characterized by 5 attributes that are 5-level discrete variables. Car profiles are taken from the conjoint analysis: Tables 13, 16, and 17 in [5]. The vehicle attributes impact the overall utility that comes from purchasing a given car. The total utility of car $j \in \{1, 2, ..., 75\}$ is the sum of partial utilities associated with the attributes of this car:

$$U_j = \sum_{n=1}^{5} PU_{j,n}, \tag{1}$$

where $PU_{j,n}$ is the partial utility of the n-th attribute of car j. These utilities were estimated also through the conjoint analysis of consumers' preferences: Table 14 in [5]. To distinguish between different types of cars, we introduce a function:

$$f(j) = \begin{cases} \text{HEV} & \text{if the } j\text{-th car is HEV}, \\ \text{PHEV} & \text{if the } j\text{-th car is PHEV}, \\ \text{BEV} & \text{if the } j\text{-th car is BEV}. \end{cases} \tag{2}$$

Following [9,11], we assume that the consumer decision-making process depends not only on the total utility of a vehicle, but also on additional external factors that account for marketing, social influence, and the availability of recharging facilities. Thus, the probability that agent $i \in \{1, 2, ..., N\}$ buys car j is expressed by the following multinomial logit model:

$$P_{i,j} = \frac{W_{i,f(j)} \cdot RFE_{f(j)} \cdot \exp(U_j)}{\sum_{j=1}^{75} W_{i,f(j)} \cdot RFE_{f(j)} \cdot \exp(U_j)}, \tag{3}$$

where $W_{i,f(j)}$ is the willingness of agent i to buy a car of a given type, which captures the impact of marketing and social influence, whereas $RFE_{f(j)}$ is the refueling effect, which reflects the availability of recharging facilities for a given car type. The refueling effect accounts for agents' concerns related with low ranges of some AFVs, such as PHEVs and BEVs. We include this effect in a functional form that has already appeared in previous studies [9–11]:

$$RFE_{f(j)} = \begin{cases} 1 & \text{if the } j\text{-th car is HEV,} \\ 1 - DPe^{-\alpha_{\text{PHEV}} N_{\text{PHEV}}/N} & \text{if the } j\text{-th car is PHEV,} \\ 1 - DPe^{-\alpha_{\text{BEV}} N_{\text{BEV}}/N} & \text{if the } j\text{-th car is BEV,} \end{cases} \tag{4}$$

where DP is a driving pattern characterized by the society, N_{PHEV} and N_{BEV} are the numbers of agents that have already adopted PHEVs and BEVs, respectively, whereas α_{PHEV} and α_{BEV} are scaling parameters used to calibrate the model.

The novelty of our model is the formula describing the willingness of agent i to buy a car of type $f(j)$:

$$W_{i,f(j)} = \underbrace{h_{f(j)}}_{\text{marketing}} + \underbrace{p_l k_{i,f(j)}/k}_{\text{local influence}} + \underbrace{p_g N_{f(j)}/N}_{\text{global influence}} + \underbrace{1}_{\text{independence}}, \tag{5}$$

where $h_{f(j)}$ reflects the effectiveness of marketing for vehicles of a given type, p_l is the strength of local social influence, p_g is the strength of global social influence, $k_{i,f(j)}$ is the number of neighbors of agent i that already possess vehicles of a given type, $k = 4$ is the total number of neighbors of an agent, and $N_{f(j)}$ is the total number of agents in the system that have vehicles of a given type. The first term of formula (5) captures not only the influence of advertisements and promotions, but also the effectiveness of various policies, benefits, and advantages related to AFVs, such as subsidies, tax releases, or free parking spaces. We assume that all vehicles with the same engine type are described by the same value of parameter $h_{f(j)}$, and thus it takes only three values: h_{HEV}, h_{PHEV}, and h_{BEV}.

Regarding social influence, we distinguish between local and global one, just like in [8]. The local influence (word-of-mouth), the second term of Eq. (5), is proportional to the fraction of neighbors with cars of the same type as the considered car. Similarly, the global influence, the third term of Eq. (5), is proportional to the fraction of all agents in the system having such cars.

A pseudo-code to simulate our model is as follows:

0. Set parameters of the model: L, DP, α_{PHEV}, α_{BEV}, h_{HEV}, h_{PHEV}, h_{BEV}, p_l, and p_g as well as the time horizon of the simulation, T. Initialize the system. Set time $t = 0$.

1. Count the number of agents in the system that have cars of each type, i.e., N_{HEV}, N_{PHEV}, and N_{BEV}.
2. Calculate the refueling effect from Eq. (4) for PHEVs and BEVs, i.e., RFE_{PHEV}, and RFE_{BEV}.
3. Draw number i from discrete uniform distribution $U\{1, N\}$. Agent i is selected to buy a car.
4. Count the number of neighbors of agent i that have cars of each type, i.e., $k_{i,\text{HEV}}$, $k_{i,\text{PHEV}}$, and $k_{i,\text{BEV}}$.
5. Calculate the willingness of agent i to buy a car of each type from Eq. (5), i.e., $W_{i,\text{HEV}}$, $W_{i,\text{PHEV}}$, and $W_{i,\text{BEV}}$.
6. For each $j \in \{1, 2, ..., 75\}$, calculate the probability that agent i buys car j, i.e., $P_{i,j}$ from Eq. (3).
7. Draw number u from continuous uniform distribution $U[0, 1]$.
8. Find index m such that $\sum_{j=1}^{m-1} P_{i,j} \leq u < \sum_{j=1}^{m} P_{i,j}$. Agent i buys car m.
9. Update time $t \to t + 1/N$. If $t < T$, go to point 1.

3 Results

To calibrate the model, we first run simulations without any marketing and social influence ($h_{\text{HEV}} = 0$, $h_{\text{PHEV}} = 0$, $h_{\text{BEV}} = 0$, $p_l = 0$, and $p_g = 0$), and tune the values of α_{PHEV} and α_{BEV} so that the stationary adoption levels of HEVs, PHEVs, and BEVs correspond to those estimated based on the survey conducted in [5]. We focus on the stationary state of the system, since it is difficult to establish a general relationship between the Monte Carlo steps and the real time. In the survey, 48.8% of the respondents declared that they would buy HEV, 32% PHEV and 19.3% BEV. We obtain similar levels of adoption for $\alpha_{\text{PHEV}} = 2.6$ and $\alpha_{\text{BEV}} = 0.05$, see Fig. 1(a). Without having data for Poland, we set $DP = 0.49$, which approximates the aggregate driving pattern for Germany [10]. All the results that we present are averaged over 40 independent simulations.

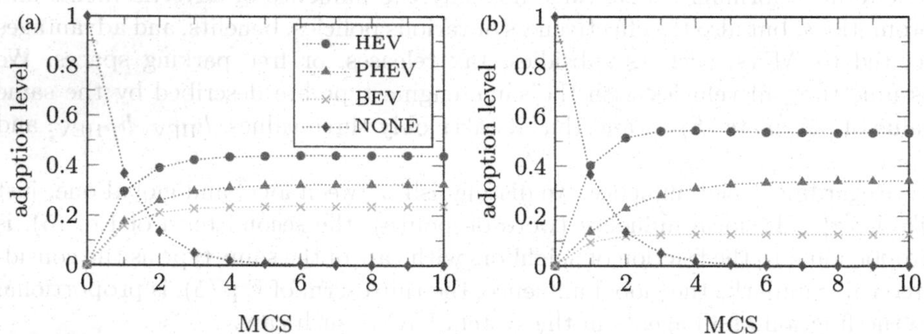

Fig. 1. Adoption levels of AFVs in a system without marketing ($h_{\text{HEV}} = h_{\text{PHEV}} = h_{\text{BEV}} = 0$) and social influence ($p_l = p_g = 0$) as a function of time measured in Monte Carlo steps: (a) $DP = 0.49$ (driving pattern for Germany [10]) and (b) $DP = 0.78$ (driving pattern for Iceland [11]). NONE represents the fraction of agents without a car.

After calibrating the model, we want to check how the driving pattern DP and different policies impact the behavior of the model. To verify the former, we simulate the system with the same values of parameters that were obtained within the calibration for $DP = 0.49$, but this time with $DP = 0.78$. This value characterizes countries with longer average daily driven distances, such as Iceland [11]. In Fig. 1(b) it is seen that increasing DP leads to a higher adoption level of HEVs at the expense of BEVs.

Next, we investigate how marketing campaigns targeting only one type of AFVs impact their stationary adoption levels. Figures 2, 3 and 4 present the results for the systems where only HEVs, PHEVs, and BEVs are advertised, respectively. Under stronger social influence, marketing targeted at HEVs leads to their higher adoption level. The opposite happens in the case of PHEVs and BEVs. However, for HEVs, stronger social influence causes smaller gains in the adoption level that result from the increase of the advertising strength, see Fig 2.

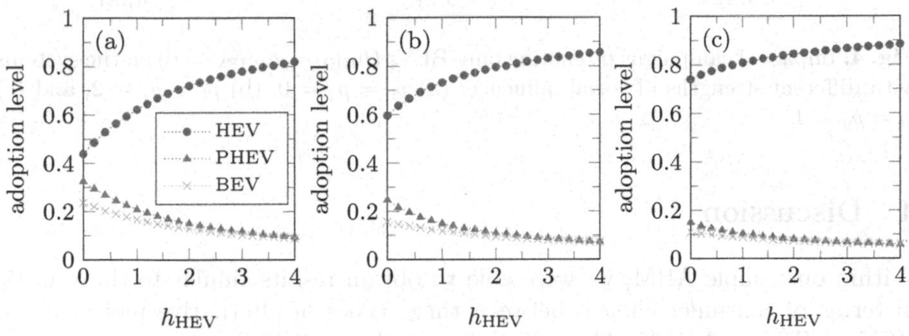

Fig. 2. Impact of campaigns promoting only HEVs ($h_{\text{PHEV}} = h_{\text{BEV}} = 0$) on the systems with different strengths of social influence: (a) $p_l = p_g = 0$, (b) $p_l = p_g = 2$, and (c) $p_l = p_g = 4$.

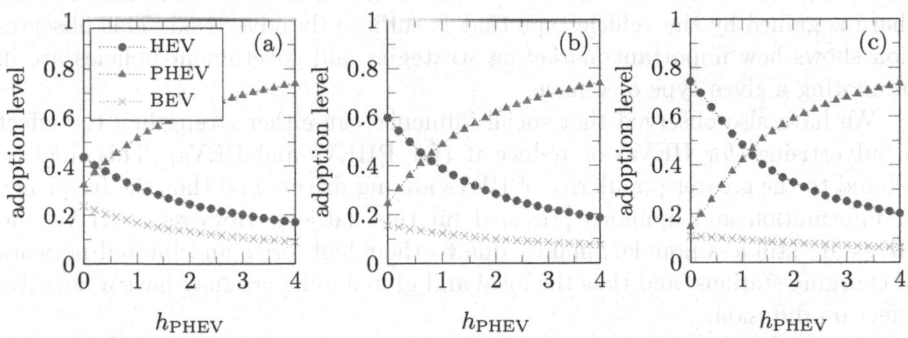

Fig. 3. Impact of campaigns promoting only PHEVs ($h_{\text{HEV}} = h_{\text{BEV}} = 0$) on the systems with different strengths of social influence: (a) $p_l = p_g = 0$, (b) $p_l = p_g = 2$, and (c) $p_l = p_g = 4$.

This diminishing effectiveness of marketing is related to the high initial adoption of HEVs. In contrast, we can achieve a considerable increase of PHEV adoption for a low intensity of advertisements when the social influence is strong enough, see Fig. 3(c).

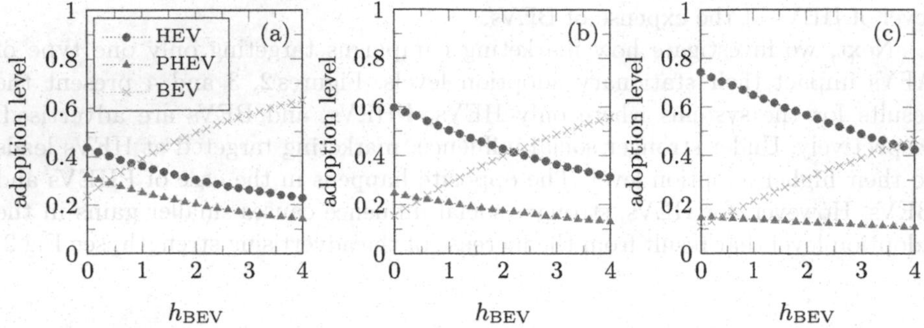

Fig. 4. Impact of campaigns promoting only BEVs ($h_{\text{HEV}} = h_{\text{PHEV}} = 0$) on the systems with different strengths of social influence: (a) $p_l = p_g = 0$, (b) $p_l = p_g = 2$, and (c) $p_l = p_g = 4$.

4 Discussion

Within our simple ABM, we were able to obtain results similar to those in [5] in terms of consumer choices between three types of alternative fuel vehicles: HEV, PHEV, and BEV. The calibration of the model allowed us not only to reflect the current sentiments on the market but also to show that the local and global impact is not always conducive to the spread of new solutions. That is why advertising is needed, whose strength must depend on the strength of social interaction to be effective. The results have revealed that the largest market share is gained by the vehicle type that is sufficiently advertised. This observation shows how important marketing strategies and government policies are in promoting a given type of vehicle.

We have also observed that social influence can either strengthen the effect of advertising (for HEVs) or reduce it (for PHEVs and BEVs). This may be related to the greater popularity of HEVs among drivers and thus the frequency of information and opinions provided on this subject. However, PHEVs and BEVs are still less popular, mainly due to their high price and limited network of charging stations, and thus the local and global influence may have a negative effect on diffusion.

Our model also allows us to observe the impact of the value of the driving pattern, DP, on the level of adoption and diffusion of vehicles. Comparison of results for medium and high values of DP indicates that BEV adoption is higher in countries more densely populated where the average distances covered are shorter and the charging station network is denser. Being aware of the weaknesses of our model (including homogeneous agents, simple network topology, lack of negative marketing, etc.), we believe that the model can be easily developed further to capture more realistic assumptions [12]. The available survey data [5] also restricted our model setup. First, we could not take into account other common means of transport, such as internal combustion engine vehicles, fuel cell electric vehicles, or public transportation. Second, not knowing how respondents had been impacted by marketing and social influence, we performed a model calibration without these factors. Taking these aspects into account would result in a more realistic setup; however, this requires more tailored empirical data, which we are missing.

References

1. Byun, H., Shin, J., Lee, C.Y.: Using a discrete choice experiment to predict the penetration possibility of environmentally friendly vehicles. Energy **144**, 312–321 (2018). https://doi.org/10.1016/j.energy.2017.12.035
2. Chang, D.S., Chen, S.H., Hsu, C.W., Hu, A.H., Tzeng, G.H.: Evaluation framework for alternative fuel vehicles: sustainable development perspective. Sustainability **7**(9), 11570–11594 (2015). https://doi.org/10.3390/su70911570
3. Huang, X., Lin, Y., Zhou, F., Lim, M.K., Chen, S.: Agent-based modelling for market acceptance of electric vehicles: evidence from China. Sustain. Prod. Consumption **28**, 206–217 (2021). https://doi.org/10.1016/j.spc.2021.04.007
4. Kangur, A., Jager, W., Verbrugge, R., Bockarjova, M.: An agent-based model for diffusion of electric vehicles. J. Environ. Psychol. **52**, 166–182 (2017). https://doi.org/10.1016/j.jenvp.2017.01.002
5. Kowalska-Pyzalska, A., Michalski, R., Kott, M., Skowrońska-Szmer, A., Kott, J.: Consumer preferences towards alternative fuel vehicles. Results from the conjoint analysis. Renew. Sustain. Energy Rev. **155**, 111776 (2022). https://doi.org/10.1016/j.rser.2021.111776
6. Lin, B., Wu, W.: Why people want to buy electric vehicle: an empirical study in first-tier cities of China. Energy Policy **112**, 233–241 (2018). https://doi.org/10.1016/j.enpol.2017.10.026
7. Mazza, S., Aiello, D., Macario, A., De Luca, P.: Vehicular emission: estimate of air pollutants to guide local political choices. A case study. Environments **7**(5), 37 (2020). https://doi.org/10.3390/environments7050037
8. McCoy, D., Lyons, S.: Consumer preferences and the influence of networks in electric vehicle diffusion: an agent-based microsimulation in Ireland. Energy Res. Soc. Sci. **3**, 89–101 (2014). https://doi.org/10.1016/j.erss.2014.07.008
9. Noori, M., Tatari, O.: Development of an agent-based model for regional market penetration projections of electric vehicles in the United States. Energy **96**, 215–230 (2016). https://doi.org/10.1016/j.energy.2015.12.018
10. Schwoon, M.: Simulating the adoption of fuel cell vehicles. J. Evol. Econ. **16**(4), 435–472 (2006). https://doi.org/10.1007/s00191-006-0026-4
11. Shafiei, E., Thorkelsson, H., Ásgeirsson, E.I., Davidsdottir, B., Raberto, M., Stefansson, H.: An agent-based modeling approach to predict the evolution of market share of electric vehicles: a case study from Iceland. Technol. Forecast. Soc. Change **79**(9), 1638–1653 (2012). https://doi.org/10.1016/j.techfore.2012.05.011

12. Sobkowicz, P.: Modelling opinion formation with physics tools: call for closer link with reality. J. Artif. Soc. Soc. Simul. **12**(1), 11 (2009). https://www.jasss.org/12/1/11.html
13. Wang, F.P., Yu, J.L., Yang, P., Miao, L.X., Ye, B.: Analysis of the barriers to widespread adoption of electric vehicles in Shenzhen China. Sustainability **9**(4), 522 (2017). https://doi.org/10.3390/su9040522

Author Index

Printed in the United States
by Baker & Taylor Publisher Services

Printed in the United States
by Baker & Taylor Publisher Services